HERSCHEL SIDRANSKY, M.D.
DEPT. OF PATHOLOGY
UNIV. OF SOUTH FLORIDA
TAMPA, FLA. 33620

HERSCHEL SIDRANSKY, M.D.
DEPT. OF PATHOLOGY
UNIV. OF SOUTH FLORIDA
TAMPA, FLA. 33620

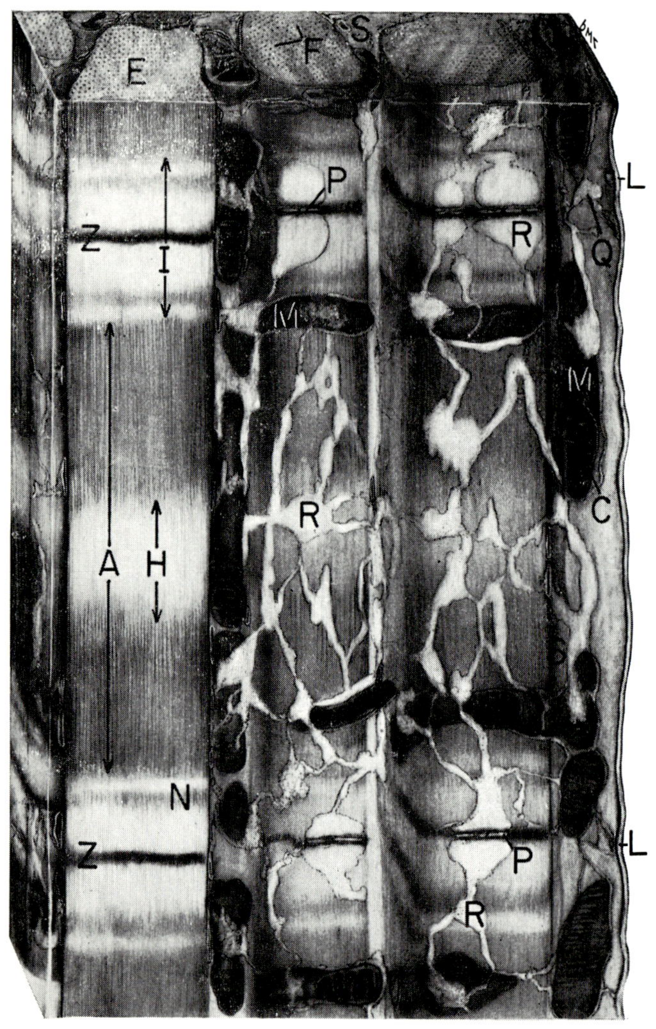

Block of striated muscle showing relationships of three myofibrils to sarcolemma (C), sarcoplasmic reticulum (R), and mitochondria (dark masses, M). Elements of sarcoplasmic reticulum are attached to cell membrane (L) near the Z bands and entwine around the myofibrils, with a special relation (P) at the Z band level. E, myofibril (cross-section surface). F, dots represent myofilaments within myofibril. A, anisotropic band of myosin filaments. I, isotropic band. R, cisternae of T system. H, H band. C, cristae of mitochondria. S, sarcous substance. N, band in I sector. Q, surface opening of T system. (From Bennett HS: Neurology, (Minneap) 8:66, 1958)

DISEASES OF MUSCLE
A STUDY IN PATHOLOGY

THIRD EDITION

Raymond D. Adams, B.A., M.A., M.D.

M.A. (Hon., Harvard); M.D. (Hon., Ghent); M.D. (Hon., Lausanne); DSc (Hon., Newcastle)
Bullard Professor of Neuropathology, Harvard University, Cambridge;
Chief, Neurology Service and Director of The John F. Kennedy, Jr. Memorial
      Laboratories, Massachusetts General Hospital, Boston;
Director, Eunice K. Shriver Laboratories, Fernald School, Waltham, Massachusetts.
Médecin Adjoint de L'Hôpital Cantonal, Lausanne, Switzerland.
Lecturer in Neurology, American University of Beirut, Lebanon.

MEDICAL DEPARTMENT
HARPER & ROW, PUBLISHERS
HAGERSTOWN, MARYLAND
NEW YORK, EVANSTON, SAN FRANCISCO, LONDON

**DISEASES OF MUSCLE: A STUDY IN PATHOLOGY,
THIRD EDITION.** Copyright © 1975 by Harper & Row,
Publishers, Inc. All rights reserved. No part of this book
may be used or reproduced in any manner whatsoever
without written permission except in the case of brief
quotations embodied in critical articles and reviews.
Printed in the United States of America. For information address Medical Department, Harper & Row, Publishers, Inc., 2350 Virginia Avenue, Hagerstown, Maryland
21740

Library of Congress Cataloging in Publication Data
Adams, Raymond Delacy, 1911–
    Diseases of muscle: a study in pathology.
    First-2d editions by R. D. Adams, D. Denny-Brown, and
C. M. Pearson.
    Includes bibliographies and index.
    1. Muscles—Diseases. 2. Muscle. I. Title.
[DNLM: 1. Muscles—Pathology. WE500 A216d]
RC925.A3 1975 616.7′4 74–34141

CONTENTS

part **I** **THE NATURE OF SKELETAL MUSCLE**

chapter **1** **EMBRYOLOGY AND HISTOLOGY OF SKELETAL MUSCLE** 2

Embryology 2 • *Muscle Fibers 2* • *Connective Tissue 7* • *Neurologic Development 11*
Anatomy and Histology 13 • *Structural Details of Mature Muscle Fiber 15* • *Histochemical Detection of Muscle Fiber Constituents 39*
Tendinous Attachments of Muscle Fiber 45 • *Supporting Tissues of Muscle 46*

chapter **2** **INNERVATION AND BLOOD VESSELS OF MUSCLE** 54

Innervation 54 • *Motor Nerve Fibers, Motor End-Plates, and Motor Units 54* • *Sensory Nerve Endings 67*
Vascular Supply and Lymphatics 71 • *Blood Vessels 71* • *Lymphatics 76*
Chemistry by Carl M. Pearson and Nirmal C. Kar 77 • *Chemical Basis of Muscular Structure and Function 77*
Contractile Proteins 77 • *Myosin 78* • *Actin 78* • *Actomyosin 78* • *Tropomyosin or Tropomyosin B 79* • *Troponin 79* • *α-Actinin and β-Actinin 79* • *Sliding Filament Theory of Contractions 80*
Electrolytes 80 • *Carbohydrates 80* • *Lipids 81* • *Nonprotein Nitrogenous Components 82* • *Enzymes in Muscle 82* • *Energy Mechanisms in Muscle Fibers 82* • *Myoglobin 85*
The Neuronal Control of Muscular Contraction 86 • *Local Contractions 89* • *Neuromuscular Transmission 90* • *"All or None" Phenomenon 93* • *Fatigue 94* • *Fibrillation 94* • *Delay in Relaxation 96* • *Spasm 97* • *Fasciculation 98* • *Duration of Action Potential 98* • *Tendon Reflex 99* • *Hypertrophy and Disuse Atrophy 100*

PREFACE TO THIRD EDITION

The authorship of this third edition of Diseases of Muscle has been altered by force of circumstance. Professor Derek Denny-Brown, with whom I have been privileged to collaborate for many years, has withdrawn because of his desire to pursue other lines of research. In order to assume important university affairs, Professor Carl Pearson has been unable to spare the time for the tedious tasks of revision. Thus it was left for the remaining author to undertake the review of the burgeoning literature on muscle disease. The reader who is familiar with the earlier editions will, upon perusing this one, realize to what extent the ideas of these two colleagues have been carried forward. Any success which this third edition may achieve must in considerable measure be attributed to their scholarship and research.

The exposition of disease encompassed in this edition should commend itself to the attention and good will of that segment of the medical public interested in the theoretic aspects of muscle diseases: The word *pathology* appearing in the subtitle is used in its broadest signification—the study of the causes and mechanisms of disease and their functional effects. Thus the book assays more than a mere description of the structural changes in the muscle fiber wrought by disease and the manner in which the morbid processes evince morphologic changes. The notes on the functional effects of these pathologic processes should be useful to those who seek an understanding of *clinical myology*. While the emphasis throughout the book is on the composite of morphologic changes, i.e., the lesion, this is considered always as a step in a dynamic mechanism. Since the visible lesions prove to be one of the most reliable and consistent attributes of disease, great pains have been taken to illustrate them profusely, which should be of particular aid to the pathologist who must serve as referee in many difficult clinical analyses.

In collecting and assembling illustrations many that were used in the first and second editions are retained in this third edition. Some of the most valuable are those that were originally prepared by Dr. Derek Denny-Brown, which are: in Chapter 1—Figures 10, 11, 13, 14, 15, 20, 21 and 22; in Chapter 2—Figures 1 through 15, 19, 22, 23, 24 and 25; in Chapter 3—Figures 1 through 40, and 54 through 58; in Chapter 4—Figure 9; in Chapter 6—Figures 1 through 12 and 19, 20, 21 and 23; in Chapter 7—Figures 25 and 26; in Chapter 10—Figures 1, 2, 3, 7, 10, 13 and 17; in Chapter 11—Figure 10; in Chapter 12—Figures 1, 2, 8, 9, 10, 11, 12 and 13.

Some fifty new illustrations, including many electron micrographs, have been added and care was taken to obtain photographs of a quality that matches the illustrative material of previous editions. New light microscopic photographs have been added where it was necessary to replace older ones of less quality or to illustrate some new aspect of the disease process. Since much of the contemporary work on the pathology of muscle has utilized histochemical preparations and electron micrographs, many of the new illustrations are of this type. Here the author has again benefited from the donations of generous colleagues. Particularly is he indebted to Dr. H. deForest Webster for his contributions of a number of lovely illustrations of normal muscle and to Dr. W. R. Kennedy for those of a human muscle spindle. Dr. Andrew G. Engel and Dr. Michael H. Brooke permitted the reproduction of several beautiful illustrations which they had used in previous publications, and Dr. Karl Åström contributed a number of valuable electron micrographs, of nemaline myopathy. Their generosity is gratefully acknowledged.

The assistance of Dr. Carl M. Pearson and Dr. Nirmal C. Kar in the preparation of the chemistry section of Chapter 2 is gratefully acknowledged.

Finally, for all the labor involved in the tedious task of collating references and of retyping the manuscript many times, while contending with heavy administrative duties for the Neurology Service of the Massachusetts General Hospital, I wish to express my sincere thanks to Mrs. Freda Foster. Also, I wish to thank the editorial staff of the Journal of Neuropathology and Experimental Neurology, Excerpta Medica and Williams and Wilkins Co. who kindly consented to the republication of a number of illustrations that had appeared in their journals and books.

R. D. A.

PREFACE TO SECOND EDITION

The enthusiastic reception of the First Edition of this monograph strengthens our conviction that such an account was greatly needed by the clinician and pathologist. Its appearance coincided with a renaissance of interest in the histology and histopathology of striated muscle brought about by the perfection of techniques of electronmicroscopy and of specific histochemical methods. In addition the greater use of muscle biopsy in the last ten years has led to unprecedented advances in knowledge of the entire subject. It is now difficult to keep abreast of the mass of new information that collects each year. The three authors have each enlarged their special interest in this field. Despite the fact that the authors are no longer working together in the Neurological Unit of Boston City Hospital, they have endeavored to provide a revision of the text that will fairly represent the present status of muscle pathology for those concerned in these diseases. The authors are particularly indebted to Dr. Stanley Bennett for permission to use his illustration for the new Frontispiece, to Drs. Grob and Hakannson, and to the journals *Acta Physiologica Scandanavica, Journal of Clinical Investigation,* and *Neurology,* for permission to use their published illustrations. Dr. F. Romanul kindly gave us the electronmicrograph shown in Figure 13. Illustrations made from sections generously lent by friends are acknowledged in their legends.

R. D. A.
D. D. -B.
C. M. P.

PREFACE TO FIRST EDITION

Diseases of skeletal muscle do not fall within the province of any one department of medicine. Probably for this reason alone no comprehensive account of the myopathic diseases can be found in any current American or British textbook of medicine or neurology. Perhaps neurologists have manifested a more consistent interest in this subject than others, for they are constantly engaged in the evaluation of disorders that affect motility. Similarly, in the field of pathology it has been the neuropathologist who has paid most attention to the morbid anatomy of muscle because the examination of this tissue is part of the routine study of peripheral nerve and spinal cord diseases. Being more familiar with the normal histology than the general pathologist, his opinion may be sought on all problems of muscle disease.

Our interest in the subject has arisen in just this way, at first from a curiosity about the changes in muscle which follow denervation and later from the necessity of having to make arbitrary decisions as to the nature of many types of muscle disease. Any competence which we possess in this field, which lies partly outside the limits of neuropathology, has been acquired from experience with the combined rich materials of the Neurological Unit and the Mallory Institute of Pathology of the Boston City Hospital.

This monograph was originally planned as a syllabus of the pathologic changes in skeletal muscle, to aid in the teaching of graduate students of neurology and pathology. It was to accompany a group of microscopic sections that were chosen to illustrate the principal classes of muscle disease. In reviewing current and past writings on this subject the task assumed proportions far greater than had been anticipated. The vast amount of factual information about the structural organization of muscle, muscular contraction, and the reactions of muscle in natural and experimental disease which had accumulated in the past fifty years was found in widely scattered references. It is for this reason that we have undertaken to write a systematic account of the subject.

Throughout the text our aim has been to aid the student of morphology to an understanding of this whole class of diseases. A survey of anatomy and physiology has been included as an orientation and background for this study. This is followed by a review of the numerous experimental studies, including much hitherto unpublished material of our own which may serve to elucidate more fully the pathologic reactions of muscle. The reader may find these sections long in proportion to the rest of the book, but in our experience the problems presented by muscular diseases constantly lead us back to basic anatomic and physiologic considerations. The second half of the book is devoted to an exposition of the pathologic changes in skeletal muscle, and we have avoided wherever possible undue speculation as to the cause and pathogenesis of lesions. Our aim throughout has been to reach an integrated conception of the mechanism of the various affections of muscle. For this reason we have stressed the details of pathology and have added only sufficient clinical data to define the different entities. In the last chapter are the methods of procedure for biopsy and autopsy and certain staining techniques which we have found most helpful in the study of diseased muscle.

During the preparation of this monograph we have received constant encouragement and generous aid from our colleagues in pathology and in particular Dr. Frederic Parker, Director of the Mallory Institute, and Dr. G. Kenneth Mallory, who have made available to us numerous cases that had accumulated in the files of the Institute. One of us (D. D.-B.) wishes to acknowledge facilities extended to him some years ago for the staining of nerve endings in normal and dystrophic muscle by the Laboratories of Physiology of Oxford University and of the National Hospital for Nervous Diseases, London. We also wish to acknowledge the assistance of Mr. L. Goodman, in preparing many of the more difficult photomicrographs.

We wish to thank the *American Journal of Pathology,* the *Proceedings of the Society of Experimental Biology and Medicine,* the *Journal of Biological Chemistry,* and the *Anatomical Record* for permission to reproduce special figures.

Neurological Unit	**R. D. A.**
Boston City Hospital	**D. D. -B.**
Boston	**C. M. P.**

DISEASES OF MUSCLE

part **I**

THE NATURE OF SKELETAL MUSCLE

EMBRYOLOGY AND HISTOLOGY OF SKELETAL MUSCLE

A knowledge of embryology and anatomy of striated muscles is necessary for studying the pathology of muscle disease. The excellent monographs of Tello,[179, 180] Kelly and Zacks,[108] Boyd,[21] and Allen[1] on the embryology of muscle are highly recommended. The best of the classic anatomy and histology presentations are to be found in the writings of Ranvier,[152] Schäfer,[162] Heidenhain,[85] Jordan,[107] Meigs,[131] and Cohn.[37] The articles of Barer,[8] Jones and Barer,[105] Huxley and Hanson,[99] Hodge,[90] Bennett,[15] and Porter and Palade[148] offer the most authoritative accounts of the ultrastructure of the muscle fiber.

EMBRYOLOGY

MUSCLE FIBERS

Skeletal muscles, so called because they produce motion of parts of the body by reason of their attachment to the skeleton, arise in the embryo from the mesodermic somites. When the embryo is but 2–3 weeks old (the embryonic disk stage), masses of cells, the somites, begin to form along each side of the axial line. Some parts of these somites differentiate into smooth muscle and various mesenchymal structures; other parts in the dorsomedial portion of the somites give rise to primitive muscle cells. The latter are present in most of the body segments and are later transformed into muscle fibers.

FOURTH AND FIFTH WEEKS

The developing myoblasts, metamerically grouped in the form of myotomes, of which there are approximately 38 (3 occipital, 8 cervical, 12 thoracic, 5 lumbar, 5 sacral, and 5 coccygeal), become united with the laterally placed dermatomes during the fourth and fifth weeks of fetal life to form the dermatomyotomes. All of the skeletal muscles and the dermal and subcutaneous structures of the trunk develop from the dermatomyotomes. Each myotome receives a spinal nerve. The myotomes are at first arranged metamerically along each side of the spinal column. Later these segments grow dorsally to come into relation with the vertebrae, and ventrally around the body to meet in the midline. Except in the intercostal muscles and a few others, this pattern of muscle segments is soon lost. Individual skeletal muscles are formed from the myotomes by a process of fusion, by splitting in either the vertical or the tangential plane, and by reorientation of the direction of the muscle fibers. In the limbs the muscles are formed by fusion of two or more myotomes. Once the muscle is formed, its relations to the embryo may be altered; it may be passively shifted by other structures. The plurisegmental nerve supply of each muscle then remains as the only evidence of its multiple metameric derivation.

The muscles of the head and neck differ from those of the trunk in that they originate from the mesenchyme of the five branchial arches of the pharynx and therefore belong to the splanchnic rather than the somatic mesoderm. Their histologic structure, however, is

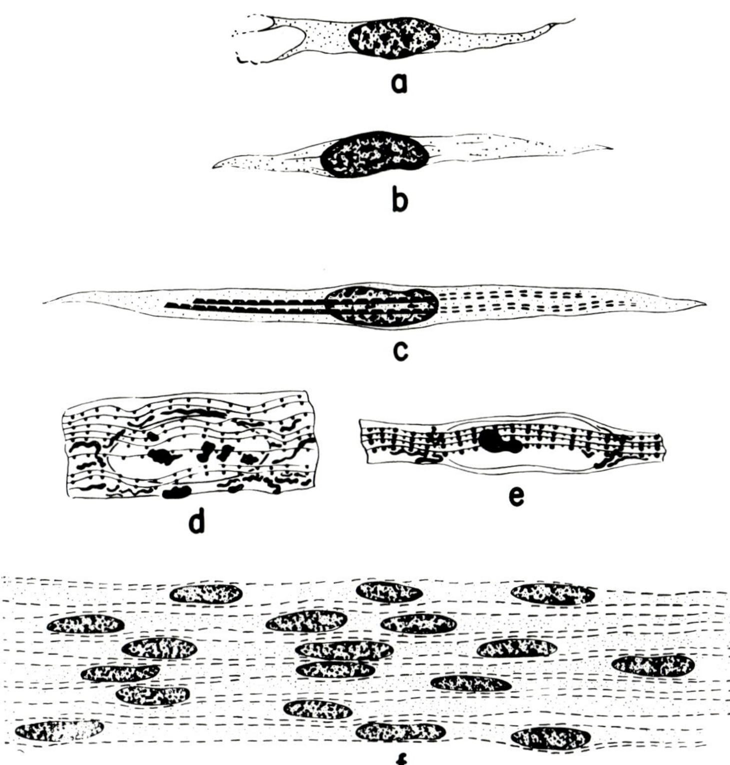

Fig. 1–1. Development of striped muscle fibers. (A) Myoblast with granular cytoplasm from a 13-month-old sheep embryo. (B) Myoblast from 1-mm guinea pig embryo showing homogeneous fibrils. (C) Myoblast with segmented myofibrils. (D) Myoblast with Z dots appearing on the myofibrils and chondriosomes near the nucleus.[56] (E) Myoblast showing the Q dots, as well as Z dots.[56] (F) Cylindrical muscle fibers with multiple nuclei and peripheral myofibrils. (After Godlewski[73])

the same. The pharynx and the upper part of the esophagus are the only parts of the alimentary canal in which striated muscle develops.

The most complete and authoritative descriptions of the histogenesis of the muscle fibers, of their tendons and investing sheaths, and of the motor end-plates and sensory organs may be found in the publications of Bardeen,[6] Lewis,[119, 120] Tello,[179, 180] Cuajunco,[43, 44] Stockdale and Holtzer,[176] Carlson,[31] Fischman,[66] Kelly,[109] Kelly and Zacks,[108] Hay,[82] and Boyd.[21] From these it is learned that during the first weeks of embryonic life the cell masses destined to form the myotomes consist of closely spaced undifferentiated cells of fusiform shape. As development proceeds, two types of cells can be distinguished: one acquiring the typical morphology of primitive branching connective tissue cells and the other having the more heavily granulated nuclei of the primitive muscle cells. The latter cells, to which Godlewski[73] gave the name myoblast (Fig. 1–1), multiply at first by mitotic division. Later they elongate and fuse to become multinucleated myocytes.[39]

The cell membrane of the embryonic muscle cell is extremely delicate, so that for a long time it was difficult to determine whether each multinucleated muscle cell is derived from a single myoblast (as proposed by Prevost and Lebert,[149] Remak,[153] and Duesberg[56]) or from the fusion of several myoblasts (as postulated by Valentin,[184] Schwann,[167] and Godlewski[73]). The former view has been largely abandoned in favor of the fusion theory. It is generally agreed that once the myoblast has formed myofilaments and fuses with other cells, becoming a myocyte, nuclear division then ceases as in all highly differentiated cells.*

* Nomenclature in the recent literature follows that of older writings, which define the *myoblast* as an undifferentiated, elongated, mononuclear cell with potentiality for development into a muscle fiber. Once there is indication of myofilament and myofibril formation, the cell becomes a *myocyte*. Fused myocytes with central nuclei and peripheral myofibrils are designated as a myotube. Some writers disagree on these points, as can be seen in Figure 1–1,[56, 73] and apply the term myoblast to all mononucleated muscle cells, regardless of whether myofibrils are visible in their cytoplasm.

SEVENTH TO NINTH WEEKS

During this period the syncytium of myocytes elongate to form primitive myotubes. The latter are multinucleated cells with a row of centrally placed ellipsoidal nuclei set in a seemingly hollow tube with only a thin border of cytoplasm. By the tenth week large protofibrils (sarcostyles*) appear in the peripheral part of the muscle fiber.

These steps of cell fusion and myofibril formation have been followed in electron microscopic preparations by Van Breeman[185] and others. The process of fusion was studied in the intercostal muscle of embryonic rat, but it was difficult to visualize because of the fragility of the cytoplasmic membranes. Yet there is no doubt that the cells do align themselves and soon thereafter are observed in a linear fused state, confirming the theory of embryogenesis proposed by Schwann[167] in 1839. Cell recognition seems to be afforded by specific phospholipids in the cell membranes of myoblasts. Further, myotubes, each surrounded by a basement membrane, continue to form for some time, the newest ones always being smaller than the older ones. The latter seem to enlarge by incorporating other undifferentiated myoblasts. These variations in size give the cross sections of fetal muscle a checkerboard pattern. New muscle fibers are still forming in rats 5 days after birth and probably longer. Kelly and Zacks[108] were unable to find evidence of a numerical increase of muscle fibers by longitudinal splitting, as had been postulated earlier from light microscopic studies by Pogogeff and Murray.[149] Presumably all these cellular events are duplicated in human myogenesis.

The electron microscope has yielded interesting information concerning the beginning of myofilament formation and the steps in the genesis of the myofibril. Holtzer,[92] Stockdale and Holtzer,[176] and Allen[1] have seen (in chick embryos) filaments in mononucleated myo-

blasts and proved their myogenic nature by the fact that they conjugate with antimyosin fluorescent antibodies. However, the synthesis of contractile proteins was clearly more vigorous in the multinucleated myotubes at later stages of development.

The actin filaments seem to form first and aggregate in small fascicles. Soon thereafter single thick myosin filaments are formed from polyribosomes, though there is some uncertainty on this point because of species variations. The first myofibrils are nonstriated. The integration of actin and myosin filaments into striated myofibrils appears initially to take place beneath the sarcolemma. Transverse tubules, presumably the first sign of the T system, then appear at intervals of 1–5 μ.* Myofibril formation next involves fusion of Z bodies to which actin filaments are attached, an alignment that depends on the sarcoplasmic reticulum and T system. The process of myofibril increase shown originally to be by longitudinal fission, a process illustrated by Heidenhain[86] in the trout embryo, seems to be in doubt. Godlewski[73] described this as arising through linear aggregation and fusion of randomly arranged granules (probably polyribosomes) and leading to continuous synthesis of actin and myosin filaments. There has been no evidence that preformed myofibrils[109] split. In mammalian muscle the striation of the myofibrils also first becomes visible just beneath the sarcolemma, though there is some indication that new myofibrils continue to be formed in the central part of the myotubes, presumably from the polyribosomes. At first the more central fibrils stain very faintly (Fig. 1–2). Sarcoplasmic reticulum and the transverse tubular system develop after the myofilaments are organized as nonstriated myofibrils. The muscle nuclei, which remain centrally located in some lower vertebrates,[164] migrate later to the periphery of the fiber in mammalian

---

* The term sarcostyle is used to designate either a group of fibrils or an individual fibril within a muscle fiber. It is now obsolete, and most anatomists prefer the term myofibril.

* Part of the difficulty in interpretation is that many types of developing cells, chondroblasts, fibroblasts, etc., contain scattered microtubules of about 100 Å width. According to Kelly,[109] these have no relationship to the formation of the thin (actin) and thick (myosin) filaments.

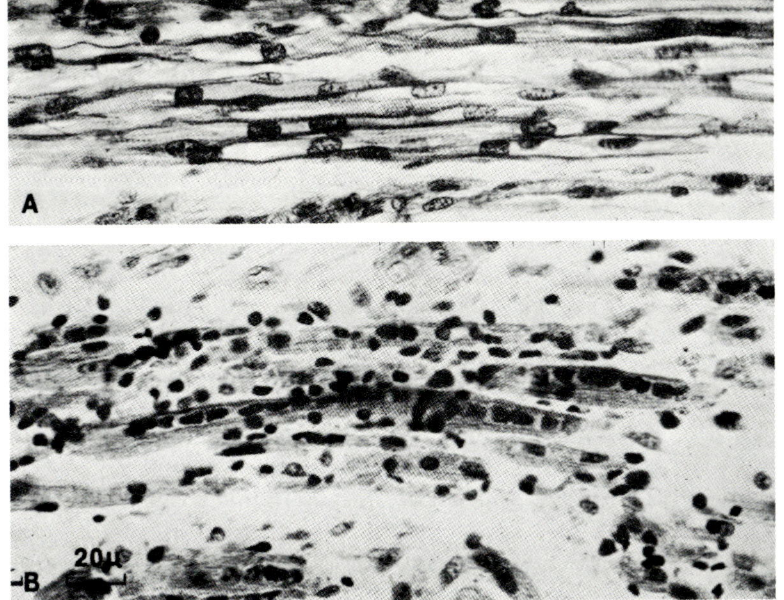

Fig. 1–2. Longitudinal section of leg muscle from a human embryo 10–11 weeks old. (A) Myotubes with central nuclei and peripheral striated myofibrils. (B) Pyknosis of sarcolemmal nuclei as an example of "retrogressive metamorphosis."[6] (phloxine, methylene blue)

species and assume a position just beneath the investing sarcolemmal membrane. As the muscle fiber becomes more densely packed with myofibrils, the nuclei become flattened against the cell membrane.

In the myotube phase degeneration of some of the muscle fibers, to make way for the connective tissue sheaths, was described by Bardeen[6] in the pig and by several workers in the tadpole. Bardeen, who refers to this process as a retrograde metamorphosis, points out that some of the larger muscle fibers swell and that their nuclei become pyknotic and irregular in outline, eventually coming to lie with their long axes transverse to the long axis of the fiber. The sarcoplasm acquires an eosinophilic quality, and finally the fiber breaks into irregular masses of protoplasm. Asai[2] and Tello[180] deny that this "physiologic degeneration" is an obligatory phase in the development of vertebrate muscle, and Cuajunco[44] was unable to detect it in his series of human fetuses. We found retrogressive change[44] of the type that had been illustrated by Bardeen in pigs in an 11-week-old human fetus (Fig. 1–2). However, it was impossible to exclude an agonal process or a reaction to some intervening disease. This problem requires further study. In

none of the more recent electron microscopic studies of muscle embryogenesis has degeneration been mentioned.

TENTH TO FIFTEENTH WEEKS

Cuajunco[44] presented one of the most useful descriptions of the development of human muscle from the tenth week of intrauterine life to term. The individual skeletal muscles, identifiable as early as the seventh week, are small but well formed by the tenth week. However, the histologic structure retains its embryonic appearance insofar as the muscle cells are still elongated, tube-like structures (myotubes) with centrally placed nuclei. The true sarcolemma is invisible, and thin myofibrils can be discerned just beneath the outer limits of the cell. Transverse striations are faintly visible in the neck, trunk, and shoulder muscles but not in the distal muscles of the arms or legs.

In a human embryo of 11 weeks the diameter of the muscle fibers has increased, the myofibrils are more abundant and fill the cells more completely as in mature muscle, and the transverse striations have become more evident in the proximal arm and trunk muscles. The undifferentiated central core of the fiber is therefore smaller, and the central nuclei appear to be crowded and compressed by the newly formed myofibrils. At this stage the distal muscles in the hand are not more developed than were the proximal muscles of the arm at 10 weeks. Similarly in the 12-week embryo the muscle fibers are still thicker, and myofibrils are more abundant and more clearly striated. Undifferentiated connective tissue cells, with more darkly staining protoplasm and nuclei, are aggregated around blood vessels and along nerve fibers on the surfaces of the individual muscle fibers. These cells later participate in the development of the motor plate.

During the 13th week the myofibrils are thicker and more darkly stained, and the anisotropic and Z disks stand out clearly. The muscle nuclei at this stage begin to change position, moving from the central clear zone of the fiber toward one side, and the myofibrils become compressed into compact columns. In humans as in animals the formation of new myotubes continues at this stage, imparting the pattern of thick and thin muscle fibers in random array. During the 14th week the myofibrils continue to increase in number and size and are often grouped into distinct bundles (Kölliker's columns or sarcostyles), which on cross section may give the appearance of Cohnheim's fields. Striations have now become prominent.

Two sizes of muscle fibers are seen during the 15th week. The larger ones have very little clear sarcoplasm in the center; the striated myofibrils are prominent and the nuclei peripherally placed. The small fibers, scattered among the larger ones, possess few myofibrils, and the sarcolemmal nuclei occupy a central position. At 16 weeks the muscle nuclei are more numerous and elongated, and all have taken a position beneath the sarcolemma.

SIXTEENTH WEEK TO BIRTH

From the 16th week until birth the muscle fibers enlarge, both in length and width, and the striations become better developed. No other histologic changes occur. Attempts to measure the cross-sectional diameter of muscle fibers at different ages of fetal life have not yielded uniform results. Cuajunco's measurements of the width of fibers (formalin-fixed) showed a gradual change of 0.009–0.016 mm for the largest fibers, and 0.005–0.007 mm for the smallest, from the 10th to the 36th week. Very small fibers (0.002 mm) were seen in the 15-week embryo by Cuajunco and appeared to have been recently formed. Such very small fibers were not seen after the 24th week. The largest fibers remain larger and more fully developed than the small fibers throughout fetal life[180] and can still be recognized by their size and distribution during early infancy (Fig. 1–3). These are the "B fibers" of Wohlfart.[192] Schäfer[162] estimated the enlargement of the muscle fibers from midfetal life until birth to be 200% and in some instances 300–400%; between birth and adult life there is a fivefold increase fiber size (Figs. 1–4 through 1–7).

It has been assumed then that the enlargement of muscles during the second half of intrauterine life is due to an increase in the size of the individual fibers and not to an increase in number. There is still some uncertainty as to whether new fibers continue to form during late fetal life and after birth. Morpurgo[136] found in muscles of the white rat that new fibers continued to form until after birth, and Tello[180] stated that newly forming fibers can be seen in the human fetus at term. MacCallum[124] observed in the human sartorius muscle a 20-fold increase in number between the third or fourth fetal month and term. Cuajunco[43, 44] was uncertain of recent formation of fibers beyond fetal life, but deReuck and Adams[50] noted an increase of 100% from birth to adulthood.

The mechanism of numerical fiber increase —long a subject of controversy[43, 44, 73]—seems now to have been settled. Couteaux[41] showed that an augmentation in the number of fibers is always due to recruitment of still undifferentiated interstitial cells. As the latter multiply by mitotic division they continue to fuse, forming new fibers, until the full adult complement is attained. This process was corroborated in embryonal rat muscle by Kelly and Zacks.[108] MacConnachie *et al.*[125] demonstrated that as rats age there is a steady increase of recently divided nuclei within muscle fibers. Using colchicine, they found 1–2% of the muscle nuclei to have divided mitotically and capable of becoming labeled with radioactive thymidine.

Whatever the mode of development may be, the picture of human muscle during fetal life corresponds closely to the descriptions given by Wohlfart.[192] As shown in Figure 1–3A, there are two populations of muscle fibers in the neonate. In each primary bundle scattered thick B fibers are surrounded by large numbers of thinner A fibers. This appearance can be discerned as early as the fourth fetal month and persists in the sartorius muscle well into the second year of postnatal life.

Our own attempt at enumeration and fiber measurement, the details of which are presented below, raise questions about the accuracy of some of the above statements. Our figures on the fiber sizes of ten muscles of thigh and leg in three human fetuses correspond to those of Cuajunco.[44] There is a steady growth of the fibers in transverse as well as longitudinal dimensions throughout childhood and adolescence (Figs. 1–4 through 1–7). However, the numbers in the superior rectus (of the eye), sternomastoid, biceps brachii, sartorius, and gastrocnemius also increased slightly throughout childhood. Moreover, there appears to be a fairly wide variation in the final number of fibers and in the size of muscle from one individual to another.

In sum, the formation of skeletal muscle may be subdivided into five overlapping stages, as follows: (1) conversion of undifferentiated mesenchymal cells into mononucleated myoblasts; (2) fusion of myoblasts into multinucleated syncytial myocytes and myotubes containing thin and thick myofibrils; (3) continuing synthesis of myofilaments and integration into myofibrils, and the formation of sarcoplasmic reticulum and the transverse tubular system; (4) growth of myotubes by assimilation of new myoblasts and formation of additional myofibrils; and (5) innervation and functional organization of contractile activity. The details of each stage in human embryonal muscle are not known; much of the above information comes from chicken and rat muscle.

CONNECTIVE TISSUE

During the first weeks of embryonic life it is difficult to distinguish between myoblasts, connective tissue cells, capillary endothelial cells, and other primitive mesenchymal elements. The connective tissue, which acquires recognizable form at about the time the myoblasts are converted to muscle cells, becomes arranged within the muscles so as to divide them into primary and secondary groups of fibers. These muscle fasciculi share a common tendon of origin and insertion. The endomysial connective tissue extends between the individual muscle fibers late in fetal life. The amount of connective tissue between the muscle fibers increases with age. The larger blood vessels are contained in the connective tissue septa. In addition to fibroblasts there are large aggregates of mononuclear cells around the blood vessels in the perimysium. Some of these cells can be identified as histiocytes; others, representing foci of erythropoiesis, are nucleated red cells, erythroblasts, myelocytes, and stem cells. These persist in scattered foci throughout fetal life (Fig. 1–8). Apparently the presence of hematopoietic centers in the muscular tissue in infants has not been well recognized. Plum,[146] in his review of extramedullary blood production, mentioned such sites as the prostate gland and the sole of the foot, but he did not include the muscle foci. Whitby and Britton[191] remarked that such foci may occur, especially in the psoas muscle, but gave no further details. Our observations seem to indicate that these foci are more likely present in the distal muscles of the extremities and that a few are also found in the proximal limb and trunk musculature.

The mode of formation of the tendon fibrils at the end of the muscle fiber is not settled as yet. In embryonic development the long processes of the muscle fibers end between connective tissue cells. These terminal processes of the muscle cells are at first slender and without nuclei, but later they enlarge and become more cellular than the rest of the fiber. The development of the tendon fibril is believed to originate from specialized connective tissue

Fig. 1–3. (A) Transverse section of normal thigh muscle from a 7-month-old fetus showing A and B fibers. (B) Longitudinal section of same. (A, phosphotungstic hematoxylin; B, phloxine, methylene blue)

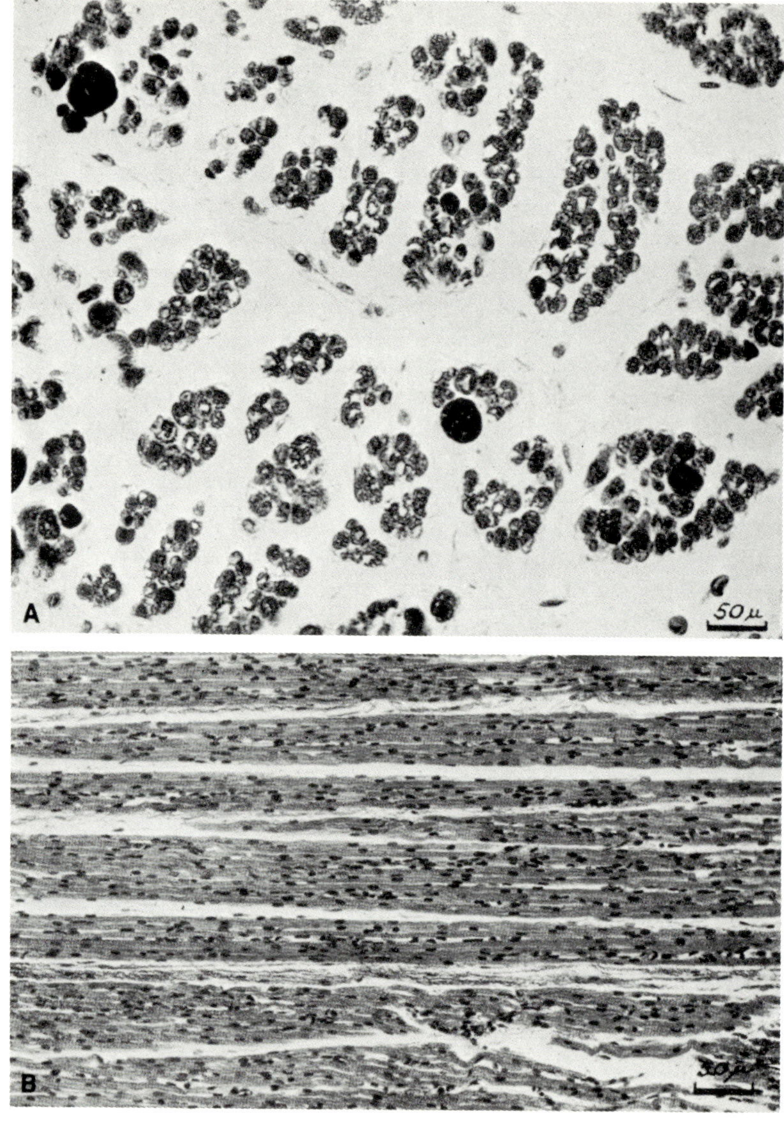

Fig. 1–4. Transverse diameters of gastrocnemius muscle from birth to old age. Each point is the average of 500 fibers in a specimen taken from near the equator of the muscle. No difference noted between male and female. (celloidin-embedded, H&E)

Fig. 1–5. Histograms of gastrocnemius muscle showing frequency distribution of narrow fiber diameters throughout life. Bimodal and trimodal curves are observed from adolescence through adult years.

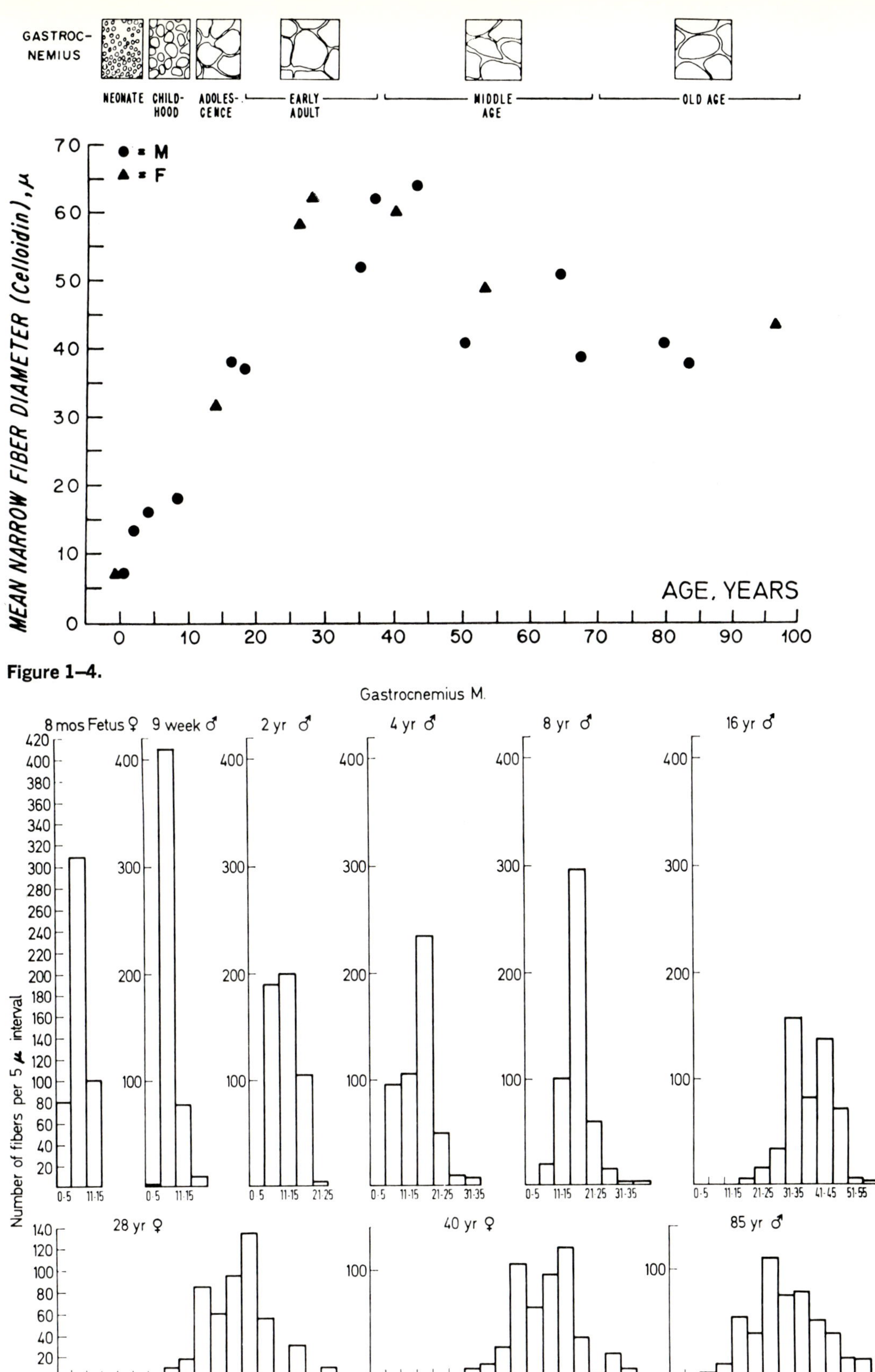

Figure 1–4.

Figure 1–5.

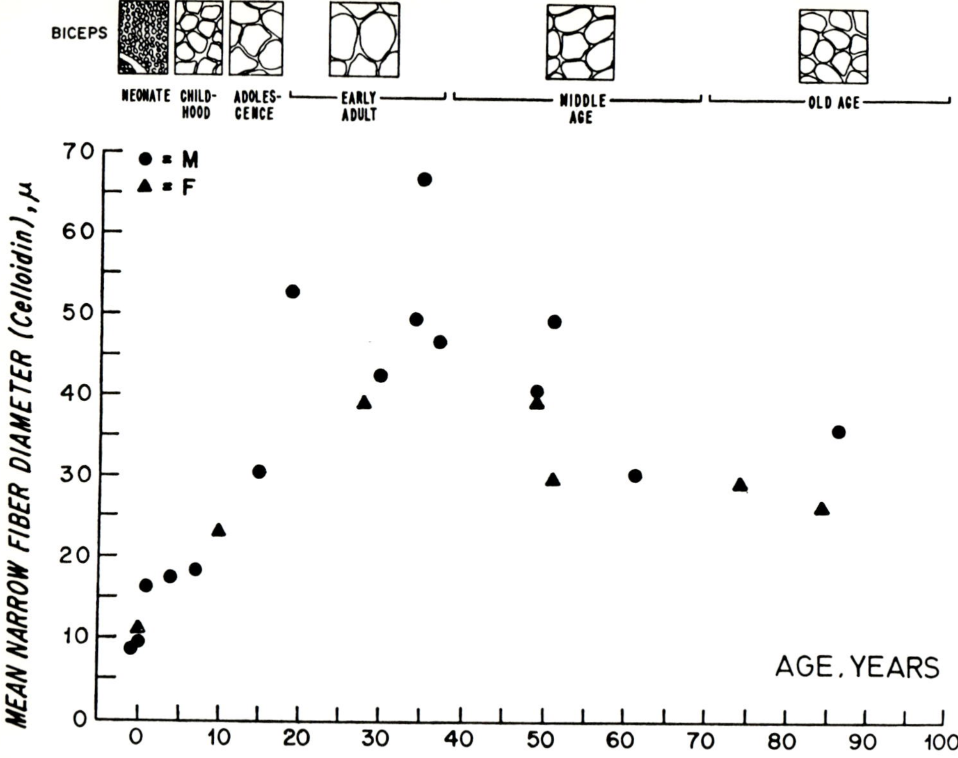

Fig. 1–6. Transverse diameters of biceps muscle from birth to old age. Each point is the average of 500 fibers in a section taken from the equator of the muscle. Note that the figures for adolescent and young adult males exceed those for females of comparable age. (celloidin-embedded, H&E)

Fig. 1–7. Transverse diameters of fibers of superior rectus muscle of the eye from birth to old age. No essential difference in size between male and female.

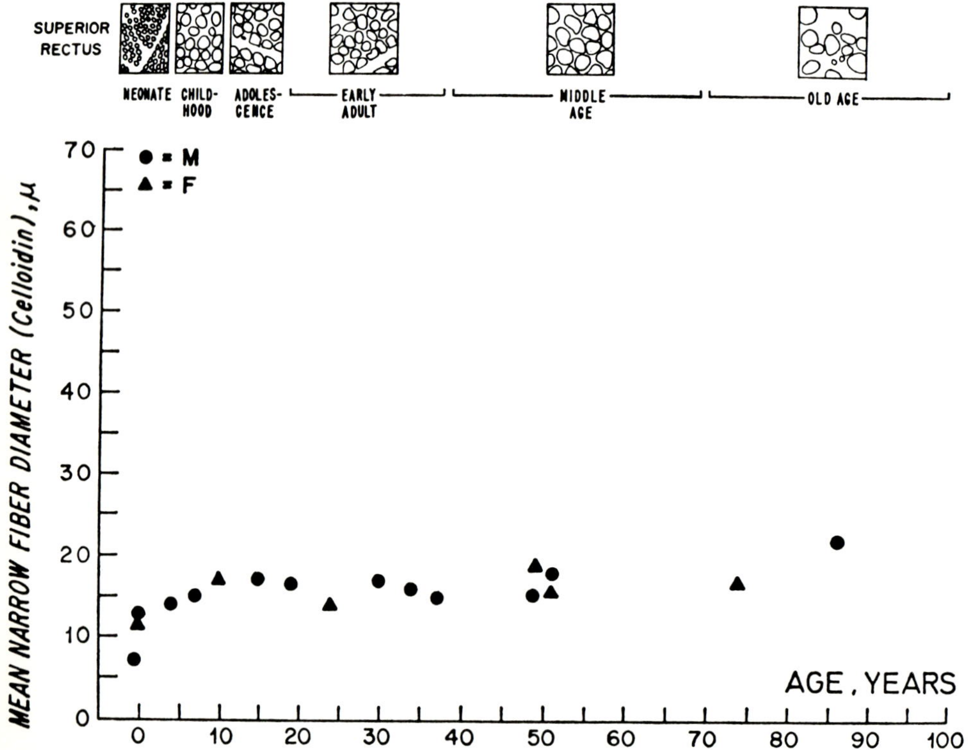

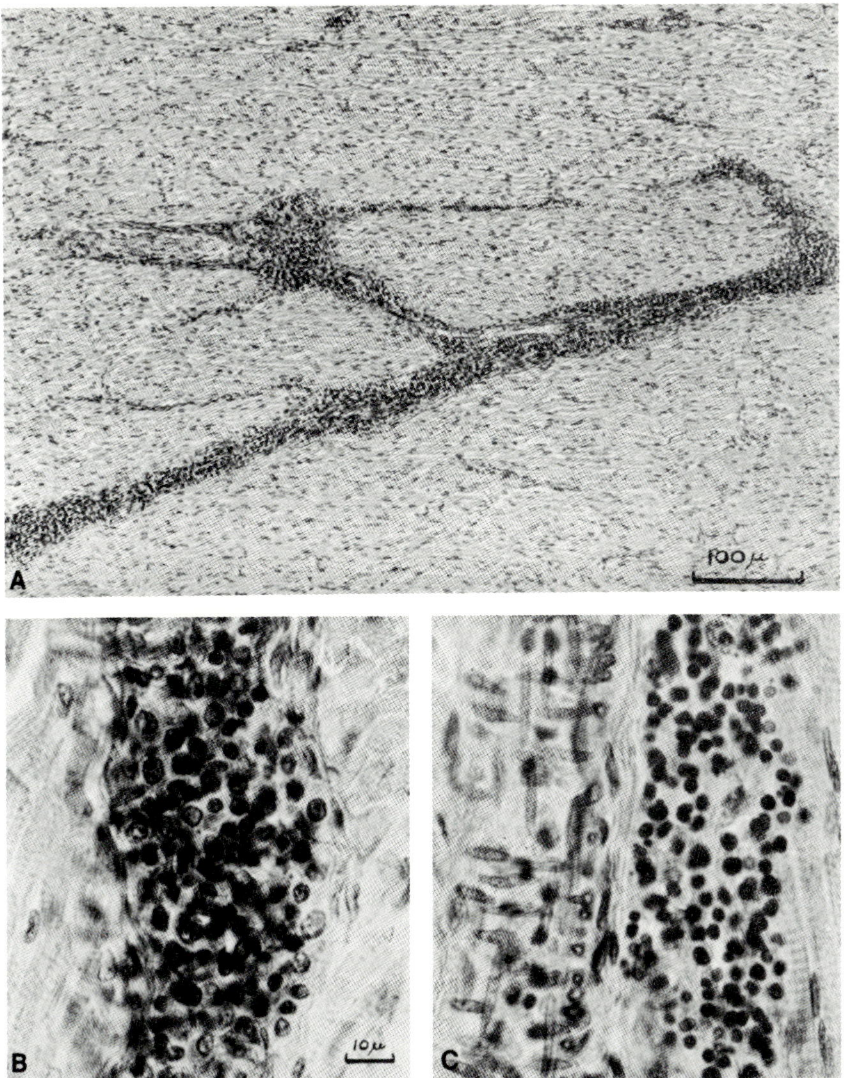

Fig. 1–8. (A) Hematopoiesis in leg muscle of normal term human fetus. (B) High-power view of one of the cellular aggregates, which is composed of nucleated red cells and myelocytes. (C) Small artery (left) and a collection of normoblasts. (H&E)

nuclei, which accumulate around the end of the muscle fiber.

NEUROLOGIC DEVELOPMENT

The embryogenesis of the motor end-plate and of the sensory organs of skeletal muscle was studied in great detail by Tello and Cuajunco.

Nerve fibers can be observed in the connective tissue during the 10th week, and by the 11th week the terminal twigs of the nerves come into contact with the muscle fibers. The growth of the nerve fibers is closely followed by migration of connective tissue cells, some of which belong to the sheath of Schwann and others probably to the specialized endoneural sheath of Henle. The end of the nerve fibril branches to form the telodendria, which are closely applied to the surface of the muscle fiber. Connective tissue cells cover the telodendria. At 13 weeks the protoplasm of this group of cells fuses with a group of sarcolemmal nuclei to form an elevation known as Doyère's eminence. From the 14th to the 24th week of fetal life these motor nerve endings are reorganized. The network of telodendria becomes more complex, but later many of these terminal nerve fibrils disappear. The telodendria develop terminal enlargements and are shaped into multiform figures during the period of fetal life extending from the 24th to the 36th week.

As nerve contacts muscle the diffusely localized cholinesterase in the sarcolemma of the embryonic fiber becomes concentrated at the neuromuscular junction. It is maintained at this point either by the continuous release of acetylcholine transmitter substance or by some peptide specific to nerve tissue.

Tello[180] provided beautiful illustrations of the embryogenesis of the muscle spindle in the chicken, and Cuajunco[43] traced its development in the human fetus. The earliest stage of formation of the muscle spindle is the 11th week of intrauterine life, when it can be recognized only by the characteristic spiral nerve endings around one or several myoblasts. By the 12th week the enclosed muscle fibers develop numerous nuclei, which produce an enlargement of the fiber under the region of the ending. The capsule, which originates from connective tissue cells, begins to form by the 12th week and rapidly surrounds the segment of the bundle of muscle fibers that receives the nerve endings. A slit-like lymph space between the capsule and the axial bundle of muscle fibers appears at the 14th week. At this time the larger nerve fibers extending through the capsule begin to myelinate; by the 24th week many are thickly myelinated. Motor fibers form a motor end-plate on each muscle fiber of the spindle at the same time that the remaining muscle fibers become innervated. By the 14th week of fetal life the muscular spindle has acquired all its essential components, and by the 24th to 31st week it has attained its mature form. The only change thereafter is an increase in its dimensions.

The formation of the tendon organs, the Pacinian corpuscles, and the free nerve endings in muscle has been followed in chick embryos, cats, and other animals by Tello,[180] but no special studies have been made in the human fetus. The modes and times of origin and innervation are similar to those of the muscular spindle.

Although the innervation of muscle takes place during early fetal life, it plays no part in the morphogenesis of muscle. This was first demonstrated by Harrison,[81] who resected the medullary tube and the neural crest of the tadpole before either nerves or muscles had differentiated. It was possible to show that the muscle fibers evolved in the usual manner in the complete absence of any formative stimuli from the nervous system. Also Bardeen[6] pointed out that in the pig embryo muscle differentiates to a considerable degree before nerve connections are established. Later, however, muscle becomes dependent on its nerve supply. In exogastrulae, in which muscle fibers develop but are not innervated, they soon disappear.[98] The fate of muscle fibers in mammals, if innervation does not occur, is unknown. Under certain circumstances, as in teratomas, striated muscle fibers survive as recognizable structures without apparent innervation for long periods. It is remarkable, moreover, that muscles which by an error in development (such as absence of the fibula) take origin from and are inserted into the same bone may still retain mature structure in later life.[172]

Muscular development and movement are orderly and predictable. The muscles of the neck and trunk are the first to develop, followed by the lingual, facial, and proximal and distal appendicular musculature. This developmental progression from cephalic to caudal, and in the limbs from proximal to distal, is manifested by the size of the muscle fibers, the maturity of innervation at different phases of fetal life, and in the organization of behavior. As Gesell[70] points out, "the fetus tends to react with the trunk and shoulder before the hands and fingers are activated. He retracts his head before he retracts and purses his lips." Close correlations between the development of muscle and its motor and sensory innervation with fetal movement have been established. Coghill[36] cited isolated observations of movement of an arm in fetuses less than 10 weeks old (crown–rump length less than 38.0 mm). Inasmuch as muscle masses are well formed but have not been innervated at this age, it must be assumed, if these observations are correct, that muscle fibers can contract prior to receipt of the motor or sensory nerve fibers. However, this seems unlikely, and further observations are needed to verify this point. The earliest reflex movements that could be evoked by Bolaffio and Artom[19] were in human embryos of 12 weeks (crown–rump length 55 mm), which coincides with the development of innervation. Strong fetal movements, which can first be perceived by the parturient woman during the fourth month of pregnancy, occur when organization of the neuromuscular apparatus is more mature.

A more recent reference on this subject is that of Dubowitz,[52, 54] who compared the histologic and histochemical status of muscle (in man and in several animal species) before and after it received its nerve supply during fetal life.

ANATOMY AND HISTOLOGY

The muscular flesh constitutes an important body tissue. In the human there are no less than 434 muscles,[130] these make up an estimated 25% of the total body weight at birth and 40–45% in adult man. The skeletal muscles are composed of a large number of muscle fibers. According to Elftmann's[58] calculations, there are about 250 million individual striated muscle fibers in the body. Each muscle fiber is a large multinucleated cell which varies in width and length from one muscle to another. Muscles vary in fiber size according to the degree to which they have been exercised and the age of the subject. In the normal adult the ocular muscles average $17.5–20.0\mu$ in diameter, and it is exceptional to find one above 25.0μ. In the frontalis and temporalis muscles it is unusual to see fibers above 80μ, but as pointed out by Halban,[78] who collected statisti-

cal data on the diameters of fibers of different muscles, groups of fibers of similar size ranging $10–60 \mu$ may be found. For the gluteus maximus Halban gives an average figure of 87.5μ in the normal adult, and for the gastrocnemius, tibialis anticus, and vastus lateralis 62.5μ but in athletic men we have found the fibers to be much larger, often exceeding 100μ. The largest and smallest fiber sizes deviate equally far from the average. In 10 subjects without neuromuscular disease Greenfield *et al.*[75] observed single fibers of the vastus lateralis muscle as small as $10–15 \mu$ and large ones as much as 100μ. The single small fibers could be explained by the fact that the plane of section passed through the tapering ends of fibers. Groups of fibers below 40μ were found only in patients with neuromuscular disease.

Comparative data on fiber sizes in infants and children were presented by Wohlfart,[192] Banker *et al.*[5] and Moore *et al.*[135] In the sartorius muscle of the newborn Wohlfart found the A fibers to have an average diameter of about 12μ and the B fibers 20μ. Banker *et al.* using formalin-fixed material measured the fibers in the gluteus maximus, hamstring, and anterior tibial muscles and obtained a figure of $9.5–11.5 \mu$ for A fibers and 20μ for B fibers. At 3 years of age the averages of both had risen; the range of A fibers was $5–25 \mu$, and that of B fibers $20–30 \mu$; by the 11th year the A fibers were $10–30 \mu$ and the B fibers $30–50 \mu$. Beyond 15 years of age the distinction between A and B fibers could not be made, and the limits of variation of the fibers at 20 years of age were 20 and 70μ, respectively.

Our own data on the size of muscle fibers at different ages are recorded in Figures 1–4 through 1–7. The growth characteristics—as ascertained in autopsy material from individuals known to have no primary disease of nerve or muscle nor to suffer from prolonged cachexia and inactivity—follow certain regular patterns. A steady but lessening increase in size occurs during the maturing years of childhood and adolescence, generally reaching a peak in early adult life. Size remains relatively constant throughout adult life only to decline slightly in old age. During the latter period there is another notable feature: a loss of muscle fibers

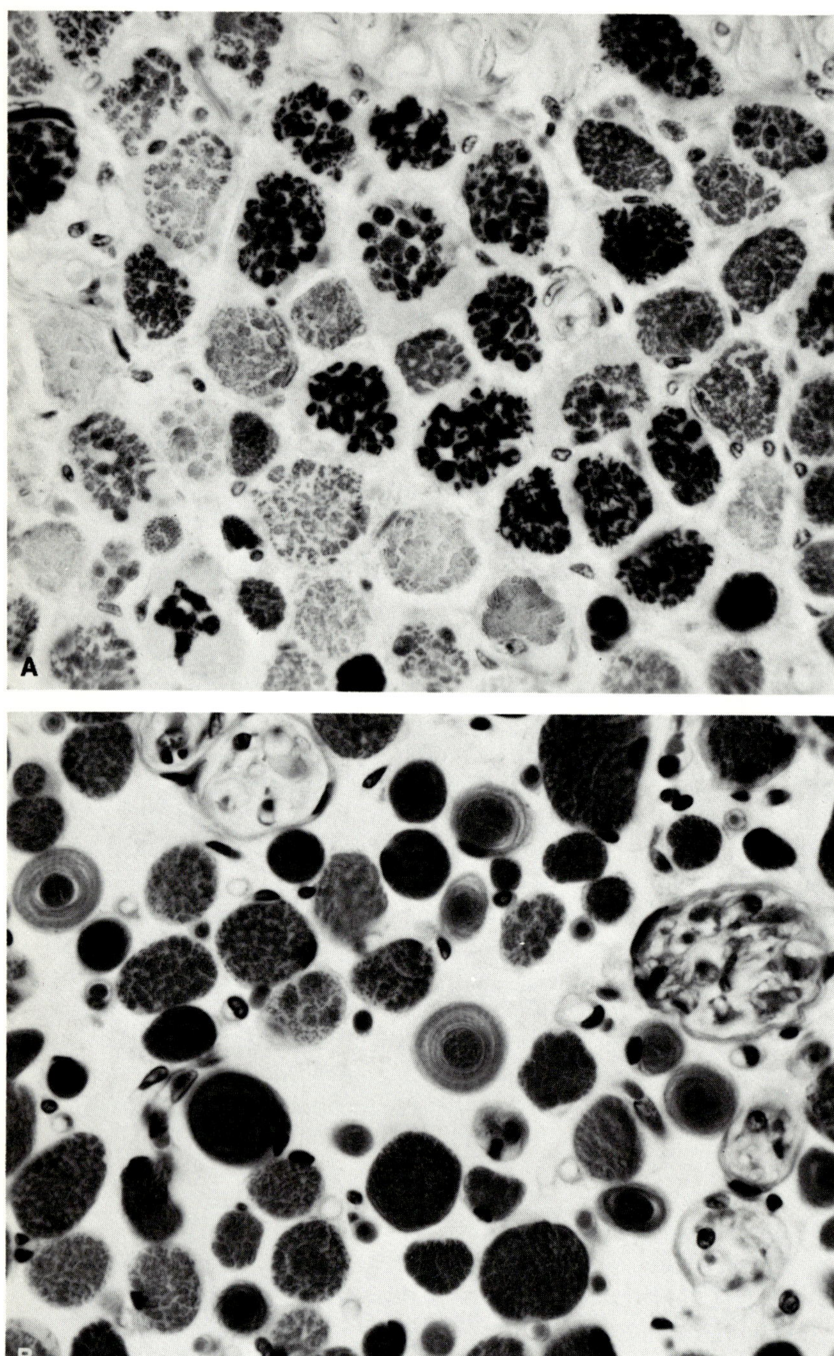

Fig. 1–9. Transverse section of superior rectus muscle of the eye. (A) This is from a 30-year-old male. Variations in intensity of staining are artifactual, but two fiber types are seen, one with heavily clustered and the other with finely dispersed myofibrils. Note relative uniformity of size. (B) This is from a 75-year-old male. Note thinning out of fibers, greater variation in size (some larger, others smaller than average); four ringbinden are seen, and several nerve twigs are present. (PTH stain, ×700)

and group atrophy signifying denervation; and another population of residual muscle fibers undergoes hypertrophy (Fig. 1–9).*

The length of fibers range from a few to over 30 mm. Huber[94] showed that within a given fasciculus there is considerable variation in the length of the fibers. For example, in a fasciculus from an adductor muscle of the thigh which measured 40 mm the fibers ranged from 9 to 30.4 mm. Lockhart and Brandt[122] isolated a fiber 34 cm long from a human sartorius 52 cm in length. In the eye muscles the fibers extend the entire length of the muscle but in most trunk and limb muscles they are joined end to end by endomysial connective tissue. As a rule the fibers do not branch or anastomose, though in the facial and lingual muscles they may divide before being inserted into skin or mucous membrane. At the tendinous end of a muscle the sarcolemma merges into a tendinous fibril which joins with the fibrils of other muscle fibers to form the tendon aponeurosis, i.e., a tendon is formed by the fusion of aponeuroses.

Certain muscles such as those of the face, tongue, larynx and eye show striking differences in their anatomy. The eye muscle fibers, for example, are not only smaller but more variable in size and more rounded in transverse sections. Many have central nuclei and the intramuscular connective tissue is more abundant. Also the fibers in the center are larger than those in the periphery. The packets of myofibrils are unusually coarse (see Fig. 1–9) and are separated by zones of clear sarcoplasm rich in mitochondria. These and certain features of other cranial muscles would be judged pathologic by the accepted standards for limb muscles.

The muscle fibers are arranged in bundles of variable thickness called fasciculi. Within a given fasciculus the fibers are parallel. The

fasciculi seldom extend from one end of the muscle to the other. In some muscles, e.g., the rectus abdominis and digastricus, the fleshy parts are interrupted by transverse tendinous bands into which the muscle fasciculi insert. The fasciculi also run parallel, though at the ends of the muscle they converge toward their tendons. Each fasciculus forms a prismatic figure and on transverse section has an angular shape (Figs. 1–10 and 1–11). The number of fibers in a fasciculus varies greatly in the same muscle, and some muscles have uniformly large fasciculi. The large ones may be subdivided into smaller bundles. The coarse or fine texture of the muscle depends on the size of the fasciculi. The arrangement of the fasciculi within the muscle varies considerably. If the tendon begins abruptly at the ends of the muscle, the fasciculi may extend the full length of the muscle; if the tendon advances into the substance of the muscle, the fasciculi insert obliquely into the sides of it. Thus a long muscle may be composed of relatively short fasciculi, or a short muscle of long ones.

In the fetus and newborn the sarcolemmal nuclei in the slender muscle fibers are relatively more numerous than in later life (Fig. 1–3B), resembling somewhat the fibers of an atrophied muscle. In infants the larger, more mature B fibers of Wohlfart[192] are often seen in the center of a group of the smaller A fibers (Fig. 1–3A). The A/B fiber ratio is about 6:1. In the adult this pattern is absent, for A and B fibers are no longer distinguishable. However, the B fibers may become visible once again during the course of muscle atrophy in the adult, according to Wohlfart.

STRUCTURAL DETAILS OF MATURE MUSCLE FIBER

Muscle fiber may be defined as a multinucleated cell that contains a large number of myofibrils embedded in a matrix of undifferentiated protoplasm, all enclosed within a fine sheath, the sarcolemma. The remarkable uniformity of structure, the ease with which the muscle fiber can be isolated, and its obvious and measure-

---

* Measurements are known to vary with type of fixative and embedding material. In comparing the unfixed frozen, formalin-fixed frozen, formalin-fixed celloidin, and formalin-fixed paraffin preparations we found the ratio of fiber sizes to be approximately 10:9:8:7, respectively. The white, or phosphorylase-rich, fast fibers are slightly (though not consistently) larger than the intermediate and also the red, oxidative and mitochondrial-rich, slow fibers.[60]

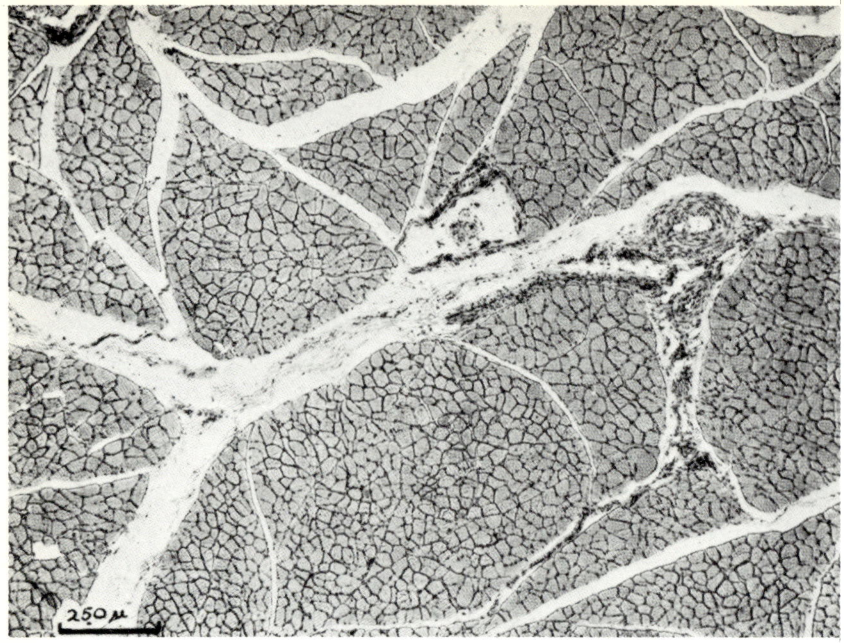

Fig. 1–10. Cross section of a normal human muscle showing arrangement of fibers. The fibers have a polygonal shape and fit together as in a mosaic. Note the spindle in the perimysial connective tissue above the center. (hematoxylin, Van Gieson)

Fig. 1–11. (A) Normal muscle in transverse section; higher magnification than in Figure 10. (B) Longitudinal section of normal muscle. (H&E)

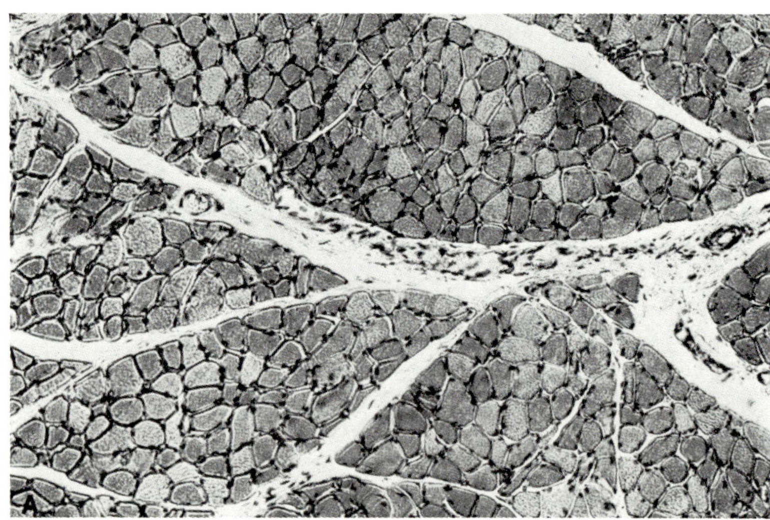

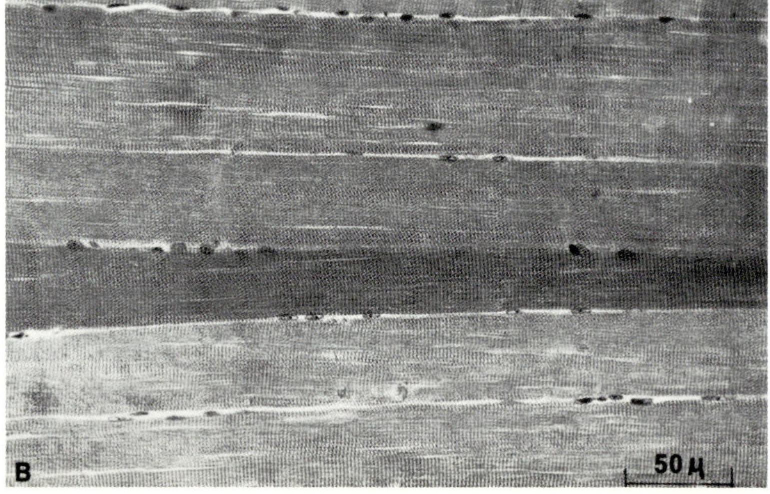

able contractile activity have made it especially suitable for investigations of cellular form and function.

SARCOLEMMA AND SARCOLEMMAL NUCLEI

The thin transparent membrane that envelops the muscle substance is called the sarcolemma. This is invisible in normal muscle fibers but can be seen clearly when a fiber is ruptured. Indeed, Bowman[20] demonstrated in 1840 that the simplest way of studying this membrane is by teasing apart the fibers in a fresh muscle preparation. This usually results in tearing some of the fibers; at the sites of injury the contents of the fiber retract to form a granular mass, known as the retraction cap. As these clots of sarcous substance separate, they leave an empty space within the sarcolemma and enable the latter to be clearly visualized. Another method is to induce intense contraction of the fiber by stimulating it with galvanic current or by treating it with caffeine or nicotine at the moment of fixation. The sarcolemma may then appear to be separated from the contracted muscle substance. When viewed under these circumstances in transmitted light, the sarcolemma appears to be a thin homogeneous membrane. By microdissection of the living muscle fiber, Kite[110] found the sarcolemma to be elastic, continuous, and resistant to the finest needle. It is adherent to the surface of the whole muscle cell and can be vitally stained by isamin blue[110] and janus green.[41]

When the sarcolemma is examined under the electron microscope with magnifications as high as $\times 30,000$[105, 144, 159] it appears to be a true membrane that has no resolvable structure. Jones and Barer observed within its substance numerous small dots 400–1000 Å in diameter arranged in a fairly regular pattern. Where the membrane was folded on itself so that it could be seen in profile, the dots were elevated above the surface of the fiber. When the fiber was treated with 3% citric acid the dots disappeared, but aside from this nothing more is known of their constitution. Jones and Barer suggested that they may represent the points of attachment of reticular fibers of the endomysium to the sarcolemma. Contrary to the contention

of Bairata[4] the structure of the sarcolemma itself is not fibrous, but Rozsa et al.[159] remarked that in their preparations numerous long threads were seen. These threads were not an integral part of the fiber but were attached to its surface and formed part of a system of fibrils surrounding the muscle fibers. These are part of the endomysium. The most recent studies with the electron microscope place the thickness of the sarcolemma at 70 Å (7 mμ).

In the studies of the sarcolemma by Bennett and Porter,[18] Bennett,[17] Ruska,[160] Robertson,[157] Porter and Palade,[148] and Fawcett and Selby,[62] higher resolution has shown two peaks of density each about 25 Å thick and separated by a band of lesser density of the same thickness. Robertson finds this trilaminate pattern characteristic of all cellular membranes and speaks of it as a unit membrane structure. Another feature is also observed—that the surface of the sarcolemma is marked by tunnel-like invaginations called caveolae intracellulares, associated with vesicles which lie close to the fiber surface. These caveolae and vesicles ramify inwardly and branch; they are about 600 Å in diameter. Their function is believed to be the transport of ions, sugars, and proteins into (and metabolites out of) the muscle fibers.

Another component of the surface membrane of the multinucleated muscle fiber is an extracellular plasma layer (external lamina or basement membrane, glycolemma) that lies external to the trilaminate membrane, or true sarcolemma (Fig. 1–12). This is a basement membrane similar to that which invests epithelial and Schwann cells. The coating may have a thickness that reaches 500 Å but is thinner in most places. Its periodic acid-Schiff (PAS) staining reaction indicates its rich polysaccharide content, reminiscent of the walls of plant cells. It appears essentially structureless even at high resolution under the electron microscope. Some of the basement substances become enclosed in caveolae and vesicles. Bennett[16] suggests that this membrane, which is rich in polysaccharides, may be capable of binding and concentrating glucose or other molecules close to the sarcolemma.

The sarcolemma must not be identical in structure at all points along its surface for there are differences in response to electrical

Fig. 1–12. Satellite cell. Note the combined basement and trilaminate membranes of the sarcolemma at either end of the cell. The basement membrane separates from the true sarcolemma to cover the external surface of the cell, and the true sarcolemma the internal surface. (Courtesy of Dr. Takeshi Sato, Institute for Neurosciences, Tokyo) (EM, ×13,600)

stimuli and pharmacologic agents. As shown by Huxley and Taylor[103] small foci near the Z band or at junctions of A and I bands (see below) are excitable to electrical stimuli which at these points are capable of inducing contraction of a single sarcomere (region between two Z bands). These authors found the region of the neuromuscular junction to be electrically inexcitable but responsive to acetylcholine, the latter substance having no effect on the rest of the fiber. In Robertson's[156] observations of myoneural junctions, the sarcolemma and basement membrane are thicker here than elsewhere. The sarcolemma according to Bennett[15] is also thicker and denser in the regions of tendinous insertion.

Beneath the true sarcolemma are the slender, flattened sarcolemmal nuclei of the muscle fiber (also called subsarcolemmal nuclei or muscle nuclei), which are oriented parallel to the long axis of the fiber (Fig. 1–13). These nuclei are invisible in the living muscle fiber. In fixed and stained sections where they are readily visualized they vary from 1 to 3 μ in width and from 5 to 12 μ in length, and each contains fine dust-like chromatin throughout the nucleo-plasm as well as one or more larger clumps that may consist either of true (Feulgen-negative) nucleoli or of chromatin (Feulgen-positive) particles (Figs. 1–11B and 1–14A). Each fiber has many such nuclei, numbering up to several thousand, all lying beneath the sarcolemma. In transverse sections 5–10 μ thick it is customary to see four to eight nuclei per fiber. Near the tendinous insertion where the fiber is of smaller diameter a few of these nuclei may be embedded between the columns of myofibrils. Otherwise, except during fetal life and in the course of muscle disease, these sarcolemmal or muscle nuclei are rarely seen in the central parts of the fiber. In this respect they differ from the nuclei of cardiac muscle. Greenfield *et al.*[75] estimated that not more than one centrally placed nucleus is seen in every 100 fibers and that more than two or three per 100 represents a pathologic change. The sarcolemmal nuclei are distributed fairly evenly along the muscle fibers.

These sarcolemmal nuclei have been the subject of a comprehensive study by Schiefferdecker.[163, 164] He pointed out that in mammals and birds the nuclei always occupy a peripheral

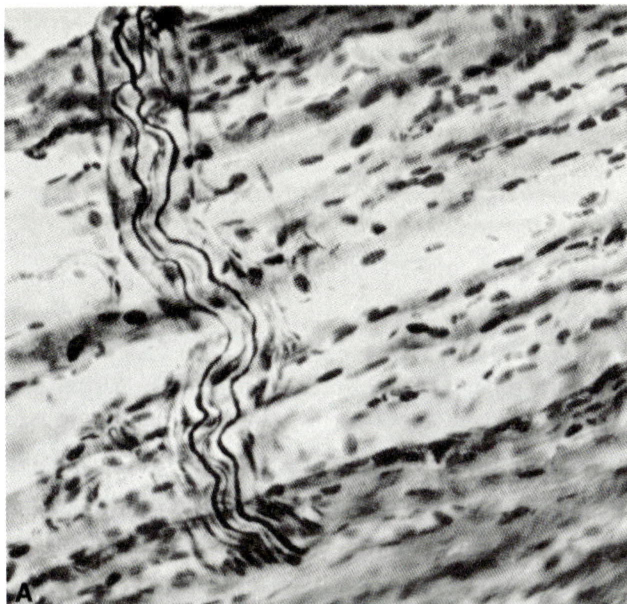

Fig. 1–13. (A) Human muscle. Henle sheath surrounding a small intramuscular nerve to a spindle and connective tissue nuclei are shown. Sarcolemmal nuclei are long and narrow. (B) Ocular muscle of cat showing argentophil network of the endomysium and some coarser elastic fibers of endomysium. Nuclei are not stained. (Bielschowsky silver method)

position, whereas in fish and amphibians they are distributed throughout the muscle fiber. He also made an elaborate attempt to establish a relationship between the function of the muscle fiber and the number and size of its nuclei, but his methods of calculation have been criticized by a number of workers in this field.

Electron microscopic study reveals two types of nuclei within the muscle fiber: one the typical muscle fiber nucleus already described, the other type lying beneath the basement membrane but external to the trilaminate sarcolemma. The latter cell, called the *Satellite cell* by Mauro,[129] comprises 4–10% of the intramuscular nuclei in frog, fowl, bat, guinea pig, mouse, rat, and human muscle. However, others[155] question their presence in normal muscle. While the satellite cell cannot be re-

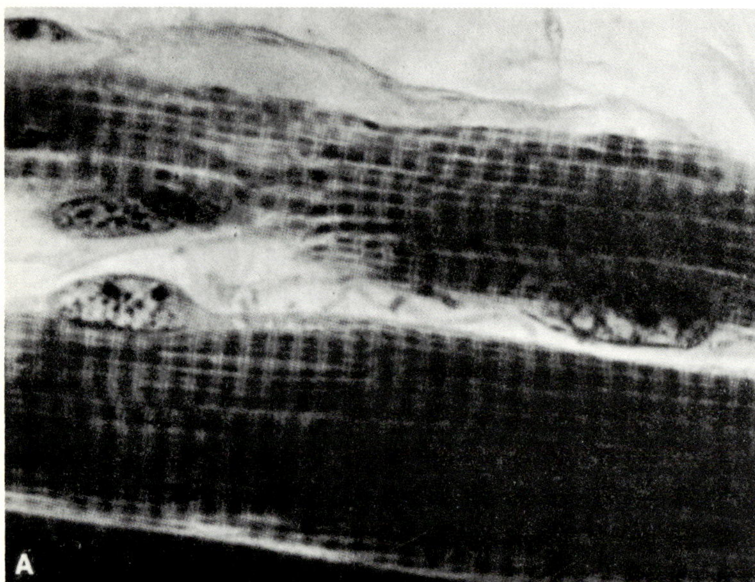

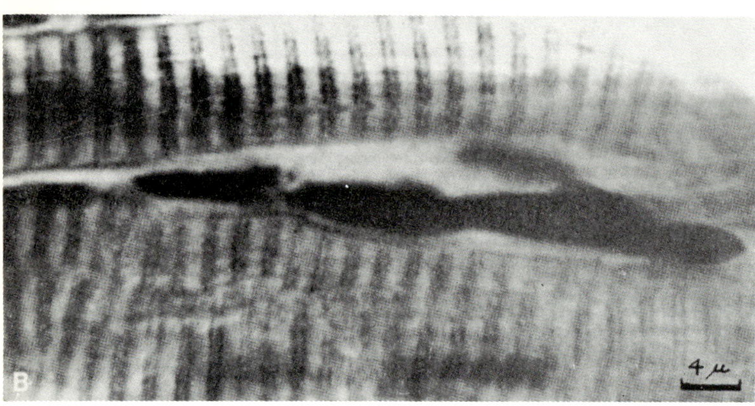

Fig. 1–14. (A) High power view of cross striations from gracilis muscle of cat, showing myofibrils. Note sarcolemmal nuclei and sarcolemma lifted from upper fiber. (B) Silver impregnation. (A, phosphotungstic hematoxylin: B, Bielschowsky method)

liably distinguished from the muscle nucleus, there are some differences useful in its identification (Figs. 1–11, 1–12, and 1–15). The satellite cells are more fusiform than true muscle nuclei; their nuclear chromatin is less evenly dispersed, nucleoli are lacking, and a clear border of sarcoplasm devoid of myofibrils separates them from the rest of the fiber. Ultrastructural characteristics include: the presence of centrioles, which are never seen in muscle nuclei; a few mitochondria; no glycogen; a small but well developed Golgi complex at the end of the muscle nucleus; and sparse endoplasmic reticulum. The space which separates the inner membrane of satellite nucleus from the outer membrane of the muscle fiber is about 150 Å wide. In places the basement membrane extends for a short distance within the cleft that separates them from the sarcolemma. Shafiq[170] observed a mitotic division in a satellite cell from a 3-week-old rat.

The nature of the satellite cell has thus been a subject of controversy. In his original report Mauro[129] suggested two possibilities: (1) that in healthy muscle there is a small proportion of muscle nuclei being split from the muscle fiber and for a time lying external to the sarcolemma but within the basement membrane, and (2) that satellite cells are survivors of the population of primitive myoblasts of the embryonic period and are being continually incorporated into preformed myotubes, a phenomenon observed by Kelly and Zacks.[108] After injury from disease the satellite cells are

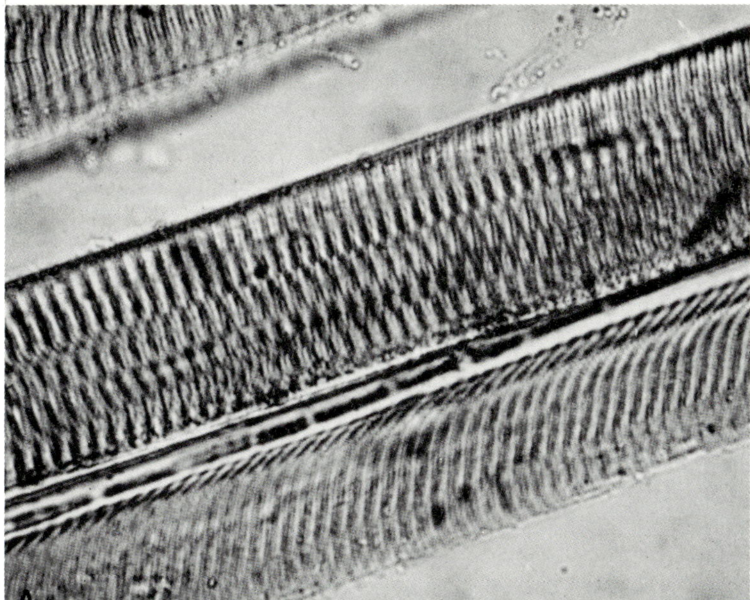

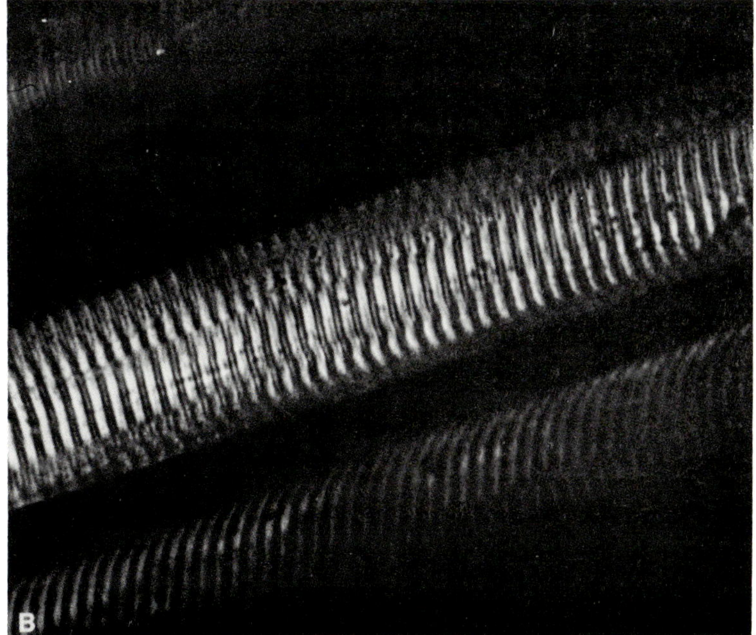

Fig. 1–15. (A) Unstained normal fibers viewed in transmitted light. (B) Same fibers under polaroid.

believed to be activated, to undergo mitosis, and then to fuse, forming new myotubes. Mauro[129] favors the theory that satellite cells represent a population of embryonal myoblasts whose function is to regenerate portions of the fiber after injury; Reznik[155] believes them to be no more than activated muscle nuclei.

Outside the true cell membrane of the muscle fiber is another closely investing layer, which is colored by stains for connective tissue and reticulum. It consists of a network of close fibrillar structure (Fig. 1–13B) and contains numbers of fibroblastic nuclei. These nuclei are more plump than the sarcolemmal nuclei, and their long axis is not strictly parallel with the muscle fiber (Fig. 1–13A). They stain more easily with silver impregnation methods than do the sarcolemmal nuclei and are particularly numerous near blood vessels.

Much confusion has arisen because the term sarcolemma was applied to each of these membranes. Most anatomists, however, now reserve the term for the true cell membrane, separable from the basement membrane or glycolemma, and they designate the fibrous structure external to the basement membrane as the endomysial sheath, a terminology followed here.

The functions of these two membranes, the sarcolemma and the endomysial connective tissue sheath, are different. The cell membrane of the sarcolemma is the active interface between the potassium-rich interior of the fiber and the sodium- and chloride-rich external environment. In this respect there is no reason to believe that it differs from the membrane of any other cell. Electrical stimulation of the surface of the sarcolemma, more particularly at points of origin of the transverse endoplasmic reticulum,[103] induces a contraction wave that rapidly diffuses over the whole fiber and is associated with the well known action potential. Electrical stimulation of the contents of the fiber by a microelectrode introduced under this membrane induces only a localized contraction that spreads very slowly or not at all and is not accompanied by electrical change.[67, 68] The cell membrane therefore appears to act as a transmitter that can "excite" the entire sarcous substance and is the means by which contractile tension is transmitted to the tendon. The myofibril has direct continuity with a tendon fibril at its end, possibly piercing the sarcolemma.

There are only a few observations of the mechanical properties of the sarcolemma, and these are conflicting. The living muscle fiber is under some slight tension even at its shortest natural length,[173] as can be demonstrated by the shortening of a muscle when its tendon or tendons are severed. From all indications this resting tension is not due to the muscle substance (the myofibrils) but to the sarcolemma and the elastic and collagenous intramuscular connective tissue. Ramsey and Street,[151] after producing retraction clots, found that the empty sarcolemmal sheath would stretch to the same degree as the normal muscle fiber.

Sichel[171] was unable to confirm their results. In his experiments the sarcolemmal sheaths would elongate 2.2 times as much as would sound portions of the fiber.

The consistency of a fully relaxed muscle, as felt in the clinic, is directly related to the size of the muscle and hence to the size of the muscle fibers. The firm muscles of athletes contain large fibers, each distending its fibrous envelope. The flabby feel of disused or atrophic muscles reflects the lack of tension of the connective tissue sheaths.

MUSCLE FIBER STRIATIONS

The histology of striated muscle fiber has been investigated by a variety of techniques, all of which confirm the existence of several interesting structures. The principal and unique cytoplasmic constituent of the muscle fiber is the myofibril. Each muscle fiber contains a large number of myofibrils, all of which are oriented in a longitudinal direction (Fig. 1–14) and run the full length of the fiber parallel to one another. Some of the more superficial ones are deflected slightly by the sarcolemmal nuclei. In especially well stained sections these myofibrils can be followed to the very end of the fiber, where they are usually more tightly packed together. Here, even though the muscle fiber often tapers and assumes a conical shape the myofibrils remain parallel up to the sarcolemma. In extremely tapered fibers those myofibrils situated in the periphery of the muscle fiber may not reach the end of it. The ends of some muscle fibers may be divided by an insertion of connective tissue, and in these there is some tendency for the myofibrils to be drawn into small groups.

It is not settled whether the myofibrils insert in the sarcolemma, pass through it and join the tendon fibrils, or stop short of it. In longitudinal sections the myofibrils are closely packed in the muscle fiber, and only small amounts of sarcoplasm can be seen between them and around the sarcolemmal nuclei (Fig. 1–14). In transverse sections there is a tendency for the myofibrils to be grouped into bundles separated from one another by a network of sarcoplasm. The map-like arrangement of these bundles of myofibrils is called Cohnheim's areas or fields, after Cohnheim[38] who first observed them in 1865. Some workers[140] believe that this compartmentation is an artifact, for it is more prominent after freezing.

The myofibril of vertebrate muscle is 0.5–1.0 μ in width. It is smaller than that of insect muscle, with a range of 1.4–2.0 μ in width, and is more difficult to dissect in the living state. The myofibril is of fairly even thickness throughout its length. It consists of a row of elongated, rod-like, cylindrical, dark particles joined end to end by a lighter substance. Some investigators (e.g., Hurthle[95]) appreciated from early times that the microscopic myofibril consists of a great many ultramicroscopic ones. This idea received support from the electron microscopic studies of Hall *et al.,*[79] who found that the myofibril is made up of filaments resembling myosin with a diameter of 0.005–0.025 μ. Also Rozsa *et al.,*[159] employing the same methods, corroborated the filamentous nature of the myofibril. According to their calculations the filaments were 100 Å in width; the length could not be ascertained because of the method of preparing the tissue (i.e., mechanical disruption), but some were at least 6–8 μ thus proving that they are continuous through the sarcomeres. When viewed in ordinary reflected light, even under the low power of the microscope, the clear pellucid muscle fiber possesses a very obvious cross striation (Fig. 1–15A) intimately related to the myofibrils. Each fiber is marked by regularly alternating light and dark bands of approximately equal length (3.0 μ). By changing the depth of focus of the microscope, it can be perceived that this striation extends throughout the substance of the fiber and is not just a surface phenomenon. These light and dark bands were noted by all early microscopists and are well illustrated in Bowman's important 1840 paper.[20] The dark band* was believed to be the principal striation and was referred to as

---

* Note that the dark appearance observed by ordinary light is due to the different refractive index, related to but not identical with the reaction to polarized light, discussed later in this chapter. By suitable alteration of the angle of light incidence, the lightness and darkness in ordinary light can be reversed.

the Q (*Querscheibe*) or transverse band, and the light one as the J band (a designation chosen by Rollett for a subdivision of the isotropic segment). When the fiber is deeply focused, the J band is found to be bisected by a thin line. This line (Figs. 1–14 and 1–15), first discovered by Dobie,[52] because of its position in the middle of each light band is termed the intermediate disk, *disque mince,* or more commonly the Z (*Zwischenscheibe*) line. Another lighter line or band may be seen in the middle of each Q band, especially when the muscle fiber is slightly extended. This is called the line of Hensen, the H band, and the M line (*Mittelscheibe*). The dark band of each myofibril in a single muscle fiber is normally opposite the dark bands of all other fibrils (Fig. 1–14A). This alignment is responsible for the cross striation of the entire fiber.[132, 133]

When the muscle fiber is examined after staining with iron hematoxylin or phosphotungstic acid hematoxylin, the striation is seen in exquisite detail (Fig. 13A). The iron-alum hematoxylin method of Heidenhain has been widely used for this purpose. The dark Q band and Z disk, evident in the unstained fiber, are well stained by hematoxylin; the J band and M line stain poorly or not at all. The reactions to hematoxylin doubtless depend on certain chemical properties of the myofibrils. Further insight into these chemical reactions was provided by Dempsey *et al.*[46] by exposing muscle to a number of different fixatives and then to acid and basic dyes. After Zenker fixation the majority of the fibers stain evenly with acidic stains such as eosin or phloxine. Along the periphery of the section, however, a few of the fibers react with a basic stain such as methylene blue. In these the Q band and Z disk are the parts where the basic dye is most heavily concentrated. After fixation in Bouin's fluid, the cross striations stain more intensely with methylene blue than after Zenker's or basic lead acetate fixatives (Figs. 1–16 and 1–17). Again the Q band and the Z disk are stained and the remainder of the J disk unstained. By modifying the pH of the stain, Dempsey *et al.*[46] were able to show that the intensity of basophila varied directly, and the acidophilia inversely, with the pH of the stain. At high pH the myofibril stained rather evenly with methylene blue. Cross striations were poorly differentiated at pH above 7. At intermediate pH the J bands of the myofibrils exclusive of the Z disk failed to stain, thus producing good differentiation of the cross striation. At pH below 5.5 the striations became dim, and below 4.0 the entire fiber took only an acidic stain.

When the muscle fiber is impregnated with metallic ions, as in the silver stains or by Bodian protargol technique, reduction of silver and consequent blackening is seen in the J band or segment and not in the Q band (Figs. 1–14B and 1–16), thus presenting the reverse appearance to that shown by hematoxylin. These argyrophilic zones are located on either side of the Z disks.

When the unstained muscle fiber is examined under ultraviolet light, it fluoresces bluish white. The fiber appears homogeneous; no cross striations are visible. According to Dempsey et al.,[46] this property of fluorescence is not altered by treating the tissue with acetone and alcohol.

Visualization of muscle fiber under polarized

Fig. 1–16. Companion to Figure 1–17. In each of the four pairs of drawings the left-hand figure against the white background represents the muscle fiber as seen with the ordinary microscope. The right-hand figure against the dark background represents the same fiber as seen between the crossed prisms of the polarizing microscope. Stained segment of the left figure may be identified as the isotropic (dark) or anisotropic (light) band of the right figure. (1) Acidophilic staining of the isotropic segments and the M disks of rat muscle. The section was stained with a dilute solution of orange G at pH 5.3 after fixation in Zenker's acetic fluid. (2) Basophilic staining of anisotropic segments and Z disks of salamander muscle. Tissue was fixed in Bouin's fluid and section stained in 5 × 10⁻⁴ M methylene blue at pH 6.3 (3) Smith-Dietrich reaction for phospholipids in rabbit muscle. Positive reaction consists of dark bands, which by comparison with right-hand figure prove to correspond to isotropic segments. (4) Glucose-1-phosphatase reaction in rat muscle. Section was incubated for 72 hours at pH 7.0. Yellowish color denotes a diffuse reaction throughout fiber, and darker bands indicate greater enzymatic activity in isotropic segments. (From Dempsey et al.[46]) (95× objective and an 18× ocular)

Fig. 1–17. Companion to Figure 1–16. In each pair of illustrations the one to the right is viewed through crossed prisms. (1) Iron alum hematoxylin stain of human muscle. Darkly stained bands in the left-hand figure correspond to the anisotropic segments when compared with the right-hand representation. (2) Argyrophilia of the Z disks in rat muscle. Tissue was fixed in Bouin's fluid and section stained by Pap's ammoniacal silver method. (3) Argyrophilia produced by Bodian's reaction in relaxed rat muscle. The entire isotropic segment is silvered, but an especially intense reaction occurs at the margins of the isotropic bands. Fixation in Bouin's fluid. Compare with 4. (4) Argyrophilia produced by Bodian's reaction in contracted rat muscle. The isotropic segments have contracted and occupy less space than in the relaxed state. The anisotropic segments are relatively unchanged in length. Fixation in Bouin's fluid. (5) Lipoidal granules in rat muscle. Frozen section of tissue fixed in formalin, stained by Sudan black. Discrete lipoidal droplets, mostly situated in the isotropic bands, are visible. These droplets were readily extractable in alcohol and acetone. The isotropic segments stain a diffuse gray with Sudan black. The sudanophilia is abolished only after prolonged extraction with hot alcohol. (6) Increase in birefringence of the isotropic segments of rat muscle after extraction of section with hot alcohol. The left-hand photograph represents the cross striations as seen with the polarizing microscope in an ordinary frozen section after formalin fixation. The right-hand figure illustrates the appearance of a similar section after 3 hours' extraction in boiling alcohol. The isotropic segments are more strongly birefringent than before extraction, approaching in brightness the intensity of the anisotropic bands. (From Dempsey et al.[46])

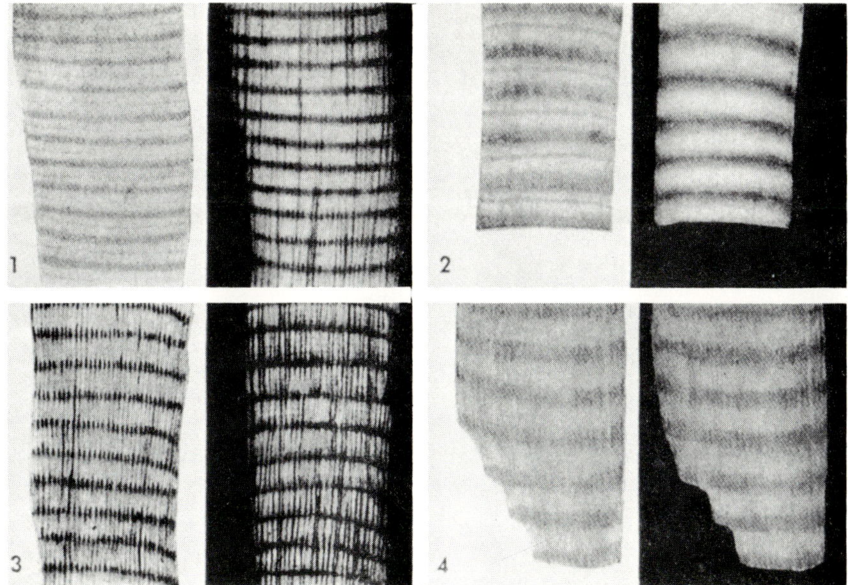

Figure 1–16.

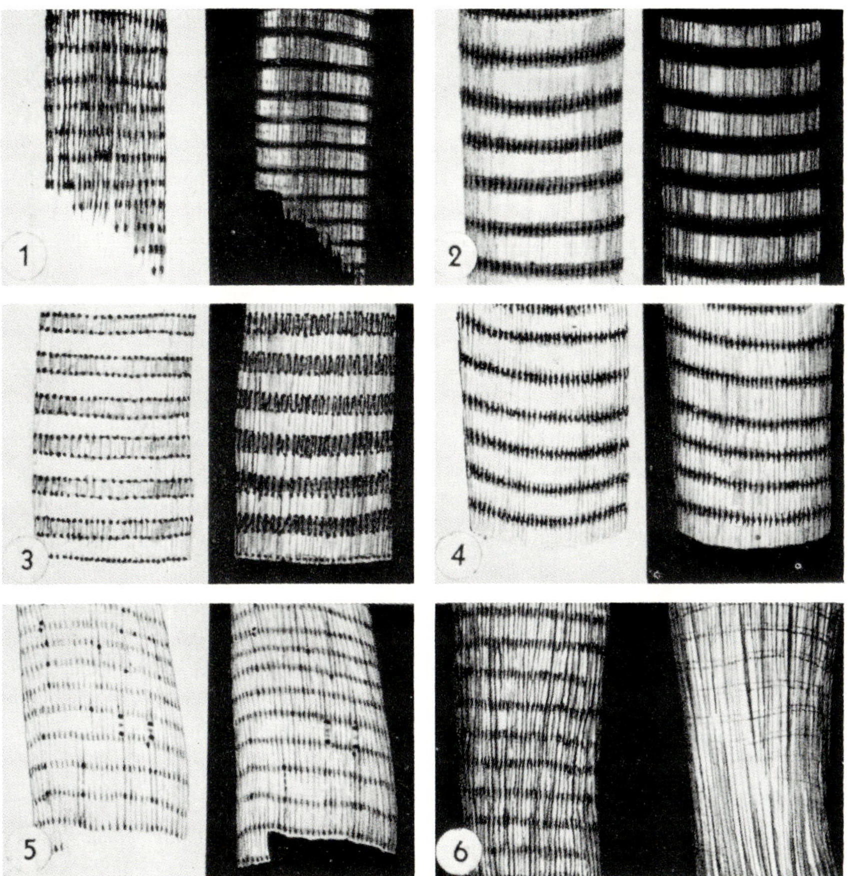

Figure 1–17.

light* has been a common method of study and has provided valuable data. Brücke[25] was the first to point out that even under polarized light the muscle fiber retains its striated appearance. Ranvier[152] made a detailed study of the birefringence of muscle. Under crossed Nicol prisms or polaroid the fiber is marked by broad bars of light anisotropic bands alternating with dark isotropic bands; the anisotropic band is known as the A band, and the isotropic as I. When compared to the stained fiber or the unstained fiber under ordinary light, the Q band is seen to be anisotropic and the J band isotropic. In other words, the striation under polarized light is the reverse of that seen in the ordinary transmitted light. Moreover, the Z disk is also anisotropic. However, it is not certain that the stainable substance of the Q band is absolutely identical with the anisotropic or A band. In most modern descriptions, however, the striation is referred to as A and I bands, whether evidenced by staining or ordinary or polarized light; the lighter band of variable width in the center of the A band is called the H band.

In a number of recent electron microscopic studies of striated muscle fiber [90, 96, 99] attempts were made to obtain more information concerning the structural details of the myofila-

---

* A polarizing microscope is one in which a pair of polaroid disks or Nicol prisms are interposed in the beam of light. One is placed between the light source and the object under study, and the other is usually placed over the microscope ocular. The latter is called the analyzer. The first disk of polaroid material or Nicol prism permits passage of light vibrating in a single plane, and when the second prism or analyzer is in the optically identical position the light which passes the first polarizer also passes through the second and the microscopic field is bright. When one of the prisms, usually the analyzer, is rotated 90° on its long axis relative to the other, the light which passes the first polarizer cannot pass the second, and the field is dark. In this field substances which themselves rotate the plane of polarized light are seen to glow, for they alter the path of light sufficiently to make it visible through the analyzer. Such substances are referred to as anisotropic or doubly refractile, or are said to show birefringence. Many crystals and some protein structures, among which are membranes (nerve sheaths, red cell membranes, etc.) and skeletal muscle show this phenomenon. For additional discussion see Schmitt,[165] Lichtman and McDonald,[121] and Fischer.[65]

ments of the myofibril originally described by Hall *et al.*[79] It now seems fairly clear that the striations of the myofibrils (Fig. 1–18 and 1–19) are based on the special molecular organization of the myofilaments. In longitudinal sections a series of nodosities occurring every 250–400 Å were observed on the surface of the myofilaments, like a string of beads, forming a series of bridges that extend transversely from one filament to another. In cross sections the myofibrils, solidly packed with myofilaments, are arranged in hexagons with spacings of about 300 Å between their centers. In addition to the larger myofilaments, averaging 100–110 Å in diameter, there are smaller ones (40 Å) appearing like satellites (Figs. 1–20D,E). These have also been observed in human muscle.[186]

The controversy over the arrangement of the myofilaments and their role in muscle contraction has now been settled. The classic model espoused by Hodge,[90] which affirmed the existence of a continuous array of myofilaments believed to be made up of actin molecules and said to extend through all the bands of the sarcomeres, proved to be incorrect. The model proposed by Huxley[96] and by Huxley and Hanson[99] in which there are two interdigitating sets of filaments is now generally accepted. The thick myosin filaments are found to be confined to the A zone; the thinner filaments of actin are solely found in the I band, on either side of the Z line, and extend into the A band as far as the H band. Thus the H band is relatively light because it contains only the thick filaments of myosin, and the remainder of the A band is dark because of the interdigitating myosin and actin filaments. The I bands are pale, having only thin actin filaments (Fig. 1–13). Muscle contraction is effected by means of a reaction between transverse bridges on the actin and myosin molecules, as a consequence of which myosin filaments slide into the I bands. The latter then lose their isotropy. With relaxation the filaments move back to their original positions, accounting for the several different geometric arrangements of myofilaments shown in Figures 1–19B and 1–20. Although some electron microscopists have found fault with Huxley's model, it best accounts for the

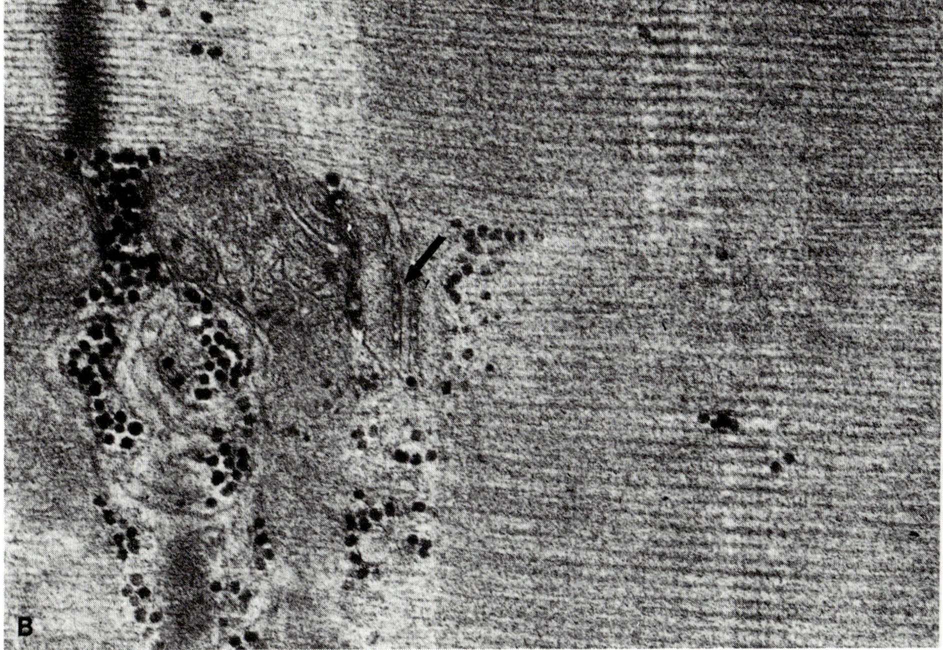

Fig 1–18. Electron micrograph of longitudinally sectioned human gastrocnemius muscle that clearly shows the parallel array of myofibrillar sarcomeres. The I and A bands are bisected respectively by Z lines and M bands, the insertion zones of the thin and thick filaments. At higher magnification the dimensions of the thick and thin filaments are readily compared. Triads (arrow) include elements of the sarcoplasmic reticulum and the transverse tubules, and are located at the A-I junction. (Courtesy of Dr. Henry deF. Webster) (*Top*, ×27,000; *bottom*, ×75,000)

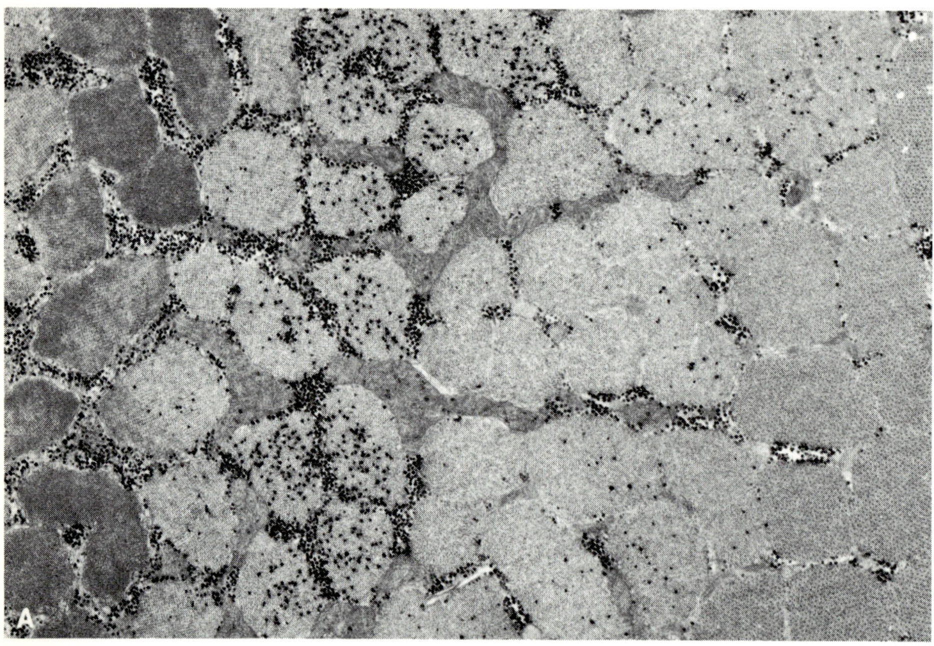

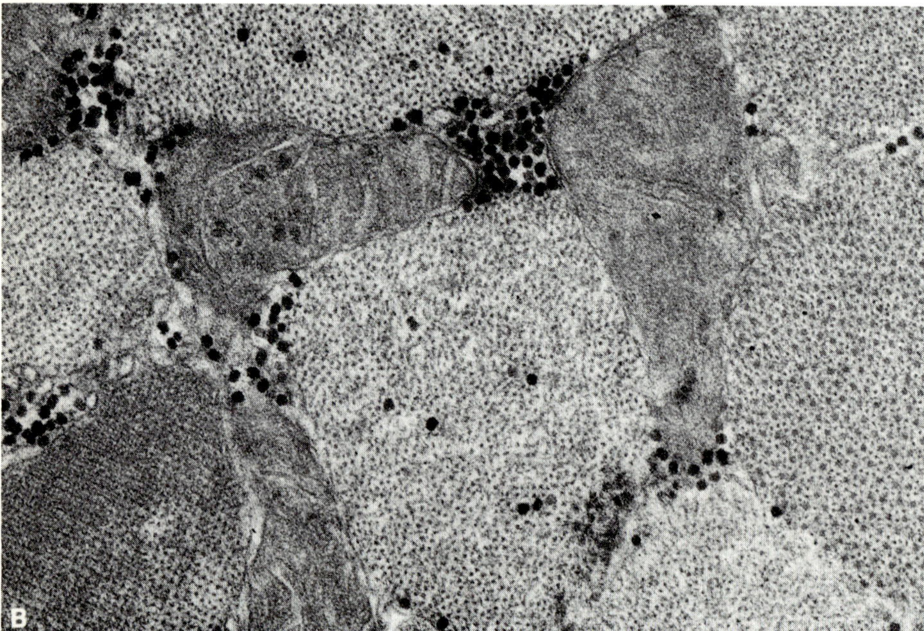

Fig. 1–19. Electron micrograph of transversely sectioned human gastrocnemius muscle. The diameter of myofibrils and the organelles between them vary along the sarcomere. On the left, numerous glycogen granules surround fibrils sectioned at the Z band level. Mitochondria and glycogen are most prominent at the I band (center) and are very sparse at the A band level shown to the right. At higher magnification differences in myofibrillar filamentous structure are also apparent. In the Z band to the left the periodicity of the paracrystalline array is about 160 Å. The thin, actin filaments in the I band (center) measure approximately 60 Å in diameter and they surround thicker 110-Å myosin filaments in the A band shown at the right. (Courtesy of Dr. Henry deF. Webster) (*Top,* ×27,000; *bottom,* ×70,000)

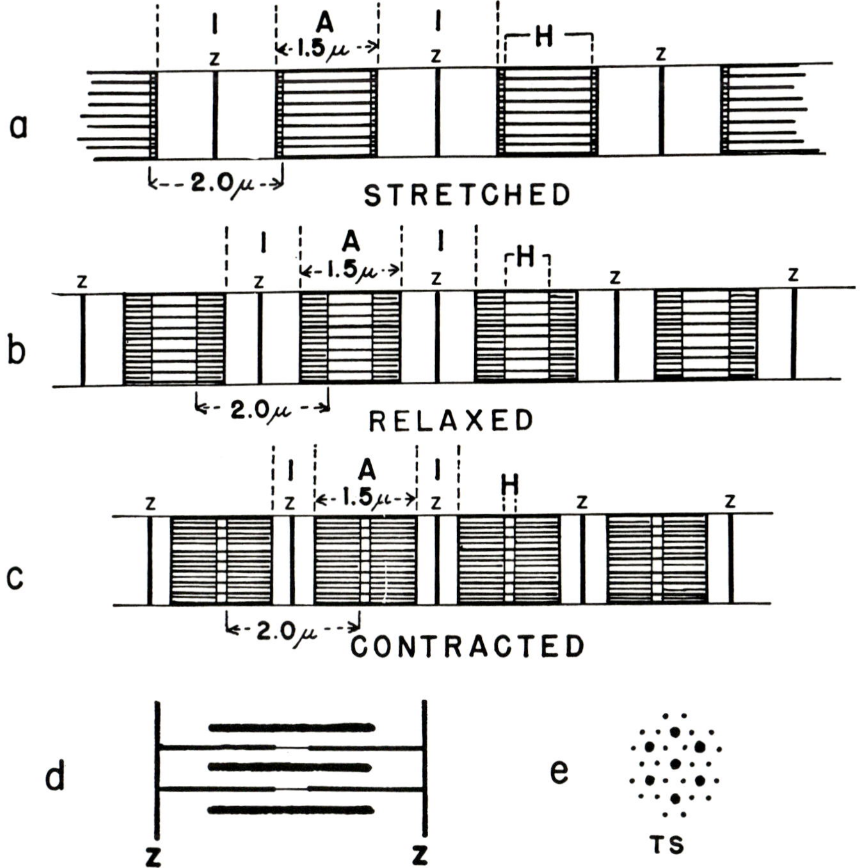

Fig. 1–20. Dimensions of muscle striation that remain constant when muscle is stretched. (A–C) In comparing the stretched, relaxed, and isometrically contracted states it is seen that width of the A band and distance between two M bands are unchanged, although the M band widens. (D) Schema of movement of the filaments within a myofibril; the thin rods of actin are attached to Z line and interdigitate with the thicker rods of myosin. Fixed length of actin rods, sliding between myosin rods, is the explanation of the relationship of striation shown above. (E) Electron microscopic appearance of overlapping zone in transverse section (TS), each rod of myosin is surrounded by six filaments of actin. (Data from Huxley[96, 99])

changes in the muscle fiber during contraction and passive stretching (Fig. 1–20).

Ruska and Edwards[161] described a unique paranuclear arrangement of sarcoplasmic substance between the sarcolemmal nucleus and the adjacent sarcolemmal membrane. Although no myofibrils are observed in this region, there are structures in the sarcoplasm that have the same "period" as the A, I, and Z bands of the myofibril. It is suggested that myofibrils grow in length or width by the addition of material initially laid down by the nucleus in this region, and that each nucleus generates myofibrillar substance in its own area and elaborates it by way of the endoplasmic reticulum.

There has been much speculation concerning the ultimate unit of the muscle fiber. In 1840 Bowman[20] described and illustrated the myofibrils and their striations and noted also the tendency of muscle fibers that had been exposed to acids, alkalis, and alcohol to fracture transversely into disks at points corresponding to the cross striations. He supposed therefore that the muscle substance was composed of a series of particles joined side by side to form disks and united end to end to form myofibrils. In Bowman's opinion the basic unit of muscle structure was the disk. Kölliker[114] took exception to this view, being impressed with the fact that muscle substance more frequently fragments into fibrils than disks. He concluded on this basis that the myofibrils rather than Bowman's disks were the actual elements of muscle and that their striation was an optical property probably unrelated to variations in chemical constitution.

Krause[115] introduced a new conception of the structure of muscle. He proposed that the Z line was not merely a striation of myofibrils but a membrane (the membrane of Krause) dividing the fiber into small compartments, each of which contained a dark Q (A) band (the sarcous element of Bowman) in the center and half of a J (I) band on either side. These segments (Fig. 1–20) were called sarcomeres. The studies of Heidenhain[85] again placed emphasis on segmentation of the whole muscle fiber into compartments by the membrane of Krause or Z line. In his terminology the membrane of Krause was called the *Telophragma* and the sarcomere was designated as the *Inokomma*. The modern version of this view was presented by Jordan,[107] whose studies elucidated a number of controversial points in relation to insect and heart muscle.

It is indeed extraordinary that in most muscle fibers the Z lines and bands are in exact alignment in all the myofibrils. Furthermore, there is usually a perfect alignment of the Z lines of all the muscle fibers in any microscopic field. This striation is of the same essential pattern in heart and voluntary muscle of all vertebrates and invertebrates and clearly must reflect some fundamental structural organization. It is not certain, however, that the whole fiber is segmented, and in fact several observations argue against the existence of a Krause membrane or Telophragma at the Z disc. For example, Kühne[116] watched a minute living nematode worm move freely up and down a muscle fiber without impediment, and Carey[29] noted the floating movement of intracellular air tubules back and forth through the extent of three to five striations in living insect muscle. Then too during observations of the process of freezing of muscle fibers under the microscope Chambers and Hale[34] noted that the columns of ice crystals advanced unhindered between the bundles of myofibrils. Also Chambers (cited by Speidel[174]) was able to manipulate a droplet of oil along the fiber under the sarcolemma without obstruction. Finally, the simple experiment of Bowman in which a frog muscle was teased apart under the microscope and its contents seen to rupture and retract some distance within the sarcolemma before the latter finally tore can be confirmed by any observer.

Single myofibrils cannot be seen clearly in the living unstained muscle fiber, and yet there is little doubt as to their existence in living tissue. Jordan[107] stated that under high magnification they could be seen in the tail of the young frog and salamander tadpoles, and Speidel[174] could see them in "irritated fibers" in the tail of the tadpole. Høncke[93] remarked that under darkfield illumination the fibrillar structure of the living muscle fiber is easily seen. The living contractile myofibrils of insect muscle can be teased apart.[162] Other observations such as the extension of longitudinal columns of ice crystals along the fiber,[34] mechanical shattering of the muscle fiber into pieces about the size of the visible myofibril,[79] and the separate contraction of groups of myofibrils[7, 80, 174] afford substantial proof that the myofibrils are the ultimate contractile elements of the muscle fiber.

Yet the matter is not so easily settled, for an extension of the Z line from an apparent attachment to the sarcolemma into the sarcoplasm has been pictured in vertebrate muscle, though less often than in heart muscle and crustacean muscle, and festoonings of the sarcolemma and separation from the fibril bundles at intervals corresponding to sarcomeres with constrictions at the Z lines has been described frequently, especially in muscle fixed during contraction.[32] When it is realized that the anisotropic segment swells during contraction and during some types of fixation and that pressure of fresh muscle against soft celloidin leaves an imprint of cross striation,[83] it appears possible that the festooned appearance is a similar imprint in the plastic sarcoplasm and sarcolemma perpetuated by fixation. Moreover, electron microscopy shows structural modifications of the sarcoplasm at the Z, M, and N lines, which may serve to bind the myofibrils into bundles at these points. We are inclined to the view based on the work of Huxley and Taylor[103] that the ultimate contractile unit of the muscle fiber is the sarcomere and that the Z bands are the bases against which contraction is effected. Moreover, the experiment of Bowman on sarcoplasmic constituents characterizes many of the pathologic reactions of striped muscle. It is possible to damage the entire contents of the fiber independently of the sarcolemma and the muscle nuclei.

The view of Krause's membrane or Z disk as a spiral or helicoidal structure[152] (which had been advanced by Heidenhain,[85] Tiegs, [181, 182] D'Ancona,[45] Aurell and Wohlfart,[3] and more recently by Ruska and Edwards[161] and Matthaei and Tiegs[128]) is clearly refuted by electron micrographs, which reveal its strictly transverse alignment and crystalline structure (Fig. 1–21). These writers were referring to the commonly observed misalignment of the striations of some myofibrils, being displaced slightly toward one or other end of the fiber. This disalignment resembles a vernier and has been called the vernier effect (Figs. 1–14 and 1–15). Such a dislocation of a bundle of myofibrils can manifestly arise from unequal tension on the muscle during fixation, as is evident in Figure 1–14. The proponents of the helicoidal hypothesis, the most convincing of whom are Aurell and Wohlfart,[3] point out, however, that the vernier effect seldom extends over more than a few sarcomeres, that in the remainder of the muscle fiber the alignment is exact and that the venier may be in two opposite directions. By focusing up and down on a striated muscle fiber it is claimed that the spiral arrangement of the Z disk is clearly seen. Speidel's[173, 175] observations on the living muscle in the tail of the tadpole also reaffirm that the arrangement of these structures is strictly transverse and particularly that the terminal Z line occurs at each end of the fiber. The latter investigator and also Barer[8] observed the formation of vernier effects and other dislocations of bundles of myofibrils in

Fig. 1–21. Sartorius muscle of cat showing bands of "extreme contraction" in one muscle fiber due to contraction at fixed length during fixation. (phosphotungstic hematoxylin)

the living fiber as a result of unequal tensions on the myofibrils in the process of their clotting.

Matthaei and Tiegs[128] made direct serial photographic recordings of the extremely slow contractile waves that occur in arthropod muscle under certain conditions. They believe that the slow waves arise by retardation of the normal rapid twitch wave, being able to observe them when an isolated muscle was placed in saline or when the animal was in a moribund condition. In some fibers the wave became so retarded that it ceased to advance and remained as a permanent or "fixed" contraction wave. Careful study of photographs led these investigators to conclude that the contractile wave initially spreads from the nerve ending *across* the fiber before moving along it and that during the sequence of contraction the front of the wave moves from one side of the fiber to the other and back again. In places a clear vernier effect was observed. They felt that such evidence would argue against primary conduction along the fibrils or the sarcolemma. Rather, they suggested the presence of cross membranes which transmitted the excitation within the interior of the fiber. These cross membranes along which the excitation wave passed correspond to the later discovered T system.

Attempts to determine the chemical composition of the myofibrils were made recently. Utilizing various chemical extraction methods, Huxley and Hanson[99] were able to demonstrate that myosin is located entirely in the A band when the fiber is in the relaxed state. Further extraction also removes actin, and then only the Z lines are visible. Since the extractive procedures are rather harsh and may remove soluble material in addition to actin and myosin, Finck *et al.*[64] employed another more precise method for the localization of myosin, the fluorescent antibody method. Specific antisera prepared in the rabbit against chicken breast muscle myosin and labeled with a fluorescent dye showed clear-cut localization of the antiserum in the A bands when viewed in the fluorescence microscope. Others recently confirmed this localization of myosin, e.g., Klatzo *et al.*[111] who likewise noted the lack of species specificity between myosin in the human, cat, rabbit, mouse, and rat. Hence two distinct

methods clearly demonstrated myosin to be present only in the A bands, while actin is probably present in both the A and I bands.

Various other substances have been localized to parts of the sarcomere. By micro-incineration of fixed striated muscle fibers Scott[168] and Scott and Packer[169] demonstrated a heavy white ash in the A bands and a light bluish ash in the I bands. Chemical analyses of the ash in the A band show it to contain calcium and magnesium, and there is some indication that potassium occurs in greatest concentration in the I band.

Caspersson and Thorell[33] adduced evidence from ultraviolet microscopic studies that adenosine compounds are concentrated in the isotropic band; Hoagland[89] confirmed this finding. Under the conditions of these experiments there was a strong absorption in ultraviolet light of wavelengths between 2400 and 2950 Å. These workers stated that the only substances in muscle that show strong absorption at this wave frequency are the purine derivatives (of which adenylic acid and adenosine triphosphate, ATP, made up nearly the total amount) and proteins. After strong contraction this ultraviolet absorbing material diffuses into the A bands.

Dempsey *et al.*,[46] in their previously quoted histochemical analysis of the muscle fiber, found the basophilic substance of the A band to differ from nucleoprotein in its dissociation curve at differing pH values, its negative Feulgen reaction, and its resistance to the action of ribonuclease. Moreover, they concluded that it was probably not a mucoprotein such as is found in mast cells, cartilage, and mucus. The I bands were found to stain heavily with Sudan black and by the Smith-Dietrich method for phospholipids. After extraction with boiling alcohol for 3 hours this staining reaction was lost and the I bands were more birefringent. It is suggested therefore that the lack of birefringence of the I segment is due to a negative birefringence of lipids.

The embryogenesis of the myofibrils and their relationship to the contractile process have also been of particular interest to biologists.[2, 56, 73] As was stated above, the first fibrils visible in the myoblast are long, unsegmented protofibrils (Fig. 1–1B). Lewis,[118, 119]

Lewis and Lewis,[120] deRenyi and Hogue,[49] and Copers[39] observed that isolated myoblastic cells migrated from fetal muscle in tissue culture, and that multinucleated bands contracted spontaneously at rates of 30–120 times a minute; yet few showed any cross striation while they were living. Spindle-shaped bodies were commonly seen after fixation and staining, and mitochondria were abundant. It is likely that the regular arrangement and characteristic striation of the submicroscopic filaments composing the myofibril become visible with ordinary microscopic stains only when sufficiently numerous threads have become aggregated into a spindle and then a myofibril. Hence it is probable that the contractile process concerns the elements of the myofibril (myofilaments) before they can be visualized as such.

The rearrangement of the muscle fiber ultrastructure during the process of contraction has claimed the attention of a number of investigators. There have been numerous attempts to measure the sarcomeres and their constituents during the resting state, during passive stretching, and in the process of contraction. Some of the reported discrepancies in the results are attributable to the different methods of preparing the muscle fiber for examination and others to technical difficulties.

An alteration in the stainable Q substance, which many of the early microscopists recorded in localized knots or bands of contraction, is commonly seen in muscle fixed in alcohol. Jordan[107] and others who studied these "contraction bands" pointed out that in stained sections the striations are certainly reversed. As the Q bands are traced from the extended to the contracted parts of the fiber they divide at their center, exposing the unstained H band, and each half of the Q band moves toward the neighboring Z disk, thus obliterating the unstained J band (Fig. 1–22). The striation is then a poorly stained H band alternating with a QZQ complex (Fig. 1–20C). Under these conditions the contraction of the whole fiber lessens the birefringence of the muscle, but the H band remains more completely anisotropic— whether the muscle is allowed to shorten,[184] left extended,[137] or prevented from thickening.[22] Thus the stainable striation is reversed

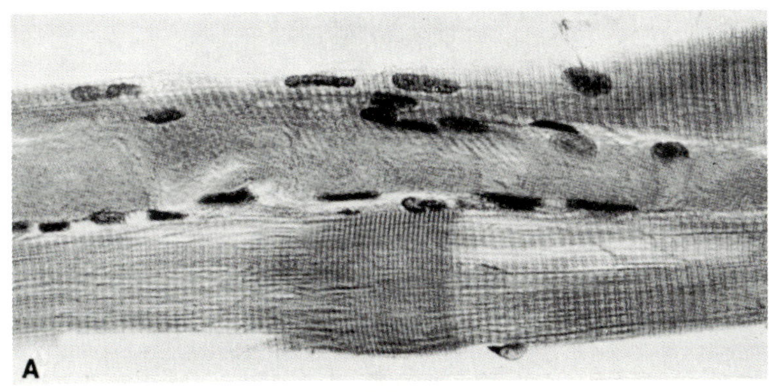

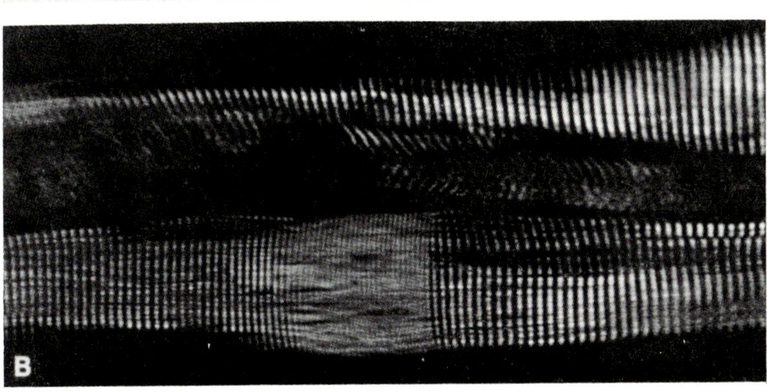

Fig. 1–22. (A) Lower fiber presents a contraction band. Fiber above it presents partial disruption by "extreme contraction." (B) The same, in polarized light. (H&E, ×600)

in contraction, though the light band in polarized light does not change. The stainable substance of the Q band therefore does not exactly correspond to the A substance.

Speidel[174] observed this reversal of striation in living tadpole muscle. If a tetanic contraction of the whole living fiber was induced, the obliteration of J bands occurred only when the muscle was allowed to shorten by 40% or more of its resting length. If less than 25% of shortening was permitted, only a relative narrowing of the J segment occurred. With a single twitch the region of the Q segment blurred for an instant. Speidel also observed that the last band at either end of the living muscle fiber was the Z line and that half a contraction band formed at this line in the contracted state.

Nageotte[140] found that if muscle was fixed while contracting in the stretched state the bands were not changed. If shortening to 70% of the resting length was allowed, the distance between the Z lines shortened proportionately and the Q substance moved as noted above. If, however, a piece of muscle was removed and subjected to every intense faradic stimulation, it could shorten to less than 60% of its length. Fixation at this instant, followed by hematoxylin staining, revealed irregular areas in which QZQ bands were so tightly packed together as to stain uniformly dark (Fig. 1–21). In this state of extreme contraction the whole of these dark areas was strongly anisotropic as compared to the remainder of the fiber. Frog muscle fibers thrown into such violent contraction can be stretched out again without structural damage, but after several repetitions of such unnatural shortening and restretching they cease to contract further. They were then found to have a high content of lactic acid.

The quantitative measurement of the living muscle fiber at rest and during contraction has yielded slightly different results. Buchthal *et al.*[26] found that in the relaxed muscle fiber the A band was 1.37 μ in width and the I band 0.81 μ. During isometric contraction the corresponding measurements were 1.13 μ and 1.05 μ. Thus although the length of the sarcomere had remained constant, the A band had shortened by 18% and the I band had lengthened by 28%. Høncke[93] confirmed these

findings. These earlier studies suffered from difficulty in obtaining good resolution of the striation with high magnification of the living fiber. Though both bands are shortened in the fixed contraction band (Fig. 1–22), contraction of this type is clearly abnormal, corresponding to reduction to 60% of length in the segments concerned. Using a special interference microscope, Huxley and Neidergerke[100, 101] found that when the living fiber was stretched within physiologic limits, or released and allowed to shorten passively, the A bands stayed at constant width. The A bands also remained at a constant width (1.5 μ) during contraction, provided that shortening was limited. Later Huxley and Hanson[99] showed in the isolated myofibril a constant width in the A band when stretched or induced to contract by ATP. It was also found that when a fibril was stretched the H band increased in width by the same amount as did the separation of adjacent Z lines (Fig. 1–20). Thus the I band increases in width at the expense of the darker segment of A band, strongly suggesting the movement of rodlets out of the A band into the I band; this finding was suggested by Huxley's electron microscopic studies and formed the basis of his model. This is also presumed to be the basis of the movement of "Q substance." It should be emphasized that these changes relate only to change in length of the fiber, whether active or passive, and are not seen in isometric contraction. In more recent studies[100, 101] these findings have been extended to show that the length of the A band was not altered in stretching the sacromere length from 2.0 to 4.5 μ (frog single muscle fiber) or in isotonic contraction from 3.2 to 2.5 μ. If shortening proceeded to less than 1.9 μ, dense lines appeared at the center of each A band; and at 1.7 μ a second set of dense lines appeared at the Z line. This last change was associated with contraction bands. Huxley and associates concluded that contraction bands develop when the A band filaments meet the Z lines.

Huxley and Hanson[99] also showed that when myofibrils are treated with solutions able to extract myosin a large amount of material is removed from the A band, and only material stretching from the Z band to the former edge of the H zone is left. It was found by electron microscopy that the rodlets had disappeared, leaving only thin filaments from the Z line to the H boundary.[99] If, in addition to myosin, actin was also extracted from the fiber, all visible material disappeared except the Z lines. On the basis of such observations and on the basis of the constancy of the 415-Å x-ray periodicity of the fibrils in stretched and contracted states, Huxley maintained that the movement of rodlets in stretching represents the activity of cross links between actin and myosin rather than the lengthening or shortening of actomyosin filaments themselves.[97] Huxley and Peachey[102] have now shown that the ability of the isolated fiber to contract is lost when it is stretched to the point at which the sacromere length is 3.6 μ, whereas it can still shorten greatly at 3.5 μ. This failure of the contractile process at maximal length had long been known from the tension-length relationship for twitch contraction in whole muscle.[151] This new evidence favors the Huxley model and the overlap between actin and myosin filaments as the site of the essential contractile mechanism, although Carlsen and Knappeis[30] recently disputed the exact coincidence of the two phenomena.

In homogenates of pigeon, rabbit, and rat muscle Harman and Montague[80] were able to study living myofibrils showing spontaneous rhythmic contractions. In each such contraction the I segments were obliterated by movement of the A segment toward the Z line, and in the relaxation phase the broad M (H) band appeared in A segment, as Huxley and Hanson described. After 1–2 hours of such contractions the fibers shortened irreversibly by 50–60%, corresponding to the "delta state" of Ramsay and Street,[151] owing to depletion of substrate and failure of oxidative phosphorylation. If exposed to excess ATP and magnesium, such fibers would then further shrivel to 10% of their initial length.

It is therefore concluded that the phenomenon of the contraction band is related to the "extreme contraction" of Nageotte and the delta state of Ramsay, and must be considered pathologic. The process of natural contraction involves much less shortening of the fiber and implies a less complex and more rapidly reversible molecular change than is exemplified by the shortening of actomysin threads induced by ATP (as would be implied, for example, by the monomeric to dimeric transformation found by Chi and Tsao[35]). The giant complexes or polymers commonly postulated are more likely to be associated with slow reversal of contraction.

SARCOPLASM AND GRANULARITY

The sarcoplasm of the muscle fiber is the undifferentiated protoplasm in which the myofibrils are embedded. It fills the spaces between the myofibrils and accumulates about the sarcolemmal and end-plate nuclei. Although difficult to see in microscopic preparations, its presence is inferred from the fact that in transverse sections the myofibrils do not completely fill the fiber. Barer,[8] who attempted to isolate myofibrils by microdissection, found the sarcoplasm to be a sticky substance that caused adherence of the myofibrils.

The amount of sarcoplasm varies in different animals and in different muscles of the same animal. In frog muscle Buchthal is said[8] to have estimated the sarcoplasm at about 40% of the total volume of the muscle fiber. In general the most constantly active muscles (ocular, masticatory, and respiratory) possess the greatest amount of sarcoplasm,[23] whereas muscles that contract quickly and fatigue readily contain the least. Knoll,[112] who carried out the most thorough investigation of the sarcoplasm in different animals, classified muscles into those rich and those poor in sarcoplasm. He observed that the former tended to be darker but that—as was pointed out later—the actual color of the muscle fiber, whether red or white, depended more on its content of myoglobin, fat, and other chemical substances within the sarcoplasm than on the amount of sarcoplasm.

Within the sarcoplasm there is a diversity of granules of different sizes. These may be seen when the fresh tissue is examined in aqueous tissue juices[14] but are more readily visible if the fiber has been cleared in 1–5% potassium hydroxide. These granules appear as more or less refractile droplets and are soluble in alcohol and ether. The larger droplets are easily stained with Sudan (Fig. 1–23), Scharlach R, osmic acid, and Nile blue, and the finer ones less so. The granules disappear rapidly after death and in certain fixing agents.

Electron microscopic studies clearly differentiated the presence of four types of inclusions in the sarcoplasm: very small particles (glycogen?), mitochondria, the interstitial granules of Kölliker (which range from 150 to 350 Å in diameter), lipid granules, and sarcoplasmic reticulum. The *mitochondria* (sarcosomes), which do not differ from those in other cells,

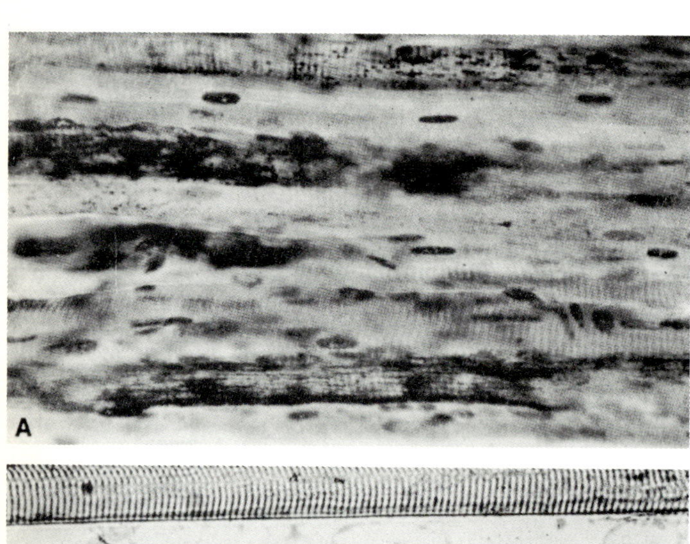

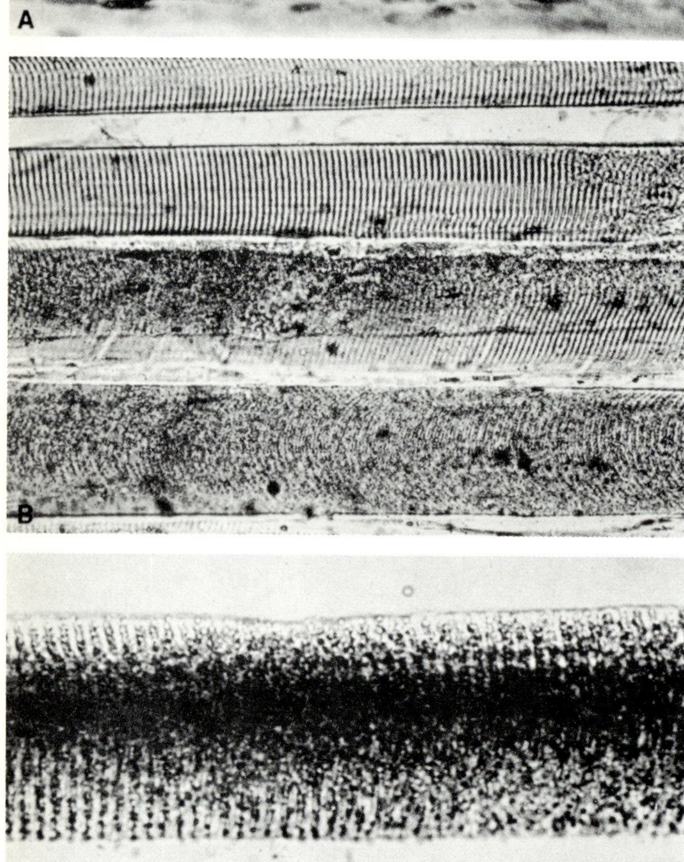

Fig. 1–23. (A) Normal human muscle. (B) Interstitial fat in muscle fibers of a 48-year-old obese woman. (A, silver stain for granules; B, Sudan stain) (C) Higher magnification of one of fibers in (B)

present a characteristic structure of little pouches with cristae in electron micrographs. In conventional sections they give a positive staining reaction with Janus green. They are irregular in size and distribution and are more numerous in some muscle fibers than in others. They are particularly abundant in human muscle,[186] and are most readily visualized by histochemical stains for oxidative enzymes (Fig. 1–24). Mitochondria are more numerous in type I red than type II pale fibers, and Z bands are thicker. Thus the two fiber types can be distinguished in electromyographs.

Kölliker[114] described fat granules first in the muscle of the winter frog and later in man and other mammals. Knoll[112] pursued this study,

being particularly interested in the distribution of fat granules in different animals. He, and later Holmgren,[91] Bell,[14] and Bullard,[27] confirmed Kölliker's findings. On the basis of the staining reactions it was assumed that the larger granules are composed of a complex lipoprotein. The distribution of lipid granules within the fiber does not follow a uniform pattern, but usually a large granule lies opposite each half of the isotropic J segment. This corresponds to the J granule of Holmgren. Other granules may be located opposite the Q segment, and many of them are irregularly dispersed. Thus the granules are not only arranged transversely at the J or isotropic line but also in longitudinal rows.[57, 126] The

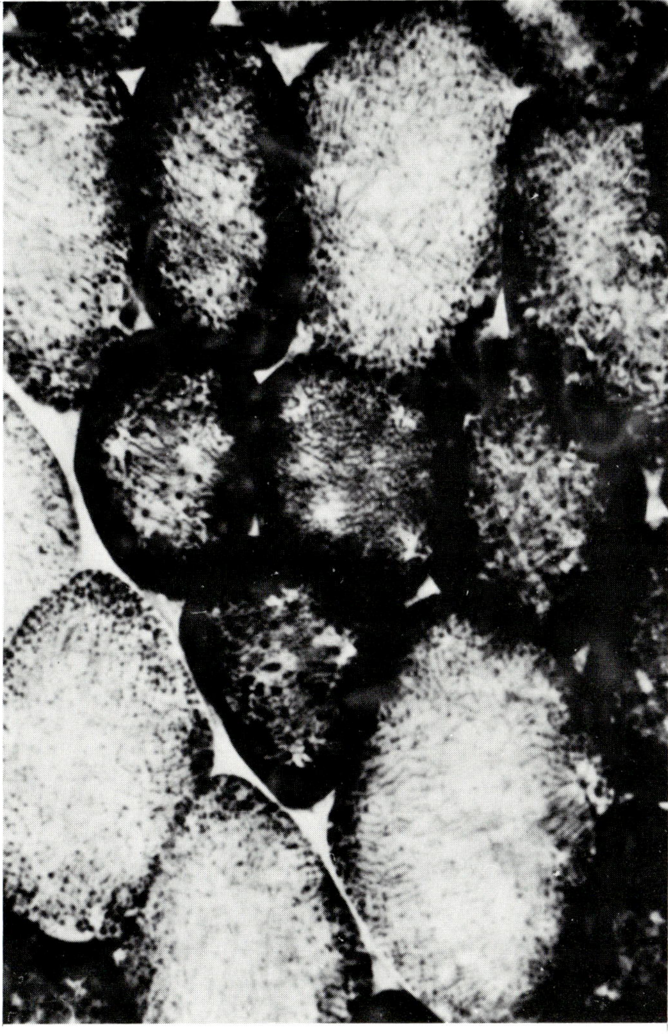

Fig. 1–24. Mitochondria of small type I and larger type II fiber. The mitochondria appear as black dots. (Courtesy Dr. C. M. Pearson) (succinic dehydrogenase stain, ×1200)

granules are not located at every J segment, and in any given longitudinal row extend only for a short distance before disappearing. From the work of Kölliker,[114] Retzius,[154] and Holmgren[91] it is quite clear that these granules lie between the myofibrils and probably represent special modifications of mitochondria. Jordan[106, 107] clearly demonstrated the difference between the granular alternation and myofibril alternation in insect muscle. Some fibers (usually type I) are laden with granules throughout their length and others are entirely free of fat (Figs. 1–18 and 1–23). The granules are usually associated with the mitochondria, according to Price.[150] Knoll and Hauer[113] showed that such fatty granules are reduced in starvation before there is any other muscle change, a fact confirmed by Bell,[14] Bullard,[27] and Denny-Brown.[47] A considerable quantity of lipid may remain, however, even in the most emaciated individual.

The muscle fibers which contain many fatty granules appear darker when examined unstained with hematoxylin. The darker fibers stain still more deeply with hematoxylin or gold impregnation, even when deprived of their fat reserves; but the staining is diffuse throughout the sarcoplasm and in both the light and dark bands, and is due chiefly to mitochondria. Hematoxylin and silver stains occasionally make clear this fundamental difference between light and dark fibers, and these same differences may be emphasized in any human muscle as a chance variant of the staining method. There is some relationship between the general pigmentation or redness of a muscle and the proportion of dark to light fibers, for the red slow muscles of certain animals (e.g., the soleus of the rabbit) are composed almost wholly of dark granular fatty fibers, whereas a pale muscle from the same animal contains less than 50% of such dark fibers. Denny-Brown[47] showed that these differences between individual muscle fibers are difficult to demonstrate at birth but become greatly accentuated during the first week after birth in the kitten, presumably reflecting the development of a specialized metabolic function. In the deep red pectoral muscle of the pigeon the contrast between light and dark fibers is very great, for all of the light fibers are large and the dark ones

uniformly small. The light fibers contract more rapidly than the dark ones (see below). The depletion of fat granules by inanition does not alter the speed of the contractile process.

The specific proteins of the sarcoplasm have proved difficult to localize in the muscle fiber. Weber[190] stated that myogen, the albumin fraction isolated from whole muscle, is confined to the sarcoplasm, but he obtained only indirect evidence of this. More likely the myogen fraction is a composite of a number of proteins with somewhat similar properties. Recent investigators[40, 61, 87] purified several fractions from myogen that are assumed to be identical with certain specific enzymes such as zymohexase, phosphorylase, dehydrogenase, and others. If this is true, and it appears likely, the sarcoplasm must certainly play a much more vital metabolic role, as suggested previously by Embden and Lange,[59] than merely that of a storehouse for foodstuff for the muscle fiber. The recent work of Hill[88] demonstrated that adenine nucleotide (including ATP) is concentrated in the I band, near the border of the A band.

There have been many other attempts to identify and localize specific chemical substances within the muscle fiber. In addition to fat in the form of granules and phospholipids in the isotropic segments, glycogen can also be demonstrated in scattered granules within the muscle fiber and, like the fat, has no direct relation to the contractile process.[62, 148] The localization of glycogen granules will be discussed below.

The protein content increases in true hypertrophy, and the addition is then clearly in the form of new myofibrils. Recent investigations show clearly that the increase in muscle size produced by growth hormone is not associated with an increase in contraction power and may indeed impair the efficiency of muscle. The ability of muscle protein to increase without an increase in the number of myofibrils must therefore also be admitted. The amount of glycogen, the amount of lipid, and the amount of protein can each vary independently of the others, the speed and power of twitch contraction remaining relatively constant. These data indicate, therefore, that muscle tissue is involved in several different metabolic functions independent of those underlying muscular contraction.[47]

In passing, it is of interest to note that the idea has been advanced from time to time that it is the sarcoplasm and not the myofibril which is the element actively engaged in contraction. Kühne in his Croonian lecture[117] in 1888 presented this hypothesis and proposed that the myofibrils were purely elastic supportive elements. This is in accord with the observation that in rather primitive and larval muscle cells sarcoplasmic contractility does exist before myofibrillar structure is recognizable.[48, 49, 120] Direct evidence for sarcoplasmal contractility in the adult vertebrate striated muscle fiber, however, is entirely lacking, and only the isolated living myofibril is contractile. Barer[8] reviewed this subject in 1948.

SARCOPLASMIC RETICULUM

The sarcoplasmic reticulum is a complex canalicular system with nonparticular contents. Although studied for the first time under the electron microscope by Bennett and Porter,[18] it was recognized by Venable[187] as early as 1902, and Heidenhain[84] illustrated a "cross-fiber network" in his 1911 book.[85] Light microscopic demonstration depended on the use of silver staining, introduced by Golgi and Cajal, which showed strands of reticulum extending transversely from the region of Z or J bands to the sarcolemma, accounting in all probability for the aforementioned granules in the latter membrane. Bennett[17] suggests that these points of contact correspond to the excitable regions of sarcolemma found by Huxley and Taylor.[104] Other parts of the reticular network extend longitudinally between the myofibrils.

According to the definitive studies of Porter and Palade,[148] the sarcoplasmic reticulum is arranged in a series of longitudinally oriented, irregular tubules and in transverse networks which form a kind of bracelet around each myofibril (Fig. 1–18). Each of the latter is located near the Z band or a short distance from it in the I band. The transverse systems are connected to the sarcolemma. Frequently subsarcolemmal vesicles and caveolae are observed at these points of juncture. In single pictures the tubules appear separated from the

inner surface of the sarcolemma, but many observers believe them to be in contact and even open onto the fiber surface. This would make the transverse tubules an inner extension of the sarcolemma. At the inner extensions of the transverse system there is a pair of blind vesicles or cisternae; the vesicles which separate them were called the "triad" of the transverse sarcoplasmic reticulum by Porter and Palade.[148] The finger-like projections of the vesicles of the transverse (T) system penetrate each myofibril, insinuating themselves between its outer myofilaments. They have a diameter of about 200 Å. Other parts of the sarcoplasmic reticulum extend longitudinally between myofibrils.

Thus the SR is believed to constitute a membrane-bound fluid compartment extending from the surface of each sarcomere into its contractile parts, the myofilaments. Its function is believed to serve as a transmitting mechanism that conveys the excitatory impulse from the sensitive sarcolemma to the interior of each contractile segment. The relationships of the transverse reticulum to the Z line and to the cell membrane at this level are shown at P, Q, and L in the frontispiece. This system of canaliculi is probably also involved in electrolyte exchange, and its distention occurs at an early stage in the vacuolar formation in certain types of edema such as periodic paralysis (Chapter 12).

See Chapter 5 for discussion of ultrastructure of the Z band.

HISTOCHEMICAL DETECTION OF MUSCLE FIBER CONSTITUENTS

During recent years the fields of histochemistry and cytochemistry have opened up new avenues for the detection and localization of a wide range of substances within tissues and cells.[10] Theoretically methods can be developed for the quantitation and localization of any metabolite or enzyme, but only a few reliable techniques are available at present. Localization of the constituents of the muscle cell poses special and difficult problems. The procedures used to prepare fresh tissue for histochemical stains (e.g., freezing or fixation in strong acidic or alcoholic reagents) often cause severe contraction and hence displacement of these constituents; and crushed fibers and staining artifacts are sources of confusion in the illustrations of several pertinent articles.

Of the many substances in muscle, glycogen and neutral fat were the first to be demonstrated. Recent improvements in techniques have provided more elegant pictures and have now settled the localization of these substances within the fiber. Methods for histochemical demonstration of fat reveal, of course, only the visible fraction of fat, always less in amount than what can be obtained by extraction methods as pointed out below. The fat is randomly dispersed within muscle fibers in large droplets and in many minute droplets which seem to be aligned in the isotropic bands.[46] (Fig. 1–23).

Glycogen, best preserved by strong fixatives such as Rossman's fluid (9 parts absolute alcohol saturated with picric acid and 1 part formalin) or Bouin's fixative modified by the addition of glacial acetic acid, followed by the McManus PAS technique[12] or the lead tetra-acetate-Schiff method,[13, 178] is stained along with other polysaccharides and glycoproteins. Previous exposure of the sections to diastase or saliva removes glycogen and leaves most of the other polysaccharides intact. Under normal conditions the amount and distribution of glycogen in muscle vary considerably. Two general patterns of distribution have been noted: (1) that of aggregates of particles of various sizes with some predilection for the perinuclear regions, and (2) that of fine granules tending to lie in relation to cross striations. At 1–12 hours after death the glycogen becomes aggregated into large masses (Figs. 1–19, 1–26); after 12–48 hours it disappears, presumably as a result of continued glycogenolysis. Malnutrition, starvation, obesity, and other abnormal physiologic conditions may alter the content of both fat and glycogen within the muscle. Further details concerning the glycogen deposits in the glycogen storage diseases are discussed in Chapter 11.

Of the many enzymes assumed to be active within the muscle fiber, only a few have been investigated by histochemical techniques. One of the most carefully studied enzymes is cholinesterase; its localization in the motor end-plate region is described in Chapter 2. The many enzymes (e.g., cytochrome oxidase and succinic dehydrogenase) involved in the oxidation substrates have been localized in mitochondria (Figs. 1–24 and 1–25). Each enzyme seems to vary in distribution within the fiber; and the intensity of staining reaction of each enzyme varies independently of the other. If muscle fibers have been destroyed in the course of disease there appears to be a concomitant decline in the quantity of these two enzymes. Phosphorylase and branching enzyme, both involved in anaerobic pathway of glycolysis, were investigated in normal muscle by Pearson[134, 143] using the Takeuchi method[178] and were shown to be evenly distributed in all fibers from several muscles. The normal muscle fiber is said not to contain alkaline phosphatase, whereas acid phosphatase was present in moderate amounts both in muscle fibers and interstitial connective tissue.[59] In diseased muscle a small amount of alkaline phosphatase appears in the fibers, and the concentration of acid phosphatase is increased.

Lipase, the fat-splitting enzyme, has been found in skeletal and heart muscle fibers of several birds.[70] In the parakeet and kite there was a uniform distribution of the enzyme throughout the pectoralis major muscle, whereas in the pigeon and other related species the enzyme was found only in the narrow red fibers that contain plentiful lipid inclusions. The reasons for this discrepancy are unknown. Evidently lipases have not been studied in human muscle. The enzyme 5-nucleotidase is not found in normal human muscle or interstitial connective tissue, but it has been reported[11] to be present in considerable quantity in connective tissue and in atrophic muscle fibers, especially in muscular dystrophy.

The presence of sulfydryl (SH) groups in muscle was demonstrated by Barnett and Seligman[9] and by Beckett and Bourne.[10] The latter workers also studied the distribution of protein-bound primary amino groups and found both of these types of compounds to be richly represented in human and several other vertebrate muscle fibers. Each seemed to be chiefly localized in the myofibrils, and no

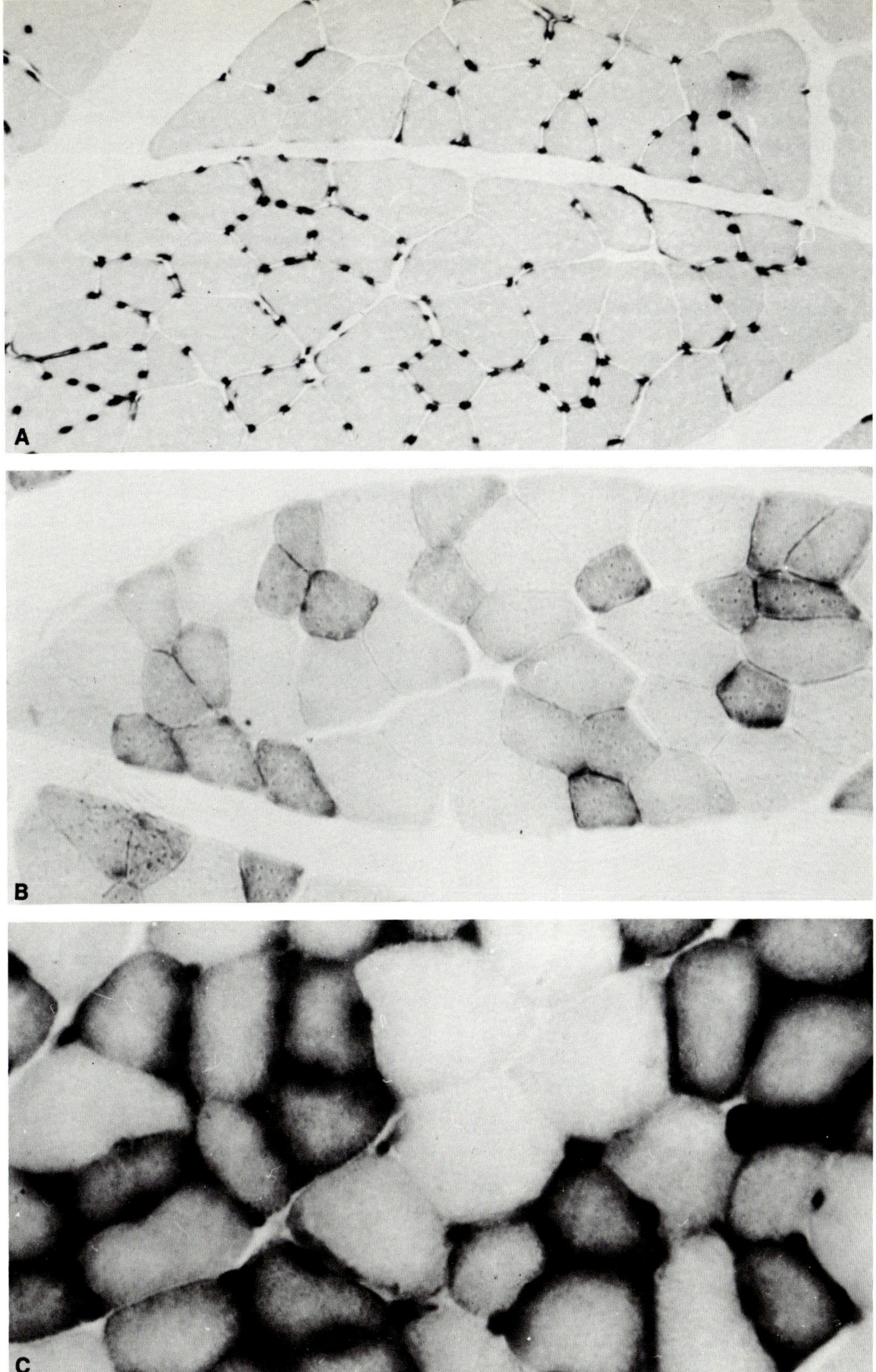

Fig. 1–25. (A–C) Transverse section of gastrocnemius muscle in rat. Note the darker and slightly smaller (red) type I oxidase-rich fibers and the larger, pale type II fibers. (A, alkaline phosphatase stain; B and C, cytochrome oxidase stain)

Fig. 1–26. Electron microscopic preparations of longitudinal section of human muscle. The black Z lines are flanked on either side by the I bands of thin myofilaments. The latter interdigitate with the thick myofilaments of A bands. Large oval cisternae of endoplasmic reticulum, some parts of longitudinal endoplasmic reticulum, and black dots of glycogen are also visible between myofibrils. (Courtesy Dr. Takeshi Sato, Institute of Neurosciences, Tokyo) (A, ×17,000; B, ×51,000)

significant variation was noted in concentration or distribution in pathologic muscle. Such is the state of knowledge at the time of this writing.

Many new cytochemical procedures have been introduced since the early 1950s, and they revealed a number of interesting differences among fiber types in many mammalian muscles[55, 60, 69, 138, 141, 142, 158, 189] (Figs. 1–22 through 1–28). In applying these methods to diseased muscle, Engel and his associates observed that the localization and indeed even the patho-

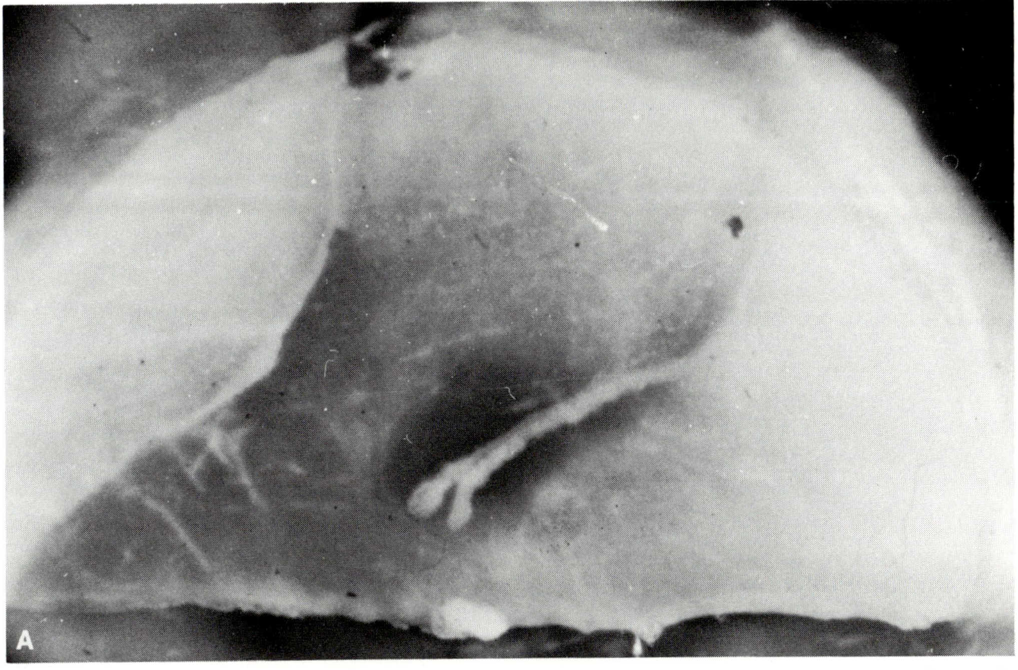

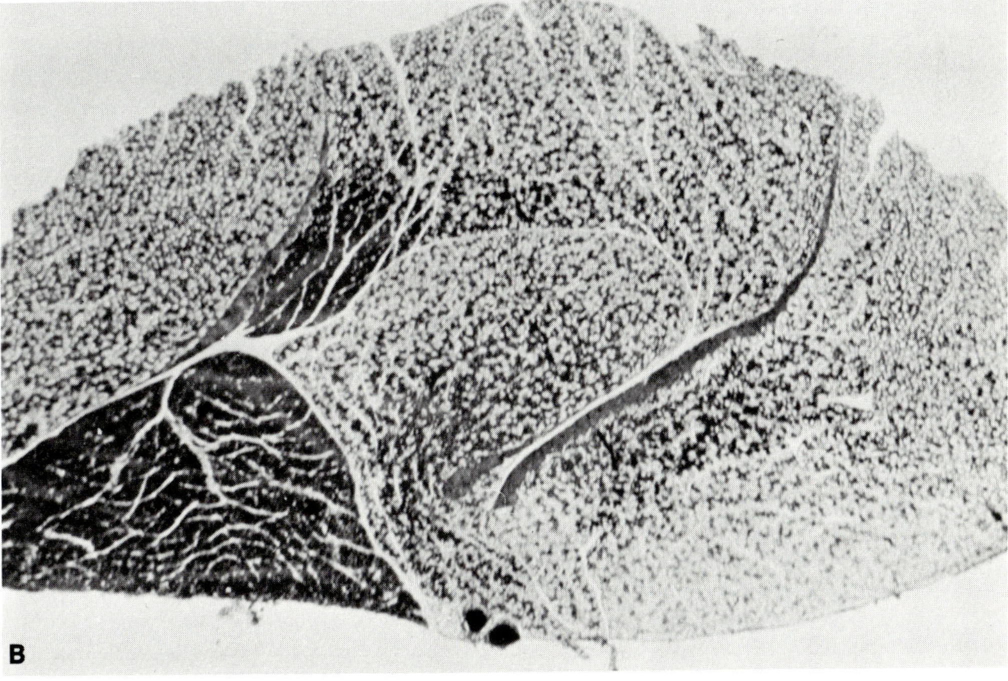

Fig. 1–27. Gastrocnemius, soleus, and plantaris muscles of rat. (A) The fresh muscle has been transected. The dark (red) soleus is seen in lower left and the (white) gastrocnemius and plantaris (with two bands of red fibers) surround the soleus. (B) The heavily stained soleus is easily seen; the plantaris lies in the center and varies in concentration of light and dark fibers; one small part of the pale gastrocnemius is dark. (B, Cytochrome oxidase stain) (Courtesy Dr. F. Romanul)

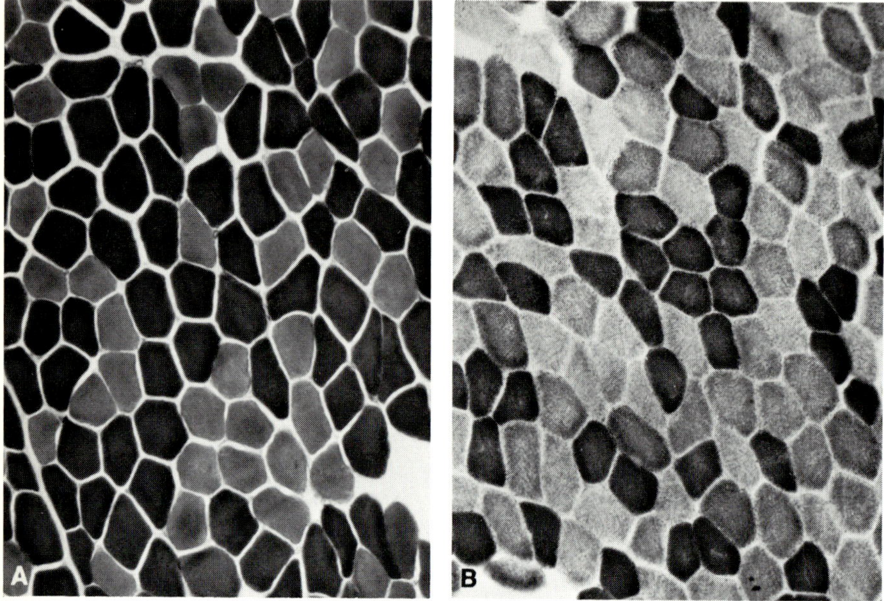

Fig. 1–28. (A) ATPase stain at pH 9.3 to show light type I (red) fibers and dark type II fibers. (B) In a succinic dehydrogenase stain the reactions are the opposite, viz. the type I fibers are dark and type II are pale. (Courtesy Dr. C. M. Pearson)

genesis of disease may be reflected in some measure by the manner in which it affects the enzymatic fiber types.

The use of size, form, position, ultrastructure, and enzymatic activity of mitochondria as the basis of a new classification of muscle is not new. Red and white muscles in animals were known to exhibit striking differences in their staining reactions. Since the first publication concerning this subject by Denny-Brown,[47] with methods which revealed oxidative and other enzymes, the red fibers are known to be rich in mitochondria (in proportion to myofibrillar mass), rich in myoglobin, and poor in myofibrillar adenosine triphosphatase (ATPase). In contrast, the white fibers contain fewer mitochondria but much phosphorylase and myofibrillar ATPase. ATPase and phosphorylase staining gives a picture the converse of that of the oxidative enzymes; the type II or white fibers are darker than type I or red fibers. According to Engel[60] there are qualitative differences between the mitochondria of type I and II fibers; the mitochondria of type I fibers, for example, are stained darkly by the method for succinic dehydrogenase, and those

of the type II fibers lightly—in contrast to the α-glycerophosphate dehydrogenase reaction, which is strong in the mitochondria of both type II and type I fibers. The staining of the transverse sarcoplasmic reticulum also is said to differ in the two fiber types. In the ATPase reaction, the one most widely used in fiber typing, the type I fibers at pH 9.4 are pale whereas the type II fibers at this pH stain darkly. (Fig. 28) Brooke,[24] who studied the resistance of type II fibers to preincubation at pH 4.3–10.4, subdivided them into types IIa, IIb, and IIc on the basis of their differing reactions at the low pH values.

The differentiation of muscle fiber types occurs postnatally at different times in various species. It is already evident in human muscle at birth according to Dubowitz[53] but continues to become more specialized as the motor units are engaged in neural mechanism.

The three or more histochemical fiber types are randomly dispersed in most muscles, as are glycogen-rich fibers. There is now fairly reliable physiologic evidence that all the fibers within a given motor unit (viz., all supplied by one neuron) are of the same type. Motor units subserving postural mechanisms tend always to be composed of type I, "slow twitch," tonic, fatigue resistant red fibers; those motor units involved in quick, phasic activities form a spectrum of white fibers, the IIa,b, and c ones of Brooke, Burke, Levine and Zajac find quite precise correlations between histochemical fiber types and their physiological properties [Burke RE, Levine DN & Zajac FE. Physiological histochemical correlations in 3 histochemical types in cat gastrochemical muscle Science 174: 709–712, 1971]. Five fiber types have been identified in eye muscle. Cross-innervation experiments[13, 158] in which a nerve to a white muscle is joined to the red muscle reverses its fiber type. Changing the physiologic activity of a muscle also alters its fiber types.

TENDINOUS ATTACHMENTS OF MUSCLE FIBER

The nature of the attachment of the muscle fiber to tendon has been a matter of dispute. There are two schools of thought: one contending that the myofibrils are attached to the inner surface of the sarcolemma at the end of the muscle fiber quite separate from the tendon fibrils, which are attached to the outer surface,[74, 123] and the other maintaining that the myofibrils are continuous with the tendon fibrils through the end of the fiber.[28, 32, 174, 177] Schultze[166] has taken an intermediate position by stating that the sarcolemma at the end of the fiber has a sieve-like arrangement through which the myofibrils join tendon fibrils. Häggquist[76] argues that muscle tension is transmitted to the sarcolemma through the Z disks rather than through the ends of the fiber.

Bennett and Porter[18] provided a further description, based on electron microscopic observations, noting that the myofibrils stop a short distance from the sarcolemma at the ends of fibers, being separated from them by a substance of unknown nature. The tendon fibrils are attached on the external surface of the sarcolemma. Bennett and Porter saw lateral attachments of Z bands to the inner surface of the sarcolemma; instead sarcoplasm always

intervenes. It must be concluded, therefore, that the transmission of force by contractile elements is at the ends and not at the sides of the fibers.

It seems impossible to settle this controversy by present methods. Barer[8] by microdissection and electron microscopic examination was unable to confirm the attachment of any part of the contents of the fiber to the sarcolemma. In our opinion, however, Carr[32] conclusively demonstrated that during the stage of development the myofibrils run directly into the sarcolemmal cap, the external surface of which attaches to the elementary fibril of the tendon. The formation of this terminal cone of the sarcolemma occurs later. This view received support from the experiments of Speidel,[173, 174] in which myoblastic differentiation provided the fibrils which connected the muscle fibers of one myotome to the next. He detected broken threads of myofibrils still attached to the terminal sarcolemma when the contents of the living fiber had retracted because of injury. Speidel noted that the myoblasts at the end of the fiber differentiate to form the tendon fibrils in living tadpole tail. This is probably the natural mode of transformation of muscle into aponeurosis. The factors governing the transformation are not known, but in general the proximal end of the muscle fiber allows an increase in length during growth and the distal end tends to form tendon aponeurosis. Commonly some central nuclei persist near each end of the fiber.

We studied fixed and stained mature muscle and are impressed by the fact that there is a distinct boundary between the muscle fiber and its attachment to tendon, as was originally shown by Péterfi,[145] but this is probably the result of the more complete differentiation of the two tissues. It also tends to be overemphasized by certain stains. The boundary is most clearly shown by the Mallory aniline blue stain, in which the deep blue of the connective tissue contrasts sharply with the orange of the muscle fiber (Fig. 1–29), and by silver impregnation of connective tissue, which demonstrates the fine reticular argentophilic fibrils. By these methods in combination with the phosphotungstic acid hematoxylin stains it can be seen that intact sarcolemma encloses the end of the

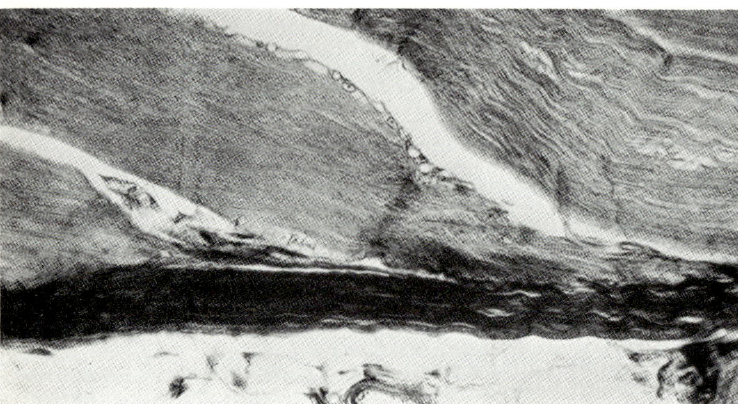

Fig. 1–29. Attachment of normal human muscle fibers to tendon. (phosphotungstic hematoxylin)

muscle fiber.[123] The shape of the ends of the fibers varies; some are tapering and conical, whereas others are rounded, blunt, truncated, or split. Argentophilic connective tissue fibrils form a cup for the end of each fiber. These fibrils then combine into a more compact bundle and after following a wavy course like any collagenous bundle reach the tendon with which they merge. The length of this terminal connective tissue strand—the distance between the end of the muscle fiber and the tendon—is 20–100 μ. If the distance is considerable, these connective tissue fibrils lose their argentophilic properties and become collagenous. Practically all of the muscle fibers meet the tendon at an angle, sometimes acute and sometimes wide. Since most muscles are four or five times as thick as their tendon, this is the only way by which muscle fibers can attach to the tendons without changing direction.[74]

One peculiarity of the musculotendinous junction that is not fully understood is its high content of cholinesterase. First observed by Couteaux in 1953 in fish and submammalian forms, it was soon confirmed by Gerebtzoff[71] in a variety of animals including man. This enzymatic localization becomes apparent midway through fetal life and persists. At first the cholinesterase is in the form of a series of dots that marginate the end of the fiber; these then fuse to form a kind of cup with finger-like projections into the muscle substance. These concentrations of cholinesterase provide the most satisfactory marker of the musculotendinous junction or the ends of the muscle fibers. Their functional significance is unknown.

SUPPORTING TISSUES OF MUSCLE

The muscle fibers are united by connective tissue sheaths which surround each individual fiber and groups of fibers. In this way muscle fibers up to several hundred in number are combined to form compact primary bundles, and several of these bundles merge to form secondary and even tertiary ones. The organization of the fibers within the muscle may follow any one of several patterns. In muscles designed to pull in one direction, all the fibers and fiber bundles are roughly parallel and the shape of the muscle is fusiform. In muscles whose action is exerted in more than one direction the fibers have a uni- or bipennate arrangement and vary accordingly in shape. Naturally the total strength of the muscle depends on the number of fibers, the direction of their action, and the manner of their origin and insertion.

The connective tissue investment of the muscle is called the *epimysium*. From this structure many bands of connective tissue project into the spaces between the primary, secondary, and tertiary muscle bundles. These form the *perimysium* and like the epimysium are composed of varying amounts of collagenous, reticular, and elastic fibers, together with fat cells. Blood vessels, lymphatics, and nerves lie within these connective sheaths. A delicate network of fibrous tissue extends from the perimysium into the primary muscle bundles and surrounds each fiber. In this, the *endomysium,* are the capillaries, nerve filaments, fibroblasts, histiocytes, and mast cells.

The endomysial network of connective tissue

as seen in a reduced silver stain is remarkably complex. It consists of fine reticular fibrils closely applied to the muscle fiber and of thicker fibrils which run between them and follow an independent, tortuous course (Fig. 1–13B). The latter arise by the coalescence of small groups of transverse or circular fibrils and terminate in the larger collagenous bundles of the endomysium and perimysium. Nagel's investigations[139] show that the collagenous fibril bundles wind in spirals around the individual muscle fibers before ramifying and joining the endomysial sheath. The course of these fibrils and the distance between them vary according to the degree of stretch or contraction of the muscle. In the contracted state they are almost at right angles to the axis of the fiber and close together, while in the extended state they are nearly parallel to the muscle fibers. Owing to this arrangement the connective tissue of muscles permits relatively easy displacement of muscle fibers and offers increasing resistance to extreme degrees of stretching. Feneis[63] emphasized the necessity of a connective tissue arrangement providing for an independent mobility of fascicles of muscle fibers. He mentions that the internal resistance of muscle connective tissue sheaths is reduced to a minimum by "displacement membranes," which are inserted between fascicles. These membranes are composed of connective tissue and often contain fat cells.

The amount of elastic tissue in the interstitial connective tissue appears to vary with the functional activity of the muscle. Continuously active muscles, such as those which move the eye, tongue, and diaphragm, have an abundance of elastic fibrils in the endomysium, whereas in the muscles of the extremities the elastic fibers are more exclusively limited to the septa between fasciculi in the perimysium.[163]

The tendons to which the muscle fibers are attached are glistening white cords of connective tissue that possess to a striking degree the properties of toughness and pliability. Their remarkable inelasticity has frequently been commented upon, for it has been observed that the bones may break asunder or the muscle may rupture before a healthy tendon yields to the strongest force. The constituent elements of tendon are arranged according to a simple plan which corresponds to the mechanical requirements of muscle, i.e., that the energy of muscle contraction be transmitted efficiently to the moving parts of the skeleton.

The collagenous connective tissue and the fibroblasts of which tendons are composed are recognizable by the time the muscle cells are well differentiated. The growth and repair of the tendon has been studied by Mason and Shearon,[127] Haines,[77] and Crawford,[42] who showed that an increase in the length of the tendon occurs chiefly during late fetal life and thereafter in the parts nearest the muscle fiber. During growth and in the repair that follows injury there is a proliferation of fibroblasts and the formation of new collagen. The closely apposed, parallel bundles of collagen normally have an undulating course and are bound together compactly. The bundles show a distinct longitudinal striation and fuse with one another at acute angles. Closely applied to the collagen fibrils are the tendon cells which produce them. These are fibroblasts that occur in rows and adapt themselves to prismatic spaces between the bundles. The nuclei of these cells, which are slender and oval, are easily seen, whereas their cytoplasm is indistinct, staining only lightly with basic dyes. When a clear view is obtained, the cells are found to be connected end to end. The primary tendon bundles also contain a small amount of elastic tissue in the form of a loose network, especially near the cells. The primary bundles are grouped into secondary bundles surrounded by septa of looser connective tissue. There are few blood vessels and nerves within these septa. The lymphatic vessels for the most part are confined to the connective tissue that surrounds the entire tendon with which the septa are continuous. A fibrous sheath, the vagina fibrosa, surrounds the tendon and may contain a cavity filled with a viscid mucoid substance similar to that of the joint cavities. The tendons and aponeuroses of muscle are connected with bones, cartilage, ligaments, and fibrous membranes. The ligaments have a structure similar to that of the tendons, except that the arrangement of the collagenous bundles is less regular.

There are four kinds of cells in the areolar connective tissue of the muscles: (1) Flattened lamellar cells (fibroblasts), which may be branched or unbranched. The branched ones have clear cytoplasm and oval nuclei, and the processes of these cells join to form an open network, as in the loose reticular endomysium. The unbranched ones are united end to end like epithelial cells. The tendon cells are examples of the unbranched variety. (2) Histiocytes or clasmatocytes, which are large, irregular cells characterized by the presence of granules or vacuoles in their cytoplasm. They possess a marked affinity for vital dyes and upon becoming phagocytic are called macrophages. (3) Granule or mast cells, which are ovoid or spherical in shape and contain within their cytoplasm large basophilic granules. (4) Plasma cells of Waldeyer and lymphocytes, which are spherical with small, dark nuclei and very little cytoplasm. These latter cells always reflect a pathologic process the nature of which may be obscure. In addition there are numerous fat cells (lipocytes), which lie between muscle fascicles and rarely between individual fibers, and of course the endothelial cells and pericytes of blood vessels; the latter are a source of histiocytes. The lipocytes occur in the perimysial and epimyseal connective tissue. Their localization is predominantly perivascular. Usually no lipocytes are seen in normal fetal or infantile muscle. Their number increases with age. Their is little or no correlation between the number of intramuscular fat cells and the amount of subcutaneous fat. Venable,[187, 188] who calculated the proportions of the main cell type in the mature mouse, finds the proportion of muscle nuclei, endothelial cells, fibroblasts, and pericytes to be 60:20: 15:5, respectively. Under the electron microscope 2% of the muscle cells qualified as satellite cells.*

Other undifferentiated or primitive mesenchymal cells (presumably with potentiality for conversion to fat, osteoblastic, or fibroblastic cells) reside in the areolar tissue of muscles. Immature cells of the red and white blood cell series can also be seen in the muscles of the fetus at term.

---

* The proportions were not altered during the atrophy that resulted from castration of the male or the hypertrophy following testosterone administration.

REFERENCES

1. ALLEN ER: Immunological and ultrastructural studies of myogenesis. The Striated Muscle. Monographs in Pathology. Edited by CM Pearson, FK Mostofi. Baltimore, Williams & Wilkins, 1973, Chap IV, pp 40–57

2. ASAI T: Beiträge zur Histologie und Histogenese der quergestreiften muskulatur der säugetiere. Arch Mikrosk Anat 86:8–68, 1915

3. AURELL G, WOHLFART G: Studien über den mikroskopischen bau der quergestreiften Muskulatur. Z Mikrosk Anat Forsch 40:402–444, 1936

4. BAIRATA A: Struttura e proprieta fische del sarcolemma. Z Zellforsch Mikrosk Anat 27:100–124, 1937

5. BANKER BQ, VICTOR M, ADAMS RD: Arthrogryposis multiplex due to congenital muscular dystrophy. Brain 80:319–334, 1957

6. BARDEEN CR: The development of the musculature of the body wall in the pig, including its histogenesis and its relations to the myotomes and to the skeletal and nervous apparatus. Johns Hopkins Rep 9:367–400, 1900

7. BARER R: The contractility of myofibrils. Proceedings of the 17th International Physiological Congress, 1947, p. 105

8. BARER R: The structure of the striated muscle fibre. Biol Rev 23:159–200, 1948

9. BARNETT RJ, SELIGMAN AM: Histochemical demonstration of protein-bound sulfhydryl groups. Science 116:323–327, 1952

10. BECKETT EB, BOURNE GH: Some histochemical observations on normal and diseased human skeletal muscle. J Histochem Cytochem 6:13–21, 1958

11. BECKETT EB, BOURNE GH: 5-Nucleotidase in normal and diseased human skeletal muscle. J Neuropathol Exp Neurol 17:199–204, 1958

12. BECKETT EB, BOURNE GH: Some observations on normal and pathological human muscle, using a combined McManus periodic acid-Schiff reaction and Sudan black on gelatin sections. Acta Anat (Basel) 34:111–124, 1958

13. BECKETT EB, BOURNE GH: Some studies on the "glycogen" of normal and diseased human skeletal muscle, using the lead tetraacetate-Schiff and the periodic acid-Schiff techniques. Acta Anat (Basel) 34:235–248, 1958

14. BELL ET: The interstitial granules of striated muscle and their relation to nutrition. Int Monatsschr Anat Physiol 28:297–347, 1911

15. BENNETT HS: Modern concepts of structure of striated muscle. Am J Phys Med 34:46–68, 1955

16. BENNETT HS: The sarcoplasmic reticulum of striped muscle. J Biophys Biochem Cytol 3 (Suppl):171–175, 1957

17. BENNETT HS: The structure of striated muscle. Structure and Function of Muscle. Edited by GH Bourne. New York, Academic Press, 1960

18. BENNETT HS, PORTER KR: An electron microscope study of section of breast muscle of the domestic fowl. Am J Anat 93:51, 1953

19. BOLAFFIO M, ARTOM G: Ricerche sulla fisiologia del sistema nervosa del feto umano. Arch Sci Biol (Bologna) 5:457–487, 1923–1924

20. BOWMAN W: On the minute structure and movements of voluntary muscle. Philos Trans Soc (Part II) 130:457–501, 1840

21. BOYD JD: Development of striated muscle. The Structure and Function of Muscle. Edited by GH Bourne. New York, Academic Press, 1960, Vol i, pp 63–85

22. BOZLER E, COTTRELL CL: The birefringence of muscle and its variation during contraction. J Cell Comp Physiol 10:165–182, 1937

23. BREMER JL, WEATHERFORD HL: A Textbook of Histology. Philadelphia, Blakiston, 1944, pp 154–168

24. BROOKE MH: The pathologic interpretation of muscle histochemistry. The Striated Muscle. Monographs in Pathology. Edited by CM Pearson, FK Mostofi. Baltimore, Williams & Wilkins, 1973, Chap VIII, pp 86–122

25. BRÜCKE E: Untersuchungen über den bau der muskelfasern mit hilfe der polarisierten lichtes. Denkschr Kaiserl Akad Wissensch Wien Math-Naturw Klin 15:69–84, 1858

26. BUCHTHAL F, KNAPPEIS GG, LINDHARD J: Struktur der quergestreiften lebenden Muskelfaser des Frosches. Acta Physiol Scand 73:162–198, 1936

27. BULLARD HH: On the interstitial granules and fat droplets of striated muscle. Am J Anat 14:1–46, 1912

28. BUTCHER EO: The development of striated muscle and tendon from the caudal myotomes in the albino rat, etc. Am J Anat 58:259–311, 1933

29. CAREY EJ: Studies on the wave-mechanics of protoplasmic motion. Am J Anat 61:159–201, 1937

30. CARLSEN F, KNAPPEIS GG: Ultrastructure and mechanical tension in striated muscle. Acta Physiol Scand (Suppl) 50:27–28, 175, 1960

31. CARLSON BM: The regeneration of mammalian muscle. The Regeneration of Striated Muscle and Myogenesis. Edited by A Mauro, SA Shafiq, AT Milhorat. New York, Academic Press, 1970

32. CARR RW: Muscle-tendon attachment in the striated muscles of fetal pig: demonstration of the sarcolemma by electric stimulation. Am J Anat 49:1–42, 1931

33. CASPERSSON T, THORELL B: The localization of the adenylic acids in striated muscle fibers. Acta Physiol Scand 4:97–117, 1942

34. CHAMBERS R, HALE HP: The formation of ice in protoplasm. Proc R Soc Lond [Biol] 110:336–352, 1932

35. CHI CW, TSAO TC: Chinese J Physiol 19:235, 1954. Quoted by Bailey K: Muscle proteins. Br Med Bull 12:183–187, 1956

36. COGHILL GE: The early development of behavior in amblystoma and in man. Arch Neurol 21:989–1009, 1929

37. COHN AE: Cardiac muscle. Cowdry Special Cytology. Second edition. New York, Hoeber, 1932, Vol 2, pp 1128–1172

38. COHNHEIM J: Ueber den feineren bau der quergestreiften muskelfaser. Virchows Arch [Pathol Anat] 34:606–622, 1865

39. COPERS CR: Multinucleation of skeletal muscle in vitro. J Biophys Biochem Cytol 7:559–566, 1960

40. CORI GT: 114th Meeting of the American Chemical Society, 1948, p. 27c

41. COUTEAUX R: Contribution a l'etude de la synapse myoneurale; Buisson de Kuhne et plaque motrice. Rev Can Biol 6:563–711, 1947

42. CRAWFORD CNC: An experimental study of tendon growth in the rabbit. J Bone Joint Surg [Br] 32:234–243, 1950

43. CUAJUNCO F: Development of the human motor end plate. Carnegie Inst Wash Publ 541, Embryology 30:127–152, 1942

44. CUAJUNCO F: Development of the neuromuscular spindle in human fetuses. Carnegie Inst Wash Publ 518, Embryology 28:95–128, 1940

45. D'ANCONA U: Contributo a una revisione della nostre conoscenze sulla morfologia della fibra muscolare striata. Protoplasma 10:179–250, 1930

46. DEMPSEY EW, WISLOCKI GB, SINGER M: Some observations on the chemical cytology of striated muscle. Anat Rec 96:221–248, 1946

47. DENNY-BROWN D: The histological features of striped muscle in relation to its function activity. Proc R Soc Lond [Biol] 104:371–411, 1929

48. DE RENYI GS, HOGUE MJ: Studies on skeletal muscle grown in tissue cultures. Arch Exp Zellforsch 16:167–186, 1934

49. DE RENYI GS, HOGUE MJ: Studies on cardiac muscle cells grown in tissue culture. Anat Rec 70:441–449, 1938

50. DEREUCK J, ADAMS RD: The number of muscle fibers in human striated muscles and their physiological and pathological determinants. In press

51. DOBIE WM: Observations on the minute structure and mode of contraction of voluntary muscular fibre. Royal Medical Society of Edinburgh. Ann Mag Nat Hist [Lond] 3:109–119, 1849

52. DUBOWITZ V: Enzyme histochemistry of skeletal muscle. J Neurol Neurosurg Psychiatry 29:23–28, 1966

53. DUBOWITZ V: Cross-innervation of fast and slow muscle: histochemical, physiological and biochemical studies. Exploratory Concepts in Muscular Dystrophy and Related Disorders. Edited by AT Milhorat. Amsterdam, Excerpta Medica, 1967, pp 164–168

54. DUBOWITZ V: Developing and Diseased Muscle, A Histochemical Study. Edited by W Heinennan. London, Medical Books, Ltd, 1968

55. DUBOWITZ V, PEARSE AGE: A comparative histochemical study of oxidative enzymes and phosphorylase activity in skeletal muscle. Histochemie 2:105–117, 1960

56. DUESBERG J: Les chondriosomes des cellules embryonnaires du poulet, et leur role dans le genese des myofibrilles avec quelques observations sur le developpement des fibres musculaires striees. Arch Zelforsch Mikrosk Anat 4:602–671, 1909

57. EBNER V: Von Untersuchungen über die Ursachen der Anisotropie Organisotropie Organisierter Substanzen. Leipzig, 1882, pp 243–244

58. ELFTMANN H: Quoted by Gelfan S: Functional activity of muscle. Textbook of Physiology. Edited by JF Fulton. Philadelphia, Saunders, 1949, p 110

59. EMBDEN G, LANGE H: Muskelatmung und sarcoplasma. Hoppe Seylers Z Physiol Chem 125:258–283, 1923

60. ENGEL WK: The essentiality of histo- and cytochemical studies of skeletal muscles in the investigation of neuromuscular disease. Neurology (Minneap) 12:778–791, 1962

61. ENGELHARDT WA: Enzymatic and mechanical properties of muscle proteins. Yale J Biol Med 15:21–38, 1942

62. FAWCETT DW, SELBY CC: Observations on the fine structure of the turtle atrium. J Biophys Biochem Cytol 4:63–73, 1958

63. FENEIS H: Ueber die anordnung und die bedentung des bindegewebes fur die mechanik der skelettmuskulatur. Morphol Jahrb 76:161–202, 1935

64. FINCK H, HOLTZER H, MARSHALL JM: An immunological study of the distribution of myosin in glycerol extracted muscle. J Biophys Biochem Cytol 2 (Suppl):175–178, 1956

65. FISCHER E: Birefringence and ultrastructure of muscle. Ann N Engl Acad Sci 47:783–797, 1947

66. FISCHMAN DA: An electron miscroscopic study of myofibril formation in embryonic chick skeletal muscle. J. Cell Biol 32:557–575, 1967

67. GELFAN S: The submaximal responses of the single muscle fibre. J Physiol 80:285–295, 1933

68. GELFAN S, BISHOP GH: Conducted contractures without action potentials in single muscle fibers. Am J Physiol 103:236–243, 1933

69. GEORGE JC, NAIK RM: Studies on the structure and physiology of flight muscles in bats. N Anim Morph Physiol 4:96–101, 1957

70. GEORGE JC, SCARIA KS: Histochemistry of muscle lipase. J Anim Morphol Physiol 5:43–48, 1958

71. GEREBTZOFF MA: Extraits. Am Histochem 1:26, 1956

72. GESELL A: The Embryology of Behavior. New York, Harper, 1945

73. GODLEWSKI E: Die entwicklung des skelet- und herz-muskelgewebes der saugetiere. Arch Mikrosk Anat 60:111–156, 1902

74. GOSS CM: The attachment of skeletal muscle fibers. Am J Anat 74:259–290, 1944

75. GREENFIELD DG, SHY GM, ALVORD EC, ET AL: An Atlas of Muscle Pathology in Neuromuscular Disease. London, Livingstone, 1957

76. HAGGQUIST G: Gewebe und Systeme der Muskulatur. Handbuch der Mikroskopischen Anatomie des Menschen. Edited by Von Molliendrof. Berlin, Springer, 1931, Vol 3

77. HAINES RW: The laws of muscle and tendon growth. J Anat 66:578–585, 1932

78. HALBAN J: Die dicke der quergestreiften muskelfasern und ihre bedeutung. Anat Hefte 3:267–308, 1894

79. HALL CE, JAKUS MA, SCHMITT FO: An investigation of cross striations and myosin filaments in muscle. Biol Bull 90:32–50, 1946

80. HARMAN JW, MONTAGUE HC: A study of contractions of skeletal muscle myofibrils by phase microscopy: motion picture demonstration. Am J Phys Med 35:182, 1955

81. HARRISON RG: An experimental study of the relation of the nervous system to the developing musculature in the embryo of the frog. Am J Anat 3:197–220, 1904

82. HAY ED: Regeneration of muscle in the amputated amphibian limb. Regeneration of Striated Muscle and Myogenesis. Edited by A Mauro, SA Shafiq, AT Milhorat. Amsterdam, Academic Press, 1970, Chap 1

83. HAYCRAFT JB: On the minute structure of striped muscle, with special reference to a new method of investigation by means of impressions stamped in collodion. Proc Roy Soc 49:287–303, 1891

84. HEIDENHAIN M: Handbuch der Anatomie des Menschen. Edited by KV Bardeleben. Jena, Fischer, 1911, Vol 8

85. HEIDENHAIN M: Plasma und Zelle. Jena, Fischer, 1911

86. HEIDENHAIN M: Ueber die enstehung der quergestreiften muskelsubstanz bei der forelle. Arch Mikrosk Anat 83:427–447, 1913

87. HERBERT D, GORDON H, SUBRAHMANYAN V, ET AL: Zymohexase. Biochem J 34:1108–1123, 1940

88. HILL DK: Autoradiographic localization of adenine nucleotide in frog's striated muscle. J Physiol 145:132–174, 1959

89. HOAGLAND CL: States of altered metabolism in diseases of muscle. Adv Enzymol 6:193–230, 1946

90. HODGE AJ: The fine structures of striated muscle. J Biophys Biochem Cytol 3 (Suppl): 131–143, 1957

91. HOLMGREN E: Von den Q- and J-Körnern der quergestreiften muskelfaser. Anat Anz 44:225–240, 1913

92. HOLTZER H: Mutually exclusive activities during myogenesis. Exploratory Concepts of Muscular Dystrophy and Related Disorders. Edited by AT Milhorat. Amsterdam, Excerpta Medica, 1967, pp 57–63

93. HONCKE P: Investigations on the structure and function of living isolated cross striated muscle fibres of mammals. Acta Physiol Scand [Suppl 48] 15:1–230, 1947

94. HUBER GC: On the form and arrangement in fasciculi of striated voluntary muscle fibers. Anat Rec 11:149–168, 1916

95. HURTHLE K: Zur kenntnis der struktur des ruhenden und tätigen froschmuskels. 3. Ueber die verteilung von wasser und fester substanz in der muskelfaser und uber den submikroskopischen bau der fibrillen. Pfluegers Arch 227:610–636, 1931

96. HUXLEY HE: Electron microscope studies on the organization of the filaments in striated muscle. Biochem Biophys Acta 12:387–394, 1953

97. HUXLEY HE: The double array of filaments in cross-striated muscle. J Biophys Biochem Cytol 3:631–648, 1957

98. HUXLEY JS, DEBEER GR: The Elements of Experimental Embryology. London, Macmillan, 1934

99. HUXLEY HE, HANSON J: Changes in the cross-striations of muscle during contraction and stretch and their structural interpretation. Nature (Lond) 173:973–976, 1954

100. HUXLEY AF, NIEDERGERNE R: Structural changes in muscle during contraction. Nature (Lond) 173:971–973, 1954

101. HUXLEY AF, NIEDERGERKE R: Measurement of the striations of isolated muscle fibers with the interference microscope. J Physiol 144:403–425, 1958

102. HUXLEY AF, PEACHEY LD: The maximum length for contraction in striated muscle. J Physiol 145:39P–40P, 1959

103. HUXLEY AF, TAYLOR RE: Activation of a single sarcomere. J Physiol (Lond) 130:49p, 1955

104. HUXLEY AF, TAYLOR RE: Function of Krause's membrane. Nature (Lond) 176:1068, 1955

105. JONES WM, BARER R: Electron microscopy of the sarcolemma. Nature (Lond) 161:1012, 1947

106. JORDAN HE: Studies on the striped muscle structure. VI. The comparative histology of the leg and wing muscle of the wasp. Am J Anat 27:1–67, 1920

107. JORDAN HE: The structural changes in striped muscle during contraction. Physiol Rev 13:301–324, 1933

108. KELLY AM, ZACKS SI: The histogenesis of rat intercostal muscle. J Cell Biol 42:135–153, 1969

109. KELLY DE: Myofibrillogenesis and Z band differentiation. Anat Rec 163:403–426, 1969

110. KITE GL: Studies on the physical properties of protoplasm. Am J Physiol 32:146–164, 1913

111. KLATZO I, HORVATH B, EMMART EW: Demonstration of myosin in human striated muscle by fluorescent antibody. Proc Soc Exp Biol Med 97:135–140, 1958

112. KNOLL P: Ueber protoplasmaarmen und protoplasmareiche muskulatur. Denkschr Kaiserl Akad Wissensch Wien Math-Naturw Klin 58:633–700, 1891

113. KNOLL P, HAUER A: Ueber das verhalten der protoplasmaarmen und protoplasmareichen, quergestreiften muskelfasern unter pathologischen verhältnissen. Sitzungsb Kaiserl Akad Wissensch Wien Math-Naturw Klin 101:315–348, 1892

114. KÖLLIKER A: Mikrosckopische Anatomie. Leipzig, W Engehmann, 1851

115. KRAUSE W: Ueber den bau der quergestreighten muskelfaser. Z Rat Med 33:265–270, 1868

116. KÜHNE W: Eine lebende nematode in einer lebender muskelfaser beobachtet. Virchows Arch [Pathol Anat] 26:222–224, 1863

117. KÜHNE W: On the origin and causation of vital movement. Proc Roy Soc 44:427–448, 1888

118. LEWIS WH: Development of the arm in man. Am J Anat 1:145–183, 1902

119. LEWIS WH: The development of the muscular system. Manual of Human Embryology. Edited by Keibel and Mall. Philadelphia, 1910, Vol 1, pp 454–522

120. LEWIS WH, LEWIS MR: Behavior of cross-striated muscle in tissue cultures. Am J Anat 22:169–194, 1917

121. LICHTMAN AL, MCDONALD JR: Birefringence in tissues. Arch Pathol 42:69–80, 1946

122. LOCKHART RD, BRANDT W: Length of striated muscle fibres. J Anat 72:470, 1938

123. LONG ME: Development of the muscle-tendon attachment in the rat. Am J Anat 81:159–198, 1947

124. MACCALLUM JB: On the histogenesis of the striated muscle and the growth of the human

sartorius muscle. Johns Hopkins Hosp Bull 9:208–215, 1898

125. MACCONNACHIE HF, ENESCO M, LEBLOND CP: The mode of increase in the number of skeletal muscle nuclei in the postnatal rat. Am J Anat 114:245–253, 1964

126. MARCUS H: Weitere untersuchungen über den bau quergestreitter muskeln. Anat Auz 55:475–497, 1922

127. MASON ML, SHEARON CC: The process of tendon repair: experimental study of tendon suture and tendon graft. Arch Surg 25:615–692, 1932

128. MATTAEI E, TIEGS OW: The path of the slow contractile wave in arthropod muscle fibre. Philos Trans R Soc 238B:349–359, 1955

129. MAURO A: Satellite cell of skeletal muscle fibers. J Biophys Biochem Cytol 9:493–495, 1961

130. MCKENZIE WC: The Action of Muscles, Including Muscle Rest and Muscle Re-education. New York, Hoeber, 1921, Chap 1

131. MEIGS EB: Striated and Smooth Muscle. Special Cytology. Second edition. Edited by Cowdry. New York, Hoeber, 1932, Vol 2, pp 1089–1125

132. MERKEL F: Die quergestreifte muskel. Arch Mikrosk Anat 8:81, 1872

133. MEYENBURG HV: Die quergestreifte muskulatur. Handbuch Spez Path Anatomie Histol. Berlin, Springer, 1929, Vol 9, pp 299–509

134. MOMMAERTS WFHM, ILLINGWORTH B, PEARSON CM, ET AL: A functional disorder of muscle associated with the absence of phosphorylase. Proc Natl Acad Sci USA 54:791–797, 1959

135. MOORE MJ, REBEIZ JJ, HOLDEN M, ET AL: Biometric analyses of normal skeletal muscle. Acta Neuropathol (Berl) 19:51–69, 1971

136. MORPURGO B: Ueber die postembryonale entwicklung der quergestreiften muskeln von weissen ratten. Anat Anz 15:200–206, 1898

137. MURALT AV: Ueber das verhalten der doppelbrechung des quergestreiften muskels während de kontraktion. Pfluegers Arch 230:299–326, 1932

138. NACHMIAS VT, PADYKULA HA: A histochemical study of normal and denervated red and white muscles of the rat. J Biophys Biochem Cytol 4:47–59, 1958

139. NAGEL A: Die mechanischen eigenschaften von perimysium internum und sarkolemm bei den quergestreiften muskelfasern. Z Zellforsch Mikr Anat 22:694–706, 1935

140. NAGEOTTE J: Sur la contraction extrême des muscles squelettiques des les vertébrés. Z Zellforsch Mikr Anat 26:603–624, 1937

141. OGATA T: A histochemical study of the red and white muscle fibers. Acta Med Okayama 12:216, 1958

142. PADYKULA HA, GAUTHIER GF: Morphological and cytochemical characteristics of fiber types in normal mammalian skeletal muscle. Anat Rec 157:296, 1967

143. PEARSON CM: Histopathological features of muscle in the preclinical stages of muscular dystrophy. Brain 85:109–120, 1960

144. PEASE DC, BAKER RF: The line structure of mammalian skeletal muscle. Am J Anat 84:175–200, 1949

145. PETERFI T: Untersuchungen über die beziehungen der myofibrillen zu den schnenfibrillen. Arch Mikr Anat 83:1–42, 1913

146. PLUM CM: Extramedullary blood production. George R Minot Symposium on Hematology. Edited by W Dameshek, Taylor. New York, Grune, 1949, pp 832–839

147. POGOGEFF IA, MURRAY MR: Form and behavior of adult mammalian skeletal muscle in vitro. Anat Rec 95:321–335, 1946

148. PORTER KR, PALADE GE: III. Studies of endoplasmic reticulum. J Biophys Biochem Cytol 3 (Suppl):269–300, 1957

149. PREVOST, LEBERT: 1844, quoted in Bardeen[6]

150. PRICE HM: Ultrastructural pathologic characteristics of the skeletal muscle fiber. The Striated Muscle. Monographs in Pathology. Edited by CM Pearson, FK Mostofi. Baltimore, Williams & Wilkins, 1973, Chap IX, pp 144–184

151. RAMSEY RW, STREET S: The isometric length-tension diagram of isolated skeletal muscle fibres of the frog. J Cell Comp Physiol 15:11–34, 1940

152. RANVIER L: Leçones d'Anatomie générale sur les Système Musculaire. Paris, Delahaye, 1880

153. REMAK: 1845, quoted in Bardeen[6]

154. RETZIUS G: Muskelfibrille und Sarkoplasma. Biologische Untersuchungen. Stockholm, Neue Folge, 1890, Vol 1, pp 81–88

155. REZNIK M: Satellite cells, myoblasts and skeletal muscle regeneration. Regeneration of Striated Muscle and Myogenesis. Edited by A Mauro, SA Shafiq, AT Milhorat. Amsterdam, Excerpta Medica, 1970, pp 133–157

156. ROBERTSON JD: New observations on the ultrastructure of the membranes of frog peripheral nerve fibers. J Biophys Biochem Cytol 3:1043–1048, 1957

157. ROBERTSON JD: The cell membrane concept. J Physiol (Lond) 140:58P, 1957

158 ROMANUL FCA: Enzymes in muscle. I. Histochemical differentiation of enzymes in individual muscle fibers. Arch Neurol 11:335–369, 1964

159. ROZSA G, SZENT-GYÖRGYI A, WYCKOFF RWG: The fine structure of myofibrils. Exp Cell Res 1:194–205, 1950

160. RUSKA H: Elektronenmikroskopischer beitrag zur histologie des skelettmuskels. Z Naturforsch [B] 9B:358, 1954

161. RUSKA A, EDWARDS CA: A new cytoplasmic pattern in striated muscle fibers and its possible relation to growth. Growth 21:73–88, 1957

162. SCHÄFER EA: Textbook of Microscopic Anatomy. New York, Longmans, 1912, pp 172–194

163. SCHIEFFERDECKER P: Muskelm und Muskelkerne. Leipzig, Ambros Barth, 1909

164. SCHIEFFERDECKER P: Untersuchung einer anzahl von baumuskeln des menschen und einiger säugetiere in bezug auf ihren bau und ihre kernerverhaltnisse nebst einer korrektur meiner herzarbeit. Pfluegers Arch 173:265–384, 1919

165. SCHMITT FO: The ultrastructure of protoplasmic constituents. Physiol Rev 19:270–302, 1939

166. SCHULTZE O: Ueber den direkten zusammenhang von muskelfibrillen und schenfibrillen. Arch Mikrosc Anat 79:307–328, 1912

167. SCHWANN: 1893, quoted by Bardeen[6]

168. SCOTT GH: Distribution of mineral ash in striated muscle cell. Proc Soc Exp Biol Med 29:349–351, 1932

169. SCOTT GH, PACKER DM: An electron microscope study of magnesium and calcium in striated muscle. Anat Rec 74:31–45, 1939

170. SHAFIQ SA: Satellite cells and muscle fiber regeneration. Regeneration of Striated Muscle and Myogenesis. Edited by A Mauro, SA Shafiq, AT Milhorat. Amsterdam, Excerpta Medica, 1970, pp 122–133

171. SICHEL FJM: The relative elasticity of the sarcolemma and of the entire skeletal muscle fiber. Am J Physiol 133:446–447P, 1941

172. SMITH NR: A note on the so-called disuse atrophy of muscle. J Anat 62:238–248, 1928

173. SPEIDEL CC: Studies on living muscles. I. Growth, injury and repair of striated muscle, as revealed by prolonged observations of individual fibers in living frog tadpoles. Am J Anat 62:179–235, 1938

174. SPEIDEL CC: Studies on living muscles. II. Histological changes in single fibers of striated muscle during contraction and clotting. Am J Anat 65:471–528, 1939

175. SPEIDEL CC: Fundamental transverse arrangement of cross striae in myofibrils of striated muscle. Anat Rec 100:91–100, 1948

176. STOCKDALE FE, HOLTZER H: DNA synthesis and myogenesis. Exp Cell Res 24:508–520, 1961

177. STUDNICKA FK: Ueber die bezichungen zwischen muskelfasern und bindegewebsfibrillen. Z Zellforsch Mikr Anat 26:36–114, 1937

178. TAKEUCHI T: Histochemical demonstration of branching enzyme (amylo-1, 4-1,6 transglucosidase) in animal tissues. J Histochem 6:208–216, 1958

179. TELLO JF: Génesis de las terminaciones nerviosas notrices y sensitivas; en el sistema locomotor de las vertebrodos superiores: histogenesis-muscular. Trab Lab Invest Biol (Madrid) 15:101–199, 1917

180. TELLO JF: Die entstehung der motorischen und sensiblen nervendigungen. Z Anat Entwicklung 64:384–440, 1922

181. TIEGS OW: On the arrangement of the structure of voluntary muscle fibres in double spirals. Trans R Soc South Aust 46:222–224, 1922

182. TIEGS OW: The structure and action of "striated" muscle fibre. Trans R Soc South Aust 47:142–152, 1923

183. VALENTIN G: 1835, quoted by Bardeen[6]

184. VALENTIN G: Die untersuchung der verkuzungserscheinungen der muskelfasern in polarisirtem lichte. Pfluegers Arch 21:307–327, 1880

185. VAN BREEMAN VL: Myofibril development observed with the electron microscope. Anat Rec 113:179–195, 1952

186. VAN BREEMAN VL: Ultrastructure of human muscle. I. Observations on normal striated muscle fibers. Am J Pathol 37:215–229, 1960

187. VENABLE JH: Constant cell populations in muscle. Am J Anat 119:263–269, 1966

188. VENABLE JH: Morphology of cells of normal testosterone-deprived and testosterone stimulated levator ani muscles. Am J Anat 119:271–297, 1966

189. WACHSTEIN M, MEISEL E: The distribution of histochemically demonstrable succinic dehydrogenase and mitochondria in tongue and skeletal muscle. J Biophys Biochem Cytol 1:483–489, 1955

190. WEBER HH: Muscle proteins and properties of the muscle. Proc R Soc Lond [Biol] B127:27–28, 1939

191. WHITBY LEH, BRITTON CJC: Disorders of the Blood. Philadelphia, Blakiston, 1950, pp 10, 397

192. WOHLFART G: Ueber das vorkommen verschiedener arten von muskelfasern in der skelettmusculatur der menschen und einiger säugetiere. Acta Psychiatr Neurol 12 (Suppl):119, 1937

INNERVATION AND BLOOD VESSELS OF MUSCLE

with brief consideration of the chemistry and physiology of muscular contraction

INNERVATION

MOTOR NERVE FIBERS, MOTOR END-PLATES, AND MOTOR UNITS

Both motor and sensory nerve fibers can be traced from the cerebrospinal nerves into the skeletal muscles. The muscular nerves are for the most part medullated and contain between 30 and 50% of sensory fibers. After entering the sheaths of the muscle, the nerve breaks up into bundles of fibers which subdivide until a single motor nerve fiber joins each muscle fiber (Figs. 2–1 and 2–2). Some long muscle fibers may receive two nerve endings, but this is rare, except in the panniculus carnosus of animals. The axon of a single motor anterior horn cell, by repeated branching, terminates on a group of muscle fibers averaging 100–200 in number, of which the physiologic function, in terms of the "all-or none" law, is indivisible. Each such group of structures is therefore termed a "motor unit" by physiologists. The coarse twitching of motor units in amyotrophic lateral sclerosis in very large muscles such as the deltoid and biceps indicates that some of the motor units are very large[74]—they are estimated by some to include more than 1000 muscle fibers[34, 95]—whereas in small muscles such as those attached to the eye, they may comprise only 10–15 muscle fibers. In conditions in which partial destruction of motor nerve fibers has occurred, a cross section of muscle commonly shows groups of muscle fibers in bundles of 10–20 distinguished by presence or absence of atrophy, and therefore it was reasonable to assume that a single motor unit must be made up of several such bundles, which, in some muscles at least, have a longitudinal arrangement.

This traditional view of the distribution of the muscle fibers of a given motor unit has been revised in the light of recent physiological and biochemical studies. There is now general agreement that all of the muscle fibers within the domain of an alpha motor neuron are more diffusely scattered within a given region and further, that each motor unit is of one histochemical type, e.g., phosphorylase-rich (type II), oxidase-rich (type I), or intermediate. Edström and Kugelberg[88] found that by stimulating the axon of a single alpha motor neuron they could deplete all of the muscle fibers of that motor unit of their glycogen, and in this way they were obviously distributed in a more or less random checker-board pattern, as illustrated in Figure 1–27. In the histochemical preparations of normal muscle only a few

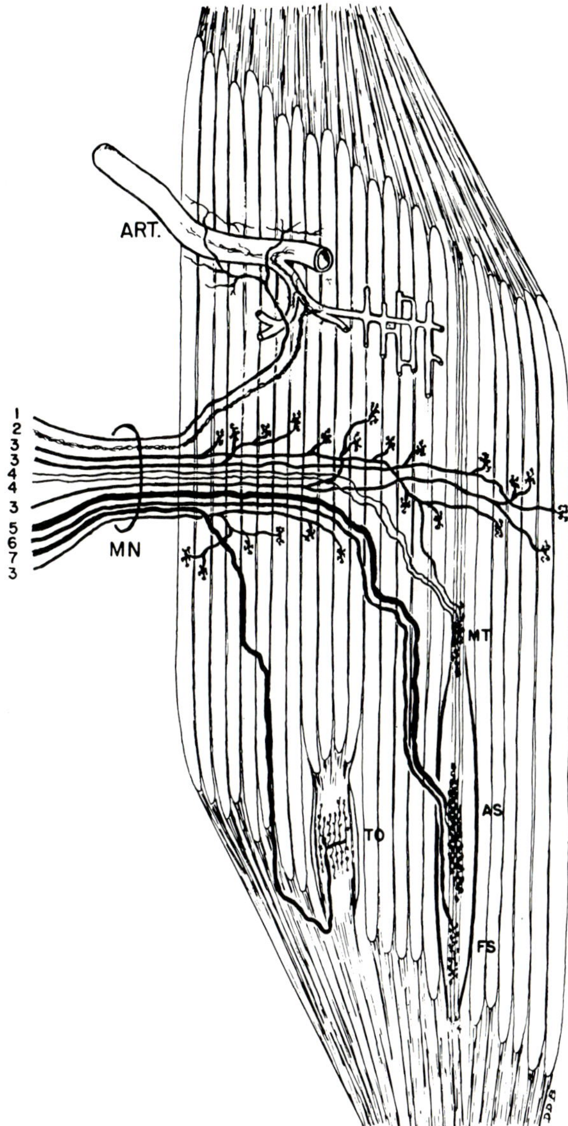

Fig. 2–1. Distribution of nerve fibers to a striped muscle. Muscular nerve (MN) contains approximately 50% of fibers derived from anterior roots. Medium-sized fibers (3, 3, 3) are distributed to motor end-plates, and small fibers (4, 4) to endplates of muscle spindles. Of fibers derived from sensory nerve roots, largest (5, 6, 7) are distributed to muscle spindles [to annulospiral endings (AS) and flower-spray ending (FS)] and to tendon organs (TO). Small sensory fibers (1), often nonmedullated in their peripheral part, are distributed in connective tissue surrounding blood vessels. Fibers derived from sympathetic nervous system (2) are distributed to muscular coats of arterioles and smaller arteries.

muscle fibers of one type are contiguous. These investigators went on to postulate that the groupings of "subunits" of atrophic fibers which feature denervative states are the result of secondary denervative atrophy of motor units after they had already collaterally reinnervated some of the isolated muscle fibers of units denervated at an earlier period.

The medullated nerve fibers remaining in a muscular nerve after degeneration of all sensory fibers fall into two groups.[85] About two-thirds of the motor fibers are of a fairly uniform diameter, between 8 and 14 μ. These are the nerve fibers of alpha motor neurons. The remaining third, the "small-nerve fibers," are nearly all 2–7 μ in size and are distributed exclusively to the specialized muscle fibers of the spindles.[147] In the frog, excitation of the small-nerve fibers contributes a small fraction to muscular tension, but in Mammalia no appreciable muscular contraction accompanies their activation. Their function will be further discussed in relation to the muscle spindles.

As each nerve fiber reaches a muscle fiber, its connective tissue sheath of Henle blends with the endomysium, and its myelin sheath terminates abruptly, as shown in the diagram in Figures 2–2 and 2–5. Upon the muscle fiber the axis cylinder of the nerve terminates in an arborization closely applied to an accumulation of muscle nuclei and sarcoplasm (Figs. 2–5 to 2–7). This was first observed by von Doyère in 1840 and later called "the motor end-plate" by Kühne,[151] who provided a complete description of it. The protoplasm of the end-plate is an expansion of the outer layer of

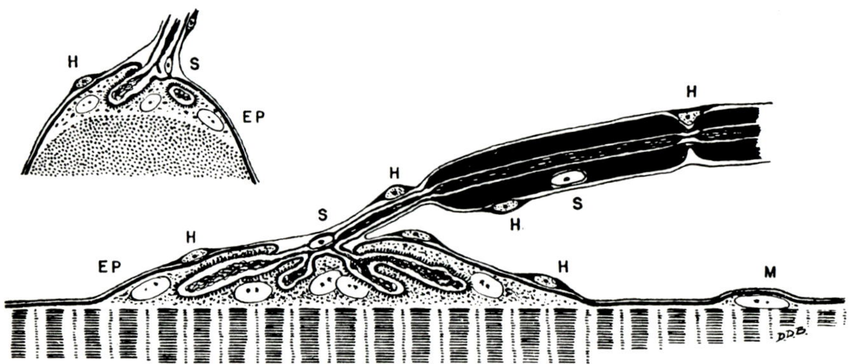

Fig. 2–2. Motor end-plate showing relationship between various structures in nerve and muscle. Last segment of myelin, with Schwann nucleus (S), terminates abruptly, leaving axis cylinder covered by sheaths of Schwann and Henle. End-plate nuclei (EP) of muscle fiber lie embedded in sarcoplasm and have same staining reactions as sarcolemmal nuclei (M). Ramifications of axis cylinder (telodendria) lie in grooves or pouches in granular sarcoplasm, each lined by spiny "subneural apparatus" of Couteaux, which is continuous with membranous sarcolemma and also Schwann membrane. Nucleus (S) of sheath of Schwann commonly lies near point of branching of axon. Sheath of Henle has small nuclei (H) and fuses with endomysial sheath of muscle fiber.

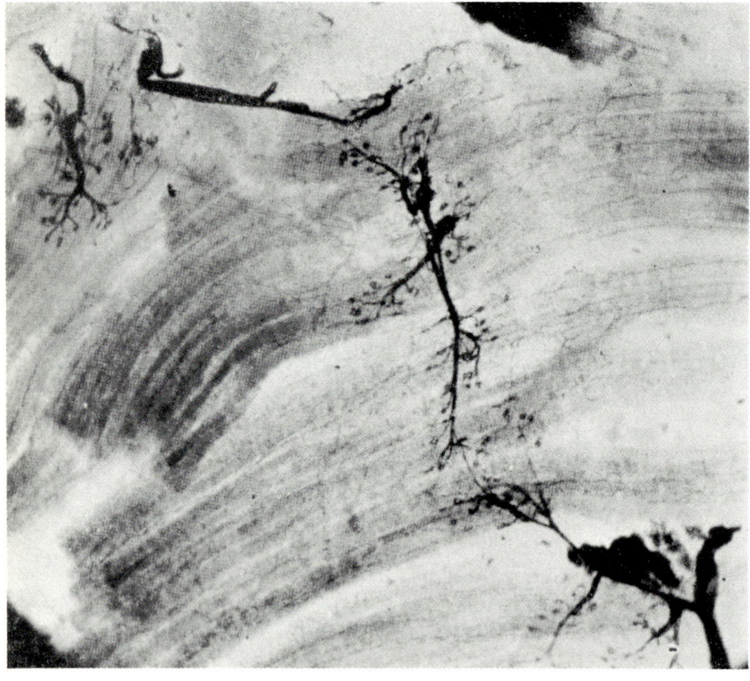

Fig. 2–3. Motor nerve endings in intercostal muscle of rat. (Ranvier gold chloride method)

sarcoplasm and consists of a mass of granular material, sometimes raised into a hillock (Doyère's hillock) within which lie the nuclei of the end-plate (Figs. 2–4 and 2–7). These motor end-plates are about 50 μ in diameter and are recognized by the collection of clear oval nuclei, which underlie the branching ramifications of nerve fibers (Figs. 2–5 to 2–7). The axis cylinder usually enters the end-plate as a single thread but occasionally divides before entering it. The pattern of ramification within the plate is greatly variable, but each branch ultimately ends as a neurofibrillar net or loop sometimes called the telodendrion, which has a more delicate network of neurofibrils with silver impregnation methods (Cajal, Bielschowsky) than with gold chloride (Ranvier) technics.

For a long period most histologists believed the axis cylinder to be naked after leaving the most distal myelin and to pierce and lie under the sarcolemma at the motor plate, thus provid-

Fig. 2–4. Motor end-plates of sheep. (Bielschowsky method)

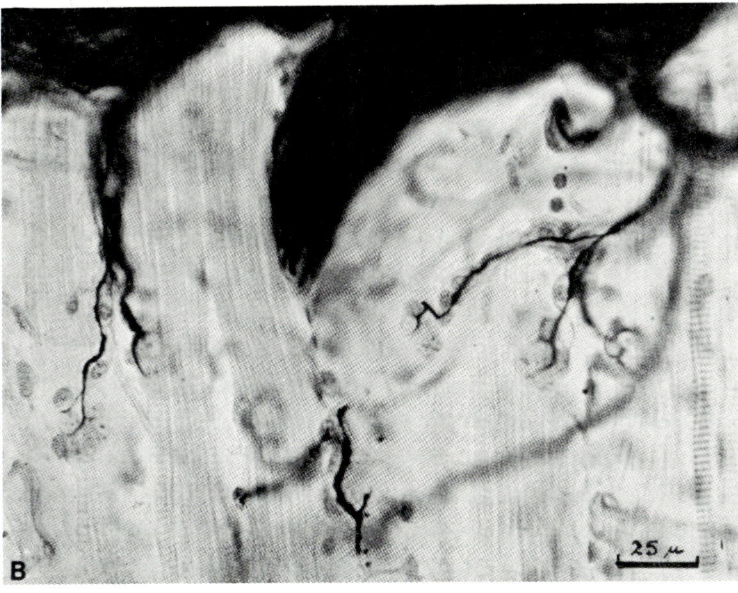

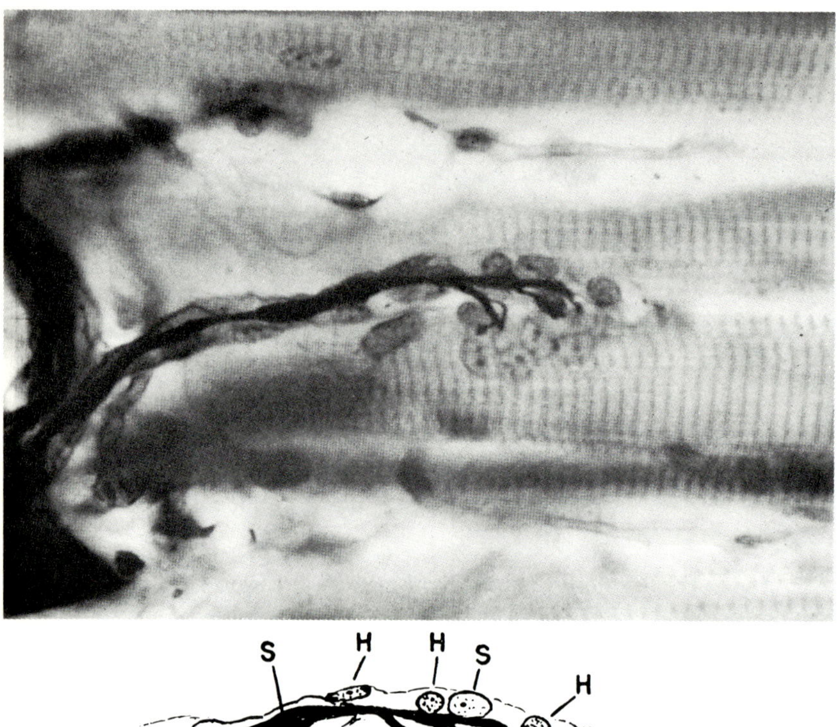

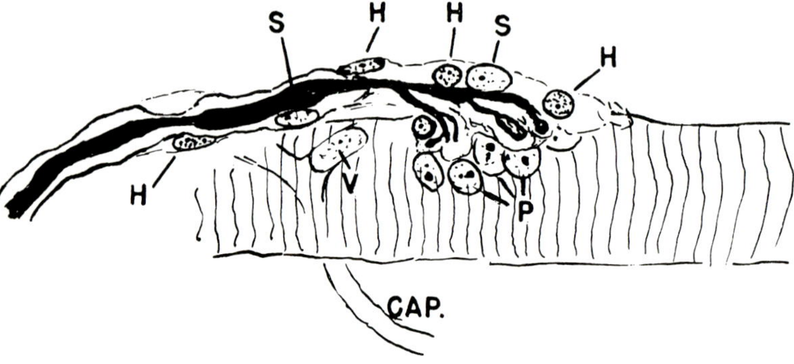

Fig. 2–5. Motor nerve endings of sheep. Diagram below distinguishes nuclei of sheath of Henle (H), sheath of Schwann (S), and of end-plate (P). A capillary nearby has an endothelial nucleus at V. (Bielschowsky method, counterstained with thionin)

ing direct continuity between neuroplasm and sarcoplasm. Boeke[23] claimed to have found a delicate network in Bielschowsky preparations, stretching from the neurofibrillar branches to make intimate contact with the myofibrils ("periterminal network"), the whole structure being under the sarcolemma (hypolemmal). In this way a clear distinction was drawn between motor and sensory endings on muscle fibers, for the latter are clearly epilemmal. Cajal[44, 45] on the other hand observed that a layer of neuroplasm always surrounded the neurofibrillar expansion, separating it from the sarcoplasm. The studies of Kühne[150, 151] supported Cajal's[43] view

that the motor ending is epilemmal and that a clear space or "matrix" covered the neurofibrillar termination of the motor axis cylinder and separated it from the sarcoplasm of the sole-plate at all points. This clear space can be seen clearly in Bielschowsky preparations, as in Figure 2–7B. Kühne observed the small dark nuclei of the sheath of Henle, which after gold inpregnation are seen to lie superficial to the terminal arborization, distinct from the nuclei of the sarcoplasm of the sole-plate.[151] The Henle nuclei may be distinguished by their round shape (Figs. 2–4 to 2–6). The matrix with the contained telodendria is embedded in

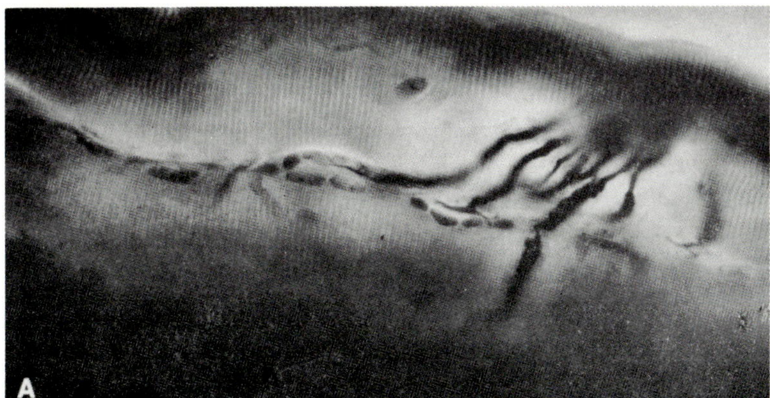

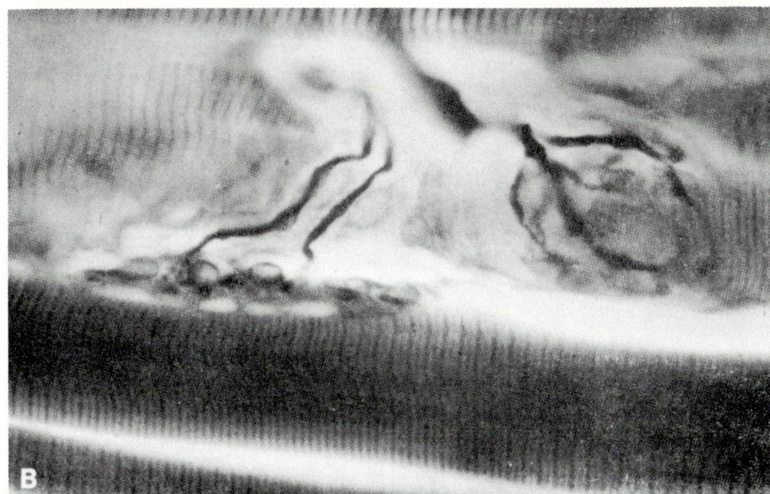

Fig. 2–6. (A) Motor endings of cat showing variations of staining of muscle nuclei (Bielschowsky method). (B) Muscle and Schwann nuclei are unstained, and Henle nuclei appear in outline.

grooves in the sole-plate. The ending was, according to Kühne, situated under the telolemma, which is the fusion of the sheath of Henle with the endomysial sheath of the muscle fiber. The sheath of Henle is a thin endoneural sheath of connective tissue which surrounds each motor nerve fiber and is best seen after silver impregnation by the Bielschowsky method (Figs. 2–6B and 2–8). The sheath of Schwann was considered to have ended with the last segment of myelin.

In sections stained by Cajal methods, or by methylene blue, the motor end-plate certainly appears to be embedded in the sole-plate, under an outer limiting membrane which is continuous with the sarcolemma, as in Figure 2–6A. When the sarcoplasm is well stained, however, the clear space or "matrix" of Kühne is clearly visible, and in a lateral view the neurofibrillar nets are separated from the limits of the sarco-plasm by a sharp line which is probably the true muscle cell membrane. The periterminal network is the result of one type of staining of the sarcoplasm of the sole-plate, and in places it may appear to be continuous with neurofibrils, as in Figure 2–7A, though a clear space may be seen to intervene if the staining is critical.

Fresh light has been thrown on the subject by Couteaux,[61, 62] whose studies of the development of the small superficial nuclei led him to the conclusion that these were Schwann nuclei, which with an associated prolongation of their cytoplasm ensheathed the neurofibrillar network to its limit, thus separating the neurone from the muscle cell. This ultimate sheath of the axis cylinder termination Couteaux calls the teloglia; it appears to correspond to the "matrix" of Kühne. Boeke[24, 25] and Tello[235] have also come to the conclusion that the more

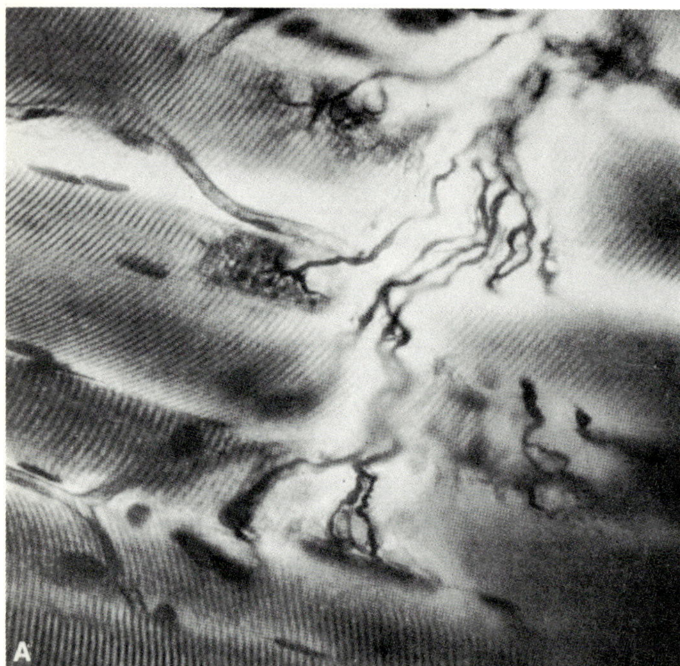

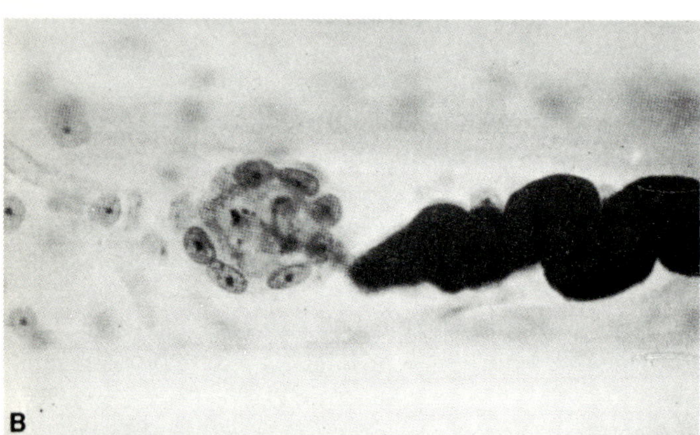

Fig. 2–7. (A) Intercostal muscle of cat. Periterminal network of Boeke (Bielschowsky method). (B) Human motor nerve ending showing granular material of sole plate and nuclei, and clear spaces around actual termination. Myelin is stained. (Bielschowsky method)

superficial group of end-plate nuclei are derived from the sheath of Schwann. In our own preparations we find that most of the superficial nuclei of the mammalian end-plate have the characteristics of those of the sheath of Henle and are so depicted in our diagram. The distinction is particularly clear in the degenerated motor plate, as in Figure 2–5, in which the small round dark nuclei of Henle are sharply differentiated from the larger vesicular muscle nuclei, on the one hand, and the elongated nuclei of the sheath of Schwann, on the other. One nucleus of Schwann appears to preside over the last segment of "teloglia," being commonly situated at the first point of branching (Fig. 2–5).

Couteaux[63] further found that brief supravital staining by Janus Green resulted in light staining of the sarcolemma and, if continued, of a curious "subneural apparatus" in each motor ending which outlines the "teloglia," as if invaginated by the arborization of the axon. The latter therefore lies external to the sarcolemma. This subneural structure covers each branch of the terminal neurofibrillar network and therefore has a branched fernleaf appear-

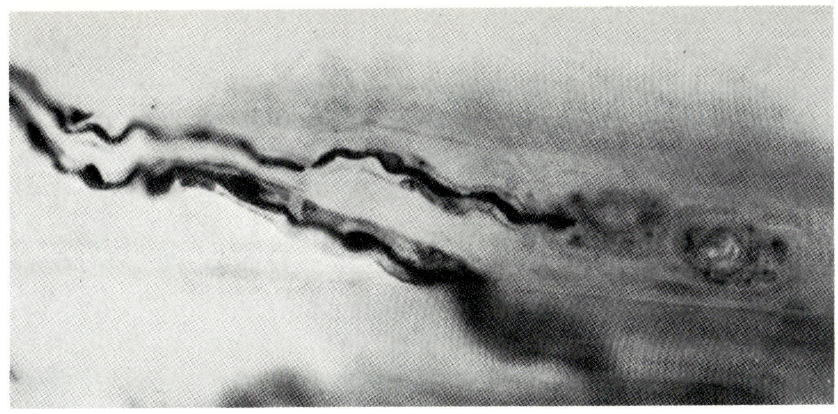

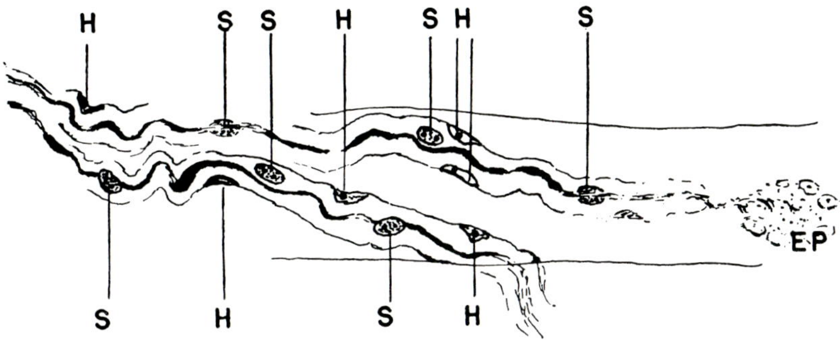

Fig. 2–8. Nerve fibers approaching end-plate showing Henle's sheath and nuclei (H) and their relation to Schwann nuclei (S).

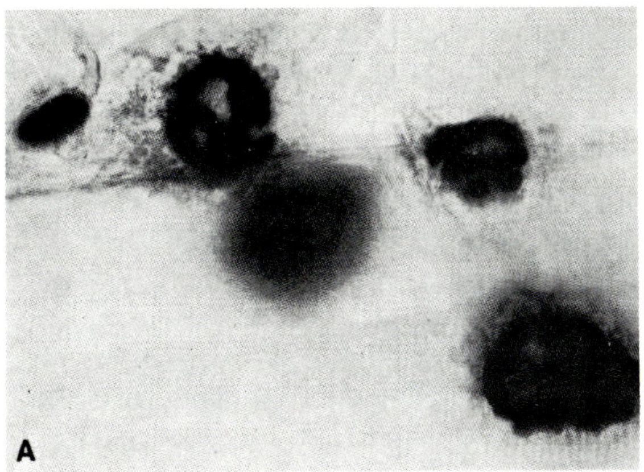

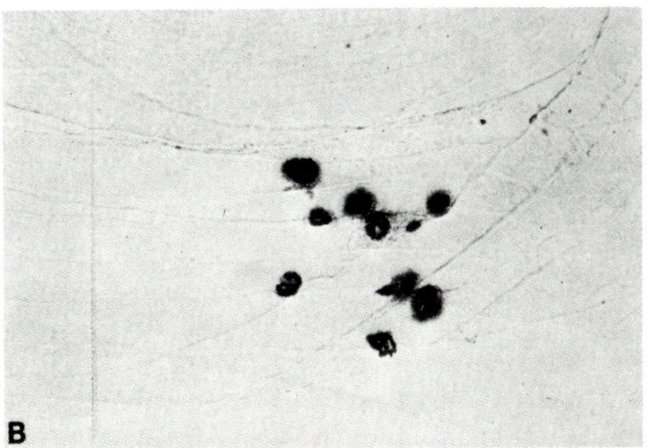

Fig. 2–9. Cholinesterase outlining "subneural apparatus" at motor end-plate. (Courtesy of Dr. Koelle)

ance, corresponding to the pattern of arborization. By means of such supravital staining, it was demonstrated that the whole neurofibrillar ramification (telodendria) is embedded in grooves in the sarcoplasm, covered throughout its extent by a layer of Schwann cytoplasm ("teloglia") and this in turn by a layer which stains with Janus Green. This last layer is present only where the teloglia is embedded in the grooves of the end-plate and does not extend onto the entering nerve fiber. The membranous interface between axis cylinder and the muscle cytoplasm is extremely complex since with the Janus Green method, a series of spinous processes called "organites" protrude into the sarcoplasm of the end-plate as if the "teloglia" was covered by short hairs or spines. The "subneural" formation of "organites" remains after degeneration of the nerve but has then lost its regular pattern. These observations and those of Iwanaga[130] Nöel,[183] and Nöel and Pomme[184] all indicate that the nerve fiber lies in grooves on the surface of the muscle fiber (epilemmal) and does not penetrate the sarcolemma. The ultrastructure of the infoldings of the sarcolemma has been studied by Reger,[205] who found evidence of three layers in the infolded pouches, and by Robertson[209] who observed five distinct layers separating axoplasm from sarcoplasm in the reptilian myoneural junction.

The most complete descriptions of the motor end-plate or neuromuscular junctions, as seen through the electron microscope, are to be found in the article by Coërs and his associates[53] and in the monograph by Zachs.[254] All these studies show the terminal axoplasm of nerve to be in immediate contact with and tightly adherent to sarcoplasm, without intervening collagenous or endothelial membranes. Moreover the plasma membranes of the motor nerve fiber and muscle fiber at their points of junction exhibit characteristics not found elsewhere—"a structural originality in its pre- and postsynaptic components," to use the words of Couteaux.[49] Some of these latter derive from interactions between axoplasm and sarcoplasm (Figs. 2–10 and 2–11).

Koelle and Friedenwald[137] have successfully stained the "subneural" apparatus of Couteaux by a histochemical method which is considered to be specific for cholinesterase activity (Fig. 2–9). Acetylthiocholine iodide was used as a substrate with copper glycinate and copper thiosulfate in teased preparations and frozen sections of rat muscle. The original method did not permit differentiation between the several different acetylcholine-splitting enzymes, though this has been possible by an improved technic.[138] The appearance is identical to that of the fernleaf appearance produced by the Janus Green method. It is therefore probable that the "subneural apparatus" is a relatively localized accumulation of cholinesterase in the portion of the membrane of the muscle cell which forms the interface between nerve and muscle. If so, it is remarkable that it should remain relatively intact following degeneration of the nerve. The collection of granules in the subneural apparatus, revealed by the Janus Green stain, prove to be accumulations of mitochondria.

The teloglia and the sheath of Henle, which attaches the collagenous layer of the nerve fiber to the outer endomysial layer of the muscle fiber, also resist degeneration after nerve section. The terminal ramification of the axis cylinder can therefore be said to lie between the endomysial connective tissue sheath and the sarcolemma, indenting the latter over the sarcoplasm of the sole-plate. A large, oval, darkly staining nucleus of the sheath of Schwann may lie over the end-plate and is accompanied by the smaller nuclei of the sheath of Henle. In human end-plates the muscle nuclei are pale and are often grouped in a circle around the edge of the sarcoplasm.

In a well-stained preparation of a small muscle it is clear that each muscle fiber receives only one motor end-plate (Fig. 2–2). Reports to the contrary are in relation to double end-plates on the muscle fibers of muscle spindles (Agduhr[2, 3]) and on very long muscle fibers such as are found in amphibia and in the panniculus carnosus of the hedgehog (Garven[99]). Sections of muscle, particularly when stained by silver impregnation, sometimes give an impression of two or even three endings on one muscle fiber, as in Figure 2–6A, but careful focusing under oil immersion shows that the endings belong to overlapping, different fibers. In muscles such as tenuissimus, in which muscle

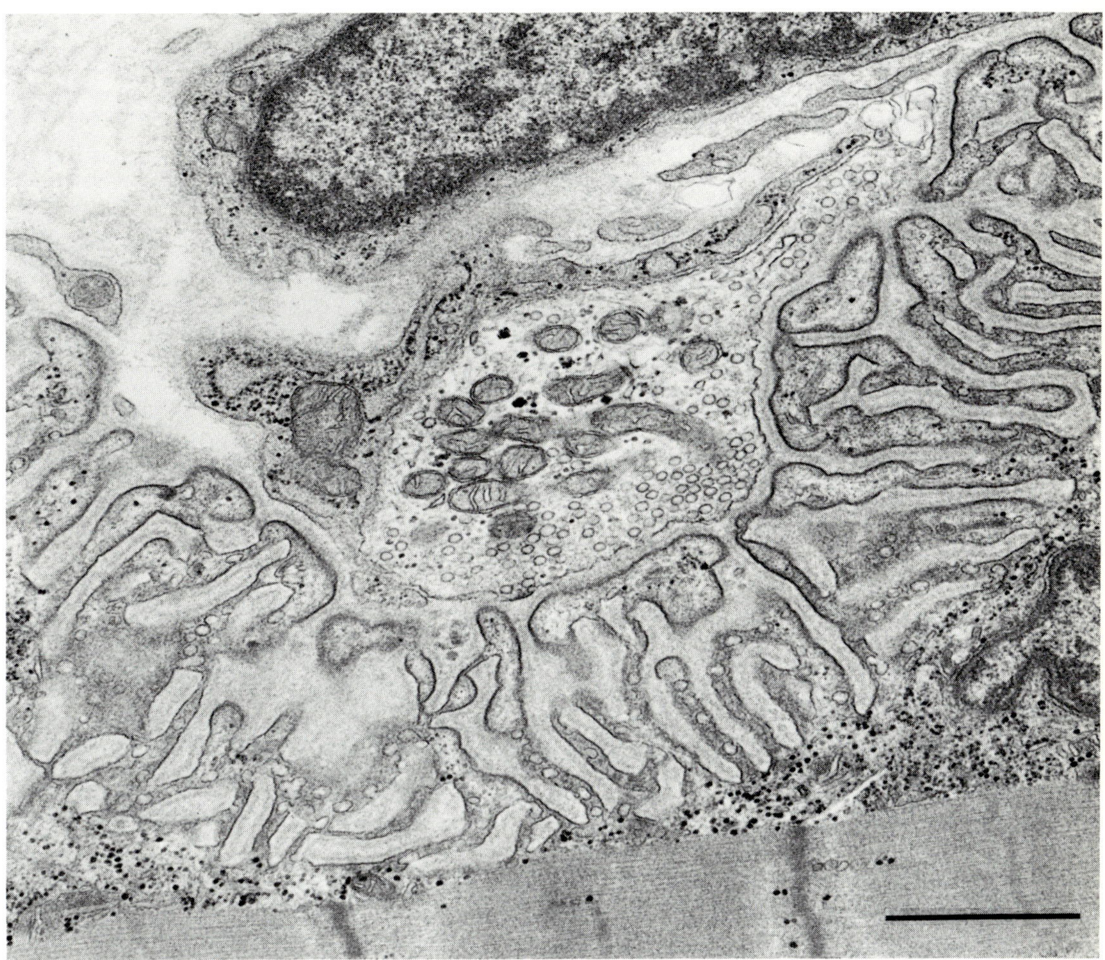

Fig. 2–10. Normal human end-plate. The axon with its presnaptic vesicles is seen in upper part of picture. The adjacent nucleus is that of a fibroblast. The muscle fiber is below. The axon indents the muscle fiber whose surface is unfolded. A homogenous material lies between the axolemma and sarcolemma. The sarcoplasmic folds are finely branched. The functional sarcoplasm contains microtubules and ribosomes. The line is 1 μ in length. (Illustrations kindly lent by Dr. A. G. Engel) (EM ×30,000)

fibers are dispersed in chains attached end to end, the individual fibers in a chain are innervated by the same parent nerve fiber, so that all in one chain must contract together.

Carey[47] has argued that morphology of the axonal termination in the end-plate reflects the functional activity of the motor unit. However, considering the great variability of morphology of end-plates, even in a single muscle,[76] and the fact that all of Carey's work was done on fixed and stained material, it is doubtful whether the changes which he reported are significant. Barer[12] has been unable to detect any special activity of the end-plate in the course of prolonged observations on the living motor nerve endings in insect larvae under phase-contrast microscopy. He noted that the end-plates move passively during muscular contraction and manifest an ameboid motion. In our own experience, only in the last stage of degeneration of the motor nerve fiber, or in phases of myopathy or myositis when the muscle fiber concerned already shows evidence of degeneration, do the telodendria show changes which are sufficiently distinctive to be significant.

In the muscles of the frog and other Am-

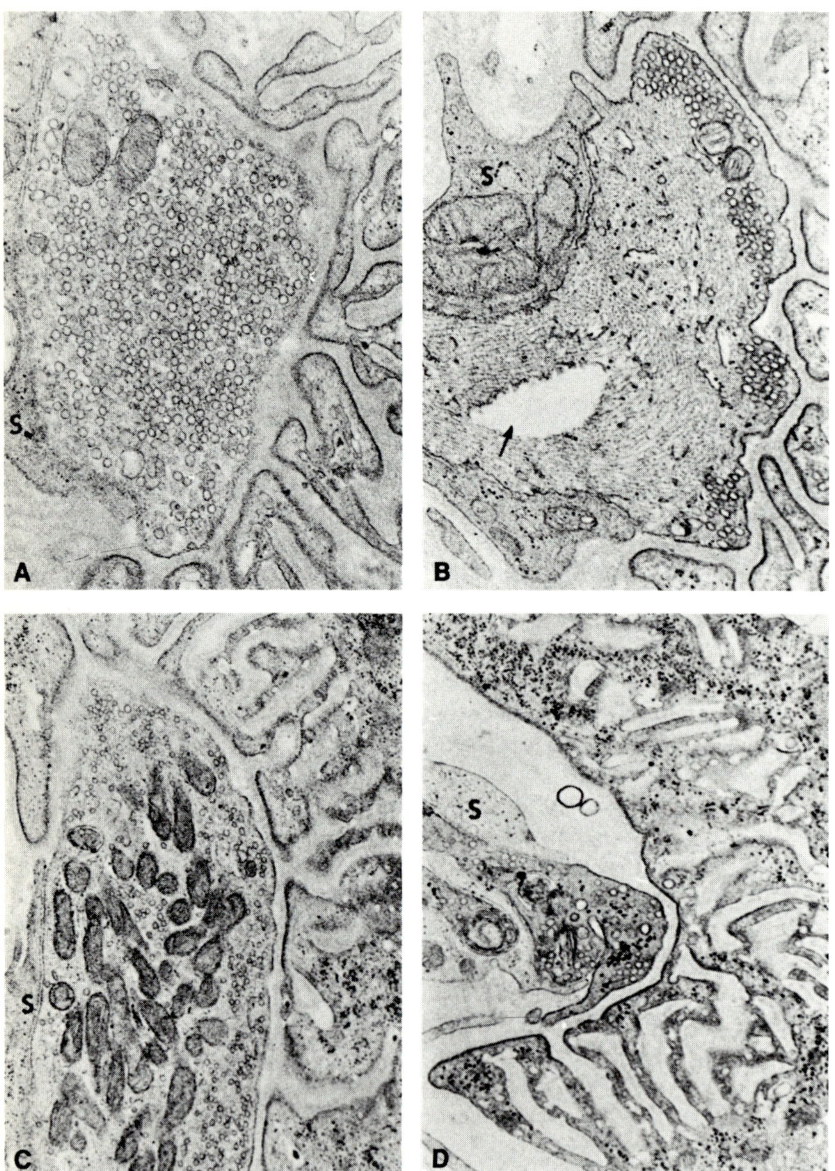

Fig. 2–11. Four normal human endplates. In A the synaptic vesicles are seen to fill the terminal part of the axon (cut transversely) (×26,800). In B the synaptic vesicles are sparse. The clear zone in the axon terminal is an artefact (×30,600). In C there are both mitochondria and synaptic vesicles in the axon terminal (×20,300). In D the nerve terminal is cut longitudinally along one side so that it seems to be small. It contains one myelin figure and dark glycogen granules and a part of it is covered by Schwann cell cytoplasm (×21,400). The primary and secondary synaptic clefts and postsynaptic membrane are seen on the right side of each figure along with glycogen granules and mitochondria of sarcoplasm. (Illustrations kindly lent by A. G. Engel)

phibia the motor end plates take the form of forked terminal branchings of the axon, each branch lying in a small groove in the sarcolemma,[63] without any Doyère eminence. In some muscles of the frog, such as the ideofibularis, some of the fibers have an even more simple arrangement, a thread of terminal axis cylinder running a considerable length of the fiber. These are evidently the muscle fibers that Kuffler[148] found to contract and relax slowly and with low excitability. They have a coarse embryonic type of myofibril.[142] In Mammalia corresponding simplified forms of nerve ending have been described only in the intrafusal fibers of the muscle spindle.[26, 53, 187]

Since 1952 Coërs[52, 53] has elaborated and perfected a method of staining nerve endings in muscle biopsies with intravital myethylene blue. By successive refinements of preliminary stimulation of the exposed muscle he has been able to localize the end-plate zone in the exposed muscle to within 2 mm. This method enables not only a high number of endings to be obtained from a small sample of muscle, but also electron microscope study. The beautiful pictures of human end-plates obtained by Coërs are typified by Figure 2–12 (compare with Fig. 2–3). The telodendria are clearly visualized, as are the subterminal branching of the axis cylinders as they leave the nerve bundle. In the figure it will be noted that the subterminal axon seldom branches; about 1 in 10 supplies two end-plates. Each muscle fiber has only one end-plate, and the terminal ramification of the telodendria, though varied, is compact and multiple. In parallel staining of the subneural apparatus by the Koelle method, it was found that this apparatus in man normally has a series of separate elements or units in each end-plate. Each unit corresponds to one piece of the telodendrion.

In a series of papers Coërs and his collaborators have explored the changes in motor nerve endings in a variety of conditions, summarized in the recent monograph by Coërs and Woolf.[53] In infantile muscular atrophy a simplified fetal type of end-plate, with a single large unit of subneural apparatus and corresponding axon expansion, was found to persist more than 1 year after birth, when the

endings normally become more complex.[51, 53] The fragmentation of endings in the first days after nerve section and progressive changes of the same kind in polyneuritis are described. Hypertrophied muscle fibers, whether residual in poliomyelitis or muscular dystrophy, or in states such as myotonia, uniformly presented an expanded large end-plate with elaborate ramifications. Small muscle fibers in states of dystrophy or myositis had simplified terminal ramifications, sometimes with enlargements or bulbs that perhaps represent a state of change.[53] In all these changes the pattern of the subneural apparatus followed the general pattern of the expansions of the exon, strongly suggesting that the end-plate is a plastic structure subject to anatomic modification according to the needs of the muscle fiber. The striking observation of long unbranched single telodendrion and subneural apparatus in myasthenia gravis, even in muscles not clinically affected by the disease, must be viewed in this light. What feature of the disease leads to this change has not been determined.

The method of Coërs also enables observations to be made on the subterminal branches of the motor axon. The branching of axons that enables one motor fiber to innervate many muscle fibers (the motor unit) normally takes place in the nerve bundles. Degeneration of some motor nerve fibers for any reason is followed by sprouting from neighboring sound fibers, and reinnervation of some of the denervated muscle fibers, as first observed by Hoffman[123] and by Edds.[87] This phenomenon, demonstrated in diseases such as poliomyelitis and amyotrophic lateral sclerosis by Wohlfart,[249] is particularly well shown by the Coërs method.[53] The new branching takes place chiefly in the subterminal portion of the motor axon. As a result such nerve fibers are seen to innervate 2–9 muscle fibers by repeated branching, whereas normally only 1 fiber in 10 shows more than 1 branch (Fig. 2–12). This state is reflected (Coërs and Woolf[53]) in the "absolute terminal innervation ratio," which expresses the number of end-plates in relation to the number of axons. It is normally 1.09:1. Values of higher than 1.27:1 are considered abnormal. A single subterminal

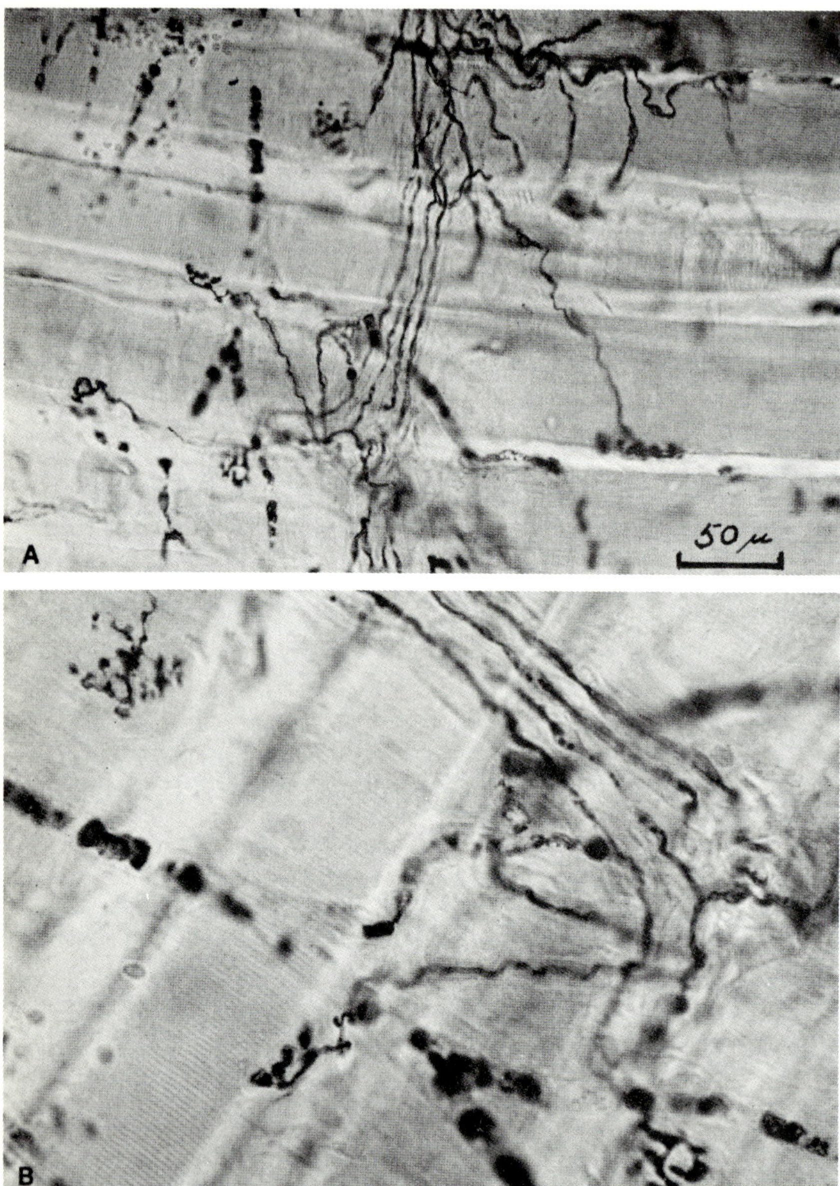

Fig. 2–12. Normal human motor end-plates and subterminal nerve fibers stained with vital methylene blue in biopsy material. Erythrocytes in capillaries also stain a dark blue. (From specimen kindly lent by D. Coërs)

fiber was observed to supply as many as 8 end-plates following poliomyelitis, and 19 after peripheral neuropathy. In primary muscular diseases such as muscular dystrophy and polymyositis, there is also branching of subterminal fibers, probably of similar type, as well as many very simple and distorted end-plate patterns.

The certainty of the Coërs method has greatly increased interest in the morphology of the end-plate and subneural apparatus but not much more information is to be expected in the future. The method also allows some observations to be made of changes in the nerve bundles.

SENSORY NERVE ENDINGS

There are four types of sensory nerve endings in muscle: 1) the muscle spindle, 2) the Golgi tendon organ, 3) corpuscles of the Golgi–Mazzoni type, and 4) free nerve endings. The general relationship of these to the muscle fibers is shown in Figure 2–1.

MUSCLE SPINDLES

The muscle spindles are highly complex structures found at the edge of a fasciculus or in the perimysium of the fleshy portions of most muscles. They were first observed by Kolliker in 1862 and Kühne[149] in 1863 and studied in more detail by Cajal[43] and Ruffini.[214] At first they were believed to represent growth centers, but in 1894 Sherrington[220] proved them to be sensory receptors. Other key references on the anatomy and physiology of this structure are those of Cuajunco,[64] Hinsey,[119] Tiegs,[238] Granit,[108] Barker,[13] and Cooper.[56]

The muscle spindles are evenly distributed in all striated muscles, but they are of simple type and difficult to recognize in the extrinsic muscles of the eye, face, mouth, and throat, in which encapsulation is often seen to be incomplete.[56] Each limb muscle contains 70–100 of these organs.[14] As the name suggests, the spindle consists of a fusiform connective tissue capsule within which a specialized region of one or more striated muscle fibers is embraced by a complex sensory receptor. The size of a spindle varies considerably; from pole to pole it may measure 0.5–3.0 mm., and two or more may be bound together in one complex. The lamellated connective tissue sheath which forms the capsule runs parallel to the muscle fibers. The specialized muscle fibers are of very small diameter and form a bundle, of which the "intrafusal" portion runs through the center of the spindle. They emerge to lie parallel with the

other muscle fibers which surround them and, like them, usually reach the aponeuroses of origin and insertion of the muscle. They may, however, terminate prematurely in a slip attached to the sheath of a neighboring muscle fiber.[13] This "extrafusal" portion of the fibers maintains a small diameter. Cross striations are easily visible in both intrafusal and extrafusal portions of the fibers.* The proximal and distal poles of the spindle are indistinguishable. There is a space (periaxial space) between the intrafusal portion of the muscle fibers and the capsule, which is traversed by connective tissue septa (Figs. 2–13 and 2–14). According to Sherrington[220] and others, this space is injectable from the subperineural channels in the nerve. At the polar regions of the spindle, the intrafusal muscle fibers have elongated, oval, sarcolemmal nuclei at widely spaced intervals, whereas in the equatorial region muscle nuclei becomes progressively more numerous in the center of the fibers. At one point they form a conglomeration which fills and distends the body of the muscle fiber, and this is the region of the annulospiral ending (Fig. 2–15). Each such nuclear bag has 40–50 nuclei of spherical shape. The number of intrafusal nuclear bags corresponds to the number of intrafusal fibers. The striated substance is displaced to the periphery of the fiber by the nuclei and in some spindles appears to be absent at the level of greatest nuclear aggregation. In addition there are smaller muscle fibers within spindles whose nuclei are not agglomerated but scattered; these are called "nuclear chain fibers." Cooper and Daniel[57] and Boyd[26] were first to distinguish these two types of muscle fibers. The nuclear bag fibers are 20–30 μ in diameter and 4–8 mm long; the nuclear chain fibers are 10–15 μ in diameter and 2–4 mm long. The difference allows nuclear bag fibers to extend out beyond the ends of the capsule. Physiological differ-

---

* Some authors (e.g., Cuajunco[64]) refer to all muscle fibers not involved in muscle spindles as "extra-fusal" fibers. Since the total number of muscle fibers concerned in muscle spindles in any given muscle is small, and since they are not totally enclosed in the spindle-shaped capsule, this use of the term appears undesirable, and we adhere to the terms used in the classic description of Sherrington.[220]

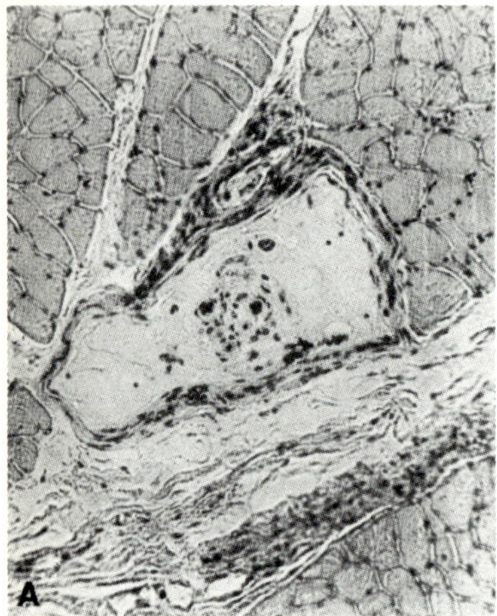
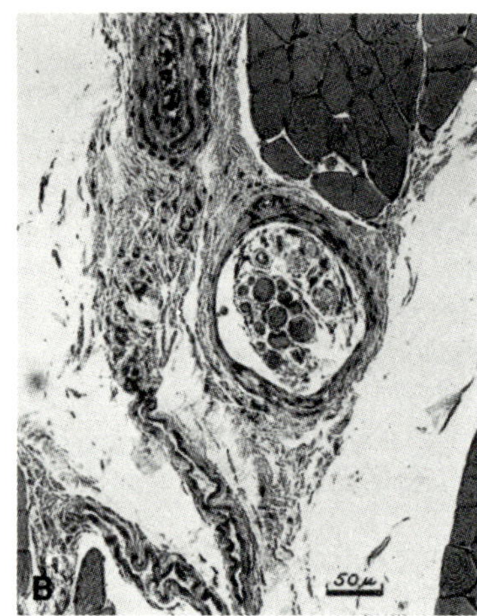

Fig. 2–13. Muscle spindles of human subject. Plane of section passes through equatorial region in (A) where one can see two large nuclear bags and several smaller nuclear chain fibers. (hematoxylin, Van Gieson)

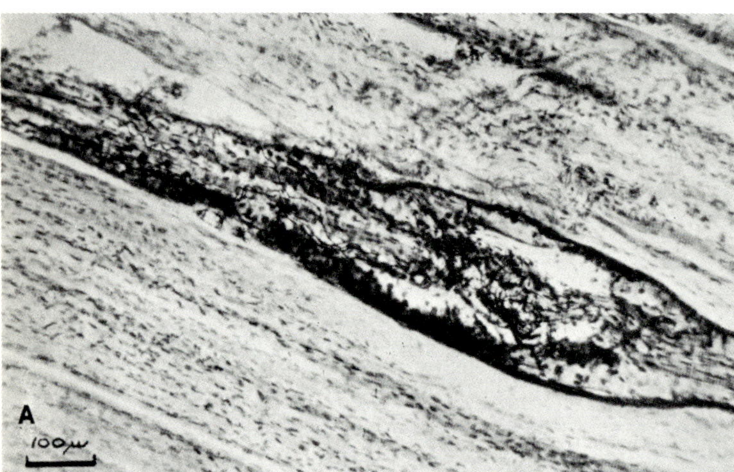
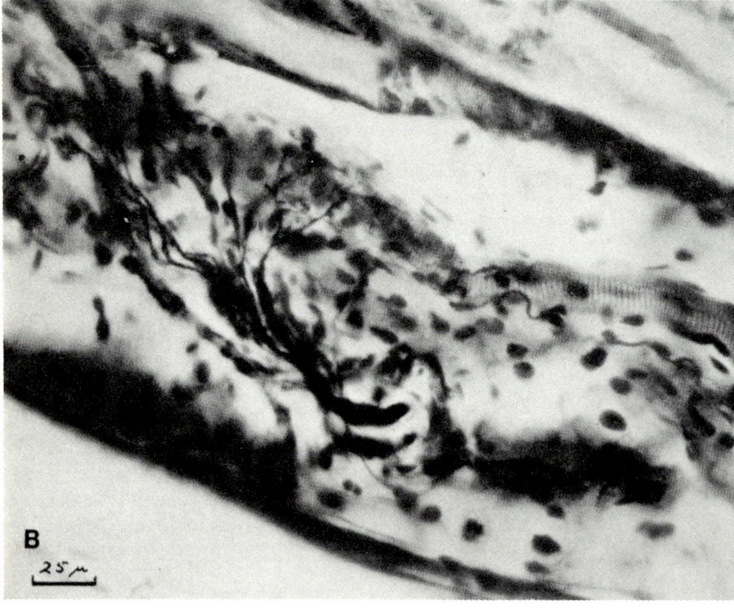

Fig. 2–14. (A) Muscle spindle of human (Bielschowsky method). (B) Higher magnification of equatorial region of (A).

ences between bag and chain fibers have been ascertained. Electron microscopic studies of the nerve endings have recently been reported by Robertson.[210]

The most complete electron microscopic studies of the muscle spindle are to be found in the writings of Gruner[111] and Banker and Girvin.[11] The capsule appears as a membrane consisting of 8–13 concentric sheets of specialized fibroblasts lined by an inner basement membrane and interspersed collagen fibrils. The capsule is traversed by channels between successive lamellae and is continuous at its apices with endomysial connective fibers (13–25 μ in diameter). The equatorial regions contain an incomplete shell of peripheral myofilaments whereas in the distal parts the arrangement of myofilaments is the same as in extrafusal fibers. In both the nuclear bag and nuclear chain fibers, Banker and Girvin[11] found arrangements of myofilaments that resemble the structure of slow striated or smooth muscle, with poor organization of I filaments and Z bands and sparse sarcoplasmic reticulum. Many satellite cells are seen near the central regions.

The average number of intrafusal muscle fibers varies with the type of spindle and the species. In the rabbit the average number is 4, in the cat 6, and in man about 10, according to Cooper.

INNERVATION OF MUSCLE SPINDLES

The description of muscle spindles is not complete without reference to their nerve fibers, on the basis of which the spindles may be classified as simple, intermediate, or complex [Ruffini[214]]. Simple spindles are short and contain less than the average number of intrafusal fibers, the majority of which are of nuclear bag "type." Its sensory innervation is equatorial (i.e., at the nuclear bag region) and spiral in form. The motor endings are small. In the intermediate spindles Cooper observes small, diffuse, fine motor endings that are visible near the poles of the muscle fibers. The complex spindles, composed of clain and bag fibers, always have two secondary sensory endings in addition to the central annulospiral primary endings.

The motor innervation of the spindle has been examined in greater detail recently by Barker[13] and Cooper.[57] They noted that each muscle fiber has typical motor end-plates in a number approximately double the number of intrafusal fibers. In a complex spindle these end-plates are distributed equally to the proximal and distal muscular regions, as shown in the beautiful illustrations of Ruffini.[214] In simple spindles the motor innervation is only at one end (see Fig. 2–15A). Some motor fibers enter the capsule independently of the sensory bundle and pass obliquely to the distal pole; those to the proximal pole usually enter with the sensory fibers. In its general features the relationship of motor fiber and the motor end-plate of an intrafusal fiber is like that of an ordinary muscle fiber, but the sole-plate is more elaborately ramified. In Reptilia, in which the spindles are very simple structures, the motor innervation is by a branch ("ultraterminal fiber") from the motor ending of a neighboring ordinary muscle fiber. In Mammalia this seldom appears to be the case, and the motor innervation is chiefly derived from the small efferent gamma fibers of the motor roots, coming from independent small nerve cells in the anterior horns of the spinal cord.[156] A single fiber of the "small-nerve fiber" system appears to innervate one or more intrafusal fibers of one or more spindles in the same muscle.[147]

The sensory innervation of the muscle spindle has been investigated by a number of workers. Ruffini,[214] in a study of the muscles of the cat, distinguished two types of sensory endings, a primary or annulospiral ending and a secondary or flower-spray ending. All the annulospiral endings in any one capsule are derived from a single very large myelinated nerve fiber. The ending consists of a series of neurofibrillar coils and spirals which are applied to the external surface of the sarcoplasmic membrane of the muscle fibers in the region of each nuclear bag (Fig. 2–15B). The secondary or "flower-spray" ending differs in minor ways. Each such ending has a separate nerve fiber and consists of a branched network which enfolds the myotube between the polar and equatorial part of the intrafusal fiber. There may be several such secondary endings, or none.[14] The parent fiber is of smaller diameter than those of the

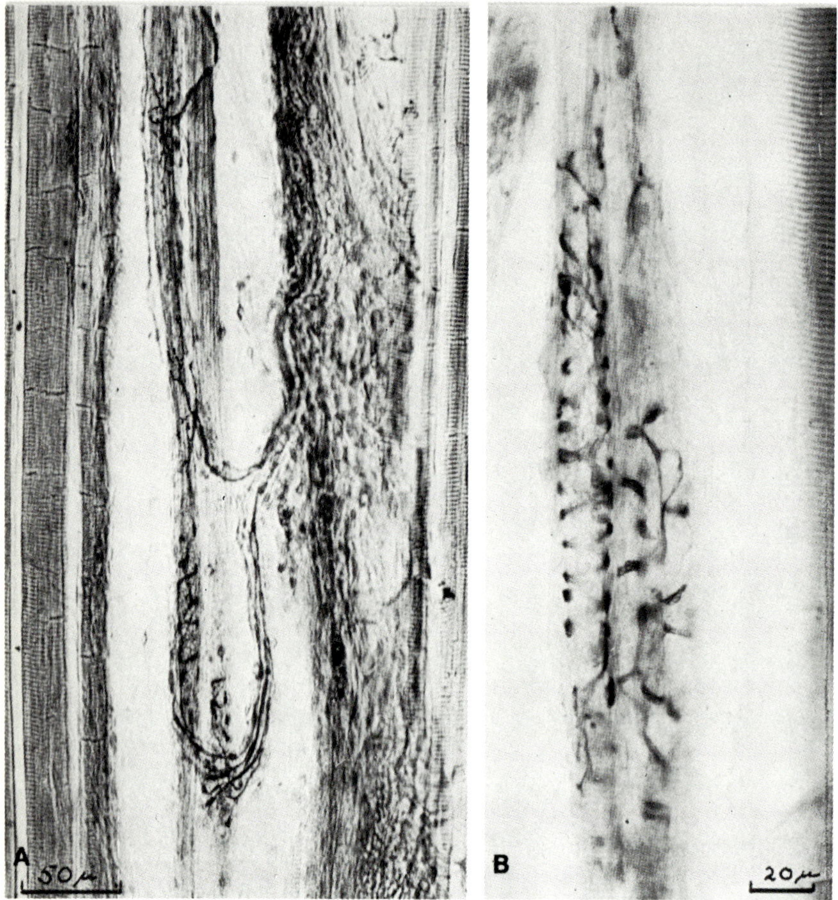

Fig. 2–15. (A) Simple muscle spindle from sartorius of cat showing relationship of annulospiral ending (below), flower-spray ending (center), and motor ending (above) (Cajal method). (B) Muscle spindle from sartorius of cat showing details of annulospiral ending. (Cajal method)

primary annulospiral ending. Both types of sensory fiber reach the capsule of the spindle in a single loose sheath of Henle and can be recognized for a considerable distance back in their course through the muscle by this feature (Fig. 2–15 and 2–17).

The arrangement of the sensory endings on the spindle suggests that mechanical distortion of the intrafusal muscle fiber excites them. From the work of Matthews[168] it would seem that muscle stretching is an effective stimulus for the largest afferent fiber (annulospiral ending). In addition, Matthews observed that strong excitation of the motor nerve may be accompanied by discharges from the spindles in muscles whose Golgi tendon organs are re-

moved by dissecting off the tendon or aponeuroses. He supposed in this instance that the intrafusal muscle fibers contract and activate the "flower-spray" ending. The work of Kuffler et al.[147] indicates that the small-fiber motor system can adjust the intensity of discharge of both annulospiral and flower-spray endings to tension over a great range in length. At any given length, activation of the small motor fibers could thus greatly intensify propriocep-tive effects and could discharge both types of spindle afferents. Both static and dynamic mechanisms are subserved (chain fibers static, nuclear bag fibers dynamic).

The muscle spindle is peculiarly resistant to muscular diseases,[15] though they are not indemnified as shown in the report of Grossiord *et al.*[110] who have provided a catalog of their pathologic reactions. The characteristic encapsulated muscle fibers are prominent in sections of atrophic muscle, whether from neuronal atrophy or muscular dystrophy, when all surrounding muscle fibers have been reduced to thin sarcoplasmic threads.

Following section of the ventral nerve roots, or of the muscular nerve,[239] atrophic changes in the muscle fibers of the spindle are greatly delayed, more so in the nuclear bag than chain fibers. After 1 or more years, when the other muscle fibers have completely degenerated, some recognizable muscle fibers may be found in residual spindles (see Figs. 10–19 and 13–13). Apart from degeneration of the sensory-nerve ending, loss of sensory nerve fibers does not affect the structure of the spindle.

GOLGI TENDON ORGANS

The Golgi tendon organs are specialized encapsulated structures (Fig. 2–16 and 2–17), each of which surrounds a bundle of tendinous fibrils which receives the attachments to 15–30 muscle fibers. They are situated where the muscle fibers join an aponeurosis and are not actually in the tendon. Each organ gives rise to a single nerve fiber. These structures are so placed as to be responsive to tension on the attached muscle fiber group; the work of Matthews[168] indicates that the threshold of tension is high. The tension produced by active contraction is much more effective than that produced by passive stretch. Their function is inhibitory under conditions of strong muscle contraction.

GOLGI–MAZZONI CORPUSCLES

The Golgi–Mazzoni corpuscles are small encapsulated endings, similar to the end bulbs of Krause and Pacini of the pleura, pericardium, and mesentery and the deeper layers of the dermal tissue. Their thin lamellar capsule encloses a branched and coiled central nerve expansion in the core. These structures are found in most tendon sheaths, aponeuroses, and intramuscular septa (Fig. 2–18), as well as in periosteum. They also are thought to respond to mechanical deformation. They are probably position sense receptors.

FREE NERVE ENDINGS

The free nerve endings are pain receptors and are similar in structure to those elsewhere in the body. In the muscles they are found chiefly in networks in the adventitia of blood vessels (Figs. 2–18 and 2–19B) or lying free in the aponeuroses (Fig. 2–19A). When they are closely related to muscle fibers they ramify in the endomysium. Autonomic endings in muscle accompany the blood vessels and terminate in their walls, particularly those of arterioles, as shown diagrammatically in Figure 2–1.

It should be mentioned that all these nerve endings require special techniques for their demonstration.

VASCULAR SUPPLY AND LYMPHATICS

BLOOD VESSELS

Skeletal muscles have a rich blood supply, the details of which have been described most completely by Spalteholz.[227] Arteries enter the muscle and branch repeatedly before reaching capillary size. A plexus of these capillaries forms a network which surrounds every muscle fiber. Veins draining the capillaries tend to run parallel to the arteries.

The total capacity of the capillary bed in the muscles is so large that vascular adjustments in this part of the circulatory system are of the utmost importance in determining the efficiency of the blood supply to many organs. Krogh[141] estimates that in a transverse section of 0.5 mm^2 of muscle, which would include about 200 muscle fibers, approximately 700 parallel capillaries are cut across. These capillaries surround the muscle fibers in such a way that each

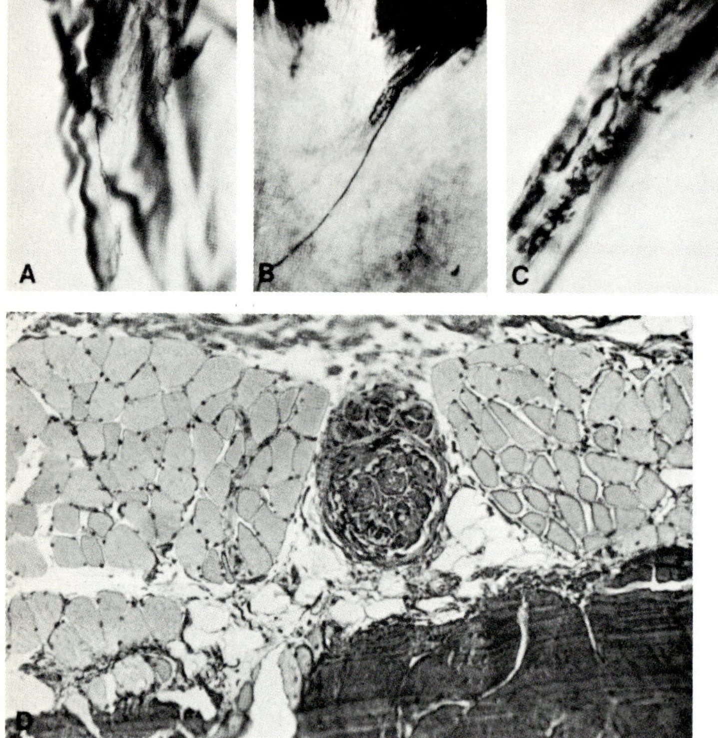

Fig. 2–16. Sensory endings in muscle. (A) Tendon organ from intercostal muscle of cat (Bielschowsky method). (B and C) Tendon organ from intercostal muscle of rat. Dark structures at top are muscle fibers (gold chloride). (D) Tendon organ in section from human subject. Tendon is below, and encapsulated organ lies between two smaller muscular fasciculi. (H&E)

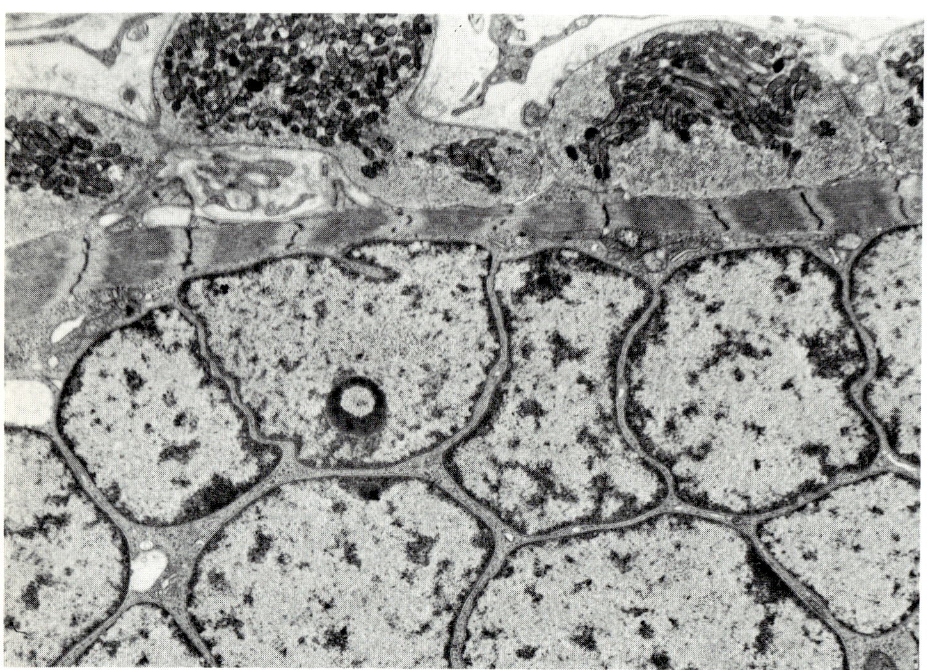

Fig. 2–17. Primary sensory ending of the IA afferent type on a normal human intrafusal nuclear bag muscle fiber, sectioned longitudinally. Myofibrils are arranged on the periphery of the fiber around the central core of nuclei. Basement membrane of the muscle extends over the endings but does intervene between nerve and muscle at the four contact areas shown. Numerous mitochondria cisternae, vesicles, and a fine fibrillar matrix fill the endings. Then cell processes of the inner capsule cells appear above the sensory endings. (Illustration kindly lent by W. R. Kennedy) ($\times$5340)

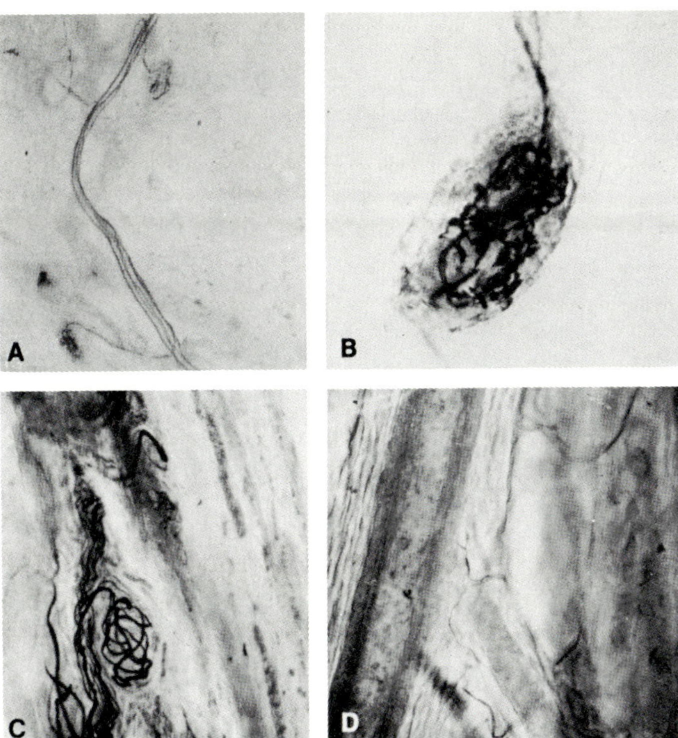

Fig. 2–18. (A and B) Golgi-Mazzoni corpuscles in perimysium of gastrocnemius of cat (gold chloride method). (C) Golgi-Mazzoni corpuscle in endomysium of gastrocnemius of child (Bielschowsky method). (D) Sensory ending at crotch of division of artery in sartorius of cat. (Cajal method)

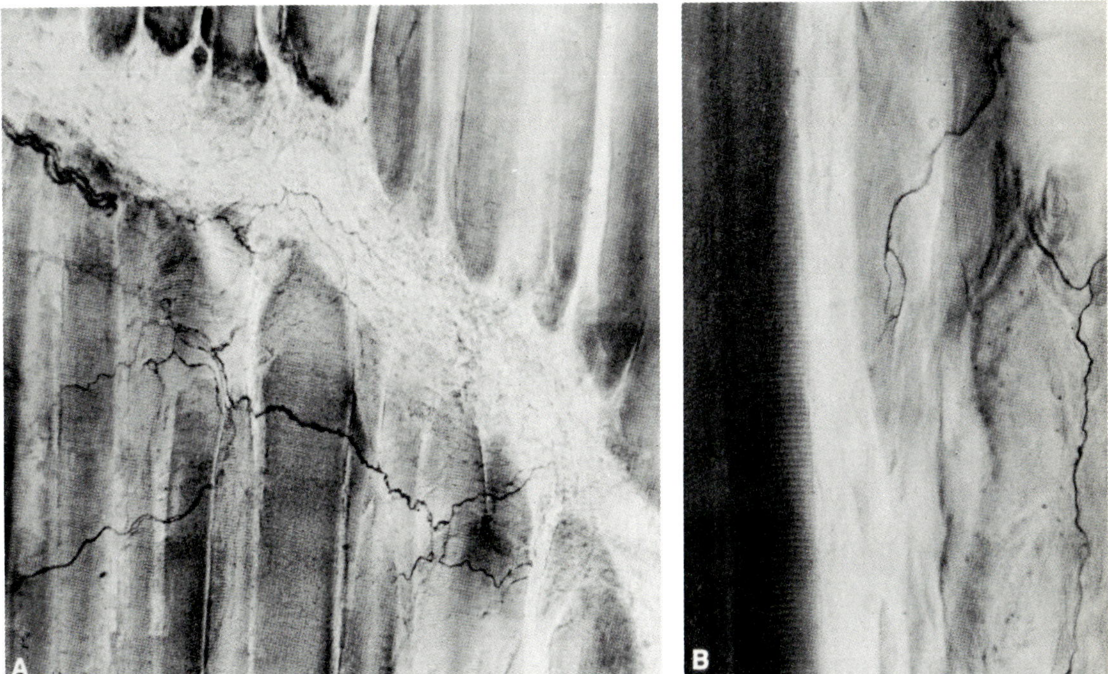

Fig. 2–19. (A) Rectus abdominis showing free endings in muscle at either side of tendinous intersection. (B) Free endings in endomysium from cat sartorius. (Cajal method)

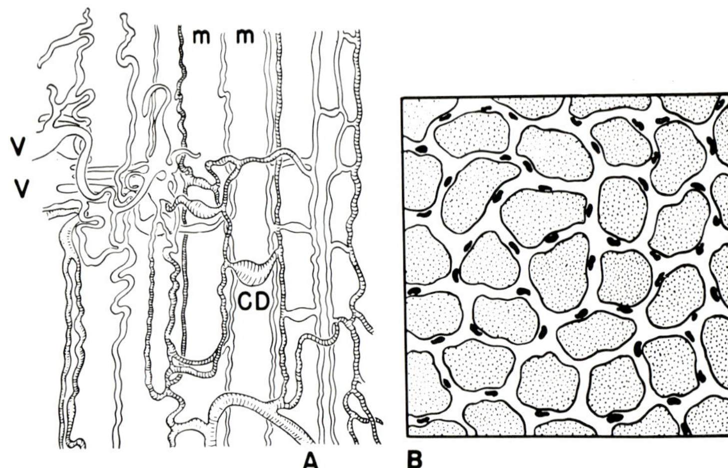

Fig. 2–20. Vascular supply of muscle. (A) Arrangement of arterioles (A), capillaries, and venules (V), showing dilated transverse capillary (CD). Spaces m and m are occupied by single muscle fibers (Modified from Ranvier[151]). (B) Cross section after injection of capillary system with India ink under pressure. Black dots are capillaries. (Modified by Krogh[110])

muscle fiber is placed in relation to four or five capillary vessels (Figs. 2–20B and 2–21). The oxidative-rich (type I) fibers of red muscles have a more abundant capillary network than the phosphorylase-rich (type II) fibers of white muscles, which explains their capacity for prolonged tonic contraction without oxygen deficit. Most of these capillaries run parallel to the long axis of the muscle fiber (Figs. 2–20A and 2–21). In the muscle mass of adult man it is calculated that the total length of all the capillaries in muscle is about 62,000 miles. Their total surface area, approximately 6300 m², allows for easy exchange of oxygen and electrolytes between blood and muscle substance.

Natural contractions of moderate degree do not use all the muscle fibers at once, and a great increase of capillary circulation results.[70] Strong contractions sufficient to induce fatigue have been shown by Merton[175, 176] to be associated with obstruction of capillary flow. The small capillary dilations in red muscle observed by Ranvier[204] (Fig. 2–20) and thought by him to represent reservoirs for blood during contraction, are too small to be of any value for this purpose. However, the increase in surface area that they afford may facilitate gaseous interchange between hemoglobin and myoglobin.

Krogh[140] and Petrén et al.[195] observed an increase of 45% in the number of patent capillaries in guinea pig muscles during exercise and therefore concluded that during inactivity many of the capillaries are closed. The blood supply is adjusted in this way to the metabolic demands of the muscle fibers. Petrén et al.[195] suggested that the increase in the bulk of a highly trained muscle is due to an increase in the number of patent capillaries. While this may contribute to the enlargement and increased firmness of such muscles, it seems to us that an overall increase in the diameter of the individual fibers is a more important factor.

Very little has been written on the intramuscular vascular patterns in man. Except for several isolated reports, information on this subject was almost lacking until Blomfield[21] made a radiographic study of a number of limb muscles in postmortem material by the injection of a mixture of barium and collodion into the main vessels. On the basis of these studies he concluded that the vascular patterns of muscle could be divided into four main categories, as follows:

1. A pattern of longitudinal anastomoses formed either from a single nutrient artery as in the gastrocnemius, or from a number of separate arteries entering the muscle from different sources, as in the soleus and peroneus longus. It is obvious that the former variety is more vulnerable to local injury; this is borne out by clinical experience, for in severe wounds involving the calf muscles, the gastrocnemius may undergo extensive necrosis while the soleus may remain relatively intact.

2. A radiating pattern of collaterals arising from a single main vessel which enters the

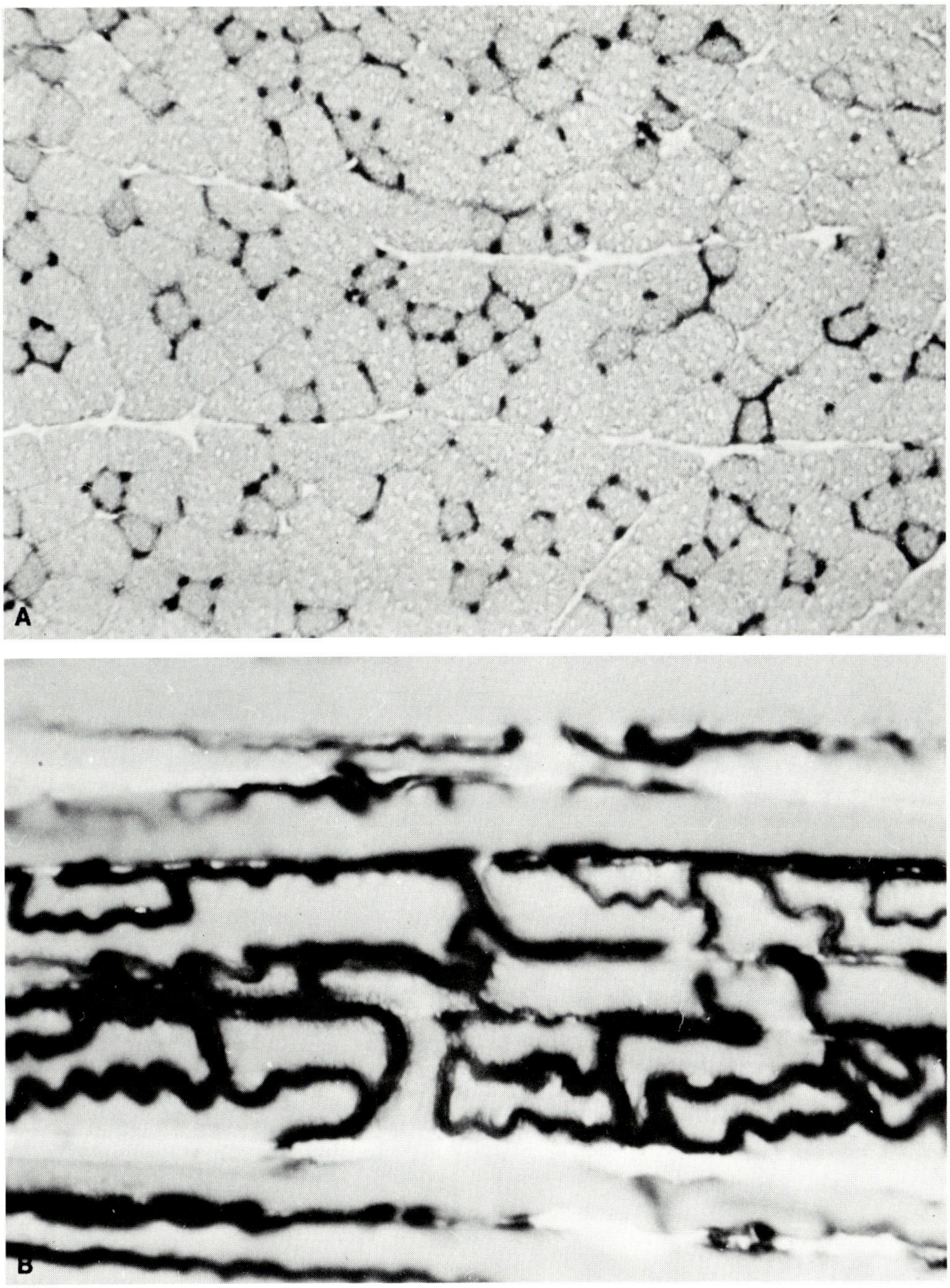

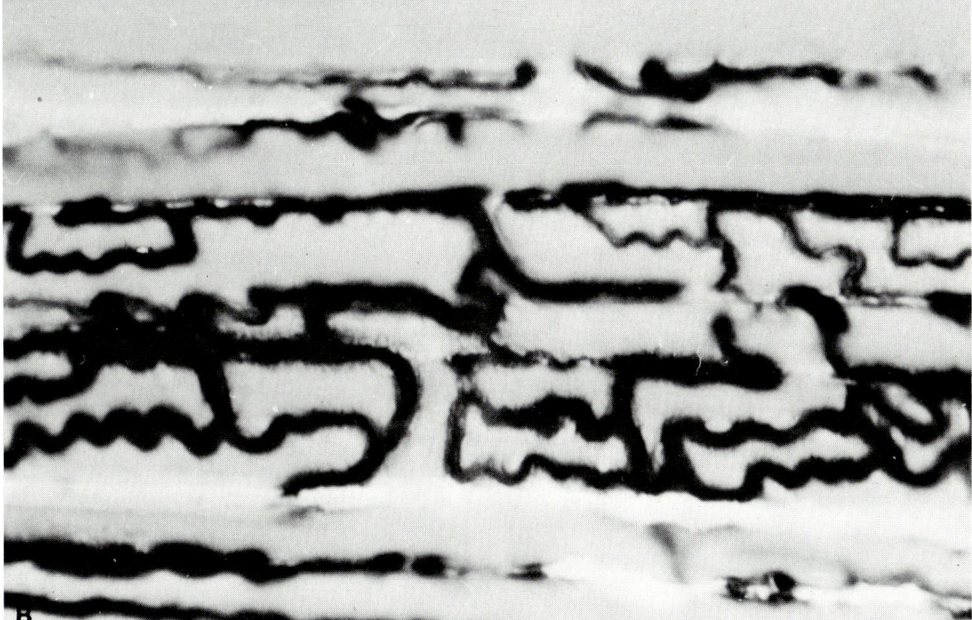

Fig. 2–21. Gastrocnemius muscle of rat. (A) Transverse section in alkaline phosphatase stain; the type I fibers are each surrounded by several capillaries (black dots). (B) Longitudinal section in alkaline phosphatase stain to demonstrate pattern of capillaries and their relationship to muscle fibers.

middle part of the muscle. This type is found in the biceps brachii, and its clinical significance was demonstrated in a case reported by Power[200] in which a massive necrosis of the long head of the biceps brachii occurred as the result of the severance of the single artery to the biceps by a shell fragment.

3. A pattern of anastomosing arcades formed by a series of vessels entering the muscle along its length. This type is found in most of the long flexors and extensors of the leg and in general appears to be the most efficient form of vascularization of muscle, although its efficiency depends partly on whether the supplying vessels are derived from the same source. An interesting contrast in this connection is shown between the tibialis anterior and the extensor digitorum longus muscles. Both muscles show the same type of intramuscular vascular supply, but whereas the former derives its entire blood supply from the anterior tibial vessel, the latter obtains part of its supply from the perforating arteries which traverse the interosseous membrane from the posterior portion of the leg. In severe injuries involving the anterior tibial vessels the tibialis anterior muscle becomes grossly necrotic, whereas the extensor digitorum longus very frequently is spared.

4. A rectangular pattern of anastomoses formed by a series of entering vessels, similar to the last category but less efficient mechanically. This type is found in the extensor hallucis longus.

The veins in striated muscle accompany the arteries and are in no way remarkable.

It has often been assumed that all muscles are provided with a rich network of intramuscular vascular anastomoses and that partial interruption of a vessel of supply to a muscle is of little practical significance. Clinical experience as well as scanty experimental evidence have not borne out these assumptions. Campbell and Pennefather[46] at the close of World War I, pointed out the relative frequency with which certain muscles were involved by gas gangrene infection in war injuries and related this incidence of infection to the vascular supply of the individual muscles. They also stressed the importance of an accurate knowledge of the vascular arrangements of a muscle before its surgical treatment is attempted, particularly after injury to the muscle.

COLLATERAL CIRCULATION

The establishment of collateral circulation in muscle is evidently very slow. For example, Power[200] has shown in humans that necrosis of small portions of muscle can occur after interruption of a small supplying vessel. These observations were confirmed experimentally by Clark and Blomfield,[49] who have observed that in certain muscles, such as the tibialis anterior in rabbits, ligation of one of its supplying vessels resulted in devascularization with reduction in the functional efficiency of the capillary circulation. This was demonstrated by the lack of coloration of the devascularized portion of muscle when a highly diffusible dye, bromphenol blue, was injected into the ear vein of the rabbit. On the other hand, the gracilis muscle in rabbits was not readily devascularized because of the plurality of its blood supply.

LYMPHATICS

Although there is an abundant supply of lymphatic vessels in the connective tissue sheaths and tendons, as far as is known there are none within skeletal muscle. These vessels serve the purpose of collecting and conveying the lymph from the muscular tissue. However, the manner in which this fluid reaches the lymphatic vessels of the perimysium and epimysium is not known. Some anatomists have suggested that it is by means of intercommunicating spaces which extend along the fascial planes between the muscle fibers.

Aagard[1] was successful in demonstrating lymphatics in many muscles by injecting colored materials into the tendons. He is of the opinion that the principal lymph channels are between fascicles of fibers. More recently Shdanow[219] observed that muscle lymphatics, though difficult to demonstrate, by injection could be seen to have the form of a wide-mesh network in the perimysium and a fine network surrounding small fascicles, and possibly even the individual muscle fibers in some instances. We have seen lymph follicles in the epimysial fat but none in the perimysial or endomysial connective tissue (see Fig. 13–5).

CHEMISTRY

CHEMICAL BASIS OF MUSCULAR STRUCTURE AND FUNCTION

Muscle is a unique organ in that it is capable of converting chemical into mechanical energy, thereby permitting movement of the organism and accomplishment of work. Inasmuch as these activities of muscle lend themselves to measurement, the energy transformations in living muscle are particularly adaptable to study. The investigations of these transformations have proceeded along two general lines: (1) the source of the chemical energy and the form in which it is made available, and (2) the manner in which the energy is utilized, i.e., organization of the contractile system.

In the development of our knowledge of muscle chemistry, four different periods can be distinguished. These were outlined by Von Muralt[244] as the prelactic acid era, the lactic acid era, the period of phosphorylations, and the myosin era. Among the prominent names associated with these eras are Fletcher, Meyerhof and Hill, Lundegaard, and Szent-Gyorgyi.

Chemical analysis of muscle fibers was first attempted more than 50 years ago.[65, 114, 243] It was reported in these and later studies that about 75% of muscle is water; and that of the solid parts the proteins make up about 80%, and the carbohydrates (chiefly glycogen), organic phosphates (phosphocreatine, adenylic acid, adenosine triphosphate, and the like), fatty acids, inorganic salts, and other dissolved substances the remaining 20%. The proteins of muscle can be separated from one another by suitable extraction methods. These have permitted isolation of a protein fraction, *myogen,* which is readily soluble in water and neutral salt solutions, and another fraction, *myosin,* which has the properties of a globulinlike protein. The latter is soluble in strong neutral salt solutions but precipitates on dilution of these solutions and is insoluble in water.

The myogens constitute the major proteins of sarcoplasm, and later studies indicate that these fractions contain many of the enzymes necessary for muscle metabolism such as phosphorylase, phosphoglucomutase, aldolase, and triosephosphate dehydrogenase. The myosin fractions, on the other hand, are believed to form the contractile elements of the muscle, the myofibrils.

Most of the earlier studies on the biochemistry of muscle contraction were carried out by Szent-Györgyi and his colleagues.[232, 233] Present knowledge about the structure and function of contractile proteins and the molecular basis of muscular contraction is described in several articles and reviews.[16, 84, 152, 248] An attempt is made to summarize current ideas on this subject in the following pages, but the reader must be reminded that the entire field is in a state of constant flux and new information is being rapidly accumulated. Therefore it is not possible to formulate any theory of muscle contraction that remains consistent with the observed facts for long.

CONTRACTILE PROTEINS

Proteins comprise about 20% of the fresh weight of muscle fibers. Roughly two-thirds of this protein fraction is in the form of the contractile proteins: myosin, actin, and a third component tropomyosin[7, 8] whose function and position within the muscle fiber is yet to be fully determined. Bundles of the contractile proteins are linked together in the myofibrils. During the past few years Ebashi[83] and his colleagues discovered several regulatory proteins that are probably involved in the assembly of myofibrillar structure. Table 2–1 shows the relative contents and the localizations of the main contractile and regulatory proteins of rabbit skeletal muscle.

The remaining third of the protein fraction

TABLE 2–1. Main Contractile and Regulatory Proteins of Rabbit Skeletal Muscle

Protein	Localization	Contents (mg/g)
Myosin	A filament	58
Actin	I filament	20
Tropomyosin	I filament	5
Troponin	I filament	3
L-Actinin (6S)	Z band	2
10S L-Actinin	?	6
β-Actinin	I filament	1
M-protein	M line	1

Data from Maruyama.[167]

is composed of myoglobin, other pigmented proteins, and many other complex soluble proteins, including all of the enzymes. Of these, myoglobin is dealt with in detail in a later section.

MYOSIN

Myosin, the largest representative of the contractile protein group, assumes a position of prime importance in the contraction scheme. It can be extracted from muscle with neutral salt solutions of high ionic strength at about pH 6.5 with minimal contamination of actin. It has interesting and unusual properties, including flow birefringence, which signifies a high degree of asymmetrical molecular orientation and the ability to serve as an enzyme[74] in the partial degradation of the energy-rich compound adenosine triphosphate (ATP). Myosin thus possesses ATPase activity, which shows two pH optima (at about pH 6 and 9) and are dependent on the presence of sulfhydryl groups in the protein.

The protein myosin is approximately 1600 Å in length, with a molecular weight of about 500,000. Further insight into the structure was obtained by studies involving its treatment with proteolytic enzymes. Brief digestion with trypsin or chymotrypsin produces two types of fragments—the so-called heavy (H-) meromyosin with a molecular weight of about 350,000 and

light (L-) meromyosin with a molecular weight of about 150,000. There is general agreement that L-meromyosin is highly helical and about 20 Å in diameter, and provides the backbone structure of the organized thick filaments of the A band. H-meromyosin is more complex, and its detailed structure is not known. It is probably globular and retains the ATPase activity of the parent molecule.[103, 234] It consists of the head (wider) portion of the myosin molecule with a variable amount of the tail region, whereas the L-meromyosin derives from the tail. Further enzymic degradation by trypsin yields the so-called subfragment I, which possesses all the enzymatic and actin-binding properties of the parent molecule.[180, 252]

ACTIN

Actin, the other main contractile protein, is a water-soluble globular (G-actin) protein with a molecular weight of about 60,000. Each molecule of G-actin contains one molecule of ATP very firmly bound to it.[230] If the ionic strength of a solution of G-actin is raised by adding salts, G-actin molecules polymerize to a high molecular weight fibrous protein (F-actin). This F-actin is a double-stranded helix which resembles the thin filament of the I bands of a myofibril. The polymerization is related in vitro to dephosphorylation of ATP,[178a, 230] to the extent that for each actin molecule that enters the polymer one molecule of ATP is split to ADP and inorganic phosphate. Whether the transformation from the G to the F form, and the reverse, occurs as a part of the natural contraction process in the living muscle is still open to question. It is possible that it represents the less easily reversible type of extreme contraction and that natural contraction involves a more rapidly reversible process of monomerization.

ACTOMYOSIN

By combining myosin with F-actin, a complex known as actomyosin is formed. The actin/

myosin ratio in this combination is about 1:3. The sulfhydryl groups of myosin are necessary for a strong linkage of these two proteins.[9] Actomyosin threads have several unique physicochemical properties which make them resemble an intact muscle fiber: (1) birefringence; (2) exquisite sensitivity to certain concentrations of ions, especially potassium; and (3) an ability to "contract" and shorten by as much as 44% upon the addition of a 0.1% solution of ATP. This shortening of actomyosin threads or superprecipitation can be readily observed under the microscope in extracts of muscle, and it may represent in a crude and oversimplified way the basic phenomenon of muscular contraction.[231] By lowering the temperature to 0°C or in the presence of magnesium ion by properly adjusting the pH and ATP concentration, it is possible to reverse or prevent the superprecipitation of actin and myosin.[102] The change in actomyosin suspension in the latter case has been termed clearing,[229] which action may be analogous to relaxation.

Both myosin and actomyosin exhibit ATPase activity that is markedly influenced in vitro by pH, ionic strength, and the concentrations of potassium, ammonium, calcium, and magnesium ions. Most significant is the fact that magnesium ion stimulates actomyosin ATPase activity, in contrast to inhibition of the myosin enzyme.[231] Calcium, however, stimulates the ATPase activities of both proteins.

It is now recognized that activation of contraction is intimately related to the binding of calcium to myofibrils.[80, 246] Both the magnesium activated ATPase and contraction of isolated myofibrils (natural actomyosin) are inhibited by chelating agents such as ethylenediaminetetraacetic acid (EDTA) or its analogue 1,2-bis (2-dicarboxylmethylaminoethoxy)-ethane (EGTA); the latter has a much higher affinity for calcium than magnesium at neutral pH. On the other hand, the magnesium-activated ATPase and superprecipitation properties of reconstituted actomyosin (prepared from isolated actin plus myosin) are not inhibited by calcium-binding agents such as EGTA.[82, 193] Recent studies demonstrated that this calcium ion dependence stems from the presence of protein factors other than actin and myosin.[83]

TROPOMYOSIN OR TROPOMYOSIN B

Isolated and characterized first by Bailey,[7] tropomyosin can be extracted from ethanol-treated muscle preparations. In the absence of salt, tropomyosin aggregates to form a viscous gel-like solution. It is notable for its high α-helix content and in this respect is very similar to the α-meromyosin fraction of myosin. It is the only fibrous muscle protein known to form true crystals. Recent studies[250] show that tropomyosin consists of two subunits, each having a molecular weight of about 30,000. In muscle, tropomyosin is observed in the thin filaments of the I band with fluorescent antibody staining technic.[190] Tropomyosin is known to interact with actin and with the troponins. The 400 Å axial periodicity visible in the filaments of vertebrate striated muscle appears to be due to the interaction of F-actin and tropomyosin.[186]

A similar protein[153] known as tropomyosin A or paramyosin is a prominent constituent of the "catch" muscles of mollusks.

TROPONIN

The original factor termed native tropomyosin by Ebashi was found later to be a complex of the tropomyosin B of Bailey and the new protein troponin.[83, 84] Native tropomyosin or troponin-tropomyosin complex is now considered to be a crucial protein in the regulation of the contraction-relaxation cycle by calcium ions. Troponin strongly binds to calcium ions unaffected by the presence of magnesium ion; ATP is not required for this binding. Troponin, in association with tropomyosin, is probably localized in the thin filaments.

α-ACTININ AND β-ACTININ

Both these proteins, discovered by Ebashi and his colleagues,[83, 167] have amino acid compositions similar to actin. Their precise function and locations in the muscle fiber have not yet been fully determined.

SLIDING FILAMENT THEORY OF CONTRACTION

The orientation of the contractile proteins within the muscle fiber was established largely through the electron microscopic and other optical studies of Huxley and Huxley and their associates[125, 126, 128] and of Hodge.[120] The electron microscopic data obtained by these authors support their hypothesis that long myosin rods of constant length are positioned in parallel arrangement throughout the A bands. Actin filaments, on the other hand, are interposed between the myosin rods and insert on either side of the Z line, from which they extend through half of the I band and into the A band, terminating at the margins of the H band in the so-called S filaments. According to this theory, the force of contraction is developed by the cross bridges in the overlap region, and active shortening is caused by movement of the cross bridges, which causes one filament to slide over the other. The molecular basis for such movement of actin and myosin filaments across each other, with subsequent bonding, is not known.

ELECTROLYTES

The muscle fibers, as do all other bodily cells, maintain a fluid internal environment that is distinctly different from the surrounding interstitial medium. It had long been recognized that the main intracellular constituents of muscle are potassium (K), magnesium (Mg), and phosphorus (P), whereas those outside the cell membrane are sodium (Na), calcium (Ca), and chlorine (Cl). The values in Table 2–2, taken from Conway's[55] analysis of rat leg muscle fibers, are believed to be representative. Analysis of normal human muscle by Nichols[136] have given essentially similar values, except for slightly higher values of Na and P.

It is virtually certain that in the normal muscle fiber the intracellular concentrations of these various electrolytes are maintained within a narrow range. The optimal biochemical functioning of the normal cell is dependent upon precise regulation of the ionic content. Since the

TABLE 2–2. Analysis of Rat Leg Muscle Fibers

Ion	Fiber water (mM)	Plasma water (mM)
Potassium	152.0	6.4
Sodium	3.0	150.0
Calcium	1.9	3.1
Magnesium	16.1	1.5
Chloride	5.0	119.0
Phosphorus	73.0	2.3

Data from Conway.[55]

sarcolemmal membrane is to some extent permeable to all these ions, a mechanism must obviously be available whereby certain ions can be expelled from the cell during excitation and recovery. Sodium is the ion transmitted in greatest quantity. It is believed to be actively and continuously expelled from muscle by some form of pump mechanism. Indeed, the operation of the Na pump has been observed in nerve and red blood cells, and it has been shown that Na transport depends on an ATPase localized in the membrane.[199, 223] Its activity shows an interesting dependence on the Na and K concentration. Recently enzymes of this type were demonstrated in cardiac muscle[217] and in the skeletal muscle of rat[194] and human.[174]

The influx of K may be intimately linked with the excretion of Na, and this active transport mechanism is an essential process in maintaining the excitability of muscle, which depends on the presence of an ionic gradient between the inside of the muscle cell and the outside medium. In certain diseases these ionic relationships are often seriously disturbed.[20] However, it is not yet possible to determine whether ionic disturbances are the cause or effect of the disturbed muscle function. Recent findings indicate that a process of active Ca transport may be involved in the mechanism of action of the relaxing factor.[81]

CARBOHYDRATES

Muscle fibers contain all the enzymes necessary to complete the Embden-Meyerhof glycolytic series. The 10 steps in this series, each requiring a particular enzyme, are capable of converting glycogen to pyruvate and lactate in the absence of oxygen and thereby can provide an energy source for muscle work in the anoxic or hypoxic state. In this metabolic scheme the abundant carbohydrate is glycogen, a complex branched polysaccharide molecule composed entirely of glucose units. When carbohydrate is needed in the contraction process, phosphorylated glucose units are split from glycogen by the action of the phosphorylase enzyme system[60] and the debranching enzyme (α-1,6-glycosidase) and become available in the Embden-Meyerhof chain of metabolic events. Glycogenolysis involves a series of reactions and is an important control point in regulating the glycolytic flux. During the resting state the glycogen stores are replenished from blood glucose by glycogen synthetase that converts glucose-1-phosphate to glycogen via uridine diphosphate glucose (UDPG), discovered in liver by Leloir and Cardini[157] and shown to be the pathway for glycogen synthesis in muscle by Villar-Palasi and Lamar[181] and by Robbins *et al.*[206] The final elaboration of glycogen in muscle, as in liver, requires the branching enzyme (amylo-1,4-α-1,6-transglucosylase), since the polysaccharide formed with UDPG has only partial branching. Healthy human muscle contains about 20 g glycogen per kilogram of intracellular water.[182] Other glycolytic intermediates in the glycolytic series—glucose-6-phosphate, phosphoglycerate, and lactate—are present in small quantities in muscle.

Muscle and liver contains most of the glycogen stores in the body. The enzyme systems in these two types of cells differ in a number of respects, the most notable of which is that muscle does not contain glucose-6-phosphatase and hence cannot dephosphorylate glucose-6-phosphate to glucose for diffusion into the blood and transport to other bodily cells and tissues.

Under normal circumstances the end-product of glycolysis is pyruvate; some of this continues to be used in the oxidative Kreb's cycle or tricarboxylic acid (TCA) cycle, and the rest is converted to lactate and released into the plasma. The lactate is carried via the bloodstream to the liver, which in turn transforms it by a complex series of reactions into glucose and glycogen. The "Cori cycle"[56] is completed when this glycogen again releases its glucose molecules into the circulation (glycogenolysis) for use in the bodily cells, especially muscle.

The exact location of the carbohydrate within the skeletal muscle fiber is not clear. Special staining for glycogen after alcohol fixation reveals a random scattering of glycogen droplets throughout the fiber (see Chapter 1). Some fibers contain many more droplets than do their neighbors, and even within a single fiber there are segmental variations in concentration. Nothing is known of the location of any of the carbohydrate intermediates. Defects in the debranching enzymes and acid α-glucosidase (acid maltase) lead to glycogen storage diseases of muscle. Absence of phosphorylase leads to McArdle's disease.

LIPIDS

Muscle is rich in lipids, largely in the form of phospholipid, cholesterol, free fatty acids, and triglycerides. Only a small part of this fat is in liquid form, capable of dissolving the dyestuffs of the Sudan group. The fat in the various forms mentioned above can be extracted with mixtures of lipid solvents. In mammalian muscle these lipid complexes comprise about 4 g/100 fresh muscle.[22] The functions of these lipid molecules within muscle are uncertain. However, during recent years it has become clear that the oxidation of fatty acids is an important source of energy in both skeletal and cardiac muscle.[97] Removal of two carbon atoms at a time from the carboxyl end of the molecule (i.e., β-oxidation) is the principal method by which fatty acids are oxidized.[109, 166] Several enzymes known collectively as the fatty acid oxidase complex, are localized in the mitochrondria in close association with the enzymes of the respiratory chain. This enzyme complex catalyzes the oxidation of fatty acids to acetyl coenzyme A (acetyl-CoA), the system being coupled with the phosphorylation of ADP to ATP. The acetyl-CoA can be further oxidized to carbon dioxide (CO_2) and water via the TCA cycle. In diseased muscle the accumulation of stainable fat droplets is a common finding, as is also the unrelated replacement of muscle fibers by what seem to be mature lipocytes.

NONPROTEIN NITROGENOUS COMPONENTS

Muscle probably contains all the α-amino acids, but the one in most abundant quantity is creatine. The latter is loosely linked with phosphocreatine (PC), one of the storehouses of energy in muscle. Plasma creatine levels and tolerance tests were used in the past as laboratory procedures to study various types of muscle disease. However, these determinations are subject to so many variables (i.e., state of liver function, muscle mass, physical activity, basal metabolic rate, dietary intake) they provide relatively little exact information concerning the state of muscle. The concentration of free amino acids in muscle is more than 10 times as great as that in plasma or extracellular fluid, and their intracellular distribution also differs considerably from that in the plasma. The role of these amino acids in muscle is not completely settled. It is possible that they may enter the TCA cycle under some circumstances (i.e., starvation and the like) to provide a source of energy. More likely, however, they are utilized in the metabolism of muscle protein, for a number of transaminases such as glutamic oxalacetic transaminase (GOT) and glutamic pyruvic transaminase (GPT) have been extracted from muscle, suggesting that an active and continuous rearrangement of amino acids is in process. Measurements of serum levels of GOT and GPT, which are capable of diffusing into the blood from diseased muscle, are valuable diagnostic and prognostic aids in some disease states, notably muscular dystrophy and myositis.[188]

In addition to the amino acids mentioned before, two unusual peptides—anserine (β-alanyl-1-methyl-L-histidine) and carnosine (β-alanyl-L-histidine)—and carnitine have been isolated from skeletal muscle. The functions of anserine and carnosine are unknown. Carnitine plays a significant role in lipid metabolism.

The other principal nonprotein nitrogenous constituents of muscle are the nucleotides. Of these, ATP is the best known and perhaps quantitatively the largest representative, but the diphosphate (ADP) and monophosphate (AMP) derivatives are said also to occur in muscle during some phases of muscle activity.

ENZYMES IN MUSCLE

Most of the earlier work on intracellular localization of enzymes in the muscle cell was carried out using relatively crude separation procedures. It is known that the glycolytic enzymes are present within the sarcoplasmic fractions of muscle homogenates, whereas the oxidative enzymes of the TCA cycle are localized within the mitochondria.[224] Although it is generally agreed that all glycolytic enzymes are soluble, recent studies of the ultrastructure of the muscle[4] suggest that the extramitochondrial sarcoplasm may be compartmentalized and the glycolytic enzymes may in part be loosely bound within certain of these compartments. They are rendered soluble by the homogenization process. It was first shown by Robbins et al.[206] that glycogen synthetase is associated with a high-speed, glycogen-rich particle fraction from rat skeletal muscle. There is also some evidence that a major part of phosphofructokinase activity is bound to the sarcoplasmic reticulum fraction.[4] The mitochondria lie close to the myofibrils, perhaps in direct contact. They are therefore well suited to supply ATP to the myofibril, and this is their main function in muscle. It is known that glocolytic enzymes in human dystrophic muscle tend to be more depleted than mitochondrial oxidative enzymes.

It is now recognized that a number of enzymes can exist in multiple molecular forms called isoenzymes, easily separable by electrophoretic methods. The most extensively studied isoenzymes in human neuromuscular disorders are those of lactic dehydrogenase (LDH). LDH exists in five molecular forms, each of which is a tetramer of the different subunits.[6, 41] One type is characteristic of skeletal muscle. (LDH-5), another of cardiac muscle (LDH-1).[28] During development the cardiac form predominates in the skeletal muscles of chickens[132] and humans.[133] In certain forms of animal and human muscular dystrophies and related disorders[30,132,246] there is a preferential loss of LDH-5 in certain skeletal muscles, and as a result the LDH isoenzyme pattern resembles that of fetal muscle.[77] It appears, however, that this change in isoenzyme pattern is completely nonspecific and is more likely to be a response to muscle injury.[66,189] Similar results for aldolase isoenzymes in diseased muscle were recently obtained by Tzvetanova.[240] A variety of other isoenzymes, especially those of GOT[134] and creatine kinase[107,135] were also studied in normal and diseased muscle, but none has provided any additional significant information.

ENERGY MECHANISMS IN MUSCLE FIBERS

Muscle fibers manifest variable levels of energy output, depending upon their state of activation or relaxation. They are never completely inactive, for in vitro studies of isolated incubated muscle such as rat diaphragm show a level of basal oxygen consumption indicating some energy production even in this condition. Relatively little is known about this basal or resting metabolic state. It is supposed that a minimal energy requirement is needed for the transport of metabolites across the cell membrane and the maintenance of concentration gradients of various electrolytes between the intra- and extracellular environments. Probably the purely endogenous activities such as protein synthesis and turnover, replenishment of PC, and synthesis of ATP and other complex molecules contribute to the maintenance of this basal state. That muscles differ in their basal metabolic requirements was demonstrated by Smelser,[225] who showed that guinea pig extraocular muscles used 50% more oxygen than did the diaphragm and almost 200% more than the lattissimus dorsi. Andres *et al.*[5] calculated that intact resting forearm muscle in man uses oxygen at the rate of 12 μmol/hour/g muscle.

Muscle metabolism may be considered also in terms of the energy it expends in contracting and producing work. This in turn can be divided into two poorly defined categories, i.e., that required during actual work, and that needed after work has been completed or until energy supplies have been completely restored. The latter is inseparable from basal metabolism. The energy for contraction is derived by an explosive release from the high-energy stores in the fiber. These endogenous sources are largely ATP, PC, and perhaps others which release energy through dephosphorylation and transphosphorylation. In the resting state muscle has four to six times as much PC as ATP. Mammalian skeletal muscle contains large amounts of the enzyme creatine kinase,[143,164] which catalyzes the reversible transport of phosphate between ADP and PC. Measurement of serum levels of creatine kinase, which is capable of diffusing into the blood from diseased muscle, is the best diagnostic and prognostic aid so far known in muscular dystrophies.[79,177] It has proved useful even in detecting some female carriers of Duchenne muscular dystrophy.[177]

Another enzyme present in muscle, myokinase[54] or adenylate kinase, catalyzes the reversible formation of one molecule of ATP and one molecule of AMP from two molecules of ADP. This reaction provides additional available energy in the form of ATP. The AMP is deaminated to form inosinic acid (IMP) by a highly specific enzyme, AMP deaminase, which is present in large amounts in muscle.[155,216] The precise role of this enzyme in normal contractile activity is not known.

Replenishment of the energy debt incurred during active contraction to restore muscle to its state of resting equilibrium must be supplied from extracellular sources, largely by the metabolism of carbohydrates and fats. During aerobic submaximal work oxidative processes supply much of the energy through the catabolism of glucose to CO_2 and water. Since a much lower yield of energy is derived from the glycolytic breakdown of glucose to pyruvate than by its subsequent oxidation to CO_2 and water via the TCA cycle, the latter process supplies the former with sufficient energy to "recharge" the phosphorylated compounds. The energy derived from various uses of foodstuffs is as follows: (1) glycolysis from glycogen to lactate yields 3 moles net of ATP per hexose unit (moles of glucose); (2) glycolysis from blood glucose to lactate yields 2 moles net of ATP per hexose unit; and (3) total oxidation of blood glucose to CO_2 and water yields 36 moles of ATP per hexose unit. In general about 6 moles of ATP are formed per mole of oxygen consumed in the oxidation of either carbohydrates or fats. During heavy work or under conditions of hypoxia, oxygen debt is incurred since glycogen and other storage substances are degraded only to lactic acid, with a relatively low yield of energy. Lactic acid accumulates and may enter the blood. Energy stores must then be replenished during the postwork rest period. Following a period of intense or prolonged exercise the restitution may extend over several hours.

Metabolites of carbohydrate, fat, and protein all contribute to the TCA cycle and oxidative phosphorylation from which their energy is derived. During heavy work quantitatively the most important source is carboydrate. Under conditions of light work it is likely that serum lipids may provide a sizable fraction of the energy supply. The same may be true during the basal state. Evidence for this comes from several indirect measurements including: (1) studies with ^{14}C-labeled glucose in man showing only 20% of expired CO_2 results from metabolism of glucose[10]; (2) numerous observations that the basal respiratory quotient (RQ) approximates 0.7 (which suggests that energy is derived largely from breakdown of long-chain fatty acids), and during strenuous exercise the RQ quickly rises to nearly 1.0 indicating dependence on carbohydrate at this time;[89] and (3) studies of Bing and co-workers indicating that cardiac muscle depends largely on blood lipids for its energy supply.[18]

CHEMICAL CHANGES DURING MUSCULAR CONTRACTION AND RELAXATION

Of the several physical changes that accompany contraction of the muscle fiber, the velocity of shortening has been measured, as have changes in stiffness, elasticity, stretch resistance, etc. Heat production during various phases of contraction and shortening was extensively studied by Hill over many years.[116, 118] Following the discovery by Fenn[96] that the greater the work, the greater the amount of heat produced in muscle (the Fenn effect), heat production in muscle was found to be separable into three phases: heat of activation, heat of shortening, and heat of work done. The sum total of energy required for these three phases is equivalent to the total quantity of energy released during the working of muscle. There has been interest in these three forms of heat production because of the general importance of the phenomena themselves and the several separate chemical reactions that subserve them. The processes from which these energy outputs are derived are the three main catabolic reactions: (1) glycogen to lactic acid, (2) PC to creatine, and (3) closest to the contractile proteins, ATP to ADP. It is known from in vitro studies on isolated myofibrils that muscle contraction leads to the hydrolysis of ATP by actomyosin ATPase. The ATP is rapidly renegerated from ADP and PC by the action of creatine kinase so that no actual fall in ATP concentration has ever been detected as a result of a single contraction. However, it was recently shown by Cain and Davies[42] that creatine kinase can be selectively inhibited by 2,4-dinitrofluorobenzene (DNFB); and in DNFB-poisoned frog rectus abdominis and sartorius muscles, which contract apparently normally for a few twitches, ATP is indeed broken down while PC is not.

The Ca ion is now considered to be the physiological activator of the contractile system. The presence of native tropomyosin (i.e., a complex of troponin and tropomyosin) is essential for the responsiveness of the actomyosin system to Ca ion.[84] Troponin is the only Ca-receptive protein in muscle and together with tropomyosin is responsible for the contraction-triggering action of Ca ion, whose concentration available to the contractile system is under nervous system control. When nervous stimulation ceases the sarcoplasmic reticulum takes up Ca ion from the surrounding medium by an ATP-dependent active transport process.[81, 245] With decline in Ca ion concentration the resting state is reestablished. Although there is yet much to be learned about the molecular mechanisms of excitation-contraction coupling, the currently accepted sequence of events may be listed as follows: (1) the excitatory stimulus generates an action potential which travels along the surface membrane and spreads inward through the transverse tubular system (t-system); (2) Ca ion is released from the sarcoplasmic reticulum into the myofilament space by an unknown mechanism; (3) the increased Ca ion concentration inhibits the troponin-tropomyosin complex, resulting in the activation of actomyosin ATPase and of the contractile process; (4) relaxation occurs when Ca ion from the myofilament space is pumped back into the sarcoplasmic reticulum.

Only a few of the many aspects of the complex chemical anatomy and physiology of muscle have been sketched in the foregoing pages. Many details of the various metabolic pathways available to the muscle fiber were purposely eliminated here. Likewise, the complexities of electrolyte balance and distribution and other matters are beyond the scope of this section. For the reader who seeks further information, a number of excellent reviews are recommended, including monographs by Szent-Györgyi,[232, 233] Mommaerts,[178] Dubisson,[78] Gergely,[104] Bendall,[16] and Laki,[153] and short articles by Lilienthal and Zierler,[160] Wilkie,[248] Weber,[245] Ebashi and Kondo,[84] and Young,[253] The technics used in the chemical analysis of muscular tissues were summarized by Mommaerts.[178b]

MYOGLOBIN

A discussion of the chemistry of muscle is incomplete without some reference to myoglobin. This pigment is an iron-protein compound present in the sarcoplasm of striated muscle cells and accounting for most of the red brown coloration of muscle. Of the total body hematin compounds, about 25% is in muscle and the remainder in the red corpuscles and other cells.

Theorell's classic isolation and crystallization of myoglobin[236] from horse heart led to numerous studies of this important heme protein. Human myoglobin was later obtained in crystalline form by Theorell and de Duve[237] and by Rose-Fanelli.[212] The three-dimensional structure of myoglobin was elucidated by Kendrew.[136]

The general properties of myoglobin are similar to those of blood hemoglobin, as would be expected from the fact that both are combinations of protoporphyrin, iron, and a specific globulin. Hemoglobin is a much larger molecule, however, having a molecular weight of 68,000 and four iron atoms per molecule, as compared with myoglobin, which has a molecular weight of about 17,500 and contains only one iron atom. These substances have the common properties of combining in a loose, reversible manner with oxygen and CO_2, of being oxidizable to "met" compounds, and of possessing the same prosthetic group ("heme").

Myoglobin is synthesized within the muscle cell,[19] and the turnover rate of its iron porphyrin is much slower than that of hemoglobin or even of cytochromes. The so-called fetal muscle heme protein is identified as fetal myoglobin.[191] Earlier claims about the presence of a fetal form of myoglobin differing from adult myoglobin[222] are disputed by later investigators.[191] The concentration of myoglobin is decreased in muscle in certain human muscular dystrophies, a finding which might explain the observed pallor of the dystrophic muscle. Perkoff *et al.*[192] first pointed out the heterogeneity of human myoglobin, and the preponderance of one fraction in some children with muscular dystrophy. However, Kossman *et al.*[139] and Rowland *et al.*[213] in their detailed electrophoretic chromatographic, and immunologic studies failed to detect any significant abnormalities in myoglobin in any of the dystrophies.

The chief function of myoglobin is to carry oxygen. Its dissociation curve is hyperbolic, and it has a greater affinity for oxygen than does hemoglobin but less than that of the oxidases. Myoglobin presumably acts as a storage place or oxygen in muscle.[177] As a short-term oxygen store it tides the muscle over from one contraction to the next by releasing oxygen to the oxidase systems within the muscle fiber. The rate of reaction of myoglobin with oxygen is of the same order as hemoglobin, i.e., about 0.001 sec. Commenting on the oxygen storage function of myoglobin, Irving[129] calculated that in a 70-kg man the myoglobin capacity for oxygen, when saturated, is about 335 cc, or about 10% of the oxygen store of the entire body. Theorell[236] found that the concentration of myoglobin in the pressed juice of the muscles of seals is about 10 times that of other mammals, a finding that possibly explains why these and other diving animals can remain submerged for such long periods of time. Also Vaughan and Pace[241] showed that continuous exposure to high altitude results in a considerable increase of myoglobin content. At 12,500 feet myoglobin increases 50%, as compared to a 25% increase of hemoglobin.

Myoglobin assays have not been carried out on all muscles in the human body. In man all normal muscles are dark red and vary less in color than those of quadripeds. Their lightness and darkness is related not only to their content of myoglobin but also to their opacity (i.e., content of lipids). Almost all fibers in the red muscles are opaque, whereas less than half of those in pale muscles have this appearance. The more rapidly contracting flexor muscles in most mammals tend to be paler than the extensors, which usually have a superficially placed, rapidly contracting pale component (or head) and a deep, slowly acting dark component.[69] In the cat and rabbit, for example, the pale gastrocnemius contrasts strongly with the underlying dark red slowly contracting soleus, and the pale long head of the triceps with the darker, more deeply placed medial short head. Less obvious differences in color and speed of contraction are found in the deep and superficial parts of mammalian flexor muscles[106] and in cardiac muscle.[40] These differences are present in the corresponding human muscles, but are even less distinct. Denny-Brown[69] pointed out that the loss of pigment and lipid in superficial muscles of the cat during the first weeks of life coincides with the development of greater speed of contraction. This strongly suggests that the pale, rapidly contracting muscle fiber is a specialized development.

It is of interest that a considerable amount of cytochrome and relatively little myoglobin has been found in muscles which regularly contract more than two or three times per second, i.e., wing muscles of flying insects, hearts of birds, small mammals, and the like. Also, muscles called upon for intense but infrequent contractions, such as the leg muscles of the frog, contain much cytochrome but little or no myoglobin. It should be noted, however, that all active muscles have a considerable amount of cytochrome regardless of their myoglobin store and that cytochrome contributes somewhat to the red coloration of muscle. Lawrie[154] has pointed out that red muscle has a much richer supply of mitochondria than does white muscle. Since mitochondria contain the enzymes for respiratory (Kreb's cycle) metabolism, this observation again emphasizes the greater capacity of red muscle for oxidative metabolism and the higher capability of white muscle for the anaerobic glycolytic process.

The breakdown products of myoglobin and their mode of excretion are unknown. deLange[67] suggested that some cases of porphyria are probably myoporphyria. He believes that some of the catabolic products of myoglobin may be coproporphyrin or possibly bilirubin or urobilin. Others failed to find spectroscopic evidence of urinary porphyrins in association with myoglobinuria.

In summary then, myoglobin serves as an oxygen transport unit and a very temporary oxygen storehouse in the human and in most air-breathing animals, except those adapted for underwater existence. It is not absolutely essential to muscular function since other of the respiratory hematin compounds (i.e., the cytochromes) can assume their roles in the presence of an adequate circulatory supply of oxygen. The most important action of myoglobin is probably to reduce the need for glycolytic processes during brief periods of relative circulatory insufficiency; as it releases its contained oxygen it allows a continuance of a much more efficient oxidative breakdown of lactate, pyruvate, and similar metabolites.

THE NEURONAL CONTROL OF MUSCULAR CONTRACTION

The response of striated muscle fiber to the receipt of a single nerve impulse is a twitch lasting approximately 0.1 sec. The onset of this mechanical tension is preceded by an electrical variation in the form of a negative potential, which begins about 0.001 sec before the contraction and lasts 10–15 msec depending on placement of the electrical leads. The duration of the twitch from its onset to the peak of contraction varies greatly in different muscles, being as brief as 7.5 msec in the extraocular muscles, 23–40 msec in pale muscles such as gastrocnemius or semitendinosus, and 90–120 msec in red muscles such as soleus, in the cat. The electrical event is therefore completed long before the peak of contraction is reached, an interval correspondingly shorter in the more rapidly contracting muscle. A brief refractory period follows the onset of the electrical potential, after which another excitation may be set up, its contractile process fusing with the previous one.

Thus repeated nerve impulses cause a summated mechanical contraction or "tetanus" in which the twitches are partially or completely fused according to the rate of stimulus. In red muscle a completely fused contraction results from rates of excitation as low as 15/sec, and in pale muscle, 100/sec; but in extraocular muscles a rate as high as 300/sec is required. Since each motor unit is innervated by one anterior horn cell, which manifests its activity by a rhythmic discharge of nerve impulses, the corresponding series of electrical rhythms (Fig. 2–22) in the muscle are readily recorded by appropriate electrodes and recording equipment, as the electromyogram. Since each motor unit is comprised of a relatively small group of muscle fibers, interdigitated with adjacent motor units,[36, 88] the size of the action potential varies with the number of muscle fibers within the unit and its distance from the recording electrode. According to the calculations of Buchthal and his associates[35] the receptive field of most types of needle electrode has a radius of 2–10 mm of the tip. Motor units at a greater distance are recorded as "blunt spikes" or undulations of the base line.[74] This applies, however, only to small units in mild and moderate contractions. With maximal effort there are always a small number of motor units with action potentials greatly in excess of the modal values, and their number increases in conditions such as old poliomyelitis,

amyotrophic lateral sclerosis, and spinal root damage. The use of two discrete electrodes in the same needle, the shaft of the needle being grounded, allows accurate recording of approximately 30 muscle fibers.

The motor unit potential is the electrical sign of excitation of the group of fibers comprising a motor unit. Since this group of muscle fibers has a transverse and longitudinal extent in the muscle it can be sampled by one or more electrodes at different points within, or in some instances (extensor hallucis brevis) even from the surface of, the muscle. The shape of the potential depends on the location of the needle electrode in relation to the axis of the active fiber group and the position of the motor end-plates (the starting point of the potential). Since the motor end-plates of a single motor unit are dispersed, often over a considerable area, the electrical change resulting from one nerve impulse is spread over a period of 8–15 millisec as compared to 2–4 millisec for one muscle fiber. The motor unit potential frequently has a polyphasic form for the same reason, and a single electrode records a characteristic shape of potential for each motor unit in its neighborhood (Fig. 2–22). The activity of a single muscle cell (Fig. 2–22) may thus be recognized by brevity as well as the small amplitude of its potentials. Conditions that reduce the number of fibers within a motor unit (e.g., muscular dystrophy or polymyositis) reduce the size and average duration of action potentials; this change has been used as a diagnostic feature.[27] Collateral innervation of denervated motor units increases the size and duration of the action potentials and renders increasing numbers of them polyphasic, features diagnostic of peripheral nerve or spinal cord disease.

The number of motor units within given muscles has been estimated in various ways. Counts of the fiber population of muscular nerves (50% of large fibers are motor according to the calculations of Feinstein *et al.*[95] Cooper,[57] and McComas *et al.*,[170] through Boyd and Davey[27] find variation from one muscle to another) divided by the number of extrafusal muscle fibers has been the method most widely used. McComas *et al.*[170] estimates the number of motor units in the small muscles of

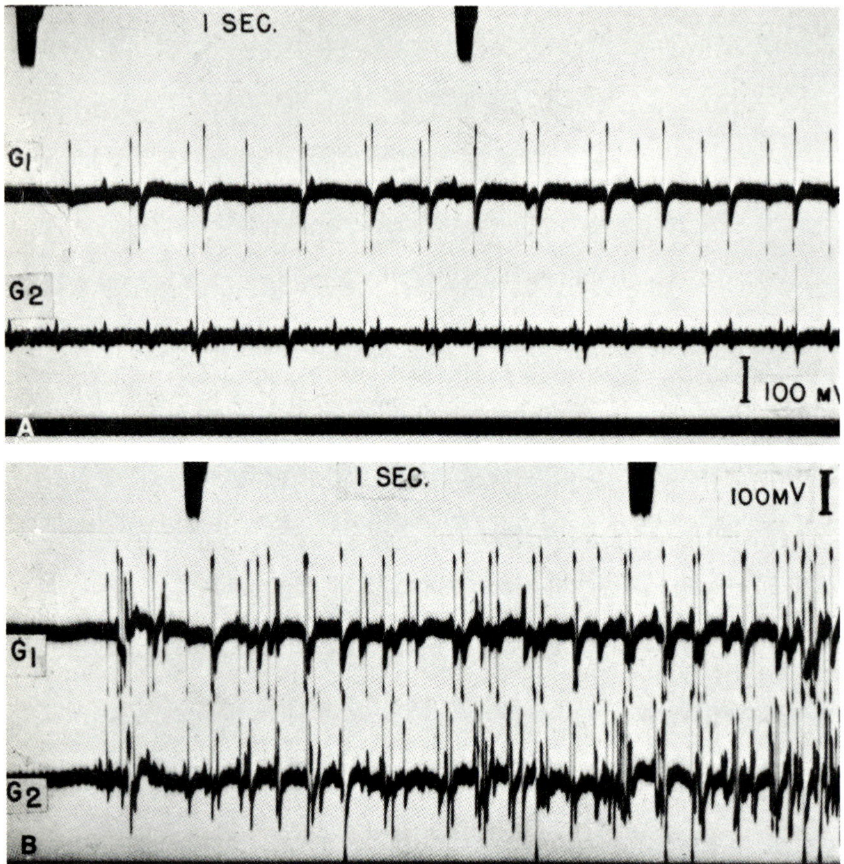

Fig. 2–22. (A) Electromyogram of onset of voluntary plantar flexion of the ankle, recorded from two needle electrodes in gastrocnemius. These and subsequent electromyograms read from left to right. Contraction begins with a regular discharge of impulses with small blunt spikes in G_1. Then three units, each with characteristic wave form, begin beating rhythmically in G_1. Finally a further unit appears in G_2. Discharge of these five units then continues. (B) Subject is asked to make same movement more forcibly. Some units begin discharge as in (A), but are rapidly joined by three more units in G_1, and five more in G_2, giving a more confused ("interference") pattern. (Calibration, 100 μV)

the foot by an electromyographic technique of measuring the electrical activity of what they assumed to be single units and determining the number required for maximal compound muscle action potential. In the human extensor hallucis brevis he estimates the total number of motor units to be 180–200; in the first lumbrical muscle of the foot there are an estimated 78 motor units, in the first dorsal interossious of the hand 99, and in the tibialis anterior muscle 370. Others find this method inadequate, failing to pick up small, large, and overlapping units. The contraction times of motor units which, as was said, range from 35–100 millisec tend to be distributed around two modes; one represents the twitches of fast groups (78–98 millisec) the other of slow groups (15–30 m/sec). Similar slow and fast motor units, identified in cat's gastrocnemius and triceps surae by Wuerker *et al.*[251] and Burke[37] and in the intercostal muscle of rat by Close[50] correspond to the type I and II histochemical fiber types seen by Brooke and Engel.[31] The type I muscle fibers have an average diameter of 45.4 μ and the type II, 49.6 μ. The rate of contraction, its duration, and the refractory periods of motor units also tend to fall roughly into two groups, which is consistent with the demonstration of at least two histochemical types of muscle fibers, each with different properties of excitation.[36, 36A, 50]

The rate of discharge of anterior horn cells and the corresponding series of potentials in the motor units varies to some extent with the intensity of innervation. They are found to be 5–25/sec in postural contraction[70, 74] and 5–50/sec in voluntary muscular contraction.[218] Since the twitch in human limb muscles lasts about 0.1 sec, it follows that the muscles are seldom in completely fused tetanic activity in these kinds of contractions. In sudden impulsive effort, as with a jump or startle response, the discharge rate may momentarily increase to 250/sec. The proof obtained by Denny-Brown[70] that the "motor unit potential" was in fact the electrical sign of discharge of an anterior horn cell was the demonstration that the corresponding reflex isometric-twitch-contraction tension was approximately 10 g in the soleus muscle of the cat. Eccles and Sherrington[85] estimated an average value of 34 g of tension for full contraction of a motor unit of cat gastrocnemius. Allowing for the negligible contribution of the small fiber system[156] this value should be increased by approximately 30%. The values for man are thought to be of a similar order.[203] The action of individual motor units is usually not detectable but in certain conditions involving small muscles of fingers or toes twitches or tremors in a large motor unit can induce a corresponding movement.[74] It is therefore assumed that natural contractions must be smoothed out by the activation of small units in small movements and large ones in phasic or powerful tonic contractions.

There is no evidence that differences in pigmentation or size of fat granules in the fiber are related to the purely mechanical aspect of the contraction process in mammalian muscle. The "slow" muscle fibers of amphibians are represented in mammals only by the relatively primitive muscle fibers of the muscle spindles. Various deductions have been made regarding the presence or absence of dark pigmentation in the wing muscles of birds with various types

of flight,[208] but it has been shown that the dark wing muscle of the pigeon has the same twitch characteristics as a pale leg muscle of the same bird.[69] In the same manner, deductions regarding the relationship of the abundant sarcoplasm of the wing muscle of certain insects to their speed of flight are misleading. Rollett[211] showed that insect muscle with flat sarcostyles as in the water beetle *Dysticus* had a twitch duration of 0.33 sec, and a fused tetanus was produced at a rate of stimulus of 9.54/sec. It is therefore unlikely that the muscles of the wing of the fly, which closely resemble those of *Dysticus* histologically, can contract and relax at the rate of the beat of the wing, which approximates 300 beats per second. Pringle[202] showed the probability that a physical resonant vibrating mechanism accounts for the rapid wing movement of insects when the innervation rate is only 5–40/sec.

In man postural contraction is invariably accompanied by action potentials in the effective parts of the muscle. Since the active muscle fibers in small postural contractions of quadriceps, for example, may be in the deeply situated crureus muscle, these are not always easy to demonstrate. In complete relaxation there are no action potentials,[74] and differences in consistency (firmness or flabbiness) in such inactive muscle appear to relate only to differences in development, i.e., the size of the muscle fiber in relation to the size of the endomysial tubes. There is no evidence that the autonomic nervous system has any influence on such consistency or can modify postural reflexes when they are present.

LOCAL CONTRACTIONS

It is possible to set up a localized contraction in the contents of a muscle fiber by stimulation with microelectrodes,[100, 101] and that such a localized contraction band is not accompanied by an action potential. Speidel[228] has observed such bands involving 5–20 sarcomeres in the intact muscle of the tail of a tadpole. They could be set up by irritation with a needle and traveled slowly in both directions at a rate of 70 μ/sec. They were quite different from a muscular twitch, which is propagated at a rate of 3 meters/sec and under the microscope appears to tighten the fiber momentarily. Such localized contraction bands can be set up by a variety of stimuli, including chemical agents or percussion (myoedema), and there is no doubt that the contraction band seen by the histologist is the result of the same phenomenon initiated in the process of fixation.

Percussion of a piece of recently excised muscle sets up an intense local contraction, the "idiomuscular contraction" of Schiff,[215] from which smaller contraction bands under some conditions travel slowly to each end, and even back again. Such bands were once thought to be due to contraction of the sarcoplasm and not the myofibrils, but are clearly only another aspect of the self-propagating contraction that can be evoked in teased myofibrils. Schiff's phenomenon can be observed in the muscles of hypothyroid patients and sometimes in those with cachexia e.g., following a pinch or percussion of a superficial muscle such as the pectoralis major. In one such patient Denny-Brown and Pennybacker[71] recorded waves traveling along the fasciculus at a rate of 2 cm/sec and lasting 0.7 sec at any one point. There was no accompanying action potential. Recently, however, Meadows[173] in a study of the response of the vastus medialis in adult man to percussion, was able to record by more refined techniques a small amount of electromyographic (EMG) activity that is propagated along the muscle fibers at a velocity of 4 meters/sec. The EMG activity lasts longer than that produced by a single electrical stimulus to muscle fibers. Nerve blockade had no effect on the percussion response. Percussion near the motor point sometimes caused delayed fasciculation due to activation of intramuscular nerve fibers.

Thus it appears that muscle possesses a basic property of repetitive discharge when stimulated by mechanical means. Such localized contractions are easily demonstrated in the muscles of crustaceans, with the difference that in these animals they can be excited by nerve impulses reaching the end-plate.[146]

NEUROMUSCULAR TRANSMISSION

The cell membrane of the muscle fiber is seen in electron micrographs as a discrete structure 70 Å (7 mμ) in thickness, comprising three distinct layers.[209] The living membrane has a high resistance to direct current and low impedance to alternating current and allows only very slow exchange of labeled ions, though once within the cell the diffusion of such ions is unrestrained.[122] The muscle membrane, like that of the nerve fiber, is concerned in two special functions, first in excluding sodium ions from the cell at rest with maintenance of a correspondingly high internal concentration of potassium ions, and secondly in transmitting the excitatory process rapidly over the entire surface of the fiber. The mechanism by which sodium is excluded from the cell is evidently enzymatic for it can be stopped by certain metabolic inhibitors.

The effect of the high internal concentration of potassium is to endow the contents of the muscle cell with a negative charge relative to the exterior. This potential difference across the cell membrane (the resting potential) may be measured by introducing a microelectrode into the fiber, which demonstrates a potential of −90 mv relative to an indifferent electrode outside the fiber. If the fiber is cut across, the same potential is recorded from the cut surface relative to any point on the intact surface of the fiber and is then called the injury potential.

Any stimulus that sufficiently rapidly breaks down this polarization of the cell membrane at any point to a value less than about −50 mv renders the polarization at that point unstable, so that breakdown continues automatically and rapidly, producing a brief positive total change of 110–130 mv (20–40 mv positive inside at that point). This potential is then rapidly propagated along the muscle fiber in both directions from its point of origin, detonating a brief contraction in the sarcomeres as it passes. When recorded at any one point by an internal microelectrode, the sharp wave is seen to be followed by a slow restitution of the original negative polarization (Fig. 2–23[112, 113] on isolated frog fiber). During the first, steeper part of the wave of depolarization, it is not yet possible for a second stimulus to

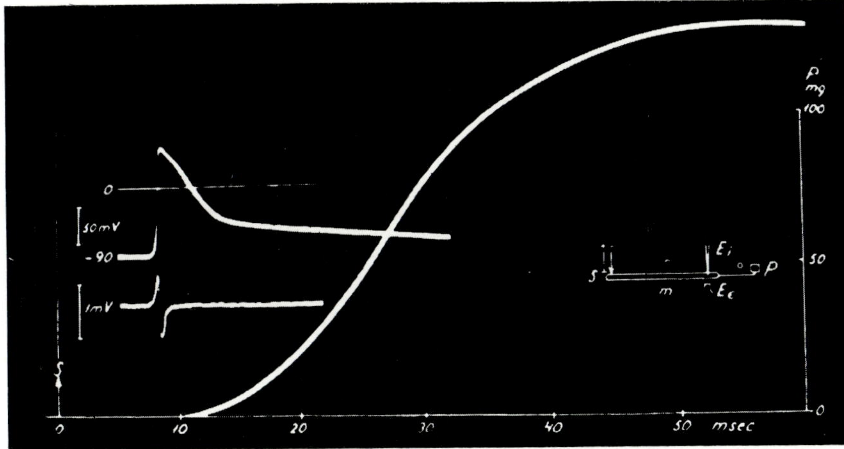

Fig. 2–23. Relationship between intracellular potential (monophasic, above), cellular potential (diphasic, to right of calibration mark 1 mV), and rise of tension resulting contraction to single stimulus in single isolated frog muscle fiber. S is time of stimulus. Time scale in milliseconds. (From Hakansson[113])

set up a second excitation (refractory period). The rising phase of the potential is due to a momentary increase of Na permeability, and the falling phase with a slower increase of K permeability of the muscle membrane.[94, 121] The net loss and gain of these ions is replaced only very slowly by the mechanism that maintains polarization. Electrodes outside the muscle fiber record only a much briefer diphasic spike, a rising phase followed by a negative phase related to the approach and passing of the electrical disturbance (action potential). The external potential is much smaller because it reflects only part of the surface event, the remainder being short-circuited by the surrounding medium.

The mechanism of transference of the excitatory process from the cell membrane to the myofibrils has been the subject of much speculation. The studies of Huxley and Taylor[127] showed that when the muscle membrane is excited opposite the Z line the related internal structure is thrown into contraction. The contraction produced by a microelectrode thus applied shortens both halves of the I band adjacent to the excited Z line. It is suggested that the transverse sarcoplasmic reticulum (T system) is the conducting structure, particularly since this tubular formation has a peculiar and special relationship to the Z line in

amphibian muscle.[198] The response of the contractile mechanism to one excitatory process is the rise of internal tension (twitch) lasting about 0.1 sec in most mammalian muscles. By means of sudden stretches and other physical analyses the contraction process has been shown to consist of an initial very brief rigidity followed by an active state exerted through elastic structure, with a rate of decay approximately corresponding to the relaxation curve of a twitch.[196] By sufficiently rapid repetitive excitation the active states of successive excitations summate with each other to produce a fused (tetanic) contraction.

The region of the membrane that is included in the motor end-plate is polarized to the same degree as the remainder but has a special sensitivity to acetylcholine (ACh), 1000 times greater than that of any other part of the membrane.[145] A nerve impulse arriving at the end-plate liberates a small quantity of ACh which has the effect of rendering this portion of the muscle membrane momentarily permeable to Na and K ions and thus depolarized, setting up a small potential, called the end-plate potential[144] (e.p.p.). Near its peak the e.p.p. forms an electrical leak sufficient to initiate a propagating spike potential in the neighboring muscle membrane, thus causing a twitch contraction. Acetylcholine is rapidly hydrolyzed to choline and acetate by cholinesterase in the end-plate, and the muscle fiber becomes repolarized. An excess of ACh, when applied to the nerve endings or injected intraarterially, or a delayed destruction of it resulting from the presence of an anticholinesterase compound such as physostigmine (Eserine), neostigmine (Prostigmin), pyridostigmine (Mestinon), edrophonium (Tensilon), and diisopropyl fluorophosphate (DFP) results in continued depolarization of the end-plate and consequent block of subsequently arriving nerve impulses for some seconds or minutes. Decamethonium has the same effect lasting for 30 min or more. Drugs that have this action of inactivating or combining irreversibly with acetylcholinesterase are called depolarizing agents, though it is the excess ACh that causes the depolarization. A cathodal current applied to the region of the motor end-plates can also depolarize them.[39] The mechanism of action of ACh and the precise manner of its production are unknown. It has been shown, how-

ever, that it is being continually released by the telodendria in very small amounts (quanta),[94] giving rise to an e.p.p. too small to set up a muscle membrane potential. Prostigmine greatly facilitates the release of ACh, resulting in spontaneous contraction of the muscle fiber, and deficiency in ionized Ca depresses the release of ACh. Botulinum toxin blocks the release of ACh from the telodendria. The sequence of events in the process of excitation and the corresponding effects of pharmacologic antagonists are summarized in Figure 2–24.[160]

Curare and its active constituent *d*-tubocurarine, pentamethonium, or an anodal current at the region of the end-plates, counteract the effect of ACh, reducing the size of the e.p.p. so that it can no longer prepare or set in readiness the neighboring muscle membrane.[39, 146] The action of curare appears to be to depress the sensitivity of the receptor substance of the end-plate to ACh, without completely eliminating it. In this sense the action of curare

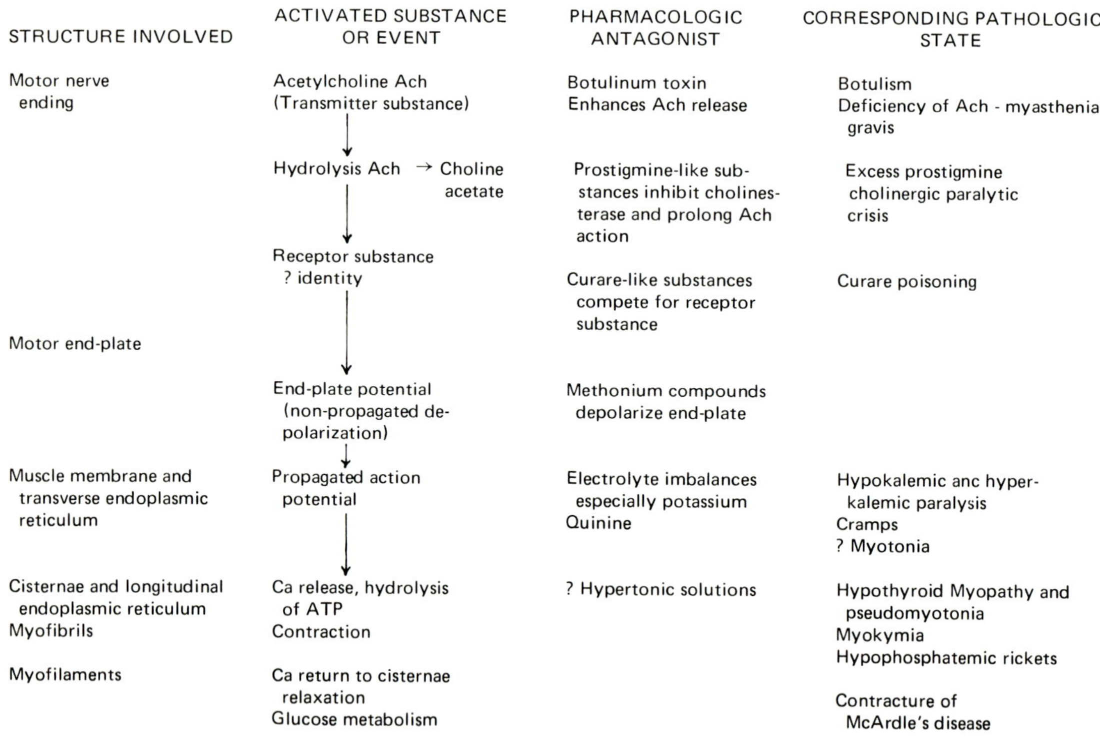

Fig. 2–24. Succession of events in excitation of muscle fiber by motor nerve. (Modified from Lilienthal and Zierler[160])

STRUCTURE INVOLVED	ACTIVATED SUBSTANCE OR EVENT	PHARMACOLOGIC ANTAGONIST	CORRESPONDING PATHOLOGIC STATE
Motor nerve ending	Acetylcholine Ach (Transmitter substance)	Botulinum toxin Enhances Ach release	Botulism Deficiency of Ach - myasthenia gravis
	Hydrolysis Ach → Choline acetate	Prostigmine-like substances inhibit cholinesterase and prolong Ach action	Excess prostigmine cholinergic paralytic crisis
	Receptor substance ? identity	Curare-like substances compete for receptor substance	Curare poisoning
Motor end-plate	End-plate potential (non-propagated depolarization)	Methonium compounds depolarize end-plate	
Muscle membrane and transverse endoplasmic reticulum	Propagated action potential	Electrolyte imbalances especially potassium Quinine	Hypokalemic anc hyperkalemic paralysis Cramps ? Myotonia
Cisternae and longitudinal endoplasmic reticulum Myofibrils	Ca release, hydrolysis of ATP Contraction	? Hypertonic solutions	Hypothyroid Myopathy and pseudomyotonia Myokymia Hypophosphatemic rickets
Myofilaments	Ca return to cisternae relaxation Glucose metabolism		Contracture of McArdle's disease

competes with ACh as a substrate for the receptor enzyme.[94] Substances of this type are known as competitive blocking agents. Prostigmine can reverse the early stages of curarization, presumably by prolonging the presence of ACh. Prostigmine has also a weak, curarizing (blocking) action manifest when present in excess amounts. Prostigmine also weakly depolarizes the muscle fiber.

Jewell and Zaimis[131] studied differences in reactivity of several muscles to various blocking agents. They found that soleus, a red slow muscle, is more resistant to decamethonium and slightly more sensitive to curare than pale muscles. In hypertrophy when the muscle is redder and slower in contraction these differences become exaggerated, whereas in disuse atrophy and denervation the differences are lessened. These changes are attributed to differences in the characteristics of the membrane of the muscle cells. Muscles that contract rapidly, such as the extraocular and laryngeal muscles, may be expected to show even greater exaggeration of such characteristics.

From the work of Castillo and Katz[48] on random miniature end-plate potentials, it became evident that the nerve impulse does not normally activate all the branches of the telodendrion in an end-plate. The second of two impulses separated by a short interval under adverse circumstances (e.g., mild curarization) can activate more of the end-plate and set up a larger e.p.p. than the first (potentiation). In some amphibian muscles particularly in winter months, there are muscle fibers whose sarcolemmal membranes are incapable of spreading spike potentials.[148] These muscle fibers are small, and their motor fiber is thin and terminates over a large surface area of the muscle fiber. A nerve impulse activates these fibers directly by end-plate potential, and summation of local potentials is needed to activate the whole contractile mechanism. The only comparable muscle fibers and endings in mammalian muscle are the intrafusal fibers of the muscle spindles.[187]

Potassum ions are essential in several aspects of the process of excitation, but most importantly for polarization of the muscle membrane. In disorders of intracellular potassium, such as periodic paralysis, propagation of the excitatory process along the muscle fiber and into the transverse sarcoplasmic reticulum fails, though impulses are still produced near the end-plate (Chapter 12). The mechanism by which substances like quinine act is complex, but there is some evidence that it affects the sarcolemma.

In the analysis of the process of muscle shortening (the mechanical twitch which follows the electrical action potential) the steps that appear to be important in regulating the active state are: (1) the availability and rate of release of Ca^{++} ions from the membrane or sarcoplasmic reticulum; (2) the rate of hydrolysis of ATP; (3) the rate of sarcomere shortening; (4) the viscosity of the elastic elements of the muscle fiber and surrounding connective tissues; and (5) the rate of uptake of intrafiber Ca ions by the sarcoplasmic reticulum. Parameters (1) and (2) presumably affect the latency of onset of the twitch (i.e., interval between the action potential and beginning of contraction); (3) and (4) determine the velocity of rise to peak tension at a given length; and (5) determines the duration of the twitch (viz., the period of full tetanic tension attained during the twitch). The total twitch time is the sum of all the above plus the time required for the twitch to be transmitted to the tendon. All these chemical events are shown diagrammatically in Figure 2–24. Little is known of the action of pharmacologic agents on each of these steps (see Chapter 11, thyroid disease).

"ALL OR NONE" PHENOMENON

In mammals the "all or none" phenomenon is a general property of nerve fibers. A single nerve excitation produces a twitch in all the muscle fibers supplied by that nerve fiber, or none. The result is not modifiable except by repetition. Increase in size of twitch contraction or tetanus at a constant frequency of excitation can occur only by increasing the number of nerve fibers excited.[165, 201] This relationship is disturbed in man in myasthenia gravis when continued excitation of the muscle results in periodic failure of contraction of some fibers of the motor unit, and eventually of all the fibers.[161] The muscles can still be excited by direct application of a faradic current, so that excitability of the sarcolemma is maintained. Botulinum toxin has a very specific action in blocking terminal branches of the motor nerve while leaving the end-plates excitable to direct stimulus.[32] As the toxin begins to take effect, parts of the motor unit are paralyzed, leaving other fibers still active. Nerve fibers are particularly vulnerable to blocking agents at their points of branching, but it is not known whether the most branched fibers are affected first. This action of botulinum toxin demonstrates a vulnerable point in motor nerve terminals that may be affected in peripheral fasciculations and muscle cramps, for which there is not as yet any adequate explanation but which originate at some point near the end-plate.

FATIGUE

The phenomena of fatigue are complex and of several different origins. It is clear that as muscular failure develops the individual strives to sustain his performance by using different groups of muscles to effect the same objective.[74] He varies the posture of the wrist and elbow, for example, in his efforts to maintain an effective hand grip. If this alternation of muscles is prevented, he uses more muscle fibers in the same muscle, producing a more complex action potential rhythm for the same amount of contraction.[89] Ultimately performance fails relatively suddenly, and the power of contraction reaches a low level, which can then be continued for a considerable time. Merton[175, 176] showed that the mechanism of such failure of sustained contraction is ischemia, the level of residual contraction being that at which the circulation of the muscle is not impeded by contraction. If the circulation has been occluded by a proximal limb cuff, the failure is complete. At this level of failure the action potentials of the muscle fibers are unaffected, and yet no contraction occurs. It is not possible to excite contraction in such fa-

tigued muscle fibers by direct stimulation. The failure must be in the myofibrils or in transmission of membrane potential to them. Although potassium is lost by the muscle in ischemic fatigue and ACh production by the end-plates is diminished, these are not factors in ordinary fatigue. Nevertheless, in states in which there is retention of muscle potassium (periodic paralysis) or defective effect of ACh (myasthenia gravis), these factors become operative in facilitating the production of symptoms. In a similar manner defects in muscle enzymes such as phosphorylase (Chapter 12) result in failure of contraction, though such phenomena are not ordinarily included under the term fatigue.

FIBRILLATION

Muscle atrophy following denervation is accompanied by spontaneous fibrillary contractions after the fifth day. These take the form of a series of twitches in individual muscle fibers, each accompanied by a small rapid action potential, at intervals of 2–10 sec (Fig. 2–25). In the following days the rate of twitch becomes a little more rapid, at intervals of 0.5 sec in some fibers.[71] Such twitching then continues as long as any recognizable muscle fiber remains but is difficult to demonstrate after 1 year. Since fibrillation occurs independently in each muscle fiber, it is quite different from the coarse fascicular twitching of motor neuron disease (amyotrophic lateral sclerosis), which is the result of single impulses arising spontaneously in motor units, with correspondingly larger and more complex action potentials (Fig. 2–26).

The spontaneous twitching of the muscle fiber deprived of its nerve supply is associated with a change in its pharmacologic reactions, the most noteworthy being an extreme sensitivity to ACh which in very small amounts provokes a great increase in fibrillary twitching and potentials. In larger amounts ACh evokes a maintained contraction (contracture) without action potentials. Decamethonium has the same effect.[207] However, curare will not

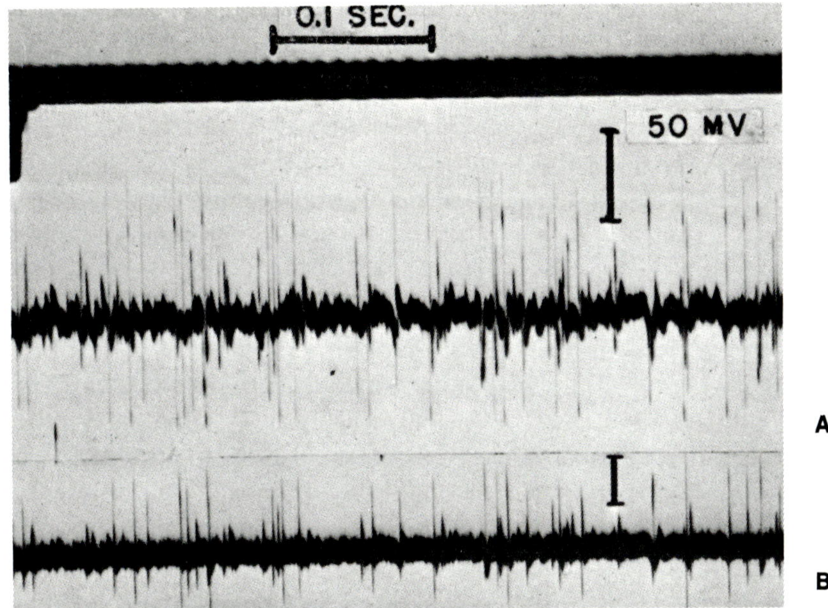

Fig. 2–25. Electromyogram of sartorius muscle of cat showing fibrillation resulting from neural atrophy. Both tracings are from same needle electrode, but amplification in (A) is approximately twice that in (B). Note smaller potential and greater speed of spikes, as compared with unit potentials recorded in Figure 2–22.

Fig. 2–26. Electromyogram from four independent needle electrodes in same muscle (gastrocnemius) in patient suffering from amyotrophic lateral sclerosis. Electrical potentials of spontaneous twitches are numbered from 1 to 8, according to their characteristic shape. Note irregular repetition of each spontaneous spike.

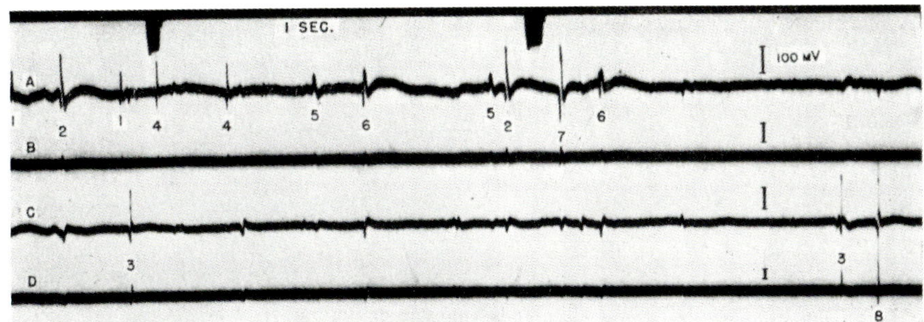

abolish such potentials. It appears therefore that besides its trophic effect on muscles, nerve in some way suppresses the spontaneous excitation of the sarcolemma. The spontaneous twitching of the outgrowths and single muscle cells derived from tissue culture of striped muscle, which were observed by Lewis[158] to occur at rates of from once every 3 sec to once

every 0.5 sec, is another aspect of the same phenomenon and indicates that fibrillation is an inherent property of the living muscle cell.

The ease with which fibrillation can be demonstrated in muscle undergoing neural atrophy is extremely variable. Fibrillation is greatly increased by mechanical stimulation, so that a burst of such potentials follows insertion or movement of a needle electrode. It can continue without such stimulation and is always visible when the surface of the muscle is exposed and viewed by reflected light. In animals such as the rat, the vibrissae show continued twitching after section of the facial nerve. Yet following diseases such as poliomyelitis with undoubted neural atrophy, one may have to search in the muscle with a needle electrode to demonstrate regular fibrillation, apart from transient "insertion" potentials. One explanation is that fibrillation is dependent upon the presence of free ACh in the circulation, for if the individual is emotionally excited the paralyzed muscle shows intense fibrillation and at times even mechanical shortening, enough to raise the paralyzed eyelid or to change the expressionless face.[17] However, there is other evidence marshaled by Shy *et al.*[221] which suggests that ACh is not responsible for the fibrillation of denervation. Instead, there is a postulated *intrinsic* hyperexcitability of cellular membranes. Shy *et al.*[221] find slower positive potentials in intracellular recordings of denervated muscle.

In work on degeneration of muscle (Chapter 3), it has been observed that the portions of muscle fibers separated from the region of motor end-plate by a zone of coagulation also show active fibrillation after the fifth to seventh day. These portions of muscle fibers in effect have been denervated. The occasional finding of foci of fibrillation in purely muscular diseases such as myositis[185] may therefore indicate only that segmental interruption of muscle fibers by disease is occurring, resulting in the presence of fragments of muscle fiber without nerve ending and not necessarily in damage to the motor nerve. Since degenerative fragmentation of muscle fibers is also found in progressing stages of muscular dystrophy, it is not surprising that isolated fibrillating fibers can sometimes be demonstrated in that disease.[105] Such muscle fibers are also sensitive to mechanical deformation and show bursts of "insertion potentials" that continue for many seconds.[172] These trains of impulses in single muscle fibers must be clearly distinguished from the phenomenon of myotonia, which is also sensitive to mechanical deformation. In true myotonia the trains of impulses are set off by muscular contraction with a characteristic delay and a decrement in successive contractions.

Fibrillation thus represents the tendency to breakdown of polarization of the isolated muscle cell. The primitive slow muscle fibers of the frog are reported to show spontaneous fibrillation,[38] even when their simple nerve ending is intact. In the normal mammalian end-plate very small spontaneous end-plate potentials are continually present, indicating release of very small quanta of ACh[48] by the nerve ending in the absence of nerve impulses, but these do not set up action potentials or contraction in the muscle fiber unless the Ca/Mg ratio is abnormal.[159]

DELAY IN RELAXATION

The later stages of relaxation from a muscular twitch produced by a single stimulus to a motor nerve may be delayed by action of the drug veratrine. This delay is the result of continued repetitive excitations in a few of the muscle fibers initially excited. A series of small electrical potentials, at rates of 5–50/sec, accompany the disturbance, and their small size indicates that single muscle fibers are contracting independently at this rate.[90] The excitation is distal to the neuromyal junction, and the impulses are not progagated back in the nerve fibers, though when intense they may set up a parallel discharge from the proprioceptive end-organs.[91] A number of substances possess a similar effect, notably the substance 2,4-dichlorphenoxyacetate (2,4-D), which in addition greatly enhances the sensitivity of muscle to mechanical stimulation.[93] The similarity of such disturbances to that existing in myotonia has been emphasized. In myotonia in man a certain number of excitations must reach the muscle before the muscle fibers begin to contract spontaneously and independently, and the phenomenon is not observed with a single twitch.[73] The substances referred to also tend to produce repetitive firing in nerve, which has not been demonstrated in myotonia. The subject is discussed further in the section dealing with congenital myotonia (Chapter 12).

SPASM

Muscular spasms associated with repetitive firing in the motor nerve following one initial excitation are more intense than those in myotonia and are associated with repetitive action potentials of the size and duration of those of motor units. Such discharges are seen in tetany, in peripheral facial spasm, and in some cases of damage to the proximal nerve roots and plexuses. In true muscular cramps parts of a contracting muscle will begin to show bursts of discharge at rates as high as 300/sec (Fig. 2–27), and the disordered type

Fig. 2–27. (A) Upper tracing is electromyogram from a small muscle in the foot (SMF) and from two leads in gastrocnemius (G_1, G_2) in a patient liable to muscle cramps. Record begins during a voluntary plantar flexion of foot. The mark above the first time mark at the top of the tracing recorded the patient's signal that a muscle cramp was beginning. Bursts of high-frequency discharge are appearing in G_2 at this time, examples being marked at X, Y, and Z. Lower tracing shows the same records; (B) at the height of a cramp; (C) at its termination. Note rhythmic high-frequency bursts. (Denny-Brown and Foley[72])

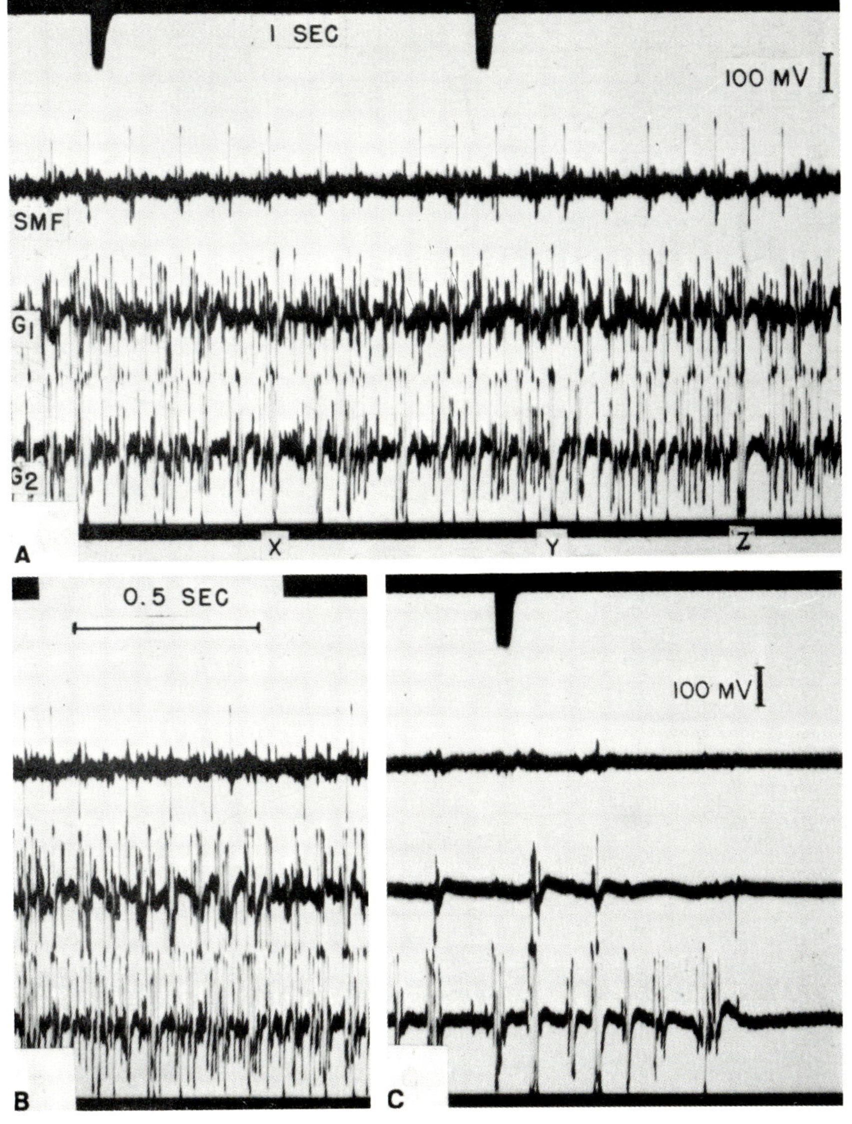

of excitation rapidly spreads to affect more and more of the muscle fibers. This is also a repetitive firing of motor units, but of such kind that a disordered unit excites its neighbors.[72] In such muscles spontaneous twitching of one or more fasciculi is common during repose, and the variable action potential of such twitches indicates that an irritable focus is present in the terminal branching of a motor unit yet involves a varying extent of its network. The associated states in which muscular cramps occur suggest that disequilibrium of electrolytes is the common factor.

In contracture such as results from chemical toxins (caffeine and ryonadine), or in metabolic disorders such as the phosphorylase deficiency of McArdle, muscle shortening is maintained without repetitive action potentials. Indeed this electrical difference between cramp with its high frequency action potentials and contracture with its electrical silence is basic.

FASCICULATION

Spontaneous twitch contractions in motor units (termed "fasciculation" by Denny-Brown and Pennybacker[71]) in distinction to the smaller and more rapid contractions of single muscle fibers (fibrillation), are a prominent feature of amyotrophic lateral sclerosis and related diseases. They also indicate an abnormal excitability in an otherwise intact motor unit. The focus of spontaneous discharge appears to be in any part of the nerve cell or its processes,[74] and the occurrence of normal unit potentials, each repeating spontaneously at irregular, slow rates—1/5 sec to 2/sec (Fig. 2–26)—clearly differentiates this condition from fibrillation. The presence of atrophic weakness and the absence of muscular cramps and other repetitive discharges usually differentiate the condition from purely peripheral hyperexcitability of the neural apparatus, such as occurs in salt deprivation, benign fasciculation (myokymia), and Prostigmin fasciculation.

In muscle that has partially lost its motor innervation rhythmic twitching of a coarse fasciculus is commonly observed during mild contraction (contraction fasciculation)[74] but ceases when the muscle is completely relaxed. In the electromyogram such contraction fasciculation corresponds to a very large polyphasic action potential. Opinion differs as to the origin of this phenomenon, some proposing that it represents a large but normal motor unit that normally enters contraction when smaller units are already active;[74] others propose that it represents a synchronization of the discharge potential of two or more units;[29, 33] yet another explanation is that following degeneration of some motor units the terminal branches of intact remaining arborizations give rise to buds that take over some neighboring denervated muscle fibers.[87, 123] Buchthal showed that the fasciculating units are still capable of participating in voluntary and postural reactions, a point which favors the latter theory. Biopsy study using histochemical stains lends further support,[249] though it is clear that such adoption of more muscle fibers by a motor unit must be a limited process, for otherwise partial neural paralysis could not persist. No significant improvement of motor disability is expected from such hyperinnervation.

DURATION OF ACTION POTENTIAL

The normal action potentials of voluntary contraction vary greatly in shape and duration according to the number of muscle fibers that make up the recorded unit and their spatial relationship to the electrode. A polyphasic unit, presenting multiple spikes, is assumed to result from a scattered disposition of the unit and variation in the distance between the electrode and the motor end-plates. Primary muscular diseases such as muscular dystrophy and some forms of polymyositis reduce the number of active muscle fibers in the motor units and thus reduce the size of the unit of action potential and its complexity.[74] There is a corresponding reduction in duration, the action potential becoming more like that of a single muscle fiber. Pinelli and Buchthal[197] measured the change in duration in a number of such cases and plotted the shift in average duration. Attempts to record single fiber potentials by the use of electrodes with a very small "pore," or active surface, in man have met with little success, for with active electrode tips below 50 μ in human muscle movement artifact becomes magnified out of proportion to the potential registered. Buchthal *et al.*,[35] however, succeeded in making very accurate measurements with multiple leads on the side of one needle. Even in advanced cases of these diseases the range of measured durations of action potentials may overlap with the wide range in normal muscle, but the mean duration differs, being much reduced, so that this electrical change has assumed considerable diagnostic importance in muscular dystrophy.

TENDON REFLEX

Close analysis of the reflex discharge of the tendon reflex indicates that it is the result of a volley of single impulses in that group of motor units in a muscle which most readily responds to sustained stretch of the same muscle. The tendon jerk is a relatively synchronous single efferent volley because the tap on the tendon produces a single discharge from each tension afferent end-organ in the muscle. For an interval of time varying from 0.04 to 0.20 sec after the efferent volley of the tendon jerk, the muscle fails to show action potentials. According to Fulton and Pi-Suñer[98] this interval during which there is no discharge indicates that the afferent end-organ in the muscle, lying between the muscle fibers ("in parallel" as opposed to an end-organ "in series"), was slackened by the tendon jerk and was again stretched as the twitch subsided. Denny-Brown showed, however, that the "silent period,"[68] was present when little mechanical movement of the muscle was possible. During the silent period the motor neurons of the muscle, and of neighboring muscles of related function, are refractory to the transmission of further impulses. This state of depression must be related to a state of inhibition produced by the active contraction of the tendon reflex,[68] and its cumulative effect as a result of a sustained stretch reflex is seen as the sudden cessation of discharge when tension is suddenly released.[70] The stretch reflex then must be recruited afresh

at the shorter length (shortening reaction). The lengthening reaction is the readjustment of the stretch reflex to a new equilibrium of excitation and inhibition after stimulation of a sudden increase in length.

There is thus evidence that both passive and active stretch on muscles reflexly excite their own motoneurons by one set of proprioceptives, and that active contraction of the muscle in addition reflexly exerts a modulating inhibition on the same motoneurons. Both influences travel centripetally by very rapidly conducting proprioceptives[108, 162, 163, 171] (Ia and Ib, respectively, in the standard nerve fiber classification) and are independent of the small-fibered nociceptive innervation of muscle. It is therefore likely that the muscle spindle and tendon organ represent the source of proprioceptive excitation and inhibition.

Matthews[168] found no proprioceptive organ that responded solely to active contraction, though some endings (B type of Matthews) discharge with both active or passive tension, and others discharge only during passive stretch (A_1 endings) but not during active contraction. A special variety of the latter (A_2 endings) discharge during passive tension and also during active tension if the stimulus is supramaximal. From the position of these endings within the muscle Matthews concluded that the A endings were muscle spindles, and the B endings Golgi tendon organs. Granit[108] reaffirmed the presence of autogenetic inhibition and its part as an automatic "governor" of muscular innervation, which under conditions of heavy loadings would otherwise become self-propelled to the point of rupture of the muscle.

Further analysis of the A discharge of muscle by Kuffler and his associates has shown clearly that the tension receptor in muscle is subject to large adjustments of its response at different lengths, according to the state of activity of the small-fiber "gamma" motor system.[147] This labile response is certainly the function of the muscle spindle. It is known that the small-fiber motor system can be activated reflexly by stimulating the skin of the feet.[124] It is therefore evident that the stretch reflex can be profoundly modified by such stimulation. More recently it was shown that a secondary afferent effect from the muscle spindles, probably from the flower-spray endings, excites the antagonistic muscles. The effect of 1% procaine on the muscular nerve was thought to block the efferent discharge in the small-fibered gamma efferents to the muscle spindles, thus abolishing the postural proprioceptive activity of the "gamma loop," yet leaving direct "alpha" innervation relatively unimpaired.[169] Such selective blockage is rarely possible, however. The muscle spindles are clearly important as self-energizers and regulators of reflex contractions.

HYPERTROPHY AND DISUSE ATROPHY

Athletic training for activities such as running or swimming that require rapid alternation of movement chiefly involves habituation to severe oxygen debt. Long distance running requires the greatest possible efficiency in respiratory and circulatory function. Such training results in only slight increase in muscle strength and size. Training for sustained efforts such as weight lifting, on the contrary, result in increased size of muscles and maintenance of contraction. It has long been recognized that near-maximal contractions of muscles are necessary to induce hypertrophy. Hettinger and Müller[115] found that maximal contraction against resistance for as little as 6 sec/day induces progressive hypertrophy of a muscle at the rate of about 5% increase in strength per week (up to 22% for some large muscles, such as hip flexors). Longer or more frequent contraction made little difference in the rate of hypertrophy. For each individual there was an upper limit of development beyond which further effort resulted in pain or other evidence of muscle damage. This is attained at about three times normal strength. If training was discontinued with 6–12 weeks the muscle strength lessened progressively at about 5% per week until it attained its original level. If training continued for longer than 3 months, it was remarkable that the hypertrophy of the muscle then persisted for many months without further maximal exercise.

The early experiments of Morpurgo[179] used dogs exercised in running wheels for 2-month periods. Biopsy of one sartorius muscle before exercise compared with counts of the other showed no increase in the number of muscle fibers. An increase in cross-sectional area of 53–55% was accounted for by an increase in cross-sectional area of individual fibers. A distribution curve of fiber diameters showed that the change was in the nature of a shift of the small- and medium-sized fibers of normal muscle to the large-fiber group. There were very few fibers in hypertrophied muscle larger than the largest normal fiber. Counts of myofibrils showed no increase in the average number of myofibrils. The size of myofibrils remained the same. Morpurgo therefore concluded that physiologic hypertrophy consisted of an increase in the amount of sarcoplasm. Some experiments conducted by Denny-Brown[75] in exercising cat soleus muscle after the gastrocnemius and biceps contribution to the Achilles tendon had been sectioned on one side 3 and 4 months earlier confirmed in general the conclusions of Morpurgo. In this case, however—particularly after 4 months of hypertrophy in a young animal—the hypertrophied muscle showed a shift to the extreme represented by only very few fibers of the largest normal group (Fig. 2–28). In addition, there was a substantial increase in the number of myofibrils (1100–1400 in normal muscle compared with 1600–2000 in hypertrophied muscle). It was also noted that small fibers staining darkly in cross section with phosphotungstic acid hematoxylin in normal muscle were no longer present in hypertrophied muscle. Dark staining of this type is related to closely packed myofibrils, these often in tightly knit groups of six to eight. We therefore concluded that after particularly prolonged heavy exercise an increase in the number of myofibrils, as well as an increase in sarcoplasm, occurs. It is assumed that this change is that which Hettinger and Müller found slow to reverse by the absence of exercise.

In disuse atrophy the first change is evidently the opposite of that produced by brief training, i.e., a loss of sarcoplasm in the medium- and large-sized fiber groups. The percentage of small dark fibers in phosphotung-

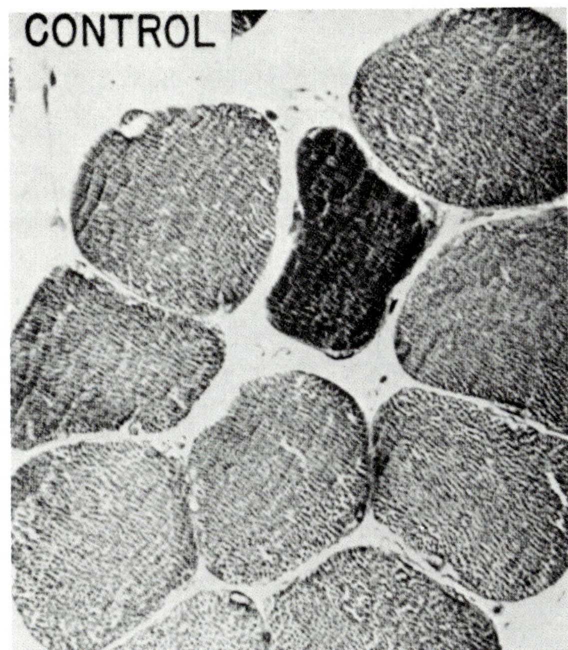

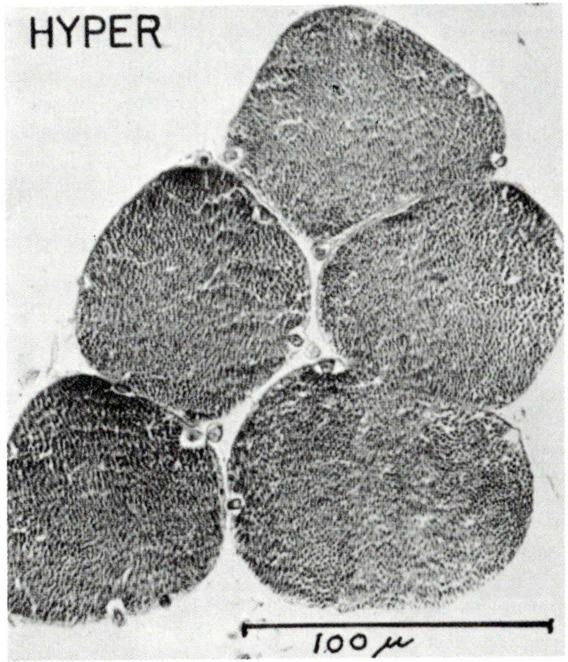

Fig. 2–28. Experimental hypertrophy in soleus muscle of cat. (phosphotungstic hematoxylin)

stic acid hematoxylin-stained sections increased. (These dark fibers should not be confused with those whose darkness is due to large lipid content.) In states of prolonged cachexia due to diseases such as carcinoma or tuberculosis the muscles lose considerable bulk though retaining remarkable powers of brief contraction. The number of small dark fibers is then very greatly increased, and the smallest may have diameters of 5–10 μ (Fig. 2–29). The myofibrils in such fibers are grouped in dense bundles, difficult to count, but are clearly greatly reduced in number. The interpretation of this pathologic picture in

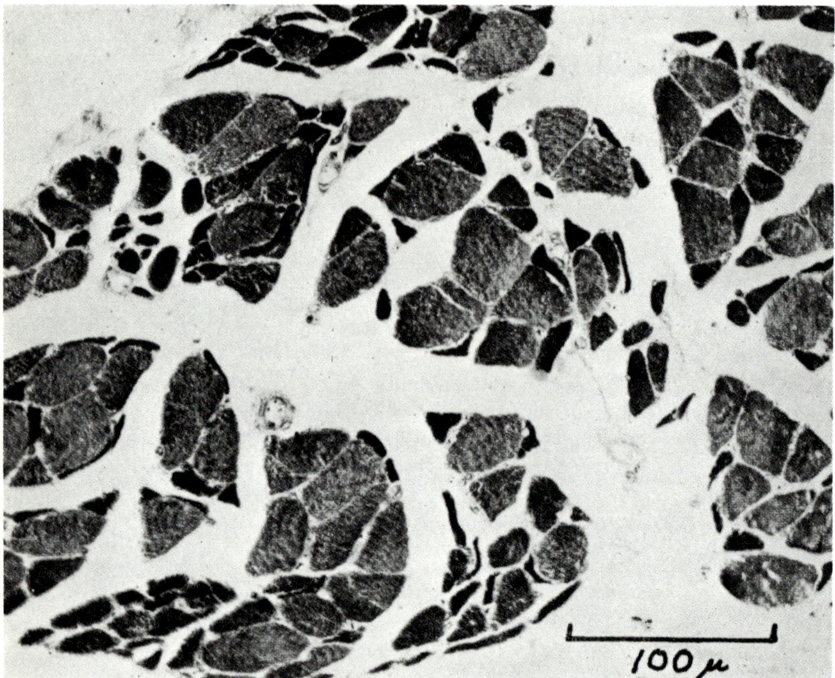

Fig. 2–29. Atrophy of muscle fibers in cachexia due to carcinoma of pylorus. (phosphotungstic hematoxylin)

disuse atrophy is difficult, for varying degrees of denervation are associated with many cachetic states (carcinomatosis, uremia, hepatic failure, vitamin deficiency, etc.). Nachmias and Padykula[181] found the type I red fibers to show greater atrophy and greater loss of cytochrome oxidase and myoglobin than type II white fibers.

Jewell and Zaimis[131] reported that in disuse atrophy the pharmacologic characteristics of muscle tend to revert to those of the undifferentiated type. Eccles[86] also found that in disuse atrophy the tetanus/twitch tension ratio lessened, indicating less power of summation of "active state" in contraction. Solandt and Magladery[226] found some increase in the sensitivity of muscle to ACh in disuse atrophy. These findings are explicable by a reversion of the membrane characteristics to those of more primitive muscle fibers. In this respect it is of interest that the distribution of the dark fibers described above is completely haphazard in muscle cross section, leading to the conclusion that every motor unit must contain some

muscle fibers that are more highly differentiated and some that are less so. This in turn may explain why some muscle fibers in each motor unit are first affected by primary muscle disorders such as myasthenia gravis, muscular dystrophy, myositis, and myoglobinuria.

REFERENCES

1. AAGARD OC: Ueber die Imphefasse der zunge des quergestreiften muskelgewebes und der speicheldrusen des menschen. Anat Hefte 47:493–648, 1912–1913

2. AGDUHR E: Ueber die plurisegmentelle innervation der einzelnen quergestreiften muskelfastern. Anat Anz 52:273–291, 1919

3. AGDUHR E: On the innervation of cross-striated muscle fibers. Upsala Läkaref-Förh 45:399–418, 1939

4. ALOISI M, MARGARETH A: The question of glycolytic enzymes localization in the skeletal muscle fiber and its bearing upon certain aspects of muscle disease. *Exploratory Concepts in Muscular Dystrophy and Related Disorders*. Edited by AT

Milhorat. Amsterdam, Excerpta Medica, 1967, pp 305–317

5. ANDRES R, CADER G, ZIERLER KL: Metabolic exchange of human muscle in situ. Am J Phys Med 34:286–290, 1955

6. APELLA E, MARKET CL: Dissociation of lactate dehydrogenase into subunits with guanidine hydrochloride. Biochem Biophys Res Commun 6:171–176, 1961

7. BAILEY K: Tropomysin: A new enzymatic protein component of the muscle fibril. Biochem J 43:271–281, 1948

8. BAILEY K: Structural proteins. II. Muscle. The Proteins. Edited by H Neurath and K Bailey. New York, Academic Press, 1954, Vol II, pp 951–1055

9. BAILEY K, PERRY SV: The role of sulfhydryl groups in the interaction of myosin and actin. Biochim Biophys Acta 1:506–516, 1947

10. BAKER N, SCHREEVE WW, SHIPLEY RA, et al: C14 studies in carbohydrate metabolism; oxidation of glucose in normal human subjects. J Biol Chem 211:575–592, 1954

11. BANKER BQ, GIRVIN JP: The ultrastructural features of the mammalian muscle spindle. J Neuropath & Exper Neurol 30:155–195, 1971

12. BARER D: The structure of the striated muscle fibre. Biol Rev 23:159–200, 1948

13. BARKER D: The innervation of the muscle-spindle. Quart J Micro Sci 89:143–186, 1948

14. BARKER D, CHIN NK: The number and distribution of muscle-spindles in certain muscles of the cat. J Anat 94:473–486, 1960 (see also J Physiol 153:8P, 28P, 1960)

15. BATTEN FE: The muscle-spindle under pathological conditions. Brain 20:138–179, 1897

16. BENDALL JR: Muscles, Molecules and Movement. An essay in the contraction of muscles. New York, American Elsevier, 1969

17. BENDER MB: Fright and drug reactions in denervated and ocular muscles in monkeys. Am J Physiol 121:609–619, 1938

18. BING RJ, SEIGEL A, UNGAR I, et al: Metabolism of the human heart; studies on fat, ketone and amino acid metabolism. Am J Med 16:504–515, 1954

19. BIÖRCK G: On myoglobin and its occurrence in man. Acta Med Scandinav 133 (Suppl 226):1–216, 1949

20. BLAHD WH, BAUER FK, LIBBY RL, et al: Radioisotope studies in neuromuscular disease: Studies in muscular dystrophy and myotonia dystrophica with sodium22 and potassium42. Neurology (Minneap)5:201–207, 1955

21. BLOMFIELD LB: Intramuscular vascular patterns in man. Proc Roy Soc Med 38:617–618, 1945

22. BLOOR WR: Distribution of unsaturated fatty acids in tissues; voluntary muscle of beef. J Biol Chem 72:327–343, 1927

23. BOEKE J: Nerve endings, motor and sensory, Cytology and Cellular Pathology of the Nervous System. Edited by W Penfield. New York, Hoeber, 1932, Vol 1, Sec VI

24. BOEKE J: The significance of the interstitial cells in the development of motor end-plates in mammals. Proc Nederl Akad Wetensch 45:444–448, 1942

25. BOEKE J: Die entwickelung der motorischen endplatten bei der säugetieren, mit besonderer berücksichtigung der kernverhältnisse und der interstitiellen elemente in der nervösen endoformation. Acta Neerlandica Morph Norm & Path 5:189–212, 1944

26. BOYD JA: The innervation of mammalian neuromuscular spindles. J Physiol 140:14P–15P, 1958

27. BOYD IA, DAVEY MR: Composition of Peripheral Nerves. Edinburgh, Livingstone, Baltimore, Williams and Wilkins, 1968

28. BRAND L, EVERSE J, KAPLAN NO: Structural characteristics of dehydrogenases. Biochemistry 1:423–434, 1962

29. BRAZIER MAB, WATKINS AL, SCHWAB RS: Electromyographic studies of muscle dysfunction in infectious polyneuritis and poliomyelitis. New England J Med 230:185–189, 1944

30. BRODY IA: The significance of lactate dehydrogenase isozymes in abnormal human skeletal muscle. Neurology (Minneap)14:1091–1100, 1964

31. BROOKE MH, ENGEL WK: The histographic analysis of human muscle biopsies with regard to fiber types. Neurology 19:221–233, 1969

32. BROOKS VB: The action of botulinum toxin on motor-nerve filaments. J Physiol 123:501–515, 1954; 134:264–277, 1956

33. BUCHTHAL F, CLEMMENSON S: On the differentiation of muscle atrophy by electromyography. Acta Psychiat & Neurol 18:377–387, 1943

34. BUCHTHAL F, GULD C, ROSENFALCK P: Volume conduction of the spike of the motor unit potential investigated with a new type of multielectrode. Acta Physiol Scandinav 38:331–354, 1957

35. BUCHTHAL F, GULD C, ROSENFALCK P: Multielectrode study of the territory of a motor unit. Acta Physiol Scandinav 39:83–104, 1957

36. BUCHTHAL F: The general concept of the motor unit. Res Pub Assoc Res Nerv & Ment Dis 38:3–30, 1960

37. BURKE RE: Motor unit types of cat triceps surae muscle. J Physiol 193:141–160, 1967

38. BURKE W: Spontaneous potentials in slow muscle fibers of the frog. J Physiol 135:511–521, 1957

39. BURNS BD, PATON WDM: Depolarization of the

motor end-plate by decamethonium and acetylcholine. J Physiol 115:41–73, 1951

40. BURRIDGE W: Ranine ventricular dimyophysitism. J Physiol (Lond) 107:P3 ,1947

41. CAHN RJ, KAPLAN NO, LEVINE L, *et al:* Nature and development of lactic dehydrogenase. Science 136:962–969, 1962

42. CAIN RF, DOVIRS RE: Breakdown of adenosine triphosphate during a single contraction of working muscle. Biochem Biophys Res Commun 8:361–366, 1962

43. CAJAL S, RAMON Y: Genesis de las fibras nerviosas en el embrion y observaciones contarias a la teoria catenaria. Trab Lab Invest Biol Madrid 4:227–294, 1905

44. CAJAL S, RAMON Y: Quelques remarques sur plaques motrices de la langue des mammiferes. Trab Lab Invest Biol Madrid 23:1–11, 1925

45. CAJAL S, RAMON Y: Les preuves objectives de l'unite anatomique des cellules nerveuses. Trab Rec Biol Madrid 29:1–137, 1934

46. CAMPBELL J, PENNEFATHER CM: The blood supply of muscles; with special reference to war surgery. Lancet 1:294–296, 1919

47. CAREY EJ: Studies on ameboid motion of motor end-plates. Am J Path 18:237–290, 1942

48. CASTILLO J DEL, KATZ B: Quantal components of the end-plate potential. J Physiol 124:560–573, 1954

49. CLARK WE LEGROS, BLOMFIELD LB: The efficiency of intramuscular anastomoses, with observations on the regeneration of devascularized muscle. J Anat 79:15–32, 1945

50. CLOSE R: Dynamic properties of fast and slow muscles of the rat during development. J Physiol 173:74–95, 1964

51. COËRS C: Etude histologique et histochemique de la jonction neuromusculaire dans les amyotonies congenitales. Acta Neurol & Psychiat Belg 54:69–77, 1954

52. COËRS C: Les variations structurelles normales et pathologiques de la jonction neuromusculaire. Acta Neurol & Psychiat Belg 55:741–866, 1955

53. COËRS C, WOOLF AL: The Innervation of Muscle: A Biopsy Study. Springfield, Ill, Thomas, 1959

54. COLOWICK SP, KALCKAR HM: The role of myokinase in transphosphorylations I. The enzymatic phosphorylations of hexones by Adenyl. J Biol Chem 48:117, 1943

55. CONWAY EJ: Nature and significance of concentration relations of potassium and sodium ions in skeletal muscle. Physiol Rev 37:84–132, 1957

56. COOPER S, DANIEL PM: Muscle spindles in human intrinsic eye muscles. Brain 72:1–24, 1949

57. COOPER S, DANIEL PM: Human muscle spindles. J Physiol Proc Phys Soc 133:1P–2P, 1956

58. COOPER S: Muscle Spindles and Motor Units in Control and Innervation of Skeletal Muscle. Edited by BL Andrew. Livingstone, Edinburgh, 1966

59. CORI GT: 114th Meeting Amer Chem Soc 1948, p 27C

60. CORI GT, CORI CF: The enzymatic conversion of phosphorylase a to b. J Biol Chem 158:321–332, 1945

61. COUTEAUX R: Sur l'origine de la sole des plaques motrices. Compt Rend Soc Biol 127:218–224, 1938

62. COUTEAUX R: Recherches sur l'histogenese du muscle striee et la formation des plaques motrices. Bull Biol 75:101–239, 1941

63. COUTEAUX R: Contribution a l'etude de la synapse myoneurale: Buisson de Kühne et plaque motrice. Rev Canad Biol 6:563–711, 1947

64. CUAJUNCO F: Development of the neuromuscular spindle in human fetuses. Carnegie Inst Wash Publ 518, Contrib to Embryol 28:95–128, 1940

65. DANILEVSKY A: Myosin seine Dorstellung, Eigenschaften, Unwandlung in Syntonin und Rückbildung aus demselben. Z Physiol Chem 5:158–184, 1881

66. DAWSON DM, KAPLAN NO: Factors influencing concentration of enzymes in various muscles. J Biol Chem 240:3215–3221, 1965

67. DELANGE CD: Myoglobin and myoglobinuria. Acta Med Scand 124:213–226, 1946

68. DENNY-BROWN D: On inhibition as a reflex accompaniment of the tendon jerk and other forms of active muscular response. Proc R Soc London, ser B 103:331–336, 1928

69. DENNY-BROWN D: The histological features of striped muscle in relation to its functional activity. Proc R Soc London [Biol], 104:371–411, 1929

70. DENNY-BROWN D: On the nature of postural reflexes. Proc R Soc London, ser B 104:252–301, 1929

71. DENNY-BROWN D, PENNYBACKER JB: Fibrillation and fasciculation in voluntary muscle. Brain 61:311–334, 1938

72. DENNY-BROWN D, FOLEY JM: Myokymia and benign fasciculation of muscular cramps. Trans Assoc Amer Physicians 61:88–96, 1948

73. DENNY-BROWN D, FOLEY JM: Evidence of a chemical mediator in myotonia. Trans Assoc Amer Physicians 62:187–191, 1949

74. DENNY-BROWN D: Interpretation of the electromyogram. Arch Neurol Psychiat 61:99–118, 1949

75. DENNY-BROWN D: Experimental studies pertaining to hypertrophy, regeneration and degeneration. Res Publ Assoc Nerve & Ment Dis 38:147–196, 1961

76. DENZ FA: Myoneural junctions and toxic agents. J Path Bact 68:235–247, 1951

77. DREYFUS JC, DEMOS J, SCHAPIRA F, *et al:* La lactico deshydrogenase musculaire chez le myopathe: persistence apparente du type fetal. C R Acad Sci 254:4384–4386, 1962

78. DUBUISSON M: Muscular Contraction. Springfield, Ill, Thomas, 1954

79. EBASHI S, TOYOKURA J, MOMOI H, *et al:* High creatine phosphokinase activity of sera of progressive muscular dystrophy patients. J Biochem 46:103–104, 1959

80. EBASHI S: Calcium binding activity of vesicular relaxing factor. J Biochem 50:236–244, 1961

81. EBASHI S, LIPMANN F: Adenosine triphosphate linked concentration of calcium ions in a particulate faction of rabbit muscle. J Cell Biol 14:389–400, 1962

82. EBASHI S: Third component participating in the superprecipitation of "natural actomyosin." Nature (London)200:1010, 1963

83. EBASHI S: Structural proteins controlling the interaction between actin and myosin. Progressive Muskeldrophie, Myotonie. Myasthenie. Edited by E Kuhn, Berlin, Springer-Verlag, 1966, pp 506–573

84. EBASHI S, KONDO M: Calcium ion and muscle contraction. Progress in Biophysics and Molecular Biology. Edited by JAV Butler, D Noble. Oxford, Pergamon Press, 1968, Vol 18, 123–183

85. ECCLES JC, SHERRINGTON CS: Numbers of contraction values of individual motor-units examined in some muscles of the limb. Proc R Soc London [Lond]106:326–357, 1930

86. ECCLES JC: Disuse atrophy of skeletal muscle. Med J Australia 28:160–164, 1941

87. EDDS MV: Collateral regeneration of residual motor axons in partially denervated muscles. J Exper Zool 113:507–552, 1950

88. EDSTRÖM L, KUGELBERG E: Histochemical composition, distribution of fibers and fatigability of single motor units. J Neurol Neurosurg & Psychiat 31:424–433, 1968

89. EDWARDS RG, LIPPOLD OCJ: The relation between force and integrated electrical activity in fatigued muscle. J Physiol 132:677–681, 1956

90. EICHLER W: Veratrinkontraktur und Endplattenrhythmik. Z Biol 99:243–265, 1938

91. EICHLER W: Ist die Muskelendplatte vom motorischen Spinalnerven durch eine Synapse tetreunt? Z Biol 99:266–276, 1938

92. ENGELHARDT WA, LIUBIMOWA MN: Myosin and adenosine triphosphatase. Nature (Lond) 144:668–669, 1939

93. EYZAGUIRRE C, FOLK BP, ZIERLER KL, *et al:* Experimental myotonia and repetitive phenomena: The veratrinic effects of 2,4 dichlorphenoxyacetate (2,4,D) in the rat. Am J Physiol 155:69–77, 1948

94. FATT P: Biophysics of junctional transmission. Physiol Rev 34:674–710, 1954

95. FEINSTEIN B, LINDEGARD B, NYMAN E, *et al:* Morphological studies on motor units in normal human muscles. Acta Anat 23:127–142, 1955

96. FENN WO: A quantitative comparison between the energy liberated and the work performed by the isolated sartorius muscle of the frog. J Physiol (Lond) 58:175–203, 373–395, 1923–1924

97. FRITZ IB, DAVIS DG, HOLTROP RH, *et al:* Fatty acid oxidation by skeletal muscle during rest and activity. Am J Physiol 194:379–386, 1958

98. FULTON JF, PI-SUNER J: A note concerning the probable function of various afferent end-organs in skeletal muscle. Am J Physiol 83:554–562, 1928

99. GARVEN HSD: The nerve-endings in the panniculus carnosus of the hedgehog with special reference to the sympathetic innervation and striated muscle. Brain 48:380–441, 1925

100. GELFAN S: The submaximal responses of the single muscle fibre. J Physiol 80:285–295, 1933

101. GELFAN S, BISHOP GH: Conducted contractures without action potentials in single muscle fibers. Am J Physiol 103:236–243, 1933

102. GERGELY J, SPICER SS: On factors affecting the reversible superprecipitation of actomyosin. Biochim Biophys Acta 6:456–460, 1951

103. GERGELY J, GONVEA MA, KARIBIAN D: Fragmentation of myosin by chymotrypsin. J Biol Chem 212:165–177, 1955

104. GERGELY J, editor: Biochemistry of Muscular Contraction. Boston, Little Brown, 1964

105. GOODGOLD J, ARCHIBALD KC: Occurrence of so-called "myotonic discharges" in electromyograph. AMA Arch Int Med 39:20–22, 1958

106. GORDON G, PHILLIPS CG: Slow and rapid components in a flexor muscle. J Physiol Proc Physiol Soc 110:6P–7P, 1949

107. GOTO I, NAGAMINE M, KATSUKI S: Creatine phosphokinase isozymes in muscles: human fetus and patients. Arch Neurol 20:422–429, 1969

108. GRANIT R: Reflex self-regulation of muscle-contraction and autogenetic inhibition. J Neurophysiol 13:351–372, 1950

109. GREEN DE: Fatty acid oxidation in soluble systems of animal tissues. Biol Rev 29:330–366, 1954

110. GROSSIORD A, LAPRESLE J, LACERT P, *et al:* À propos d'une forme localisée de névrite hypertrophique. Rev Neurol (Paris) 119:248–252, 1968

111. GRUNER JE: Structure fine du fusean neuromusculature humain. Rev Neurol 104:490–507, 1961

112. HÄKANSSON GH: Action potentials recorded intra- and extra-cellularly from the isolated frog muscle in Ringer's solution and in air. Acta Physiol Scandinav 39:291–312, 1957

113. HÄKANSSON GH: Action potential and mechanical response of isolated cross striated muscle fibres at different degrees of stretch. Acta Physiol Scandinav 41:1–18, 1957

114. HALLIBURTON WD: On muscle-plasma. J Physiol (Lond) 8:133–302, 1887

115. HETTINGER TH, MÜLLER EA: Muskelleistung und Muskel-training. Arbeitsphysiologie 15:111–126, 1953

116. HILL AV: The abrupt transition from rest to activity in muscle. Proc R Soc London (Biol) 136:399–420, 1949

117. HILL AV: A challenge to biochemists. Biochim Biophys Acta 4:4–11, 1950

118. HILL AV: The thermodynamics of muscle. Brit M Bull 12:174–176, 1956

119. HINSEY JC: Some observations on the innervation of skeletal muscle of the cat. J Comp Neurol 44:87, 1927

120. HODGE AJ: The fine structure of striated muscle. A comparison of insect flight muscle with vertebrate and invertebrate muscle. J Biophys Biochem Cytol 1:361–380, 1955; 2 (Suppl):131, 1956

121. HODGKIN AL: Ionic basis of electrical activity in nerve and muscle. Biol Rev 26:339–409, 1951

122. HODGKIN AL, HUXLEY AF: Currents carried by sodium and potassium ions through the membranes of the giant axon of *Loligo*. J Physiol 116:449–482, 1952

123. HOFFMAN H: Local reinnervation in partially denervated muscle. Aust J Exp Biol Med Sci 28:383–397, 1950

124. HUNT CC: The reflex activity of mammalian small-nerve fibers. J Physiol 115:456–469, 1951

125. HUXLEY AF: Interpretation of muscle striation —evidence from visible light microscopy. Brit M Bull 12:167–169, 1956

126. HUXLEY AF, NEIDERGERKE: Measurement of the striations of isolated muscle fibres with the interference microscope. J Physiol 144:403–425, 1958

127. HUXLEY AF, TAYLOR RE: Local activation of striated muscle fibres. J Physiol 144:426–441, 1958

128. HUXLEY HE: The ultrastructure of striated muscle. Brit M Bull 12:171–173, 1956

129. IRVING L: Respiration in diving mammals. Physiol Rev 19:112–134, 1939

130. IWANAGA I: Studien über der motorischen Nervenendigungen: I. Ihre Histogenese. Mitteil allg Pathol Anat 2:257–342, 1925

131. JEWELL PA, ZAIMIS EJ: A differentiation between red and white muscle in the cat based on responses to neuromuscular blocking agents

132. KAPLAN NO, CAHN RD: Lactic dehydrogenase and muscular dystrophy in the chicken. Proc Nat Acad Sci USA 48:2123–2130, 1962

133. KAR NC, PEARSON CM: Developmental changes and heterogeneity of lactic and malic dehydrogenases of human skeletal muscles and other organs. Proc Nat Acad Sci USA 50:995–1002, 1963

134. KAR NC, PEARSON CM: Glutamic oxalacetic transaminase iosenzymes in human myopathics. Proc Soc Exp Biol Med 116:733–735, 1964

135. KAR NC, PEARSON CM: Creatine phosphokinase isoenzymes in muscle in human myopathies. Am J Clin Path 43:207–209, 1965

136. KENDREW JC: Myoglobin and the structure of proteins. Science 139:1259–1266, 1963

137. KOLLE GB, FRIEDENWALD JS: A histochemical method for localizing cholinesterase activity. Proc Soc Exper Biol Med 70:617–622, 1949

138. KOELLE GB: The histochemical differentiation of types of cholinesterases and their localizations in tissues of the cat. J Pharmacol Exper Therap 100:158–179, 1950

139. KOSSMAN RJ, FAINER DC, BOYER SH: A study of myoglobin in disease with comments concerning the myoglobin minor components. Cold Spr Harb, Vol 29, 1964 pp 375–385

140. KROGH A: The number of distribution of capillaries in muscles with calculations of the oxygen pressure head necessary for supplying the tissue. J Physiol 52:409–415, 1918–1919

141. KROGH A: The Anatomy and Physiology of Capillaries. New Haven, Yale University Press, 1922, pp. 5–11

142. KRUGER P: Tetanus and Tonus der Quergestreiften Skellettmuskeln der Wirbeltiere und des Menschen. Leipzig, Geest and Portig, KG 1952

143. KUBY SA, NODA L, LARDY HA: Adenosine triphosphate creatine transphosphorylase I. Isolation of the crystalline enzyme from rabbit muscle. J Biol Chem 209:191–201, 1954

144. KUFFLER SW: Electric potential changes at isolated nerve-muscle junction. J Neurophysiol 5:18–26, 1942

145. KUFFLER SW: Specific excitability of the end-plate region in normal and denervated muscle. J Neurophysiol 6:99–110, 1943

146. KUFFLER SW: Membrane changes during excitation and inhibition of the contractile mechanism. Am New York Acad Sci 47:767–779, 1947

147. KUFFLER SW, HUNT CC, QUILLIAM JP: Function of medullated small-nerve fibers in mammalian ventral roots: Efferent muscle spindle innervation. J Neurophysiol 14:29–54, 1951

148. KUFFLER SW, VAUGHAN-WILLIAMS EM: Small nerve junctional potentials. Distribution of small motor nerves to frog skeletal muscle, and the membrane characteristics of the fibres they innervate. J Phyiol 121:289–317, 318–340, 1953

149. KÜHNE W: Die Muskelspindeln. Virchow's Arch Path Anat 28:528–538, 1863

150. KÜHNE W: Ueber die verbindung der Nervenscheiden mit dem Sarkolemma. Z Biol 19:501, 1883

151. KÜHNE W: Neue Untersuchungen über motorische Nervenendigung. Z Biol 23:1–148, 1886

152. LAKI K, editor: Contractile Proteins and Muscle. New York, Marcel Dekker, 1971

153. LAKI K: Tropomyosin A. Contractile Proteins and Muscle. Edited by K. Laki, New York, Marcel Dekker, 1971, pp 273–288

154. LAWRI RA: Relation of the energy-rich phosphate in muscle to myoglobin and to cytochrome-oxidase activity. Biochem J 55:305–309, 1953

155. LEE YP: 5-Adenylic acid deaminase, I. Isolation of the crystalline enzyme from rabbit skeletal muscle. J Biol Chem 227:987–998, 1957

156. LESKELL L: The action potential and excitatory effects of the small ventral root fibres to skeletal muscle. Acta Physiol Scandinav 10 (Suppl 31): 1–81, 1945

157. LELOIR LF, CARDIN CE: Biosynthesis of glycogen from uridine diphosphate glucose. J Amer Chem Soc 79:6340–6341, 1957

158. LEWIS MR: Rhythmical contraction of the skeletal muscle tissue observed in tissue cultures. Am J Physiol 38:153–161, 1915

159. LI C–L, SHY GM, WELLS J: Properties of mammalian skeletal muscle fibres with particular reference to fibrillation potentials. J. Physiol 135:522–535, 1957

160. LILIENTHAL JL, ZIERLER KL: Diseases of muscle, Biochemical Disorders in Human Disease. Edited by AHS Thompson and EJ King. New York, Academic Press, 1957, pp 445–493

161. LINDSLEY DB: Myographic and electromyographic studies of myasthenia gravis. Brain 58: 470–482, 1935

162. LLOYD DPC: Neuron patterns controlling transmission of ipsilateral hind limb reflexes in cat. J Neurophysiol 6:293–315, 1943

163. LLOYD DPC: Conduction and synaptic transmission of reflex response to stretch in spinal cats. J Neurophysiol 6:317–326, 1943

164. LOHMAN K: Uber die enzymatische Aufspaltung der Kreatin-phosphorsaure: Zugleich ein Beitrag zum Ehemismus der Muskelcontraktion. Biochem Z 271:264–277, 1934

165. LUCAS K: The "all or none" contraction of the amphibian skeletal muscle fibre. J Physiol 38: 113–133, 1909

166. LYNEN F, OCHOA S: Enzymes of fatty acid metabolism. Biochim Biophys Acta 12:299–314, 1953

167. MARUYAMA K: Regulatory proteins. Contractile Proteins and Muscle. Edited by K. Laki, New York, Marcel Dekker, 1971, pp 289–310

168. MATTHEWS BHC: Nerve endings in mammalian muscle. J Physiol 78:1–53, 1933

169. MATTHEWS PBC, RUSHWORTH G: The relative sensitivity of muscle nerve fibres to procaine. J Physiol 135:245–269, 1957

170. MCCOMAS AJ, FAWCETT PRW, CAMPBELL MJ, et al: Electrophysiological estimation of number of motor units within a human muscle. J Neurol Neurosurg Psychiat 34:121–131, 1971

171. MCCOUGH GP, DEERING ID, STEWART WB: Inhibition of knee jerks from tendon spindles of crureus. J Neurophysiol 13:343–350, 1950

172. MCINTYRE AR, BENNETT AL, HINMAN JS: Fibrillation in dystrophic muscles. The Physiologist 1:59, 1957

173. MEADOWS JC: Observations on the responses of muscle to mechanical and electrical stimuli. J Neurol Neurosurg Psychiat 34:57–67, 1971

174. MEISTER A: Biochemistry of the amino acids. New York, Academic Press, 1957

175. MERTON PA: Voluntary strength and fatigue. J Physiol 123:553–564, 1954

176. MERTON PA: Problems of muscular fatigue. Brit M Bull 12:219–221, 1956

177. MILLIKAN GA: Muscle hemoglobin. Physiol Rev 19:503–523, 1939

178. MOMMAERTS WFHM: Muscular Contraction, A Topic in Molecular Physiology. New York, Interscience Publishers, 1950

178a. MOMMAERTS WFHM: The molecular transformation of actin. III. The participation of nucleotides. J Biol Chem 198:469–475, 1952

178b. MOMMAERTS WFHM: Chemical investigation of muscular tissues, Methods in Medical Research. Chicago, Year Book Pub, 1958, Vol 7, pp 1–59

179. MORPURGO B: Ueber Activitats-Hypertrophie der willkürliehen Muskeln. Virchow's Arch Path Anat 150:522–554, 1897

180. MUELLER H, PERRY SV: The degradation of heavy meromyosin by trypsin. Biochem J 85: 431–439, 1962

181. NACHMIAS VT, PADYKULA HA: A histochemical study of normal and denervated and white muscles of the rat. J Biophys Biochem Cytol 4:47–59, 1958

182. NICHOLS N: Intracellular glycogen and electrolyte concentrations in human skeletal muscle. Proc Soc Exp Med Biol 97:363–366, 1958

183. NÖEL R: La structure de la substance proto-plasmique dans les plaques motrices de vertébrés. Bull Histol Appl Physiol 2:124–133, 1925

184. NÖEL R, POMME B: Zone de jonction myoneurale, ou plaque motrice, a l'état normal et dans quelques cas pathologiques. Ann d'Anat Pathol 12:621–642, 1935

185. O'LEARY PA, LAMBERT EH, SAYRE GP: Muscle studies in cutaneous disease. J Invest Dermatol 24:301–310, 1955

186. OHTUSKI I, MASAKI T, NONONURA Y, et al: Periodic distribution of troponin along the thin filament. J Biochem 61:817–819, 1967

187. PASCOE JE: Two types of efferent fibres to the muscle spindles of the rabbit. J Physiol 140:29P, 1958

188. PEARSON CM: Serum enzymes in muscular diseases. I. Serum glutamic oxalactic transaminase. N Eng J Med 256:1069–1075, 1975

189. PEARSON CM, KAR NC: Isoenzymes: general considerations and alterations in human and animal myopathies. Ann N Y Acad Sci 138:293–303, 1966

190. PEPE FA: Some aspects of structural organization of the myofibril as revealed by antibody-staining methods. J Cell Biol 28:505–525, 1966

191. PERKOFF GT: Further consideration of fetal muscle heme protein. J Lab Clin Med 71:610–613, 1968

192. PERKOFF GT, HILL RL, BROWN DM, et al: The characterization of adult human myoglobin. J Biol Chem 237:2820–2827, 1962

193. PERRY SV, GREY TC: A study of the effects of the substrate concentration and certain relaxing factors on magnesium-activated myofibrillar adenosine triphosphatase. Biochem 64:184–192, 1956

194. PETER JB: A (Na+ K+) ATPase of sarcolemma from skeletal muscle. Biochem Biophys Res Commun 40:1362–1367, 1970

195. PETRÉN T, SJOSTRAND T, SYLVEN B: Der Einfluss des trainings auf die Heufigkeit der Capillaren in Herz und Skeletmuskulatur. Arbeitsphysiologie 9:376–386, 1936

196. Physiology of Voluntary Muscle. Brit M Bull 12, Spec No (3), 1956

197. PINELLI P, BUCHTHAL F: Muscle action potentials in myopathies. Neurology 3:347–359, 1953

198. PORTER KR, PALADE GE: Studies on the endoplasmic reticulum. III. Its form and distribution in striated muscle cells. J Biophys Biochem Cytol 3:269–300, 1957

199. POST RL, MERRITT CR, KINSOLVING CR, et al: Membrane adenosine triphosphatase as a participant in the active transport of sodium and potassium in the human erythrocyte. J Biol Chem 235:1796–1802, 1960

200. POWER RW: Gas gangrene: With special reference to vascularization of muscles. Brit M J 1:656–658, 1945

201. PRATT FH, EISENBERGER JP: The quantal phenomena in muscle: Methods with further evidence of the all-or-none principle for the skeletal fiber. Am J Physiol 49:1–54, 1919

202. PRINGLE JWS: The excitation and contraction of the flight muscles of insects. J Physiol 108:226–232, 1949

203. RALSTON HJ, POLISSAR MJ, INMAN VT, et al: Dynamic features of human isolated voluntary muscle in isometric and free contractions. J Appl Physiol 1:526–533, 1949

204. RANVIER L: Lecons d'Anatomie Genérale sur le Systeme Musculaire. Paris, Delahaye, 1880

205. REGER JF: The ultrastructure of normal and denervated neuromuscular synapses in mouse gastrocnemius. Exp Cell Res 12:662–665, 1957

206. ROBBINS PW, TRAUT RR, LIPMANN F: Glycogen synthesis from glucose-6-phosphate, and uridine diphosphate glucose in muscle preparations. Proc Nat Acad Sci USA 45:6–12, 1959

207. ROBERTS DV, THESLEFF S: Neuromuscular transmission in vivo and the actions of decamethonium: A microelectrode study. Acta Anesth Scand 9:165, 1965

208. ROBERTS F: Degeneration of muscle following nerve injury. Brain 39: 296–347, 1916

209. ROBERTSON JD: The ultrastructure of a reptilian myoneural junction. J Biophys Biochem Cytol 2:381–394, 1956

210. ROBERTSON JD: Electron microscopy of the motor end-plate and the neuromuscular spindle. Am J Phys Med 39:1–43, 1960

211. ROLLETT A: Beiträge zur Physiologie der Muskeln. Denkschr Kaizerl Akad Wissensch Wien Math Naturw Kl 53:1–64, 1887

212. ROSSI-FANNELLI A: Crystalline human myoglobin: some physicochemical properties and chemical composition. Science 108:15–16, 1948

213. ROWLAND LP, DUNNE PB, PENN AS, et al: Myoglobin and muscular dystrophy. Arch Neurol 18:141–150, 1968

214. RUFFINI A: On the minute anatomy of the neuromuscular spindles of the cat and on their physiological significance. J Physiol 23:190–208, 1898

215. SCHIFF JM: Muskel und nerven-physiologie, Cyclus. Edited by Schauenberg, Lahr, 1858, Vol 1, Part 9

216. SCHMIDT G: Uber Fermentative Desaminierung in Muskel.

217. SCHWARTZ A: A sodium and potassium stimulated adenosine triphosphatase from cardiac tissues. I. Preparation and properties. Biochem Biophys Res Commun 9:301–306, 1962

218. SEYFFARTH H: The behavior of motor units in healthy and paretic muscles in man. Acta Psychiat Neurol 16:79–109, 261–278, 1941

219. SHDANOW DA: Die Kollaterallymphwege der Brushthohle des Menschen. Anat Anz 82:417–440, 1936

220. SHERRINGTON CS: On the anatomical constitution of nerves of skeletal muscles: With remarks on recurrent fibres in the ventral spinal nerve-root. J Physiol 17:211–258, 1894

221. SHY GM, MCLEAN L, DIGIACOMO N, SAUNDERS CL: Trophic effects of nerve on the muscle cell, Exploratory Concepts of Muscular Dystrophy and Related Disorders. Edited by AT Milhorat. Amsterdam, 1968, pp 357–377

222. SINGER K, ANGELOPUOLOS B, RAMOT B: Studies in human myoglobin. II. Fetal myoglobin: Its identification and its replacement by adult myoglobin during infancy. Blood 10:987–998, 1955

223. SKOU JC: The influence of some cations on an adenosine triphosphotase from peripheral nerves. Biochim Biophys Acta 23:394–401, 1957

224. SLATER EC: Biochemistry of Sarcomes. The Structure and Function of Muscle, Edited by GH Bourne, New York, Academic Press, 1960, Vol II, pp 105–140

225. SMELSER GK: Oxygen consumption of the eye muscles of thyroidectomized and thyroxin-injected guinea pigs. Am J Physiol 142:396–401, 1944

226. SOLANDT DY, MAGLADERY JW: A. comparison of effects of upper and lower motor neurone lesions on skeletal muscle. J Neurophysiol 5:373–380, 1942

227. SPALTEHOLZ W: Die Vertheilung der Blutgefässe im Muskel. Abhandl, Math Phys Cl Gesellsch d Wissensch 14:507–534, 1888

228. SPEIDEL CC: Studies on living muscles. II. Histological changes in single fibers of striated muscles during contraction and clotting. Am J Anat 65:471–528, 1939

229. SPICER SS: The clearing response of actomyosin to adenosine triphosphate. J Biol Chem 199:289, 1952

230. STRAUB F, FEUER G: Adenosine triphosphate: the functional group of actin. Biochim Biophys Acta 4:455–470, 1950

231. SZENT-GYÖRGYI A: Studies on muscle. Acta Physiol Scand (Suppl 25) 9:11, 1945

232. SZENT-GYÖRGYI A: Nature of Life: A Study on Muscle. New York, Academic Press, 1948

233. SZENT-GYÖRGYI A: Chemical Physiology of Contraction in Body and Heart Muscle. New York, Academic Press, 1953

234. SZENT-GYÖRGYI A: Meromyosins: The subunits of myosin. Arch Biochem 42:305–320, 1953

235. TELLO JF: Sobre una vaina que convuele toda la ramificacion del axon en los terminociones motrices de las muscolos estriados. Trab Inst Cajal 36:1–62, 1944

236. THEORELL H: Kristallinisches myoglobin: H-V. Biochem Ztschr 268:46–82, 1934

237. THEORELL H, DE DUVE C: Crystalline human myoglobin from heart muscle and urine. Arch Biochem 12:113–124, 1947

238. TIEGS WW: The flight muscles of insects—their anatomy and histology; with some observations on the structure of striated muscle in general. Philos Trans R Soc (London) 238:221, 1955

239. TOWER SS: Atrophy and degeneration in the muscle spindle. Brain 55:77–90, 1932

240. TVETANOVA E: Adolase isoenzymes in patients with progressive muscle dystrophy and in human fetuses. Clin Chem 17:926–930, 1971

241. VAUGHAN BE, PACE N: Changes in myoglobin content of the high altitude acclimatized rat. Am J Physiol 185:549–556, 1956

242. VILLAR-PALASI C, LARNER J: A uridine coenzyme-linked pathway of glycogen synthesis in muscles. Biochim Biophys Acta 30:449, 1959

243. FURTH O VON: Ueber die Eiweisskörper des Muskelplasmas. Arch Exper Path u. Pharmakol 36:231–274, 1895

244. MURALT A VON: The development of muscle-chemistry, a lesson in neurophysiology. Biochim Biophys Acta 4:126–129, 1950

245. WEBER A: Energized calcium transport and relaxing factors. Current Topics in Bioenergetics. Edited by DR Sanadi, New York, Academic Press, 1966, Vol 1, pp 203–254

246. WEBER A, WINICUR S: The role of calcium in the superprecipitation of actomyosin. J Biol Chem 236:8198–3202, 1961

247. WEIME RJ, HERPOL JE: Origin of the lactate dehydrogenase isoenzyme pattern found in the serum of patients having primary muscular dystrophy, Nature (Lond) 194:287–288, 1962

248. WILKIE DR: Facts and theories about muscle. Progr Biophys 4:288–324, 1954

249. WOHLFART G: Collateral regeneration from residual motor nerve fibers in amyotrophic lateral sclerosis. Neurology 7:124–134, 1957

250. WOODS EF: Molecular weight and subunit structure of tropomyosin B. J Biol Chem 242:2849–2871, 1967

251. WUERKER RB, MCPHEDRAN AM, HENNEMAN E: Properties of motor units in a heterogeneous pale muscle. J Neurophysiol 28:85–99, 1965

252. YOUNG DM, HIMMELFARB S, HARRINTON WE: On the structural assembly of the polypeptide chains of heavy meromyosin. J Biol Chem 240:2423–2436, 1965

253. YOUNG M: The molecular basis of muscle contraction. Ann Rev Biochem 38:913–950, 1969

254. ZACKS SI: The Motor Endplate. Philadelphia, Saunders, 1964

part II

PATHOLOGIC REACTIONS OF SKELETAL MUSCLE

chapter 3

EXPERIMENTAL PATHOLOGY

Many of the pathologic processes that occur naturally in the skeletal muscles of man have been the subject of experimental inquiry. These scientific studies are widely scattered in anatomic and pathologic journals and are often overlooked by the casual student. Our intention here was to collect these experimental data and to add certain observations of our own which may afford the reader a more complete understanding of some of the processes at work in human muscle disease.

DENERVATION ATROPHY

EFFECT OF DENERVATION ON MUSCLE

The general effect of denervation on muscle is well known. After severance of a nerve the distal portion maintains its excitability until the axis cylinders begin to break up into fragments after about 48–72 hours. Though the muscle cannot then participate in any central nervous activity, the region of the motor end-plate still retains its full electrical excitability for 5–10 days after degeneration of the distal portion of the motor nerve has commenced. It is during this period that the last fragments of axis cylinder in the motor end-plate disappear. The excitability of this region then abruptly changes to approximately that of the remainder of the muscle fiber. From this time onward a gradual loss of excitability occurs, though some trace of contraction can usually be obtained by very strong stimulation as long as any striate substance remains (1–3 years).

Since muscular excitability has a prominent time factor that requires a longer duration of stimulus the less excitable the tissue, the abrupt fall in excitability that follows degeneration of the motor nerve is revealed in the failure of rapid electrical fluctuation (faradic current) to excite the muscle when applied to the overlying skin, while slow electrical change (galvanic current) is still effective.

REACTION OF DEGENERATION

The *reaction of degeneration,* as originally defined by Erb[64] in 1868, refers to the aforementioned changes in excitability of denervated muscle in response to the galvanic stimulus, which though seldom now used in the clinic are of considerable theoretical interest. These changes were tested by Pollock and his associates,[156, 157] who reaffirmed Erb's original finding of a rise of galvanic excitability (less amperage is necessary to excite) and increased efficiency of an anodal closing stimulus. The initial increase in galvanic excitability runs parallel with an increased sensitivity of the muscle fiber to chemical stimulation, particularly by acetylcholine, and appearance of spontaneous fibrillary twitching, both of which begin when the last vestige of axis cylinder substance and acetylcholnesterase disappears from the motor end-plates. The index of excitability (chronaxie), when measured in terms of the most brief interval required to excite the muscle fibers, does not show this change and reflects instead only a very slow decline. Also, if reinnervation occurs, change in chronaxie is delayed for a period after motor power returns presumably because regenerated nerve fibers are at first very small and are not excited by rapid electrical change. On the other hand, a galvanic stimulus begins to have less effect as soon as regenerating nerve fibers reach the muscle, before any muscular contraction can be mediated by the nerve.

In man a sudden increase in rheobase (minimal voltage necessary to excite the muscle when applied for infinite time—1 sec or longer) antedates the first recovery of willed movement by many days, often many weeks.[156] In this instance galvanic excitability may have considerable prognostic value. It should also be emphasized that the use of the faradic stimulus in determining the reaction of degeneration in man is based on the relatively large size of human muscles in relation to the size of the electrode and thickness of the skin. If the muscle fibers are exposed or even stimulated through the skin in a small animal, a faradic stimulus excites the atrophic muscle as long as any muscle fiber remains, but the intensity of shock must then be greater than for normal muscle.

Lapique's formula for chronaxie therefore fails to reveal the most significant changes in excitability in denervated muscle, which are better shown by strength-duration curves. Chronaxie is a better indicator of excitability change of nerve than of muscle. The phenomenon of fibrillation, one aspect of the change in excitability, can be demonstrated with certainty by electromyography (EMG), even in very small and deeply situated muscles such as the cricoarytenoids.[213] The onset of fibrillation on the fifth to seventh day after nerve section and its abrupt cessation days or weeks before the first return of neural contraction are thus clinical criteria of the same phenomenon shown by change in galvanic excitability.

Intracellular recording of denervated fibers reveals, in addition to the spontaneous fibrillation potentials: (1) a fall in the resting membrane potential; (2) increased resistance to an intracellular electrical stimulus (raised voltage); and (3) a change in the character of the action potential—the spontaneous fibrillation potential is followed by a period of hyperpolarization of ± 100 millisecs. The site of origin of the fibrillation potentials, according to

Belmar and Eyzaguirre,[11] is in or near the motor end-plate. Thesleff[191] believes the changes in the muscle fiber membrane to be due to a reduction in resting K conductance and a more prolonged K current during action potential. However, modifications of organelles in the atrophic fiber may account for some of its contractility characteristics.

These changes in muscular function, as expressed in terms of excitability, are inseparable from the process of atrophy exhibited by all mature skeletal muscle when it is anatomically isolated from motor nerve fibers. There is no evidence of any special trophic innervation of muscle, except that derived through continuity of the motor nerve fiber. Tower[197, 198] isolated lumbar segments of the spinal cord in young dogs by section of the spinal cord above and below, and of the dorsal nerve roots of the surviving segments. The muscles still innervated by the surviving isolated segments were quiescent and suffered moderate loss of bulk. Some increase in fibrous tissue and contracture developed in the more peripheral musculature. Such changes can be accounted for by inactivity alone since Brooks[21] and others failed to demonstrate by surface and intracellular recordings any change in membrane excitability.

In studies of nerve block produced by the compression of the motor nerve, Denny-Brown and Brenner[42] found that fibrillation or change in size of muscle fiber did not occur, though the conduction of the nerve was abolished for as long as 3 weeks. The axis cylinders and nerve endings remained intact, and it is therefore assumed that anatomic continuity is more important in preventing atrophy than the reception of nerve impulses. Section of the dorsal nerve roots alone or autonomic nervous supply does not induce histologic change or any alteration of the response to excitation by the motor nerve. Section of the ventral roots produces the same changes as interruption of the peripheral muscular nerve.

HISTOLOGIC CHANGES AFTER DENERVATION

The histologic changes which follow muscle denervation have been the subject of many

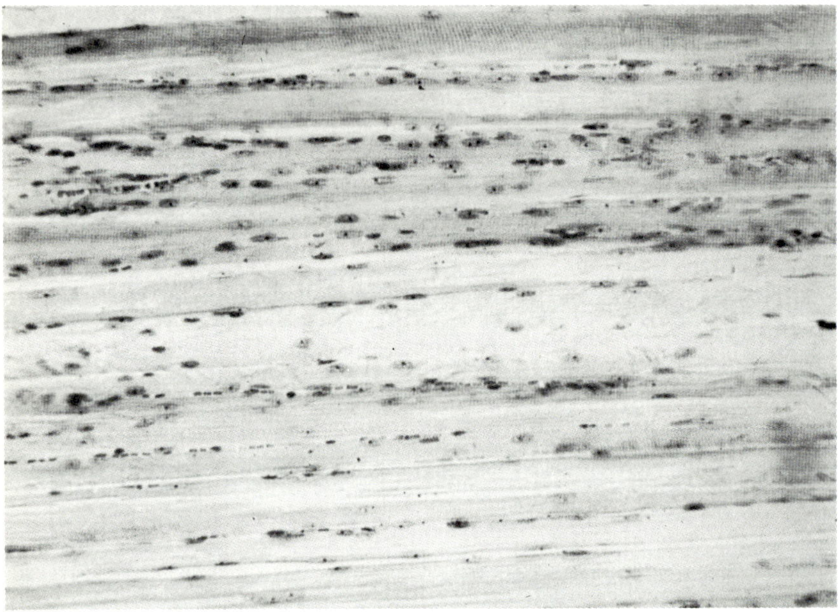

Fig. 3–1. Sartorius muscle of cat 2 weeks after nerve section. (methylene blue, eosin)

careful studies. Those of Ricker and Ellenbeck,[168] Morpurgo,[144] deBuck and deMoor,[38] Tower,[195] and Chor *et al*.[28] were complemented more recently by Gutmann and Zelena,[90] Pellegrino and Franzini,[154] Miledi and Slater,[140] and Stonnington and Engel.[187] The following account is from our own experience in experimental neural atrophy in the cat and man.

After section of either the muscular nerves or the motor nerve roots, the first evidence of histologic change in the muscle fibers is in the sarcolemmal nuclei. At the end of the first week many of these nuclei lose their flattened, compressed appearance and become rounded and plump. The nucleoli become more prominent. These changes occur earlier in some muscle fibers than in others. Tower[196, 199] found the earliest change in the vicinity of the motor end-plates. By the end of the second week after denervation (Fig. 3–1), many of the sarcolemmal nuclei have lost their elongated narrow shape and are oval or round. Other evidence of activity such as enlargement of the nucleolus and hyperchromatism appear, and some nuclei take up a position among the myofibrils in the more central parts of the muscle fiber. Small accumulations of sarcoplasm appear around some of the nuclei, giving the impression of a halo or cleft. The sarcoplasm of the end-plate stains darkly, and the end-plate nuclei appear to be increased in number. The small dark nuclei of the sheath of Henle stand out in strong contrast to the larger, less heavily chromatinized sarcolemmal nuclei and Schwann cells (Fig. 3–2). Little evidence of numerical increase in muscle nuclei is seen, but in places the sarcolemmal nuclei are so close together as to touch one another. During the first 2 weeks no definite change in the remaining structure is seen. It is hard to show reduction in size of individual muscle fibers, but beginning loss of weight and bulk of the whole muscle indicates a lessening in size of each fiber that can be determined only by careful measurement. Transverse and longitudinal striations are well preserved; the sarcomeres are in midcontraction and of the usual length. The endomysial connective tissue is more prominent, but it is doubtful that there is any increase in the number and size of fibroblasts. Histiocytes or clasmatocytes and occasional small round cells become more prominent.

At the end of a month there is a perceptible

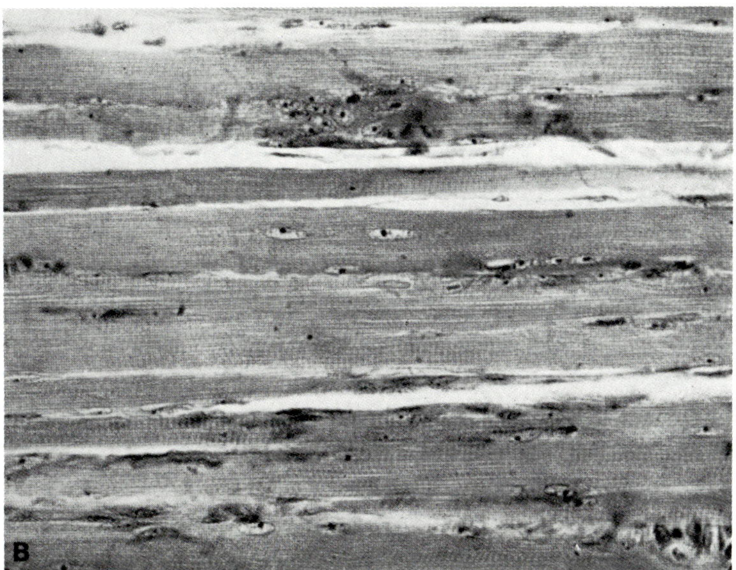

Fig. 3–2. Neural atrophy in cat. (A) Sartorius muscle, nerve to which was sectioned 2 weeks previously. Two upper fibers show degenerating motor end-plates with darkly stained cytoplasm and nuclei of Henle. Several fibers show centrally placed nuclei. (B) Biceps brachii, nerve to which was sectioned 4 months previously. End-plate nuclei are at top. (methylene blue, eosin)

reduction in the diameter of muscle fibers, which now appear rounded instead of polygonal in transverse section. The striations, both transverse and longitudinal, the length of sarcomeres, and the relative proportions of dark and light bands are unchanged. The longitudinal striations are more evident, as if many clefts had appeared in the columns of myofibrils; and through loss of myofibrils there is less uniformity in the diameters of the muscle fibers than is normally found. Subsarcolemmal nuclei are enlarged and vesicular, but there are also scattered smaller, darkly staining trigroid nuclei that appear to be regressing. The nuclei are more evenly scattered, and pairs are no longer so evident as previously. It is quite doubtful if nuclear proliferation is occurring. There is disagreement concerning the apparent increase in sarcolemmal nuclei. Willard and Graw[217] and Sunderland and Ray,[189] by careful counts in longitudinal and transverse sections, found no increase in the number of nuclei in the opossum and mouse. Others[19, 28] have reported a numerical increase. Mitoses are not

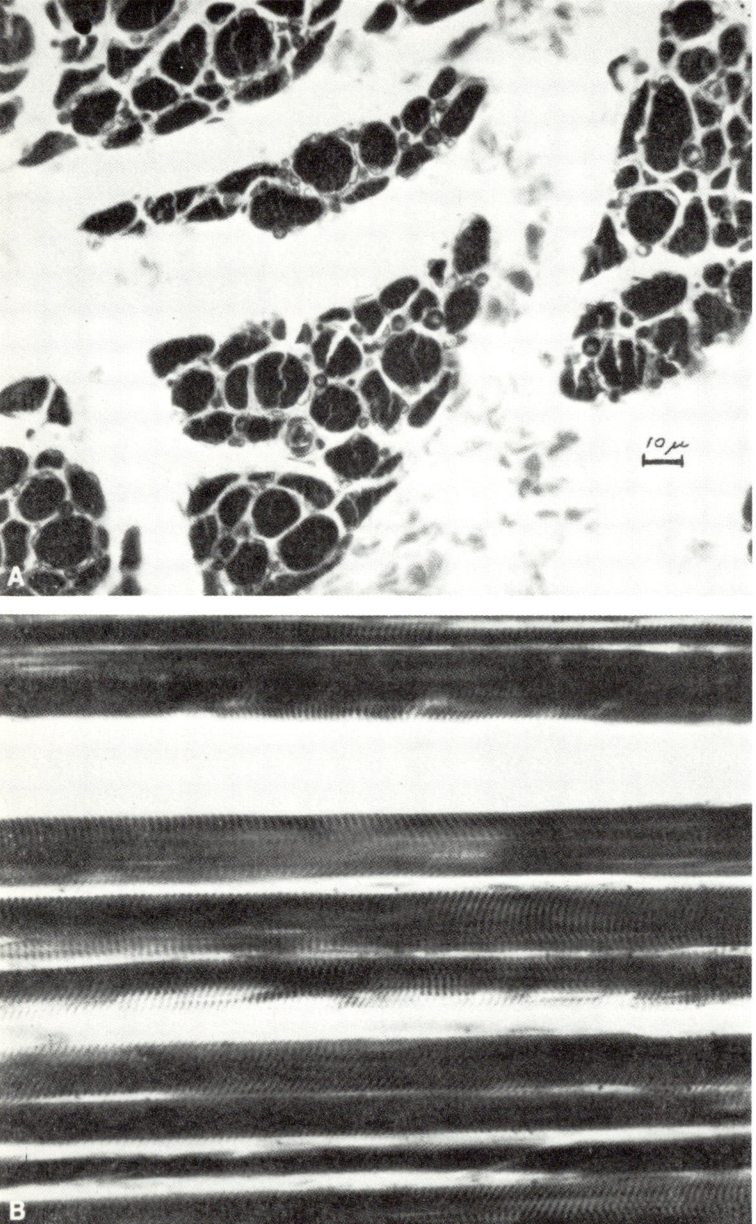

Fig. 3–3. (A) Sartorius of cat, 6 weeks after nerve section. (B) Same. Note preservation of striation. (phosphotungstic hematoxylin)

seen. Altschul[5] attributes an increase in size of sarcolemmal nuclei to a reduction in interstitial pressure. The connective tissue is more evident, and a few scattered lymphocytes and histiocytes are found.

At the end of 2 months the decrease in diameter of all the muscle fibers is far advanced and more pronounced in some fascicles than others. Most muscle fibers are reduced to less than half their normal diameters, though some always remain larger than others. It is mainly the myofibrils that share in this atrophy. The A substance is less sharply defined, and the sarcomeres are of more variable length than normal. The striations are intact, and though there is obviously reduction in the number of myofibrils, they still fill the muscle fibers (Fig. 3–3). An occasional fiber shows an altered staining reaction and the beginning of fragmentation, a change termed degeneration, but

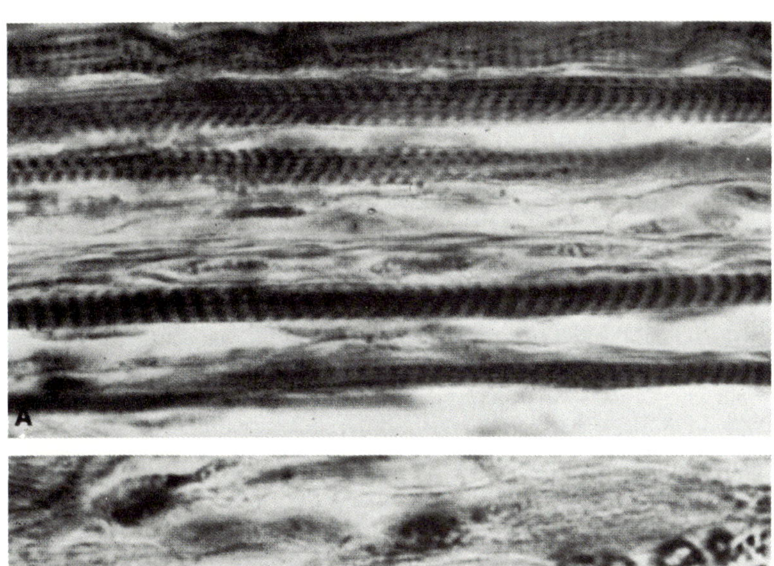

Fig. 3–4. Late atrophic changes in myofibrils. Sartorius muscle from cat denervated for 8 months. (A) Least degenerated region of muscle shows reduced numbers of otherwise intact myofibrils in simple atrophic fibers. At center, one fiber shows myofibrils that have lost transverse striation. (B) Region showing terminal changes. Very thin fiber in center shows a few intact myofibrils; another at lower border has lost striation. At top, a fiber with clumped "tigroid" nuclei shows faint residual striation between clumps. (phosphotungstic hematoxylin)

this varies from one species of animal to another.* The differences between red "pigmented" and white "pale" muscle fibers are no longer evident; all fibers contain some fat granules and pigment, as in the fetus. Knoll and Hauer[115] found that the great difference between the pale large and dark small muscle fibers of the pigeon had vanished by the 35th day.

Four months after denervation, atrophy is progressing less rapidly. The striations are still present. In some fibers the sarcolemmal nuclei are now becoming clumped into groups, and the degeneration of isolated fibers is more prominent. These latter fibers first become granular and basophilic with rows of nuclei, then become thinned at several points, and finally disintegrate into smaller clumps or tubes of muscle nuclei with very scanty cytoplasm. Some nuclei stain progressively more darkly and become shrunken (Figs. 3–2 through 3–6). Otherwise, the age of the process is indicated only by the thinness of the fiber and the condensation of the endomysium. It is still doubtful if any increase in the number of fibroblastic nuclei has occurred. The number of myofibrils progressively lessens, and surviving ones are thinner ($< 1\ \mu$). At a late stage transverse striation is lost (Fig. 3–4). There is no evidence of fatty change in the muscle fibers. By almost imperceptible degrees the contractile tissue shrinks to small tubes containing rows of nuclei and some poorly staining fibrils. Such tubes then break up into fragments of varying length, some still keeping a few fibrils (Fig. 3–4). Finally, only a string of darkly staining rounded muscle nuclei remains. At this stage it

---

* The terms atrophy and degeneration refer to essentially different processes. From the histologic point of view, atrophy consists in reduction in size of a cell or tissue while its form is conserved. Degeneration, on the other hand, means the destruction of cells or tissue or a change in appearance of such degree that recognition is no longer possible. In regard to muscle, atrophy is probably reversible in all but its terminal stages, whereas degeneration is probably irreversible from the moment of its onset.

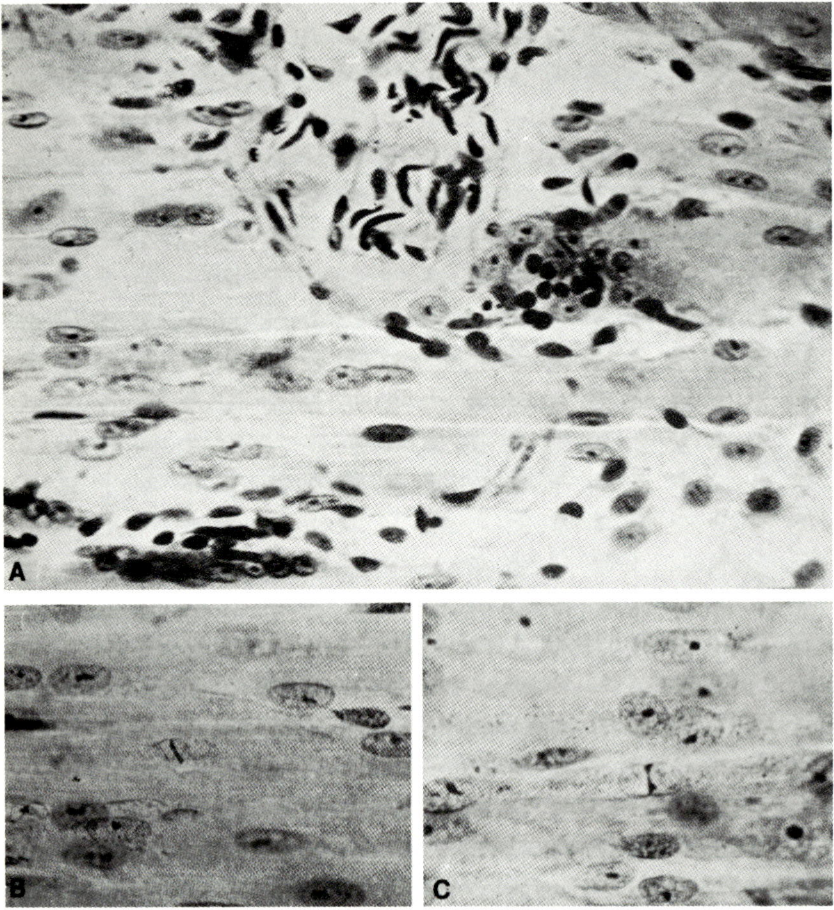

Fig. 3–5. (A) Sartorius of cat, denervated 6 weeks. Motor end-plates. Note Schwann cells in degenerated nerve bundle. Sheath of Henle nuclei (black) and muscle nuclei of the end-plate with their abundant sarcoplasm. (B and C) Sartorius after neural atrophy of 6 weeks. (methylene blue, eosin)

is often difficult to define any sarcolemmal membrane, and the nuclear remnants appear as dark, shrunken, homogeneous disks lying free in strands of fibrous tissue. The fibrous endomysium is prominent but probably not more than might be expected from condensation of a preexistent fibrous supporting network. An occasional macrophage is found in the interstitial tissue.

The degree of reduction in size and weight of muscle after denervation has been computed by several workers using the rabbit, rat, cat, monkey, and opossum as experimental animals. This subject is reviewed in the report of Sunderland and Ray.[189] The latter conducted chronic experiments in the Australian opossum

and followed the atrophic process for 485 days. There was a 30% loss of weight during the first 29 days and a 50–60% loss in 60 days. Thereafter the atrophy progressed much more slowly, reaching a fairly constant value of more than 80% at 120 days and beyond. Reduction in the size of the fibers proceeded apace; the atrophy was rapid during the first 30 days, and by 60 days the cross-sectional area was about 70% of normal. Since it is estimated that the connective tissue of muscle accounts for 10–25% of its weight,[116, 214] it can be seen that the muscle fibers are reduced markedly in size. The rates of denervation atrophy were approximately the same in three different muscles of the opossum (flexor carpi ulnaris, flexor carpi radialis, and

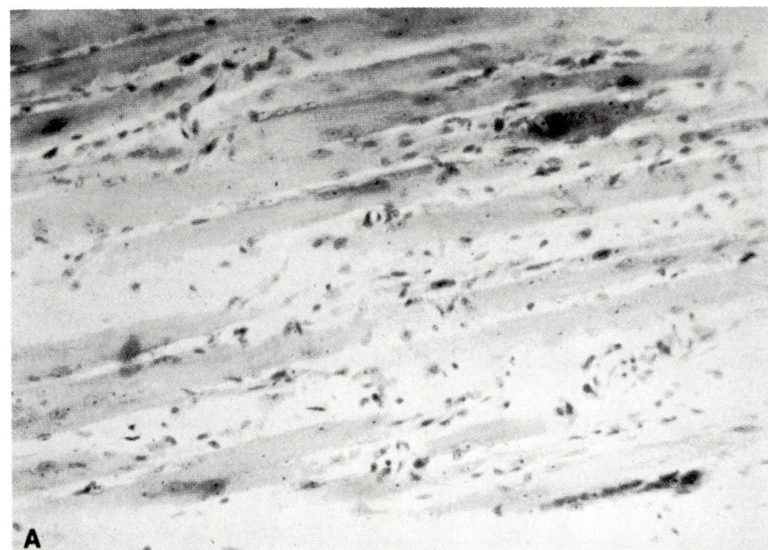

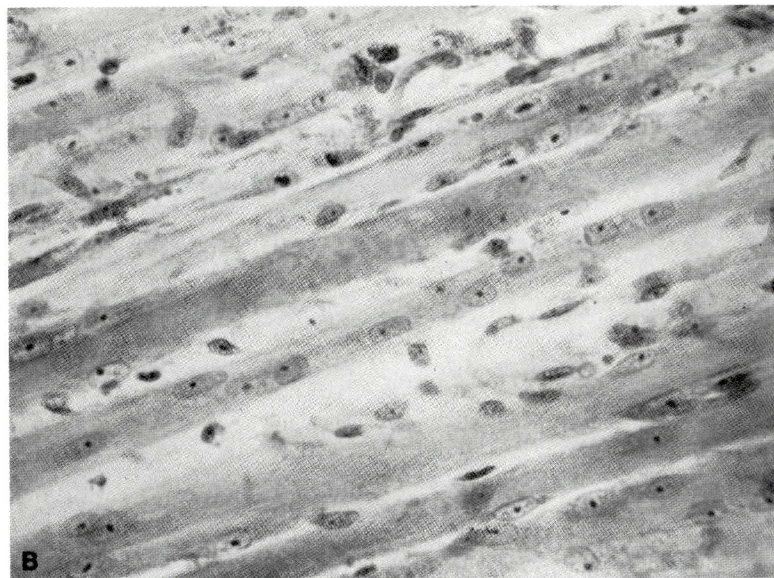

Fig. 3–6. (A) Sartorius muscle of cat 6 weeks after section of femoral nerve, showing in places unusually precocious degenerative changes in isolated muscle fibers, with basophilia and fragmentation into pyknotic mononucleated cells. (B) Note degenerate daughter myoblasts of recent cell division. (methylene blue, eosin)

pronator terres) and were in fairly close agreement with those determined by Roberts,[169] Langley,[121] Langley and Hashimoto,[122] and Altschul.[4] Sunderland and Ray observed that although the muscles were slightly more pale, they retained their red color.

Several electron microscopic studies of the earlier phases of denervation atrophy of rat muscle[154, 187, 212] and frog muscle[140, 148] serve to clarify the intramuscular events. As the muscle fiber undergoes atrophy there is a parallel myofibrillar atrophy in both red and white fibers. Stonnington and Engel[187] found this to be expressed morphometrically as a reduction in the total myofibrillar content of the fiber. It had decreased in 28 days to less than 25% of the original volume while the total volume of the fiber decreased to 20%. The process begins in the periphery of the fiber as a degeneration of myofilaments and is first observable at the Z lines spreading transversely from one myofilament to the next, according to Pellagrino and Franzini.[154] A single fiber shows this change at several points along its length, at first with normal intervening sarcomeres. Loss of peripherally placed myofilaments within a myofibril reduces the diameter of the latter from 1.0 to 0.2 μ, and there is an increasing variability of diameters. The degenerating myofilaments leave no trace, as though the breakdown process had been an immediate lysis.

The mitochondria in both the red and white fibers at first decrease in size (for 24 hours) then increase during the second and third days. Thereafter there is an absolute and relative decrease in mitochondrial mass proportionate to reduction in fiber size. The mitochondria elongate and lay in more strict longitudinal orientation. The sarcotubular surface increases absolutely and relatively during the first days and then decreases but always to a lesser extent than myofilaments.

The surface of the atrophic fibers becomes irregular and displays small papillary projections.[1] Irregular Z disks are observed where myofilament degeneration is most active. There are progressively fewer glycogen granules. Lipofuscin granules and lysosomes increase, as do other cytoplasmic degeneration products. Acid phosphatase and β-glucuronidase (two of the lysosomal enzymes) also increase. Focal dilatations of both the transverse sarcoplasmic tubules and longitudinal sarcoplasmic reticulum are often seen, and the spatial arrangements of the sarcotubular system are less regular. In the perinuclear regions there are increased amounts of rough endoplasmic reticulum and of free ribosomes during the first few days of denervation. Mitochondrial degeneration (formation of myeloid structures and entrapments of autophagic vacuoles) are observed after the first week of denervation. Central nucleation becomes slightly more frequent than in the normal fiber.

From their observations Pellagrino and Franzini postulate two types of change in denervation atrophy—the first a degeneration of areas of sarcous substance, the second a depletion within myofibrils of their peripheral myofilaments. Early volumetric reduction is attributed to the first process, and a slower, later diminution in fiber size to the second. The number of myofibrils falls from a normal level of 1200–1900 per muscle fiber in gastrocnemius and 1900–3000 per fiber in rat soleus to a few hundred after 2 weeks; the second atrophic process

reduces the size of the residual myofibrils. Red and white muscle fibers come more to resemble one another in their atrophic state, i.e., they dedifferentiate.

It is of interest that although Chor and Dokeart[27] mentioned isolated instances of degenerative change in denervated muscle, Wechsler and Hager[212] and Stonnington and Engel[187] saw only the peripheral loss of myofilaments. This discrepency leaves unsettled the validity of the early degenerated areas. Other studies are needed to exclude the possibility of artifacts from imperfect fixation.

The published electron microscopic observations of early atrophy correlate well with biochemical data that show a decrease in protein content,[68, 104] contractile material,[69] uptake of labeled ammoacids,[174, 175] and protein synthesis.[221] Also mitochondrial enzymes (e.g., cytochrome oxidase and succinic dehydrogenase) diminish, as has been shown by both direct chemical analyses and histochemical procedures,[9, 149] while lysosomal enzymes and adenine triphosphatase (ATP) increase. Nachmias and Padykula[149] also report a decrease in sudanophilia and in glycogen content in denervated muscle.

The foregoing account applies to the early course of events observed following denervation of any large muscle in the limbs, such as soleus, gastrocnemius, or biceps brachii, and follows in general the account given by Tower.[195, 198] There are, however, significant deviations from this process as exemplified by the small muscles of the larynx, which reach a state of advanced atrophy within 2 weeks[38] following section of the recurrent laryngeal nerve. The striped muscle of the external sphincter of the anus, on the contrary, is extremely resistant to atrophy following denervation and the diaphragm may actually hypertrophy. Denny-Brown found, following section of the femoral nerve in the cat, that in some animals the sartorius muscle can begin to show isolated foci of fragmentation of muscle fibers after 6 weeks (Fig. 3–6), whereas the vastus and rectus divisions of the quadriceps muscle in the same animal have reached only the stage pictured in Figure 3–1. Further, the medial part of the sartorius muscle 6 weeks after denervation may present a degree of nuclear proliferation and fragmentation of muscle fibers seldom seen at any one stage in larger muscles. In the latter muscle nuclear division is proceeding in all the muscle fibers and examples of cellular division are not infrequent, even in motor end-plates (Fig. 3–5). In such a muscle the processes of atrophy and degeneration are both accelerated, and in this state it is clear that fragmentation is the ultimate fate of all the muscle fibers. In other animals following an identical procedure the degeneration of sartorius was less advanced 6 weeks after nerve section, though this muscle still showed more nuclear change than the quadriceps.

Chronic denervation has been studied little in the experimental animal, and there are many discrepancies in the accounts of different authors. Fibers of normal form, nuclear distribution, and striations have been observed by Sunderland and Ray[189] in the muscles of the opossum up to 485 days after denervation. Whatever may be the reason for acceleration of atrophy in some muscles, it appears that the more rare degenerative changes in isolated fibers of larger muscles must represent a similar precocious terminal change not yet evident in the bulk of the muscle. Degenerative changes are frequently found in all fibers of the larger muscles 6 months following section of the motor nerve. Those who recently described experimental neural atrophy did not report a terminal stage in which small chains of darkly staining pyknotic nuclei alone represent the muscle fibers. Yet such chains are commonly found 12 months or more after neural atrophy in man and were noticed in the early experiments of Stier,[186] Ricker and Ellenbeck[168] and deBuck and deMoor[38] in animals.

Following section of the femoral nerve in the cat, with transposition of the proximal stump into the abdomen to prevent regeneration, the process of degeneration of the various femoral muscles 8 months later was found by Denny-Brown to be especially instructive with respect to the chronic phases of atrophy. For such studies it is of course important to exclude not only the possibility of neural regeneration but also the effects of sepsis, trauma, or other muscle disease. For this reason, proximal muscles were chosen to avoid the trophic lesions of the feet that complicated Tower's long-term experiments. Adequate controls of symmetrical muscles of the other limb that retained their innervation were always used, with identical fixation and staining (Fig. 3–7). It was noted that, in spite of histologic and electrical proof of absence of nerve regeneration, various parts of the muscles were in greatly different phases of atrophy and degeneration. In the lateral part of sartorius, for example, the majority of muscle fibers presented an atrophy only slightly advanced beyond that observed at the end of the second month (Fig. 3–7A). In the medial third of the muscle no intact muscle fibers remained, and in intermediate parts a mixture of atrophic and degenerated fibers was found (Fig. 3–7C). The lateral aspect of the vastus lateralis was severely degenerated, whereas the medial portion (shorter head) still showed only moderately severe atrophy. Foci of recent nuclear proliferation, loss of myofibrils, and clumping of muscle nuclei could be found in the areas presenting moderate atrophy (Fig. 3–8). Fat cells also began to appear in large numbers at this time. In the degenerating areas each muscle fiber preserved for a long time its own endomysial tube (Fig. 3–9), but the myofibrils stained poorly and disappeared and the sarcolemma became indistinguishable. Strands of muscle nuclei, many showing regressive changes, lay in vacuolated spaces beside others which are connected by short threads of sarcoplasm (Fig. 3–10A). A few short bundles of myofibrils remained in some of these sarcoplasmic threads (Fig. 3–10B), but most appeared to have cytoplasm with only faint remnants of myofibrils or none at all. In places columns of dark shrunken muscle nuclei lay free in the tissue, as in the lower part of Figure 3–10A.

At a later stage fat cells accumulated in long rows, which broke up the cellular remnants of the muscle fibers into columns and clumps (Figs. 3–10 and 3–11). The nuclei of the fat cells had the appearance and staining characteristics of normal fat cells, and the clumps of residual muscle nuclei, though often sandwiched between fat cells, showed no evidence of participating in the formation of fatty or fibrous tissue. The residual endomysial nuclei

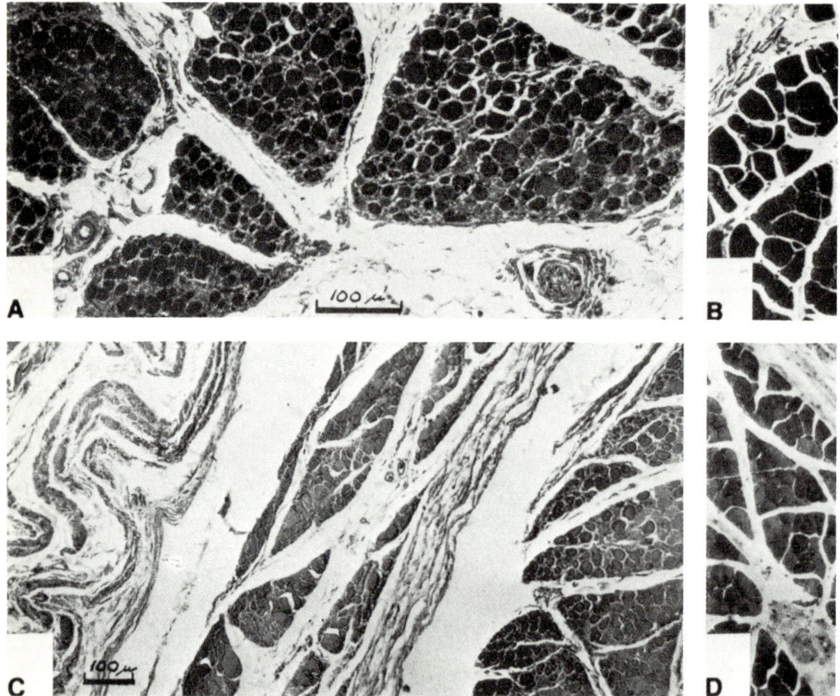

Fig. 3–7. Sartorius muscle of cat denervated 8 months showing different rates of degeneration in its different parts. (A) Transverse section of thicker lateral parts of sartorius. (B) Control section, opposite sartorius. (C) Section of all three parts of sartorius: medial, intermediate, and lateral, from left to right. (D) Control, intermediate part. (A, B, phosphotungstic hematoxylin; C, methylene blue, eosin)

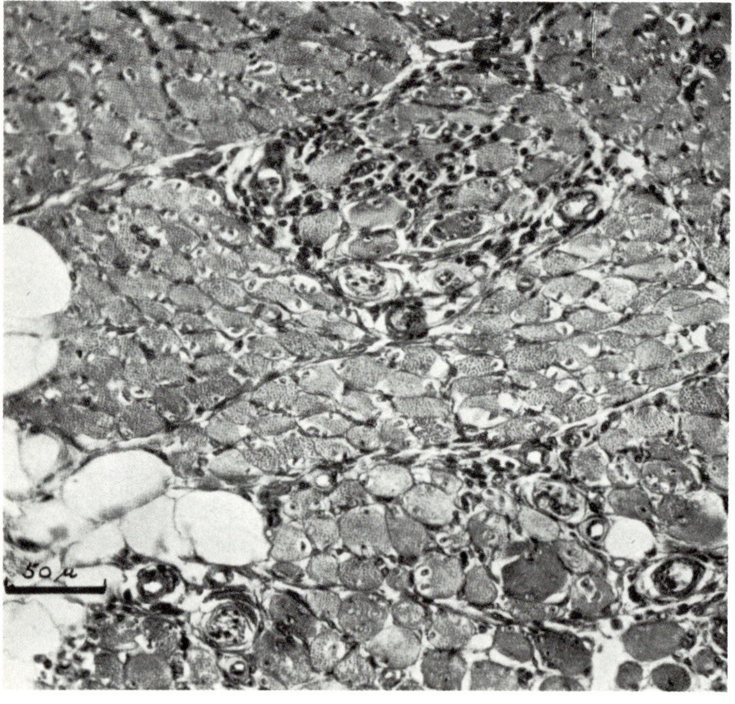

Fig. 3–8. Vastus medialis of cat denervated 8 months showing a small focus of degeneration with cellular response and thickening of endomysium. In the lower part of figure there are also scattered degenerating fibers. (methylene blue, eosin)

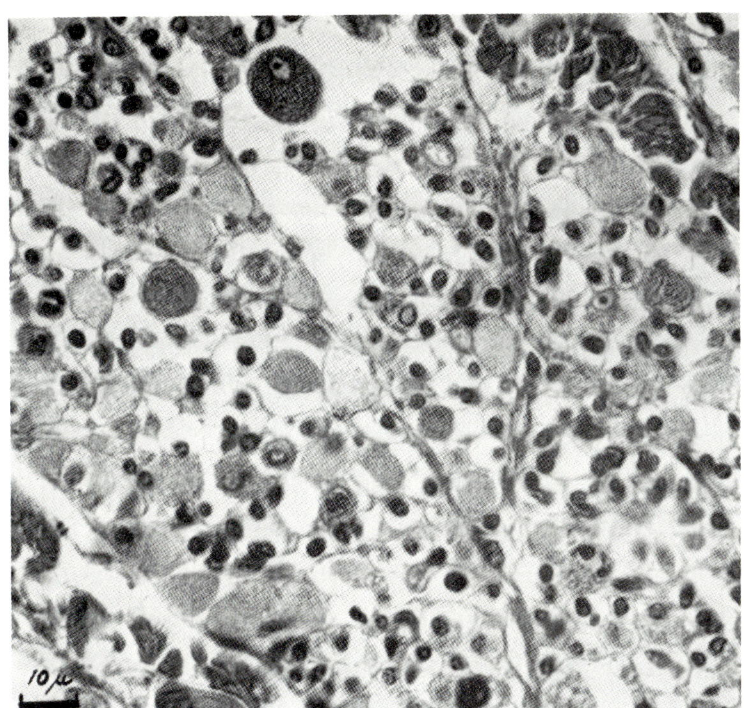

Fig. 3–9. Vastus lateralis of cat denervated 8 months; transverse section of moderately degenerated region. (methylene blue, eosin)

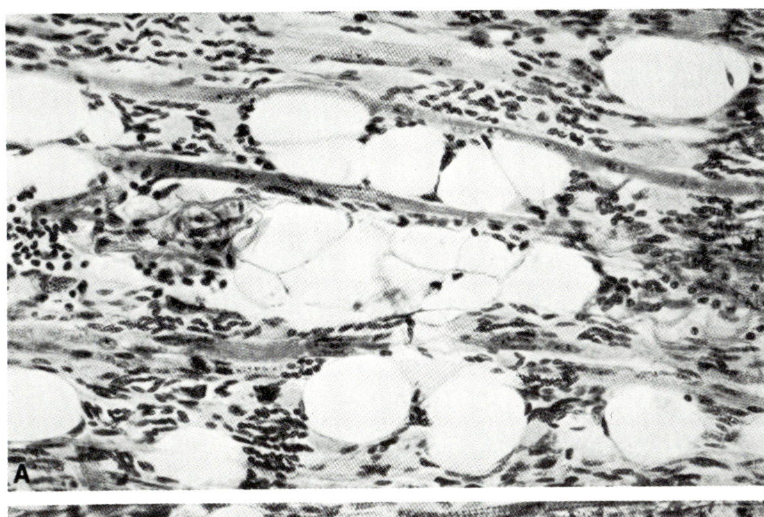

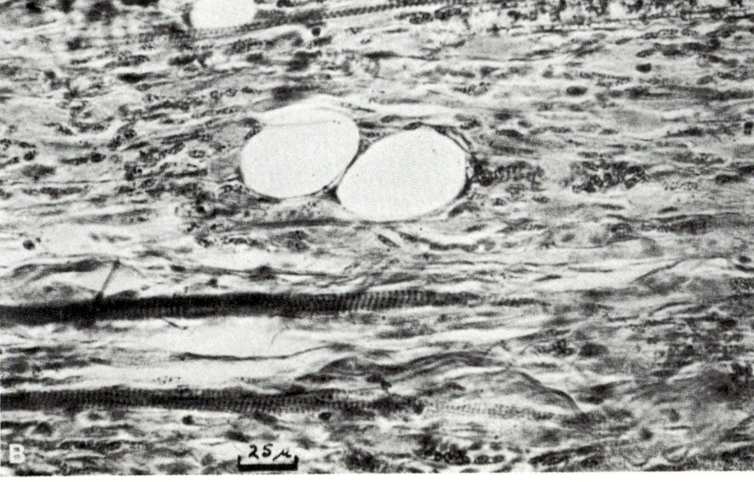

Fig. 3–10. Same muscle as in Figure 3–9, longitudinal section. (A, methylene blue, eosin; B, phosphotungstic hematoxylin)

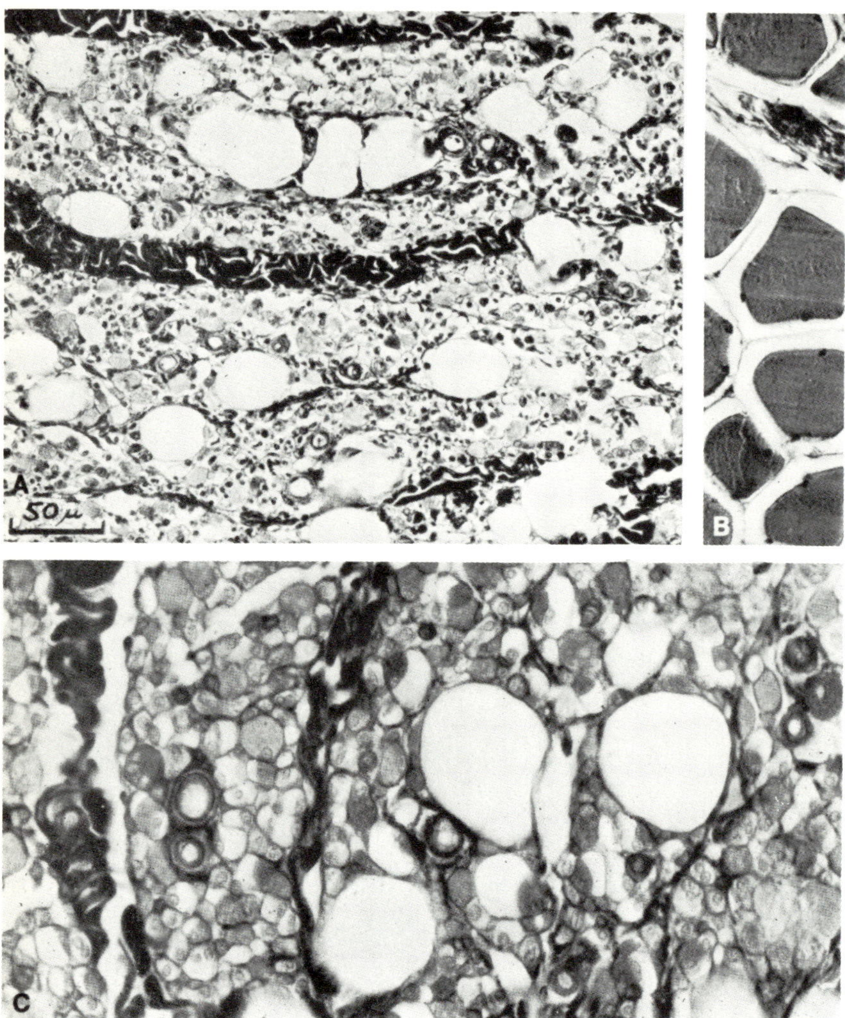

Fig. 3–11. (A) Same muscle as in Figure 3–9, showing thickened interfascicular septa and scattered collagenous threads in the endomysium. Note also thickening of walls of small blood vessels and presence of fat cells. (B) Control from opposite side, same magnification. (C) Endomysial and perimysial septa stained light blue (reticulin), and elastic tissue is represented only in condensed bundles of aponeuroses and walls of blood vessels. (A, B, phosphotungstic hematoxylin; C, orcein-aniline blue and orange G method of Margolena and Dolnick[130])

appeared inactive. Stains which differentiate between reticulin, elastin, and collagen (e.g., as the orcein-silver-picro-fast green blue method of Lewis and Jones[125] revealed an accumulation of collagen in the aponeurotic septa, around blood vessels, and in the substance of the intramuscular nerves at this stage. The endomysium presented only a condensation of its natural reticulin and occasional elas-

tic fibers (Fig. 3–11). There was no evidence of collagen deposition in the endomysium. In the areas of most advanced degeneration the former muscle was represented only by columns of fat cells interspersed with strands of inert fibrous tissue (Fig. 3–12) in which lay clumps or short chains of former muscle nuclei. They were recognizable by their shape but were evidently only an inert chemical resi-

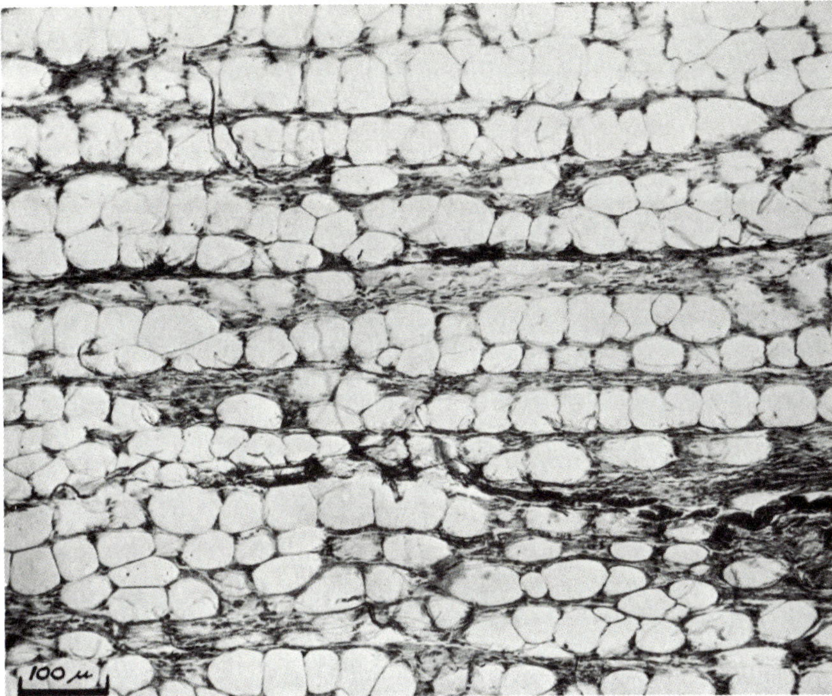

Fig. 3–12. Vastus lateralis of cat denervated 8 months. Severely degenerated portion of muscle showing ultimate replacement of muscle fibers by fatty tissue. Two dark strands of fibrous perimysium cross the figure. Only scattered muscle filaments and residue of muscle nuclei remain in bands of endomysium. (phosphotungstic hematoxylin)

due that often appeared to be calcified. Here and there a muscle spindle could be recognized by its capsule and some structurally intact intrafusal muscle fibers with a nuclear bag. In serial section it was occasionally possible to follow the extrafusal portion for long distances, and it was clear that all of the muscle fibers of the muscle spindle were well preserved for many months after nerve section, with darkly staining myofibrils in sharp contrast to the atrophy, poor staining reaction, and degeneration of surrounding muscle fibers (Fig. 3–13). The study by Denny-Brown convinced him that fragmentation is a stage in the process of neural atrophy, and that the small chains of pyknotic nuclei in late atrophy represent not whole muscle fibers but the remnants of the larger fragments of the final stage (Fig. 3–14). This explains why such chains, or nuclear bags, are never of any considerable length. However, these findings in

certain leg muscles of the cat are still open to the criticism that some aberrant inflammatory or traumatic disease has supervened, adding itself to the atrophic process.

The rate of the chronic phases of denervation atrophy in different species was studied by Knowlton and Hines.[116] They found that the speed of atrophy in a given species is related to the growth rate and life span of the animal: An increased rate of atrophy is associated with a shorter life span. Bowden and Gutmann[19] and others have observed some identifiable muscle tissue in biopsy specimens of human muscle that had been denervated for as long as 26 years.

There can be no doubt, however, that at some stage of neural atrophy the atrophic process gives way to changes of degenerative kind, which occur much earlier in some animals and some muscles than others. This is evidently the process of "metamorphosis" discussed by Durante,[49, 50] though we find no

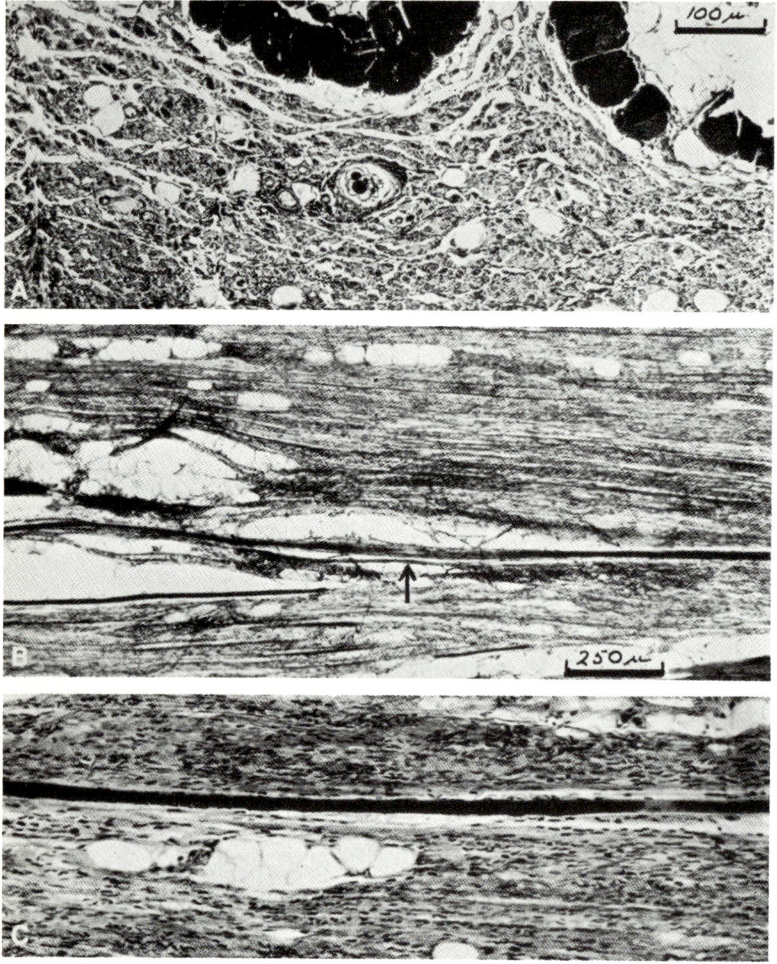

Fig. 3–13. Vastus lateralis of cat denervated 8 months showing intact muscle spindles. No regeneration has occurred. (A) Transverse section, near an aponeurosis. (B) Longitudinal section. Equatorial region with nuclear bag is indicated by arrow. Part of muscle fiber belonging to another spindle lies immediately below the capsule. (C) Muscle fiber of spindle, 1 cm away from region shown in (B), surrounded by degenerating fibers.

Fig. 3–14. Human muscle with neural atrophy of more than 10 years' duration. Residual muscle fiber remnants are seen in connective tissue. At top and center are small muscle fibers with a few striated myofibrils. Other fibers show fragmentation with clumping. (methylene blue, eosin)

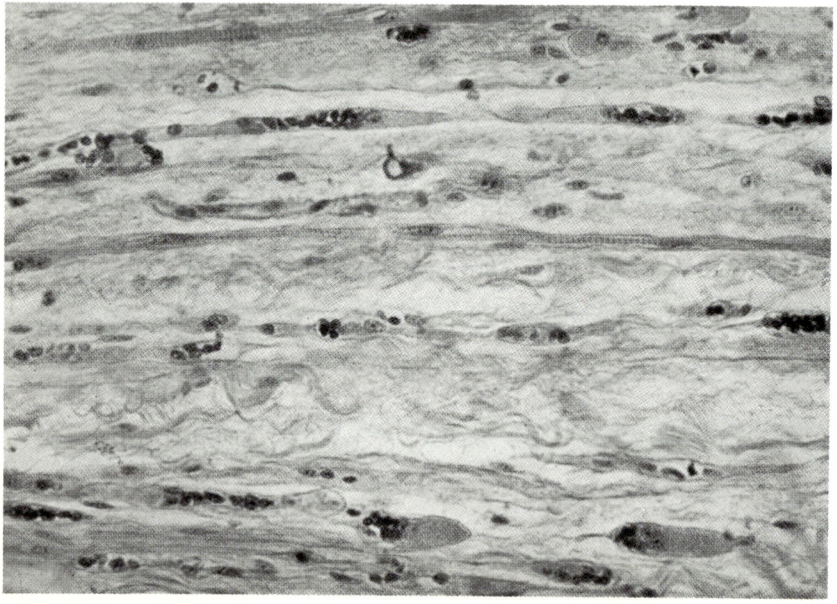

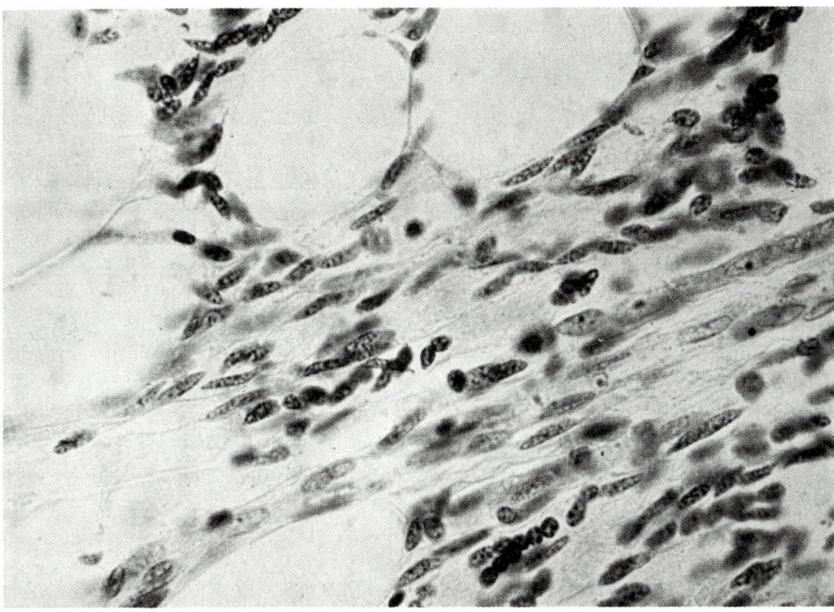

Fig. 3–15. Cat sartorius denervated 8 months. Residual muscle fibers running obliquely from middle right to lower left, with chains and clumps of oat-shaped tigroid nuclear clumps, chains of shrunken round dark muscle nuclei (terminal stage), and fat cells. Each fat cell has a single dark round or oval nucleus, of which only two are shown here. (methylene blue, eosin)

evidence in support of his contention that it represents adipose or fibrous transformation. Of the possible factors responsible for the degenerative phase, trauma appears the most likely. The large range of stretch in double-joint muscles such as sartorius must impose unequal stresses on their individual muscle fibers and offers an explanation for their special susceptibility in this regard.

Tower[196] described transformation of muscle fibers to fibrous tissue, which she designated as fibrotic dedifferentiation. She states, "By progressive fading of the cross striation to the point of disappearance with conservation of the fibrils, and by progressive slimming of the nuclei, large regions of the muscle tissue finally lose their specific character as such, and become transformed into fibrous tissue." Altschul[4] described a similar change in denervated cat and rabbit muscle. He also was convinced of the replacement of muscle fibers by adipose tissue and expressed the opinion that both fibrous and fatty tissue were the result of the metaplasia of muscle fibers. In our own prep-

arations the last stage of disappearance of muscle fibers seems to be disintegration of the sarcolemma, leaving clumps of darkly staining disks that have the aspect of nonviable aggregations of the sarcolemmal nuclear materials (referred to above). A histiocytic response is excited by degeneration.[25, 26] In this last stage the absolute quantity of fibrous tissue does not appear to be increased (Fig. 3–12).

We have not observed any transitional stage between muscle fiber and either fibroblast or fat cell. Indeed, both fat cells and increase in collagenous tissue first appears before the sarcolemma disintegrates. The progressive replacement by adipose tissue that ultimately occurs appears to us to result from the differentiation of mesenchymal cells. This transformation is difficult to observe, for it is evidently a very slow process. The process of degeneration is associated with small irregular collections of round mononuclears in perivascular spaces. Such a clump is seen in Figure 3–10A (at left). The terminal stage of the muscle nucleus is dark, elongated, or

Fig. 3–16. Various foci of degeneration after denervation. (A) Cat sartorius 6 weeks after nerve section. (B and C) Cat biceps brachii 4 months after nerve section. Spindle-shaped cells are splitting from fragmented fiber. Arrow in (B) indicates a spindle containing a mitotic figure.

round, and many are seen near small fat cells (Fig. 3–15); proof of direct transformation is lacking. The nucleolus is no longer present.

In summary, denervation atrophy of muscle exhibits several distinct phenomena: (1) atrophy with reduction in size of fibers but preservation of cross and longitudinal striations; (2) clumping of muscle cell nuclei, then fragmentation of the fiber, degeneration of scattered but progressively numerous muscle fibers; (3) ultimate replacement by fat cells, with a variable increase in the amount of reticulin fibrils in the endomysium and collagen in the perimysium and aponeuroses. Figures 3–16 through 3–18 show the later stages of the process.

All these changes can be observed in human muscles. There is occasional simple fragmentation of muscle fibers without cellular reaction, probably the result of mechanical trauma during the process of excision or of fixing, embedding, and sectioning the tissue. The fragments tend to assume the form of ovoid, granular masses, and the nuclei are dark and pyknotic. Vacuolar degeneration of fibers is rarely seen.[19] The nuclear changes and loss of

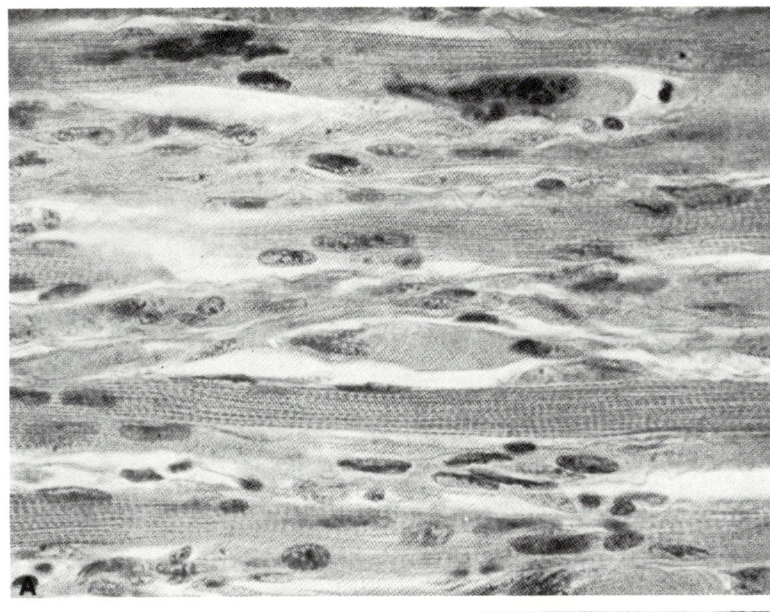

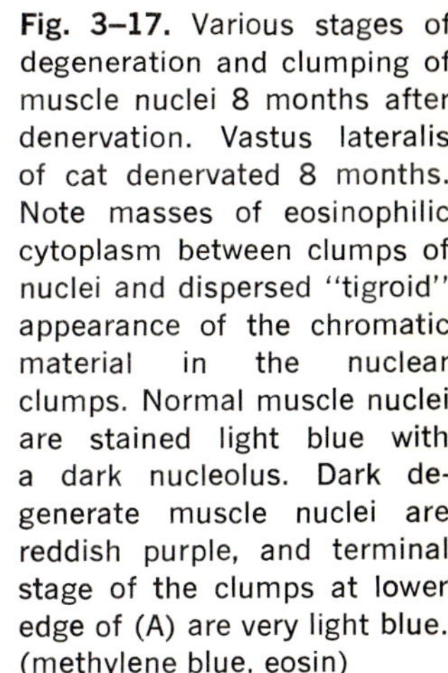

Fig. 3–17. Various stages of degeneration and clumping of muscle nuclei 8 months after denervation. Vastus lateralis of cat denervated 8 months. Note masses of eosinophilic cytoplasm between clumps of nuclei and dispersed "tigroid" appearance of the chromatic material in the nuclear clumps. Normal muscle nuclei are stained light blue with a dark nucleolus. Dark degenerate muscle nuclei are reddish purple, and terminal stage of the clumps at lower edge of (A) are very light blue. (methylene blue, eosin)

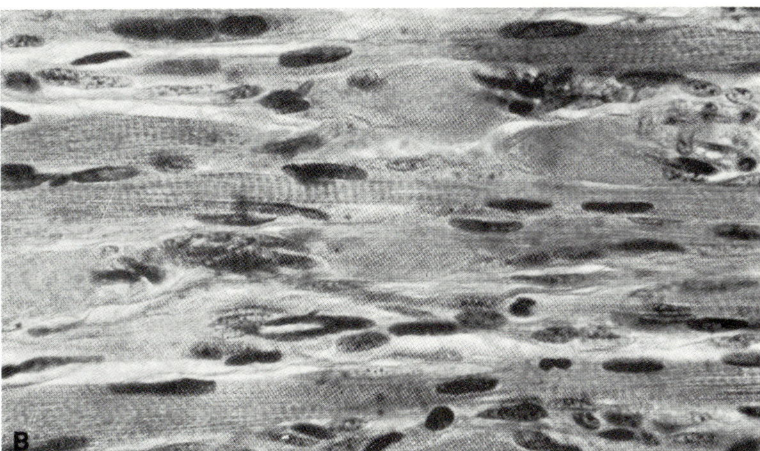

diameter occur as described above. In the late stages of atrophy, darkly staining nuclei are found in rows within the thin sarcolemmal tubes. In places even the tube is absent, and small groups of dark nuclei appear to lie within the fibrous tissue. These are seldom observed within 12 months and are associated with the appearance of fat cells. Bowden and Gutmann[19] could not find evidence of fibrotic dedifferentiation. In our own material it is surprising how little fibrous tissue may separate the thin and tightly packed thread-like remnants of muscle fibers even years after the muscle has become atrophic (old poliomyelitis, etc.).

The small muscle fibers of the muscle spindles seem peculiarly resistant to the process of denervation atrophy, for in cross section of very atrophic muscle the spindle capsules, with intrafusal fibers, stand out prominently.[43] The difference is more apparent than real during the first few months, for there is a reduction of all muscle fibers to a variably small diameter. After this the spindle more obviously escapes atrophy. The fibers of the muscle spindle, being already small and fetal in type, demonstrate a resistance to atrophy characteristic of the fetal muscle fiber. Tower[195] claims to have observed fibrous transformation of the intrafusal muscle fibers with loss of the nucleated structure if the

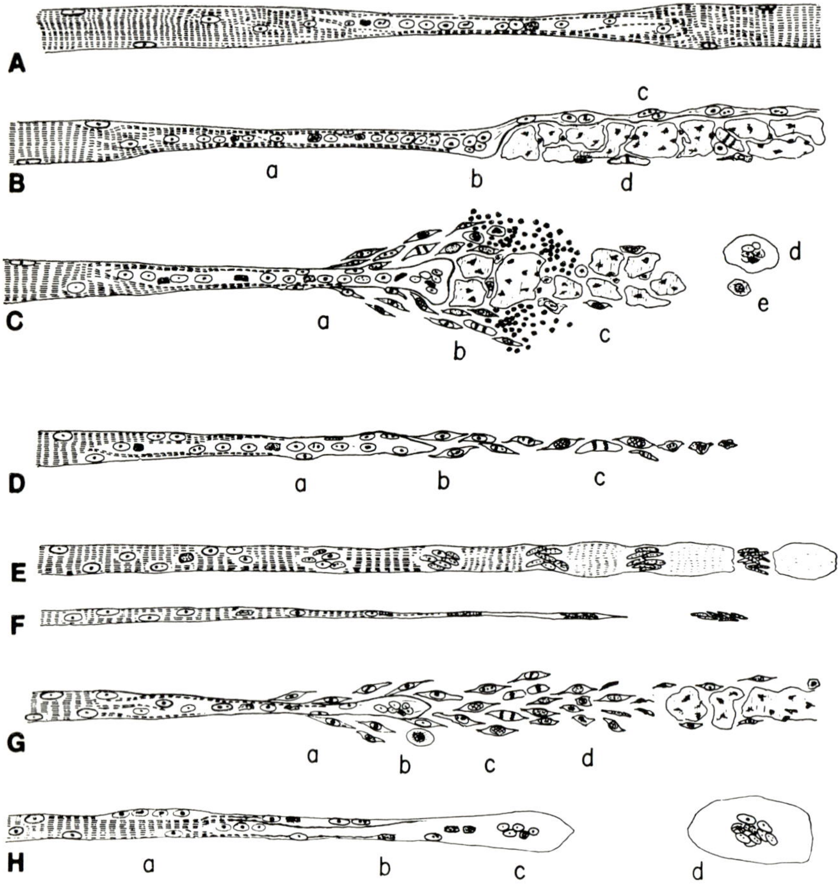

Fig. 3–18. Proliferative reactions of muscle nuclei. In (A) response of a muscle fiber to mild trauma is shown as loss of myofibrils and nuclear proliferation in affected segments. Some nuclei are shrunken and stain darkly. (B) Necrosis of part of the fiber, with partial damage to neighboring segment (c–d), is associated with regenerative budding at b, with surviving spindle cells (D) and histiocytes in necrotic clumps. (C) Regeneration is obstructed by hemorrhage with resulting spindle fragmentation of partially damaged segment. Spindle cells show mitosis and degenerative changes, ending in rounded forms (a, b). (D) Spontaneous fragmentation of denervated muscle is represented. (E and F) Phenomenon of "clumping" fragmentation. (G) Response of denervated muscle to coagulation of one segment, resulting in degenerate clubs (b), and "giant cells" (H, d).

sensory nerve fiber is destroyed in cats. This did not occur in our material, though degenerative changes may be found in some of the intrafusal fibers 6 months or more after section of the whole nerve. In human muscle some intrafusal structures may be preserved for at least 1 year and probably longer after poliomyelitis and in chronic peripheral neuropathy. The remaining intrafusal fibers are then very small, and some may have disappeared. The nuclei in the nuclear bag become very

shrunken even after 3 months of denervation.

Of course it is not unusual in many pathologic conditions for muscle fibers to undergo destruction, and in most of these there are associated reparative changes. However, the process of neural atrophy is of particular interest in that it presents a simple type of degeneration of muscle fiber uncomplicated in its more advanced stage by regeneration. Thus experimental studies to ascertain the ultimate fate of the muscle fiber after denervation for

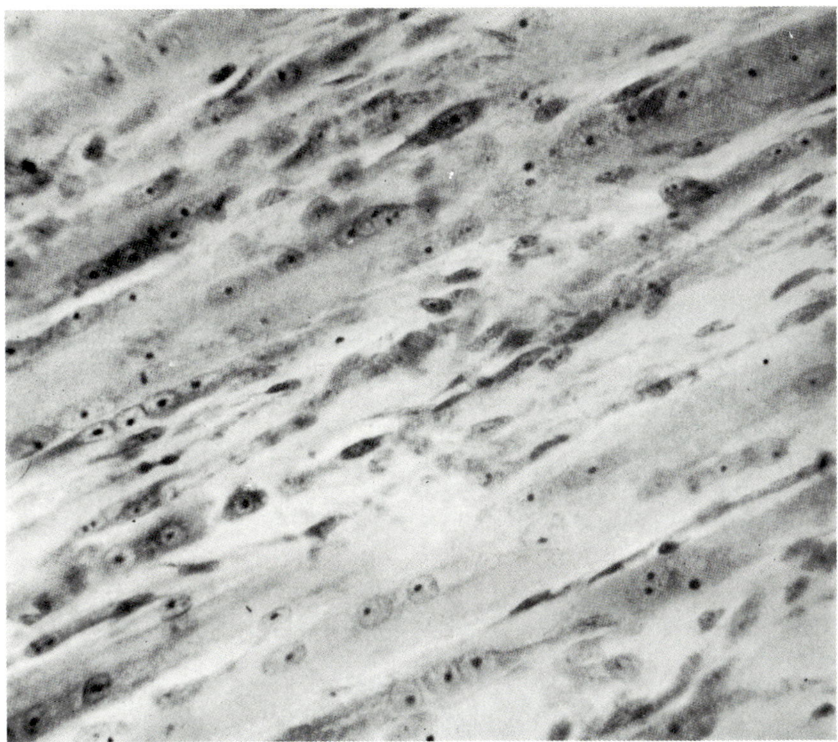

Fig. 3–19. Sartorius muscle of cat showing accelerated degeneration.. Femoral nerve was sectioned 24 days earlier. Segment of muscle from which section was taken was subjected to heat coagulation by application of a metal bar heated to 85° C for 1 min 10 days before sacrifice. Remainder of muscle resembled that in Figure 3–2B.

long periods are of importance in assessing the significance of purely degenerative changes in other circumstances.

EFFECT OF DENERVATION ON REACTION TO INJURY

A number of studies have been designed to investigate the effect of denervation on degeneration and regeneration after various types of injury. For example, it was found[40] that if a heated metal bar at a temperature of 85° C or higher was applied to the surface of a denervated muscle for 1 min the degenerative process was greatly intensified in the region immediately surrounding the surface fibers that are coagulated by such treatment. This type of injury of normal muscle is followed by intense regeneration in all portions of the muscle

fiber that surround the coagulated area. If the muscle has been denervated for 4 weeks or less, it is certainly difficult to detect any defect in the process of regeneration.* In cat muscle denervated for a longer period, Denny-Brown found that the number of fibers showing failure of regeneration increased. The zone of heat coagulation or crush injury in a muscle injured in this way during the third and fourth months following denervation shows an increasing proportion of degenerative change, with corresponding lessening of regenerative development of new muscle fibers. In medial sartorius of the cat it may be found as early as 3 weeks after denervation (Fig. 3–19). After 4 months of denervation in a muscle that had

* Kirby[113] studied this reaction at 2 weeks, Denny-Brown at intervals up to 2 months, Walton and Adams[207] at intervals up to 6 weeks, and Saunders and Sisson[173] at 3 weeks.

Fig. 3–20. (A) Sartorius muscle of cat showing the reaction 10 days after heat coagulation and 5 weeks 3 days after nerve section. (B) Biceps of cat showing reaction 10 days after heat coagulation and 2 months 10 days following nerve section. (C) Sartorius muscle of cat showing abortive fibrils without striation 10 days after heat coagulation and 3 weeks after nerve section. (A, B, C, methylene blue, eosin)

not developed more than one spontaneously degenerating fiber per low power field, a zone treated by heat and examined 2–3 weeks later showed general fragmentation of muscle fibers bordering the coagulated zone, and complete absence of true regeneration (Fig. 3–20B).

Such universal fragmentation results in a brief phase of intense cellular activity. The rapid synthesis of nucleic acids is reflected in a basophilia evident on methylene blue staining. Fragmentation of the ends of undamaged parts of the fibers into sheaves of degenerating spindle cells is then a remarkable feature (Figs. 3–21 and 3–22) in which all the characteristics we described as simple neural degeneration are exaggerated. All stages of degenerating spindle cells may be seen. Active and abortive mitosis is frequent. (Fig. 3–23). The zone of partial damage extending 2–3 mm around the coagulated area becomes intensely cellular between the 7th and 14th days after injury, owing to the enormous numbers of spindle myoblasts, each with dispersed nuclear chromatin. By the 14th day most of these daughter myoblasts show shrinkage and pyknosis (Fig.

Fig. 3–21. Budding of spindle cells from ends of partially damaged fibers in biceps muscle of cat denervated 4 months. Zone coagulated by heat 11 days earlier lies approximately 2 mm to right of these fibers. (methylene blue, eosin)

3–22B). By the 21st day the damaged area shows only a few fibroblasts and cellular fragments. Such complete disintegration of the muscle substance is followed by fatty replacement of the damaged zone, the remainder of the muscle continuing to show only the slower change of neural atrophy and ultimate patchy fragmentation. At the transition between the two areas the surviving muscle fibers end bluntly with a terminal clump of degenerating nuclei. The same failure can be observed following gentle application of a clamp or hemostat across the fibers of a denervated muscle (Fig. 3–22A).

Some have questioned the identification of muscle spindle cells, and particularly of mitosis in such cells. There is in fact little difficulty in differentiating myoblast and histiocyte, for the former is nearly four times the size of the histiocyte in the motile stage. The granular amphophilic cytoplasm of the histiocytic phagocyte lends further distinction in stains based on methylene blue or hematoxylin. Even in the granule cell stage of the histiocyte the mitotic figure is approximately only one-half the diameter of the very obvious myoblast spindle in mitosis (Fig. 3–23A). The rounded forms of degenerating myoblasts present a nucleus

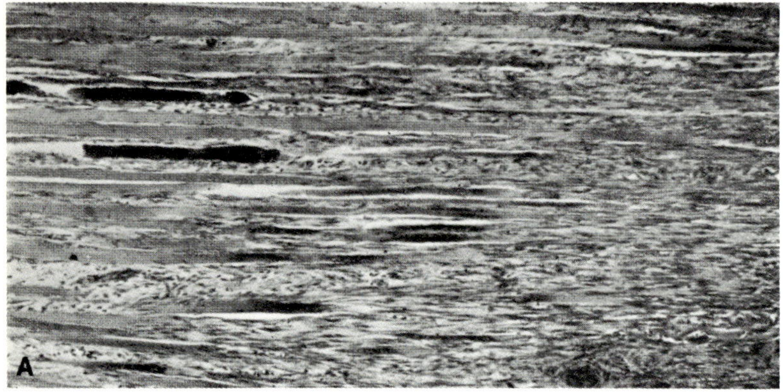

Fig. 3–22. (A) Partially damaged ends of muscle fibers bordering a zone of crushing injury 10 days earlier (at right). Biceps of cat denervated 4 months. Note proliferation of spindle cells and intense basophilia of damaged and partially damaged segments. (B) Various types of degenerating spindle cells near zone of heat coagulation in same muscle shown in Figure 3–21. (methylene blue, eosin)

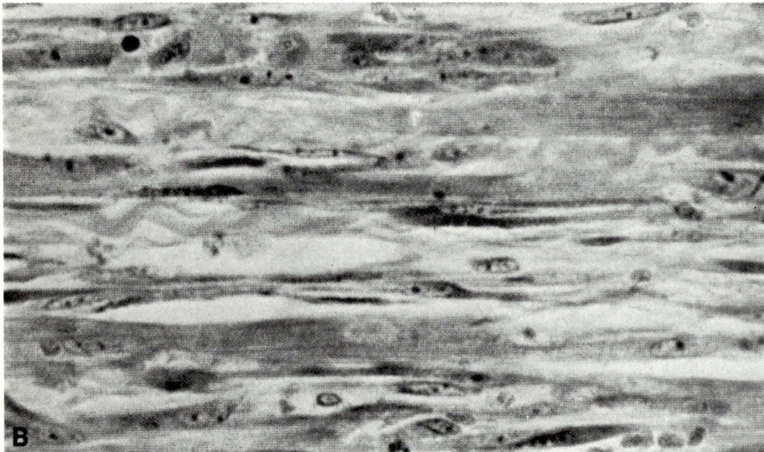

50% larger than that of the histiocyte, and the cytoplasm stains uniformly dark with methylene blue or hematoxylin. Differentiation between myoblast and fibroblast is often more difficult, particularly with methylene blue stain. The fibroblast nucleus tends to be narrower than that of the myoblast, its cytoplasm is amphophilic, and the nucleolus is not as prominent as it is in myoblasts (note those at the top of Figure 3–23A). The processes of the myoblast are more plump, and with the phosphotungstic acid method the cytoplasmic fibrils stream singly, whereas the fibrils of the fibroblast are bifid. Mitosis in the fibroblast is small and insignificant. In shrunken cell remnants all these criteria tend to be obscured, but the fibroblast is more resistant to the types of damage we described.

The reaction of the myofibrils in "neural degeneration" also presents some special features that are accentuated by trauma. In the area bordering coagulation of muscle denervated for 4 months, in which regressive changes and fragmentation are occurring, the least affected muscle fibers show severe loss of their content of myofibrils, compared with undamaged parts of the same muscle. The remaining myofibrils, though staining intensely, are grouped in bundles, often of two or four, and tend to lie peripheral to a zone of afibrillar central cytoplasm that is basophilic with methylene blue (Fig. 3–23B). In such a region most of the muscle nuclei are large and vesicular with prominent nucleoli but poorly developed peripheral nuclear chromatin. Other muscle nuclei in the same fiber are shrunken and square-ended and stain darkly, documenting the essentially degenerative nature of the process. The cytoplasm shows no trace of the protofibrils that precede true regeneration.

If the reparative process is sampled in this way, the failure of the denervated muscle cell

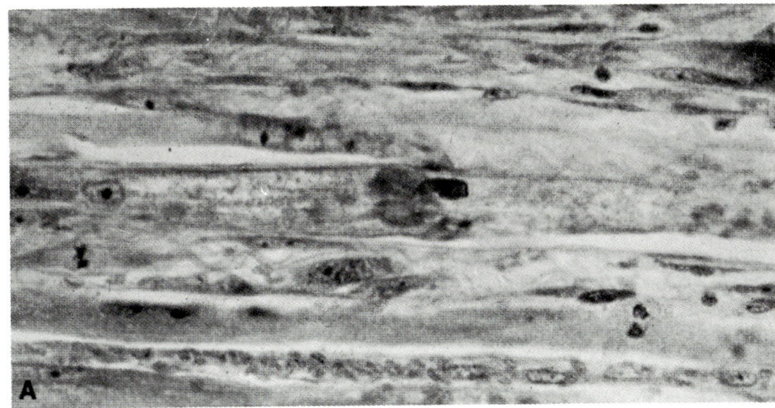

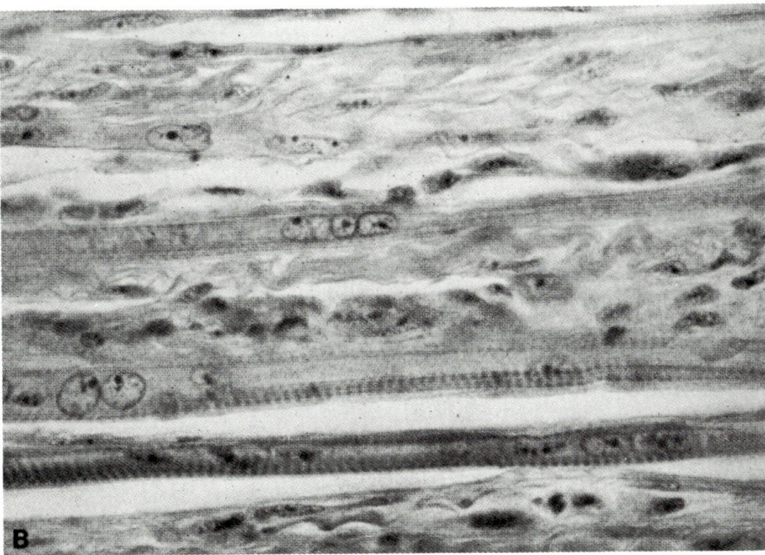

Fig. 3–23. (A) Zone of thermal coagulation on 11th day of biceps muscle denervated for 4 months. Contrast fibroblasts (thin chain in upper part of figure), a spindle cell in mitosis (at left), and a histiocyte showing mitosis (lower right). (B) Various degrees of myofibril loss in degenerating muscle fibers. Note regressive changes in muscle nuclei. (A, methylene blue, eosin; B, phosphotungstic hematoxylin)

to respond to trauma by regeneration is found to be gradual. It is present in some fibers in the domestic cat as early as the fourth week after denervation and is complete after 2 months. In Figure 3–18 the reactions to injury of the normal and denervated fiber are compared in a diagram taken from the article by Denny-Brown.[41]

It is concluded that the fate of denervated muscle fibers is fragmentation and degeneration. Trauma accelerates this process and betrays some defect in the multicellular cohesion of the muscle fiber and failure of regeneration. An essential early part of the defective reaction is nuclear and cytoplasmic proliferation. This phase of the process appears to be similar to, if not identical with, the multiplication that precedes regeneration. Many investigators in-

cluding this author regard nuclear proliferation in muscle as essentially a regenerative process. It is the primary reaction of muscle to injury, and in healthy muscle it leads to full restoration of the fiber. In the denervated one the inevitable result is fragmentation and degeneration. The most important signs of the degenerative process are the nuclear changes.

We view the final stage of neural atrophy as a *dystrophic process* in which some factor supplied to muscle by its junction with the motor nerve fiber is lacking.[40] This factor is essential to the preservation of myofibrils, and its absence is reflected in the gradual loss of tensile structures in the contents of the fiber. Muscular tension, instead of stimulating the production of new myofibrils (as in hypertrophy), appears at least in the late stages to accelerate the dystrophic process. Proliferation of muscle nuclei and increased sarcoplasm, seen early, are explained as only an ineffective response of the muscle fiber to the stimulus of tension or trauma. The nature of the factor supplied by nerve is unknown. There is some evidence[45] that it is acetylcholine, though Miledi[139] and Eccles[53, 54] postulate some other trophic substance such as a peptide specific to nervous tissue. It is not cyclic AMP.

It is remarkable that those who have studied tissue culture preparations of skeletal muscle[82, 83, 155] have found it possible to maintain subcultures of such preparations, many with some degree of development of striated fibrils, for "many months." This is certainly true of muscle fibers from sickly rats.[82] In a state of less clear differentiation, Geiger and Garvin[78] maintained subcultures of adult human muscle for as long as 1 year. In comparable terms of the reaction of denervated muscle in vivo, the *maximal* survival time of some muscle fibers or cells within a denervated muscle is not known, but our own observations indicate that the proportion of surviving cells steadily diminishes with the passage of time. As the months pass, the degree of cell loss increases steadily, and our impression is that this also occurs in tissue culture.

The difficulties surrounding the correct interpretation of the late effects of denervation in the experimental animal are not inconsiderable. One of the most troublesome inconsistencies has been the variability of the findings in different animals and in different muscles within a single animal. All investigators seem agreed that the commonest change, requiring 3–4 months of denervation, is the reduction of many fibers to a variably small diameter with relative increase in sarcolemmal nuclei, some of which are dark. This pathologic picture may persist for 1–2 years, as shown by Sunderland and Ray, [189] and it may be the only change seen in denervated human muscle after severe paralytic poliomyelitis when the muscle is examined several years after the onset of the paralysis. The histologist must be cautioned against the interpretation of clustering of dark sarcolemmal nuclei and short segmentation as fragmentation, for this can result from the tendency of thin fibers to become

tortuous (a wavy fiber cut in the longitudinal plane appears to be segmented). We are not entirely certain that basophilia or actual disintegration of sarcoplasm or loss of muscle fibers can result from simple denervation. In some muscles foci of degenerative change with pyknosis of sarcolemmal nuclei, disintegration of sarcoplasm, and later proliferation of sarcolemmal nuclei and basophilia of sarcoplasm in what might be interpreted as regenerating fibers may be found within 6 weeks (Figs. 3–6 and 3–8). Here it is difficult to determine whether the denervated muscle had later suffered an injury or if there was some intercurrent disease process—so frequently found in all our colonies of experimental animals—that had been superimposed on the denervation atrophy.

CHEMICAL AND NEURAL ACCOMPANIMENTS OF MUSCLE DENERVATION

Physiologists have had some success in isolating and studying the various chemical and neural accompaniment of muscle denervation. Although it is difficult to establish the presence of any change during the first 3–7 days until the fragments of axis cylinder have disappeared from the end-plates, Fishback and Fishback[70, 71] and more recently Goldberg[84] recorded a decrease in the incorporation of amino acids for the synthesis of protein. Tower,[198] who reviewed this subject at some length, states that one of the more striking alterations is the increase of calcium content, which is doubled by the end of the first month of atrophy.[100] The content of chloride also mounts steadily. Potassium, after a slight increase during the first 3 days, may progressively decline in concentration,[66] although this is debated by Eichelberger et al.[57] The phosphorus content declines rapidly, particularly the phosphocreatine (PC) and ATP. The former may reach 30% of the control value after 28 days.[98] Some increase in inorganic phosphate is probably related to the increase in calcium content. Phosphocreatine and glycogen fall very rapidly with the onset of fibrillation, and creatine is slowly and progressively lost after the 15th day. The decrease in phosphate compounds, creatine, and potassium is in proportion to the loss of muscle substance, and the increase in collagen, fat, and chlorides to the relatively greater amount of nonmuscular tissue.[97, 99]

The findings of Eichelberger *et al.*[57] suggest that the chemical events that occur in denervation atrophy are virtually the reverse of those seen during growth of muscle. In growth, the increase in cell solids parallels the increase in numbers of cells, and muscular development is characterized by the displacement of extracellular fluid with an increase in tissue cells. In atrophy there is a decrease in cell solids due to a loss of sarcoplasm, with a corresponding relative increase in connective tissue mass and extracellular fluid. The sarcoplasm that is lost is of normal composition with respect to several electrolytes, and hence that which remains has normal concentrations of these elements. In support of this reasoning, these authors found that after prolonged denervation in puppies the concentrations of potassium and magnesium per unit weight of muscle fiber water were similar to control values.

Hoagland[104] compared the alterations in muscle chemistry which result from motor denervation with disuse atrophy produced by tenotomy or immobilization by plaster cast. He notes that the lactic acid and glutathione content and collagen are increased both in denervated and inactivated muscles. The total protein content, the amount of precipitable myosin, creatine, glycogen, phospholipids, oxygen, glucose, carbon dioxide, and succinodehydrogenase activity are decreased in both groups. There are no significant differences between the values obtained in denervated and inactivated muscles. These changes in the chemical constituents of muscle are believed to be an expression of a relative loss of muscle cell content and a relative increase in the proportion of connective tissue. A decrease in myosin and actomyosin was also reported by Fischer and Ramsey[69] and Fischer.[68] They found that the decrease in these muscle proteins as well as in ATP activity corresponds to the weight loss of muscle during denervation atrophy.

The first demonstrable loss in glucose and PC coincides with the onset of fibrillation, on the fifth or sixth day after denervation. The change may begin as early as the third day in some muscles of certain animals, e.g., the rat. The spontaneous rhythmic contractions occur at first only in isolated muscle fibers, but within a few days all muscle fibers show independent rhythmic contractions at rates of 2–120/min. The resulting quivering activity of the muscle can be clearly seen when its surface is viewed by reflected light. The muscle as a whole does not shorten owing to the lack of synchronism in the twitching of the various fibers. Further, the activity is curiously variable from one time of day to another and may be intense in one part of a muscle and absent for brief periods in another. Such twitching is identical with that observed in growing muscle tissue in tissue culture[126, 127] and appears to be an inherent property of isolated skeletal and cardiac muscle fibers.

Fibrillation is stimulated by exposure to air or cold but continues to occur in the absence of such extraneous stimuli. It is always present in some degree in the atrophic tongue (after peripheral lesions of the hypoglossal nerve, for example) and can be then directly observed through the thin mucous membrane (Chapter 2). Fibrillation therefore is but an index of the increase in excitability of the muscle fiber following denervation. If sensory nerve fibers are intact in the atrophic muscle, antidromic impulses in these can cause a small maintained contraction (Vulpian-Sherrington phenomenon) presumably due to generation of a small amount of acetylocholine.

Langley and Kato[123] and later Tower[196] ascribed the reduction in muscle fiber size to the continued fibrillary activity, which they believe causes depletion of glycogen and other materials normally stored in muscle. A contrary view was presented by Solandt,[182] who observed that the rate of atrophy in both denervated muscle and inactivated muscle with intact nerve supply was of the same speed and magnitude during the first 10 days, even though fibrillations were present only in the former. After 10 days the denervated muscle continued to atrophy, whereas muscle inactivated by splinting or severance of the tendon began to regain some of its weight and reactivity. In an earlier experiment Solandt and Magladery[183] also succeeded in arresting the fibrillation of denervated muscle in rats by the use of quinidine without preventing muscle atrophy, but these experiments were less convincing because the animals were sick and lost weight. There is some evidence indicating that lack of tension on the muscle fiber is in some way responsible for atrophy.[97]

Histochemical studies reveal other aspects of denervated muscle.[9, 63, 170] The three identifiable histochemical fiber types (Chapter 2) revert to an embryonic undifferentiated state within a few weeks of the onset of denervation. Thus white, fast-contracting muscles rich in type II phosphorylase fibers lose their heavy content of this enzyme and come to react like the weakly staining oxidative-rich type I fibers. In red muscle the converse occurs, the oxidative-rich type I fibers become indistinguishable from type II fibers with respect to these enzymes. In states of partial denervation intact motor nerve fibers, by collateral regeneration, adopt some of the denervated muscle fibers and convey to them their own capacity for slow or quick muscle contractions and their predominantly oxidative or phosphorylative metabolism. The normal checkerboard pattern is then replaced by large aggregates of fibers of uniform histochemical staining.

EFFECTS OF ELECTRICAL STIMULATION ON DENERVATIVE ATROPHY

Many attempts have been made to determine the value, if any, of repeated electrical stimulation of a muscle undergoing denervation atrophy. It appears that regularly repeated galvanic stimulation delays the loss of weight of such a muscle as compared with a control.[88, 101] The treated muscle contained more creatine and glycogen than the controls but had the same water content. The process of atrophy was, however, not prevented, and after the first 4 weeks the treated muscles had lost approximately as much weight as the controls. The speed of functional recovery following reinnervation was not significantly changed. It has also been demonstrated that fasting and inanition do not appreciably alter the rate of denervation atrophy.[102]

FIBROUS CONTRACTURE

The final phase of so-called fibrous contracture* is of such uncertain origin that we do not include it in the process of neural atrophy. In some denervated limbs after several years the muscles are reduced to areolar tissue with small isolated muscle fibers and chains of nuclear remnants. In other muscles the remaining fibers are embedded in dense fibrous tissue. In this case the joint concerned is usually greatly limited in movement, and the fibrous change is found in the group of muscles most shortened in such fixed "contracture." It appears that the supervention of contracture is an independent process that favors the appearance of fibrous tissue. Fibrous contracture therefore does not necessarily follow neural atrophy. Some muscle fibers may remain intact even in dense scar tissue,[19, 47, 196] and at least partial return of function may occur after reinnervation some years later.

REINNERVATION OF MUSCLE AFTER NERVE INJURY

The reinnervation of muscle after nerve injury was investigated by Gutmann and Young.[89] They studied the motor end-plates of muscle fibers of rabbits after denervation and during the process of regeneration. It was noted that after a nerve was interrupted by a crush close to the muscle, one nerve fiber returned to each motor plate. The axonal tip advanced at a rate of 4.4 mm/day. The nerve fiber was very thin and branched just before reaching the end-plate. There was a delay of not more than 5–7 days between arrival of the nerve fiber at the surface of the muscle fiber and its penetration into the end-plate. The end-plates on the denervated muscle fiber remained intact, even

---

* A better term is pseudocontracture to distinguish it from true contracture.

after severe atrophy, and there was no increase in the nuclei of the end-plate.

Reinnervation of muscle fibers after longer periods of atrophy (e.g., when the nerve was crushed at a distance from the muscle) was less complete. Some of the regenerating axons failed to reach the motor endplates, and new end-plates were formed where these fine fibers contacted the muscle fiber. After further section and resuturing of the nerve, reinnervation was still less complete. In this latter instance many small fibers, some sensory or sympathetic, entered the muscle and formed elaborate networks among the muscle fibers.

The number of nerve endings entering the motor end-plates was decreased during reinnervation but returned to normal about 70 days after crush of a distal nerve and about 100 or more days after the crush of a proximal one. After severance of the nerve and resuture, many of the motor nerve endings on the muscle fibers had not been established at the end of a year. The delay between the return of the motor nerve fiber and the resumption of reflex function was about 11 days when the crush was close to the muscle and as much as 25 days after severance close to the muscle. The delay was even greater if the muscle had undergone considerable atrophy. It was concluded that the success of reinnervation is prejudiced by (1) considerable atrophy, i.e., new motor end-plates are formed slowly; (2) abnormal patterns of reinnervation with many wrong connections; and (3) failure of many fibers to receive motor nerves.

In Chapter 2 we mentioned the phenomenon described independently by Hoffman and by Edds whereby surviving subterminal axons of motor nerve fibers begin to sprout and form new endings on denervated fibers following partial denervation of a muscle. Edstrom and Kugelberg[56] find that the muscle fibers adopted by the regenerating axons occupy small groups corresponding to the subunits of Buchthal. Later denervation of these subunits is responsible for the familiar pathologic picture of "group atrophy" secondary to nerve or spinal denervation. The large numbers of fine branches given off by a regenerating axon as it reaches denervated muscle fibers are further reflections of this neurobiotaxis, or attraction

phenomenon. In spreads of end-plates obtained by the method of Coërs,[32] this multiple branching of regenerated subterminal nerve fibers is very characteristic and can be used as a diagnostic criterion. The extent of enlargement of regenerated motor units resulting from such collateral branching may be considerable, as shown by the studies of Wohlfart[219] in amyotrophic lateral sclerosis, poliomyelitis, and related conditions. Some biologic limit is soon reached, however, for otherwise all partially denervated muscle fibers would be collaterally reinnervated.

After the regenerating axons reach the muscle fibers, atrophy no longer progresses and some degree of muscle function is restored. A gradual recovery of muscle mass and strength of contraction follows. Usually the bulk and strength of the reinnervated muscle are less than normal because some of the nerve fibers fail to establish contact with the muscle fibers.[46] These remain atrophic and eventually degenerate. Fibrillation ceases when the muscle fiber is reinnervated. There is then an increased content of phosphates and creatine, and the glycogen content is restored to normal.

DISUSE ATROPHY

Atrophy that results from inactivation of a muscle whose nerve supply is intact can be produced in several ways: (1) skeletal fixation or application of a cast to restrict activity; (2) cutting the tendon of a muscle; (3) an interruption of all the neural pathways by which muscle contraction, either reflex or volitional, is produced, but leaving the motor neurons intact. In the latter method, utilized by Tower,[197, 198] the spinal cord is transected at L_1, and all the posterior lumbar and sacral roots are sectioned. In this way the distal or lumbar stump of the cord is isolated from all afferent and all descending nerve impulses. After this operation the anterior horn cells of the spinal cord show little or no change. The muscles supplied by these isolated cord segments are paralyzed and completely flaccid with only rare, noncoordinated bursts of activity, usually associated with handling of the animal (traction on the spine) or the application of pressure over the isolated segment of the cord.

Eccles[52] observed a loss of 40% of the muscle bulk 3 weeks after inactivation of the leg muscles of cats by the above method of isolating the lumbosacral cord segments. No fibrillation was noted, and there was no definite change in the sarcolemmal membrane characteristics.[21] (See also the comparative chemical studies of skeletal muscle after neurotomy and tenotomy done by Hummoller *et al.*[108] and McMinn and Vrbová.[135] The atrophy could be prevented for the most part by electrical stimulation of the nerves supplying the muscles for as brief a period as 10 sec/day. The importance of maintaining muscle tension and activity in preventing muscular atrophy have been stressed by Eccles[52, 53] and Knowlton and Hines.[116] Chor *et al.*[28] applied leg casts to monkeys for variable periods of time and then determined the weight loss of these muscles as compared to controls. They found 12.8% weight loss in 2 weeks and 29.8% in 6 weeks, the limits of their experiments. These investigators also analyzed the wasted muscle and found no change in the relative amounts of water, nitrogen, and protein.

Ferguson *et al.*[67] observed the effects of applying casts to hind limbs in the rabbit. The limbs were immobilized in various positions. When fixed in a relaxed position the anterior tibial muscle gradually declined in weight whereas that of the gastrocnemius rapidly decreased. When placed in the stretched but immobilized position, the anterior tibial muscle responded with a significant (up to 65%) degree of hypertrophy. Hence tension on the muscle, even in the disuse state, appeared to protect against atrophy, at least in the anterior tibial muscle of the rabbit.

The rate of atrophy of a muscle after severance of its tendon is about equal to that following skeletal fixation; but it is slower, especially after the first 7–10 days, than that following denervation. Eccles[53] found that the ratio of the maximum tetanic contraction to the maximum twitch contraction was more impaired after tenotomy than in isolation atrophy, an indication that the contractile mechanism is more gravely affected by tenotomy than isolation

atrophy. The muscle continues to receive nerve impulses, and the significant factor is loss of tension.

Information regarding the histologic changes occurring during disuse atrophy of muscles has been supplied by the experiments of Tower,[198] Reid[163] and Brooks.[21] The essential change was a gradual reduction in size of all the muscle fibers. The sarcoplasm and myofibrils were both diminished, but the longitudinal and cross striations were preserved until very late. There was no striking increase in the size and number of the sarcolemmal nuclei, as was observed after denervation of muscle, and the elements of the motor end-plates were among the last rather than the first to disappear. These latter elements were often present as dilated fusiform structures in markedly narrowed and atrophic muscle fibers. There were none of the acute vacuolar and granular degenerative changes which Tower had observed in denervated muscle. Slow conversion into fibrous tissue, the so-called fibrotic dedifferentiation, and a proliferation of the fibrous interstitial tissue is a disputed finding we are not prepared to accept.

HYPERTROPHY

In the experimental animal hypertrophy can be induced easily by severing the fibers of part of a muscle or by cutting the tendon of one of a synergistic group of muscles. This requires the remaining fibers or muscles to carry the full burden of work, and in order to meet the demand they begin to enlarge within a day or two. Concomitantly there is a prompt increase in the rate of assimilation of amino acids and synthesis of contractile proteins.[84]

The stimulus to this type of *work* or *use hypertrophy* is the excess activity of the muscle fibers.* Controlled exercise of the muscles of the laboratory animal allows close analysis of the exact physiologic conditions which give rise to it. As to the principle derived from

---

* Stretching alone is effective, as shown by the experiments of Sola *et al.*[181] who induced hypertrophy even in recently denervated muscle.

such studies hypertrophy occurs only when a muscle contracts with a force greater than that to which it is accustomed. The contraction must be near the maximal power in a given unit of time. If, instead, the resistance against which the muscle works remains the same but the muscle is required to perform only for a longer time, no hypertrophy results. Sieber and Petow[179] confirmed this principle (that hypertrophy is related to the intensity and not the duration of effort), and in human muscle Müller and Hettinger[147] showed it to be applicable. In quantitative terms they found that not until exercise involves a force of contraction in excess of 70% of the maximal capacity did hypertrophy ensue. Interestingly, they determined that the maximal strength of contraction of a muscle is approximately three times the tension produced in everyday activity. By a rigid training schedule they observed that the greatest effort, produced for as brief a period as 6 sec/day, incited hypertrophy and permitted an increase in muscle strength of about 5% per week. Further activity had no greater effect in enlarging the muscle. The threshold stimulus was said to be a temporary oxygen deficit.

The histologic changes of work or exercise hypertrophy were precisely described by Morpurgo of Siena[144, 145] in 1897. In a study of the sartorius muscles of two dogs before and after running in exercise wheels for a 2-month period (for distances of 3218 and 1550 kilometers, respectively) the hypertrophy amounted to slightly more than 50% (53% and 55% to be exact). The number of fibers remained the same. The enlargement was due to an increase in the diameters of the individual fibers (Fig. 12–14). Each enlarged fiber contained a greater number of myofibrils. Denny-Brown[40–42] confirmed this morphologic finding by exercising cats whose gastrocnemius tendon had been severed. In the overworked soleus the number of myofibrils increased from 1679 to 2063, an increase of 500–600 over the control muscle. Isometric contraction has the same effect, according to Müeller.[146] Müeller and Hettinger[147] also observed an increased number of central nuclei in the enlarged fibers. The number of functioning capillaries around the enlarged fibers also increased, contributing to the turgor of the well conditioned muscle.

The ultrastructural changes appearing in the course of hypertrophy have never been completely described. Apart from an increase in the number of myofibrils the composition of each myofibril is probably quantitatively increased by the addition of myofilaments peripherally. Also the interfibrillar substance increases. Mitochondria are enlarged and more numerous. The arrangements of transverse and longitudinal endoplasmic reticulum remain constant. The differentiation of histochemical fiber types becomes less distinct.

Forms of hypertrophy other than those due to exercise have also been investigated in the experimental animal. Papanicolau and Falk[151] observed that the temporal and other muscles of the adult male guinea pig are larger than those of the female. If, however, the male is castrated at puberty his temporal muscles remain small and flat, like those of the female; or if the female is given injections of adequate amounts of androgenic hormone over a period of several weeks her temporal muscles become round and protuberant, like those of the adult male. If the gonads of male and female are removed before puberty, gonadotropic hormone is ineffective whereas testosterone (the secretory product of interstitial cells) produces hypertrophy in both castrated and spayed animals. Thus it appears that the androgenic hormone accounts for the superior size and strength of the adult male muscle, at least in the guinea pig. In this animal overuse may explain part of the difference, for the temporal muscles are engaged in masculine mimetic movement (chewing forms one part of the expressive movements of the male of the species).

Striking confirmation of the capacity of muscle to increase in size as a result of endocrine influence was obtained by Bigland and Jehring,[16] who studied growth hormone. The muscle fibers increase in size, presumably with increased content of myofibrils and storage of protein but without change in their contractile properties.

Not all workers are agreed that exercise hypertrophy is purely a matter of volumetric increase in individual fibers. On the basis of human biopsy and postmortem studies, some believe there is also a numerical increase in muscle fibers (Chapter 4).

GENERAL REACTIONS OF NORMAL MUSCLE TO INJURY

A variety of histologic changes is seen in muscle as a result of various types of injury, including vacuolation, hyaline changes with and without fracture of fiber into disks, and waxy degeneration. The extent to which these represent specific types of change, or alternatively merely various degrees of the same type of change, and the question as to the point at which damage inflicted upon the muscle fiber becomes irreversible can be answered only by experimental methods. After Zenker's original description[223] in 1864 of the degenerative changes in muscle in fatal cases of typhoid fever, there were numerous attempts to reproduce such changes in experimental animals by Waldeyer,[205] Weber,[211] Tschainski,[200] Gussenbauer,[87] Kraske,[118] Volkmann,[204] and others. More recently Forbus,[74, 75] Clark,[30] and Reznik[166] reviewed the subject.

Degeneration of muscle fibers can be produced by a variety of physical agents. Fishback and Fishback [70, 71] compared the effects of direct trauma, vascular occlusion, chemical and bacterial agents, and parasitic infection. The types of injury which resulted from exposure to these agents varied from minimal cloudy swelling of the sarcoplasm to total necrosis of the muscle fiber. These authors believe that an orderly series of changes can be observed in muscle undergoing necrosis regardless of the cause. The changes proceed from a slight granularity and swelling of the fiber, to edema and vacuolation, and finally to a form of waxy swelling and segmental fragmentation similar to that of Zenker's degeneration. Although, in general, we are in agreement with their conclusions, there are many complicating factors that cause considerable diversity of histologic appearance of the lesions and also materially affect the final outcome. To understand these, we need to inquire more deeply into the mechanism of damage. Necrosis of the muscle

fiber with preservation of a well marked refractile striation and a tendency of the fibers to fracture transversely across the Z line, forming disks or short segments, is common in ischemic lesions or after the injection of chemical substances into muscle. The application of extreme heat or cold or other physical trauma commonly may induce waxy or hyaline fragmentation or else a shredded appearance of the contents of the fibers. Subsequent osmotic and autolytic changes result in a floccular appearance of the contents of the fiber.

TOTAL NECROSIS WITH PRESERVATION OF STRIATION

In complete necrosis caused by arrest of the arterial blood supply, Clark[30] found the general form of the muscle fibers and supporting tissues to be fairly well preserved with recognizable cross striation, though sarcolemmal nuclei had disappeared and the whole muscle fiber had a diffuse acidophilic staining property. Fragmentation of the muscle fibers across the transverse striation and the formation of disks was generalized, but Clark raised the question as to whether this was due to the action of the fixative used. The same changes were seen in the center of muscle grafts. Such fragmentation surely occurred during life, for at the edges of the damaged region similar fragments were surrounded by phagocytes. Within a few days after production of the lesion the whole necrotic area had become edematous with sparse infiltration of neutrophilic leukocytes and histiocytes. The muscle fibers were swollen and the cross striations less evident. Vacuolation was not apparent, and the fragments of fibers gradually shrank and became more widely separated. They remained in this state until they were reached by the advancing zone of activated histiocytes and fibroblasts. If necrosis was caused by injection of a fixative such as alcohol into muscle, as in the experiments of Forbus,[74] the muscle fibers acquired a glassy appearance with highly refractile but unstaining striations even when the sarcolemmal nuclei escaped destruction. The same fragmentation and other changes proceeded as outlined above.

The relation of these changes to the chemistry of the affected muscle fibers was studied by Harmon and Gwinn.[91] With the use of ischemia caused by application of a tourniquet to the leg of rabbits and rats, it was found that the power of muscular contraction was greatly impaired after 4 hours of circulatory arrest and totally lost after 6 hours. Contractility returned after release of the tourniquet. After 8 hours of ischemia there was recovery of only 50% of contractility within the following 24 hours.

The corresponding histologic findings were as follows. After only 2 hours of ischemia the transverse striation of many fibers was found to be accentuated, and after 4 hours some fibers were beginning to show fracture into disks at the Z line in vitro, as described by Bowman,[20] with preservation of striation. After 12 hours of circulatory arrest, 90% of fibers showed this "discoid degeneration." The number of fibers presenting hyaline change and discoid fracture correlated with failure of recovery of contractility. If the ischemia was relieved and survival allowed for a day or more, Zenker's type of hyaline degeneration with homogeneous, structureless, faintly acidophilic change in the sarcoplasm was observed. The muscle nuclei were pyknotic, and the striations had disappeared. The granular necrosis described by Fishback, with deeply acidophilic cytoplasm and embedded dark nuclei, was also common. Neither the hyaline nor granular type of change was seen unless the circulation had been allowed to return. In some parts of the muscle Bowman's hyaline discoid change remained for a long period in spite of renewed circulation. The waxy and granular types of necrosis with loss of striation therefore require some autolysis of the fiber contents and are secondary to the initial coagulation.

The lactic acid content of histologically altered muscle was found by Harmon and Gwinn[91] to be raised during the first 24 hours in all types of change, indicating that glycolysis had occurred. In addition, severe damage to the biochemical mechanism was produced by ischemia before any recognizable histologic change could be demonstrated. ATP, glycogen, and PC were completely lost and returned only to about one-third of control values 24 hours after release from a relatively brief occlusion of the circulation. Degeneration and necrosis therefore represent a more serious disintegration than mere loss of metabolic reserves. As Harman and Gwinn observed, death of the fiber is not established until structural collapse occurs. Whether this corresponds with the destruction of all enzyme systems has not been determined; but even after death of the fiber alternating change in refraction of light indicates that some semblance of striations persists in the clotted framework. The stainability of the Q and Z bands, on the other hand, is reduced in proportion to the disorder of chemistry. Waxy or hyaline change must represent a further, delayed breakdown of the fundamental architecture of the myofibrils. Evidence shows that granular change, in the absence of waxy or hyaline transformation, may characterize milder and reversible degrees of damage.

We may conclude that the hyaline preservation of segments of fibers exposed to a protoplasmic precipitant such as alcohol is none other than the process of tissue fixation. If, in the course of a disease process necrosis is massive, there is a period when the destroyed fibers have not yet been exposed to autolysis and phagocytosis. This is the stage known as Zenker's hyaline degeneration. The myofibrils cease to stain except for an eosinophilic reaction and yet retain a highly refractile striation; this supposedly is due to lysis of some of the more labile basophilic features of the mechanism, leaving an intact structural myofibrillar framework. Following damage by ischemia, the striation is rapidly obscured. Early swelling and blurring of the structure, giving a waxy appearance, are then manifestations of a digestion process. An extension of the same process results in a floccular appearance. Both these methods of inducing damage affect the contents of all the muscle fibers in the region of the injury. In a more severe degree of both types of injury the sarcolemma and sarcolemmal nuclei as well as the contents of the fiber undergo necrosis. The contents of the fiber evidently become brittle and break transversely at numerous points, often across a Z line, just as a muscle fiber treated with strong acid or alkali in vitro breaks into the disks of Bowman. The waxy segments of varying length then remain unchanged until eroded by phagocytes.

TOTAL NECROSIS WITH IRREGULAR TEARING OR SHREDDING OF CONTENTS OF FIBER

EFFECT OF STIMULATION

Volkmann[203] found that in necrosis following the application of extreme heat or cold the contents of the muscle fibers presented irregular loss of structure, with obscuration of striations, which we call the "shredded" appearance. An explanation for the occurrence of this appearance was found by Nageotte.[150] He first showed that very strong faradization of a piece of excised muscle caused it to shorten to 60% of its resting length, and to become rigid in the state of "extreme contraction" (Chapter 2). If the stimulus was continued while the muscle was placed in fixative, sections stained with hematoxylin showed a series of branched, irregular dark streaks transversing the muscle fiber, interlacing with one another and separated by clear areas in which the Q lines were visible. If the stimulus was stopped before fixation, the muscle softened but remained shortened. Fixation and hematoxylin staining of the muscle in this state revealed the same interlacing bands, but they were no longer uniformly dark. They showed narrow Z lines very close together, as in contraction bands. Such dark zones had the same irregular interlacing arrangement as in the former condition but were even broader. These dark zones were anisotropic.

If the muscle was stretched after stimulation, it could gradually be brought back to its original length. Renewed stimulation could again induce the state of "extreme contraction," following which the muscle could be extended and again contracted a number of times before complete failure of the contractile mechanism occurred. In the extended state fixation and sectioning of the muscle revealed no abnormality. The state of extreme contraction was therefore a reversible physiologic phenomenon by which muscle freed from its attachments could be thrown into an abnormal degree of shortening. The meaning of the irregular pattern of the more dense bands of contraction is unknown, but it probably represents uneven permeability of the muscle membrane under these conditions of extreme excitation.

The relevance of the irregular pattern of extreme contraction in the present connection is that Nageotte proceeded to show that if a piece of muscle, freed from its attachments, is frozen and then fixed and sectioned, it shows exactly the same pattern of irregular, darkly stained anisotropic bands as does muscle in the state of extreme contraction. The mechanism of production of these bands was found to be a sudden extreme contraction of the muscle at the moment of thawing. In this case, however, the condition was irreversible and the muscle was necrotic. Irreparable damage had occurred to the vital mechanism of the muscle fiber. In such extreme contraction a large amount of lactic acid is liberated, but this also occurs in the reversible type produced by faradic current. The mechanism of irreversible damage must be an independent phenomenon. Nageotte showed that the same irreversible change occurred when a fragment of muscle was heated or fixed by certain volatile anesthetics such as chloroform or ether.

The irregular lacework of Nageotte is occasionally seen following fixation of a freshly excised, irritable piece of normal muscle in fixatives such as Zenker's solution. An example of fibers in this state as a result of immersion in a Cajal fixative containing ammonia for the study of nerve endings is shown in Figure 3–24. In the upper fiber in that figure, the light bands no longer show striation, having stretched to the breaking point. When extreme cold or heat is applied to a portion of a muscle in situ through an incision in the skin and subsequent survival of the animal is allowed for varying periods, the necrotic muscle fibers show irregular transverse bands of dense material separated from each other by lighter areas—the "shredded" appearance to which we referred. The denser areas have the same pattern as the dark bands seen by Nageotte following freezing but have torn away from each other owing to the inability of the muscle to shorten beyond 70% of its resting length without such damage. We therefore regard the shredded appearance as the result of necrosis in a state of extreme contraction.

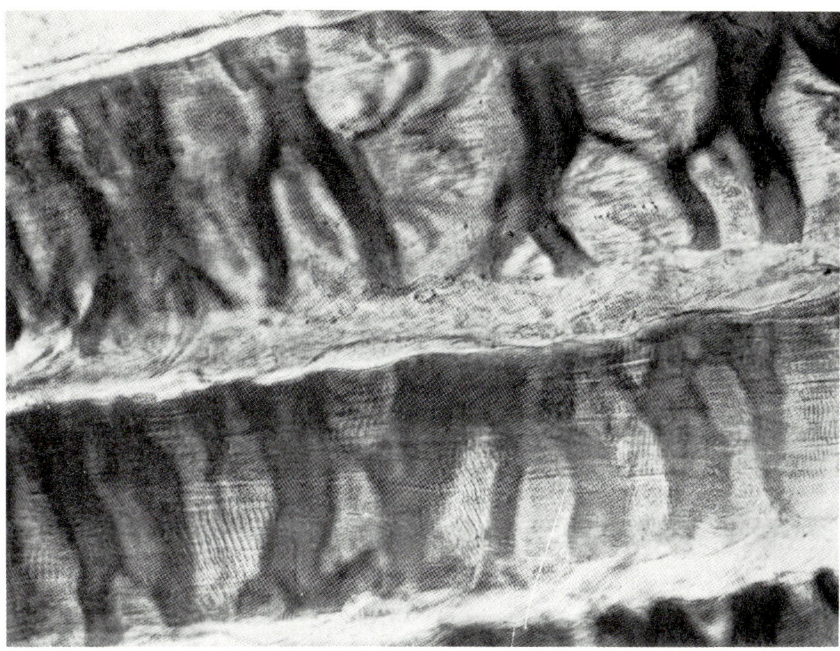

Fig. 3–24. "Shredded" appearance of muscle fixed in alcohol while still contractile. Fixative was Cajal's ammoniated alcohol.

The reaction of living muscle to types of injury involving excitation of the' muscle fibers was observed in great detail by Speidel[184, 185] in the tail of the tadpole. Whereas the types of reaction we discussed above were massive reactions in a whole segment of muscle, the observations of Speidel relate to the reactions of single muscle fibers and clearly explain phenomena commonly observed in isolated fibers in pathologic muscle, as well as some kinds of fixation artifact. We already mentioned the contraction band that results from stimulation of a localized region of the contractile contents of the fiber. A localized injury to a muscle fiber, such as a needle prick or local application of heat, sets up such a band, consisting of the close contraction of 2–50 sarcomeres.

If the injury was sufficient Speidel observed that the band remained fixed for hours or days. The consistency of this part of the fiber increased, and the striation was obscured. Upon closer study it could be seen that the sarcoplasmic constituents had clotted. Further injury resulted in the addition of more and more sarcomeres to the clot, with consequent stretching and ultimate rupture of the remainder of the contents of the fiber. A clotted contraction band (retraction cap) formed on the ends at the point of rupture and likewise absorbed a number of neighboring sarcomeres. If the injury was severe, multiple clots and retraction caps formed on the same fiber, and still more violent injury reduced it to a series of clots with no remaining normal striate substance. Such clots were produced by electric current, cutting, burning, extreme cold, ischemia, and various chemicals. They could remain localized or could progressively involve less damaged parts of the fiber, according to the nature of the injury. They could not be induced in embryonic fibers until the striate contractile substance had developed and obviously involved the myofibrillar substance, often pulling the columns of myofibrils out of alignment. Under some circumstances a "thin clot" was observed, which appeared to be identical with the reversible type of contraction described by Nageotte. Isolated clots were removed by phagocytosis or under-

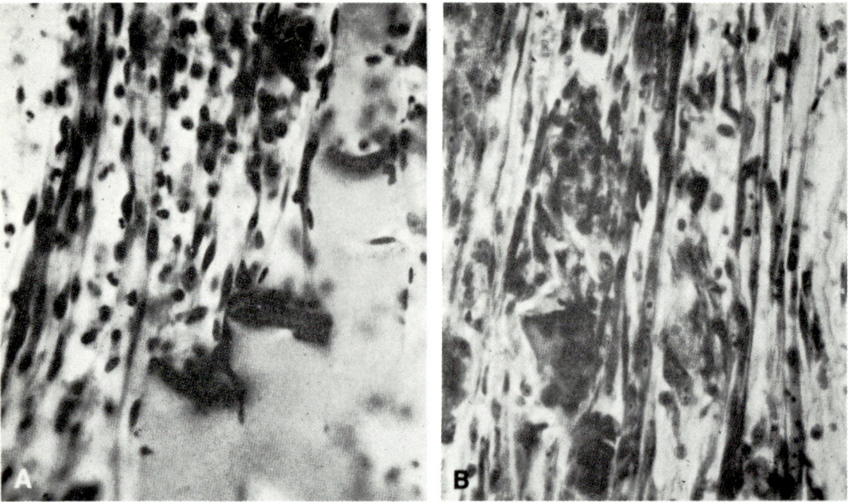

Fig. 3–25. (A) Gracilis muscle of cat showing the ends of the damaged fibers following heat coagulation of a strip of muscle by application of a heated metal bar 4 mm wide at 100° C for 1 min 4 days earlier. Tendon of muscle had been sectioned just before application of heat. (B) The same procedure was followed but tendon was left intact. Dark retraction caps have been torn off the waxy portion which lies beyond the field shown.

went autolysis, small isolated ones sometimes disappearing completely in as short a period as 24 hours. The process of formation and removal of clots was entirely independent of the fate of the sarcolemma and the muscle nuclei. If the latter survived, they were concerned in the regeneration of a new fiber but played no part in the formation or removal of clots.

We conclude therefore that necrosis of muscle fibers in the form of irregular transverse clots, either in isolated bands or generalized so as to present a shredded appearance with disruption of the internal structure of the fiber, is the result of extreme contraction at the moment of death of the affected portion of the fiber. A waxy change with fragmentation is then a secondary phenomenon and follows all types of massive necrosis. The importance of natural tension within the muscle in causing fragmentation is seen in the difference between the edge of the necrotic lesion in a muscle in which the tendon had been cut at the time of injury, and in one in which the tendon remained intact.[40] In the former case (Fig. 3–25A) the necrotic fiber ends in a waxy re-

gion with a densely staining retraction cap. If the tendon is intact and the muscle allowed to be used, the retraction cap is torn off (Fig. 3–25B), and pulled as much as 1 mm away from the nearest zone of waxy degeneration, which is broken into short segments, each widely separated from the next.

In any type of necrosis portions of the sarcolemma may survive; but usually it reacts as an elastic membrane, stretching or tearing according to the mechanism of the damage. The muscle nuclei rapidly become pyknotic and lose their staining quality unless they are adherent to a portion of surviving sarcolemma. In the latter case the nuclei begin to proliferate on the 2nd day after injury and, still attached to the sarcolemma, become separated from the necrotic fiber contents. Their subsequent behavior is discussed later in relation to regeneration. The necrotic contents commonly show patchy but diffuse staining with hematoxylin or acid fuchsin during the first 2–3 days, probably owing to their exposure to lactic acid at the time of necrosis. Their pale color when the necrosis is fresh is also probably due to the

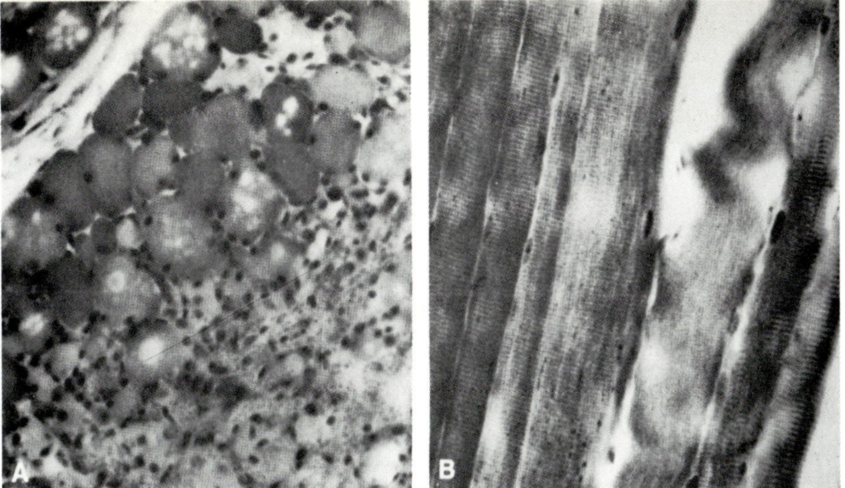

Fig. 3–26. (A) Border of area of heat coagulation showing vacuolation of partially damaged muscle fibers on 4th day. (B) Same tissue showing granular change and separation of columns of myofibrils. The coagulated zone (not shown) is to the right in each figure. (A, silver impregnation; B, Gros-Bielschowsky)

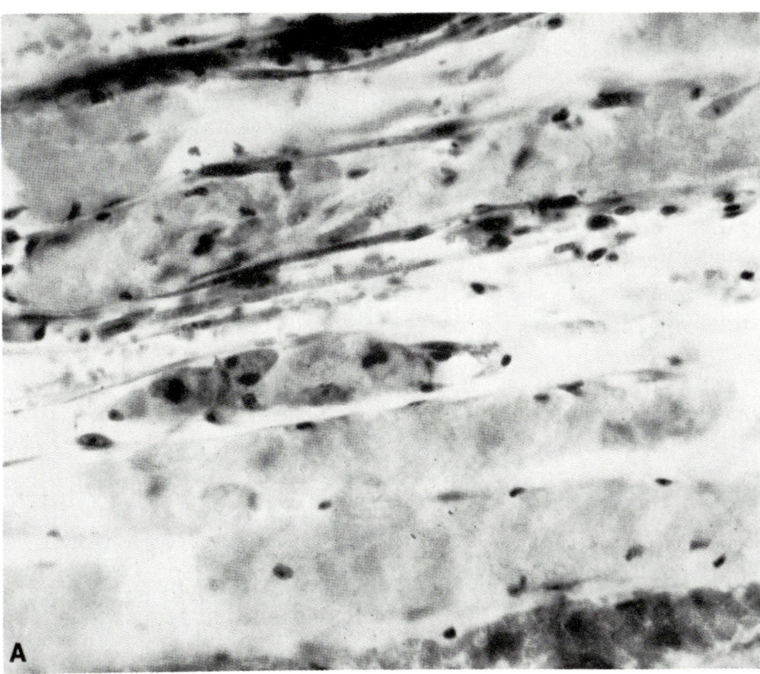

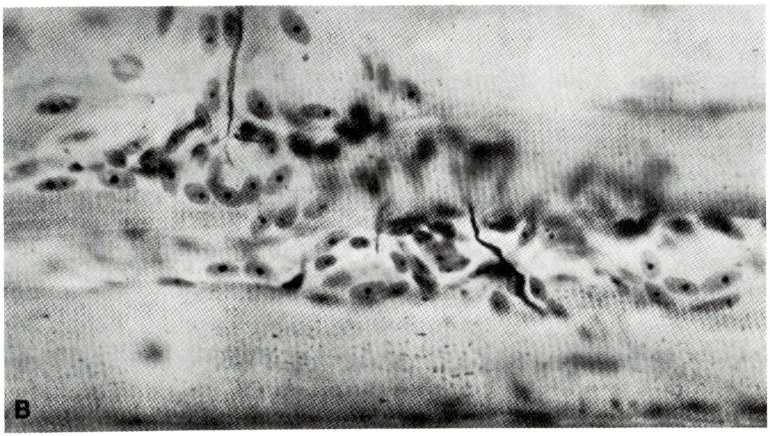

Fig. 3–27. (A) Heat coagulation on 4th day showing edge of necrotic segment where sarcolemmal nuclei and endomysium have survived. (B) Two motor end-plates bordering the zone of necrosis. Note the sarcoplasmic swelling, poor staining of telodendria, and vacuolation. (Gros-Bielschowsky)

bleaching action of lactic acid. Thereafter they remain faintly eosinophilic and floccular in appearance (Figs. 3–31B and 3–33).

TYPES OF PARTIAL INJURY AND REPARATIVE REACTIONS

THERMAL INJURY

The leisons discussed in the preceding sections are undoubtedly related to necrosis; here we are concerned with the cellular reactions to destruction of only segments of fibers (segmental necrosis) and the reparative reactions thereto in the intact parts of the fiber. We used heat as a convenient means of producing these changes, but cold is equally effective. Similar changes may be observed after other modes of physical and chemical injury.

Application of a metal bar 2–3 mm in width and at a temperature of more than 85° C immediately induces "shredded necrosis" in a zone of the muscle fibers. Irregular, transverse markings in an otherwise diffusely acidophilic sarcoplasm are all that remain of the natural striations (Fig. 3–26). After 24 hours the contents of the necrotic fibers present a cloudy "flocculated" appearance and are then only faintly acidophilic (Fig. 3–27). In the center of the zone all the sarcolemmal nuclei are shrunken and pyknotic, whereas those at the edge may reveal viability by their normal staining reactions (Fig. 3–27A). In many of the affected fibers there is a rupture of sarcoplasm between necrotic and intact segments with retraction of the healthy part. Often the sarcolemma remains intact and simply collapses in the region of rupture. The retracted segment tends to stain more deeply with hematoxylin, and its end is sometimes swollen but rarely vacuolated. The striation in this region is preserved but stains less clearly than that of the normal parts of the fiber, and the sarcoplasm is faintly granular. Where the fiber contents remain in continuity with the necrotic region, the junctional segment exhibits a hyaline swelling of the type shown in Figure 3–25A. Proliferation of sarcolemmal nuclei begins within 3 days, and many then migrate to the center of the fiber.

Lesser degrees of damage are noted in adjacent muscle fibers. Commonly such fibers exhibit an irregular vacuolation, giving an appearance of longitudinal splitting (Fig. 3–26A). All muscle fibers bordering the coagulated zone have an increased bulk that is evident as early as 24 hours after injury, and the swelling progresses for the first 4–6 days. By the fifth day these partially affected fibers begin to diminish in size, soon to less than one-fifth their normal diameter; those 2 mm away lose half their bulk. The loss of sarcous substance appears to be mainly at the expense of myofibrillar mass (Figs. 3–28 and 3–29). Some such fibers contain only a few myofibrils by the fifth day, giving them a fetal appearance (Fig. 3–32A). In other fibers just peripheral to the necrotic area the preserved striate substance stains poorly, and the intact nuclei appear to migrate into vacuoles by the sixth day. An even lesser degree of injury is reflected in a more evident longitudinal striation due to separation of the constituent myofibrils. In silver impregnation stains the sarcoplasm of such slightly damaged fibers has become highly granular (Fig. 3–29B), and in places the natural argentophilic pattern of the myofibrils is diminished or lost. Instead, a diffuse argentophilia features the necrotic segments of the fibers for many days after the injury.

Loss of substance, reduction of number of myofibrils, appearance of myofibrils clumping into coarse bundles producing a frayed aspect, granulation, altered staining quality of myofibrils, and vacuolation therefore represent successively more severe degrees of partial damage to the contents of the fiber. Within 6 days the sarcolemmal nuclei of such fibers proliferate, and the process of repair begins (Fig. 3–28). Many nuclei assume a central position in the clefts between bundles of myofibrils. Only when the myofibrils no longer stain can one declare the fiber to be necrotic. The loss of myofibrils is complete by the sixth or seventh day following injury, and thereafter any residual myofibrils become hyperchromatic with hemotoxylin. The corresponding granular appearance imparted by such myofibrils in

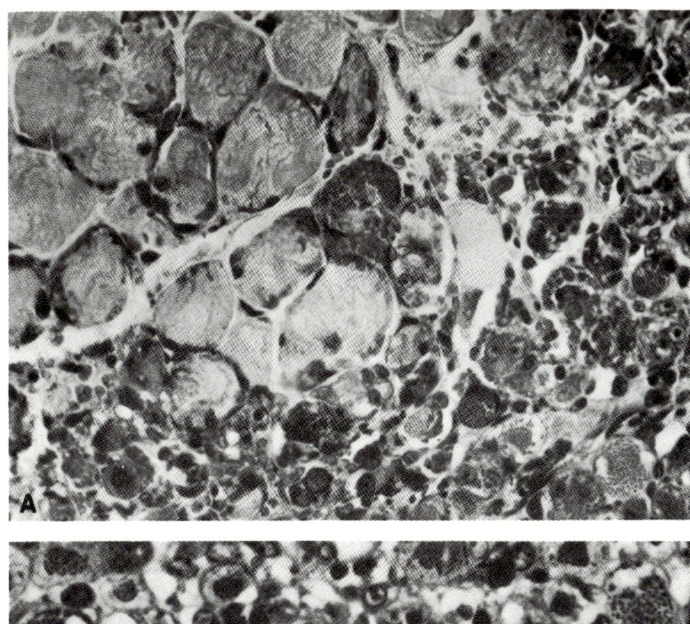

Fig. 3–28. (A) Effect of heat coagulation 11 days earlier. Necrotic zone is at top left. Regenerating and partially damaged fibers are at bottom right. (B) Muscle fibers contiguous to those in lower part of (A). (phosphotungstic hematoxylin)

transverse sections should not be interpreted as evidence of regeneration, for it is uniformly present in the area of partial damage several days before regenerative protofibrils appear and long before the myofibrils of growing buds begin to stain. Only during the third week do new myofibrils become identifiable under the light microscope.

The electron microscope has not been used to explore the status of organelles in partially damaged fibers. Fahimi and Cotran[65] studied the earliest phases of heat injury using colloidal laudanum as a tracer. This method showed increased permeability of cell membranes within 2 hours. The tracer had traversed the membrane and was concentrated around the mitochondria. Also at this stage muscle glycogen was lost, the mitochondria were swollen, the sarcoplasmic reticulum was disrupted, and contractile elements were beginning to degenerate.

Application of a block of solid carbon dioxide (Dry Ice) 5 mm in width to the surface of a muscle for 40 sec results in intense shortening of the damaged zone. Relaxation occurs in about 1 hour. Within the next few hours the frozen area is pale and swollen, demarcating it clearly from the normal muscle on either side. In most instances the band of necrosis extends completely through the muscle

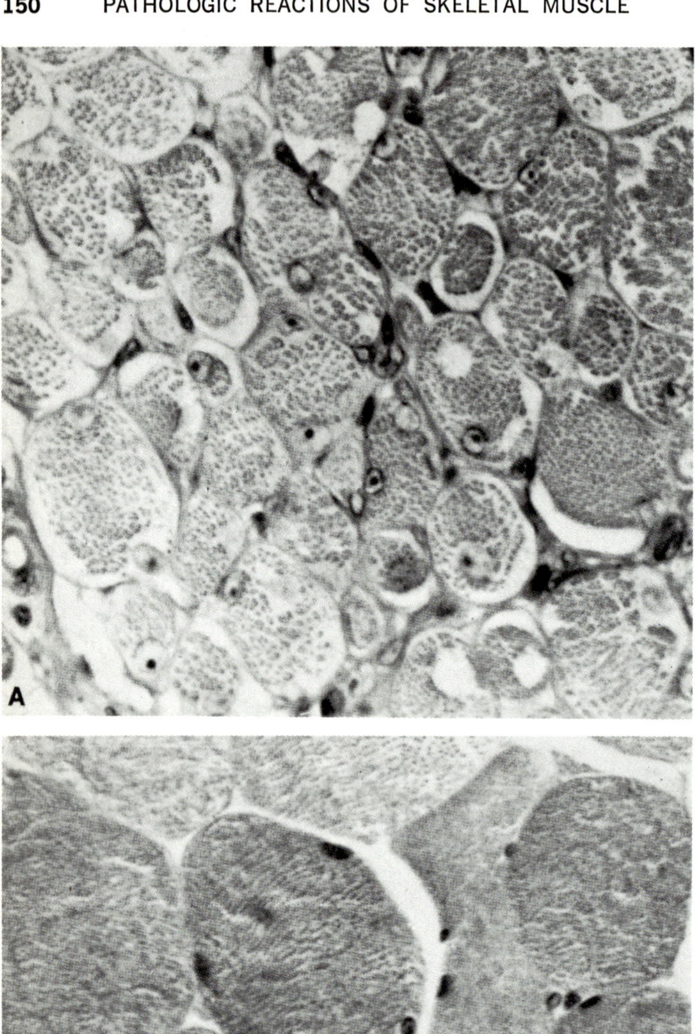

Fig. 3–29. (A) Same section as in Figure 3–28, 2 mm from zone of necrosis. (B) Muscles fibers in unaffected part of muscle 7 mm from zone of necrosis.

belly, being more widespread than that caused by heat. There occur essentially the same types of fiber alteration as were noted above. However, since the entire region is damaged, predictable types of fiber damage are usually seen in all the central and neighboring areas. These consist of a central zone of uniform hyaline destruction in which a severe contractile pattern is seen. Again the process is essentially one of coagulative necrosis. Within a few days the sarcolemmal nuclei lose their staining reactions as do the sarcous contents of the

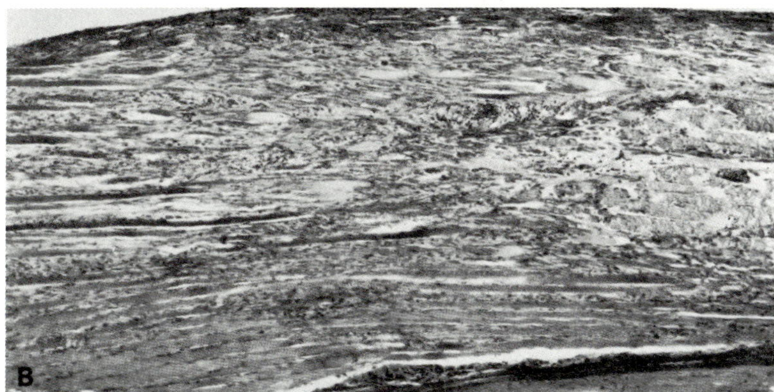

Fig. 3–30. Gracilis muscle of cat. Border of area of heat coagulation, with intact tendon 4 days after coagulation (A) and after 11 days (B). Coagulated area is at left in (A) and at right in (B). Darkly stained broken retraction caps seen in (A) have become hyaline masses in (B), in which they are separated from darkly staining intact portion of affected muscle fiber and from necrotic one. (hematoxylin, eosin)

muscle fiber, again after an initial diffuse argentophilia. The sarcolemma is variably preserved. The alterations of fibers adjacent to the necrotic zone differ in no important ways from those consequent upon heat injury. Histiocytic reactions of the type described below under *Simple Section* are prominent.

THE REPARATIVE PROCESSES

Muscle Buds and Giant Cells. The reparative processes are especially well visualized after thermal injury, especially with heat. An extensive zone of necrosis leaves a large mass of sarcoplasm to be removed. Fragments of debris filling the endomysial spaces are surrounded by histiocytes during the first 4 days (Zellenschläusche), and any remaining sarcolemmal nuclei have by that time commenced active proliferation (Figs. 3–30 through 3–32). By the 8th to 12th day the fragments of necrotic fibers have become smaller and fewer, and the previously empty portions of the sarcolemmal tubes are becoming filled by deeply staining granular sarcoplasm with chains and clumps of sarcolemmal nuclei in both peripheral and central positions (Fig. 3–33). Most of these show two nucleoli. The damaged ends of the sarcolemmal tubes are filled with a mixture of histiocytes and sarcolemmal nuclei. By the 8th day the former ends of the sarcolemmal tube have thus become transformed into buds of sarcoplasm with the nuclei clumped into irregular rows. Slender sarcoplasmic extensions into the fibrinous gaps accompanied by chains of sarcolemmal nuclei then appear. In places within the old tube the newly formed sarcoplasm has a finely fibrillar appearance, but the bud is not transversely striated. At this time one of the nucleoli of the sarcolemmal nucleus is sometimes represented by a rod-shaped structure, but we have not seen the "rhombosomes" described by Millar[142] with any constancy. Between the 8th and 14th days striated myofibrils

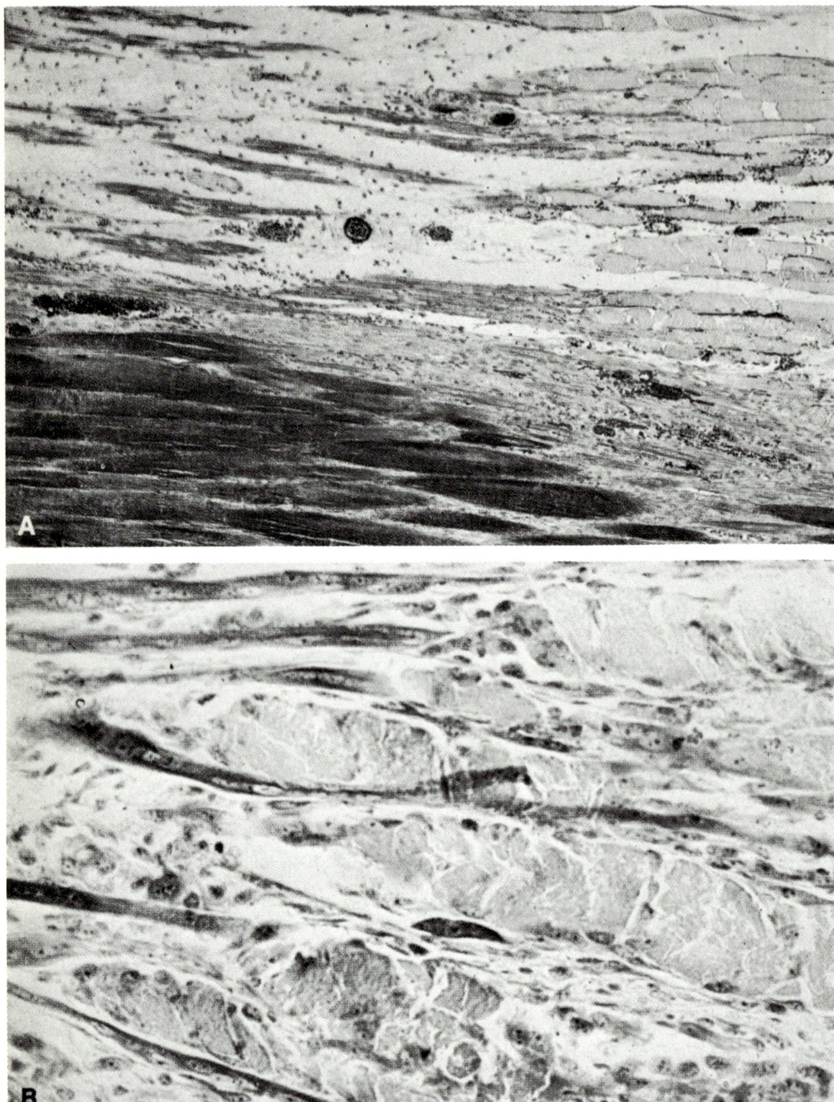

Fig. 3–31. Normal mouse biceps femoris. (A) Zone coagulated by heat 5 days earlier: right margin, separated from normal fibers; lower left, by pale zone of partial injury. Ribbons have already started to grow between coagulated fibers. (B) Section taken 10th day after injury showing ribbons growing between old sarcolemmal sheaths. Some degenerate spindles are also visible. (A, phosphotungstic hemotoxylin; B, methylene blue, eosin)

are observed in the growing stem of the muscle fiber, preceded by nonstriated cytoplasmic filaments.

The distal end of the bud advances rapidly in long-pointed projections, of which there may be two or three from one muscle bud. If the distance is short, these soon meet and fuse with corresponding spouts from the muscle fibers on the other side of the gap. If an obstruction is met (e.g., a blood clot or a dense wall of fibroblastic tissue) the sprout becomes club-shaped with a central collection of nuclei, and expansions in various lateral directions may begin. Some of these sprouts form long spindle cells with one or two nuclei and abundant granular cytoplasm, which soon has a longitudinal un-

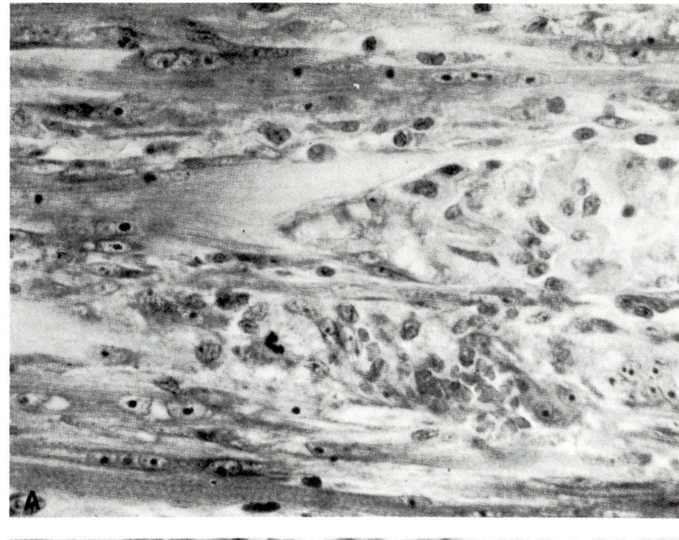

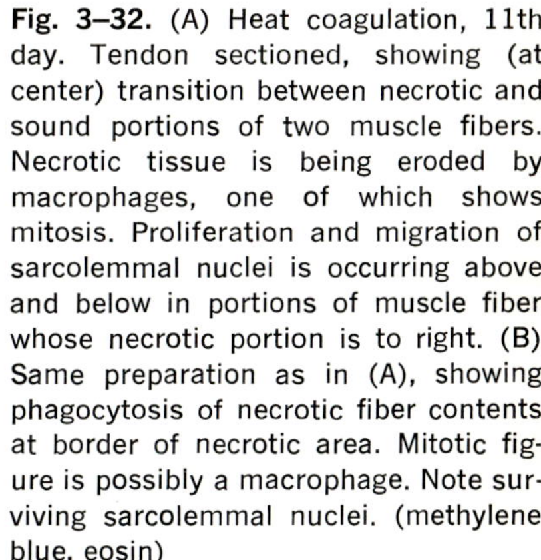

Fig. 3–32. (A) Heat coagulation, 11th day. Tendon sectioned, showing (at center) transition between necrotic and sound portions of two muscle fibers. Necrotic tissue is being eroded by macrophages, one of which shows mitosis. Proliferation and migration of sarcolemmal nuclei is occurring above and below in portions of muscle fiber whose necrotic portion is to right. (B) Same preparation as in (A), showing phagocytosis of necrotic fiber contents at border of necrotic area. Mitotic figure is possibly a macrophage. Note surviving sarcolemmal nuclei. (methylene blue, eosin)

Fig. 3–33. Changes after a clean cut in surface of muscle (upper left) 11 days earlier. (hematoxylin, eosin)

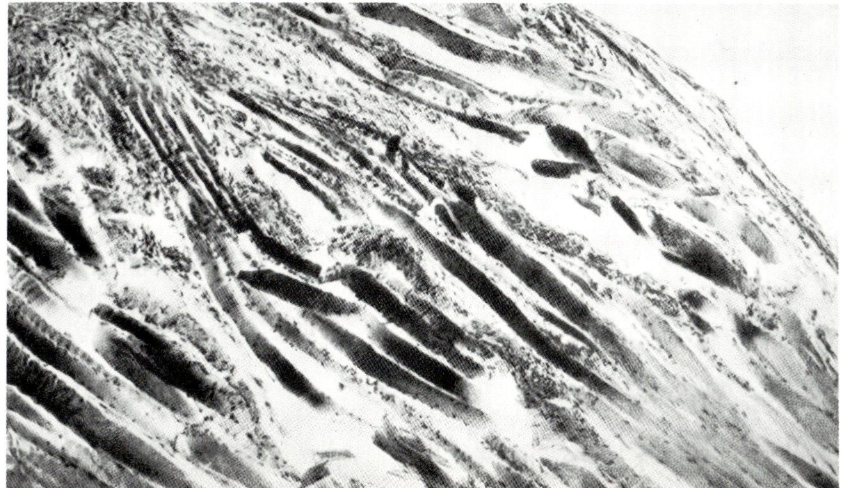

striated fibrillar appearance. These have been called myoblasts, but they are usually connected to a muscle sprout by a long filament.

Similarly, masses of cytoplasm with a central clump of muscle nuclei may seem to arise independently, presumably from isolated fragments of the original sarcolemma, and have somewhat smaller nuclei than those of the growth sprouts. When there has been little trauma such isolated "muscle giant cells" are few or absent. By the 21st day the myofibrils in the regenerated muscle columns have become deeply stained and more abundant, cross striation has appeared, and a sarcolemma has formed over the newly grown section of fiber where it has reformed a connection across the gap. During the following 2 weeks newly constituted fibers are perfected, their striations become regular, and the nuclei move to the periphery. The remaining sarcoplasm becomes less and less obvious. Small abortive outgrowths regress and atrophy. Large multicellular knobs which have met an obstruction persist for many weeks, gradually showing more nuclear and general shrinkage and vacuolation.

When the zone of segmental necrosis is narrow and particularly if retraction of the sound portion of the muscle fiber from the necrotic segment is prevented by section of the muscle tendon at the time of injury, the histologic picture is less confused (Fig. 3–30). The end of the necrotic segment is also in this instance filled by histiocytes, which often show mitoses (Fig. 3–32). At their edge, and occasionally among them, sarcolemmal nuclei (myoblasts) may be seen; but there is a sharp demarcation from the end of the sound portion of the fiber, where a zone of poorly stained striations merges into waxy swelling. From this waxy cytoplasm buds or sprouts carrying muscle nuclei with them insinuate themselves between the sarcolemma of the necrotic segments and the endomysial partitions. The growing sprouts progress deeply into this cleft by the 5th and 6th day after injury and are seen either as robust cords of clear, moderately basophilic cytoplasm or as chains of spindle-shaped cells with prominent muscle nuclei (Figs. 3–31 and 3–32). In contrast, the fibroblasts are small and insignificant.

Proximal to the budding end the remainder of the sound muscle fibers show nuclear and cytoplasmic reactions for distances of as much as 1–2 mm away from the injury. In this zone of partial damage there is intense nuclear proliferation, the nuclei being large and vesicular, with large and prominent nucleoli and collecting in rows in the center of the fiber separated by abundant granular cytoplasm. The cytoplasm in this transition zone is basophilic, staining moderately deep blue with methylene blue and to a lesser extent with hematoxylin. In some instances the whole thickness of the fiber in this region is basophilic. It seems possible that the nuclei are migrating toward the growing waxy bud. The myofibrils are reduced in number in this reactive transition zone and often stain darkly. When the insult has not been sufficient to induce coagulation of any part of a muscle fiber, nuclear proliferation and cytoplasmic accumulation can occur alone and then mark the zone of lesser injury.

This description applies to the most usual consequence of gentle application of heat to the surface of the muscle. In muscle fibers most removed from the trauma, where continuity between the coagulated and noncoagulated segments has been preserved, fewer growing buds are seen outside the sarcolemma. In this circumstance, particularly if an abundance of sarcolemmal nuclei remains in the coagulated segment, these are found linked in an irregular chain from the 4th or 5th day on. Proliferation and movement of spindle myoblasts then appear to be greatly reduced at all stages, with correspondingly less evidence of degenerative loss of myoblastic spindles during the 2nd week. When the fiber contents have been only partly coagulated or not coagulated at all, spindle cells are not seen in isolation and the regenerative process is ushered in by nuclear proliferation and basophilic sarcoplasmic aggregation in the least damaged part of fiber contents. This last type of minimal injury is prominent in the response to administration of plasmocid.

The process by which necrotic material is removed is rapid, once the fiber has been invaded by histiocytes. By the 10th or 11th day the contents of the more superficial fibers have been converted to a foamy material embraced

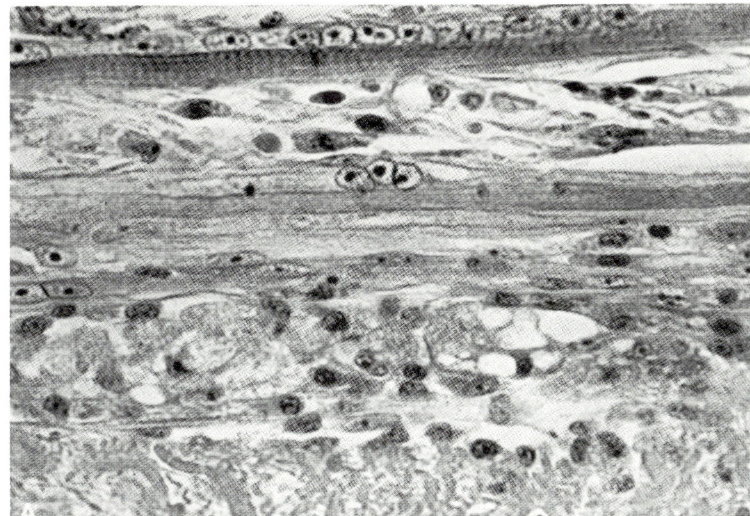

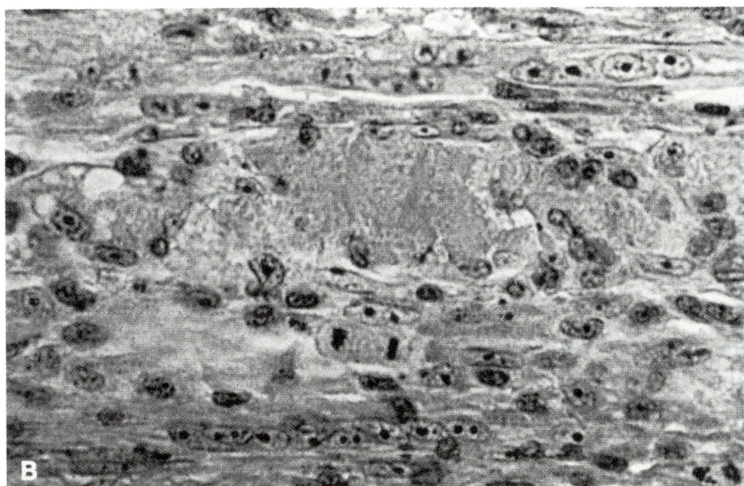

Fig. 3–34. Semitendinous muscle of cat, 11 days after coagulation of narrow zone. (A) Section at border of lesion shows partially damaged fibers (top) and two coagulated fibers, one with beginning histiocyte invasion (bottom), the other with fully developed "Zellenschläuche." (B) Further details with mitosis of a spindle (near center). (phosphotungstic hematoxylin)

by the expanded cytoplasm of histiocytes (Fig. 3–34A). Removal is evidently accomplished by lysis and diffusion, for though such foamy lipid-containing inclusions of histiocytes are commonly seen within the old sarcolemma, such cells are seldom found outside the nuclear bags. From the 15th day onward the most superficial fibers in the lesion are found to have collapsed, and only shrunken histiocytes are seen within and around them. During the 3rd week small collections of dark, round cells surounding a neighboring venule are the only representation of the earlier invasion.

In most situations in which a portion of the muscle fiber is rendered necrotic, there is a great deal of abortive proliferative change. Many of the myoblasts derived from muscle nuclei surviving within the old sarcolemma undergo degenerative changes during the second week. When a growing bud meets an obstruction such as a clot, hemorrhage, or fibrous tissue, it tends to fragment into a number of spindle cells, many of which show degenerative changes as early as the 10th day. No spindle cells remain after the 14th day. Mitosis in such spindle cells, both inside and outside the coagulated segments, is frequent and is evidently followed by dispersal of nuclear chromatin as an agonal event comparable to that described earlier in this chapter as a regular part of the degenerative cycle of the muscle cell (Fig. 3–18).

If tension upon the muscle has resulted in fragmentation and wide division at the border of the damaged region, these zones become widely separated (Fig. 3–30A). The gap between sound fiber and necrotic region is large, and the stretched parts of the tubes are filled with long spindle-shaped, multinucleated muscle cells in continuity with an actively proliferating muscle bud at the end of the sound segment of the muscle fiber. Some of these may be seen in Figure 3–25B. The hyaline retraction caps lie midway between sound and necrotic segments. In this circumstance the sarcolemma is torn, and resulting muscle spindle cells appear to link up with one another more frequently than following simple coagulation in continuity. This type of regeneration by fused spindle cells is common at the border zone following cold injury. In this case intense contraction at the time of injury has a similar effect in disrupting the border zone. Regeneration in the region of the coagulated segments then appears to be the result of extension of fingerlike processes into the space outside the old sarcolemmal sheaths.

Reconstitution of a large necrotic region proceeds slowly, being delayed until phagocytosis is complete. The endomysial partitions are reconstructed by fibroblast activity, and phagocytosis of the hyaline fragments proceeds from both ends, being closely followed by the growing muscle bud. If the endomysial architecture and sarcolemma have not been destroyed, removal of the hyaline material is more rapid, for histiocytic invasion of the sarcolemmal sheath then occurs all along the damaged region. Ultimate reconstitution of the fiber occurs in either case.

Myofibril Formation. After the 6th day the cords of sarcoplasmic buds begin to acquire striation, except at the growing tip. At first this is seen as faint and seemingly discontinuous longitudinal fibrils, as in the lowest fiber in (A), and the uppermost in (B) of Figure 3–35. Between the 10th and 14th days these "protofibrils" begin to develop alternating transverse A striation. It is remarkable that from the beginning such transverse striation is in approximate alignment in all the fibrils (Fig. 3–35). The Z dots do not become

manifest until the end of the 2nd week. The striation at first stains faintly and only toward the end of the 3rd week develops the intensity of that of unaffected myofibrils. It is striking that the development of striation under these conditions is a mass reaction. This tendency to "polymerize out," to use the analogy of Porter,[158] was also noted by Godman[83] and is in contrast to the serial development of myofibrils in the embryonic muscle fiber. It is evident that in completely isolated muscle cells in the coagulated zone, as in tissue culture, protofibrils and myofibrils may also appear but are few and rudimentary in comparison with those of the regenerating muscle sprout. In the course of repair of mammalian muscles, completely isolated muscle cells become rounded and undergo regression and degeneration during the 2nd week. Myofibrils in such cells seldom appear to develop beyond transitory protofibrils. The growing tip of the regenerating sprout retains clear cytoplasm.

At the borders of the well defined lesion produced by heat or cold coagulation, there are many small muscle fibers with reduced numbers of myofibrils and rows of centrally located vesicular nuclei. This appearance is seen both in muscle fibers that have been in part coagulated and in those deep to the lesion that have escaped coagulation in any part (Fig. 3–31). In such partially damaged muscle fibers the myofibrils stain with increased intensity from the 4th day onward. They are clumped in bundles of two and four, giving a stranded appearance (Figs 3–36 and 3–37). The nuclei of such muscle fibers become large and vesicular with prominent nucleoli, and the remaining cytoplasm becomes faintly basophilic. When undeniable regeneration is occurring, such fibers begin to show protofibrils throughout the remaining cytoplasm after the sixth day. Subsequently, slow development of transverse striation and full staining characteristics appear during the 2nd and 3rd weeks as in newly developed muscle buds (Fig. 3–36). Muscle fibers with central nuclei and deeply staining pairs or multiples of myofibrils are common in a stage of partial damage, and the nuclei may then be vesicular with prominent nucleoli. Whether regeneration occurs in such muscle fibers cannot be determined until the

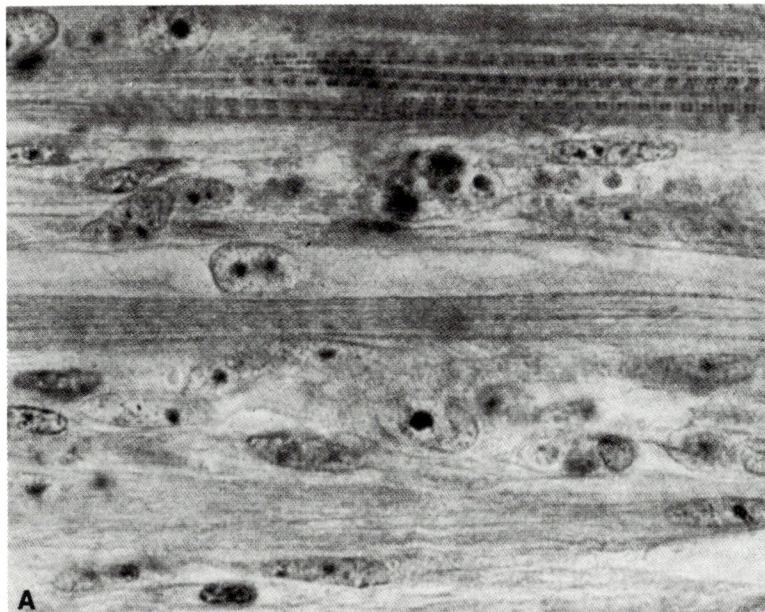

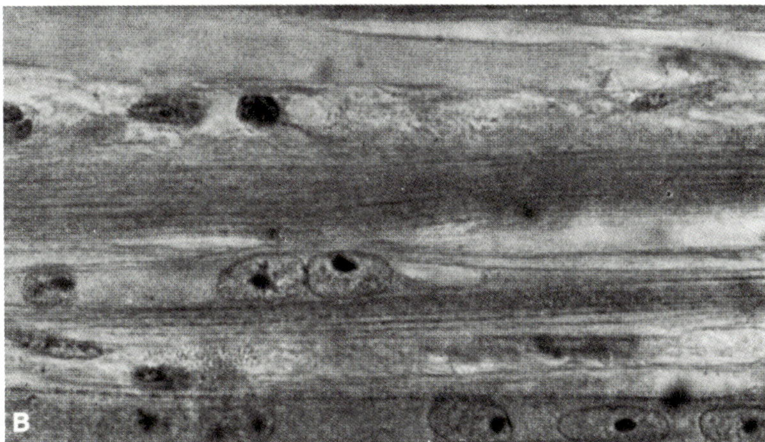

Fig. 3–35. Semitendinosis muscle of cat showing various stages in development of myofibrils from protofibrils—lower (A) and upper (B)—through first appearance of A band, and finally first Z line (upper fiber in A). (phosphotungstic hematoxylin)

fainter new striation can be seen. During the 3rd week some of the muscle nuclei of such fibers become shrunken and pyknotic or give rise to abortive muscle spindles, which we interpret as failure of regeneration.

The successfully reconstituted muscle fiber remains smaller than those in the rest of the muscle for several weeks. The increased number of centrally placed muscle nuclei distinguish the regenerated segment for at least 6 weeks. By this time all abortive buds and other signs of the regenerative process have been resorbed.

The sequence of events following any other type of necrotizing injury is similar. Contraction and tearing of the junctional region at the time of injury, as is caused by extreme heat or extreme cold, is followed by liberation of more spindle myoblasts, with delayed ribbon development. On the other hand, toxic effects that do not damage the sarcolemma at all (see *Plasmocid,* below) can be followed by extremely rapid onset of regeneration, though the ultimate complete reconstruction of the full complement of myofibrils is still long delayed. If the sarcolemma and its attached nuclei are destroyed over a considerable length of the fiber, regeneration occurs from both intact ends, but early fibroblastic invasion of the totally necrotic area tends to result in a tough,

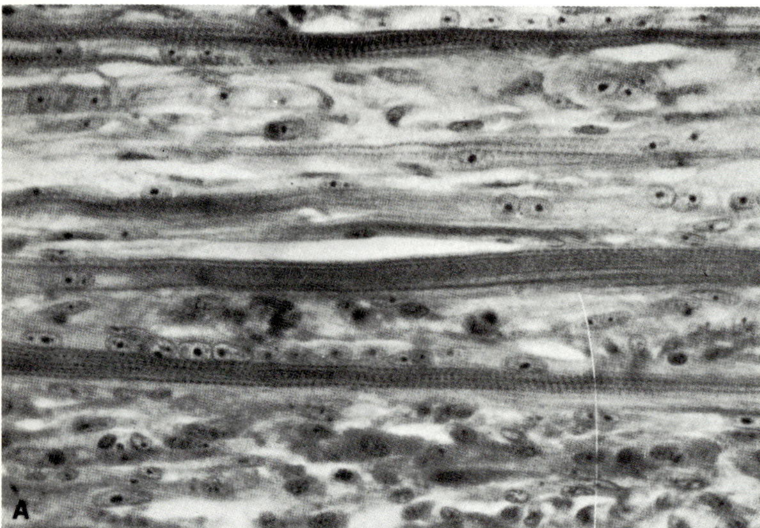

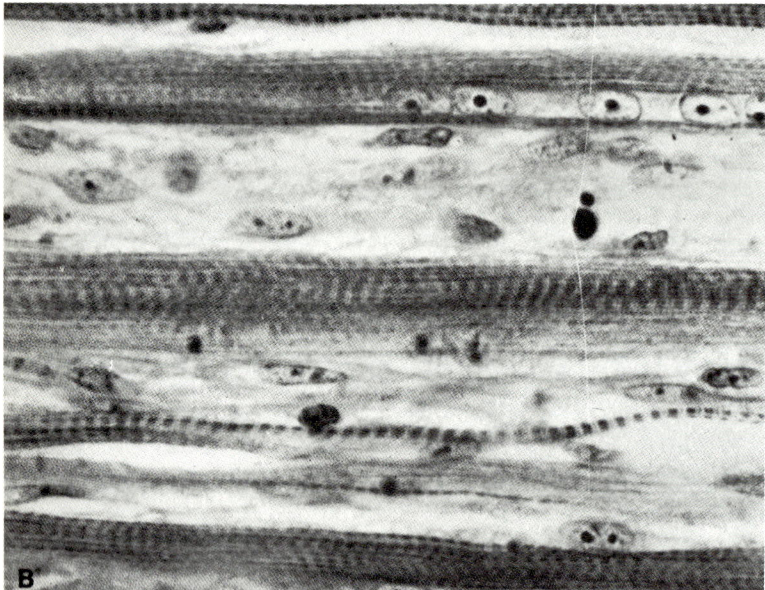

Fig. 3–36. Heat coagulation 11th day showing surviving myofibrils at the edge of the necrotic zone. In lower part of (B) a double myofibril courses alone past a muscle nucleus that is just out of focus and appears only as a dark dot. Protofibrils fill newly constituted sarcoplasm. (phosphotungstic hematoxylin)

impenetrable fibrous scar. If fragments of the sarcolemma and its nuclei remain, these form multiple growing points.

Coagulation of a segment of muscle fiber leaves one or both surviving segments without innervation, depending on whether the motor end-plate is destroyed. If it is destroyed, regeneration of nerve fibers presumably results in reinnervation of any surviving or regenerated muscle fibers near the terminal branches of the nerve bundle. In any case, some parts of the surviving muscle fibers are separated from the end-plate until regeneration connects them with an innervated segment or with a new motor ending. At 5 or 6 days after the injury the surviving portion of muscle fibers without end-plates begins to fibrillate. We have often observed and recorded such fibrillation during coagulation in cat and mouse muscle. This fibrillation extends into the region of budding, so that it is difficult to determine whether the myoblasts and other muscle fragments in that region are also fibrillating. On the basis of the behavior of such fragments in tissue culture,

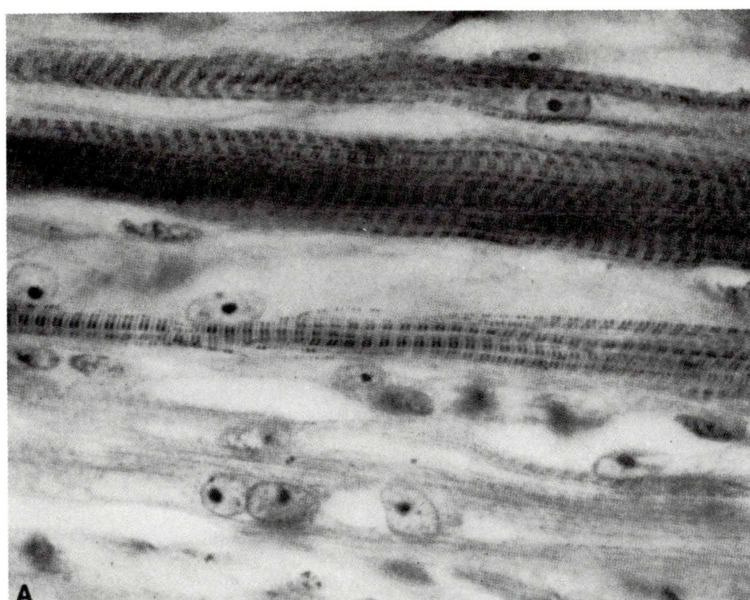

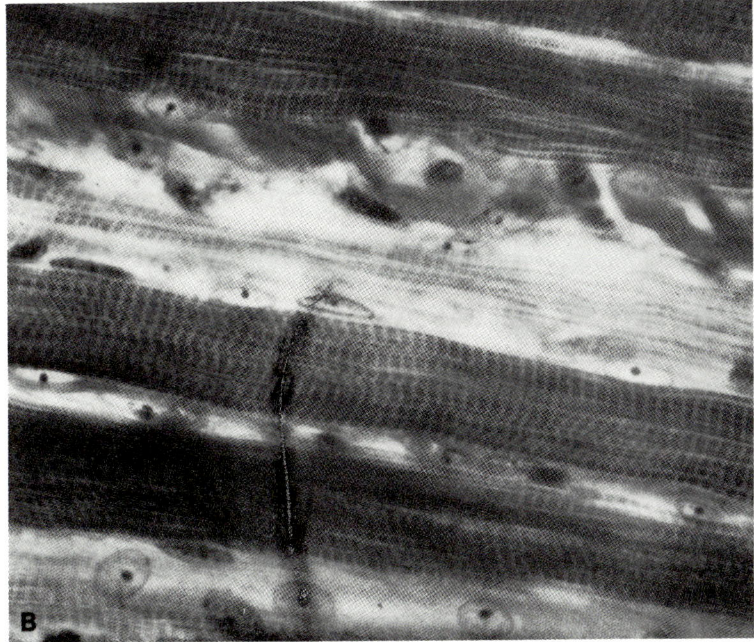

Fig. 3–37. Heat coagulation from same section as in Figure 3–36, showing partially damaged myofibrils next to necrotic zone in (A), and at a greater distance in (B). Note beginning of transverse striation on protofibrils near nucleus in lower right of (A).

such fibrillation may be expected to occur from the 6th day after injury onward until restitution of innervation occurs.

SIMPLE SECTION

The sequence of events following simple section of muscle fiber represents another well recognized type of injury. A small, clean cut of a fasciculus at the surface of a muscle, for example, is followed by retraction of the contents of the fibers within the sarcolemma, thus leaving this tube gaping and empty for a short distance. A small necrotic clot or "retraction cap" is commonly present on the end of the retracted portion, which otherwise stains deeply and diffusely. During the first 2 days polymorphonuclear leukocytes and monocytes appear in the fibrinous clot that fills the wound, and there is increasing histiocytic activity at the margin. On the 2nd day many of the sarcolemmal nuclei become rounded out for a distance of 1 mm on either side of the injury. By the 3rd day a large number of histiocytes are actively engaged in phagocytizing debris, and some have penetrated the sarcolemma if the retracted contents contain any dense clots or a "retraction cap." Any muscle nuclei included in such clots have become shrunken and pyknotic. The remaining sarcolemmal nuclei, whether remaining adherent to the empty part of the sarcolemmal tube or in the portion of the tube containing retracted muscle substance, now show striking activity, being present in chains of three to eight or more, with zones of homogeneous sarcoplasm separating them from the bundles of myofibrils. Each nucleus now presents a prominent nucleolus. Some nuclei lie between the bundles, which stain diffusely and appear somewhat condensed in the viable portions of the fiber. The portions of the muscle fiber at a greater distance from the injury are unchanged. The empty portion of the tube contains sarcolemmal nuclei as well as histiocytes. Each sarcolemmal nucleus is plump, with an intensely staining large nucleolus and has abundant basophilic cytoplasm. Some nuclei have two nucleoli, but mitotic figures are not seen. Unlike thermal injury there is intense ·fibroblastic proliferation in the endomysium, and many of the mononuclear cells are undifferentiated; fibroblasts and myoblasts are often indistinguishable.

ISCHEMIA

The damage resulting from partial ischemia presents a special circumstance, and the studies of Clark[30] have illuminated a number of obscure features of the process. Necrosis of the anterior tibial and gracilis muscles in rabbits was produced by several different methods. In one series of experiments a small portion of gracilis muscle was completely excised and immediately sutured back in place. Its vascular supply was thus completely disrupted. In another series the main vessels to the anterior tibial muscles were ligated, producing necrosis of the distal two-thirds of the muscle. Material from the devitalized muscles was examined histologically at several intervals of time.

In the ischemic areas the muscle fibers rapidly died, their nuclei disappeared, and the contractile substance fragmented. In addition the endomysial nuclei also disappeared, but many of the sarcolemmal and endomysial membranes remained intact, as did the perimysial connective tissue. Thus the general architecture of the necrotic muscle was preserved. The first sign of renewed activity was edema of the endomysial tissue followed by an invasion of the area by a large number of neutrophilic leukocytes and some histiocytes. These cellular infiltrations were first noted at the junction of a necrotic and a normal muscle. They were soon followed by massive phagocytosis of the contractile substance of the dead muscle fibers and a proliferation of fibroblasts. The fibroblasts grew into the endomysial spaces between the dead muscle fibers so that each fiber was progressively enclosed in a tube of endomysial connective tissue. Phagocytosis of the necrotic muscle fibers was progressive from each end, and the growing sprouts from the region of intact muscle closely followed the zone of phagocytosis. The newly formed endomysial tubes were of prime importance in the regenerative processes, for these channels provided guides for growing muscle fibers. The latter sprouted from the stumps of the damaged fibers that had survived at the margin of the necrotic muscle and grew into the necrotic area. The new muscle buds consisted of plasmodial outgrowths of granular sarcoplasm and parallel rows of longitudinally arranged nuclei. Occasionally longitudinal striations were visible in the sarcoplasm. The tips of the sprouts were cone-shaped except when indented by some obstruction such as a fragment of dead muscle fiber. The new muscle fibers grew rapidly down the endomysial tubes and replaced the dead ones as fast as they underwent dissolution and phagocytosis.

The longitudinal growth rate of regenerating fibers was calculated by Clark and Wajda[31] at 1–1.5 mm/day. Longitudinal striations could be seen in new muscle after the 2nd day, and cross striations began to develop on the 6th day. After 10 days to 3 weeks the average thickness of the fibers was about one-third normal, and by 3 months these fibers had increased to normal thickness.

The importance of the directive action of the endomysial tubes in the regeneration of muscle fibers is demonstrated in Clarke's experiments,[30] in which pieces of rabbit gracilis muscle were excised and resutured into place at right angles to their original position. The direction of growth within the graft was oriented in the line of the original architectural framework. At the junctions of the graft and the healthy muscle, continuous fibers were seen in turn at right angles and to follow the course of the rest of the fibers. A large blood clot interposed in the course of regenerating muscle fibers disrupted the continuity of the endomysial tubes, and the sprouting fibers then coursed in all directions. The larger the graft, the greater was the tendency to ultimate fibrosis.

CRUSHING

Crushing muscle with a clamp (e.g., a hemostat) is commonly used as a means of studying the reaction of muscle to injury.[30] The results differ from those of the methods already discussed in several respects. Firstly, the remnants of the damaged muscle fibers are usually in the form of small fragments which are rapidly removed by phagocytosis during the first 2 weeks. Secondly, unless the pressure is applied very gently, not only the muscle fibers but also the pattern of the supporting endomysial architecture of the damaged region is destroyed. The method is therefore ideal for the study of muscle buds.

The area of necrosis is invaded by histiocytes within 24 hours, and by the 2nd day fibroblasts from the intact endomysium stream into the damaged area. Proliferation of muscle nuclei at the ends of the remaining portions of muscle fiber on both sides of the injury is evident on the second day. By the 5th day there is an accumulation of sarcoplasm and nuclei at the end of the muscle stumps, which also show evidence of partial damage to their con-

tents (Figs. 3–38 and 3–39). At this time it is clear that many isolated single muscle cells are present in the damaged region, recognizable by their large nucleus and abundant cytoplasm (Fig. 3–39). Some of these soon link up in chains with extensions from the growing bud, but a few become aggregated into multinucleated clumps (muscle giant cells). Others show regressive nuclear changes and disappear. If the extent of necrosis is not too large, many of the sprouts from buds join with those from the other side of the gap so that by the end of the third week there is substantial repair of the injury. Since the guidance of the endomysial tubes has usually been lost, there is much interlacing of the restored muscle fibers (Fig. 3–40). Several branches of one bud may make reunion across the gap, and when these become mature the appearance is that of a muscle fiber split into daughter fibers. In old muscle wounds regenerated muscle fibers may also split into two or more divergent, fully matured daughter fibers. This phenomenon no doubt gave rise to the mistaken notion of regeneration of muscle fibers by longitudinal splitting of the old fiber. We have not observed longitudinal division of the sound portion of the fiber.

The successful branches of muscle buds fill with myofibrils (Fig. 3–40C). After 3 weeks many of the nuclei show degenerative changes, and it is evident that unsuccessful branches of a muscle bud become vacuolated, show nuclear pyknosis, and are resorbed (Fig. 3–40B). Such degenerative change is characterized by intense basophilia of both cytoplasm and nuclei. Any myofibrils contained in such fibers at first stain intensely and then suddenly disappear. The shrinkage of the nuclei is particularly evident, and the basophilia is much darker than in a regenerating fiber.

"REGENERATION" AFTER SEGMENTED NECROSIS

From the results of experimental investigations of muscle necrosis it can be seen that regeneration of striated muscle fibers is an active and vital process influenced by many factors. Moreover, it is evident that the process begins al-

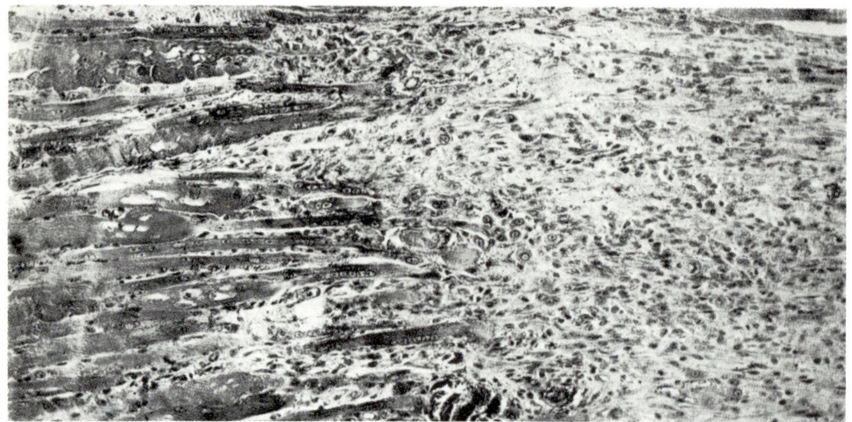

Fig. 3–38. Sartorius muscle of cat, showing border of an area crushed 5 days earlier with a hemostat. (phosphotungstic hematoxylin)

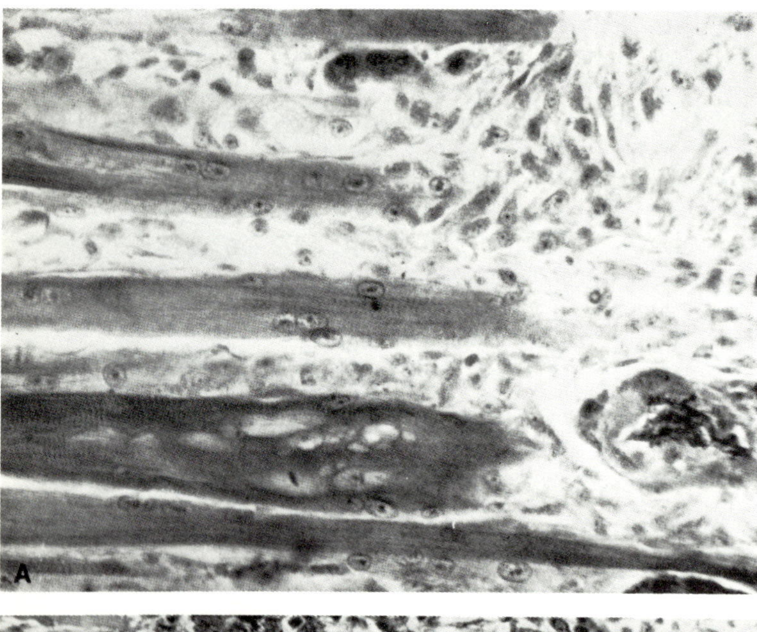

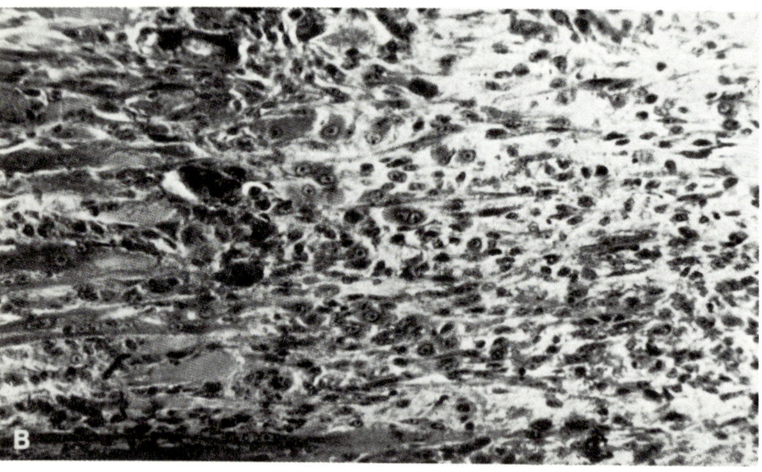

Fig. 3–39. Same muscle as in Figure 3–38 showing (A) detail of ends of surviving segments and (B) surviving isolated muscle cells with abundant cytoplasm (center).

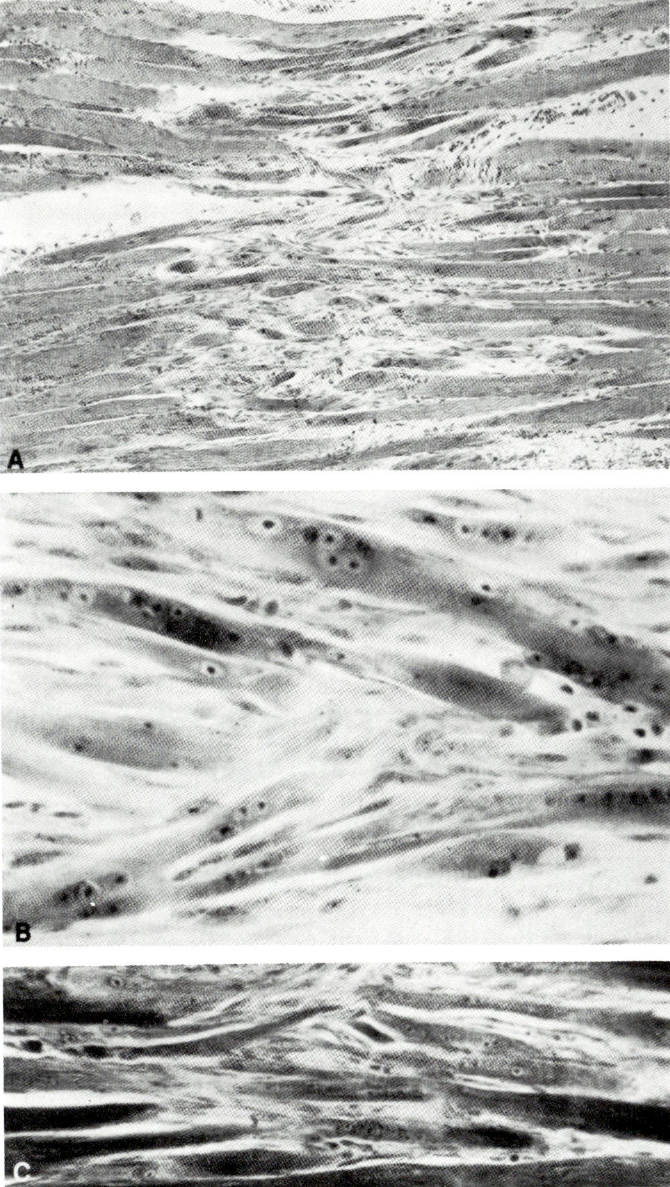

Fig. 3–40. (A) Regeneration 17 days following a localized crush injury. (B) Same muscle showing regressive nuclear changes in side branches. (C) The same. (A, B, methylene blue, eosin; C, phosphotungstic hematoxylin)

most as soon as the first cellular response to tissue necrosis, so that the two types of change are difficult to separate (Kraske[118]).

If it is to be complete, regeneration must lead to full reconstitution of the muscle fiber; and the latter must be attached at both ends to the tendons or connective tissue sheaths and contain a normal complement of myofibrils capable of contracting under the influence of nervous stimuli.[22] When analyzed under these conditions it is apparent that the so-called regenerative process comprises several types of change, occurring in sequence: (1) myoblastic, (2) myotubular or sarcoblastic, and (3) maturational and reinnervative stages.

MYOBLASTIC STAGE

Among the mélange of cells that proliferate within and along the margins of a necrotizing lesion certain ones, by virtue of their position in relation to the sarcolemma of damaged fibers, assume the characteristics of mononuclear myoblasts. Some evidently are normal sarcolemmal nuclei that have survived injury but have become isolated from the fiber of which they were a part. Others are obviously derived from healthy sarcolemmal cells of intact portions of fibers in the margin of the lesion.

The reactive sarcolemmal cell has an enlarged, rounded, hyperchromatic nucleus. As it separates from the parent fiber it comes to lie between the sarcolemmal and basement membranes, resembling at this stage the satellite cells described in frog muscle by Mauro.[134] When the cells enlarge and grow in isolation they assume a spindle shape and are difficult to distinguish from fibroblasts. Mitoses are visible in some of them, either while maintaining their satellite relationship or lying free in the lesion. Reznik[165, 166] argues, quite persuasively, that the satellite cell is no more than a reacting sarcolemmal nucleus preparing to split away from the muscle fiber. The rapid increase in the population of mononuclear myoblasts by nuclear (mitotic) division can be readily proved by their incorporation of tritium-labeled thymidine.[105]

Basophilia of sarcoplasm is an integral part of the early phase of regeneration; unfortunately, it has proved most difficult to interpret, and there is not complete agreement about its significance. In histochemical terms basophilia means that acidic elements have increased in the tissue cells. These in turn show an affinity for basic dyes, e.g., hematoxylin and methylene blue. All actively proliferating tissues show basophilia, some to a greater degree than others. The pancreatic cells show a most intense basophilia, and it is significant that this organ is at all times actively engaged in protein synthesis. A significant component of the acidic substances that account for basophilia in growing or otherwise active tissues is an increase in nucleic acid units. Desoxyribonucleic acid (DNA), localized entirely within the nucleus, is moderately increased in quantity, especially during mitotic activity. Ribonucleic acid (RNA), found in the resting muscle cell in the nucleolus and to a slight extent in the

cytoplasm especially in the paranuclear zone, increases greatly during cellular proliferation, large amounts of it being found in nucleoli as well as in cytoplasm. It has been clearly shown in recent years that the synthesis of protein molecules is closely linked with RNA function. Hence it is very likely that intense sarcoplasmic and nucleolar basophilia, as observed in routinely stained sections of regenerating muscle cells, is most often the result of an increase in RNA, which in turn is actively engaged in the reconstruction of the many sarcoplasmic proteins, including the actin and myosin of the myofibrils.

We studied the basophilia of regenerating muscle fibers by several techniques and concluded, on the basis of specific histochemical methods, that RNA is present in the sarcoplasm and nucleolus in sizable amounts during regeneration, whereas it is largely confined to the nucleolus in mature muscle. The basic metachromatic dye azure B, when applied to tissue sections* at pH 4.0, shows a specific staining affinity for RNA.[72] Muscle cells that contain RNA stain a royal purple, whereas mature fibers are faintly basophilic or unstained (Fig. 3–41A). Nuclear membranes and inflammatory cell nuclei are colored a light or dark blue and mast cell granules a deep purple. When unstained sections are first digested by the specific enzyme ribonuclease (RNAse) at pH 6.5 and then stained with azure B, all the purple coloration of muscle fibers is removed. The blueness of the nuclear membranes and the invasive cells is unchanged, and the granules of the mast cells remain purple due to the intense concentration of sulfated acid mucopolysaccharides. The lack of basophilia of regenerating muscle fibers when prior digestion

---

*Formalin-fixed and paraffin-embedded sections are used in this method. Other techniques may be used for the detection of RNA. These include: (1) the methyl-green-pyronin stain technique of Brachet,[86] (2) ultraviolet spectroscopy, and (3) the use of fluorescence to identify RNA after staining with acridine orange.[12] The staining methods have been briefly investigated and show no advantage over the simpler azure B technique. Localization of RNA was similar in all when similar tissues were studied. Azure B is a derivative of methylene blue.

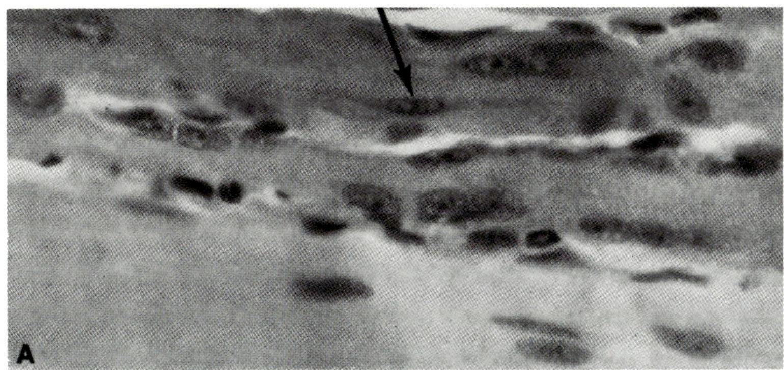

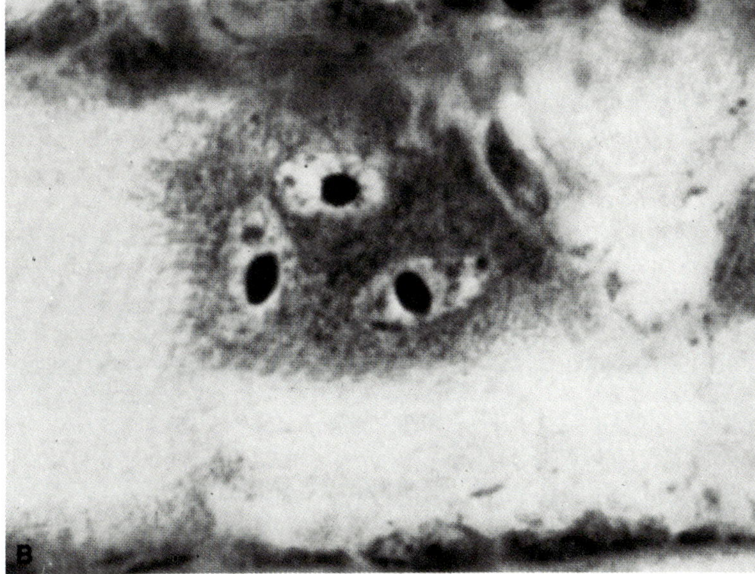

Fig. 3–41. Rat masseter muscle. (A) Spindle-shaped region of basophilia in a centrally positioned nucleus (arrow) 48 hours after plasmocid administration. (B) Granular basophilic RNA surrounding three prominent sarcolemmal nuclei. Note faint coloration of normal cross striations and their absence in basophilic zone. Rat diaphragm 36 hours after plasmocid injection. (C) Prominent sarcolemmal nuclei are embedded in short and long ribbons of basophilic RNA. Rat diaphragm 72 hours after plasmocid injection. (A, hematoxylin, eosin, ×180; B, azure B, pH 4.0, ×1875; C, azure B, pH 4.0, ×750)

by RNAse has been carried out proves that most if not all of the basophilic material present after azure B staining is due to the presence of RNA.

The azure B method, when applied to muscle tissue during various phases of regeneration after necrosis induced by plasmocid or extreme cold, reveals interesting patterns of distribution of RNA. Within 3 hours after necrosis the sacroplasm loses its faint basophilia but the nuclei and nucleoli are stained, except in plasmocid necrosis in which many nuclei are damaged and lose their basophilia. The first evidence of an increase in RNA content appears between 24 and 36 hours when preserved nuclei at the margins of the zone damaged by cold or in the centers of groups of fibers damaged by plasmocid show enlargement, reduplication, and increased staining intensity of nucleoli (Fig. 3–4A, B). At this time or soon thereafter faint granular basophilic caps of sarcoplasm appear at either end of the prominent nucleus. Between 36 and 72 hours the basophilia surrounds each nucleus, and in the activated retraction caps of cold-damaged muscle the entire club-shaped end of the fiber is basophilic, evidently because of confluence of basophilic RNA from the many nuclei in the cap. During the next 24 hours terminal sprouting of the new fiber begins. Intense sarcoplasmic basophilia persists in the growing tips and gradually fades in the more mature proximal portions of the fiber until at about 14–21 days the basophilia is again confined to the nuclear caps and the nucleoli.

The areas of basophilia originally show a course and then a finer granularity. The cross striations of the adjacent myobrils do not extend into the zones that are undergoing repair (Fig. 3–41C). During the fourth to sixth days while basophilia is still intense, coarse, longitudinal, dense, purple photofibrils are seen. These are followed by the appearance of multiple finer fibrils, described above. As the finer fibrils appear, basophilia gradually diminishes, and when cross striae develop on the 5th to 9th days most of the basophilia disappears, except for a perinuclear halo and the prominent nucleoli.

Careful study of this sequence of events during the repair of muscle suggests that each sarcolemmal nucleus controls its own area in the sarcoplasm; and if the myofibrils or sarcoplasm in that specific region are injured, the parent nucleus promptly responds by enlargement and elaboration of RNA, which diffuses outward from the nucleus (perhaps from the nucleolus). The RNA in turn actively synthesizes the necessary protein constituents for repairing the contractile units and replacing the sarcoplasmic elements, including the soluble enzymes, sarcosomes, and other structures.

The idea of control and maintenance of an area of sarcoplasm and a group of myofibrils by a nucleus is consistent with the observations of Tiegs[194] that in embryonic insect muscle each free myoblast generates a single or several fibrils and that the muscle fiber is gradually built up by a progressive addition of free myoblasts to the fibril column. Each myoblast brings with it its nucleus, fibrils, and sarcoplasm to contribute to the adult fiber. Furthermore, in their electron micrographic studies on thrush muscle, Ruska and Edwards[171] detected a paranuclear pattern of sacroplasm and myofibrils, a finding that supports this hypothesis because these newly formed materials appear to remain localized in a certain area of the muscle fiber.

MYOTUBULAR OR SARCOBLASTIC STAGE

Mononuclear myoblasts have been observed under the electron microscope to fuse, and thus to form the myotube, just as it is formed in embryonic life. Also from the ends of injured muscle fibers sarcoplasmic buds extend like pseudopodia into the lesion, and myoblasts may also attach to their surface. The myoblasts of these strap-like structures are heavily nucleolated and surrounded by sarcoplasm. On either side of the nuclei myofibrils soon become visible. Once joined to other cells no further mitotic division occurs in myoblasts and they can no longer assimilate tritium-labeled thymidine. Such intrafiber nuclei are always diploid in DNA content, never haploid or tetraploid as would be the case if they had just divided.[124] This explains why Clark,[30, 31] who attempted to arrest mitosis by colchicine, was never able to find a mitotic figure within myotubes even though they were common in free proliferating myoblasts, endomysial fibroblasts, and histiocytes. Presumably there is also migration of sarcolemmal and satellite cells from the healthy parts of the fiber into the injured segment. Some of these leave the fiber and form free myoblasts.

These observations seem to settle once and for all the question of amitotic division of sarcolemmal nuclei. What really is happening is the active fusion of postmitotic myoblasts. Moreover, the old debate as to whether muscle regeneration occurs only from the damaged ends of muscle fibers by a process of budding (i.e., giant-cell formation) or by the continuous mode as maintained by Millar[142] and Clark[30, 31] is settled: Both occur. Also the problem of whether new fibers can form from the fusion of isolated myoblasts (discontinuous mode) as argued by Godman,[82, 83] Hay,[93] Walker,[206] and Betz *et al.,*[13, 14] is now resolved by these newer histologic techniques.

We found that the formation of isolated muscle cells is favored by two circumstances: (1) a broad but mild crushing injury and (2) fragmentation of a few fibers in otherwise healthy muscle, as in plasmocid toxicity or the toxemia of infectious diseases such as typhoid fever. The sarcolemma and its nuclei are then preserved in the small segments between clumps of necrotic, hyaline masses. As Volkmann[203, 204] first showed, and Gussenbauer later demonstrated in teased preparations,[87] the isolated groups of sarcolemmal nuclei form mononuclear myoblasts, though they do not persist in this state. If necrosis of a whole mass of muscular tissue occurs as Volkmann[203, 204] showed, regeneration by cellular buds from the undamaged segment of muscle fiber is the usual mode. The latter form multiple branches that bridge the necrotic gap. Thus in reality the old controversy between the protagonists of regeneration by terminal budding[170] and those who argued for regeneration from embryonic single myoblasts was long ago settled by Volkmann. All that is new is that the so-called buds have been shown to involve also a fusion of free myoblasts, some incompletely separated from muscle fibers (satellite cells) and others multiplying mitotically outside the muscle fiber. Regeneration following fiber necrosis is essentially a process of ribbon or pseudopodial extension of sarcoplasm in surviving cell masses.

Spindle cells that do not fuse within a few days degenerate. By the 16th day after injury isolated spindle cells and multinucleated sarcoplasmic cells undergo regressive alterations in the form of intense basophilic changes in their cytoplasm with nuclear shrinkage and pyknosis.

Myosin fibrils can be demonstrated by a fluorescent antimyosin antibody and has been seen in the cytoplasm of mononuclear myoblasts. Allbrook[2] and Betz and Reznik[15] followed the formation of myofilaments and myofibrils with the electron microscope. Ribosomes line up in chains and spirals to form polyribosomes and soon thereafter thick and thin filaments, presumably corresponding to the myosin and actin filaments of Huxley. As differentiation proceeds the myofilaments become organized into myofibrils and assume a longitudinal orientation. Opaque, transverse Z bands develop across the bundles, and the latter become aligned with one another.

By light microscopy the development of striated myofibrils is evidently preceded by the presence of long, poorly organized rods within the cytoplasm. These were observed by Millar,[142] are well known to embryologists,[48, 81] and are also seen in tissue cultures.[127, 181] Myofibrils that stain darkly and sharply, commonly in pairs or multiples of two, suggest longitudinal division as was observed in the trout embryo by Heidenhain[94] and shown here in Figure 3–2. These are always continuous with the sound portion of the fiber but for several days are sparse in the region of the retracted but viable stump. New striation develops all at once in the faint precursor unstriated fibrils that appear throughout the new and damaged sarcoplasm. Lewis and Lewis[127] observed development of striated fibrils de novo in the buds in tissue cultures, but this observation and the statement that contraction could occur without myofibrils in such cultures disregards the presence of the poorly staining fibrillary precursor. Speidel,[184, 185] in his studies of muscle regeneration in the tail of the living tadpole, observed the proliferation and migration of sarcolemmal nuclei and the formation of single myoblastic cells, which, however, as in tissue culture, tended to be attracted by and linked up with neighboring myoblastic masses. Striated myofibrils appeared in groups, often in isolated cells, and joined with those of other cells when the cytoplasm fused. Single myofibrils were not seen to form, but the method of observation was not such as to establish the type of primary development of the myofibrils.

MATURATION AND REINNERVATION OF MYOTUBES

As myotubes become mature muscle fibers the sarcolemmal nuclei change their position and structure, moving to a peripheral position; the bundles of myofilaments thicken and lengthen as they form myofibrils. The mechanism of myofibril increase is believed to be the same as that which led to their initial formation. Probably it duplicates the embryologic state.

The connective tissue is adjusted to support individual fibers; but if the connective tissue sheaths have been injured as well as muscle fibers, the fibroblasts proliferate to form scars, which may obstruct muscle regeneration. Actually the pattern of connective tissue proliferation provides as much information about the nature of the injury as do the regenerating muscle fibers themselves.

The final stage of muscle regeneration is reinnervation. The terminal intramuscular nerve fibers may have been damaged by the same process that injures muscle fibers, in which instance axonal regeneration must first occur. If nerve fibers escaped, as the muscle regenerates the axons come into contact with muscle fibers at approximately the same points as were occupied originally by motor endplates. New ones may also form as revealed in the studies of Gutmann and Young[89] and Edds.[55] Many fibers may thus have more than one motor ending.

It is always difficult in muscle regeneration after injury to follow the reinnervation process. Axons can be seen invading the necrotic lesion within the 1st week, and eventually they make contact with maturing muscle fibers.[3, 15] Acetylcholinesterase stains reveal sites of developing motor end-plates in the regions where sarcolemmal nuclei are aggregated.

Without reinnervation normal morphogenesis of muscle is halted short of full maturation of fibers. While the abundant proliferation of myoblasts and myotubes proceeds without need of nerve supply, Zelena[222] showed that the growth of myotubes to mature muscle fibers is impaired.

Cytochemical methods have also been applied to the study of muscle regeneration but have shed little additional light. The early sarcoplasmic basophilic sarcoplasm due to excess RNA contains ribosomes and polyribosomes. This is pronounced in the myoblast, and both the basophilia and ribosomes progressively diminish as the myotubes mature to muscle fibers. Glycogen shows an opposite trend, being absent in myoblast and reaching a maximum in the mature fiber. Techniques for acetylcholinesterase show a diffuse staining of myotubular structure, beginning at points where they connect with intact surviving muscle fibers. In muscle buds it is the proximal cellular zone that has the highest concentration of the enzyme.[14, 62] As the myotubes develop the enzyme becomes concentrated in irregular foci along the surface of the fibers. Later the concentration is clearly confined to motor endplates and myotendinous junctions. The converse occurs in denervation where the normal concentration of cholinesterase at the motor end-plate disappears. Thus nerve contact and nerve stimuli seem to be necessary to its maintenance.

The writings of Bourne and Beckett[18] may be consulted for further data concerning the concentrations of succinic dehydrogenase, acid phosphatase, phosphorylase, 5-nucleotidase, and nucleoside phosphotransferase in regenerating muscle.

OTHER SPECIAL TYPES OF EXPERIMENTAL INJURY AND DISEASE

OCCLUSION

In contrast to localized ischemic damage to parts of a muscle, interference with the circulation of an entire extremity produces a rather more variable effect on the muscles.

ARTERIAL OCCLUSION

When the principal artery of the dog's leg is permanently obstructed by ligature, the results are variable. There may be no demonstrable structural or functional change owing to adequacy of the collateral circulation. For example, after ligature of one iliac artery small anastomoses from the mammary and deep epigastric arteries maintain the physical integrity and functional activity of leg muscles and other structures as well. Occasionally for the first few days, however, a condition of ischemic paralysis appears after continued use of the limb; with rest the muscles regained normal function. This fatigue paralysis usually disappears after 1–2 weeks. Microscopic examination of such muscles reveals no significant changes. In some cases, particularly if the collateral circulation is impeded, gangrene of the limb develops. Necrosis of skin is then nearly always more extensive than that of muscle. In such muscles there is sharply localized degeneration of muscle fibers and supporting tissues without hemorrhage, inflammatory reaction, or fibrosis.

The effects of temporary occlusion of the arterial supply to a limb are slightly different. As in the case of permanent occlusion, there is no demonstrable paralysis or tissue change, if collateral circulation is adequate. However, if the collateral circulation is restricted (e.g., by application of a tourniquet) and if the arterial obstruction exists for as long as 17 hours, there are scattered foci of necrosis with degeneration of muscle fibers or extensive necrosis of muscle and skin. The necrosis occurs despite resumption of normal pulsations in the major arteries of the limb. The pathologic change in the muscles consists of necrosis of muscle fibers and supporting tissues, with edema, hemorrhage, and a slowly developing inflammatory reaction. Fibrosis and permanent shortening of the muscle are slight.

Experiments in which the nutrient artery of a single muscle is occluded were also conducted by Brooks.[21] The rectus femoris was selected because most of its blood supply is obtained from one branch of the femoral artery. The results were similar to those which followed occlusion of the primary arteries to the extremity. The muscle tissue either remained viable and functioned quite normally or it became wholly necrotic and disappeared completely. Again, permanent shortening and diffuse fibrosis were not prominent features of this lesion.

Kaspar *et al.*[110] gave a rather complete description of the gross changes in the anterior tibial muscle of the rabbit after arterial occlusion. At 24 hours the affected muscle was swollen, soft, and hemorrhagic; by the fourth day it had become translucent, and by the sixth day its color was pale gray. These authors traced the degenerative and regenerative processes up to the 45th and 90th days, by which time the muscle had regained a normal appearance. Interestingly the connective tissue that had reacted earlier regressed as the myotubes matured.

VENOUS OCCLUSION

The results of obstruction of the vein from the rectus femoris muscle are quite different. Within an hour after operation the muscle is dark blue and swollen, and bleeds if cut. Electrical stimulation of the muscle or its nerve gives a slight contraction. After 24 hours the degree of vascular congestion is very pronounced, and there is extravasation of blood within the muscle and around it. Extensive migration of neutrophilic leukocytes is observed. At the end of 48 hours many of the muscle fibers are necrotic, and by the 4th day muscle degeneration is still more evident and is accompanied by a marked proliferation of fibroblastic connective tissue. At 10 days connective tissue is abundant, and in 2 of the 19 animals it almost completely replaced the muscle fibers. Usually some muscle tissue remains, and in it some degree of electrical excitability is preserved. The marked inflammatory reaction and the fibrosis always extends into the tissue surrounding the muscle. Shortening of the muscle is always present.

The experiments of Brooks established that the circulation disorder produced by venous occlusion is much more destructive to muscle than that following arterial occlusion. Both types of disorder vary in intensity according to the extent of the collateral circulation. The studies of Clark[30] demonstrated the astonishing ability of the remaining endomysium to regenerate the endomysial tubes following partial experimental ischemic necrosis. The fibrosis characteristic of venous occlusion therefore requires some additional factor. Brooks suggested that the stimulus to the fibrosis is a mechanical one. With the vein occluded and the artery open, the pressure in the capillaries is greatly elevated. This would damage the walls of these vessels, permitting extravasation of red corpuscles and plasma, and at the same time would interfere with nutrition of the tissues. The combination of ischemic damage and hemorrhage without complete obstruction to the blood supply is a more likely explanation of this pseudoinflammatory reaction and fibrosis.

EFFECTS OF IRRADIATION ON SKELETAL MUSCLE

Warren summarized the small amount of information available on the effects of irradiation on striated muscle. Many pathologists describe as "radiosensitive" those tissues and tumors that are severely damaged by less than 2500 r in divided therapeutic doses; as "radioresponsive" those affected by 2500–5000 r; and as "radioresistant" such tissues as do not react or react only to more than 5000 r. In the latter class are included liver, adrenal, lung, kidney, pancreas, central nervous system, and skeletal muscle. The unusual resistance of muscular tissue to irradiation is evidently due to its high degree of differentiation. In fields in which heavy irradiation has produced fibrosis and vascular damage, atrophy and other changes in the muscle fibers may develop secondarily.

Usually roentgen therapy produces only minimal atrophic changes. With light irradiation there may be no change at all or a slowly progressive diminution in fiber size, evidently due to sarcoplasmal damage. With heavy irradiation there appears vacuolation of the sarcoplasm with distortion of the myofibrils. The cross striations usually diminish in number or disappear early after heavy irradiation, and the zone between atrophic and normal fibers is sharp. Occasionally there are some hypertrophied fibers at the periphery of the atrophic zone. Interstitial fibroblastic proliferation with subsequent fibrosis follows the atrophic changes in the muscle fibers.

Interstitial irradiation—radon in amounts of 5–55 mc—produces within 40 hours well defined zones of chalky white necrosis in the lumbar muscles of rabbits. The necrotic zones are surrounded by congested capillaries and some small hemorrhages. Histologically the muscle fibers in the inner zone exhibit a loss of striations, fragmentation, and homogenization of sarcoplasm. Peripherally the muscle nuclei seem normal in spite of sarcoplasmic changes. In the inner zone the nuclei are broken apart. The amount of radon exposure determines the size of the lesion, e.g., 5 mc produces a zone of necrosis 4 mm in diameter and 55 mc produces a zone of necrosis 15 mm in diameter. The necrosis develops rapidly, and exposure beyond a critical time produces little further change.

Bade[8] exposed the skeletal muscles of dogs to doses of roentgen rays that produced permanent loss of the overlying hair. Histologic examination of the irradiated muscle tissue from 4 days to 2 months after exposure disclosed no morphologic changes. When rats were given sufficient radiation to produce skin ulceration that would eventually heal, neither the muscle fibers nor their connective tissue sheaths showed any perceptible alterations during observation periods of 4 days to 4 weeks after the last treatment; no histologic changes were detectable in the muscle. The author therefore concluded that in outlining a plan of radiation therapy injury of muscle need not be expected if the skin tolerance dose is not exceeded.

Hirschberg,[103] on the other hand, reported localized muscular atrophy in humans (lower leg in one case and posterior neck region the other) several years after radiation therapy. In one instance the radiation had been given for a sarcoma of the tibia and in the other for trichophytosis of the hair on the head. In both cases the muscles had been irradiated during childhood, and the atrophy did not become manifest until later. It seems likely, in view of experimental work, that muscle atrophy in these cases developed secondary to vascular damage and slowly progressive interstitial fibrosis. This delayed sequence of vascular injury is well known in organs such as the brain.

The effect of irradiation on myogenesis in experimental muscle disease was studied by Bade,[8] Reznik and Betz,[167] and Sloper and Pegrum.[180] Using doses of 750–1500 rads, an amount insufficient to injure muscle, Reznik and Betz[167] observed that carbon dioxide (CO_2) freezing or crushing of muscle in mice was not followed by the usual reaction. The lower dosage prevented the majority of myoblasts from entering more than one mitosis, and by the 10th day there was a diminished number of myotubes. At the higher dosage the suppression was marked, and only rare, rudimentary myotubes were ever seen. In contrast the degree of inflammatory reaction and amount of phagocytosis were unchanged. These observations suggest that myogenesis depends on local factors at the site of injury which are radiosensitive.

EXPERIMENTAL DEGENERATION INDUCED BY CHEMICAL TOXINS

Adult skeletal muscles, unlike other tissues, are resistant to almost all known parenterally administered chemical toxins. For example, carbon monoxide, cyanide, and other substances which in low concentrations readily damage brain tissue are without effect on muscle, and the same is true of carbon tetrachloride, which is harmful to the liver; and alloxan, which destroys beta cells of the pancreatic islets. It seems that the mechanism of action of these substances is either an interference with specific enzyme systems or metabolic pathways that do not exist in muscle, or that muscle can use anaerobic cycles of metabolism when tissue respiration is damaged by substances such as cyanide.

MONOIODOACETIC ACID

Lundsgaard[128] first demonstrated an inhibition of muscle metabolism by a chemical agent. He showed that when an isolated muscle was made to contract in the presence of monoiodoacetic acid it rapidly became fatigued and developed rigor. Moreover, he found that no lactic acid was produced during contraction, in contrast to normally active muscle in which considerable lactic acid accumulates. This chemical inhibition of muscle, since known as the Lundsgaard effect, was revolutionary in the field of muscle physiology. To the best of our knowledge no histopathologic study of this effect was ever carried out. When iodoacetate was administered parenterally even in quite small doses, the animals usually died rapidly. Recently it has been shown that this substance uncouples oxidative phosphorylation and damages mitochondria.

PLASMOCID

During the course of a study on the metabolic inhibitors of brain tissue, Hicks[96] discovered that plasmocid almost specifically injures skeletal muscle. Plasmocid, or 8-(3-diethylaminopropylamino)-6-methoxyquinoline dihydrochloride, a known protoplasmic toxin, was originally tested as an antimalarial compound but was discarded because of its high toxicity.

Personal observations of the effects of this substance[153] have shown that a very small amount of plasmocid (12–16 mg/kg body weight) in divided doses, an amount equivalent to only 2–3 mg in a 200-g animal, caused destruction of striated muscle tissue in rats and mice. Cardiac muscle was less often affected. If a fairly large dose was given intraperitoneally the animals became ill within an hour and died soon thereafter. Sublethal doses caused some of the animals to become weak and listless whereas others were not visibly affected. No muscular spasm or actual paralysis was observed.

Microscopic examination of the tissues of these animals demonstrated that the striated muscles which suffered the most extensive damage were those that evidently maintained the highest levels of activity, such as the diaphragm, abdominal musculature, external eye muscles, masseter, and tongue. The diaphragm was damaged in all 50 animals given lethal and

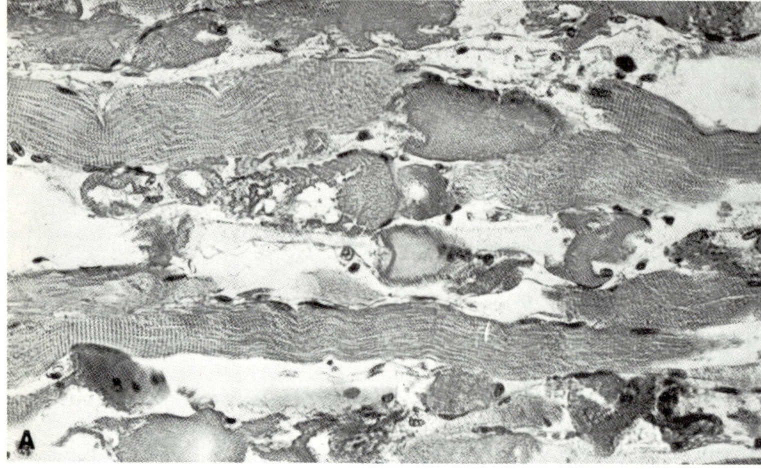

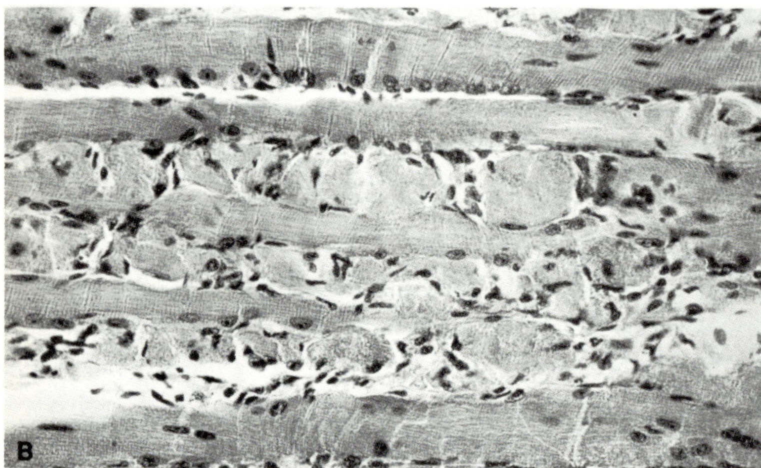

Fig. 3–42. (A) Plasmocid intoxication. Hyaline degeneration of muscle fibers 6 hours after intraperitoneal administration of plasmocid. (B) Twenty-four hours after plasmocid. Note granular degeneration, early phagocytosis, and proliferation of sarcolemmal nuclei. (hematoxylin, eosin)

sublethal amounts of plasmocid; the extrinsic eye muscles and abdominal musculature were involved approximately 80% and the tongue and masseter in about 70%. Other muscles—those of the extremities, the pharyngeal constrictors, the spinal extensors, were affected in a spotty and unpredictable fashion. It should be noted that the lesions were most pronounced in muscles whose fibers are rich in granules (mitochondria).

The first sign of muscle injury was seen about 6–8 hours after intraperitoneal administration of plasmocid. By this time some of the muscle fibers had already developed an eosinophilic hyaline or granular appearance, and the striations, both cross and longitudinal, had become indistinct or disappeared. Some fibers were swollen and of irregular width; others were of natural appearance (Fig. 3–42A).

Interstitial edema, as indicated by separation of muscle fibers, and a general looseness of the tissue was prominent.

At 12 hours destruction of muscle fibers was more obvious. The acidophilic granular and hyaline fibers were more numerous and often were fragmented with the formation of retraction caps or clots of muscle protoplasm. Most of the sarcolemmal nuclei, as judged by their form and staining reaction, remained viable even when the muscle contents were necrotic. Some, however, were shrunken and pyknotic with folded nuclear membranes. In places there were infiltrations of polymorphonuclear leukocytes and evidence of early activation of interstitial connective tissue elements. In some muscles the involvement was spotty, affecting only a solitary fiber or segment of a fiber, which was converted into rounded acidophilic

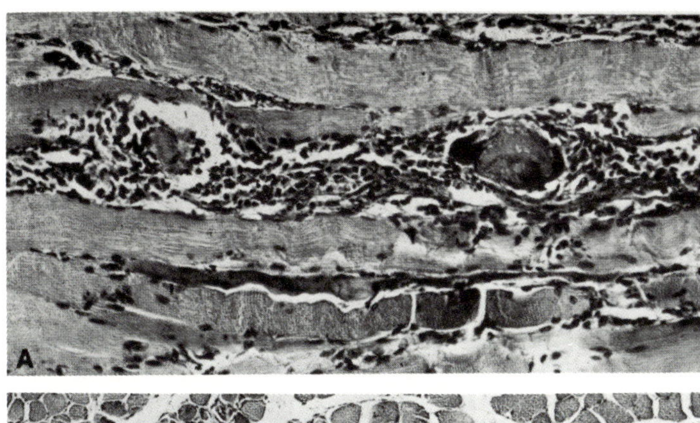

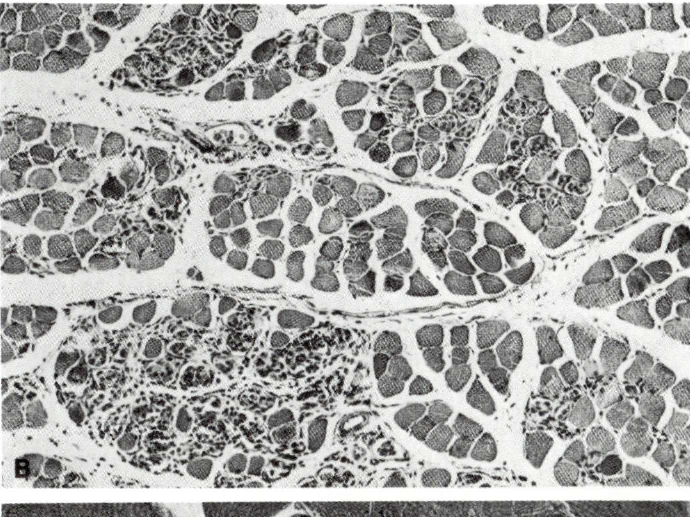

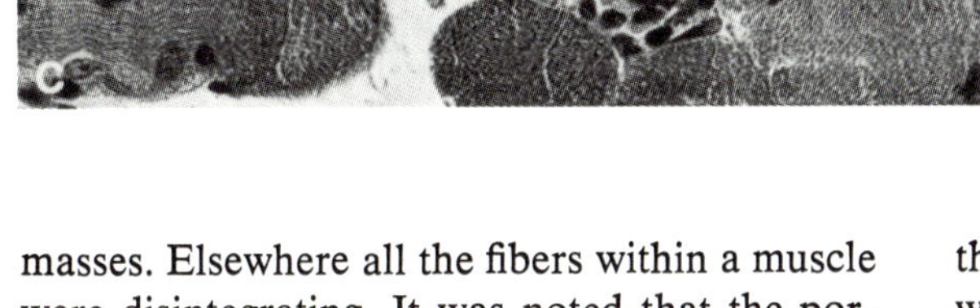

Fig. 3–43. (A) Necrosis of isolated fibers and groups of fibers in a limb muscle after plasmocid administration. (B) Cross section of focal destruction. (C) Focal destruction in two fibers. (A, ×175)

masses. Elsewhere all the fibers within a muscle were disintegrating. It was noted that the portions of muscle fibers near tendinous or fibrous attachments tended to be most vulnerable to destruction.

At 18–24 hours intense cellular activity was in evidence. There were large numbers of hyperplastic sarcolemmal nuclei, presumably developing from the remaining viable muscle nuclei, and many invading phagocytic cells (Fig. 3–42B). The sarcolemmal tubes were filled with these cells. Even in otherwise normal fibers the sarcolemmal nuclei appeared to have been activated, and many had migrated toward the center of the muscle fibers. These nuclei were enlarged and basophilic and contained one to three prominent nucleoli; they were often seen in isolated segments of empty sarcolemmal tubes, the contents of which had been removed by phagocytosis or had retracted. A striking feature in partially affected muscles was the involvement of small circumscribed groups of muscle fibers (Fig. 3–43B), as if some parts of the muscle were more susceptible to the toxin than others.

After 24–36 hours cellular activity was prolific. There was a profusion of macrophages and inflammatory cells (Fig. 3–43A), and many could be seen in the process of mitotic division. This cellular activity was indeed the most striking feature of the entire process and may indicate that plasmocid specifically stimulates such a response. No similar intense cellular response has been seen in muscle following other types of injury. In some areas all muscle fibers had disappeared and were replaced by a field of heterogeneous cells. It was usually possible to distinguish sarcolemmal nuclei from phagocytic cells. The destructive action of the toxin seemed to be directed chiefly to the con-tents of the muscle fiber. The sarcolemmal sheath usually remained intact, and many sarcolemmal nuclei were unharmed.

Between 24–48 hours after administration of plasmocid the first signs of regeneration of the muscle fibers became manifest. In the lumen of the tubes or lying flattened against the inner walls of the sarcolemmal membrane were plump, elongated nuclei which contained one to three prominent nucleoli, vesicular nucleoplasm, and heavily staining nuclear membranes (Fig. 3–44). The cytoplasm was often

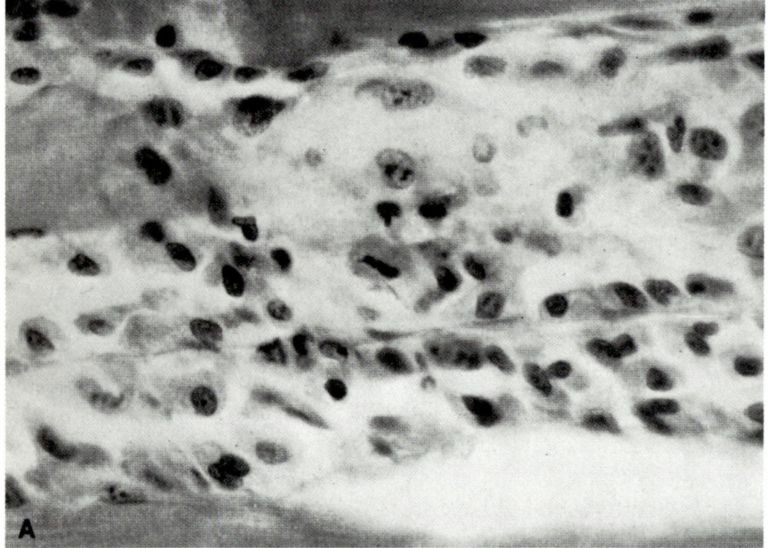

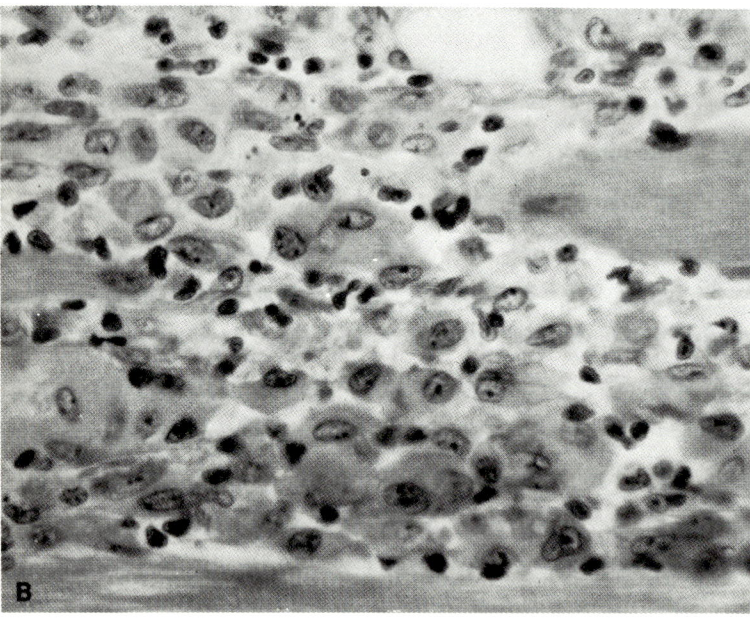

Fig. 3–44. (A) Sarcolemmal tubes containing many myoblasts whose nuclei have one to three prominent nucleoli. Macrophages are still present in these tubes. Plasmocid toxicity, 48 hours. (B) Myoblasts containing large amounts of lilac cytoplasm. Fusion has occurred in the multinucleate cell (below center), while sarcoplasm of several myoblasts in the central fiber is in the process of fusion with adjacent single cells and with the tip of the parent fiber. Note that myoblastic nuclei are identical with sarcolemmal nuclei in retracted end of intact muscle fiber. Plasmocid damage, 55 hours. (hematoxylin, eosin, ×650)

sparse, spindle-shaped or angular, and amphophilic or basophilic. These were the myoblast nuclei that were preserved as the sarcoplasmic substance was destroyed. During the next few hours there was a striking increase in these myoblasts by (1) activation of all the surviving sarcolemmal nuclei, (2) aggregation with the formation of short rows of juxtaposed nuclei in a ribbon-shaped band of cytoplasm. Mitotic division was also in evidence in isolated cells. All the myoblasts acquired a large amount of granular lilac cytoplasm that gave a positive stain for RNA. At about 55 hours after injury fusion of the cytoplasm of adjacent myoblasts began to occur (Fig. 3–44B). Some of these cells joined with the terminal clubshaped stumps of viable muscle fibers, which contained one or several nuclei identical with those in the proliferating myotubes (Fig. 3–44B). Faint longitudinal fibrils extended between the newly fused cytoplasm of some of the cells. By 72 hours there was a sudden transformation so that single or multinucleate myoblasts were difficult to find and in their

place were one or several parallel columns of closely packed or elongated nuclei embedded in a lilac granular sarcoplasm (Fig. 3–41) that showed an intense reaction for RNA. Faint longitudinal fibrils were often carefully aligned in these new muscle fibers. At 96 hours the nuclei were still centrally positioned in columns; the fibers were packed with myofibrils, most of which were cross-striated; and the basophilia of the fibers was changing to pink. By the 7th day the new fibrils were mature in every respect except that they were somewhat narrower than normal and contained an excess of nuclei, which were usually peripherally displaced in the fiber, and the sarcoplasm still retained a faint bluish tint. By 3 weeks the muscle tissue had resumed its normal appearance except for a slight increase in sarcolemmal nuclei (Fig. 3–45). At no time during the rapid evolution of this regenerative process was terminal budding from intact fibers a prominent feature of the sequence.

By the 30th day after plasmocid administration (the limits of the experiment), only mini-

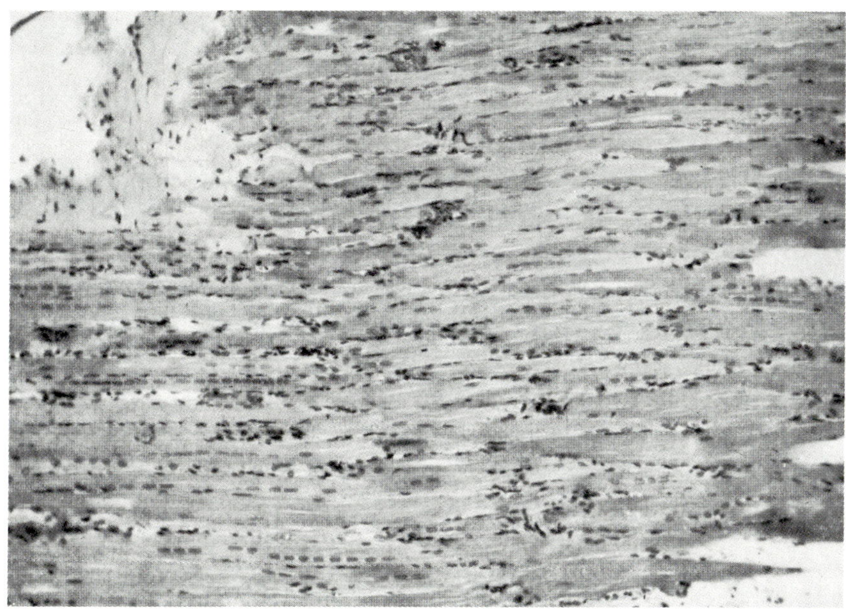

Fig. 3–45. Rat diaphragm 21 days after plasmocid necrosis. New fibers are of approximately normal width, though they are not sufficiently parallel to the plane of the section to be certain of this. Faint basophilia is still present. Short or long rows of centrally positioned nuclei are still present in many fibers.

mal evidence of the previously existing damage remained. A very rare fragment of a necrotic fiber could be seen, and some healthy-appearing fibers contained more nuclei than was normal. The muscle fibers were by this time of normal diameter. There was no evidence of connective tissue replacement of muscle, as was so often the case in other types of muscle destruction. No branched forms of muscle fibers were present, and no interlacing of fibers, as noted after crushing of a living muscle. There was no vascular or nerve damage. The status of the motor end-plates in this type of destruction has not been studied. Nerve

cells in the ganglia (such as the gasserian) were destroyed, but all other organs and tissues were unaffected.

Colchicine, a known mitotic inhibitor, was given to a few of these animals 48 hours after plasmocid and 18 hours before death. In these animals, if proliferation of fibroblasts, histiocytes, and sarcolemmal cells was pronounced, a great many atypical mitotic figures were visible in the interstitial cellular elements and occasionally within muscle fibers (Fig. 3–46). It was not possible to determine on morphologic grounds alone whether these mitoses within the fibers were actually those of muscle

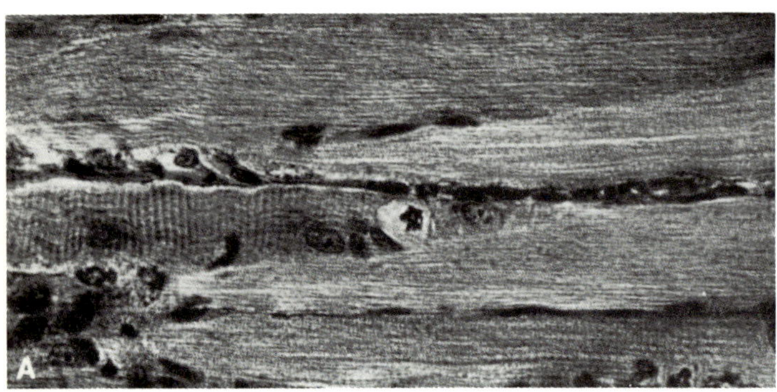
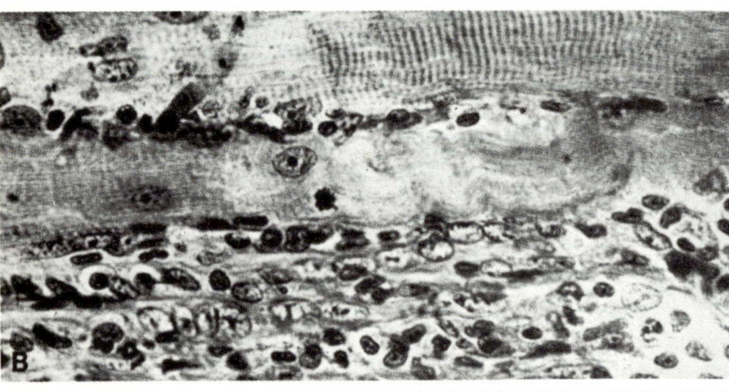
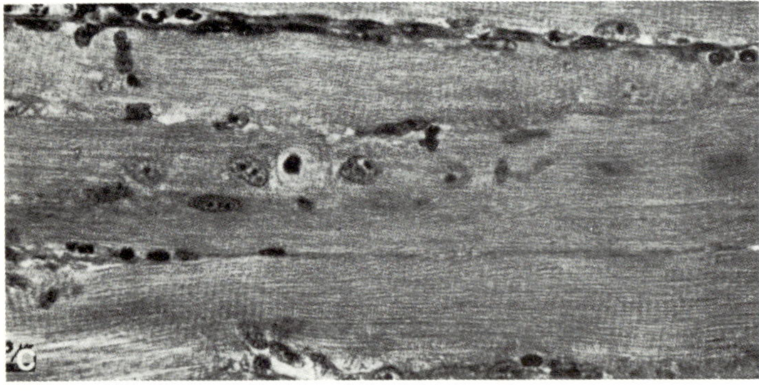

Fig. 3–46. Mitoses in muscle after plasmocid and colchicine. (×570)

nuclei or of macrophages that had invaded some of the fibers.

In a small group of animals, repeated small doses of plasmocid were given for several weeks. Cumulative effects of the drug did not occur; in fact, the animals became quite resistant to dosages that in other animals were lethal within a few hours. In one such animal a total dosage of 194 mg/kg body weight was given in increasing amounts over a 30-day period. This was about 15 times the usual lethal dose. Sections of the diaphragm and abdominal muscles demonstrated a curious mixture of muscle destruction and regeneration of many different ages. Interstitial cellular infiltrates were not prominent in these lesions. Many of the other skeletal muscles showed no damage at all.

The histologic effects of several other known poisons and metabolic inhibitors have been studied by Hicks,[95, 96] including fluoroacetate, cyanide, dinitrophenol, carbon monoxide. None on these affected skeletal muscle, and only two (fluoroacetate and dinitrophenol) occasionally damaged cardiac muscle.

It may be concluded from these studies that the quinoline compound plasmocid has an almost specific affinity for striated muscle and may be considered a myotoxic agent. The destructive effect of this substance is exerted principally upon the contents of muscle fibers and leaves intact the sarcolemmal membranes and most of the muscle nuclei, as well as the interstitial tissue and its cellular components, the nerves, and the blood vessels. The response of histiocytes to the administration of plasmocid or to the necrosis of muscle it produces is striking. Because of the specificity of this agent for muscle and the rapidity of its action, the effects can probably be ascribed to a selective interference with the metabolism of the muscle fiber, e.g., the blockage or poisoning of a muscle enzyme or enzyme system. Unfortunately, we have no knowledge of the particular action of plasmocid. A more thorough investigation of such specific myotoxic agents may provide a means of securing valuable information about the metabolic functions of the muscle fiber in vitro and about the level of metabolic activity of individual muscles in the body.

EXPERIMENTAL CHLOROQUINE MYOPATHY

Experimental chloroquine myopathy is another example of a toxic myopathy rather different from plasmocid myopathy in that vacuolation of the muscle fiber rather than segmental necrosis is the primary lesion. First observed in a patient who under prolonged treatment with chloroquine had developed severe weakness only to recover when the drug was stopped, the vacuolar and presumably reversible nature was recognized from the beginning. Rewcastle and Humphrey[164] studied other examples and emphasized that the vacuoles contained cytoplasmic degradation products. MacDonald and Engel[129] investigated the lesion in the laboratory using rats of different ages.

The excellent study of MacDonald and Engel demonstrated three types of lesions: (1) small vacuoles, either central or peripheral in an increasing proportion of fibers, appearing after 5 days of treatment and involving 70% of fibers by the 20th day; (2) fiber splitting, with the daughter fibers being of many different sizes; (3) optically dense granules measuring up to 10 μ present in an increasing proportion of the fibers (Fig. 3–47).

The vacuoles contained material that stained intensely in acid phosphatase preparations. Also periodic acid-Schiff (PAS) stains were positive even after exposure to diastase. The dense granules gave a staining reaction indicative of phospholipids.

Electron microscopy revealed newly formed membranes derived from the longitudinal sarcoplasmic reticulum and the T system in both red and white fibers. In fact all the vacuolar spaces were lined by single or double membranes. An array of tubular networks of T system origin surrounded some of the vacuoles. The optically dense granules in phase microscopy were composed of myeloid structures in places situated next to Golgi complexes, SR membranes, or T system networks. Analysis of these lesions convinced MacDonald and Engel that the vacuoles are truly autophagic in nature and the myeloid bodies were interpreted as secondary lysosomes presumably reactive to chloraquine molecules.

Whisnant *et al.*[216] reported also on the noninflammatory vacuolar myopathy in man, and

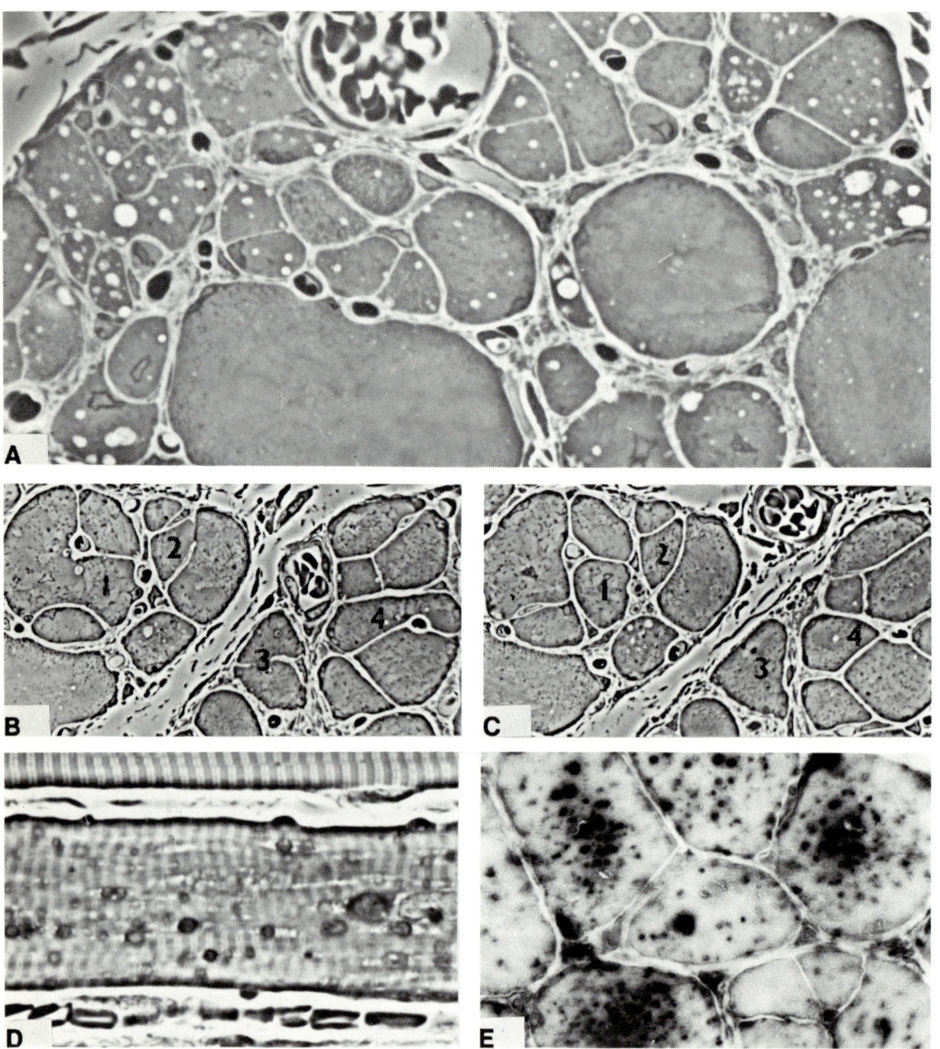

Fig. 3–47. (A) Soleus muscle from chloroquine-treated rat. Majority of fibers contain vacuoles of various sizes. Fiber diameter varies greatly, and smaller fibers are molded together. (B and C) Soleus muscle of chloroquine-treated rat. Nonconsecutive serial sections separated by 60-μ intervals. Fibers 1, 2, and 4 can be seen to split off from the parent fiber. Incomplete subdivision of fiber 3 is noted in (B). (D) Gastrocnemius muscle from rat treated with chloroquine for 100 days. Optically dense structures of different dimensions are present in fiber. Some of these are in or adjacent to small clear spaces. (E) Soleus muscle from rat treated with chloroquine for 45 days; acid phosphatase reaction. Intense enzyme activity is present in circumscribed areas of all fibers. (Courtesy of Andrew Engel[128]) (A, ×650; B, C, ×520; D, ×820; E, ×420)

the lesion was reproduced in swine by Chester and Gleiser[24] and by Eadie and Ferrest.[51] Vacuolation, necrosis, and splitting of individual fibers and the appearance of optically dense bodies were seen under the light microscope and preceded myophagocytosis. The first ultrastructural change, seen after 5 days of treatment, consisted of proliferation of membrane systems and "encirclement of small cytoplasmic areas by double membranes." Enlargement and fusion of such spaces led to the formation of larger vacuoles; acid phosphatase in the vacuoles indicated their autophagic nature. A vigorous myoblastic reaction restored the integrity of fibers within a few weeks. Lesser degrees of intoxication resulted in greater focalization of lesions and in the heart muscle fibers in perinuclear vacuolation. The peripheral nerves are also affected (neuromyopathy) and fat cells (lipodystrophy).

METALLIC POISONS

The effects of metallic poisons such as lead, arsenic, and antimony on striated muscle have been investigated with unconvincing results. Villaverde[202] carried out a series of experiments on lead intoxication, one of which was a study of the changes induced in the muscle fiber. He reports two types of lesion, one a muscular atrophy secondary to a parenchymatous lead "neuritis" and the other a peculiar fibrillary degeneration. The latter consisted of a progressive loss of transverse striations and an exaggeration of longitudinal striations. The muscle fibers tended to decompose into homogeneously colored longitudinal columns separated by clear spaces. The sarcolemmal nuclei became shrunken and dark and eventually pyknotic. No proliferation of sarcolemmal nuclei or other reactive changes were observed even in the most advanced and chronic stages of lead poisoning. Villaverde was at a loss to explain this lesion. He noted that it differed from Zenker's degeneration and did not believe it was determined by the nutrition of the animals. Changes of this type are difficult to evaluate and it is not at all clear that lead intoxication is responsible for them. Agonal changes and artifacts of fixation and staining can produce a similar appearance.

POLYMYOPATHY IN EXPERIMENTAL INFECTIONS

A group of viruses capable of producing paralysis in experimental animals was discovered in the study of human poliomyelitis. These are sometimes referred to as the "poliomyelitis group of viruses." Some are pathogenic for rodents, and others for swine, monkeys, and the like. Included in this group are the Lansing strains of poliomyelitis; the SK strain of Trask, Vignec, and Paul; the MF strain of van Rooyen; the MM strain of Jungeblut; Theiler's encephalomyelitis virus; and the Coxsackie group of viruses first described by Dalldorf. All produce infection in mice and other rodents. The characteristic lesions in the nervous system consist of necrosis of nerve cells in the anterior horns of the spinal cord and in the brain, with neuronophagia and inflammatory changes consisting of perivascular and focal infiltration of lymphocytes and other mononuclear cells. The skeletal muscles, which are deprived of motor innervation, become atrophic. Clearcut criteria for classification of this group of viruses is lacking, though antigenic differences have been demonstrated.

COXSACKIE VIRUS

More relevant to the subject of muscle disease are the reports by Dalldorf and Sickles,[36] Dalldorf,[35] Rustigian and Pappenheimer,[172] and Melnick and Godman.[137] The first of these reported on the isolation of a filtrable virus from two human cases of alleged poliomyelitis from Coxsackie, New York, which produced a severe polymyositis in suckling mice and hamsters.[36] The two patients were males aged 9 and 3½ years. The symptoms were of acute onset and consisted of nausea, headaches, and pains in the extremities. There was weakness of trunk muscles and both legs in one case, and of one leg in the other. The weakness was still present at the end of 7 months in the first case and had disappeared by the 8th month in the second. Muscle biopsies were not performed. The same virus has been isolated from other patients in whom the clinical picture was that of nonparalytic poliomyelitis, an epidemic myalgia (pleurodynia), or aseptic meningitis.

Fecal suspensions from these two cases were inoculated intracerebrally into albino mice and hamsters. Suckling mice 3–7 days of age became paralyzed, while those more than 12 days old did not. Suckling hamsters were also susceptible. After the initial isolation injection of the virus into other animals usually resulted in paralysis, "spasticity," or tremulousness within 3–5 days after inoculation. Paralysis also followed intraperitoneal and intramuscular injections. Mouse brain suspensions, filtered through a Mandler candle, still produced paralysis when injected into other animals. Neutralizing antibodies were not present in the serum during the acute phase but were found in the convalescent phase of the two human cases.

Extensive studies during the years following the original discovery by Dalldorf and Sickles[36] of the prototype Coxsackie viruses in 1948 have revealed that there are at least 19 immunological different viruses within group A and 5 immunologically different types in group B.[85] The A group causes a pharyngitis called herpangina and probably several other less well defined illnesses. The B group has been established as the etiologic agent in epidemic pleurodynia, aseptic meningitis, and some cases of acute benign pericarditis, as well as in myocarditis in the newborn.

Pathologic Changes. The pathologic changes permit separation of the two strains of the virus.[35] The A strain affects the muscle primarily. The B strain involves both the muscles and the central nervous system. Opaque, whitish streaks can be seen in some of the muscles of the limbs, trunk, face, and scalp in animals inoculated with the A strain. The microscopic pathology consists of severe and widespread degeneration of skeletal muscle fibers. There is loss of striations, swelling, hyalinization, and acidophilia of the individual fibers (Fig. 3–48). Phagocytosis of fragments of degenerated fibers is quite marked, producing intense cellular reactions within the sarcolemmal tubes. Sarcolemmal nuclei usually survive and proliferate actively, in places forming mutinucleated buds. Lymphocytes, activated histiocytes, and a few plasma cells infiltrate the endomysium but are not numerous. Chemical analysis of the affected muscle reveals a loss of potassium and creatinine. There is also a creatinuria.[79] Smooth muscle and myocardium are not involved. There are no lesions in the central or peripheral nervous systems. Animals inoculated with the B strain are more often spastic and tremulous, and at autopsy their brains

and spinal cords are soft and the fat pads usually pale. The muscle lesions are focal and less extensive. Patchy necrosis of gray and white matter of the nervous system occurs, with a loosening of the ground matrix, degeneration of neurons, and infiltration of the lesions by inflammatory and phagocytic cells. In fact, a very large area of cerebrum or cerebellum may exhibit necrosis. Aseptic necrosis of fat pads is another characteristic finding. These two groups of viruses are also serologically distinct.

Repeated recovery of the virus from the specimens of human feces and the immunologic response of the patients indicate that this virus is capable of infecting man and has the unique property of causing a severe myositis in animals. It is believed that this agent can be differentiated from the Theiler viruses, from the MM virus, the virus of encephalomyocarditis, and the Lansing mouse-adapted poliomyelitis stains.

Dalldorf[34] expressed the opinion that the muscle lesion produced by virus is similar to Zenker's degeneration and emphasized that the earliest change, a hyaline necrosis, preceded the cellular infiltration. We agree with this interpretation and conclude that this muscle lesion is more properly an acute toxic myopathy rather than a true myositis or dystrophy and that it is remarkably like that produced in suckling animals by vitamin E deficiency and by plasmocid intoxication. There are recent reports that muscle lesions of this type occur in man. The peculiar susceptibility of suckling animals of certain species to these viruses is unexplained.

OTHER VIRAL INFECTIONS OF MUSCLE

Rustigian and Pappenheimer[172] investigated this problem experimentally. They tested a mouse encephalomyelitis virus of Theiler as well as the SK virus of Jungeblut, the lymphocytic choriomeningitis virus, herpes simplex virus, Lansing strain of poliomyelitis, JHM virus, and eastern equine encephalitis viruses. From this study it became clear that certain neurotropic viruses have a marked pathogenicity for skeletal muscle in young suckling animals. The most intense lesions were pro-

Fig. 3–48. Virus myositis, showing a focus of fiber destruction.

duced by Jungeblut's SK virus. Even in high dilutions symptoms were recognized within 15 min and definite lesions were produced. There was edema of intramuscular septa between the fibers and on occasional neutrophilic leukocyte. Necrosis of the muscle with hyalinization and segmentation of fibers and leukocytic infiltration were evident within 3–12 hours. Sarcolemmal nuclei often survived, and mitoses presumably in myoblasts were seen within 1–2 days. The animals usually died in 48 hours. The peripheral and central nervous systems were not affected. The mouse encephalomyelitis virus produced similar but more chronic lesions. Many

of the muscle fibers were necrotic and had been converted into eosinophilic masses. Between the fibers there was a sparse inflammatory exudate of neutrophilic leukocytes, lymphocytes, and mononuclear histiocytes. Many of the necrotic fibers were invaded by these inflammatory cells, and phagocytosis of degenerated muscle tissue could be observed. The inflammatory reaction extended to the intramuscular fat and fibrous tissue. Regenerative activity in surviving fragments of muscle fibers began by the fourth day. The sarcolemmal nuclei that escaped destruction proliferated actively. They aligned themselves longitudinally, myofibrils appeared at their inner surface, and soon cross striations also appeared. In the late stages fibrosis was common, and calcification of necrotic muscle occurred occasionally. The nuclei of the regenerating muscle contained eosinophilic inclusions (Fig. 3–49). The lymphocytic choriomeningitis virus, when injected intramuscularly into suckling mice, gave rise to a profuse lymphocytic and monocytic infiltration of the intramuscular fat and connective tissue but caused no necrosis of muscle fibers. Similar lesions were produced by the MM virus. The other viruses have no effect on striated muscle.

More recently oncogenic viruses (polyoma and SV-40) were shown by Fogel and Defendi[73] to infect myoblasts in tissue culture. These cells later fused to form myotubes and were able to synthesize DNA before and also after fusion, unlike normal myoblasts. In contrast, multinucleated myotubes were resistant to penetration by virus particles. Yaffe and Gershon[220] found, in addition, that by 50 hours after exposure to the virus some of the myotube nuclei began to undergo mitotic division. The nuclear division, however, progressed only as far as metaphase, and thereafter the chromosomes became clumped. Kaighn et al.[109] did not succeed in corroborating the latter finding in avian muscle.

Viruses have been seen in electron micrographs of normal human eye muscle[23] and in biopsies from polymyositic patients.[29, 133, 153]

NONVIRAL INFECTIONS

The effects of bacteria and other infectious agents on skeletal muscle have not been studied systematically. Diphtheria toxin in guinea pigs has produced a fatty degeneration of the cardiac muscle and more recently similar changes were seen in the leg muscles of rabbits. Direct inoculation of bacteria into muscle has resulted in lesions that do not differ from those produced by any bacterial necrosis (*Clostridial Infections,* Chapter 7). Klinge[114] attempted to produce the lesions of rheumatic fever in rabbits by injecting purified horse serum into previously sensitized animals. In addition to proliferative lesions on the heart valves, he succeeded in producing a focal necrosis of

skeletal muscle, with waxy degeneration, phagocytosis of muscle fragments, and proliferation of sarcolemmal nuclei. Infiltrations of neutrophilic and eosinophilic leukocytes were also observed. The lesions had little resemblance, in our opinion, to Aschoff bodies.

Clostridium perfringens toxin when instilled near a muscle hydrolyzes the various phospholipids of the sarcolemma causing destruction of the fibers. Strunk et al.[188] followed the evolution of the discrete defect in the plasma membrane by the use of ferritin, which penetrates the fiber at the site of the lesion. The subsequent degeneration of the myofibrils begins at the I and Z bands. Later the sarcoplasmic reticulum, mitochondria, and nuclei degenerate.

The coyotillo fruit causes a toxic myodegeneration in goats.[44]

DEGENERATIVE CHANGES IN MUSCLE DUE TO VITAMIN DEFICIENCY

VITAMIN E

In recent years it has been established that at least in some circumstances vitamin E is a necessary factor for preservation of the integrity of skeletal muscle fibers.[59] This has been demonstrated in a number of species of animals, including mice, guinea pigs, rabbits, goats, sheep, rats, monkeys, dogs, tree kangaroos, and birds.[152] Contrary to the early reports that vitamin E was necessary for the successful completion of pregnancy, no sexual changes were observed in most of the experimental animals.

The onset of muscular lesions was usually manifested by a dragging of the hind limbs and "clenching" of the paws in rats and guinea pigs. In mice, however, no recognizable symptoms were usually exhibited. No instance of frank paralysis was observed.

The gross and microscopic changes have been dealt with in several comprehensive studies by Pappenheimer and his associates[152] and by West and Mason.[215] In some of the severely affected animals, yellowish gray streaks were evident in the muscles upon gross examination. In young mice born of

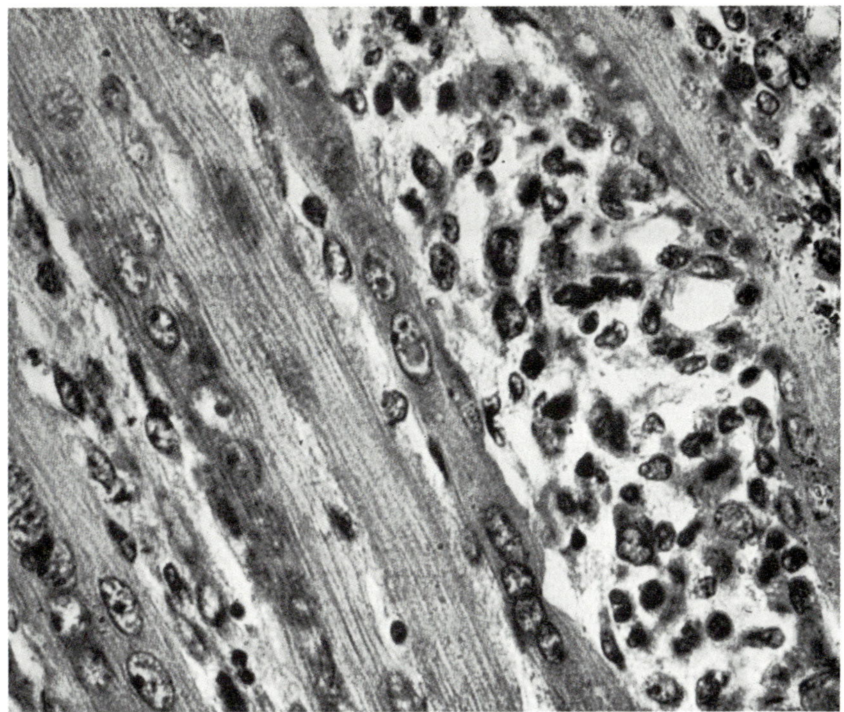

Fig. 3–49. Muscular degeneration caused experimentally by Theiler's virus. Note eosinophil inclusion body in sarcolemmal nucleus (at center).

Fig. 3–50. Vitamin E deficiency. (A) Early change in chick muscle. (B) Edge of a large area of duckling gluteus muscle in which a central necrotic zone (left lower corner) makes transition to normal fibers (right upper corner).

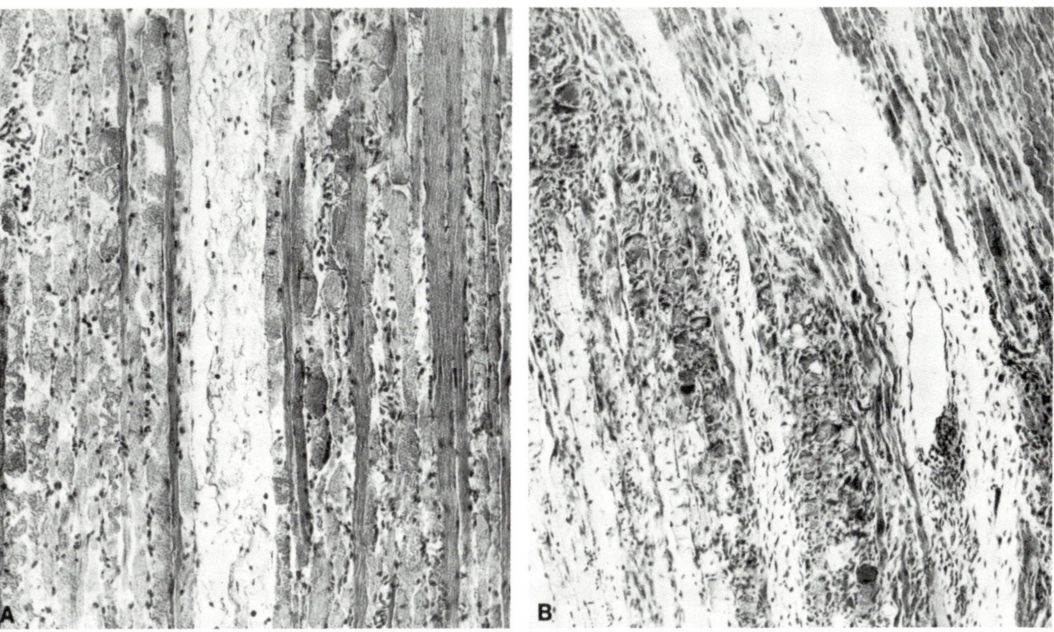

mothers deficient of vitamin E during gestation, about 50% had muscle lesions; mice sacrificed from the 2nd to the 5th day showed lesions in about one-third of the cases; the group sacrificed between the 6th and 15th day had lesions in 97%; those surviving 16 to 35 days were affected in 59% of cases; those killed at 35 to 39 days exhibited changes in about 9%. The incidence rose in succeeding generations kept on vitamin E-deficient diets.

The gray streaks in the muscles are due to a patchy hyaline necrosis that rapidly becomes intensely cellular. In the earliest stage large areas of a section of an affected muscle show fragmentation of hyaline eosinophilic muscle fibers (Fig. 3–50). The sarcolemmal nuclei stain darkly but are pyknotic only in the center of such areas, where damage to the endomysium, always of relatively minor degree, may also occur. Within the 1st week there is intense proliferation of the sarcolemmal nuclei, and the hyaline fragments each attract abundant phagocytic histiocytes and mononuclear cells (Figs. 3–50 and 3–51). An intense cellular

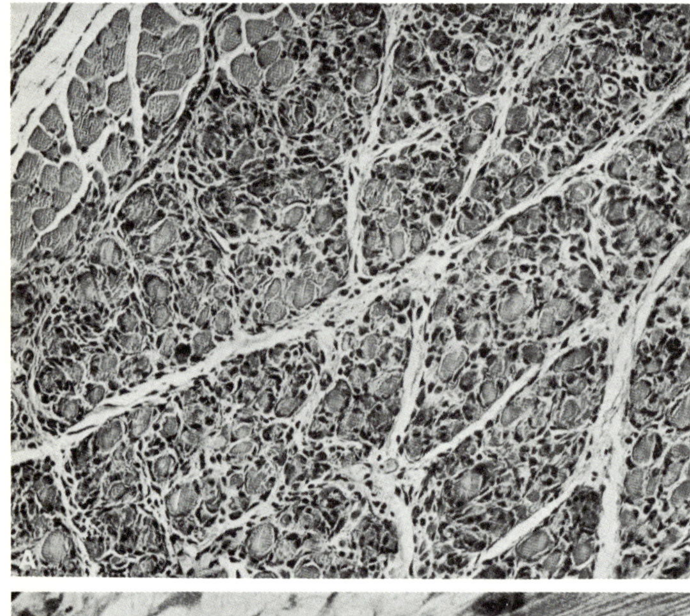

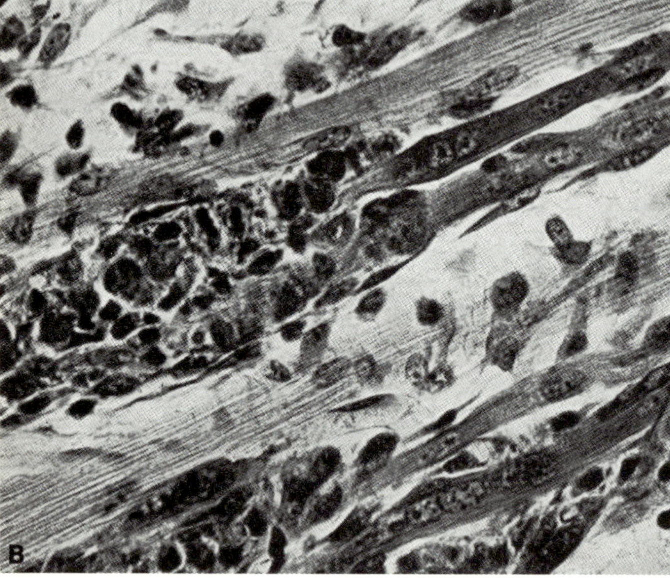

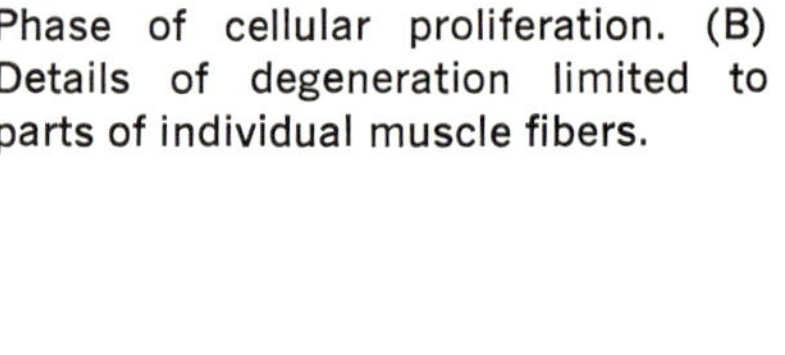

Fig. 3–51. Vitamin E deficiency. (A) Phase of cellular proliferation. (B) Details of degeneration limited to parts of individual muscle fibers.

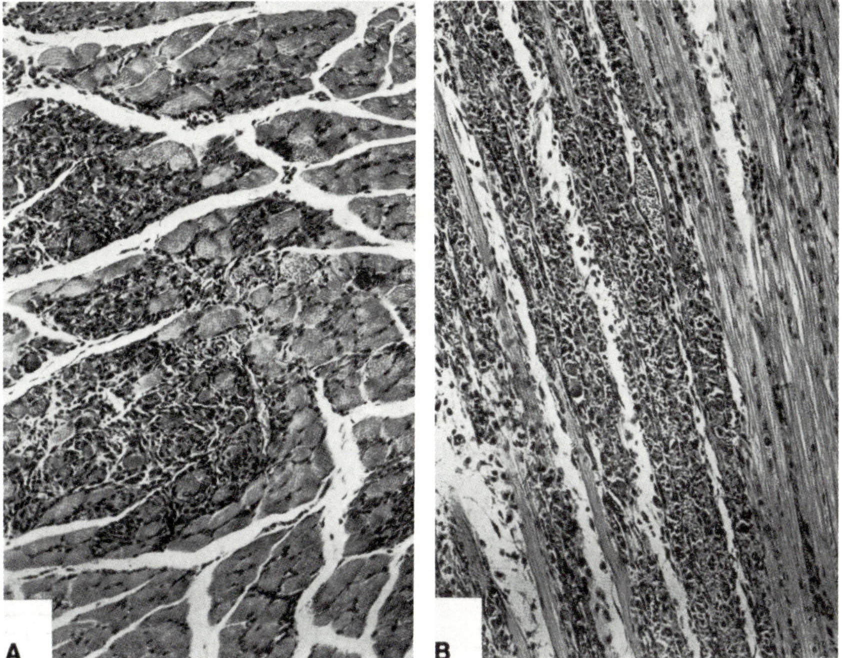

Fig. 3–52. Vitamin E deficiency. Selective effect of lesions is illustrated.

Fig. 3–53. Vitamin E deficiency. (A) Isolated muscle cells with fibroblasts and histiocytes. Architecture of the muscle has been destroyed. (B) Terminal stage, with isolated residual muscle fibers lying in adipose tissue.

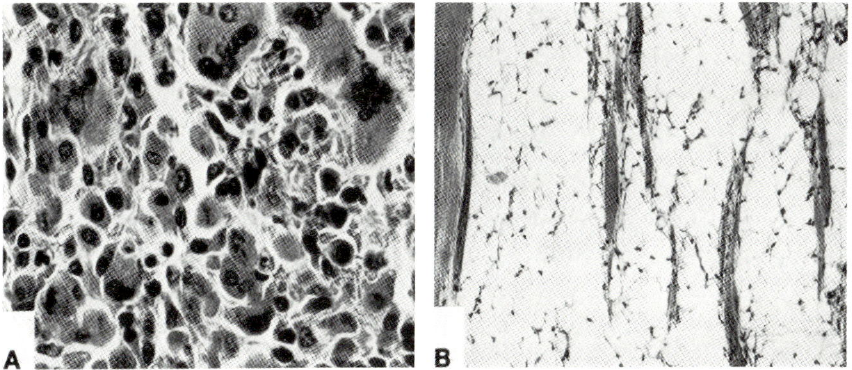

activity results which represent a diffuse mixture of concurrent phagocytic and regenerative processes. The sharp transition from cellular activity to normal muscle fibers at the edge of the damaged area emphasizes the patchy selectivity of the disorder (Figs. 3–51 and 3–52). In the center of the larger areas persistent hyaline change (Fig. 3–50) reveals the truly necrotic zone where cellular reaction is accompanied by regeneration of fibers.

When examined under the electron microscope[201] the earliest change is said to be in the mitochondria, where swelling and fragmentation of cristae are observed. Many membranous profiles soon form within mitochondria, as well as electron-dense bodies. The massive necrosis of all the contractile elements follows the mitochondrial change. Lysosomes and lipoidal deposits appear late. Van Fleet *et al.*[201] believe that a deficiency of vitamin E, which is stored in mitochondria, results in lipid perioxidation. Their work does not support the previously enunciated hypothesis that there is an increased lysosomal hydrolytic activity in this pathologic state. Howes *et al.*[107] observed under the electron microscope the formation of lipochrome (ceroid) consequent upon dietary deficiency. Thickening of the membranes of sarcoplasmic reticulum, dense bodies, and peculiar myelin-like formations in mitochondria preceded the appearance of lysosomes.

It is evident that vitamin E deficiency, like plasmocid or certain viruses, may produce large areas of damage to muscle, within which the contents of the muscle fiber are primarily, perhaps solely, damaged. In such cases regeneration of the contents of the sarcolemmal and endomysial tubes may indeed be completely restored. The elements of the process can readily be distinguished in longitudinal sections (Fig. 3–51) which represent all the changes produced by heat coagulation, crushing, and other types of moderate injury to the contents of the muscle fiber. If the architecture of the muscle is destroyed, however, a derangement of muscle cells and tissue elements results. The muscle cells derived from surviving sarcolemmal fragments then become aggregated into rounded multinucleated giant cells (Fig. 3–53A), which later will show regressive changes and disappear. After many weeks the damaged area of muscle becomes replaced by fatty tissue, in which lie occasional groups of surviving muscle fibers (Fig. 3–53B). A section of the muscle at this time shows areas which bear a remarkable resemblance to the late stage of muscular dystrophy; but as our studies of neural atrophy and degeneration have shown, an ultimate replacement by fat cells occurs in any diffuse degeneration of muscle fibers. In some cases calcified fragments surrounded by multinucleated cells and histiocytes survive the stage of intense cellular reaction.

The disease runs its course during the first month of extrauterine life and is not progressive. If the animals survive, the lesions undergo spontaneous remission, even though vitamin E is withheld. Edema of the subcutaneous and intramuscular connective tissue is found in the young animals but not in the older ones. The muscle of the myocardium is not involved, and the nervous system is intact.

A muscle paralyzed by nerve section is strikingly immune to degeneration, as also has been found in some types of virus myopathy. The lesions of intense vitamin E deficiency are so strikingly similar to those of plasmocid intoxication and of virus myopathy that the same disorder of some metabolic factor in muscular chemistry in each disease seems probable. The special susceptibility of very young animals is presumably related to the operation of this metabolic factor.

In distinction to the changes induced by vitamin E deficiency in the newborn animal, it has been found that adult rats made chronically deficient in vitamin E have developed histologic changes within 4–5 months and an evident muscular weakness of the trunk and legs within 7–8 months. This was described by Einarson and Ringsted,[59] Mason and Emmel,[132] and Einarson.[58] In the mature animals the changes in the skeletal muscles are of somewhat different character from those of the suckling animals. The muscles have a brownish coloration. Muscle fiber degeneration is not prominent; only rare hyalinized or vacuolated fibers are seen. These are occasional areas of fibrosis representing foci in which muscle fibers have degenerated. A more striking change is the occurrence of brown granules in the sarcoplasm of many fibers. These granules are arranged in rows between myofibrils, and the large ones are surrounded by vacuoles. Macrophages between the fibers contain similar pigment particles. The pigment is fuchsinophilic and resembles "wear and tear" pigment or lipofuscin but occurs in a distribution and in an amount not seen in normal animals. It does not give a dopa reaction, lacks the characteristic fluorescence of porphyrin, and does not contain hemosiderin. Evidently it is inert because many of the muscle fibers containing the pigment are otherwise healthy, suggesting that it is an abnormal product of intermediate cell metabolism, a metabolite. It shares with ceroid an acid-fast quality.

There have been numerous attempts on the part of clinicians to determine whether a deficiency of vitamin E was responsible for any of the neuromuscular diseases of humans. At one time it was postulated that amyotrophic lateral sclerosis and progressive muscular dystrophy were due to a lack of this vitamin, but subsequent studies disproved any such relationship. Nevertheless, a few investigators[141, 159] reported suggestive evidence that large amounts of vitamin E and synthetic α-tocopherol are beneficial in polymyositis and in "menopausal muscle dystrophy," which is probably a chronic form of polymyositis.

VITAMIN A

Animals made deficient in vitamin A are said to develop lesions in skeletal muscles[13, 178] that in most respects are similar to those of vitamin E deficiency. However, as was demonstrated by Krakower and Axtmeyer,[117] these lesions, at least in vitamin A-deficient animals, can be prevented by the administration of α-tocopherol. It is probable therefore that sufficient care was not taken to control the diets in these animals in order to avoid a concomitant vitamin E deficiency.

VITAMIN C

Histologic changes similar to those produced by vitamin E deficiency have been observed in guinea pigs made deficient in vitamin C. The muscle fibers became opaque and strongly eosinophilic; the striations disappeared, and the fibers were fragmented. In older lesions there was active phagocytosis of the degenerating fibers and hyperplasia of sarcolemmal nuclei, sometimes taking the form of muscle giant cells. Replacement by fibrous connective tissue was observed in some of the more chronic experiments. These changes occurred most frequently in the diaphragm, masseter, and intercostal muscles. The muscles of the pelvis and the limbs were not affected unless the animals were exercised on a rotating barrel. Florid lesions then developed, showing that the degeneration was related to the activity of the

vitamin C-deficient muscle.[34] Aschoff and Koch[6] noted hemorrhagic areas in the skeletal muscles of scorbutic humans.

CORTISONE MYOPATHY

The muscles of the trunk and extremities of rabbits were found by Ellis[60, 61] to degenerate after the administration of massive doses of cortisone (10 mg/kg day for periods of 7–21 days or longer). The myopathy appeared after 1–2 weeks. When examined at the end of this period the affected muscles were swollen and pale, and microscopically the destroyed segments of the fibers were homogeneous and occasionally impregnated with calcium salts. Necrotic sarcoplasm was being actively phagocytized by histiocytes, often with giant cell formation. This was followed by active regeneration, becoming complete in a few weeks, providing the cortisone was discontinued. No infiltrations with inflammatory cells were observed. Potassium deficiency produced a similar lesion in skeletal muscle, which differed from that of the cortisone-treated animals in that the myocardium was also affected. Glaser and Stark[80] found that prolongation of the refractory period typical of potassium deficiency is absent in cortisone myopathy.

Further investigation of the effects of glucosteroids on muscle, which were undertaken by Tice and Engel,[192] have brought to light a reversible atrophy (amyotrophy) with a specific alteration of certain organelles. Triamcinolone, a fluorinated steroid, was more potent than cortisone. After a few days of administration of the drug to rats the mitochondria began to proliferate, more so in white (gastrocnemius) than red (soleus) muscle. Then many of the interfibrillar mitochondria disappeared. There followed an atrophy of myofibrils in red muscles; the myofibril degeneration began at the Z line with disruption of sarcomeres.

A reversible myopathy of similar nature, with widespread weakness of muscle fibers in the lower limbs and later in the shoulder muscles has been seen in man following administration of cortisone derivatives.[112, 130, 218] In man also, compounds containing fluorine attached to the steroid nucleus (triamcinolone) have most frequently led to this complication. The details of the pathologic changes in muscle will be discussed in Chapter 12.

HEREDITARY MYOPATHY OF THE MOUSE

Although a genetic disease process is not experimental, the recent discovery of genetically determined muscular degenerative disease in mice[138] with many resemblances to the genetic process called muscular dystrophy in man has provided the possibility of an experimental approach to its study. The muscular lesions of sheep in the disease called scrapie, which is less clearly allied to muscular dystrophy,[17, 39, 92] and certain muscular degenerations in calves and lambs[111] have not been shown to have such genetic association and cannot be profitably discussed in this connection. A hereditary muscular abnormality was described in the domestic chicken[7, 33] that bears some resemblance to human muscular dystrophy, and the same is true of the dystrophy observed in the Syrian hamster by Homburger *et al.*[106] and in the guinea pig by Webb.[210] Mouse myopathy is considerably better known and defined. Some experimental work by Banker and Denny-Brown[10] elucidates certain aspects of this disease.

Michelson *et al.*[138] in 1955 reported a mutation in the Bar Harbor 129 strain of house mice, which they called dystrophia muscularis. This myopathic disorder occurs as a result of autosomal gene whose full recessive expression is observed to a maximal average of 27%. The animals develop a progressive weakness of proximal muscles that is recognizable 3–4 weeks after birth by dragging of the hind limbs, kyphosis, and a nodding of the head with gasping that is possibly only an expression of dyspnea. There is reduction in body weight and progressive atrophy of axial and limb muscles that is seldom compatible with survival beyond the 3rd or 4th month. Muscle tremors are seen in some of these animals, but their significance is obscure.

Small patchy lesions have been found in the muscles at birth, and general atrophy and pallor of proximal muscles become manifest by the end of the 1st month. The muscle then shows a great variation in diameter of fiber (Fig. 3–54)— from 3 to more than 85 μ, as compared with a usual range of 30–40 μ (Fig. 3–55). In some areas, possibly the site of earlier more focal degeneration, there are more small fibers than in others, but otherwise the scattering of small and large fibers without any organized pattern is identical with that of human dystrophy. The architecture of the fasciculi is not broken up by columns of fat cells as it is in severely affected muscles in human dystrophy, but this is probably only a reflection of shorter duration of fiber loss. Some fat replacement ultimately occurs in places in which loss of fiber has become very severe. There is some increase in the connective tissue, but this is generally mild in degree and related only to areas of more intense change.

Many of the larger fibers have centrally placed nuclei, but long chains of such nuclei are rare. The smaller fibers are basophilic and show irregular nuclear proliferation. Many of the nuclei of the smallest fibers are pyknotic, while others stain somewhat darkly and show slight shrinkage. The smallest fibers appear to be undergoing fragmentation, with resulting atrophic spindle cells and larger fragments with pyknotic nuclei (Figs. 3–54 and 3–56). The myofibrils stain well until the last stage of fragmentation. The appearance of such small fibers is identical with what we earlier described as degenerative change in neural muscular atrophy, but in mouse myopathy the muscular nerves are intact.

In relation to foci of more intense fiber degeneration there is an occasional necrotic large or medium-sized muscle fiber with histiocytic cellular reaction. In the neighborhood of such foci, large or medium-sized muscle fibers occasionally have a hyaline appearance in cross section. This appearance, as in the uppermost fiber in Figure 3–56, is not associated with any cellular reaction. Since such fibers commonly have a contracted appearance, or even recent rupture in longitudinal section, we regard such a change as a defect in fixation, though possibly reflecting an undue irritability. An electomyogram shows some reduction in voltage in the motor units of severely affected muscle. Occasionally, as in any degenerating muscle but more frequently than in human dystrophy, a hyperirritable muscle fiber is encountered by the electromyograph electrode, resulting in a brief train of very fast spontaneous activity. We have not encountered true myotonic discharge.

Creatine in the dystrophic muscle of the mouse was found to be consistently diminished. Expressed as a ratio of grams of creatine per gram of noncollagenous nitrogen, the muscle of myopathic mice ranged from 0.03 to 0.082, as compared with a range of 0.094–0.12 for normal muscle.[10] In denervated normal mouse muscle this creatine ratio ranged from 0.062 to 0.085. The reservoir of creatine in myopathic mouse muscle was therefore, weight for weight of muscle fiber, lower than that of denervated normal muscle. This finding is remarkable in that a great many of the muscle fibers in the myopathic muscle (of which Figure 3–54 shows a fair sample) presented a superficial appearance of normality, though nuclei were plump, numerous, and showed great variation in size. This certainly appears to indicate that the disorder of dystrophic muscle extends to more than the few muscle fibers presenting obvious histologic degeneration. Measurements of the uptake of ^{14}C-labeled creatine showed that dystrophic muscle took up slightly less creatine during a given period of time than normal muscle. The radioactive creatine/radioactive creatinine ratio remained within normal limits, showing that the creatine taken up was metabolized normally. In these mice the total serum and red blood cell creatine was elevated and the creatinine decreased, resulting in creatinuria.

Susheela *et al.*[190] found the dystrophic change to be distributed largely in areas of muscle rich in type I fibers. It was their impression that type I fibers showed a greater tendency to atrophy and type II fibers to hypertrophy. As the dystrophic process progresses, however, the type differentiation became more difficult to demonstrate. The free fatty acid content of parts of muscle, contained largely in type I fibers, had increased by 300%, and the succinic dehydrogenase activity to a lesser

degree. They assumed the free fatty acids had been esterified to triglycerides. β-Hydroxybutyrate dehydrogenase, which is responsible for the oxidation of free fatty acids, did not increase. The content of glycogen was not altered by the dystrophic process. It was the impression of these investigators that a crush injury hastened the dystrophy. These findings were not fully corroborated by Meier *et al.*[136] who noted the earliest degenerative changes in tertiary phosphorylase-rich fibers to be a disruption of myofilaments.

Banker and Denny-Brown[10] found that a striking change in fiber diameter in dystrophic muscle occurred 2–6 weeks after denervation of dystrophic muscle in the mouse (Fig. 3–55). The large and medium-sized fibers then rapidly disappeared, and the very small fibers correspondingly increased in number. At the end of 6 weeks the small fibers were considerably smaller than those of a normal muscle denervated for the same period. After denervation the creatine index of myopathic muscle dropped to extremely low levels (0.026–0.064). It was concluded that after denervation the myopathic process continued but was now evident only as progressive smallness of fiber. In addition, the presence of very large fibers is demonstrably dependent upon intact innervation. Such fibers therefore probably represent some form of compensatory reaction in the least-affected muscles.

Denny-Brown[41] found that myopathic mouse muscle reacted unfavorably to trauma in the form of heat coagulation or crushing. The damaged area of the muscle tends to become paper thin in the following 3 weeks, and the remaining nontraumatized portion of the muscle becomes smaller than the control muscle of the opposite limb. The initial reaction to infliction of a band of coagulation across the surface of the muscle closely resembles that of the same injury in the same muscle in a normal mouse. By the 5th day it was evident that the loss of substance in muscle fibers in the zone of partial damage was consistently greater than in control muscles. In this region all muscle fibers in myopathic muscle had become reduced to small tubes for a distance as great as 3 mm on either side of the lesion. Such partially damaged fibers gave rise to enormous numbers

Fig. 3–54. Hereditary myopathy of mice. (A) Appearance of biceps femoris muscle in transverse section. (B) Longitudinal section showing degenerating fibers. (A, hematoxylin, eosin; B, methylene blue, eosin)

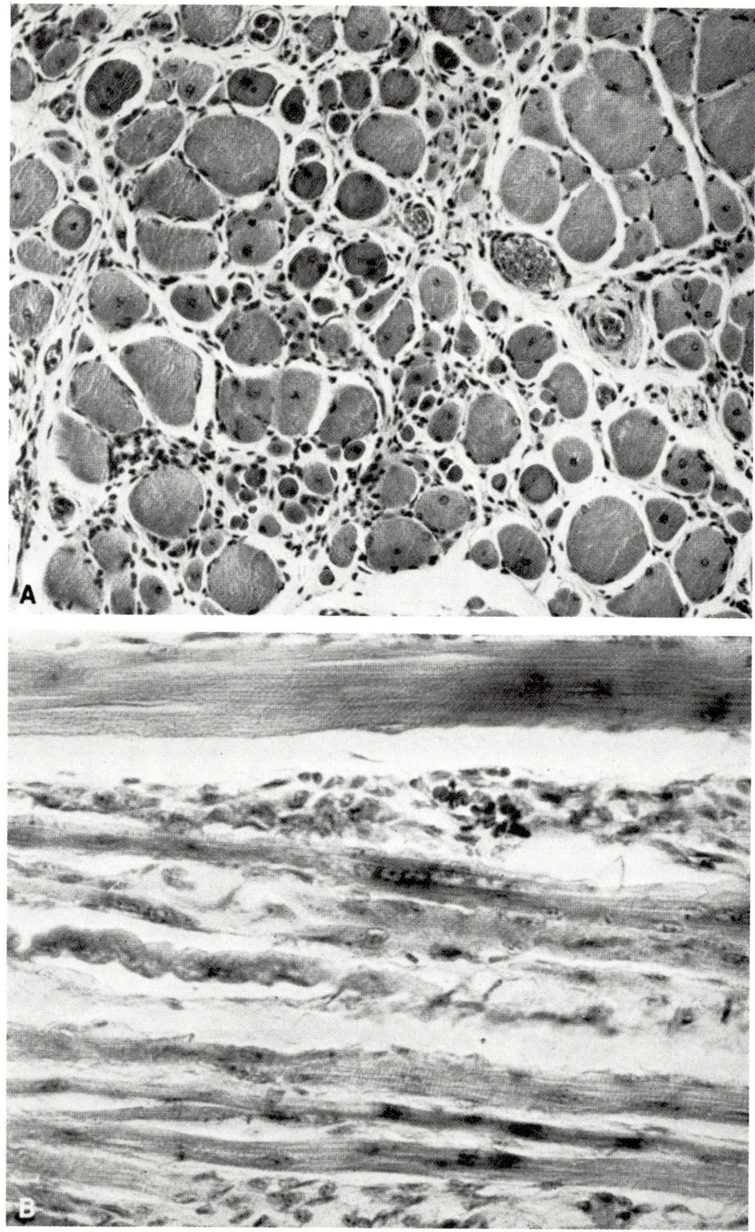

Fig. 3–55. Plot of fiber diameters of gastrocnemius muscle of myopathic mice (black dots) and normal control littermates (open circles), beginning at 2 weeks of age, as compared with denervation of same muscle for 2–6 weeks and still innervated muscle of dystrophic mouse at 6 weeks.

Fig. 3–56. Various types of fragmentation of muscle fibers into degenerating spindle cells in hereditary mouse myopathy. A mitotic figure is seen at center of lowest illustration. (methylene blue, eosin)

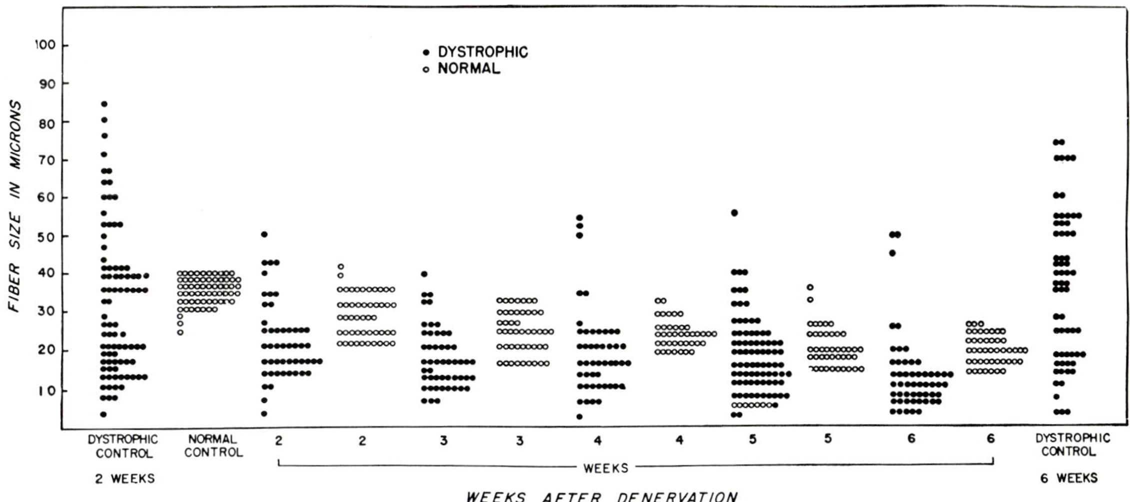

Figure 3–55.

Figure 3–56.

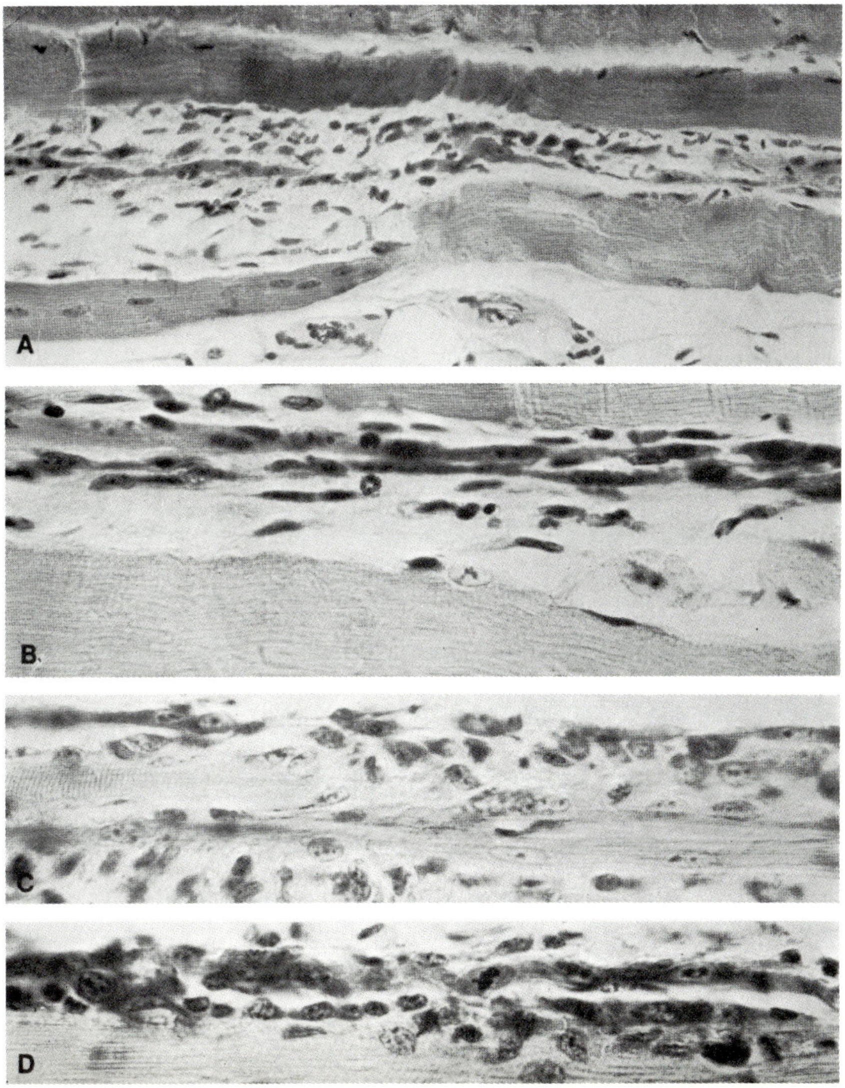

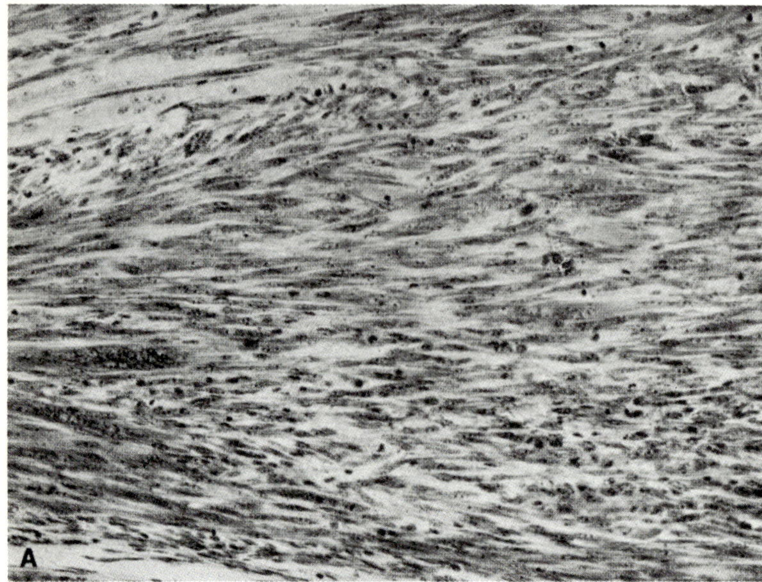

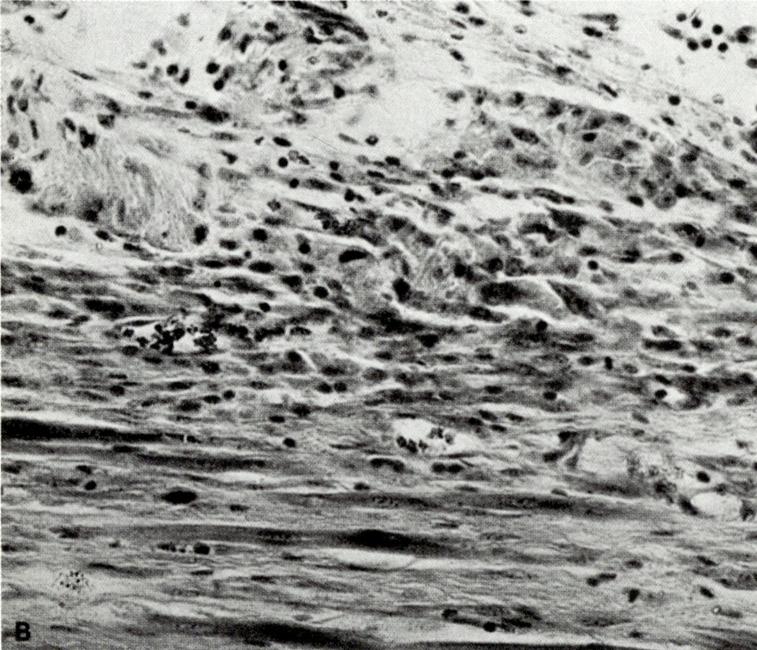

Fig. 3–57. Hereditary mouse myopathy. (A) Florid proliferation of spindle cells from ends of partially damaged fibers in mouse myopathy. Zone coagulated by heat 10 days earlier is at right. (B) Border deep to a zone coagulated 12 days earlier. Floccular coagula (at top) are surrounded by histiocytes and degenerated spindles. Partially damaged fibers (at bottom) show advanced myopathic fragmentation (though myopathic degeneration was rarely present in the remainder of this muscle). (methylene blue, eosin)

of spindle cells from the fourth to sixth day, leading to the production of a cellular mass (Fig. 3–57A) in which the nuclei showed a high frequency of regressive changes. The rapid reduction in number of myofibrils and intense staining of some of the remaining bundles of myofibrils occurred in partially damaged muscle fibers. By the 10th day after injury, the region of partially damaged muscle fibers con-

tiguous to the zone of coagulation showed evidence of degenerative change in every muscle fiber. After the 12th day dark basophilia of cytoplasm and of a high proportion of nuclei and many shrunken and deeply pyknotic nuclei were associated with patchy fragmentation of fibers and scattered, deeply stained, single muscle spindle cells (Fig. 3–57B). By the 16th day the coagulated zone showed only degen-

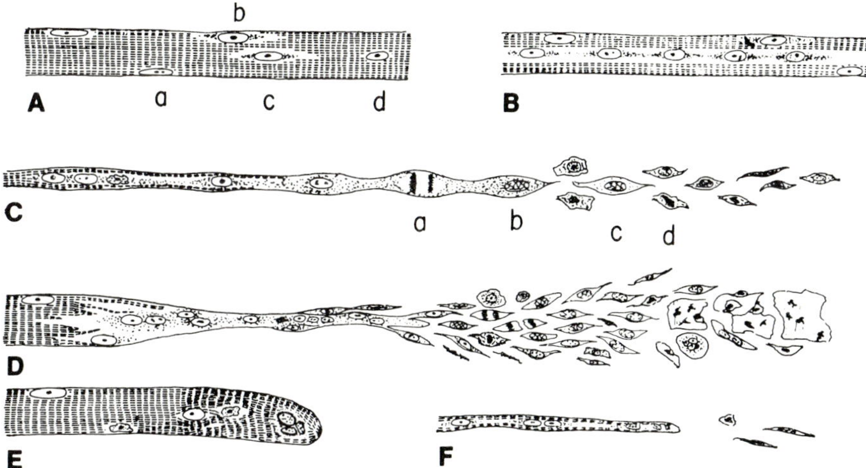

Fig. 3–58. Changes observed in hereditary mouse myopathy. The early nuclear changes (A) progress through central nuclear proliferation and thinning of the fiber (B) to spontaneous fragmentation into spindles (C). The spindles show progressive dispersion of nuclear chromatin (B–C), then shrinkage and pyknosis (D). Thermal damage to the affected muscle results in acceleration of these changes in the partially damaged fiber (D). The remainder of the fiber ends bluntly with regressive nuclear changes (E, F).

erating cells, and the region bordering that zone presented severe myopathic changes in virtually every remaining muscle fiber, though such changes were scattered and infrequent in the remainder of the muscle. Three weeks or more after the injury the muscle fibers reaching the zone of previous injury ended abruptly in a bulb or blunt end with regressing nuclear changes. The few muscle fibers traversing the border of the previous injury showed severe myopathic changes and fragmentation. It was concluded[41] that segmental thermal coagulation in mouse dystrophy provokes local intensification of a degenerative process closely resembling dystrophic change instead of the normal regenerative response. The Bar Harbor 129 strain of mice with myopathy therefore present useful material for the study of genetically determined muscular dystrophy. Since the fault underlying human muscular dystrophy is not known, it is not possible to establish absolute identity with the disease in mice. There is a remarkable absence of regenerative changes in the histology of muscles in some stages of the human disease. In the myopathy of mice this feature also constitutes the chief histologic differentiation from the intense cellular re-

sponse in vitamin E deficiency and virus disease. Also as in human dystrophy, the muscle appears capable of producing myofibrils up to a certain stage of physiologic use but thereafter presents a progressive loss (atrophy) followed by absolute failure (degeneration). Even in the hypertrophic stage a stimulus that normally excites regeneration fails to have that effect. Isolated myoblasts shrink and disappear, as also do large fragments. These reactions of the muscle fiber in mouse myopathy are represented in Figure 3–58 in which the cycle of spontaneous changes are seen in (A), (B), and (C), and the response to localized injury in (D). The terminal stage of the uninjured fiber, instead of showing regeneration, presents the appearance shown in (E or F).

Degeneration of the muscle fiber into spindle myoblasts and the subsequent degenerative changes in them are associated with intense nuclear and cytoplasmic activity. This activity is represented by basophilia as a result of staining with methylene blue and similar stains and is accompanied by intense labeling with tritiated thymidine[41] representing vigorous synthesis of nucleic acids.

Homburger and his associates[106] discovered a similar hereditary dystrophy in the Syrian hamster, and Webb[210] in the guinea pig. Since the lesions do not differ from those of murine dystrophy it would be repetitious to describe them.*

CONTRACTURE

When a muscle is excited to contract either through its nerve or by induction shocks applied directly, this state of activity is manifested by the development of tension or shortening, the liberation of heat and accumulation of catabolic products, and a change in the electrical potential. These are the measurable expressions of muscle contraction. A single momentary stimulus induces a wave of contraction to pass along the muscle, and the muscle gives a single twitch. Repeated stimuli cause a series of twitches, and if sufficiently close together they fuse into a sustained contraction or tetanus. In contrast to such tetanic shortening of a muscle there are other states in which an entire muscle shortens or maintains a tension without evidence of repeated neural excitation of the contractile mechanism. The types of shortening that are not accompanied by the usual evidence of activity of the contractile mechanism and are reversible are designated by the term *contracture*. Once contracture becomes irreversible, it is called *rigor*.

These terms, long in use by physiologists, have been assigned precise though limited meanings. However, some of them, particularly the term contracture, have been adopted by clinicians to denote all types of fixed shortening of skeletal muscle other than a tetanus produced by nerve impulses. Thus the term has been applied to conditions as different as muscle fibrosis and arthritis. In conditions such as dystonia musculorum deformans the fixed shortening in part is due to abnormal nervous discharge, and in part to fibrosis of muscle. The

relative proportion of the latter can be determined only by the degree of postural abnormality remaining under anesthesia. This confusion about the different states of contracture has led some physiologists to advise that the term be abandoned. Yet its use for a condition fixed resistance to passive stretch of muscle is so firmly embedded in medical writing that it will surely be retained, and what is therefore required is more exact definition.

TIEGEL'S CONTRACTURE AND VULPIAN-HEIDENHAIN-SHERRINGTON PHENOMENON

A review of the literature discloses a remarkably large number of physiologic studies of different types of contracture. Kronecker[119, 120] was the first to observe that an irritable frog muscle stimulated directly with successive supramaximal induction shocks to give a series of twitches did not relax completely between stimuli. The base line of each twitch becomes higher until a steady level is maintained, which continues some time after stimulation is stopped. This phenomenon is known in the literature as Tiegel's contracture after Tiegel,[193] who studied it. Also, the slow contraction of muscle deprived of its motor nerve supply upon stimulation of the sensory nerve (the Vulpian-Heidenhain-Sherrington phenomenon) has been studied intensively and was shown to be a type of acetylcholine contracture. The reversible changes produced in muscle by thermal stimulation, by mechanical stimulation (idiomuscular contraction) of Schiff,[176] and the sustained shortening which follows perfusion with chemical substances such as chloroform, alcohol, ether, acid, alkalis, veratrine, caffeine, and acetylcholine have all been the subject of a voluminous literature, which was reviewed by Gasser.[77]

There have been but few careful histologic studies of the muscle fibers in these states of pharmacologic contracture. Secher[177] remarks on a lateral shift of the myofibrils (which destroys the regular appearance of the cross striations) and a wrinkling in the sarcolemma during the contracture caused by caffeine.

---

* These and other animal dystrophies are reviewed in the monograph of Innes on Veterinary Pathology and in his article in the ARNMD volume on neuromuscular diseases.

These changes are said to be reversible by perfusion with Ringer's solution. One hesitates to attach any significance to changes of this type since they are judged by most pathologists to be unreliable. There is need in this field for close cooperation between the physiologist and the histopathologist versed in the use of the electron microscope and other modern histologic methods.

MYOSTATIC CONTRACTURE

Another condition that has never been fully elucidated is the myostatic contracture of Moll.[143] He clearly demonstrated that if the bony attachments of a skeletal muscle were approximated and immobilized for several days there occurred a fixation of the muscle at the length thus imposed upon it. If undertaken sufficiently early, the contracture can be overcome by active or passive movements, but if the shorter length is maintained for weeks, it cannot be immediately overcome without structural damage. Fröhlich and Meyer[76] showed that this fixation occurs only if the nerve supply is intact. Immobilization of the denervated muscle does not cause its length to become set. These authors maintained that section of the dorsal nerve roots alone prevented myostatic contracture, indicating that integrity of local reflexes was essential for its development. The same fixation at a shorter length occurs after sectioning the tendon of a muscle, and this too fails to occur if the nerve is also sectioned. Once it has developed, such myostatic contracture persists under deep anesthesia or after section of the motor nerve to the muscle.

The numerous investigations of this problem were reviewed by Ranson and Sams,[161] who reaffirmed the findings of others that no histologic change was demonstrable in the affected muscle fibers except possibly a slight decrease in clarity of cross striation, which gives a "glassy" appearance. There was no increase in connective tissue, even after a period of many weeks. Ranson and Sams found that muscles involved in the spasm induced by the injection of tetanus toxin, which in the cat and rat causes only a localized spasm, undergo a similar change after the spasm has persisted for some days. In the beginning the hypertonus of such "local tetanus" is reflex and is abolished by section of the motor nerve, but myostatic contracture is later superimposed. This contracture developed in the same manner if the tendon of the muscle had been sectioned and was preventable by nerve section; but whereas normal muscle lost weight, the muscle in tetanus contracture after tendon resection did not. The process of contracture is therefore independent of the mechanism of disuse atrophy. After tetanus contracture has persisted a number of weeks, some increase in the number of sarcolemmal nuclei was observed as well as a haziness of cross striation, but there was no other histologic change. The height of contraction and rate of relaxation following stimulation of the motor nerve were unchanged. It was assumed therefore that the alteration was related to a change in the ductility of the elastic elements in the muscle fiber such as to shorten its resting length by 10–40%. The length of tendon was unchanged. We are, however, not convinced that the change is necessarily in the muscle fiber and believe that a similar phenomenon is common to all true elastic tissue. The epimysium and perimysium of muscle contain much true elastic tissue.

Ramsey and Street[160] claimed that isolated muscle fibers suffered severe and irreversible changes (the "delta state") when stimulated and allowed to shorten below 65% of their resting length. This appears to be the state of "extreme contraction" which Nageotte[150] found to be reversible if sufficient time was allowed. Such shortening seldom occurs after simple tenotomy, but Eccles[52] using more exact methods than Ranson and Sams found a loss of 40% in contraction tension per unit weight after tenotomy, as compared with 20–30% in simple disuse atrophy. The loss of muscle weight after tenotomy was only about 20% as compared with the 40% loss in disuse atrophy. It therefore appears that tenotomy does impair the contractile process to some degree, but Eccles did not find it appreciably worsened nor atrophy prevented by intermittent stimulation of the tenotomized muscle. No fibrillation occurred in the inactive muscle, and there is reason to believe that the contractile mechanism is not gravely damaged. Probably there is only a change in elasticity of the fiber. Changes such as uneven intensity of staining, wavy striation, and indistinct cross striation, which were reported by Davenport *et al.,*[37] are within the limits of variations inherent in the process of fixation and can be accepted only with reservations.

PHARMACOLOGIC CONTRACTURE

Finally there is the remarkable state of persistent shortening or contraction of muscle that is independent of the nerve impulses. Electromyographically the muscle is silent, in contrast to the virtual electrical storm that emanates from the sarcolemma during tetanus and cramp. Maintenance of the shortened state does not involve an active biochemical process, oxygen utilization, or liberation of heat. This type of contracture has been much studied by pharmacologists and is now known also to characterize McArdle's phosphorylase and the phosphofructokinase deficiencies (Chapter 5), and possibly the myopathy of malignant hyperthermia (Chapter 11).

Recent reviews of the pharmacologic contracture inform us that drugs such as halothane and dinitrophenol administered to dogs cause their muscles to become rigid because of depletion of ATP and prevention of its reformation by mitochondria, as well as the uncoupling of oxidative phosphorylation and acceleration of glycolysis. Under these conditions as calcium is liberated from the sarcoplasmic reticulum into the sarcoplasm, where it combines with the troponin complex, it cannot be restored to the sarcoplasmic reticulum. Contraction cannot then be released. There are actually two proteins of the sarcoplasmic reticulum involved in the reabsorption calcium. One is magnesium-sensitive ATPase and the second calciquestrin, the latter is capable of binding up to 43 moles of calcium without ATP.

Caffeine also causes contracture, supposedly in this instance by blocking phosphodiesterase, an enzyme in the degradation pathway of adenosine monophosphate (cyclic AMP). As shown by Rasmussen,[162] either the increase in cyclic AMP or a blocking of phosphodiesterase causes both muscle rigidity and hyperthermia, a state similar to human malignant hyperthermia.

FIBROUS AND NEUROGENIC CONTRACTURE

In clinical writings the term contracture is used loosely to describe a limitation of movement at a joint caused by a fixed shortening of a muscle or group of muscles. The muscles concerned tighten abruptly when stretched to a given length and resist further attempts to lengthen them. The pain associated with passive traction on such muscles suggests that the change is primarily in the elastic supporting tissues. The muscular contracture in certain cases of Addison's disease or rheumatoid arthritis are examples, and in the latter is usually reversible by careful stretching and use of physical therapy methods. "Fibrous" contracture may also result from destruction of muscle tissue and replacement fibrosis. "Neurogenic" contractures may occur in pyramidal or extrapyramidal diseases which impose fixed postures that are maintained for long periods of time, but the condition in such instances appears to be the same as that which occurs after any other fixation of posture.

Many problems in this field need further study. The relation of some of the forms of pharmacologic contracture to clinical states in which fixed shortening of muscle occurs has not been precisely determined. The contracture and rigor of the pharmacologist appear to be clearly related to a defect in the phase of relaxation of the contractile process. The contracture of the clinician concerns primarily, and perhaps solely, changes in the supporting tissues of muscle. There is still relatively little exact information as to the cause of the shortening of muscle after simple immobilization, and the reactions of the elastic tissue of the endomysium and its content of collagen and elastin have not been explored in this respect. There is need of applying some of the newer techniques for the staining of elastic tissue (Margolena and Dolnick[13]).

REFERENCES

1. AFIFI AK, ALEU FP, GOODGOLD J, ET AL: Ultrastructure of atrophic muscle in amyotrophic lateral sclerosis. Neurology (Minneap) 16:475, 1966

2. ALLBROOK DB: An electron microscopic study of regenerating skeletal muscle. J Anat 96:137–152, 1962

3. ALLBROOK DB, AITKEN JT: Reinnervation of striated muscle after acute ischemia. J Anat 85:376, 1951

4. ALTSCHUL R: Atrophy, degeneration and metaplasia in denervated skeletal muscle. Arch Pathol 34:982–988, 1942

5. ALTSCHUL R: On nuclear division in damaged skeletal muscle. Rev Can Biol 6:485–495, 1947

6. ASCHOFF L, KOCH WA: Skorbut, einepathologischeanatomische Studie. Jena, Gustav-Fischer, 1919

7. ASMUNDSON VS, JULIAN LM: Inherited muscle abnormality in the domestic fowl. J Hered 47:248–252, 1956

8. BADE H: Lässt sich ein Einfluss von Röntgenbestrahlungen, die Toleranzgrenze de Haut nicht überschreiten, auf die quergestreifte Muskulatur nachweisen? Strahlentherapie 65:455–567, 1939

9. BAJUSC E: Succinic dehydrogenase in muscular dystrophy and on secondary changes after denervation. Exp Med Surg 23:169, 1965

10. BANKER BQ, DENNY-BROWN D: A study of denervated muscle in normal and dystrophic mice. J Neuropathol Exp Neurol 18:517–530, 1959.

11. BELMAR J, EYZAQUIRRE C: Pacemaker site of fibrillation potentials in denervated mammalian muscle. J Neurophysiol 29:425–441, 1966

12. BERTALANFFY L VON, BICKIS I: Identification of cytoplasmic basophilia (ribonucleic acid) by fluorescent microscopy. J Histochem Cytochem 4:481–493, 1956

13. BETZ H: Contribution à l'étude de la dégénérescence et de la régéneration musculaire. Arch Anat Microsc Morphol Exp 40:46–89, 1951

14. BETZ EH, FIRKET H, REZNIK M: Some aspects of muscle regeneration. Int Rev Cytol 19:203–217, 1966

15. BETZ EH, REZNIK M: La régéneration des fibres musculaires strieés dans differentes conditions expérimentales. Arch Biol (Liege) 75:567–594, 1964

16. BIGLAND B, JEHRING B: Muscle performance in rats, normal and treated with growth hormone. J Physiol (Lond) 116:129–136, 1952

17. BOSANQUET FD, DANIEL PM, PARRY HB: Myopathy in sheep; its relationship to scrapie and to dermatomyositis and muscular dystrophy. Lancet 2:737–746, 1956

18. BOURNE GH, BECKETT EB: Histochemistry of developing skeletal and cardiac muscle. Muscle. Edited by GH Bourne. New York, Academic Press, 1960, pp 89–108

19. BOWDEN REM, GUTMANN E: Denervation and reinnervation of human voluntary muscle. Brain 67:273–313, 1944

20. BOWMAN W: On the minute structure and movements of voluntary muscle. Philos Trans Soc Lond Part II 130:457–501, 1840

21. BROOKS B: Pathologic changes in muscle as a result of disturbances of circulation: an experimental study of Volkmann's ischemic paralysis. Arch Surg 5:18–216, 1922

22. CARLSON BM: The regeneration of mammalian muscle. The Regeneration of Striated Muscle and Myogenesis. Edited by A Mauro, SA Shafiq, AT Milhorat. Amsterdam, Academic Press, 1970, Chaps 1 and 2

23. CAULFIELD JB, REBEIZ J, ADAMS RD: Viral involvement of human muscle. J Pathol 96:232–234, 1968

24. CHESTER A, GLEISER B: Study of chloroquine toxicity and a drug-induced cerebrospinal lipodystrophy in swine. Am J Pathol 53:27–47, 1968

25. CHEVREMONT M: Le système histiocytaire ou réticulo-endothélial. Biol Rev 23:267–295, 1948

26. CHEVREMONT M: Le muscle squeléttique cultivé in vitro: transformation l'éléments musculaires en macrophages. Arch Biol 51:313–333, 1940

27. CHOR H, DOLKEART RE: A study of simple disuse atrophy in the monkey. Am J Physiol 117:626–630, 1936

28. CHOR H, DOLKEART RE, DAVENPORT HA: Chemical and histological changes in denervated skeletal muscle of the monkey and cat. Am J Physiol 118:580–587, 1937

29. CHOU SM: Myxovirus-like structures in a case of human chronic polymyositis. Science 158:1453–1455, 1967

30. CLARK WL: An experimental study of the regeneration of mammalian striped muscle. J Anat 80:24–36, 1946

31. CLARK WL, WAJDA HS: The growth and maturation of regenerating striated muscle fibers. J Anat 81:56–63, 1947

32. COËRS C, WOOLF AL: The Innervation of Muscle: A Biopsy Study. Springfield, Ill, Charles C Thomas, 1959

33. CORNELIUS CE, LAW GRJ, JULIAN LM, et al: Plasma aldolase and glutamic-oxalacetic transaminase activities in inherited muscular dystrophy of domestic chicken. Proc Soc Exp Biol Med 101:41–44, 1959

34. DALLDORF G: The lesions in the skeletal muscles in experimental scorbutus. Exp Med 50:293–298, 1929

35. DALLDORF G: The coxsackie viruses. Bull NY Acad Med 26:329–335, 1950

36. DALLDORF G, SICKLES BM: An unidentified, filtrable agent isolated from the feces of children with paralysis. Science 108:61–62, 1948

37. DAVENPORT HK, RANSON SW, STEVENS E: Microscopic changes of muscle in myostatic contracture caused by tetanus toxin. Arch Pathol 7:978–992, 1929

38. DE BUCK D, DE MOOR L: Morphologie de la régression musculaire. Nevraxe 5:229–262, 1903

39. DELEZ AL, GUSTAFSON DP, LUTTREL CM: Some clinical and histological observations on scrapie in sheep. J Am Vet Med Assoc 131:439–447, 1957

40. DENNY-BROWN D: The influence of tension and innervation on the regeneration of skeletal muscle. J Neuropathol Exp Neurol 10:94–96, 1951

41. DENNY-BROWN D: Experimental studies pertaining to hypertrophy regeneration and degeneration. Assoc Res Nerv Ment Dis 38:1, 1960

42. DENNY-BROWN D, BRENNER C: Paralysis of nerve induced by direct pressure and tourniquet. Arch Neurol Psychiatry 51:1–26, 1944

43. DE REUCK J, VANDER EECKEN H, ROELS H: Biometrical and histochemical comparison between extra and intra-fusal muscle fibers in denervated and reinnervated rat muscle. Acta Neuropathol 25:249–258, 1973

44. DEWAN ML, HENSON JB, DOLLAKITE D, et al: Toxic myodegeneration in goats produced by feeding mature fruits from coyotillo plant. Am J Pathol 46:215–227, 1964

45. DRACHMAN DB: Pharmacological denervation of skeletal muscle in chick embryos treated with botulinus toxin. Trans Am Neurol Assoc 90:241–243, 1965

46. DUBOWITZ V: Pathology of experimentally reinnervated skeletal muscle. J Neurol Neurosurg Psychiatry 30:99, 1967

47. DUEL AB: Clinical experiences in the surgical treatment of facial palsy by autoplastic nerve grafts. Arch Otolaryngol 16:767–788, 1932

48. DUESBERG J: Les chondriosomes des cellules embryonaires du poulet, et leur rôle dans le génèse des myofibrilles avec quelques observations sur le développment des fibres musculaires striées. Arch Zellforsch Mikr Anat 4:602–671, 1909

49. DURANTE G: Anatomie pathologique des muscles. Manuel d'Histologie Pathologique. Second edition. Edited by V Cornil, L Ranvier. Paris, Felix Alcan, 1902

50. DURANTE G: Du processus histologique de l'atrophie musculaire. Arch Med Exp 14:658–677, 1902

51. EADIE MJ, FERREST M: Chloroquine myopathy. J Neurol Neurosurg Psychiatry 29:331–337, 1968

52. ECCLES JC: Disuse atrophy of skeletal muscle. Med J Aust 2:160–164, 1941

53. ECCLES JC: Investigations on muscle atrophies arising from disuse and tenotomy. J Physiol (Lond) 103:253–266, 1944

54. ECCLES JC: Interrelationship between the nerve and muscle cell. The Effect of Use and Disuse on Neuromuscular Function. Edited by E Gutmann. Prague, Publishing House of Czechoslovakian Academy of Science, 1963

55. EDDS MV: Collateral regeneration of residual motor axons in partially denervated muscles. J Exp Zool 113:507–552, 1950

56. EDSTROM L, KUGELBERG E: Histochemical composition distribution of fibers and fatigability of single motor units. J Neurol Neurosurg Psychiatry 31:424–433, 1968

57. EICHELBERGER L, AKESON WH, ROMA M: Effects of denervation on the histochemical characterization of skeletal muscle during growth. Am J Physiol (Lond) 185:287–298, 1956

58. EINARSON L: Criticizing review of the concepts of the neuromuscular lesions in experimental vitamin E deficiency, preferably in adult rats. Acta Psychiatr Neurol Scand [Suppl] 78:76, 1952

59. EINARSON L, RINGSTED A: Effect of Chronic Vitamin E Deficiencies on the Nervous System and the Skeletal Musculature in Adult Rats. New York, Oxford University Press, 1938

60. ELLIS JT: Studies on the nature and pathogenesis of muscular degeneration in cortisone-treated rabbits. Bull NY Acad Med 29:814–817, 1953

61. ELLIS JT: Necrosis and regeneration of skeletal muscle in cortisone-treated rabbits. Am J Pathol 32:993–1013, 1956

62. ENGEL WK: The essentiality of histo- and cytochemical studies of skeletal muscles in the investigation of neuromuscular disease. Neurology (Minneap) 12:778–794, 1962

63. ENGEL WK, BROOKE MH, NELSON PG: Histochemical studies of denervated tenotomized cat muscle. Ann NY Acad Sci 138:160, 1966

64. ERB W: Zur pathologie und pathologischen anatomie peripherischer paralysen. Dtsch Arch Klin Med 4:534–578, 1868

65. FAHIMI HD, COTRAN RS: Permeability studies in heat-induced injury of skeletal muscle using lanthanum as fine structural tracer. Am J Pathol 62:143–159, 1971

66. FENN WO: Factors affecting the loss of potassium from stimulated muscles. Am J Physiol 124:213–229, 1938

67. FERGUSON AB, VAUGHAN L, WARD L: A study of disuse atrophy of skeletal muscle in the rabbit. J Bone Joint Surg [Am] 39A:583–596, 1957

68. FISCHER E: Some enzyme systems of denervated muscle. Arch Phys Med 29:291–300, 1948

69. FISCHER E, RAMSEY VW: Changes in protein content and in some physicochemical properties of the protein during muscular atrophies of various types. Am J Physiol 145:571–582, 1946

70. FISHBACK DK, FISHBACK HR: Studies of experimental muscle degeneration. I. Factors in the production of muscle degeneration. Am J Pathol 8:193–209, 1932

71. FISHBACK DK, FISHBACK HR: Studies of experimental muscle degeneration. II. Standard method of causation of degeneration and repair of the injured muscle. Am J Pathol 8:211–217, 1932

72. FLAX M, HIMES M: Microspectrophotometric analysis of metachromatic staining of nucleic acid. Physiol Zool 25:297–311, 1952

73. FOGEL, M, DEFENDI V: Infection of muscle cultures from various species with oncogenic (DNA) viruses. Proc Natl Acad Sci USA 58:967, 1967

74. FORBUS WD: Pathologic changes in voluntary muscles. I. Degeneration and regeneration of the rectus abdominis in pneumonia. Arch Pathol 2:318–339, 1926

75. FORBUS WD: Pathologic changes in voluntary muscle. II. Experimental study of degeneration and regeneration of striated muscle with vital stains. Arch Pathol 2:486–499, 1926

76. FRÖHLICH A, MEYER HH: Ueber dauerverkürzung der gestreiften harmblütermuskeln. Arch Exp Pathol Pharmakol 87:173–188, 1920

77. GASSER RS: Contractures of skeletal muscle. Physiol Rev 10:35–109, 1930

78. GEIGER RS, GARVIN JS: Pattern of regeneration of muscle from progressive muscular dystrophy patients cultivated in vitro as compared to normal human skeletal muscle. J Neuropathol Exp Neurol 16:532–543, 1957

79. GIFFORD H, DALLDORF G: Creatinine, potassium and virus content of muscles following injection with the "Coxsackie virus." Proc Soc Exp Biol Med 17:589–592, 1949

80. GLASER GH, STARK L: Excitability in experimental myopathy. II. Potassium deficiency: an initial study. Neurology (Minneap) 8:708–711, 1958

81. GODLEWSKI E: Die entwicklung des skelet- und herz-muskelgewebes der Säugetiere. Arch Mikr Anat 60:111–156, 1902

82. GODMAN GC: On the regeneration and redifferentiation of mammalian striated muscle. J Morphol 100:27–82, 1957

83. GODMAN GC: Cell transformation and differentiation in regenerating striated muscle. Frontiers in Cytology. Edited by SL Palay. New Haven, Yale University Press, 1958, Chap 13, pp 382–416

84. GOLDBERG AL: Protein turnover in skeletal muscle: protein catabolism during work-induced hypertrophy and growth-induced with growth hormone. J Biol Chem 244:3217–3222, 1969

85. GORDON RB, LENNETTE EH, SANDROCK RS: The varied clinical manifestations of Coxsackie virus infections. Arch Intern Med 103:63–75, 1959

86. GURR E: A Practical Manual of Medical & Biological Staining Techniques. New York, Interscience, 1956, pp. 336–340

87. GUSSENBAUER C: Ueber die veränderrungen des quergestreiften muskelgewebes bei der traumatischen entzündung. Arch Klin Chir 12:1011–1047, 1871

88. GUTMANN E, GUTTMANN L: Effect of electrotherapy on denervated muscles in rabbits. Lancet 1:169–170, 1942

89. GUTMANN E, YOUNG JZ: The re-innervation of muscle after various periods of atrophy. J Anat 78:15–43, 1944

90. GUTMANN E, ZELENA JR: Morphological changes in denervated muscle. The Denervated Muscle. Edited by E Gutmann. Prague, Publishing House of Czechoslovakian Academy of Science, 1962

91. HARMAN JW, GWINN RP: The recovery of skeletal muscle fibers from acute ischemia as determined by histological and chemical methods. Am J Pathol 25:741–755, 1949

92. HARTLEY WJ, DODD DC: Muscular dystrophy in New Zealand livestock. NZ Vet J 5:61–66, 1957

93. HAY ED: Electron microscopic observations of muscle dedifferentiation in regenerating amblystoma limbs. Dev Biol 1:555, 1959

94. HEIDENHAIN M: Ueber die entstehung der quergestreiften muskelsubstanz bei der forelle. Arch Mikr Anat 83: 427–447, 1913

95. HICKS SP: Brain metabolism in vivo. I. The distribution of lesions caused by azide, malonitrile, plasmocid and dinitrophenol poisoning in rats. Arch Pathol 50:545–561, 1950

96. HICKS SP: Brain metabolism in vivo. II. The distribution of lesions caused by cyanide poisoning, insulin hypoglycemia, asphyxia in nitrogen and fluoroacetate poisoning in rats. Arch Pathol 49:111–137, 1950

97. HINES HM: Neuromuscular denervation, atrophy and regeneration. Fed Proc 3:231–235, 1944

98. HINES HM, KNOWLTON GC: Changes in the skeletal muscle of the rat following denervation. Am J Physiol 104:379–391, 1933

99. HINES HM, KNOWLTON GC: Effect of thyroparathyroidectomy and thyroxin on the rate of atrophy of skeletal muscle. Proc Soc Exp Biol Med 31:1029–1030, 1934

100. HINES HM, KNOWLTON GC: Electrolyte and water changes in muscle during atrophy. Am J Physiol 120:719–723, 1937

101. HINES HM, MELVILLE E, WEHRMACHER WH: The effect of electrical stimulation on neuromuscular regeneration. Am J Physiol 144:278–283, 1945

102. HINES HM, THOMPSON JD, LAZERE B: A compara-

tive study of muscle atrophies caused by denervation and acute inanition. Am J Physiol 140: 115–118, 1943

103. HIRSCHBERG M: Dauerheiling cines Tibiasarkoms mit Röntgenstrahlen und atrophische Vorgänge in der bestrahlten Muskultur. Strahlentherapie 34:421–424, 1929

104. HOAGLAND CL: States of altered metabolism in diseases of muscle. Adv Enzymol 6:193–230, 1946

105. HOLZER H, ABBOTT J, LASH J: On the formation of multinucleated myotubes. Anat Rec 131:567, 1958

106. HOMBURGER F, BAKER JR, NIXON CW, *et al:* Primary generalized polymyopathy and cardiac necrosism in Syrian hamsters. Med Exp (Basel) 6:339, 1962

107. HOWES EL, PRICE HM, BLUMBERG JM: Effects of diet on production of lipochrome pigment (ceroid) in ultrastructure of skeletal muscle in rat. Am J Pathol 45:599–633, 1964

108. HUMMOLLER FL, GRISWOLD B, MCINTYRE AR: Comparative chemical studies in skeletal muscles following neurotomy and tenotomy. Am J Physiol 161:406, 1950

109. KAIGHN ME, EBERT JD, STOTT PM: The susceptibility of differentiating muscle cones to Roux sarcoma virus. Proc Natl Acad Sci USA 56:133, 1966

110. KASPAR W, WIESMANN U, MUMENTHALER M: Necrosis and regeneration of the tibialis anterior muscle in rabbit. Arch Neurol 21:363–372, 1969

111. KEITH TB, SCHNEIDER AP: Muscular dystrophy in calves and lambs. I. The relation of environmental conditions to incidence. J Am Vet Med Assoc 131:519–522, 1957

112. KENDALL PH, HART MF: Side-effects following triamcinolone. Br Med J 1:682–685, 1959

113. KIRBY E: Experimentelle Untersuchungen über die Regeneration des quergestreiften Muskelgewebes. Beitr Pathol 11:302–319, 1892

114. KLINGE F: Das Gewebsbild des fieberhaften Rheumatismus. XII. Zusammenfassande kritische Betrachtung zur Frage der geweblishschen Sonderstellung des rheumatischen Gewebeschaden. Virchows Arch [Pathol Anat] 286:344–388, 1932

115. KNOLL P, HAUER H: Ueber des Verhalten der protoplasmaarmen und protoplasmareichen, querquestreiften Muskelfasern unter pathologischen Verhältnissen. Sitz Akad Wissensch Wien Mathnat C1 101:315–348, 1892

116. KNOWLTON GC, HINES HM: Kinetics of muscle atrophy in different species. Proc Soc Exp Biol Med 35:394–398, 1936

117. KRAKOWER D, AXTMEYER JH: Effect of alpha-tocopherol on the lesions of skeletal muscles in rats on vitamin A deficient diets. Proc Soc Exp Biol Med 45:583–586, 1940

118. KRASKE P: Experimentelle Untersuchunger über die Regeneration quergestreiften Muskeln. Halle, Habschr, 1878

119. KRONECKER H: Monatsschr Acad. Berlin, 639, 1870. Cited in Mosso A: Ueber die Gesetze der Ermüdung: Untersuchungen an Muskeln des Menschen. Arch Anat Physiol Leipzig 89–168, 1890

120. KRONECKER H: Ueber die From des minimalen Tetanus. Arch Anat Physiol Leipzig 571–573, 1877

121. LANGLEY JN: Observations on denervated and on regenerating muscle. J Physiol (Lond) 51:377–395, 1917

122. LANGLEY JN, HASHIMOTO M: Observations on the atrophy of denervated muscle. J Physiol (Lond) 52:15–69, 1918

123. LANGLEY JN, KATO T: The rate of loss of weight in skeletal muscle after nerve section, with some observations on the effect of stimulation and other treatment. J Physiol (Lond) 49:432–440, 1915

124. LASH JW, HOLTZER H, SWIFT H: Regeneration of mature skeletal muscle. Anat Rec 128:679–702, 1957

125. LEWIS AL, JONES RS: A combined connective tissue stain for elastin, reticulin and collagen. Stain Technol 26:85–87, 1951

126. LEWIS MR: Rhythmical contraction of the skeletal muscle tissue observed in tissue cultures. Am J Anat 38:153–161, 1915

127. LEWIS WH, LEWIS MR: Behavior of cross-striated muscle in tissue cultures. Am J Anat 22:169–194, 1917

128. LUNDSGAARD E: Untersuchungen über Muskelkontraktion ohne Milchsäurebildung. Biochem Z 217:162–177, 1930

129. MACDONALD RD, ENGEL AG: Experimental chloraquine myopathy. J Neuropathol Exp Neurol 29:479–499, 1970

130. MACLEAN K, SCHURR PH: Reversible amyotrophy complicating treatment with fluorocortisone. Lancet 1:701–703, 1959

131. MARGOLENA IA, DOLNICK EH: A differential staining method for elastic fibers, collagenic fibers and keratin. Stain Technol 26:119–122, 1951 (see also Lewis and Jones[125])

132. MASON KW, EMMEL AF: Vitamin E and muscle pigment in the rat. Anat Rec 93:33–60, 1945

133. MASTAGLIA FL, WALTON JN: Coxsackie virus-like particles in skeletal muscle from a case of polymyositis. J Neurol Sci 11:593–599, 1970

134. MAURO A: Satellite cell of skeletal muscle fibers. J Biophys Biochem Cytol 9:493–495, 1961

135. MCMINN RMH, VBROVÁ G: The effect of tenotomy on the structures of fast and slow muscles in the rabbit. Q J Exp Physiol 49:424, 1964

136. MEIER H, WEST WT, HOAG WG: Preclinical histopathology of mouse muscular dystrophy. Arch Pathol 80:165, 1965

137. MELNICK JL, GODMAN GC: Pathogenesis of Coxsackie virus infection: multiplication of virus and evolution of the muscle lesion in mice. J Exp Med 93:247–266, 1951

138. MICHELSON AM, RUSSELL ES, HARMAN PJ: Dystrophia muscularis: a hereditary primary myopathy in the house mouse. Proc Natl Acad Sci USA 41:1079–1084, 1955

139. MILEDI R: Properties of regenerating neuromuscular synapses in the frog. J Physiol (Lond) 154:190–205, 1960

140. MILEDI R, SLATER CR: Electron microscopic structure of denervated skeletal muscle 174:253–269, 1969

141. MILHORAT AT, TOSCANI V, BARTELS WE: Effect of wheat germ on creatinuria in dermatomyositis and progressive muscular dystrophy. Proc Soc Exp Biol Med 58:40, 1945

142. MILLAR WG: Regeneration of skeletal muscle in young rabbits. J Pathol 38:145–151, 1934

143. MOLL A: Experimentelle untersuchungen über den anatomischen Zustand der Gelenke bei andauernder immobilisation derselben. Virchows Arch [Pathol Anat] 105:466–485, 1886

144. MORPURGO B: Sur les processus histologiques consécutifs à la névrectomie sciatique. Arch Ital Biol 17:432–445, 1892

145. MORPURGO B: Ueber activitäts-hypertrophie der willkürliehen muskeln. Virchows Arch [Pathol Anat] 150:522–524, 1897

146. MÜELLER EA: Die einflussung der muskelkraft durch isometrische kontractionen. Munch Med Wochenschr 103:341–347, 1961

147. MÜELLER EA, HETTINGER T: Ueber Unterschiede der Trainings geschwindkeit atrophierter und normalen Muskeln. Arbeitsphysiologie 15:223, 1953

148. MUSCATELLO V, MAGRETH A, ALOISI M: On the differential response of sarcoplasm and myoplasm to denervation in frog muscle. J Cell Biol 27:1–24, 1965

149. NACHMIAS VT, PADYKULA HA: A histochemical study of normal and denervated red and white muscles of the rat. J Biophys Biochem Cytol 4:47, 1958

150. NAGEOTTE J: Sur la contraction extrême des muscles squelettiques des les vertebrés. Z Zellforsch Mikr Anat 26:603–624, 1937

151. PAPANICOLAU GN, FALK EA: General muscular hypertrophy induced by androgenic hormone. Science 87:238–239, 1938

152. PAPPENHEIMER AM: On Certain Aspects of Vitamin E Deficiency. Springfield, Ill, Charles C Thomas, 1948

153. PEARSON CM, MOSTOFI FK, EDITORS: The Striated Muscle. Baltimore, Williams & Wilkins, 1973

154. PELLEGRINO C, FRANZINI C: An electron microscopic study of denervation atrophy in red and white muscles. J Cell Biol 17:327, 1963

155. POGOGEFF JA, MURRAY MR: Form and behavior of adult mammalian skeletal muscle in vitro. Anat Rec 95:321–336, 1946

156. POLLOCK LJ, GOLSETH JG, ARIEFF AJ: Strength-interval curves and repetitive stimuli in electrodiagnosis. Surg Gynecol Obstet 84:1077–1082, 1947

157. POLLOCK LJ, GOLSETH JG, ARIEFF AJ, ET AL: Reaction of degeneration in electrodiagnosis of experimental peripheral nerve lesions. War Med 7:275–283, 1945

158. PORTER K: Cell and tissue differentiation in relation to growth (animals). Dynamics of Growth Process. Edited by EJ Boell. Princeton, Princeton University Press, 1954

159. RABINOVITCH R, GIBSON WC, MCEACHERN D: Neuromuscular disorders amenable to wheat germ oil therapy. J Neurol Neurosurg Psychiatry 14:95–100, 1951

160. RAMSEY RW, STREET SF: The isometric length-tension diagram of isolated skeletal muscle fibers of the frog. J Cell Comp Physiol 15:11–34, 1940

161. RANSON SW, SAMS CF: A study of muscle in contracture; the permanent shortening of muscles caused by tenotomy and tetanus toxin. J Neurol Psychopathol 8:304–320, 1928

162. RASMUSSEN H: Cell communication, calcium ion, and cyclic adenosine monophosphate. Science 170:404–412, 1970

163. REID G: A comparison of the effects of disuse and denervation upon skeletal muscle. Med J Aust 2:165–167, 1941

164. REWCASTLE NB, HUMPHREY JG: Vacuolar myopathy, clinical histochemical and microscopic study. Arch Neurol 12:570–582, 1965

165. REZNIK M: Possible origin of myoblasts during muscle regeneration. J Cell Biol 39:110, 1968,

166. REZNIK M: Origin of myoblasts during skeletal muscle regeneration: electron microscopic observations. Lab Invest 20:353–363, 1969

167. REZNIK M, BETZ EH: Régéneration du muscle ischémié préalablement irradié. C R Soc Biol (Paris) 159:2078–2080, 1965

168. RICKER G, ELLENBECK J: Beiträge zur kenntniss der veränderungen des muskels nach der durchschneidung seines nerven. Virchows Arch [Pathol Anat] 158:199–253, 1899

169. ROBERTS F: Degeneration of muscle following nerve injury. Brain 39:297–347, 1916

170. ROMANUL F, HOGAN EL: Enzymatic changes in denervated muscle. I. Histochemical studies. Arch Neurol 13:263, 1965

171. RUSKA H, EDWARDS GA: A new cytoplasmic pattern in striated muscle fibers and its possible relation to growth. Growth 21:73–88, 1957

172. RUSTIGIAN R, PAPPENHEIMER AW: Myositis in mice following intramuscular injection of viruses of the mouse encephalomyelitis group and of certain other neurotropic viruses. J Exp 89:69–92, 1949

173. SAUNDERS JH, SISSON HA: The effect of denervation on the regeneration of skeletal muscle after injury. J Bone Joint Surg [Br] 35B:113–124, 1953

174. SCHAPIRA G: Biochimie Comparee des Muscles Atrophiques et Dystrophiques Actualites Depathologie Neuromusculaire. Paris, G Serratrice & H Roux L'Expansion, 1971, pp 45–53

175. SCHAPIRA G, DREYFUS JC: Biochemistry of Hereditary Myopathies. Springfield, Ill, Charles C Thomas, 1962

176. SCHIFF JM: Muskel und nerven-physiologie. Cyclus. Edited by Schauenberg. Lahr, Part 9, Vol 1, 1885

177. SECHER KJA: Untersuchungen über die einwirkung des caffeine auf die quergestriefte muskulatur. Arch Exp Pathol Pharmakol 77:83–121, 1914

178. SEKIZIMA K: Comparative observations on changes of skeletal muscles in experimental A, B and C avitaminoses. Mitt Med Acad Kioto 32:836–838, 1941

179. SIEBER W, PETROW W: Studien über die hypertrophie des skelettmuskels. Z Klin Med 109:427–433, 1926

180. SLOPER JC, PEGRUM GD: Regeneration of finished mammalian skeletal muscle and effects of steroids. J Pathol 93:47, 1967

181. SOLA OM, CHRISTENSEN DL, MARTIN FW: Hypertrophy and hyperplasia of adult chicken anterior latissimus dorsi muscles following stretch with and without denervation. Exp Neurol 41:76–100, 1973

182. SOLANDT DY: Atrophy in skeletal muscle. JAMA 120:511–513, 1942

183. SOLANDT DY, MAGLADERY JW: The relation of atrophy to fibrillation in denervated muscle. Brain 63:255–263, 1940

184. SPEIDEL CC: Studies on living muscles. I. Growth, injury and repair of striated muscle, as revealed by prolonged observations of individual fibers in living frog tadpoles. Am J Anat 62:179–285, 1938

185. SPEIDEL CC: Studies on living muscles. II. Histological changes in single fibers of striated muscle during contraction and clotting. Am J Anat 65:471–528, 1939

186. STIER S: Experimentelle untersuchungen über das verhalten der quergestreiften muskeln nach läsionen des nervensystems. Arch Psychiatr Nervenkr 29:249–298, 1896

187. STONNINGTON HH, ENGEL AE: Normal and denervated muscle. Neurology (Minneap) 23:714–725, 1973

188. STRUNK SW, SMITH CV, BLUMBERG JM: Ultrastructural studies of lesion produced in skeletal muscle fibers by crude type A Clostridium birefringens toxin and its purified alpha fraction. Am J Pathol 50:89–109, 1967

189. SUNDERLAND S, RAY LJ: Denervation changes in mammalian striated muscle. J Neurol Neurosurg Psychiatry 13:159–177, 1950

190. SUSHEELA AK, HUDGSON P, WALTON JN: Histological and histochemical studies of experimentally induced degeneration and regeneration in normal and dystrophic mouse. J Neurol Sci 9:423–442, 1969

191. THESLEFF S: Spontaneous electrical activity in denervated rat muscle. The Effect of Use and Disuse on Neuromuscular Functions. Edited by E Gutmann, P Hnick. Prague, Publishing House of Czechoslovakian Academy of Science, 1963

192. TICE LW, ENGEL AG: The effects of glucosteroids on red and white muscles in the rat. Am J Pathol 50:311–333, 1967

193. TIEGEL E: Ueber muskelcontractur in genesatz zu contraction. Pfluegers Arch 13:71–83, 1876

194. TIEGS OW: The flight muscles of insects: their anatomy and histology with some observations on the structure of striated muscle in general. Philos Trans R Soc 238:221–348, 1955

195. TOWER SS: Atrophy and degeneration in the muscle spindle. Brain 55:77–90, 1932

196. TOWER SS: Atrophy and degeneration in skeletal muscle. Am J Anat 56:1–43, 1935

197. TOWER SS: Function and structure in the chronically isolated lumbosacral spinal cord of the dog. J Comp Neurol 67:109–127, 1937

198. TOWER SS: Trophic control of non-nervous tissues by the nervous system: a study of muscle and bone innervated from an isolated and quiescent region of spinal cord. J Comp Neurol 67:241–261, 1937

199. TOWER SS: The reaction of muscle to denervation. Physiol Rev 19:1–48, 1939

200. TSCHAINSKI J: Ueber die entzündlichen veränderungen der muskelfasern. Stud Inst Exp Pathol Wein 1:86–94, 1870

201. VAN FLEET JF, HALL BV, SIMON J: Vitamin E deficiency. Am J Pathol 51:815–830, 1967

202. VILLAVERDE JM: Les lesions de la fibre muscu-

laire dans l'intoxication saturnine expérimentale. Trab Lab Invest Biol Madrid 27:227–248, 1931

203. VOLKMANN R: Die ischaemischen muskellähmungen und kontrakturen. Zentralbl Chir 8:801–803, 1881

204. VOLKMANN R: Ueber die regeneration des quergestreiften muskelgewebes beim menschen und saugetier. Beitr Pathol 12:233–324, 1893

205. WALDEYER W: Ueber die veränderungen der quergestreiften muskeln bei der entzündung und dem typhusprozees, sowie über die regeneration derselben nach substanzdefecten. Virchows Arch [Pathol Anat] 34:473–514, 1865

206. WALKER BE: The origin of myoblasts and the problem of dedifferentiation. Exp Cell Res 30:80, 1963

207. WALTON JN, ADAMS RD: The response of the normal, the denervated and the dystrophic muscle cell to injury. J Pathol 72:273–298, 1956

208. WARREN S: Effects of radiation on normal tissues. XIV. Effects on striated muscle. Arch Pathol 35:347–349, 1943

209. WARREN S: Histopathology of radiation lesions. Physiol Rev 24:225–238, 1944

210. WEBB JN: Naturally occurring myopathy in guinea pigs. J Pathol 100:155–159, 1970

211. WEBER O: Ueber die neubildung quergestreifter muskafasern, insbesondere die regenerative neubildung derselban nach verletzungen. Virchows Arch [Pathol Anat] 39:216–253, 1867

212. WECHSLER W, HAGER H: Electronmikroskopische befunde im atrophisen quergestreiphen skelettmuskel der Ratte nach Nervendurchtrennung. Naturwissenschaften 47:185, 1960

213. WEDDELL G, FEINSTEIN B, PATTLE RE: The electrical activity of voluntary muscle in man under normal and pathological conditions. Brain 67:178–257, 1944

214. WEISS P, EDDS MV: Spontaneous recovery of muscle following partial denervation. Am J Physiol 145:587–607, 1946

215. WEST WT, MASON KE: Degeneration and regeneration in experimental muscular dystrophy. Am J Phys Med 35:223–239, 1955

216. WHISNANT JP, ESPINOSA RE, KIERLAND RR, et al: Chloroquine neuromyopathy. Mayo Clin Proc 38:501–513, 1963

217. WILLARD WA, GRAW EC: Some histological changes in striate skeletal muscle following nerve section (abstract). Anat Rec 27:192, 1924

218. WILLIAMS RS: Triamcinolone myopathy. Lancet 1:698–701, 1959

219. WOHLFART G: Collateral regeneration from residual motor nerve fibers in amyotrophic lateral sclerosis. Neurology (Minneap) 7:124–134, 1957

220. YAFFE D, GERSHON D: Multinucleated muscle fibers, induction of DNA synthesis and mitosis by polyoma virus infection. Nature (Lond) 215:241, 1967

221. ZAK R, GUTMANN E: Lack of correlation between synthesis of nucleic acids and proteins in denervated muscle. Nature (Lond) 185:766–767, 1960

222. ZELENA J: The effect of denervation on muscle development. Denervated Muscle. Edited by E Gutmann. Prague, Publishing House of Czechoslovakian Academy of Science, 1962, pp. 103–126

223. ZENKER FA: Ueber die Veränderungen der willkürlichen Muskeln im Typhus Abdominalis. Leipzig, Vogel, 1864

chapter 4

GENERAL REACTIONS OF HUMAN MUSCLE TO DISEASE

Before proceeding to human myopathology, we might summarize the principles that emerge from studies of the anatomy and experimental pathology of the muscle fiber. Consideration will also have to be given to other principles that are enunciated because they have application to certain human diseases that have no counterpart in the animal kingdom.

The multinucleated tubular structure forming the elementary muscle fiber is presented as a complex metabolic and functional unit. Because the number of such functional units in any given skeletal muscle evidently remains fixed (so in a sense the units are irreplaceable). All the numerous tissue reactions evoked by diseases of muscle must in the end be traced to changes in the individual fiber. It is obvious that agreement has not yet been reached as to the exact significance of each of the many histologic features manifested by this complex tissue, and yet there are some well established general principles that aid in the interpretation of changes commonly observed in disease.

The most striking feature of skeletal muscle is its dependence upon innervation by the motor neuron. The process of atrophy which follows denervation is in many respects a reversion to the status of the fetal muscle fiber. The contractile process is greatly slowed and no longer coordinated with that of other muscle fibers to impart movement or perform work. In morphologic terms the gradual loss of myofibrils, reduction of all fibers in a muscle to fairly uniform small diameter, and loss of histochemical specificity of sarcoplasmic constituents also indicate a reversal of the differentiation of these various properties and a resumption of the state which exists during the later months of fetal life. Tissue culture studies and those on the embryogenesis of muscle indicate that the myoblast is primarily independent of the nervous system. The status of striped muscle during the first 6 months after denervation is a reversion to that state. Degeneration of isolated fibers in denervated muscle and the gradual fragmentation of the cytoplasm in the terminal period proceed at greatly differing rates in different muscles and in different fibers in the same muscle. These late changes are more properly classed as abiotrophic or degenerative, and of the many factors concerned in their appearance, trauma is undoubtedly prominent. Only after months or years are changes in the supporting tissue well defined, and these must therefore be regarded as secondary to the essential process.

Another equally obvious principle of myopathology is the influence of repeated or continuous contractile activity upon the structure of the fiber. Disuse from whatever cause (casting, section of tendon, joint ankylosis, or denervation) leads almost immediately to an excess of catabolic activity (over anabolic), and this is reflected in atrophy with a loss of myofibrils, a relative hypernucleation, and a reduction in power of contraction and endurance. Conversely, muscle hypertrophy can readily be brought about by exercise and is manifestly the result of individual fiber enlargement and not an increase in the number of fibers. From the multitude of tightly packed myofibrils seen in transverse section of such fibers, the conclusion is inescapable that physiologic activity can determine an increase in number of myofibrils. As to the manner in which such an increase comes about, we may suppose they are derived from polyribosomes that proliferate to form myofilaments, which are added to existing myofibrils and also integrated into new ones. In other words, it involves the same mechanism as that which produced the original myofibrils embryologically. The splitting of myofilaments postulated by McComas *et al.*[25,26] seems implausible.

Of the various organelles within the muscle fiber, it is the longitudinally oriented thin and thick filaments that stamp it as a unique cellular element. Although actin-like filaments are seen in astrocytes and other cells, organized packets of thin and thick laments that form a myofibril occur in no other cell—and of course this arrangement confers on muscle its characteristic transverse striation. That these myofibrils constitute a kind of insoluble skeleton is revealed by necrotic muscle where even after nuclei and other elements have disappeared the dead tissue still retains a cross striation that is visible as a variation in refraction and polarization.

Those who have studied the histologic changes of striped muscle during contraction emphasize the division of muscle into transverse compartments (sarcomeres, inokommas), each bounded by a transverse Z membrane (telodiaphragma) which bisects a light disk. The histologic events of contraction can be conveniently described in relation to such a unit, and it is likely that the chemical events of contraction will ultimately be elucidated in terms of such units. Muscle fragments across the region of the Z line when undergoing necrosis or when treated with acids. Yet it is doubtful if the sarcomere has a unit survival value. The dark band of one bundle of myofibrils in a muscle fiber may not always be in absolute alignment with that of another bundle, and in the process of myofibril regeneration after injury the alternation may be at complete variance between different bundles. Whatever the explanation of such a phenomenon it is also certain that *pathologic processes affect the contents of the fiber as a whole, and not individual sarcomeres.* Even following a

localized injury, such as section of a fiber, the contents retract and undergo reconstruction for a considerable distance on either side of the sarcomere. Indeed, the smaller clots of muscle fibers commonly involve 4–50 sarcomeres.

The essential living unit of the muscle fiber is the sarcolemmal or muscle nucleus, which shares a syncytial cytoplasm with its fellows, and in the resting state lies flattened against the sarcolemma by pressure from the tightly packed myofibrils. If the sarcolemmal nuclei are destroyed, regeneration does not take place. If they survive, their reaction to injury is rapid and effective. Proliferation of myoblasts derived from the activated sarcolemmal nuclei (called satellite cells when still within the basement membrane) by mitotic division, swelling, appearance of intensely staining nucleoli, and fusion to form myotubes form an invariable sequence of reaction to devastating segmental injury of all types. With some nuclear stains the nuclei of the fibroblastic supporting tissue may be mistaken for sarcolemmal nuclei. The position of sarcolemmal nuclei within the sarcolemmal tube, their usual strict alignment with the myofibrils, and their tendency to prominent nucleolation when reacting constitute their chief distinguishing characteristics. The means by which injury and nerve section excite nuclear activity are unknown.

The course of events following experimental injury to muscle depends on two factors: the fate of the sarcolemmal nuclei and the cell membrane or sarcolemma, and the fate of the contents of the sarcolemmal tube. The contents may undergo either massive coagulation necrosis or varying degrees of waxy or granular degeneration, according to the intensity of damage. In either case the coagulated material is eventually fragmented and undergoes slow dissolution. If the sarcolemma and its nuclei are also necrotic, the whole structure is treated as a foreign body. If any part of the sarcolemma and its nuclei remains, the process of attempted reconstruction begins with a proliferation of sarcolemmal nuclei and an increase in the perinuclear sarcoplasm.

The study of muscle diseases reveals, besides the well known neural atrophy and frank segmental necrosis, a number of other primary affections of muscle, including a number of disorders of neuromuscular transmissions and the well known progressive muscular dystrophies, various other types of myopathies, and polymyositis. As experimental prototypes of muscular dystrophy we have only the murine and other animal dystrophies, which may not be comparable to the human diseases. Nevertheless they share one property—that of a defect in the power of survival and of restitution of the muscle fiber after degeneration or atrophy. Loss of muscle fibers is the most obvious change, but this surely represents an end-stage lesion. There are a variety of other changes, not strictly interpretable, which presumably reflect various degrees of the dystrophic process, and although the sarcolemmal nuclei may divide and show other signs of activity, their effective reparative function is limited. In various tumors, on the other hand, the reproduction of multinucleated cytoplasmic masses and the provision of new cross striation becomes strikingly evident. This is also the case in most types of polymyositis. The nature of the fundamental dystrophic defect has not been resolved, and the striking changes are a challenge to further studies in pathogenesis.

HISTOPATHOLOGIC REACTIONS OF MUSCLE TISSUE

Compared to the complexity of striated muscle structure and the range and variety of diseases to which muscle is subject, the specific histopathologic changes of the muscle fiber are relatively limited. Perhaps this conclusion is illusory and reflects the current inadequacy of available histologic methods for showing the significant lesions. In accord with the modern conceptions of disease, the denotative pathologic features which reliably distinguish one muscle disease from another comprise (1) the character of the changes of the muscle fibers and their supporting tissues, (2) the additional data derived from ascertainment of the age and distribution of these changes within one or

many muscles, (3) the functional effects of the lesions, and (4) the presence or absence of genetic and other factors.

With reference to the changes of the muscle fibers, we have seen that they generally fall into three categories: (1) segmental necrosis (necrobiosis); (2) disfiguration, in which the preserved fiber undergoes a subtle morphologic alteration, such as vacuolation with water, glycogen, or fat accumulation or the hyperplasia of one or several organelles; and (3) changes in volume (dysvoluminal). In (1) and (3) the abnormality may be visualized with a light microscope in standard histologic preparations; in (2) the electron microscope is needed for full demonstration of the lesion, meaning that the changes are essentially ultrastructural.

SEGMENTAL NECROSIS

In a variety of experimental conditions and in human disease the primary abnormality is complete destruction of the content of the muscle fiber, leaving the supporting tissues relatively intact. The necrotizing process seldom involves the entire fiber. Instead, a few or many sarcomeres are usually affected, and the least degree of it is the central part only (a core) of a single sarcomere (Fig. 4–1). Within any given muscle in transverse sections there tends often to be a grouping of the necrotic fibers; 5–10 nearly contiguous fibers are in the same stage of degeneration.

In the earliest stage of the necrotizing process the altered sarcous substance has a highly refractile, vitreous appearance and is brightly eosinophilic or acidophilic after aniline staining. The fragility of the coagulated sarcoplasm is shown by its manifest tendency to fracture (due to muscle tension) in a transverse direction. Rupture usually occurs at the junction of normal and regenerative zones (discoid degeneration), leaving empty parts of the sarcolemmal tube. The vitreous change which antedates cellular reactions is difficult to distinguish from the artefact induced by crushing or cutting the living fibers during biopsy (Chapter 13). Within 24–48 hours an infiltration of neutrophilic leukocytes and invasion by pleomorphic histiocytes and marcophages (the reaction of myophagia) occur, providing indisputable evidence of necrobiosis (Figs. 4–1 and 4–2). Many of the sarcolemmal nuclei within the damaged segments are shriveled and pyknotic.

As the necrotic sarcoplasm with its faintly visible transverse striation disintegrates, it becomes loose and coarsely granular as though liquefying by the action of proteolytic enzymes (Fig. 4–1). Its removal involves this latter process as well as myophagia (Fig. 4–2A). The latter results in formation of a series of cellular endomysial cords and faintly visible sarcolemmal tubes.

In the intermediate region between necrotic and normal muscle a number of changes occur that are best seen with the electron microscope. Whorls of densely packed membranes derived from sarcoplasmic reticulum and mitochondria proliferate within a few days. Disruption and rearrangement of sarcoplasmic tubules lead to formation of honeycomb structures. The newly formed mitochondria contain densely packed cristae.[35] The Z lines in places enlarge and thicken.

REGENERATION

Muscle regeneration under various experimental conditions was discussed at length in Chapter 3. It was pointed out that some degree of muscle fiber regeneration is a natural sequel to many types of muscle degeneration. This reparation of muscle tissue is always the result of proliferation of sarcolemmal nuclei. Connective tissue elements do not participate in this process except to bridge the defect and to offer support for the new fibers. This regenerative process takes two forms. One, regeneration by budding from the surviving parts of the muscle fibers, occurs whenever segments of the muscle fibers and their sheaths are destroyed; the other, regeneration by proliferation of cellular bands (cellular bands of Waldeyer), occurs when many of the sarcolemmal nuclei are spared and can re-form a sarco-

plasmic band by linkage of their cytoplasmic processes. This latter form of regeneration is often referred to as an embryonal type of regeneration because of its resemblance to the original development of the muscle fiber during fetal life.

Both types of regeneration can be observed in human muscle diseases. In destructive lesions of muscle—traumatic necrosis, hemorrhage, infarction, and suppurative myositis—regeneration is by budding. The buds consist of undifferentiated plasmodial masses in which numerous myoblasts are incorporated. They arise from intact muscle fibers at the edge of the lesion and elongate and extend into the lesion. Fibrillar differentiation gradually becomes evident in the cytoplasm between these nuclei. Where the sheaths of the original muscle fibers are destroyed, fibroblasts proliferate to form support for new fibers. Due to interruption of the total architecture of the muscle, this regeneration is haphazard and may present an admixture of such diverse cell forms and nuclear activity as to be confused with neoplastic change (Fig. 4–4). Regeneration of this type can extend 3–4 mm or more into a scar or defect.

In the muscle degenerations of waxy or hyaline type in which the striated contents are destroyed but the sarcolemmal nuclei and endomysium are usually preserved, the regeneration is by proliferation of the surviving muscle nuclei as well as by extension from any intact portions of fiber. This embryonal type of regeneration is less frequent than regeneration by budding. Here the muscle nuclei proliferate in the intact sarcolemma. As macrophages remove the necrotic portions of the fiber, these nuclei elongate and divide repeatedly and fuse into myotubes. There is an accompanying increase in sarcoplasm, which is at first clear but as early as the 5th day begins to acquire a fibrillar appearance (Fig. 4–2). The longitudinal striation appears first at the two poles of the multinucleate sarcoplasmic masses. Later the medial portion undergoes fibrillar differentiation. The sarcoplasm of adjacent masses fuses to form a syncytium. The transverse striation develops later than the longitudinal. Budding may take place in one part of the fiber and the embryonal

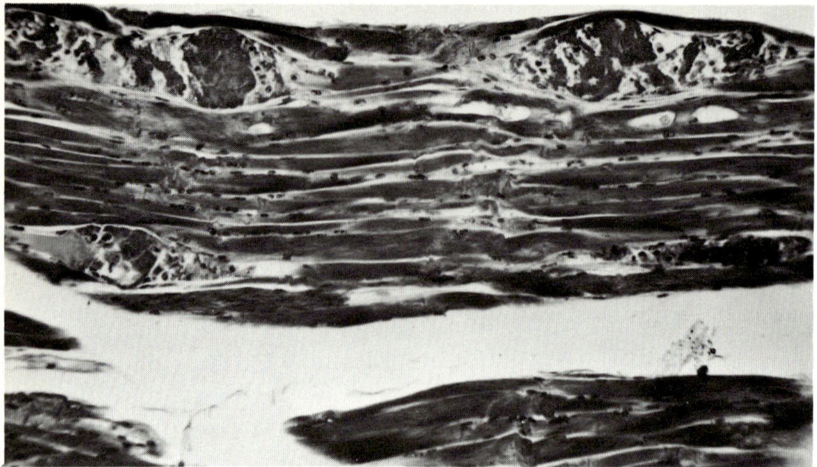

Fig. 4–1. Longitudinal section of gastrocnemius muscle from a patient with alcoholic myopathy. At top, several necrotic sarcomeres of a fiber have fragmented, swelled, and are beginning to disintegrate. (hematoxylin, eosin, ×110)

Fig. 4–2. Acute degeneration of muscle fibers. (A) Upper fiber shows a retraction cap. Central fibers have degenerated and are being invaded by macrophages. (B and C) Note rows of sarcolemmal nuclei in thin regenerating fibers. (hematoxylin, eosin)

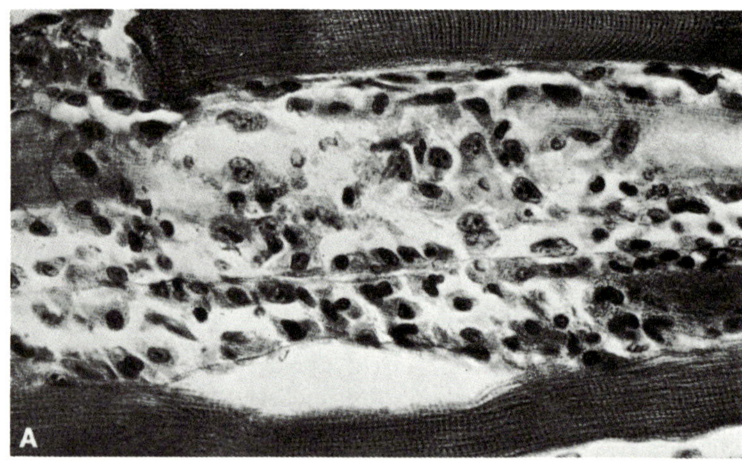

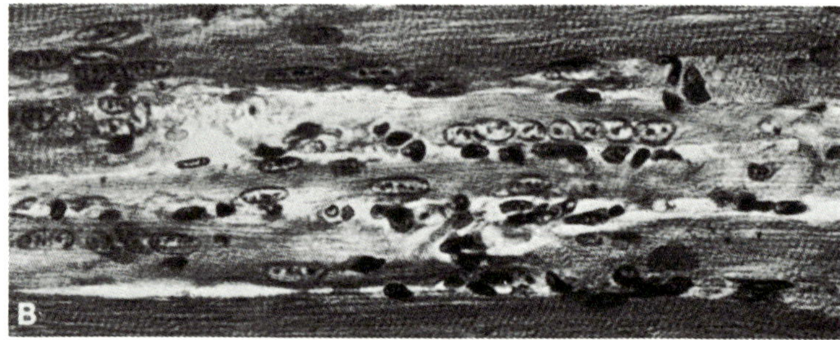

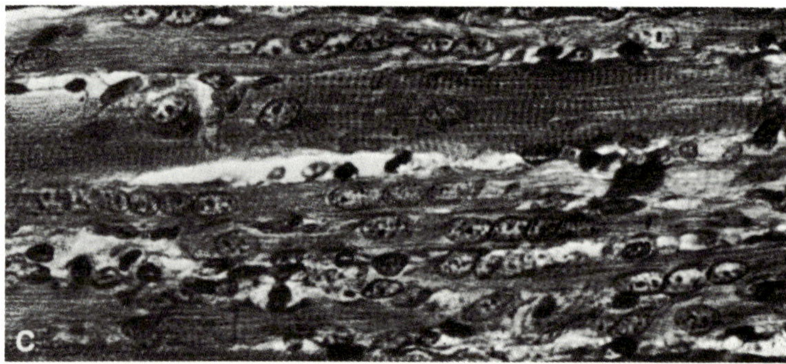

type of cellular regeneration in another. The phase of repair merges with the reaction to injury, but in general the regressive nuclear and other changes resulting from injury have disappeared in most fibers by the 12th day. By this time, unequivocal evidence of repair is obvious in the form of faint transverse striation, vesicular nuclei, and weakly basophilic cytoplasm. If the endomysial support of the fibers is not destroyed, the new fibers replace degenerated ones, and in this way long stretches of muscle may be reconstituted. If the endomysial tube has been destroyed, the growing bud becomes a multinucleated club, which may or may not be successful in sprouting branches, depending upon the amount of obstruction by fibrous tissue or hemorrhage. Isolated muscle cells tend to joint together to form cell bands or clusters.

In the examples of experimental muscle disease described in Chapter 3, several mechanisms of segmental necrosis are portrayed: viral invasion of sarcolemmal nuclei, inflammatory necrosis by invading microbes, antigen-antibody reactions in allergic polymyositis, toxic effects after plasmocid injections, and primary biochemical disorder in vitamin E deficiency. In human muscle disease segmental necrosis is the primary lesion in acute and subacute polymyositis, Zenker's degeneration, Meyer-Betz familial paroxysmal myoglobinuria, epidemic myoglobinuria and Haff disease, acute trichinosis, and toxoplasmosis. It is said to be the principal abnormality in alcoholic myopathy (Fig. 4–1), but this entity is not firmly established, or at least cannot be distinguished from some of the pannecroses of muscle tissue such as result from ischemia and trauma. Segmental necrosis also features the more rapidly advancing muscular dystrophies (Duchenne, for example). Massive necrosis is also produced by trauma (Figs. 4–3 and 4–4) and circulatory arrest.

One of the central problems in interpreting the necrobioses of muscle relate to the occurrence of minor degrees of it in normal muscle after the muscle has been exercised to an extreme degree. Presumably this accounts for single-fiber necrosis seen occasionally during examination of routine postmortem or biopsy

material from individuals not known to have muscle disease.[1, 29, 44] The differential diagnosis of the necrotizing segmental lesion depends on demonstration of the causative agent.

MUSCLE FIBER DISFIGURATIONS

The term disfigurations is suggested to refer to a series of subcellular lesions, some of which are consistently associated with certain diseases. Their delineation is not secure, for the problem of distinguishing lesion from artifact is as difficult in freeze-dried histochemical stains and electron micrographs as in the more conventional histologic preparations. Also, they must not be confused with certain minor alterations of muscle fibers that accompany systemic diseases, i.e., "cloudy swelling." The latter condition in which the entire content of the fiber is swollen and takes on a faintly granular appearance with obscuration of transverse striations was clearly described by Meryon[27] and Virchow[43] more than a century ago. It is a finding we have had trouble in distinguishing from the changes of postmortem autolysis and from the artifacts resulting from strong fixatives and extractives. Similarly we think that the traditional concept of granular and waxy degeneration in routine pathologic material is relatively useless. Here vitreous and granular change in sarcoplasm, usually with complete effacement of transverse striations, are nearly impossible to interpret because of their frequent occurrence when unfixed muscle is crushed or cut, especially if it is still in a state of active contractility. Rigor mortis is another source of confusion.

Interestingly, the entire group of muscle fiber disfigurations have been brought to light through the study of certain peculiar human diseases. The fact that they have been seen mostly in biopsy rather than autopsy material and have not been successfully reproduced in the experimental animal under the controlled conditions of the laboratory has added to our difficulties in interpretation. Moreover, their demonstration has usually required techniques such as the freezing and drying of living muscle

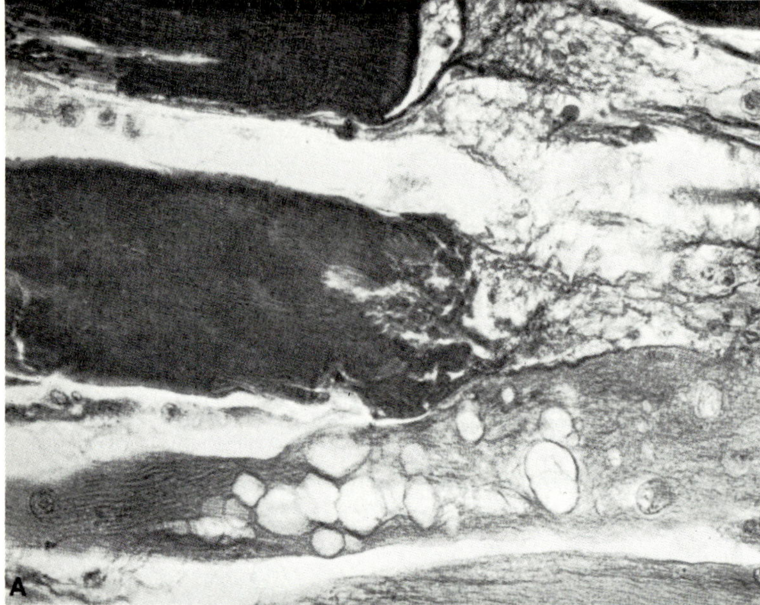

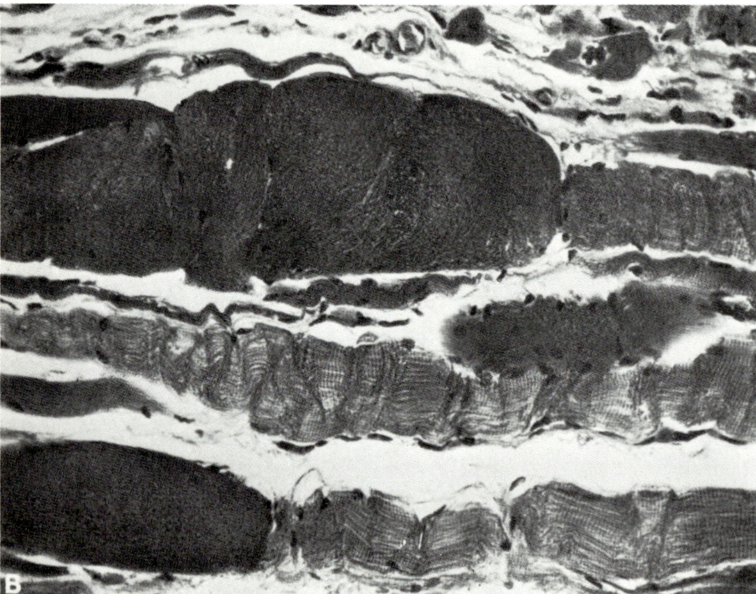

Fig. 4–3. (A) Recently traumatized muscle, showing varying effects of injury on adjacent muscle fibers. Two upper fibers are undergoing hyaline degeneration and have broken and retracted from main portion of fiber, leaving empty reticulated sarcolemmal sheaths (at right). Lower fiber is undergoing advanced cloudy swelling and early vacuolar degeneration. (B) Granular degeneration. (A, hematoxylin, eosin; B, phloxine, methylene blue)

and the use of special fixatives for phase and electron microscopy, with which most pathologists have had little experience.

The following disfigurative lesions, although subtle, appear to rest on a firm basis:

1. Dense cores near the center of each fiber with loss of mitochondria, of ATPase in the A bands, and of phosphorylase (also Z band streaming).

2. Rod bodies (also called nemaline bodies) in packets beneath the sarcolemma and cytoplasmic bodies. Electron microscopically they resemble the material of the Z band from which Gonatas *et al.*[16] and Price *et al.*[33] believe them to emanate.

3. Massive proliferation of mitochondria, some containing round, dense, osmophilic bodies.

4. Increase in number and enlargement, or abnormality, of mitochondria (megaconial—

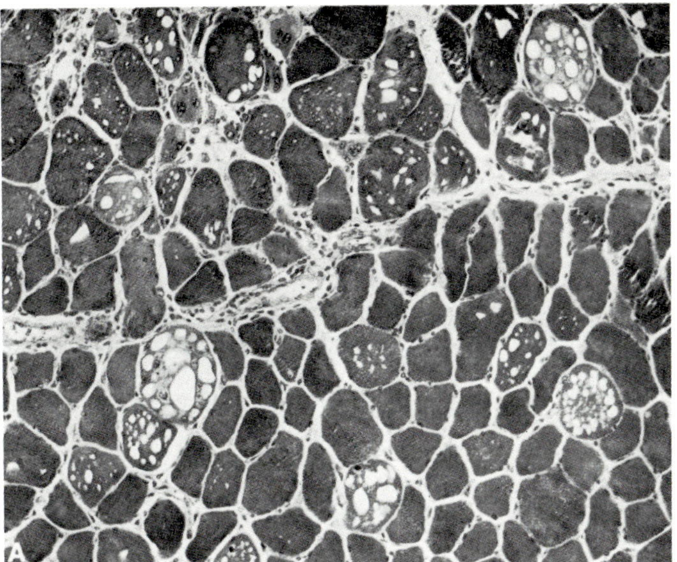

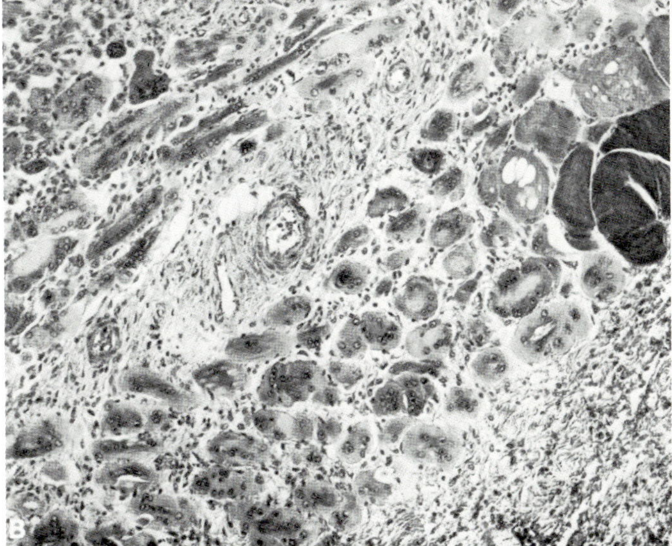

Fig. 4–4. Traumatic lesion of muscle 2 weeks after injury. (A) Many of muscle fibers are vacuolated. (B) Muscle buds (in section appearing as muscle giant cells) have formed at margin of damaged tissue. (phosphotungstic hematoxylin)

Shy; atypical mitochondria—Price). Some have rectangular bodies and spherical inclusions, excess intracellular lipid droplets mostly in type I fibers, and an excess of oxidative enzymes and mitochondria.

5. Loose spacing or lack of central myofibrils with central nuclei, and in some fibers an increase in DPNH dehydrogenase reaction with osmophilic material and myelin-like figures.

6. Gross vacuolation and dilation of endoplasmic reticulum, usually with an increase in glycogen in liposomes and lipid droplets (vacuolar myopathies).

7. Accumulation of glycogen.

8. Cytoplasmic bodies and crystalline masses, some of which are interpreted as viruses.

In our review of experimental pathology of the disfigurative changes, only the vacuolar myopathy was consistently reproduced in the laboratory animal. All the others described in Chapter 5 represent new processes—i.e., central core disease with the congenital myopathy by that name, described by Shy and Magee;[37] nemeline or rod bodies with the congenital myopathy, first reported by Shy et al.,[38] and the pleoconial and megaconial changes with the congenital myopathy by Shy,[36] Price,[32] and others; central clearing and nucleation with the congenital myotubular myopathy presented by Spiro and Kennedy;[39] the vacuolation with familial hypokalemic periodic paralysis, which has long been known; glycogen accumulation with acid maltase deficiency, otherwise known as Pompé disease. Already the specificity of these changes has been brought into question by Afifi et al.,[2] who found nemaline bodies in one member of a family with a congenital myopathy and central cores in another. Moreover, as was pointed out by Engel,[12] Mauro, Shafiq, and Milhorat,[24] Engel[13] and Price,[32] the exact significance of these alterations of organelles in relation to the internal economy of the muscle fiber remains uncertain. Presumably they portray in some instances a metabolic abnormality of which the most obvious expression is the alteration of a certain organelle; in other instances they must be regarded as an adaptive or compensatorial reaction, as when an excess of mitochondria reflects a defect in oxidation. Again the ultrastructural change may be due to an accumulation of a metabolic product resulting from an enzyme defect, e.g., glycogen; or it may represent a defective structural organization of contractile proteins, e.g., sarcoplasmic masses and abnormal configurations of myofilaments in myotonic dystrophy.

As experience with electron microscopy enlarged it was discovered that no one of the above changes is specific. Central cores are similar if not identical to some of the "target changes" of denervative diseases, according to Engel.[13] Nemaline bodies were also observed in muscle fibers after the tendon was sectioned and in dermatomyositis, Sjogren's syndrome, lupus erythematosus, pathologic cramp syndrome, etc.; Z-line streaming and deformity have been noted in a variety of diseases, as have mitochondrial abnormalities, according to Engel.[12] Thus in each of the above diseases there must be not one but a constellation of changes, and their relationship is far from certain. In several instances, if not in all, the primary abnormality is certainly something other

than a change in organelles. Finally it is appreciated that artifactual change is even more troublesome under the electron microscope than under the light microscope, since now the imperfections resulting from faults in the biopsy and fixation technique are magnified 10,000-fold (Chapter 13).

VOLUMETRIC CHANGE IN THE MUSCLE FIBER AS A BASIC ABNORMALITY

DENERVATIVE ATROPHY

The concept of the motor unit meaning the motor neuron of spinal cord and brain stem, its axon, and all the muscle fibers to which it is attached (Chapter 2), has application in myopathology, for all diseases of nerve, spinal cord, and brain stem express themselves in terms of such units. In other words, diseases of the motor nerve cell not only render muscle functionally useless (paralyzed) but impair its nutrition as well, leading to a bulk reduction of 80–90% within a few months. The latter includes a reduction in content of contractile proteins, amino acids, enzymes, and myoglobin evidenced under the electron microscope as a loss of peripheral myofilaments in each myofibril and even a total reduction in numbers of myofibrils. They are replaced by vesicular elements, glycogen, and rare mitochondria (Figs. 4–5 and 4–6).

Partial denervation permits two processes to compensate for the weakness of contractility, one the collateral regeneration of denervated muscle fibers by axonal sprouting in adjacent intact axons, the other a hypertrophy of surviving units from overwork. The first process increases the nerve/muscle fiber ratio in motor units resulting in large action potentials and polyphasic units in the electromyogram (EMG). It also results in the grouping of fiber types together, replacing the normal haphazard distribution. The second process widens the range of fiber sizes and causes a clustering of fiber sizes around two or three modes in a quantitative spectrum analysis rather than in the normal gaussian curve.

Denervation differs only in degree (and in fiber

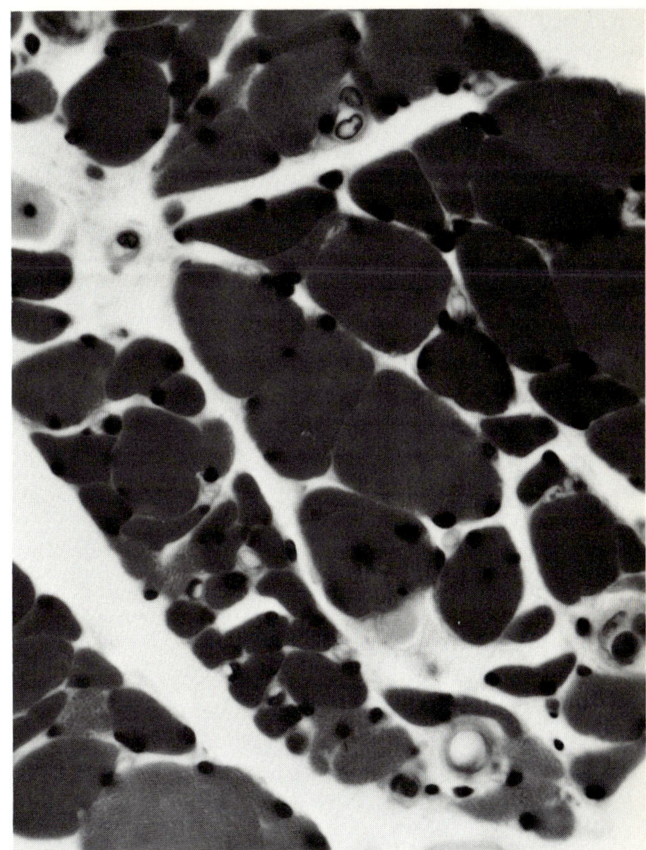

Fig. 4–5. Transverse section of gastrocnemius muscle, showing "group atrophy" due to denervation in an elderly man not known to have neuromuscular disease. (hematoxylin, eosin, ×610)

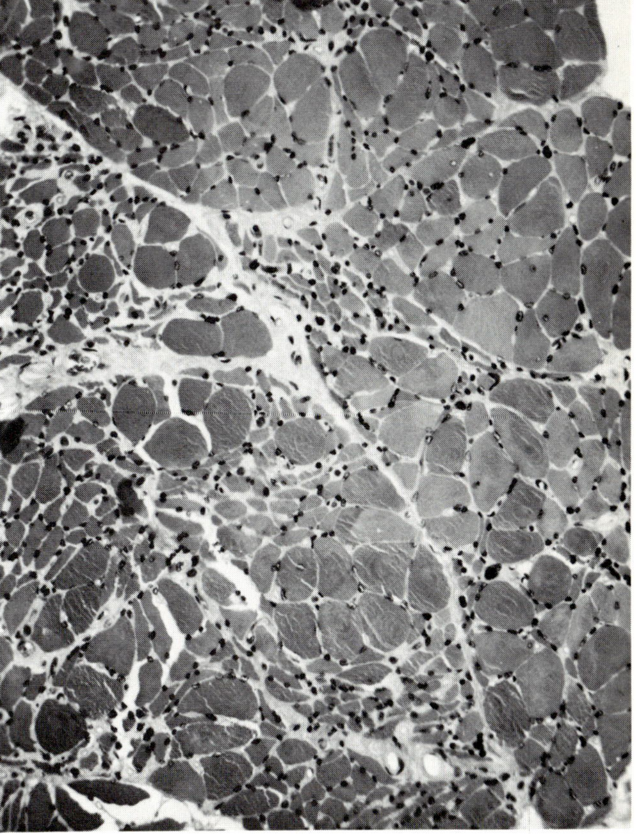

Fig. 4–6. Transverse section of gastrocnemius muscle from an elderly woman showing "group atrophy" with no known neuromuscular disease. (hematoxylin, eosin, ×220)

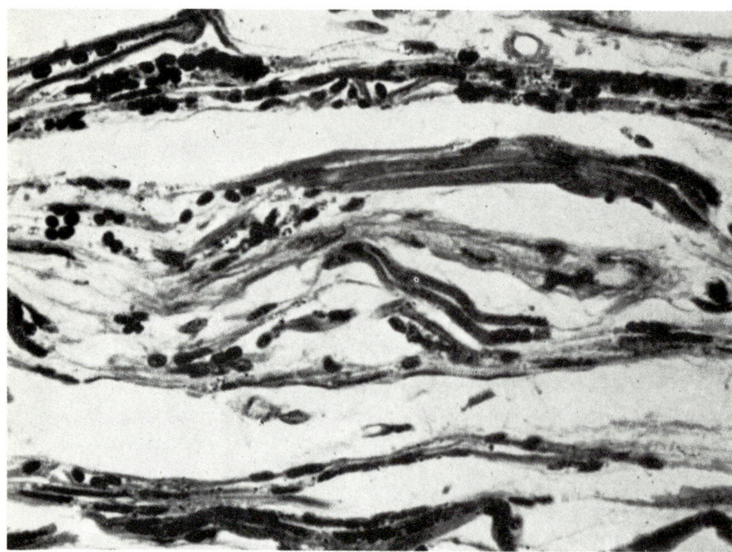

Fig. 4–7. Extreme disuse atrophy of muscle in an advanced case of rheumatoid arthritis with skeletal fixation. Muscle fibers are present as thin strands; note the rows of dark nuclei and remnants of sarcoplasm. (hematoxylin, eosin, ×700)

type selectivity) from disuse atrophy, although there are also physiologic differences between the two processes. In cachexia and disuse (Fig. 4–7) it is the white type II fibers that are more involved, whereas in denervative diseases both red and white types are usually affected.

The fate of the denervated fibers is highly controversial. Nerve is essential to the growth and natural development of the muscle fiber and to some extent its survival. After a variable interval of time, months to years, many of the muscle fibers degenerate, including some of the larger ones with sound innervation still. These late dystrophic changes have been a source of dispute among myopathologists who have mistakenly identified them with the primary pathologic process. The finding of a certain smudging and coagulation in the center of some of the surviving fibers, the so-called "target fiber" of Engel, has no explanation: It has not been reproduced experimentally and may represent some delayed secondary process unrelated to denervation.

In sum, severe diffuse atrophy and "group atrophy of single motor units" either simultaneously or successively stand as cardinal myopathologic criteria of all neural and spinal diseases, and histochemical regrouping is a valid index or reinnervation. The usefulness of these data to the myopathologist in biopsy studies is limited by the fact that a certain amount of time must elapse before these changes become manifest, and they are readily confused at an early stage with the effects of disuse and aging.

HYPERTROPHY IN HUMAN MUSCLE

In the strict sense of the word, muscular hypertrophy denotes an increase in the number or size of the constituent fibers. Any other condition which results in an augmentation of the volume of a muscle, such as fatty infiltration or lipomatosis, a tumor, or an inflammatory process, is properly termed pseudohypertrophy. In the latter the increase in fat cells and connective tissue or the infiltrations of tumor or inflammatory cells which determine the increase in bulk may actually mask a concomitant atrophy or destruction of muscle parenchyma. These distinctions between true hypertrophy and pseudohypertrophy should be recognized by the student of muscle pathology even though they may at times merge with one another, as in certain cases of muscle dystrophy.

True hypertrophy, with which we are occupied in this section, may be due to excessive use of a muscle or muscle group or may be an expression of a muscle fiber disease. Its cause then may be either physiologic or pathologic. The former results from abnormal activity.

Pathologic hypertrophy is uncommon, being encountered only in the early phases of progressive muscular dystrophy, in Thomson's disease or myotonia congenita, and as a relatively rare idiopathic affection known as hypertrophic myopathy or true muscle hypertrophy. Each of these conditions is discussed briefly in the following pages. Further details as to the morbid anatomy can be obtained from Durante,[11] who presented a well rounded discussion of the whole subject and whose general outline we have adopted.

PHYSIOLOGIC HYPERTROPHY

Whenever a healthy muscle is required to perform heavy work over any extended period of time, whether as a result of exercise or because of denervation or disease of part of a muscle, it increases in volume, and proportionally in strength and endurance (Chapter 2). This elementary fact is in part the basis of athletic training programs. Since certain muscles are apt to be engaged more than others, the hypertrophy may be limited to particular muscle groups, e.g., the heavily developed shoulder and arm muscles of the weight lifter and the enlarged gastrocnemius muscles of ballet dancers. Likewise, hypertrophy of certain muscle groups are the trademark of certain occupations, e.g., the large forearms and hands of the old-time blacksmith. However, it must not be assumed that muscle size is determined exclusively by exercise, for genetic factors also operate, some individuals being endowed with large, thick muscles as part of their body habitus. The herculean muscle development of the strong man in the circus is probably more a matter of inherited somatic form than of physiologic hypertrophy.

Muscle hypertrophy of the physiologic type may develop in the course of certain diseases and physiologic states. In hypertension and arteriosclerosis the heart muscle becomes hypertrophic; in chronic pulmonary diseases, such as fibrosis of the lungs and emphysema, the accessory muscles of respiration (intercostal, sternomastoid, and scaleni) enlarge; during pregnancy there is hypertrophy of the abdominal muscles to hold in place the heavy gravid uterus; and in those diseases of the brain that result in states of increased muscle tone and rigidity (e.g., athetosis and dystonia) muscular hypertrophy is a common sequel.

Upon gross inspection the physiologically hypertrophied muscle, apart from its greater size, has a deep red color, prominent fascicles which impart a coarse granularity to the cut surface, and a tough consistency. From the histologic point of view the muscles appear to be entirely normal, and the increase in volume is easily overlooked. The fibers are well striated and their sarcolemmal nuclei are normal in number. By careful measurement it can be demonstrated that the bulk of each fiber is augmented slightly, that the number of myofibrils is increased (Fig. 2–25), and that the sarcolemmal nuclei are either normal or slightly larger than average. The total amount and disposition of fat and connective tissue are natural. The vascular arrangements are probably expanded.[31]

Whether this physiologic hypertrophy is the result of an increase in the volume of each fiber or an increase in the number of muscle fibers has been debated at length. The problem poses numerous experimental difficulties. A satisfactory study would be one that establishes the range and average of fiber diameter and length for a given muscle in an animal species, and then comparable measurements after the induction of physiologic hypertrophy.

Our investigations of physiologic hypertrophy in the animal support the volumetric hypothesis rather than the numerical one. The experiments and observations that have been carried out were described in Chapter 2, and it is probable that the results are applicable to human muscle. Endocrine factors undoubtedly are also of importance in determining the ultimate size of skeletal muscles. The muscles of the male in most mammalian species tend to be heavier than those of the female, owing to larger size of the individual fibers.

MUSCULAR HYPERTROPHY IN PROGRESSIVE MUSCULAR DYSTROPHY

Enlargement of certain muscles is a common finding in progressive muscular dystrophy

(Chapter 6). It has been noted in all the subvarieties of dystrophy but is more conspicuous in the pseudohypertrophic form than in others. The leg muscles, particularly the gastrocnemii, are affected most frequently, but the arms, deltoid, triceps, and biceps are often involved, and even the masseters on occasion. Usually the enlarged muscles have a peculiar firmness on palpation and contract with somewhat less vigor than is normal.

As a rule, macroscopic examination alone permits the distinction between pseudohypertrophic enlargement and true hypertrophy, because in the former the muscular tissue is pale red or even yellowish whereas in the latter the natural color is preserved. In microscopic sections, however, the distinction cannot be drawn so sharply, for there exists constantly in progressive muscular dystrophy a number of more or less hypertrophic fibers. It has been assumed by many workers, since first being pointed out by Erb,[15] that enlargement of individual muscle fibers represents the first phase of the dystrophic process. Regardless of the correctness of this assumption, microscopic examination almost invariably discloses some degree of fiber enlargement. Individual fibers may attain a size of 200–250 μ. In a few exceptional cases[4] it has been demonstrated by biopsy that a general enlargement of all muscle fibers and a minimum of muscle fiber atrophy and degeneration had occurred. Later these same muscles become wasted and lose strength of contraction as well as tendon reflex and electrical excitability, thus showing that the disease followed the expected course of muscle dystrophy. The weakness in atrophic diseases usually results in some compensatory hypertrophy in unaffected muscles, and it is a moot point whether some fibers in a dystrophic muscle may hypertrophy in response to added stress owing to loss of others in the same muscle. This seems to occur in poliomyelitis.

It would appear that in progressive muscular dystrophy an increase in muscle bulk and firmness is attributable to an infiltration of fat cells between individual muscle fibers or fascicles of fibers, a pseudohypertrophy in some muscles, while hypertrophy of individual muscle fibers may add to the increase in muscle size, and

exceptionally may alone be responsible for it (true hypertrophy).

MUSCLE HYPERTROPHY IN THOMSEN'S DISEASE (MYOTONIA CONGENITA) AND IN PARAMYOTONIA

Muscle hypertrophy is one of the predominant clinical features of myotonia congenita. This fact was well documented in early publications on this subject, which note that many of the original Danish males were employed as "strong men" in circuses. The outstanding microscopic change is an increase in the volume of individual fibers, the transverse diameters often being in excess of 100 μ or even 150 μ. An increase in number of myofibrils and probably of sarcoplasm as well accounts for this enlargement. It is possible that the enlargement is a form of physiologic hypertrophy produced by the high intensities of neuronal innervation that are a secondary phenomenon in this disease. Infiltration of fat cells and hyperplasia of connective tissue do not occur.

HYPERTROPHIC MYOPATHY OR HYPERTROPHIA MUSCULORUM VERA

There have been isolated observations of true hypertrophy of muscles without accompanying atrophy or other signs of dystrophy.[3, 21, 47, 49] The strength of the hypertrophied muscles was usually increased, though in many instances there was weakness and excessive fatigue. Other tissues than muscle did not partake of this hypertrophy, thus distinguishing it from the general hypertrophy of a limb or one side of the body sometimes seen in conjunction with neurofibromatosis, angiomatosis, and other neuroectodermal disorders.

Most pathologists who studied this disorder are agreed that the only uniform change is an enlargement of the muscle fibers. When measurements were available, the fibers ranged in size from 95 to 200 μ. Similar findings are reported in the cases of congenital hypertrophy of muscles reported by Bruch,[6] de Lange,[9] Debré and Semelaigne,[8] and Hall et al.[18]

HYPOTROPHY AND HYPOPLASIA AND PREMATURE SENESCENCE

A certain delicately balanced genetic mechanism must determine the configuration of the musculature in every species and accounts for individual variations within a species (Chapter 2). Some individuals appear to be destined to have heavy, short muscles, and others long, thin ones; moreover, certain patterns of variation are obvious. The herculean muscular development of the strong man in the circus is probably more a matter of inherited somatic form than physiologic hypertrophy. Little is known of how the exact number and arrangement of fibers in each muscle is fixed, or of the manner in which the process of muscle fiber formation is naturally arrested at a particular moment. Presumably the musculature in individuals suffering from certain developmental diseases of the nervous system is deficient, but this possibility has never been investigated in quantitative terms (computation of number and size of fibers in a given muscle).

It has been assumed that all the muscle fibers an individual is to possess exist at birth, although in a number of studies with animals and humans de Reuck and Adams[10] showed that the number slowly increases after birth. The main growth of muscles postnatally, however, is due to an increase in the size of fibers already formed. The growth rate is steady all through childhood and is paralleled by an increase in strength.[26] At puberty the male muscle rapidly doubles in size and power, and by adulthood reaches a plateau. Female muscle remains volumetrically smaller and weaker after puberty, but the sex differences are much greater in some muscles than in others.

As an example of pure developmental hypoplasia only one disease has been defined—one in man called myotubular myopathy (Chapter 5). Probably the Praeder-Willi syndrome should also be included. Here all the fibers retain their fetal characteristics even to the lack of differentiation into fiber types. There are other diseases to be considered such as the congenital hypoplasia of Krabbé and congenital benign hypoplasia of Walton, but here one cannot specify the nature of the abnormality or even state whether the thinness of muscle is due to a numerical or volumetric deficiency of fibers.

Throughout life it appears that each person must conserve his population of muscle fibers. Complete destruction of a fiber, which rarely happens as a single event, results in permanent loss. In a sense the muscle fibers are like nerve cells in being postmitotic and incapable of proliferating after birth. When one speaks of regeneration of muscle it is in reality a restoration of injured segments by the uninjured sarcolemmal nuclei of that same fiber to which one refers. Significant losses of fibers consequent to disease appear to be irreparable.

During late life the aging processes of the body express themselves also in the musculature, but these have not been systematically studied. Turning to a large number of muscle specimens from a series of more than 125 individuals not known to have neuromuscular disease, Moore et al.[28] and in another series Tomlinson et al.[41] found an increasing degree of atrophy, darkening of color, and microscopic disappearance of fibers during each of the late decades of life. Some fibers shrink, become relatively hypernucleated (leaving rows or clumps of small dark nuclei) with increased amounts of lipofuscin and folded cytoplasmic and nuclear membranes. Other fibers increase in size (? work hypertrophy) so that the range of fiber diameters increases. In certain muscles (e.g., ocular muscle) there are cleared zones of sarcoplasm and a loss of peripheral myofibrils, i.e., ringbinden or spiral annulets (Fig. 4–8). The type II fibers are thinner, and there is an attendant reduction in enzymes. In the leg muscles of individuals older than 60 years "group atrophy" from loss of anterior horn cells is frequent. It is uncertain whether these changes correspond to what French neurologists have termed "myosclerosis of the aged" because of imprecision in the use of this term.

Oculopharyngeal dystrophy, a dominant hereditary disease of late life, seems to exemplify an exaggerated aging process. The lesion (Chapter 5) differed only in degree from that seen in older control subjects. This led Rebeiz et al.[34] to postulate that the disease

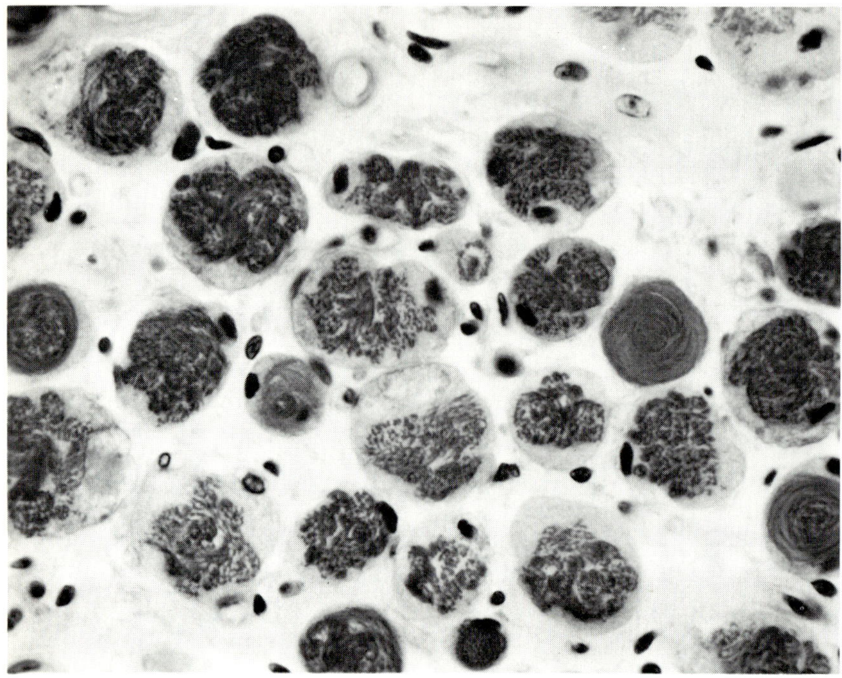

Fig. 4–8. Transverse section of superior rectus muscle in an octogenarian. Note grouping of myofibrils, cleared zones of sarcoplasm beneath sarcolemma, and ringbinden. (phosphotungstic hematoxylin, ×610)

Fig. 4–9. Localized blebs of sarcoplasm raising the sarcolemma in relation to abnormal bands of "extreme contraction," probably an artifact of fixation. In cross section these appear as zones of fiber free of myofibrils. Sarcoplasm is finely granular and stains intensely with periodic acid-Schiff. Taken from a biopsy specimen of a patient having amyotrophic lateral sclerosis with muscle cramps. (methylene blue, eosin)

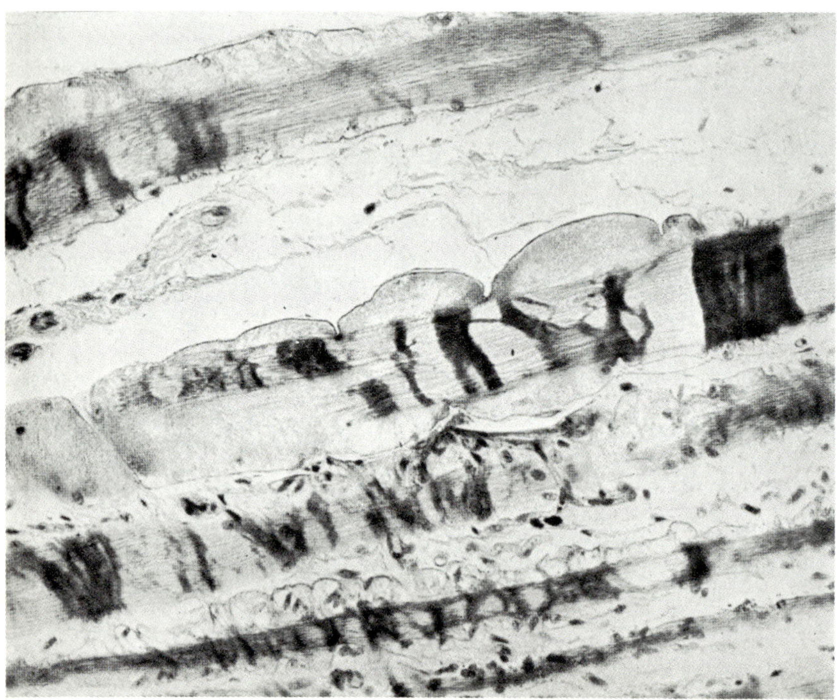

might represent a presenile myopathy analogous to the presenile neuronal degeneration of Alzheimer. The validation of this concept awaits more complete description and understanding of the natural life cycle and aging of the muscle fiber.

MYOPATHOLOGY OF DISEASES INVOLVING NEUROMUSCULAR TRANSMISSION OF NERVE IMPULSES

Here a new principle confronts the student of myopathology—that of paralysis or spasm due not to disease of the motor neuron or muscle fiber but to disorder in the conduction of innervative impulses. It results either in weakness and paralysis or in spasm and cramp. There is in this group of diseases the widest discrepancy between functional change and structural alteration, mainly because of the inadequacy of our techniques for studying morphologically the conducting apparatus (Chapter 12). In fact only in comparatively recent times has our concept of the conducting apparatus been broadened to include not only the terminal nerve twigs and the muscle endplate but also the true sarcolemma, transverse sarcoplasmic reticulum, terminal vesicles, and longitudinal sarcoplasmic reticulum. Not until Coërs and Woolf[7] applied the methylene blue method to the motor point of muscle and the Kolle method of cholinesterase and electron microscopy to the motor end-plate as well as the muscle fiber was it possible even to inspect these structures.

The pathologic alterations of the conducting apparatus responsible for some of these severe functional changes have been difficult to demonstrate, which is not unexpected in view of their transitory and reversible nature. The diseases that fall in this category are myasthenia gravis, Eaton-Lambert syndrome, periodic paralysis, myotonia congenita, tetanus, tetany, botulism, pathologic cramping, and some of the thyroid and parathyroid effects on muscle. Such data as can be assembled are fragmentary (Chapter 12).

Since the muscle in some of these states is excessively irritable and subject to contraction bands (Fig. 4–9) it is not surprising that fracture of one or more of the bundles of myofibrils within a muscle fiber, with coiled or circular arrangement of the broken bundle (Fig. 4–6), is a common finding. It is especially frequent in biopsy material and most often appears in conditions in which there is myotonia or some related abnormality of muscle excitability such as a cramp. The perfect preservation of striation and abrupt point of fracture strongly suggest that the phenomenon is usually an agonal event immediately preceding fixation. It must be conceded, however, that this phenomenon of "ringing" of muscle fibers by aberrant myofibrils which encircle the periphery or penetrate the sarcoplasm has been a highly controversial finding for a long time. As was first noted by Heidenhain[19, 20] and later confirmed by Wohlfart[48] in man and by Weiss and James[46] in amphibians, these myofibrils turn at right angles to the usual longitudinal plane of the section and wrap themselves around the outer surface of the fiber. In transverse sections the longitudinally oriented central myofibrils are surrounded by a collar of these myofibrils. Sometimes they pass through the fiber, coming to the surface on the other side. These aberrant myofibrils retain their cross striations and reorient the sarcolemmal nuclei in parallel. The distance between their Z disks is said to be the same as that of the longitudinal myofibrils.[17]

As to whether ringbinden develop during life or are the result of a biopsy procedure, their striking incidence in myotonic dystrophy, as emphasized by Wohlfart, might be in keeping with either possibility in view of the extreme irritability of the muscle. They do, however, occur in other forms of muscle disease.[30] The author is inclined to agree with Schaeffer and Erb, who stated that when peripherally placed and unassociated with other alterations of the fiber, these circular striations are probably artifacts. However, when as in myotonic dystrophy they are associated with sarcoplasmic masses and rows of sarcolemmal nuclei, it seems that Wohlfart[48] and Greenfield *et al.*[17] are right in assuming them to be a true structural defect representing a disfigurative disorganization. The process that organizes myofibrillar structure appears to have failed.

REGRESSIONS AND METAPLASIAS OF MUSCLE

There remains to be discussed only a few concepts that have been so firmly established in pathology textbooks as to merit additional comment in reference to general principles of human muscle diseases.

PIGMENTARY ATROPHY

Brown atrophy of the liver and heart and to a lesser degree of other abdominal viscera are commonplace findings during old age, and we classify them as effects of age. In most textbooks, however, they are considered as separate entities. We refer here to the yellowish brown granules of lipochrome that accumulate at either pole of the sarcolemmal nucleus of the heart muscle fiber or are dispersed all through the cytoplasm of the liver cell. Simple atrophy of the cell or fiber and an increase in interstitial connective tissue accompany this pigmentation.

Skeletal muscles seldom show brown atrophy to the same degree as heart muscle, though in advanced age small numbers of lipochrome granules may be seen.[11, 45] As in heart muscle, they accumulate at the poles of the sarcolemmal nuclei, more in the atrophic than in healthy fibers (Fig. 4–10), and are faintly sudanophilic and acid-fast. The source of this "wear and tear" pigment, as it has been called, is not known. The idea that it could be a degeneration product of myoglobin is untenable because it has the same staining reactions as the lipochrome of other cells (e.g., nerve cells) that contain no myoglobin. The possibility of it being a breakdown product of cytochrome C consequent on the action of lysosomal enzymes is more plausible but has not been confirmed, nor has its relation to the pigment in vitamin E-deficient animals (Chapter 3) been clarified. In our opinion the appearance of lipochrome

granules is always a pure aging effect and has little or no correlation with degeneration of the muscle fibers. The not infrequent association of this pigmentary change with the atrophy and degeneration of muscle fibers and the increase in fibrous tissue during advanced age has resulted in undue emphasis on pigmentary atrophy or degeneration as a special pathologic state. Denny-Brown was impressed by the frequency of a combination of granular and pigmentary atrophy in carcinomatosis of various types, but particularly in relation to ovarian carcinoma. This association points to a possible metabolic origin.

MUSCLE REGRESSION AND METAPLASIA

Other modifications of the muscle fiber designated as regressive and metaplastic are less well known. These have been described in detail by many of the most able students of muscle pathology, but because of the difficulty in tracing these various chronic transformations from beginning to end, they remain controversial. In these states the muscle fiber does not degenerate but either loses its striation slowly as the protoplasm becomes undifferentiated (the plasmodial regression of Durante[11] and others) or its muscle nuclei abandon their syncytial relationship and become individualized (cellular regression; Fig. 4–11). The isolated fibers or muscle nuclei then gradually lose their identity and are stated by some to undergo a transformation or metaplasia to fibroblasts, lipoblasts, and osteoblasts. Cellular regression is said to account for the formation of fibrous connective tissue, fat cells, and bone.

In several of the most authoritative monographs on muscle disease the regressions have either been presented as isolated pathologic events or as distinct pathologic processes. This impression is quite erroneous, for they occur predominantly in atrophic muscle and are almost invariably part of the general process of simple atrophy. Although they may be seen in muscle undergoing regeneration, they do not occur in the particular fibers that show regenerative activity.

In simple atrophy, whether from chronic

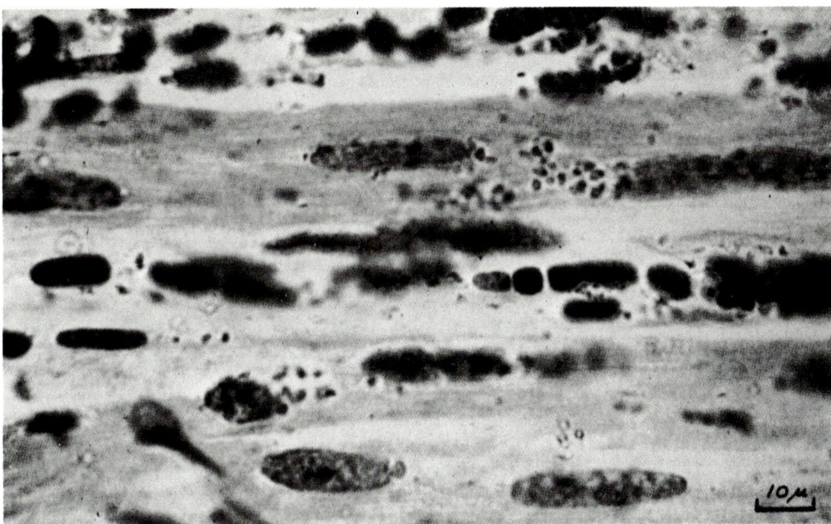

Fig. 4–10. "Pigmentary atrophy" in a biopsy specimen from a patient with rheumatoid arthritis. Note lipofuscin granules at each end of sarcolemmal nuclei. (hematoxylin, eosin)

Fig. 4–11. "Plasmodial regression" of muscle fibers several months after trauma. Isolated surviving fibers, some of which exhibit signs of abortive regeneration, are in the process of degeneration. (phloxine, methylene blue)

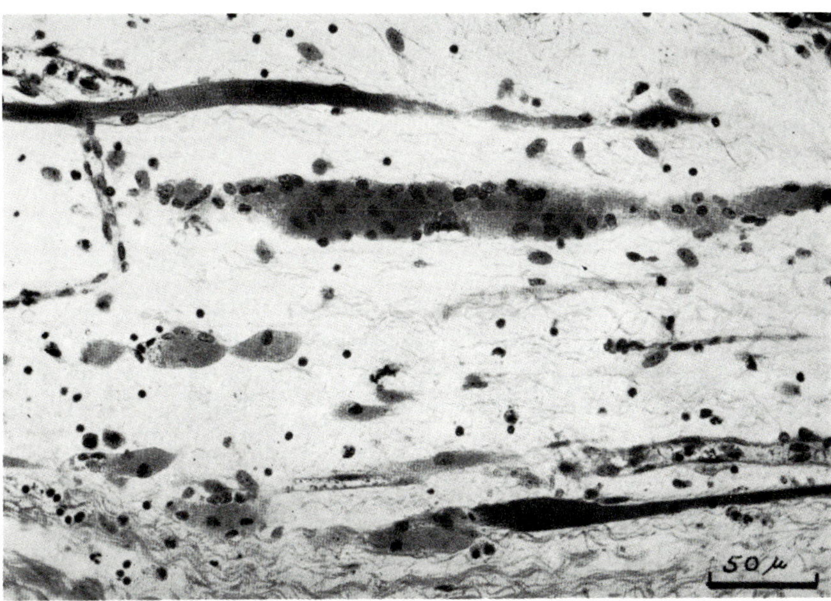

malnutrition, immobilization, or denervation, the muscle fibers gradually diminish in size and the sarcolemmal nuclei increase in number and size. The striations of the fiber are retained, the longitudinal ones often being more prominent than the transverse. The fibers are not all reduced to the same size. Those that were originally large probably require longer to atrophy; and some fibers, probably unaffected ones, even exhibit a true hypertrophy. The smaller fibers become thin ribbons of sarcoplasm with rows of small dark sarcolemmal nuclei, and eventually only the nuclei remain to mark the previous site of the muscle fiber (Fig. 4–7). The cloudy swelling and granular and waxy degeneration found in some fibers are probably of a degenerative nature. These are the standard changes of muscle atrophy (Chapter 3).

Under the influence of circumstances yet to be elucidated, regression seems to refer to particular modifications of all or parts of some of the muscle fibers. In some fibers the sarcolemmal nuclei begin to enlarge and to multiply at one point on the surface of the fiber. The sarcoplasm around them becomes clear and undifferentiated. Gradually one or several of these nuclei begin to separate from the muscle fiber and acquire a fusiform shape. This small group of cells and sometimes a few myofibrils may form an acute angle with the parent fiber and extend into the surrounding connective tissue. They may detach themselves completely and become a true exfoliation. In our material this is the result of disintegration of the parent fiber, we believe that exfoliation cannot occur from an intact fiber. The fate of the muscle exfoliate varies. In our own material (Chapter 3) we are convinced of the fact that these cells disintegrate and disappear. Some authors, particularly Krösing[23] and Durante,[11] state that the isolated muscle cells can again resume their syncytial relationship and under ideal circumstances can produce a new muscle fiber, or else they may gradually disappear or become indistinguishable from fibroblasts. Krösing,[23] who has written an interesting monograph on metaplasia of muscle fibers, looks upon this as a transformation of the muscle fiber to connective tissue and terms it status fibrosis. He points out that it is most convincingly demonstrated adjacent to a compressive muscle tumor. Tower[42] probably refers to a similar process by the term fibrous dedifferentiation. Other exfoliated cells are said to become laden with fat droplets, which eventually form a large central globule. The sarcolemmal nuclei are arranged around this globule. The identification of this pseudo-adipose cell as the remnant of a muscle fiber is possible only if there is retention of striations in the cytoplasm. Gradual transitions from this cell to a typical fat cell are described by Krösing

and Durante. The latter believes that the excess fat cells of muscle dystrophy and to a lesser extent of muscle atrophy are transformed muscle cells. He points to their occurrence in longitudinal rows with the same orientation as that of the muscle fibers as proof of this. Also some of the relatively rare intramuscular lipomas are believed to originate as a focal metaplasia of fat cells. The gradual conversion of exfoliated sarcolemmal cells to osteoblasts was also observed by Krösing, who expressed the opinion that this is the essential process underlying traumatic or idiopathic myositis ossificans.

According to Krösing,[23] similar metaplastic changes may involve the muscle fiber in its totality and not merely a small exfoliation of it; or a partial plasmodial regression may begin within the depth of the muscle fiber and result in its division into two or more parts. If there is incomplete separation, Y-shaped figures may be formed in longitudinal sections. At the points of bifurcation, the sarcolemmal nuclei are enlarged and aggregated. The newly divided fibers are surrounded by sarcolemma and acquire an endomysial sheath and capillary supply. In transverse sections these dividing fibers may, if cut in a fortuitous plane, be segmented, as pictured in Erb's monograph on muscle dystrophy[14] or may even show a central capillary. The destiny of the newly formed fiber is as variable as that of the exfoliations. Some give rise to healthy muscle fibers, whereas others are said to undergo fibrous, adipose, osseous, and even vascular transformations. The parent cell may retain a natural aspect with a strong striation, even though it is reduced in bulk. It is emphasized by many early pathologists that the small volume of fibers in atrophied muscle is due not only to simple wasting but also to these exfoliations and divisions. Moreover, longitudinal splitting is looked upon as a means either of increasing the total number of fibers in hypertrophy or of reducing the total number by disappearance of both parent and daughter fibers in the "numerical atrophy" of Klippel.[22]

These regressive changes are deserving of the most careful consideration, and further experiments should be set up to elucidate them more completely. In his own material the author can confirm longitudinal division as a change that arises from bifurcation of the sprouts of regenerating muscle fibers, and it may occur in hypertrophy but is not seen in simple muscle atrophy. Here one must be careful to avoid the optical illusion of division created by two overlapping fibers. Furthermore, there can be no doubt that muscle fibers compressed in connective tissue gradually lose their identity, coming to resemble fibroblasts or undifferentiated cells in their last identifiable stage. At this stage, however, the nuclei appear pyknotic, and it is doubtful if transformation occurs. The author has not observed a metaplasia of muscle cells to lipoblasts or osteoblasts and supposes in the light of more recent cytologic teaching that the transformation of one specialized cell into another is without precedent. However, the possibility of the muscle cell undergoing a regressive change to an undifferentiated state and then giving rise to a fat cell, fibroblast, or osteoblast can be neither affirmed nor denied from histologic observation. The alternative hypothesis—that in muscle disease undifferentiated mesenchymal cells may form fat cells or bone or connective tissue to replace the muscle fibers—seems more consistent with observed facts.

FATTY DEGENERATION

The natural fat content of striated muscle was described in detail in Chapter 1. Some muscle fibers contain finely dispersed fat and others do not. This fat, which is in the form of fine granules, occurs in the muscles of both infants and adults but tends to involve a larger proportion of the muscle fibers in the aged. When present in slight degree these granules serve to emphasize the longitudinal striation, but as the amount of fat increases the transverse striation tends to be masked and the fiber becomes more opaque.

We already expressed the view that lysosomes in muscle play no part in the contractile process but are related to certain general metabolic activities, depending upon the nutrition of the patient. Yet there is no doubt that in some muscle fibers an excessive deposition of unusually large fat granules is associated with atrophy and finally degeneration of the fiber. It is in this latter state, in which many of the fibers are of uneven caliber and some are degenerating, that the term fatty degeneration is usually applied.

There are two possible sources of the visible fat in the muscle fiber; one is the lysosomes and the other complex lipoprotein molecules which occur in muscle substance. Patients immobilized by muscular diseases are likely to store more natural fat as liposomes than individuals in normal health. In certain conditions such as diphtheria (in which the fibers of the heart frequently exhibit fatty change) and phosphorus poisoning, the liberation of bound fat appears to be one of the most important pathologic changes. In most other diseases in which fatty degeneration is commonly seen (e.g., chronic infections, in the late stages of neural and spinal amyotrophy, and in certain inflammatory diseases) it is doubtful if fatty degeneration is a primary process. The frequent conjunction with granular degeneration suggests that it is only one manifestation of a severe irreversible change which should be referred to as a granulofatty degeneration. The fat probably arises from liberation of bound fat and by coalescence of lysosomes. The formation of fat droplets may also be seen in hyaline degeneration if appropriate stains are used. One form of vacuolar myopathy also appears to be related to a disturbance of fat metabolism (Chapter 5); and the same is true of certain intoxications (e.g., chloraquine myopathy). There are autophagic vacuoles that appear to be caused by release of lysosomal enzymes.

RIGOR MORTIS

Rigor mortis is a manifestation of partial contraction of the body's skeletal musculature which develops after death. After a short agonal period of relaxation, the muscles harden and fix the joints so firmly that attempts to move the limbs passively encounter great resistance. Rigor usually begins in the jaw muscles and spreads to the trunk and extremities; it disappears in a similar order. The rigor usually appears 2–4 hours after death, reaches its height in less than 48 hours, and disappears in another 48 hours; it does not recur. When the temperature of the environment is warm, rigor occurs early and disappears early; in a cold atmosphere the rate of appearance and disappearance is retarded. It begins earlier if there has been considerable muscular activity prior to death as in tetanus, strychnine poisoning, poisoning with prussic acid, convulsions and sunstroke. Rigor also involves the heart muscle and smooth muscles, particularly those of the middle-sized and small arteries, pylorus of the stomach, urinary bladder, intestines, and eye (smooth muscle of the iris) producing constriction of the pupil. The cutis anserina of the cadaver is caused by rigor of the arrectores pilorum muscles. Rigor also may occur in intrauterine fetuses that died recently and may thus lead to difficulty in labor.

The freshly exercised muscle tissue in a state of rigor can no longer be induced to contract. According to Durante,[11] the fibers retain a clear striation. Saline solutions do not modify the birefringence and only rarely produce the false vitreous degeneration that is so easily brought about in living muscle. The fibers are slightly opaque with a fine granularity of the contractile substance and are less soft and diffluent than in the living state. After the cadaveric rigidity vanishes the muscle fibers once again became softer and less fragile. This is the ideal time for excising tissue for microscopic examination.

In the past it was believed that cadaveric rigor was due to coagulation of muscle proteins arising from postmortem accumulation of lactic acid and a corresponding hydration and swelling of the sarcoplasm. Recent chemical studies did not confirm this simple hypothesis and pointed to a more fundamental change.

It has been known for a long time that if there is extreme activity immediately preceding death, rigor mortis develops early and is pronounced. Bate-Smith and Bendall[5] investigated this phenomenon experimentally in rabbits. They found that the time of onset and degree of rigor were influenced by three factors: the glycogen reserve of the muscle, the pH of the muscle at the time of death, and the temperature of the environment. In well fed animals with high glycogen reserves there is a long delay before the onset of rigor, whereas in animals with little or no glycogen reserves the onset of rigor is immediate or occurs shortly after death. Its time of appearance is directly related to the pH of the muscle at the moment of death—being more prompt if the pH is low, as when the animal is active up to the moment of death, and if the glycogen store is poor. Raising the external temperature from 17 to 37° C resulted in a shortening of the interval between death and the onset of rigor.

Szent-Györgyi[40] in the course of his studies on muscle chemistry found that the energy-rich compound adenosine triphosphate (ATP) is normally absorbed in considerable quantities into the muscle proteins and that this state appears to be necessary for relaxation of the muscle fibers. Erdos (quoted by Szent-Györgyi) discovered that as the muscle begins to stiffen the amount of ATP is reduced, and he concludes that this ATP reduction is the chemical event that precipitates rigor.

Bate-Smith and Bendall[5] confirmed Erdos' and Szent-Györgyi's findings and extended their original observations. They determined that there was a rapid turnover of ATP in muscle immediately after death and that during the interval before the onset of rigor the breakdown of ATP is balanced by its resynthesis by the glycolytic cycle, or by its further synthesis from ADP, utilizing phosphocreatine (PC). Hence the reactions

$$\text{ATP} \xrightarrow{\text{ATPase}} \text{ADP} + \text{P}$$

and

$$\text{2ADP} \xrightarrow{\text{Myokinase}} \text{ATP} + \text{adenylic acid}$$
$$\Updownarrow$$
$$\text{glycolytic energy}$$

or

$$\text{ADP} + \text{PC} \rightarrow \text{ATP} + \text{creatine}$$

are exactly balanced. As the glycogen stores are exhausted—this occurs sooner in starved or exercised animals—the resynthesis of ATP cannot keep pace with its degradation and rigor ensues. Bate-Smith and Bendall found that rigor begins to appear when the level of ATP falls below 85% its initial value and is nearly complete when the level reaches 15%. The

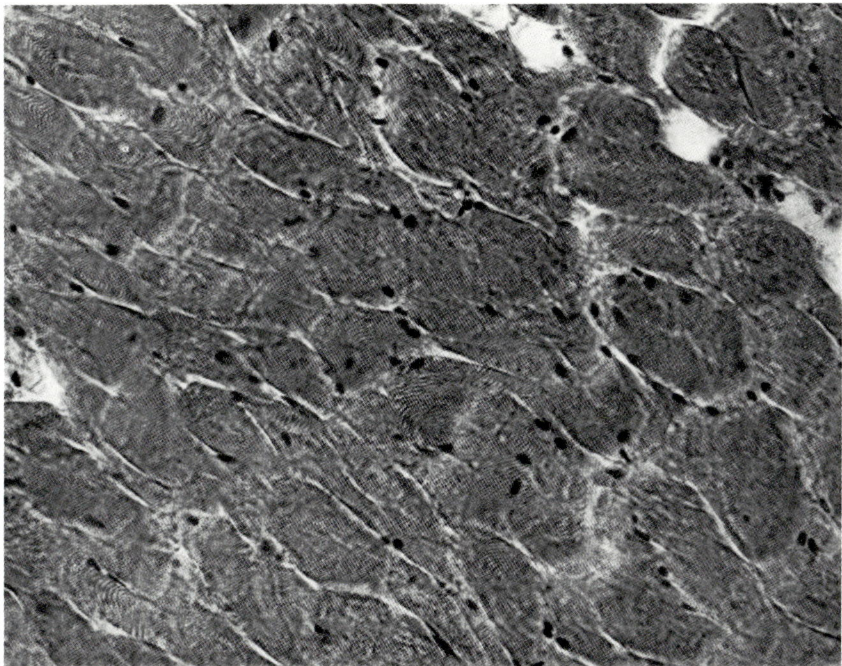

Fig. 4–12. Postmortem changes in muscle removed from body 2 weeks after death. (phloxine, methylene blue)

degree and speed of onset of rigor is determined by the initial ATP, PC, and glycogen content.

Once developed, rigor may persist for days if bacterial contamination of the muscle is prevented.[5] The onset of putrefaction has a "plasticizing" effect, restoring ductility. Other chemical processes (e.g., deamination of adenine nucleotide and accumulation of ammonia) also play some part in the disappearance of rigor.

The changes in muscle during putrefaction are evidently due to autolytic liquefaction of the muscle protoplasm. These are often marked, and on this account it is important to recognize them so as not to mistake them for vital processes. The myofibrils become less distinct, tend to dissociate, and finally disintegrate. The contents of the muscle fiber are transformed into a fine emulsion, and the individual fibers appear to swell and become compressed against their neighbors (Fig. 4–12). They stain deeply with eosin and are more brittle and easily fractured than otherwise. Sarcolemmal and other nuclei are shrunken and stain darkly and homogeneously. A later change is disappearance of the chromatin so that the nuclei are faint or invisible. The entire tissue then becomes loose and reticulated.

CLASSIFICATION OF STRIATED MUSCLE DISEASES

One of the first requisites for a discussion of muscle disease is a classification. Etiology affords the only sound basis for such a classification, but this is impossible at the present time because the causes of only a few of these diseases are known. We are compelled therefore to divide them according to their chief and distinguishing pathologic features. The limitations of a classification that depends on morbid anatomy becomes evident to the reader when he finds that some diseases of supposedly different etiology produce almost identical changes in the muscle and that other diseases generally regarded as muscular (e.g., myotonia congenita, periodic paralysis, and myasthenia gravis) are accompanied by no perceptible morphologic changes. The classification followed in the succeeding chapters has as its chief recommendation that it is based on etiology, when known, and otherwise on morbid anatomy.

REFERENCES

1. ADAMS RD, REBEIZ J: Unpublished observations

2. AFIFI AK, SMITH GW, ZELLWEGER H: Congenital nonprogressive myopathy; central core disease and nemeline myopathy in one family. Neurology (Minneap) 15:371, 1965

3. AUERBACH L: Ein Fall von Wahrer Muskelhypertrophie. Virchows Arch [Pathol Anat] 53:397–417, 1871

4. BARNES S: A myopathic family; with hypertrophic, pseudohypertrophic and atrophic terminal (distal in upper extremities) stages. Brain 55:1–46, 1932

5. BATE-SMITH EC, BENDALL JR: Factors determining the time course of rigor mortis. J Physiol (Lond) 110:47–65, 1949; Br Med Bull 12:230–235, 1956

6. BRUCH F: Ueber einen Fall von congenitaler Makroglossie combiniert mit allgemeiner, wahrer Muskelhypertrophie und Idiote. Dtsch Med Wochenschr 15:229, 1899

7. COËRS C, WOOLF A: The innervation of muscle—a biopsy study. Springfield, Ill, Charles C Thomas, 1959

8. DEBRÉ R, SEMELAIGNE G: Syndrome of diffuse muscular hypertrophy in infants causing athletic appearance. Am J Dis Child 50:1351–1361, 1935

9. DE LANGE C: Congenital hypertrophy of muscles, extrapyramidal motor disturbances and mental deficiency: a clinical entity. Am J Dis Child 48:243–268, 1934

10. DE REUCK J, ADAMS RD: Changes in muscle fibers with age. International Congress Series. Elsevier Press 1974, Amsterdam, Excerpta Medica, 1971, in press

11. DURANTE G: Anatomie pathologique des muscles. Manuel d'Histologie Pathologique. Edited by V Cornil, L Ranvier. Paris, Félix Alcan, 1902, pp 1–477

12. ENGEL AG, MACDONALD RD: Ultrastructural reactions in muscle disease and their light microscope correlates. Muscle Disorders. Proceedings of the International Congress, Milan Elsevier Press. Series Amsterdam 199, 1969, pp 71–89

13. ENGEL WK: A critique of congenital myopathies and other disorders. Exploratory Concepts in Muscular Dystrophy and Related Disorders. Edited by AT Milhorat. Amsterdam, Excerpta Medica, 1966, p. 27

14. ERB W: Ueber die "juvenile form" der progressiven Muskelatrophie ihre Beziehungen zur sogenannten Pseudohypertrophie der Muskeln. Dtsch Arch Klin Med 34:467–519, 1884

15. ERB W: Dystrophia muscularis progressiva: klinische und pathologische-anatomische Studien. Dtsch Z Nervenheilk 1:13–94, 1891

16. GONATAS NK, SHY GM, GODFREY EH: Nemeline myopathy: the origin of nemeline structure. N Engl J Med 274:3–35, 1966

17. GREENFIELD JG, SHY GM, ALVORD EC, et al: An Atlas of Muscle Pathology in Neuromuscular Diseases. Edinburgh, Livingstone, 1957

18. HALL BE, SUNDERMAN FW, GITTINGS JC: Congenital muscular hypertrophy. Am J Dis Child 52:773–783, 1936

19. HEIDENHAIN M: Ueber die Entstehung der quergestreiften Muskelsubstanz bei der Forelle. Arch Mikr Anat 83:427–447, 1913

20. HEIDENHAIN M: Ueber progressive Veränderungen der Muskulatur bei Myotonia Atrophica. Beitr Pathol 64:198, 1918

21. JEWESBURY RC: Case of generalized muscular hypertrophy in a boy, aged 10 years. Proc R Soc Med 15:34, 1922

22. KLIPPEL: Arrêt du développment du membre supérieur consécutif a un traumatisme datant de l'enfance; atrophie musculaire numérique. Presse Med 2:49, 1897

23. KRÖSING R: Ueber die Ruckbildung und Entwickelung der quergestreiften Muskelfasern. Virchows Arch [Pathol Anat] 128:445–484, 1892

24. MAURO A, SHAFIQ SA, MILHORAT AT, EDITORS: Regeneration of Striated Muscle, and Myogenesis. Amsterdam, Excerpta Medica, 1970

25. MCCOMAS AT, SICA REP, PETITO F: Muscle strength in boys of different ages. J Neurol Neurosurg Psychiatry 36:171–174, 1973

26. MCCOMAS AT, SICA REP, PETITO F: Physiological changes in aging muscles. J Neurol Neurosurg Psychiatry 36:174–183, 1973

27. MERYON E: On granular and fatty degeneration of voluntary muscles. Med Chir Trans 35:73–84, 1852

28. MOORE MJ, REBEIZ JJ, HOLDEN M, ET AL: Biometric analyses of normal skeletal muscle. Acta Neuropathol (Berl) 19:51–69, 1971

29. PEARSON CM: The incidence and type of pathologic alterations observed in muscles in a routine autopsy survey. Neurology (Minneap) 9:757–766, 1959

30. PERRY RE, SMITH AG, RENN RN: Ringbinden of skeletal muscle. Arch Pathol 61:450–455, 1956

31. PETRÉN T, SJOSTRAND T, SYLVEN B: Der Einfluss

des Trainings auf die Haufigkeit der Capillaren in Herz- und Skeletmuskulatur. Arbeitsphysiologie 9:376–386, 1936

32. PRICE HM: The skeletal muscle fiber in the light of electron microscope studies—a review. Am J Med 35:589–605, 1963

33. PRICE TML, ALLOTT EM: The stiff-man syndrome. Br Med J 2:682–685, 1958

34. REBEIZ JJ, CAULFIELD JB, ADAMS RD: Oculopharyngeal dystrophy—a presenescent myopathy. Progress in Neuro-Ophthalmology. Edited by J Brunette, A Barbeau. Amsterdam, Excerpta Medica, 1969, pp 12–31

35. REZNIK M: Muscle fiber regeneration. The Striated Muscle. Edited by CM Pearson, FK Mostofi. Baltimore, Williams & Wilkins, 1973, Chap 10, pp 185–225

36. SHY GM: Some metabolic and endocrinological aspects of disorders of striated muscle. Res Publ Assoc Res Nerv Ment Dis 38:274–317, 1960

37. SHY GM, MAGEE KR: A new congenital non-progressive myopathy. Brain 79:610–621, 1956

38. SHY GM, WANKO T, ROWLEY PT, et al: Studies in familial periodic paralysis. Exp Neurol 3:53–121, 1961

39. SPIRO AG, KENNEDY C: Hereditary occurrence of nemaline myopathy. Arch Neurol 13:155, 1965

40. SZENT-GYÖRGYI A: Chemistry of Muscular Contraction. New York, Academic Press, 1947

41. TOMLINSON BE, WALTON JN, REBEIZ JJ: Effects of aging and cachexia upon skeletal muscle. J Neurol Sci 9:321, 1969

42. TOWER SS: Atrophy and degeneration in skeletal muscle. Am J Anat 56:1–43, 1935

43. VIRCHOW R: Ueber parenchymatöse Entzündung. Virchows Arch [Pathol Anat] 4:261–324, 1852

44. WALLACE SL, LATTES R, RAGAN C: Diagnostic significance of the muscle biopsy. Am J Med 25:600–610, 1958

45. WATJEN J: Zur Kenntnis des Pigmentvorkommens in der quergestreiften Muskulatur. Frankfurt Z Pathol 54:213–220, 1940

46. WEISS P, JAMES R: Aberrant (circular) myofibrils in amphibian larvae: an example of orthogonal tissue structure. J Exp Zool 129:607–622, 1955

47. WEITZ W: Ueber einen interessanten Fall von Muskelhypertropie. Dtsch Z Nervenheilk 71:330–338, 1921

48. WOHLFART G: Ueber das vorkommen verschiedener Arten von Muskelfasern in der Skelettmuskulatur des Menschen und einiger Saügetiere. Acta Psychiatr Neurol (Kbh) 12 (Suppl):1, 1937

49. WOODS AH: Muscular hypertrophy with muscular weakness. J Nerv Ment Dis 38:532–538, 1911

part **III**

PATHOLOGY OF MUSCLE DISEASES

chapter 5

CONGENITAL NEUROMUSCULAR DEFECTS

The congenital origin of several neuromuscular diseases is more firmly substantiated than in 1962 when the second edition of this volume was issued. Many of the cases of progressive muscular dystrophy of early childhood (Duchenne pseudohypertrophic form) have been shown to originate in utero and to exist long before muscular weakness is demonstrable. Moreover, a new group of congenital myopathies, probably of dystrophic nature, have been discovered that are sometimes expressed clinically at birth as a form of arthrogryposis. Furthermore, the group of hereditary motor system diseases, which formerly included only Werdnig-Hoffman disease, has been enlarged by the addition of several more slowly evolving, less malignant forms. Finally, an entirely new class of congenital, relatively nonprogressive myopathies in which the muscle fibers are altered or disfigured rather than destroyed have come to light.

On the basis of this new information we therefore suggest the following classification of congenital neuromuscular diseases:

1. Congenital absence of muscle(s)
2. Restricted fixed palsies (absence of muscle or of its innervation), often in conjunction with other congenital anomalies
3. Congenital contractures and pseudocontractures due to primary abnormalities of muscle or spinal cord
4. Congenital dystrophies of uncertain progressivity and the intrauterine forms of the established types of dystrophy
5. Congenital myopathies of disfigurative type
   a. Central core disease
   b. Nemeline myopathy
   c. Mitochrondrial (pleoconial and megaconial) myopathies
   d. Myotubular myopathy
   e. Acid maltase deficiency
   f. Other glycogen storage diseases
6. Hereditary motor system diseases (Werdnig-Hoffmann) and polyneuropathies
7. Unclassified myopathies
   a. Oppenheim's amyotonia congenita
   b. Walton's congenital benign hypotonia
   c. Krabbe's congenital hypoplasia of muscle

Each of these groups is discussed in the following pages, except those in group 6 which are covered in Chapter 10.

All of the more severe affections of muscle, present at birth or manifested soon thereafter, can be accommodated in the above classification. However, there remains a seemingly endless series of minor peculiarities of posture and muscle action (viz., a slight turning in of one foot, walking on the side of the foot, slight rotation of the hip), the nature of which has not been determined. Browne[12] believes them to result from fetal malpositioning with restricted movement during intrauterine life; he shows, for example, how the abnormal posture of a foot coresponds to persistent pressure indentations made by it on the surface of the body. However plausible this explanation may be, one is left in doubt as to the cause of the malpositioning—whether it is due to a slight deficiency of motor neurons, nerve fibers, or muscle fibers—about which we have no information. Many of these postural abnormalities are outgrown; special shoes, splinting, and casting appear to be helpful in correcting some of the more lasting ones during critical phases of growth.

CONGENITAL ABSENCE OF MUSCLES

Some individuals are born without certain muscles, a fact well known to physicians and to geneticists. This is true not only of inconstant muscles, such as the palmaris longus, which have no important functions in human motility, but of some of the constant and important muscles as well. The most authoritative writings on these anomalies of muscle are by Le Double[64] and Bing.[5] Le Double wrote a didactic treatise of two volumes on the muscular variations in man based on his own experience in the anatomic dissecting room. He discussed the congenital absence of muscles and also variations of structure, such as differences in size, origin, and insertion. Statistics as to the relative incidence of these muscle anomalies in males and females and in different races were presented. In addition, Le Double collected data on congenital abnormalities of homologous muscles in various species of animals. Bing's article[5] described one case in full detail and contained a review of 185 cases that had accumulated in the medical literature up to 1902. Table 5–1, which summarizes his findings, brings out clearly that the pectoral muscles, particularly the sternocostal portions, are deficient more frequently than any other group and that smallness or absence of the trapezius is next. Also it appears that any muscle can occasionally be lacking.

As a general rule these muscle defects are discovered at birth or soon afterward, and the resulting disability remains stationary. The abnormalities tend to be unilateral and are often limited to a single muscle or a related group of muscles. Nevertheless, numerous cases of bilaterally symmetrical deficiencies have been reported. For example, in one case the trapezius, sternomastoid, and lower part of the pectoral muscles were absent on both sides,

TABLE 5–1. Frequency of Muscle Defect or Absence

Muscle involved	No. cases	Muscle involved	No. cases
Pectoralis	102	Platysma	2
Trapezius	18	Extensor carpi ulnaris	2
Serratus anticus major	14	Longissimus dorsi	1
Quadratus femoris	16	Supinator longus	1
Omohyoideus	8	Levator scapulae	1
Semimembranosus	7	Intercostales	1
Brachioradialis	4	Facial muscles	1
Gemelli	4	Gastrocnemius	1
Deltoideus	4	Subclavius	1
Latissimus	4	Triceps brachii	1
Sternocleidomastoideus	3	Brachialis ant.	1
Rhomboidei	3	Glutei	1
Supra- and infraspinatus	3	Extensor digitorum proprius	1
Biceps brachii	3	Flexor digitorum sublimus	1
Hand muscles	2	Stylohyoideus	1
Quadriceps femoris	2		

After Bing.[5]

and in another case there were no pectorals and no serratus magnus, biceps, or triceps on either side.

MUSCLE DEFECTS WITH OTHER CONGENITAL ANOMALIES

Several syndromes in which a defect of skeletal muscle is combined with some other abnormality have been delineated. Some of the best known of these are the following.

Congenital defect of the *mammary gland* and pectoral muscle in conjunction with syndactylism and microdactyly has been described in over 50 cases by Morley,[75] Christopher,[14] Brown and McDowell,[11] Resnick,[81] Krabbe,[62] and others. Agenesis of the pectoral muscles may be accompanied either by scoliosis and webbed fingers or underdevelopment of the ipsilateral arm and hand (Poland's syndrome).

An unusual syndrome in which there is a congenital deficiency of the *abdominal musculature* was reported by Osler,[77] Garrod and Davies,[39] McClendon,[70] Lichtenstein,[65] and more recently reviewed by Silverman and

Huang.[91] The weakened abdominal wall bulges with every respiratory effort, and both coughing and defecation are hampered. Infants so afflicted often die of respiratory infection within a few months. There may be an associated defect of ureters, bladder, and genital organs. Congenital absence of the *diaphragm* with displacement of the abdominal viscera into the thoracic cavity causing collapse of the lung is another well known condition.

RESTRICTED PALSIES WITH OR WITHOUT OTHER NERVOUS SYSTEM ABNORMALITIES

CONGENITAL PTOSIS

Probably the commonest example of a congenital weakness or palsy of a cranial muscle is that of *congenital ptosis* due to involvement of the levator palpebrae. This varies in degree from the slightest perceptible narrowing of the palpebral slit to a ptosis of the eyelids so complete as to cover the pupils. The other muscles supplied by the oculomotor nerve may not be affected. In some cases movement of the jaw is to myotonic dystrophy. Cases with a later development of myasthenia gravis have also been reported. The implication that congenital ptosis is related to muscular dystrophy is heightened if more than one member of a family is affected. In some cases movement of the jaw is accompanied by an elevation of the ptosed eyelid; this combined movement has been termed the "jaw-winking phenomenon" of Gunn,[51] who first called attention to it. The eyelid usually is elevated when the jaw is opened strongly and descends when it is closed, but the opposite may happen, or lateral movement of the jaw may be more effective than either opening or closing. The levator muscle in such cases obviously retains innervation. We know of no pathologic studies of the muscles.

Complete paralysis of all muscles innervated by the oculomotor nerve has been reported a number of times as an isolated abnormality in one person, in several generations of one family, or as one of many conjoined abnormalities. Similarly, the abducens muscles may be paralyzed from birth. Congenital inequality of pupils or immobile pupils are also frequent, and ptosis may occur as part of a congenital Horner's syndrome. Ford[38] remarks on cases with congenital absence of the pterygoid and masseter muscles. With this there is an unusual deformity of the jaw in which the mandible is very narrow; and when the molar teeth are approximated, the incisors are separated by more than 1 cm. Such bony change indicates that natural muscular stress is essential to development of the mandible during the period of growth. We observed a similar abnormality of the jaw in myotonic dystrophy.

Congenital ophthalmoplegia may be combined with retinal pigmentary degeneration, congenital heart disease, endocrine deficiency, cerebellar ataxia, and mental backwardness (Kearns-Sayre syndrome). The skeletal muscles are thin and hypotonic. The nature of the muscle lesion has not been ascertained but the extraocular muscles exhibit many "ragged red fibers" in Gomori stained sections due to aggregates of mitochondria and excessive oxidative enzyme activity.

Another deficiency of development of cranial muscles is exemplified by the congenital facial diplegia syndrome. The condition consists of facial diplegia in combination with a bilateral paralysis of ocular muscles, particularly the abducens, and has been known since Von Graefe's original publication in 1880[46] and the one by Moebius[74] a few years later. Because of the lucidity of his description, the condition is often called Moebius' syndrome. According to the count by Danis, by 1945[17] more than 70 cases had been recorded. The most complete review in English is that of Henderson.[53]

Usually the facial palsy is manifested during the first days of life by failure of the eyes to close completely during sleep and difficulty in nursing. It becomes even more obvious after a few weeks when the face remains quite smooth and impassive during crying and laughing. Often the mouth is not completely closed. The lips are prominent and the lower lip everted. In learning to talk, labial consonants are poorly formed. Food lodges in the cheeks.

This paralysis of facial muscles may be partial or complete, may involve upper or lower parts of the face separately, and may be either uni- or bilateral. It does not progress; in fact, in some cases a certain degree of movement is acquired later in life. About 20% of the cases are pure in the sense of being isolated abnormalities in an otherwise healthy child. In the remainder some other muscular palsy or congenital malformation is associated with the facial diplegia. Of these, paralysis of one or both abducens muscles with convergent strabismus is the most frequent finding, occurring in about 75% of cases. An external ophthalmoplegia is the next commonest. Deformity of the external ears; deafness, defect of pectoral muscles; paresis of tongue, soft palate, or jaw muscles; bilateral clubfoot; mental defect; and epilepsy have been observed in various combinations in a proportion of these cases (Table 5–2).

TABLE 5–2. Abnormalities Associated with Facial Diplegia

Abnormality	No. reported cases
Facial diplegia	61
Abducens palsy	45
Ophthalmoplegia externa	15
Ptosis	6
Lingual palsy	18
Mental defect	6
Clubfoot	19
Brachial malformation	13
Pectoralis muscle defect	8

From Henderson.[53]

The cause of this peculiar condition is not known. Usually it can be distinguished from facial palsy resulting from birth trauma by its bilateral occurrence and its association with ocular palsies and other malformations. In four of the reported cases in which a similar facial diplegia was observed in one or several siblings, the possibility of a hereditary determination of the disease was suggested However, most of the reported cases have been nonfamilial. The available pathologic data favor the notion that nuclear hypoplasia is responsible for the cranial nerve palsies,[82] and it may

be assumed that the muscle should show a second atrophy of the neural type. The possibility of a primary muscle defect with secondary degeneration of the nuclear structures cannot be fully dismissed. However, it is said that the nerve cells do not degenerate when there is merely an absence of muscle, as shown by the cases of Lichtenstein[65] and others. Only when the peripheral nerve is injured during intrauterine development, as in the case of Edinger,[26] does one see atrophy of the motor neurons of the spinal cord. This point is still disputed, and it is possible that some of the cases of congenital diplegia belong to the mildest type of progressive dystrophic ophthalmoplegia (Chapter 6). Ocular and facial palsies have also been reported in the Bonnevie-Ullrich-Turner syndrome.

Ullrich's ideas[98] concerning the etiology of the facial diplegia syndrome are of special interest. His hypothesis is based on the embryologic studies of Bonnevie,[6] who observed similar anomalies in a strain of experimental mice raised by Little and Bagg.[67] The animals exhibited congenital deformities of the limbs and eyes. Mutations of this type were induced by means of roentgen irradiation of the pregnant female, and their incidence was increased by inbreeding for several generations. Bonnevie, after examining over 700 live embryos of this strain, claimed that there was an excessive formation of cerebrospinal fluid in the newly formed choroid plexuses of the fourth ventricle. She concluded that the pressure of this fluid either inhibited the development of or directly damaged the motor nuclei in the floor of the fourth ventricle. This explanation lacks credibility and has not been tested.

Most of these congenital defects of muscles are hereditary. Several interesting pedigrees can be found in the publications of Gates.[40] The identical muscle may be missing in several members of one family in the same or successive generations. This inherited pattern is especially striking in some of the inconstant human muscles, e.g., palmaris longus, pyramidalis, anterior peroneal and plantaris. When a muscle defect occurs sporadically in one member of a family, the possibility of muscle atrophy secondary to a congenital disease of nerves or spinal roots must be considered.

We have had no opportunity to examine the muscle tissue or nervous system in a case of congenital deficiency or absence of skeletal muscle. There have in fact been relatively few pathologic studies of these conditions. In Bing's case part of the pectoral muscle had failed to develop, and microscopic sections showed a reduction in the number of muscle fibers, variation in the size of fibers, and in the thin fibers an increase in the sarcolemmal nuclei. The amount of interstitial connective tissue was increased. In other writings on this subject the defective muscle has consisted of a cord or sheet of connective tissue with only a few scattered small muscle fibers. There is no agreement as to the findings in the nervous system in such cases. In Bing's case no changes in the peripheral nerves or spinal cord could be demonstrated, and one must conclude therefore that the aplasia or hypoplasia of muscle tissue was primary. Other writers (e.g., Lichtenstein[65]) described an atrophy or absence of spinal motor neurons in the segmental levels corresponding to the aplastic muscle. Presumably, then, the muscle defect may be either a primary aplasia or one secondary to a neural or spinal deficiency. Pathologic studies are available in only four of the reported cases of congenital facial diplegia, those of Huebner,[55] Rainey and Fowler,[80] Spatz and Ullrich,[92] and Richter.[82] The cases were examined post mortem at 19 months, 10 weeks, 1 month, and 3 months of age, respectively. In all four papers the findings in the central nervous system were reported as being the same. The seventh and sixth cranial nerves were unusually thin and their nuclei deficient in nerve cells. Those cells which remained were smaller than normal. Unfortunately, the facial muscles were not examined adequately. Rainey and Fowler[80] were the only ones to mention the muscle tissue, and they remarked that the few fragments of muscle that could be found were normal histologically.

It was reported that in four biopsies of frontalis muscle no muscle fibers were found at all. Also observed were concretions of calcium and iron among the few remaining nerve cells in the abducens and facial nuclei, and it was argued that on the basis of these findings the process is degenerative.

In a few of the reported cases of facio-scapulo-humeral (Landouzy-Déjerine) type of muscular dystrophy there has been congenital absence of part of one pectoral or another muscle such as brachioradialis before the onset of the dystrophic disease. The relationship between such congenital defects and the later onset of muscular dystrophy is obscure. The relative incidence of congenital defects of muscle as compared with other localized anomalies in 1000 human embryos and infants at birth was determined by Mall.[69]

CONGENITAL CONTRACTURES AND PSEUDOCONTRACTURES OF MUSCLES AND JOINT DEFORMITIES*

Shortening and fibrosis of muscles are responsible for a variety of deformities in infants and children. Some of the commonest of these are congenital clubfoot, torticollis, and congenital elevation of the scapula. In all of these conditions the postural distortion is maintained by the shortened muscle. The exact cause of the shortening cannot always be determined. In some instances the shortened muscle is the normal one; and because of its excessive activity due possibly to a lack of opposition by a weakened antagonist, a constant posture is maintained. Conversely the agonist muscle may be permanently shortened (true contracture). This explanation is most apposite when an obvious structural deficiency or absence of one muscle of an apposing pair is responsible for the prevailing posture. Trauma to a muscle with resulting fibrosis during intrauterine life or at birth may be the explanation of a pseudo-contracture in some cases. In congenital torti-

---

* The reader's attention is drawn again to the distinction between contracture and pseudocontracture. The first term is used here to refer to a shortened state of the muscle not dependent on the normal contractile process, as in prolonged positioning of a muscle in a cast or in McArdle's syndrome. Pseudocontracture refers to a fixation of limb posture due to destruction of muscles, fibrosis of muscle and periarticular tissues, and shortening of ligaments. If joints are primarily affected so as to impede motion, the condition is ankylosis.

collis, for example, damage to the main muscular artery or vein supplying one sternomastoid muscle (giving rise to a sternomastoid pseudotumor) is held by some to be the causal lesion (see below). Finally, the joints themselves may not have developed properly (perhaps as one of the series of inherited congenital anomalities) so that all possibility of joint motion is prevented.

CONGENITAL CLUBFOOT (TALIPES)

Clubfoot is one of the most frequent and best known of all the congenital muscular and bony abnormalities, and yet there is relatively little precise knowledge of its pathogenesis. The feet may assume any one of several abnormal postures at birth—plantar flexion of the foot and ankle (talipes equinus), dorsiflexion of the foot and ankle (talipes calcaneus), inversion and adduction (talipes varus), or eversion and abduction (talipes valgus). A combination of these deformities is commonest; talipes equinovarus comprises about 77% of all clubfoot cases. It is easily recognized at birth, being present usually on one side and sometimes bilaterally. The feet are inverted to such a degree that the soles can be brought into contact. Attempts at passive movements in order to return the foot to its normal position are resisted by taut ligaments and shortened muscles. With growth the developmental failure of certain leg muscles (e.g., peroneal) and the unopposed action and shortening of others (e.g., gastrocnemius) become evident. The tendons are displaced, and the bones of the foot become deformed by unusual stress.

The genetic literature is replete with examples of multiple incidence in one family; Gates[40] says that the heredity is generally recessive or possibly in some families dominant but of low penetrance. Sex-linked inheritance has also been observed. Despite this, however, a hereditary factor appears to operate in only a small percentage of instances. Direct transmission from parent to child is rare. Two-thirds of the cases occur in males.

Opinions differ as to the cause of the foot deformity, and no one explanation can be applied to all cases. That misshapen feet and other skeletal abnormalities can result from unusual pressures and abnormal positions in utero has long been argued, most convincingly by Browne.[12] Another explanation, equally difficult to prove, was advanced by Thomasen,[97] who attributed the deformity to a defect in the bones of the foot and ankle. More specifically the view held by most orthopedists postulates an embryonic disturbance in the formation of the tarsal and metatarsal bones, particularly talus, os naviculare, cuboids, and os calcis. This results in a shortening of the inverter muscles and lengthening of the everters.[54] The primary defect is believed to be in the cartilaginous anlage of the talus.[58]

Another cause of foot deformity is congenital spasticity. Here the ankle is flexed with persistent toe-walking and there is a pes cavus and dorsiflexion of phalanges due to hypertonia of long flexors and extensors.

A primary defect of the spinal cord with weakening or paralysis of some muscles of the foot leaving others normally innervated to contract without opposition has been demonstrated in a limited number of cases. According to Courtillier,[16] this idea was first proposed by Morgagni. Michaud,[71] in a case of bilateral congenital clubfoot, discovered two foci of sclerosis in the spinal cord—one in the thoracic region and the other in the lumbar area. Durante and Courtillier[24] found unquestionable lesions in the spinal cord in each of four cases. These were situated at the junction of the thoracic and lumbar segments and consisted of an apparent failure of development of anterior horn cells, an atrophy or developmental deficiency of Clarke's columns, and a paucity of medullated fibers in the anterior roots. In each case the clubfoot was unilateral and the lesions in the spinal cord ipsilateral. Examples of congenital poliomyelitis and congenital syphilis of the spinal cord have also been reported.[41] Dittrich[18] suggested that damage to peripheral nerves of the leg can occur in utero as a consequence of traction, and that the foot deformity may be secondary to a neural paralysis of certain groups of muscles.

In the majority of cases of clubfoot studied pathologically, no abnormality of the spinal cord or peripheral nerves has been demonstrated,[63] and attention has therefore been directed to the musculature of the leg. Dittrich[18] examined the leg muscles of an 8-month-old fetus with bilateral equinovarus deformities of the feet and noted slight atrophy of muscle fibers and nuclear proliferation. The atrophy was most marked in the anterior muscular groups (peronei and anterior tibials). Bechtol and Mossman[3] studied 2 embryos, 12 and 16 weeks old, with foot deformities. In the first case—a possible example of amyoplasia congenita—the development of muscle fibers in the muscles whose action may have been deficient was retarded. Pyknosis of sarcolemmal nuclei and smallness of the fibers of some of the muscles were observed in the second case. Pearson reviewed the sections from this case. The striking changes in the muscle nuclei and sarcoplasm, which closely resembled the retrogressive metamorphosis of Bardeen (Chapter 1), were widespread in the muscles of the affected limb, especially in the calf muscles. Sections of many other muscles, including those of the opposite limb, showed essentially normal muscle and only an occasional degenerating

fiber. The most important observation in these cases, in our estimation, is that at an early age structural differences can be found in muscles which normally move the deformed foot.

In a study of two cases of clubfoot, both bilateral, and a normal control case of about the same age (all 25- to 30-week-old premature fetuses[100]) we were able to confirm this variation in fiber size. In the normal fetus the average diameter of the fibers in all of the leg muscles was approximately $7.5 \pm 0.2 \mu$. In the cases of clubfoot the muscles that appeared to maintain the foot in the deformed position were of the same average size as in the control case, whereas the opposite muscles whose action was presumed to be deficient were 5.0–6.0μ in diameter (Fig. 5–1). These differences were consistent and of such magnitude as to be highly significant statistically. No other microscopic abnormality was noted. In one of the clubfoot cases the anterior horn cells of the lumbar and sacral cord segments in a few random sections were present in normal number and were of normal appearance. In the other, only one section of a lower lumbar segment of

Fig. 5–1. Clubfoot, fetus aged 32–33 weeks. Foot was dorsiflexed and inverted. (A) Transverse section of soleus muscle; average diameter of fibers is 6.3 μ (range 4.7–8.5 μ). (B) Transverse section of anterior tibial muscle; average diameter of the fibers is 8.1μ (range 6.0–9.8 μ). Intervals of the scale (at bottom) are 7 μ. (hematoxylin, eosin)

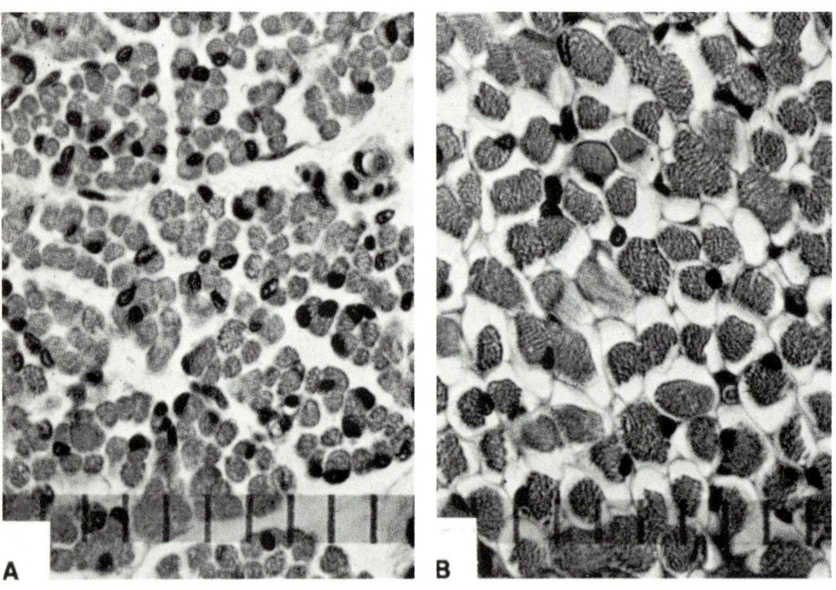

the spinal cord was available for study, and no definite abnormality was noted. In one case it was possible to demonstrate intact motor end-plates in the thin, underdeveloped muscle fibers.

Durante and Courtillier[24] described the muscles in cases of congenital clubfoot at a more advanced age. The anterior tibial and peroneal muscle swere affected to the greatest degree. In some cases the muscles were atrophic but not otherwise remarkable. In others there was an obvious excess in adipose tissue in addition to the atrophy. In a few cases no recognizable muscle tissue could be found. Histologically the majority of the fibers were of uniformly small diameter and possessed a rather indistinct cross striation. Sarcolemmal nuclei were more numerous than usual. Muscle fibers were widely separated by large numbers of fat cells in some cases. Instances of grossly atrophic muscles with normal histologic appearance have also been observed and are cited by Durante[23] as examples of the "numerical atrophy" of Klippel,[61] i.e., an atrophy due to a reduction in the number but not the size of the fibers.

While these pathologic studies are too few to permit broad generalizations as to the cause and pathogenesis of congenital clubfoot, certain facts begin to emerge. It would seem that clubfoot does not have a single cause. In some cases the foot deformity is secondary to a defect of the spinal cord which prevents proper innervation of some of the foot muscles. In others it is equally certain that no primary neurologic abnormality exists, and those muscles which maintain the deformity have fibers of average size whereas opposing muscles are composed of thin, poorly developed fibers. Whether this difference in size, which is not due to a defect in motor innervation, represents retardation of development of the muscle per se or is the result of disuse or stretching can be determined only by further observation and experiment. Replacement of the thin muscles by fat and connective tissue sheds no light on the nature of the original fault. In our opinion antenatal contracture, the important clinical feature of both clubfoot and amyoplasia congenita, is not adequately explained by small-ness of one set of muscle fibers.

CONGENITAL TORTICOLLIS (WRYNECK)

Congenital torticollis is a condition that usually begins during the first months of life. Due to an obvious shortening of the sternomastoid muscle, which seems abnormally firm and taut, the head is inclined and the occiput slightly rotated to the side of the affected muscle. Usually there are pathologic changes in the skeleton. An exostosis can be felt on the clavicle at a point where the clavicular head of the sterno-mastoid muscle attaches. The face and skull are asymmetrical (plagiocephaly). The eyebrow on the side of the shortened muscle tends to slope downward, and the face is shortened in the vertical dimension and broadened. The ipsilateral frontal eminence is flattened and the occipital bone bulges slightly, while on the other side of the skull the frontal bone bulges and the occiput is flat. These bony changes are due to misalignment of the skull and unusual tension on the bones during the growth period.

When the sternomastoid muscle is exposed during surgical operation, it is seen to be of reduced bulk; its lower half is white, glistening, and tough, contrasting with the upper half, which retains its natural color. Microscopic examination reveals partial or complete replacement of the muscle fibers by relatively acellular connective tissue. Scattered small muscle fibers of variable diameter are seen in some parts. These fibers are well striated and appear to be healthy.

The cause of this condition has been debated. A hereditary factor is excluded by lack of familial incidence in most cases. The possibility of birth injury has been affirmed by numerous writers but the exact mechanism of the injury has not been established. Many patients with congenital torticollis are known to have had a sternomastoid tumor (actually a pseudotumor) of infancy. This is a spindle-shaped swelling, occupying the position of one sternomastoid muscle, which appears during the 1st or 2nd week of life. It is firm and slightly tender and accounts for a slight inclination of the head. When surgically exposed it appears grossly as a white, glistening swelling composed of cellular fibrous tissue with scattered remnants of the original muscle fibers.

These fibers show vacuolation and granular degeneration; some are in the process of being phagocytized. It is assumed by Middleton[72] and others that this represents an early phase of injury to the sternomastoid muscle. Moreover, as the tumor is resorbed after a few months it is replaced by connective tissue, which induces fibrous contracture. Middleton produced a similar change in the sartorius muscle of the dog by ligating all the veins, and he assumed that congenital torticollis is due to an acute venous obstruction in the lower part of the muscle during labor. He confirmed Brooks' experimental findings[10] (Chapter 3) by showing that arterial occlusion and infarct necrosis of muscle caused very little fibrosis whereas venous occlusion was attended by marked fibroblastic proliferation. This theory had been suggested previously by Nové-Josserand and Viannay.[76]

Since the pathologic changes are similar to those of Volkmann's contracture, the ischemic theory is certainly the most satisfactory explanation of this condition, but there is no agreement as to whether this is due to venous or arterial occlusion. Hulbert[57] discussed the possible causes of this condition, and Lidge et al.[66] provided a complete review and bibliography on the etiology and pathology of torticollis up to 1957.

CONGENITAL ELEVATION OF THE SHOULDER (SPRENGEL'S DEFORMITY)

Although von Eulenberg is stated by Fairbank[37] to have described this deformity in 1863, some 28 years before Sprengel's publication,[95] the latter's name has become attached to it. Greig[48] has presented an excellent review of this subject. A change in the position and shape of the scapula is the dominant feature of the abnormality. The scapula is usually short in the vertical and widened in the transverse dimension, generally misshapen, lies nearer the spine than its fellow, and is retracted in movement unless fixed to the spine or ribs by a bony bridge or strand of connective tissue. There may be an exostosis on the medial border of the scapula. The affected scapula is elevated and rotated so as to bring the lower angle nearer the spine. Since the condition is usually unilateral, the shoulders are asymmetrical, the deformed side being elevated and advanced. Scoliosis is often present, and in a small proportion of cases there is torticollis. All movements of the arm are possible except abduction of the shoulder beyond 90°.

There have been few pathologic studies. One of the most careful is that of Fairbanks,[37] who gave a detailed description of a dissected specimen. In his case both scapulas lay at a high level and were fixed to the spine in an abnormal manner by bands of connective tissue or a cartilaginous bridge. The trapezius muscle had not formed below the levels of the scapular spine, being replaced by connective tissue. In this case there was also a cleft palate.

This type of abnormality is probably related, at least in some instances, to a defect in the innervation or in the development of either the trapezius or serratus magnus muscles. When accompanied by a cervical spina bifida or a Klippel-Feil deformity of the spine, the muscular atrophy may be of the secondary or spinal type.

ARTHROGRYPOSIS

The first description of multiple congenital contractures of the extremities was given by Otto in 1841 (quoted by Middleton[73]). Many other accounts appeared in German and French orthopedic and surgical journals in the latter part of the nineteenth century. Rocher[84] collected 31 cases from the literature and delineated a clinical condition which he called multiple congenital articular rigidities. Stern[96] introduced the term arthrogryposis multiplex congenita (*arthrogryposis* meaning curved joints), and later Middleton[73] designated it myodystrophica congenita deformans. Sheldon,[87] who conceived of the disease as a primary deficiency of muscle fibers or of whole muscles, suggested the shorter, and to him more appropriate name of amyoplasia congenita.

From these publications it can be learned that typical cases are recognizable at birth by deformity and rigidity of the extremities which may in some instances actually interfere with the birth process. In Heijbroek's[52] case and in one of our own, fractures of both femurs occurred during a difficult delivery. Many of the early descriptions were of cases of amyotonia congenita with contractures of the joints of some limbs. In some cases all four extremities were affected, but there are other instances in which deformities were limited to part of a single limb. Familial incidence of all forms— myelopathic, neuropathic, and myopathic—is well documented.

MYELOPATHIC ARTHROGRYPOSIS

Two different clinical syndromes and pathologic processes have been observed in our own material. One of these might be called the myelopathic form of arthrogryposis, the other the myopathic form. In the myelopathic variety, the infant has an appearance which has been likened to that of a wooden doll.[7] The extremities may be fixed in almost any position, but most often the arms are rotated inward and extended at the elbows, with forearms pronated so that the backs of the hands present forward. The hands themselves are flexed, often with ulnar deviation, and the fingers are curled into the palms. Usually the legs are flexed at the hips and externally rotated while the knees are either partly flexed or extended; the feet assume a pronounced equinovarus position. Clubhand or clubfoot are present in some cases. The erector spinae muscles may be involved, with resulting scoliosis. The muscles of the head are spared, excepting in one case of Ealing[25] that was marked by rigidity of temporomandibular joints. The affected limbs are small in circumference, and the joints by contrast appear unusually large or fusiform. Complete ankylosis of the joints is uncommon, some degree of active and passive motion being possible. Passive motion may encounter no resistance until the firm hindrance of ligamentous tissues is encountered. Muscular weakness and hypotonia are prominent features in most of the cases. The muscles are thin and sometimes cannot be felt at all; roentgenograms show them to be small. The electrical reactions of the muscle

are feeble or absent, but the reaction of degeneration is not observed. The tendon reflexes usually cannot be obtained.

The skin and subcutaneous tissues are thickened, wrinkled, and flabby, and the general nutrition tends to be subnormal. The skeleton is intact though in some cases with marked hyperextension of the knees the patellar bones may not develop. Luxations and subluxations of the joints may occur. Other congenital abnormalities such as webbed fingers, polydactyly, hydrocephalus and malformations of the skull, absence of the sarcrum, shortness of the cervical spine, and elevated scapulas have sometimes been associated with contractures. Often intellectual functions are impaired, but the special senses are intact.

Pathologic Changes. There have been only a few reports on the pathologic findings in this disease, and these are contradictory. Probably the uncertainty as to the etiology arises in part from this fact alone. We have had access to the pathologic material of only a few cases of this type, and the examination in one of these was not sufficiently complete to permit more than tentative conclusions. The following description is based on these cases and on those reported in the literature.

The muscles in the contractured limbs vary in appearance. Some are of normal size, color, and texture whereas others are greatly reduced in size, as much as a third to a sixth according to Gilmour.[42] Some muscles cannot be found at all, having either not developed or having been replaced by fat and connective tissue. The locations of the atrophic and healthy muscles and their relation to the contracture has not been specified by most writers. In one of our own cases with flexion contracture of the right hip, the flexor muscles of the thigh were of fair bulk whereas the posterior thigh and gluteal muscles were thin. It appeared that the contracture was due to the unopposed action of the psoas and iliacus muscles. The more affected muscles are pale link.

The most consistent histologic change was smallness of most of the muscle fibers (Fig. 5–2). They ranged from 3 to 5 μ in width and showed a greater variation in size than normal, though not as much as in Gilmour's case, in

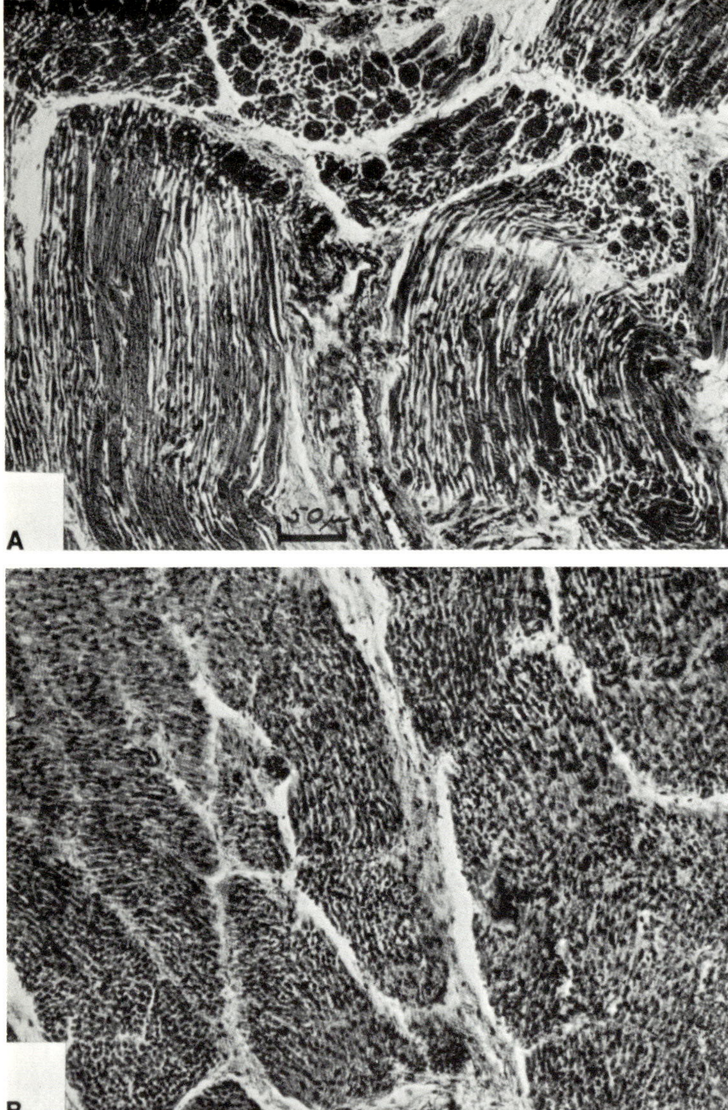

Fig. 5–2. (A) Arthrogryposis. Note large and small fibers in psoas muscle. (B) Gastrocnemius muscle from same infant. All muscle fibers are small and of fetal type. (phloxine, methylene blue, ×200)

which they varied from 3 to 24 μ, as compared with 7–10 μ in a normal infant of the same age. Most of the small fibers retained both longitudinal and transverse striation, clear in some parts and indistinct in others. In some muscles affected to a moderate degree there were groups of much large, well striated fibers amidst the atrophic ones (Fig. 5–2A). In one of our cases many of the muscle fibers—both atrophic and large—were hyalinized, but since this was constant in all sections with relatively little disintegration of the fibers or cellular reaction of any type we were inclined to regard it as being due to postmortem autolysis or secondary to the trauma of delivery. The endomysial connective tissue was increased little or not at all. Fat cells were numerous in some muscles and not in others. The muscle spindles were of normal appearance.

In one of our cases in which all four extremities were involved, the most important lesions were found in the spinal cord (Fig, 5–3). In the lumbosacral segments, all of the anterior horn cells had disappeared or had not developed. The lumbosacral anterior roots contained few or no nerve fibers and the posterior

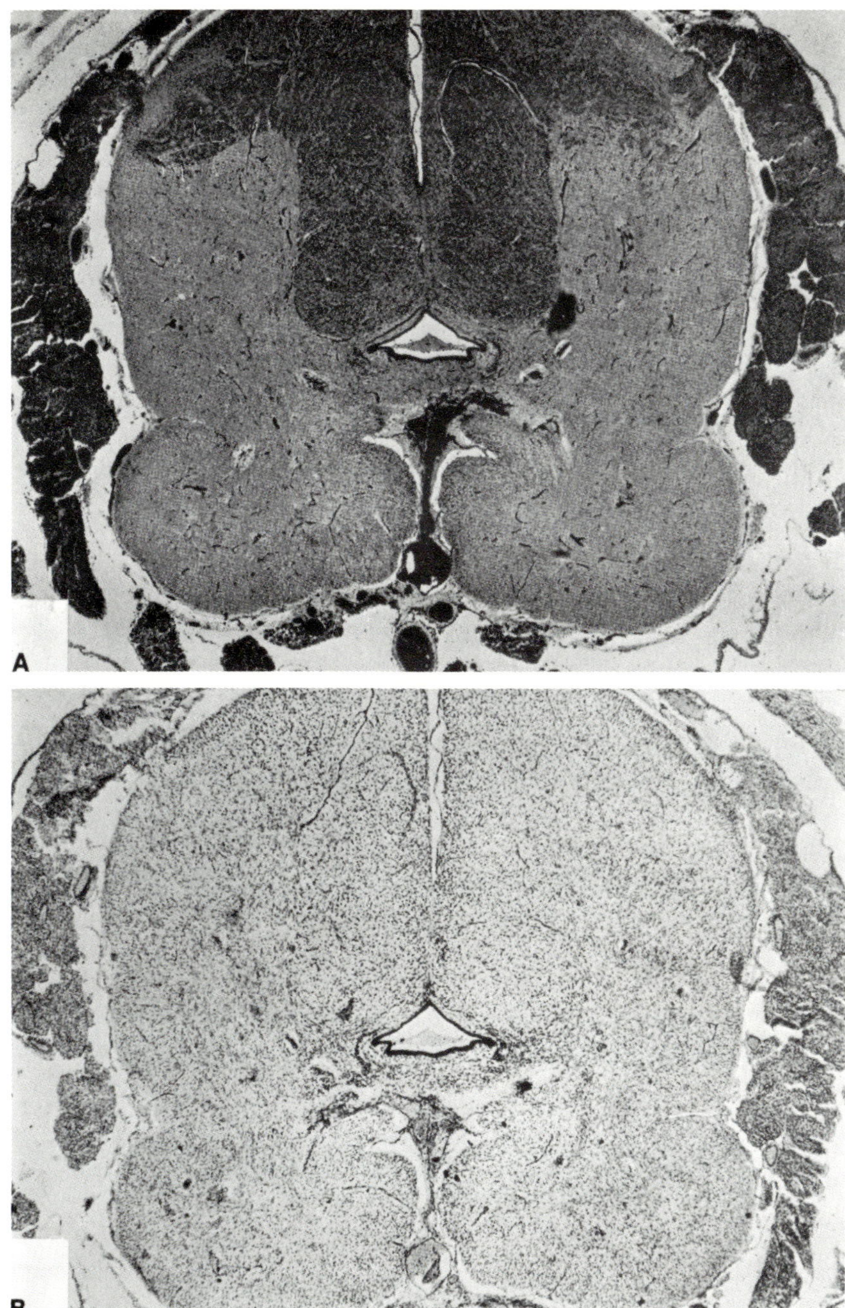

Fig. 5–3. (A) Amyoplasia congenita; S_1 segment of spinal cord. Dorsal columns and roots are dark. No medullated fibers are present in anterior and lateral columns. Anterior roots are thin and devoid of fibers. (B) Same segment showing absence of anterior horn cells. (A, Loyez myelin stain, ×20; B, Nissl, or thionine, stain)

roots the usual number. In the upper lumbar, thoracic and cervical segments the population of the anterior horn cells was greatly reduced and many of those remaining were small and darkly stained. There was no hyperplasia of microglial cells or of astrocytes in the gray matter and no infiltrations of inflammatory cells. The nerve cells in the posterior horns, the lateral horns, and in Clarke's column appeared to be normal, as did the dorsal root ganglion cells. The brain was small and undeveloped, with incomplete fissuration and large lateral ventricles. In our other case, in which only the legs were affected, there was a significant reduction in the number and size of the anterior horn cells in the lumbosacral cord segments.

Other writers found similar changes within the central nervous system. Gilmour[42] stated that the anterior horn cells in his case were reduced in number and of small size, but he gave no idea of the degree of this cell loss. Brandt[7] described a coarse granularity, vacuolation, and pyknosis of the anterior horn cells but noted no abnormalities of the nerves or roots. His illustrations, however, depicted only the common agonal or postmortem changes so commonly seen in the nerve cells of infants and children. Furthermore, the same cell alterations were noted in the cerebrum. In other case reports the central nervous system was intact.

The articular surfaces of the acetabulum and glenoid fossa were examined in Gilmour's case and were believed to be normal, and the same was true in one of our cases. However, surgeons who have operated on older children with this disease found destruction of the articular surfaces and synostosis of opposing bones.

MYOPATHIC ARTHROGRYPOSIS

In myopathic arthrogryposis the abnormality observed at birth may be a hypotonic, hyporeflexic weakness of the limbs, with or without significant contracture. Only the case with contracture falls within the group of arthrogrypotic disorders. The clinical picture may be duplicated in more than one sibling, thus establishing

with reasonable probability the existence of a hereditary factor.[2] Contractures, when present, are likely to be in the flexor muscle of the hip, knee, and elbow, i.e., flexion contractures. The affected limb muscles are weak and lack tendon reflexes. The distal limb muscles may be larger than the proximal ones. The skin over the extremities is loose and wrinkled. The trunk muscles are also weak. The head cannot be lifted from the bed and is often held to one side. Cranial nerve functions are not remarkable, although ptosis of one eyelid was noted in one of our cases. Mental reactions in general, to the extent that they can be tested, usually appear normal in this group of patients.

Pathologic Changes. The affected muscles (in our cases the glutei, thigh, trunk, shoulder girdle, and neck) are pale and thin and of firm, fibrous consistency. The lower leg and forearm muscles usually retain their natural bulk and color. The joints are not abnormal in any way, and when the limbs are dissected it can be readily demonstrated that contracture is due to shortened muscles. Under the microscope the most severely affected muscles (e.g., gluteus medius in one of our cases) may lack a large proportion of their fibers (Fig. 5–4A), which are replaced by fat cells. Residual fibers are extremely variable in size and are separated by connective tissue. The less involved muscles (Fig. 5–4B) show a great disparity of fiber size, with large and small in haphazard arrangement, and increased endomysial connective tissue. Relatively little evidence of fiber degeneration may be seen, and regenerative activity is lacking under such circumstances. Intramuscular nerves and anterior horn cells are normal (Fig. 5–5), in contrast to neuropathic arthrogryposis. There are no abnormalities of the brain.

Etiology. Our view based on these cases and reports in the medical literature is that all of the cases classified as arthrogryposis are not alike, and that this term is applied to states of contracture as well as pseudocontracture of the limbs in child or adult. In some there is a developmental defect of the spinal cord, with

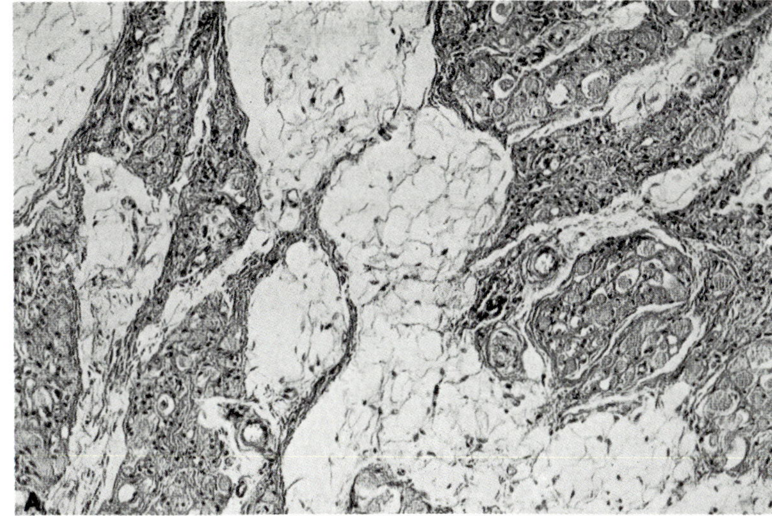

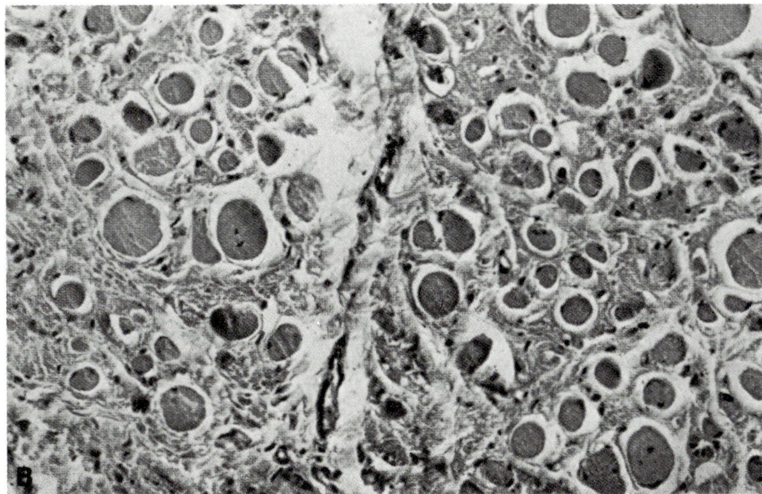

Fig. 5–4. Myopathic arthrogryposis. (A) Gluteus medius. Note loss of muscle fibers, replacement by fat cells, variation in size of residual fibers, and fibrosis. (B) Biceps femoris. Transverse section showing random distribution of small and large fibers and pronounced endomysial fibrosis. (A, B, hematoxylin, eosin; A, ×400; B, ×500)

failure of skeletal muscle innervation. In others an infantile muscular dystrophy is the cause of the contracture and is distinguished clinically by the presence of flexor rather than extensor postures.[2] A developmental defect in the joints themselves may be responsible for contractures of the limbs in other cases. Mental retardation and other congenital anomalies so often present in the more frequent myelopathic variety often overshadow the amyoplasia and combine to form a confusing clinical picture. The overlap of arthrogryposis and congenital clubfoot is noteworthy.

The etiology of arthrogryposis multiplex is obscure. A hereditary or familial incidence has been recorded in the myopathic but not in the myelopathic cases. The most likely explanation of the pseudocontracture in the latter cases such as we have observed is that there is a congenital defect in the development of the anterior horn cells or a chronic degeneration of these cells during fetal life. The tendency to contracture, which is so pronounced in arthrogryposis and so relatively slight in amyotonia congenita, may be due to the failure of muscle innervation and the maintenance of a fixed position in utero. In several of the case records it was noted that fetal movements were feeble and that either oligoamnios or hydramnios existed. Some writers have implied that reduced

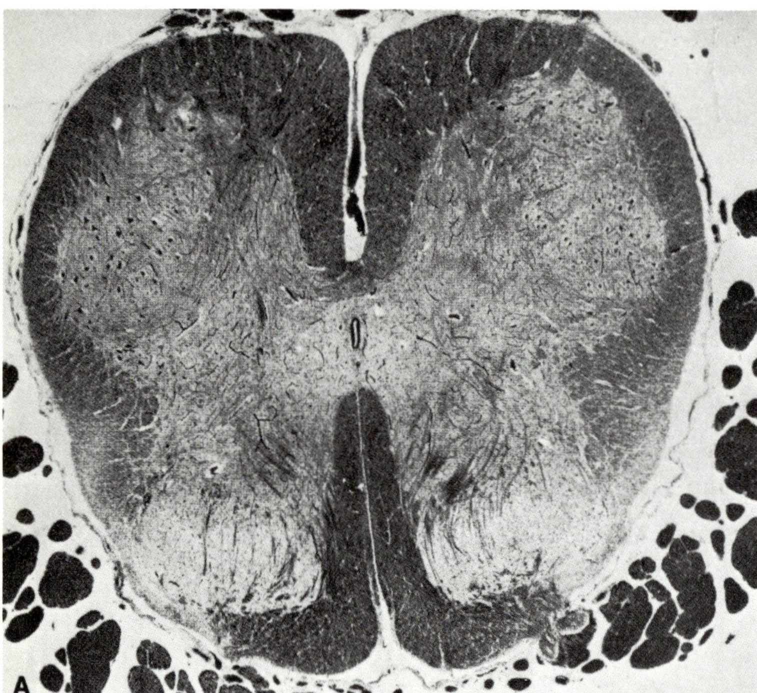

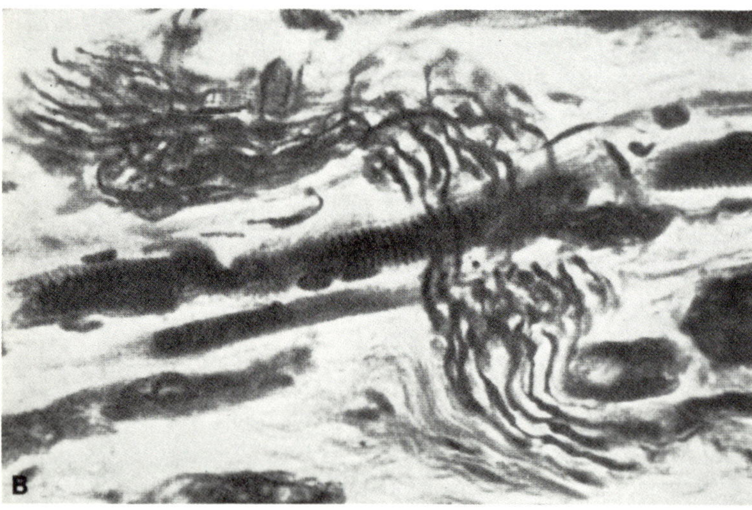

Fig. 5–5. Myopathic arthrogryposis. (A) Spinal cord L₅ showing normal complement of anterior horn cells. Myelination of anterior and posterior roots is normal. (B) Quadriceps muscle demonstrating intact intramuscular nerves. (A, Loyez, ×15; B, Glees, ×15)

amniotic fluid could interfere with fetal movements and thereby cause contractures. There is no evidence on this point, and the explanation is hardly credible. Analogy to a somewhat similar disease, though with a clear hereditary transmission, that occurs in newborn lambs is stressed in several discussions of amyoplasia congenita. This disease was described by Roberts.[83] The lambs, which are stillborn, exhibit multiple articular rigidities due to muscle contracture. When the muscles are sectioned, the limbs become freely movable. According to Middleton,[73] the muscles consist of adipose and undifferentiated myoblastic tissue and scattered small muscle fibers. In the myopathic cases the factors responsible for the development of dystrophy must be at play.

All cases of arthrogryposis cannot be reduced to these two patterns (myelopathic and myopathic) for there are also a few reported instances of these deformities with diseases of peripheral nerves and of abnormal tendon attachment.

CONGENITAL MUSCULAR DYSTROPHIES

The term myopathy covers all forms of primary muscular disease. Dystrophy, in contrast, has been given a more restricted meaning—specifying a hereditary, progressive degenerative disease of muscles. Difficulty arises in applying these definitions to individual cases where family incidence is unknown (? mutation) and the patient dies too soon to ascertain progressivity. One might expect the histopathologic changes in the muscle to settle the problem, but there is still uncertainty as to the primary lesion in the muscular dystrophies (Chapter 6).

ESTABLISHED PROGRESSIVE MUSCULAR DYSTROPHY AT BIRTH OR INFANCY

Three of the well recognized types of progressive muscular dystrophy can be identified during the neonatal period or infancy: Duchenne's pseudohypertrophic type, facioscapulohumeral type of Landouzy-Déjerine, and the myotonic dystrophy of Steinert. The onset in each instance is during intrauterine life, and by birth certain muscles may be manifestly weakened and show typical myopathic electromyographic (EMG) changes. Serum enzymes have been consistently elevated in the Duchenne type at the time of birth or even during intrauterine life. The pathologic changes in each of these diseases are described in Chapter 6.

CONGENITAL MYOPATHY (? DYSTROPHY) OF UNIDENTIFIED TYPE

There is a growing literature on congenital dystrophies that do not conform to the three aforementioned types. Gubbay *et al.*[50] and others describe unmistakable cases of progressive myopathy. The pathologic alterations of muscle are the same as those described above under myopathic arthrogryposis (Fig. 5–4).

RELATIVELY NONPROGRESSIVE CONGENITAL MYOPATHIES

Beginning in 1956 with Shy and Magee's account of a patient whose muscle fibers exhibited a peculiar central densification of sarcoplasm called cores,[88] a series of new diseases has gradually been delineated. In most instances diagnostic measures were prompted by the observation of motor developmental retardation or slackness and hypotonia and thinness of limbs. Such cases fall in the category of the floppy baby syndrome. It is noteworthy that hypotonia or floppiness is emphasized as the essential clinical state at this age because the more usual tests of strength and endurance of muscle action cannot be applied.[47]

Further study revealed that the diseases of this group are not confined to infancy. Each of the entities described below has been observed at a later age, even in the middle period of adult life; and if the disease is mild there is often no way of deciding whether it has been present since birth. Extremely slow progression, which contrasts to the more rapid pace of muscular dystrophy and Werdnig-Hoffman disease, characterizes most of them. Yet examples of more rapid progression are known, and prior to the laboratory investigations the original clinical diagnosis was muscular dystrophy in several of the recorded examples. Coincidence in other members of a family has also been established, so the clinical line of separation between this group of diseases and the muscular dystrophies remains ambiguous.

The lesions in the congenital myopathies are revealed most clearly by the systematic use of histochemical stains and in preparations for phase and electron microscopy. Some of the abnormalities are also revealed by the conventional stains used in light microscopy but are always difficult to distinguish from biopsy and fixation artifacts. Thus as a group, one might say that their discovery is a product of a new histologic technology.

CENTRAL CORE DISEASE OF SHY AND MAGEE

CENTRAL CORE MYOPATHY

Central core myopathy described first by Shy and Magee, affected 5 of 17 members in a family. The pattern of inheritance suggested a mendelian dominant. All the affected patients were delayed in learning to walk until the age of 4–5 years and for the remainder of their lives continued to have difficulty in rising from a sitting posture, climbing stairs, and running. The oldest patient was 65 years and the youngest 2 years. Muscular weakness was chiefly proximal and was most severe in the lower extremities, though the adults showed some distal weakness in the lower limbs and some weakness in shoulder musculature. The tendon reflexes were active and symmetrical in all patients.

Muscular atrophy was not a prominent feature, though one patient showed poor general muscular development, and the photograph of another suggests some flattening of the deltoids. No fasciculations or definite focal atrophy were found. No myotonia could be elicited. The electrocardiograms were normal. Except for an increase of urinary creatine and a decrease of urinary creatinine excretion, no chemical defect was detected by laboratory studies. No endocrine defect could be found.

Muscle biopsy revealed a unique type of change in all five affected patients. In all cases, in specimens fixed in formalin, Zenker's, or Bouin's fixatives the central core of myofibrils in almost every muscle fiber presented an amorphous hyaline appearance. The central zone of altered myofibrils still presented some evidence of cross striation under polarized light or in phase microscopy, and gave a positive periodic acid-Schiff (PAS) reaction. This core also stained blue with Gomori's trichrome stain, contrasting with the normal red color of the peripheral fibrils. No focal destruction of fibers was found, though there was fatty infiltration of some fasciculi. The majority of fibers measured approximately 80 μ, though some large fibers ranged to as much as 240 μ.

Dubowitz and Pearse[19] examined the histochemical characteristics of a biopsy specimen from another family suffering from this disease and found that the central core of myofibrils was devoid of oxidative enzymes and of phosphorylase activity. In their case some muscle fibers contained two or more separate cores of degeneration.

One of our own cases is illustrated in Figure 5–6, in the upper part of which is seen the characteristic hyaline change in the central part of the fiber. In the longitudinal section the most central part of the core in this case is seen in typical form surrounded by a thin layer of thickened, darkly stained myofibrils separating the hyaline core from the normal myofibrils, as in the lower part of Figure 5–6. In addition, the relative size of the "core" varied greatly, and in widely scattered parts of the section isolated muscle fibers that had become completely hyaline were seen with dark pyknotic subsarcolemmal nuclei. There was no cellular reaction to this process, even in the most completely involved muscle fibers, but remnants of hyaline fragments of muscle fibers could be found in various parts of the section (Figure 5–6, lower right corner). Our own impression is that this disease process is progressive, but only very slowly so. It differs from any other known type of degeneration of muscle fibers and therefore presumably reflects some unique metabolic faults of unknown nature.

Additional cases have been reported by Engel *et al.,*[33] Dubowitz and Pearse,[20] Seitelberger *et al.,*[85] Gonatas *et al.,*[44,45] Bethlem and Myjes,[4] Dubowitz and Platts,[21] and Dubowitz and Roy.[22] In several cases the onset of disease occurred in adult life and was limited to proximal muscles, thus simulating limb-girdle dystrophy. Atrophy of the muscles of only one shoulder girdle has also been observed early in the disease. Usually the progression is so slow as to be barely discernible. The intelligence of the patients is normal, and there are no other

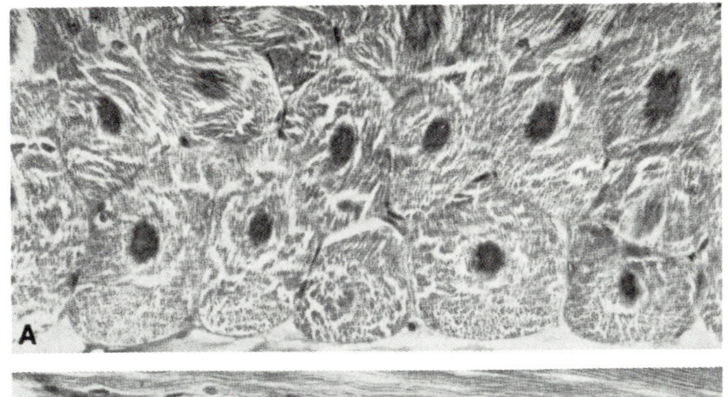

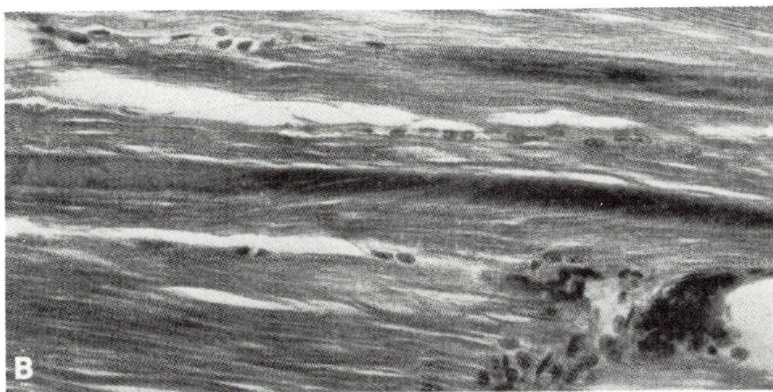

Fig. 5–6. "Central core disease." (A) Appearance of dark homogeneous core in most muscle fibers on transverse section. (B) In longitudinal section core has hyaline center and darker edge in which some of the surrounding myofibrils became enmeshed. Dark mass (lower right) is a shrunken, darkly staining fragment of totally hyalinized fiber in the process of dissolution. (Masson trichrome stain)

neurologic abnormalities. In particular nerve fibers, spinal roots and anterior horn cells are all of natural appearance.

In several reported instances of congenital myopathy with central cores, a dislocation of hips had led to the investigation of the status of the musculature. Fiber typing has shown the cores to predominate in type I fibers in some and in both types I and II in others.[32] Failure of fiber differentiation was exemplified in one of the cases of Dubowitz.

Further electron microscopic studies have shown that in the core region where mitochondria have disappeared many of the myofibrils are relatively normal; a zigzag distortion or streaming of the Z bands is the usual finding. Z bands have been lacking in other myofibrils, and in places the thin filaments of I bands are missing as well. However, as pointed out by Engel,[30] such Z band streaming is also observed in the center of "target fibers" seen in human denervated muscle. Engel therefore emphasizes the absence of mitochondria, ATPase in A bands, and phosphorylase as being the more fundamental changes.

The specificity of the cores in the disease was brought into question by the observations of Shafiq *et al.*[86] who found central cores in the muscle of one member of a family and nemaline rods in a sibling. Multicores within muscle fibers have been seen in other cases, possibly in a slightly different disease. In one of our own cases both central cores and nemaline bodies were observed.

NEMALINE (ROD-BODY) MYOPATHY

Nemaline myopathy also expresses itself as hypotonia in infancy; and during the most active growth period the ocular, facial, lingual, pharyngeal, trunk, and limb muscles remain thin and weak. The tendon reflexes are reduced or absent. EMG shows a myopathic picture, and serum enzymes are normal. As a rule intelligence is unimpaired. Both dominant and recessive inheritance have been documented. This clinical picture was noted in the original cases of Shy *et al.*,[89] as well as those of Conen *et al.*,[15] Engel *et al.*,[33, 36] Spiro and Kennedy[93] Price *et al.*,[79] Gonatas *et al.*,[43, 45] and Afifi *et al.*[1] Severely affected individuals die during infancy of inanition and respiratory infections, but the milder ones attain adulthood. Engel and Reznick[35] observed individuals who developed a proximal, progressive muscular weakness during middle age.

The characteristic lesion here is visible under the light microscope, especially in PTH stains where small packets of bacillus-like structures $1.0 \times 3.0 \mu$ are visible beneath the sarcolemma. In their staining qualities they resemble fractured myofibrils, but under the electron microscope they have more or less the same electron density as Z material (Figs. 5–5 and 5–7 through 5–9) from which they appear to arise.[79] Continuity with I filaments has also been noted by these investigators and others.[28, 45] In many of the affected fibers the subsarcolemmal muscle nuclei are slightly enlarged and displaced. Central nucleation and variations in fiber size give the muscle a myopathic aspect, but segmental necrosis, phagocytosis, and regeneration are seldom observed. Other organelles in the fiber appear to be relatively normal. Histochemical stains show marked reduction of type II fibers. Nerve fibers and neuromuscular junctions are normal.

The diagnosis of rod-body myopathy is justified only when large numbers of rods are seen in relatively normal fibers. The separation of a primary form of rod myopathy from secondary forms where rods may be formed (see below) is admittedly somewhat arbitrary, depending more on collateral data than the number of rods.

MITOCHONDRIAL MYOPATHIES

Two types of mitochondrial disease have been distinguished by Shy and his colleagues.[90] In one, called *pleoconial* because of an increase in the number of mitochondria, the patient (an 8-year-old boy) had had hypotonia and slow motor development since early life. A brother and cousin were similarly affected. Diffuse weakness was noted at the time of examination. It was reported that there was a craving for salt and that three episodes of more severe weakness rather like those of periodic paralysis

had occurred. Abundant mitochondria were noted in 20% of the muscle fibers. Within some of the mitochondria there were electron-dense inclusions. No vacuolation was seen; hence Shy believed the condition to differ from hypo- or normokalemic periodic paralysis. Instead, a disorder of mitochondria was postulated, an interpretation challenged by Engel[30, 31] who argues that such increases may signify only that the mitochondria suffered the least damage and were proliferating. To support this point of view similar mitochondrial proliferations are seen in the perinuclear zones of fibers in hypokalemic myopathy, progressive muscular dystrophy, in the sarcoplasmic masses of myotonic dystrophy and several cases of familial ophthalmoplegia (Chapter 6).

The other myopathy, called megaconial, was observed in an 8-year-old girl who had had a proximal weakness from the early months of life. Her sibling had died of a similar disease (diagnosed as infantile muscular atrophy). The muscle fibers did not show denervative atrophy but instead contained an excess of lysosomes and numerous giant mitochondria within which were rectangular structures not unlike the striped bodies ("galons") seen by Gruner[49] and Zintz[104] in the eye muscle in Von Graefe's progressive ophthalmoplegia and in the Kearns-Sayre type of ophthalmoplegia (Figs. 5–10 and 5–11). In the latter disease many of the type I fibers in Gomori trichrome stain (normal appearance is red) are ragged —hence "ragged red fibers. Their appearance is due to the presence of masses of mitochondria. Because excessive fat was found in the liver and muscles (lipid filled vacuoles and cytoplasmic inclusions), Shy postulated a generalized defect in lipid metabolism. Nerves and neuromuscular junctions are normal. Other similar cases were reported as lipid myopathies.

MYOTUBULAR (CENTRONUCLEAR) MYOPATHY

In myotubular centronuclear myopathy, a remarkable familial disease, hypotonia and feebleness of muscle action become manifest

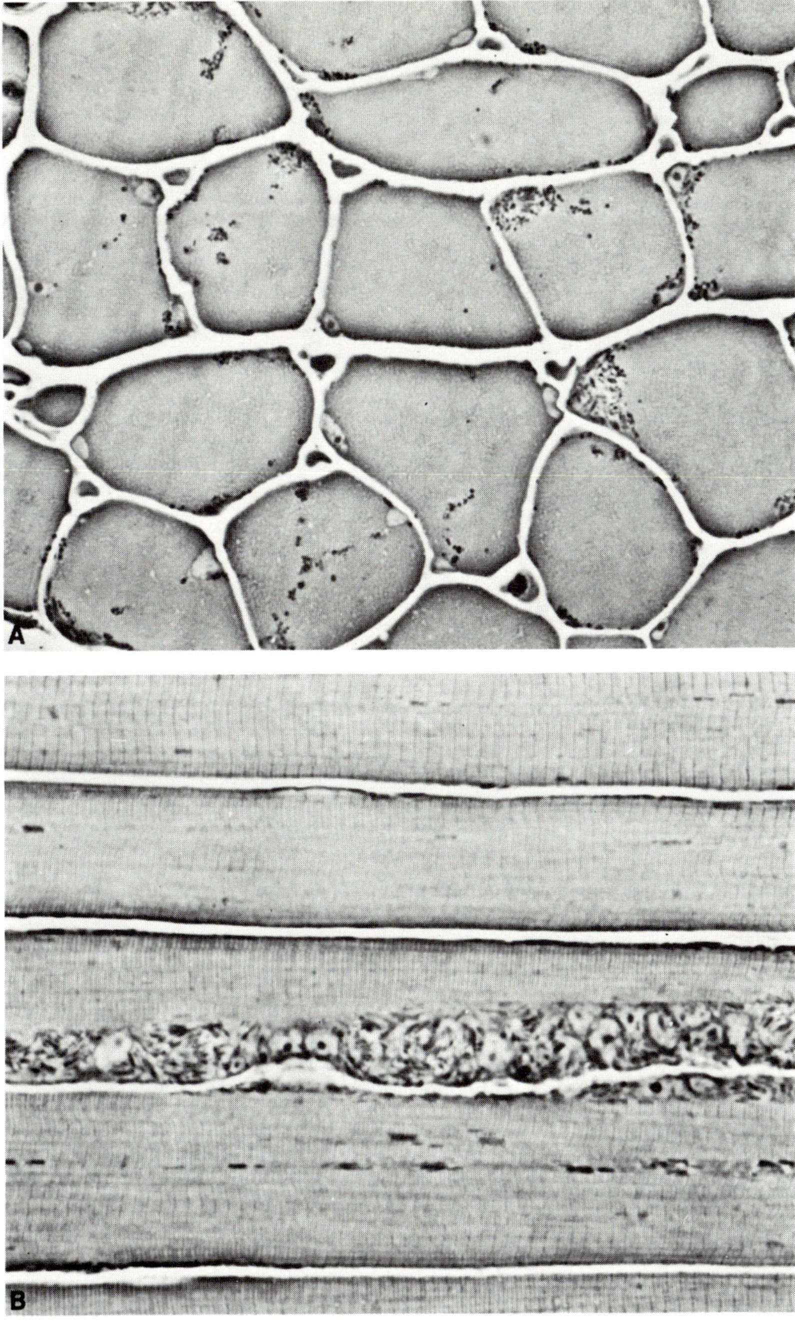

Fig. 5–7. Nemaline myopathy. (A) Transverse section. Many single dark bodies or aggregates of them are visible in the homogeneous sarcoplasm. Most lie in the periphery of the fiber on one side near the sarcolemma. One small fiber is present (at left). (B) Longitudinal section of five fibers, demonstrating an elongated mass of rod bodies, many surrounded by a clear space. (phase microscopy, ×540) (Courtesy of Dr. K. E. Astrom)

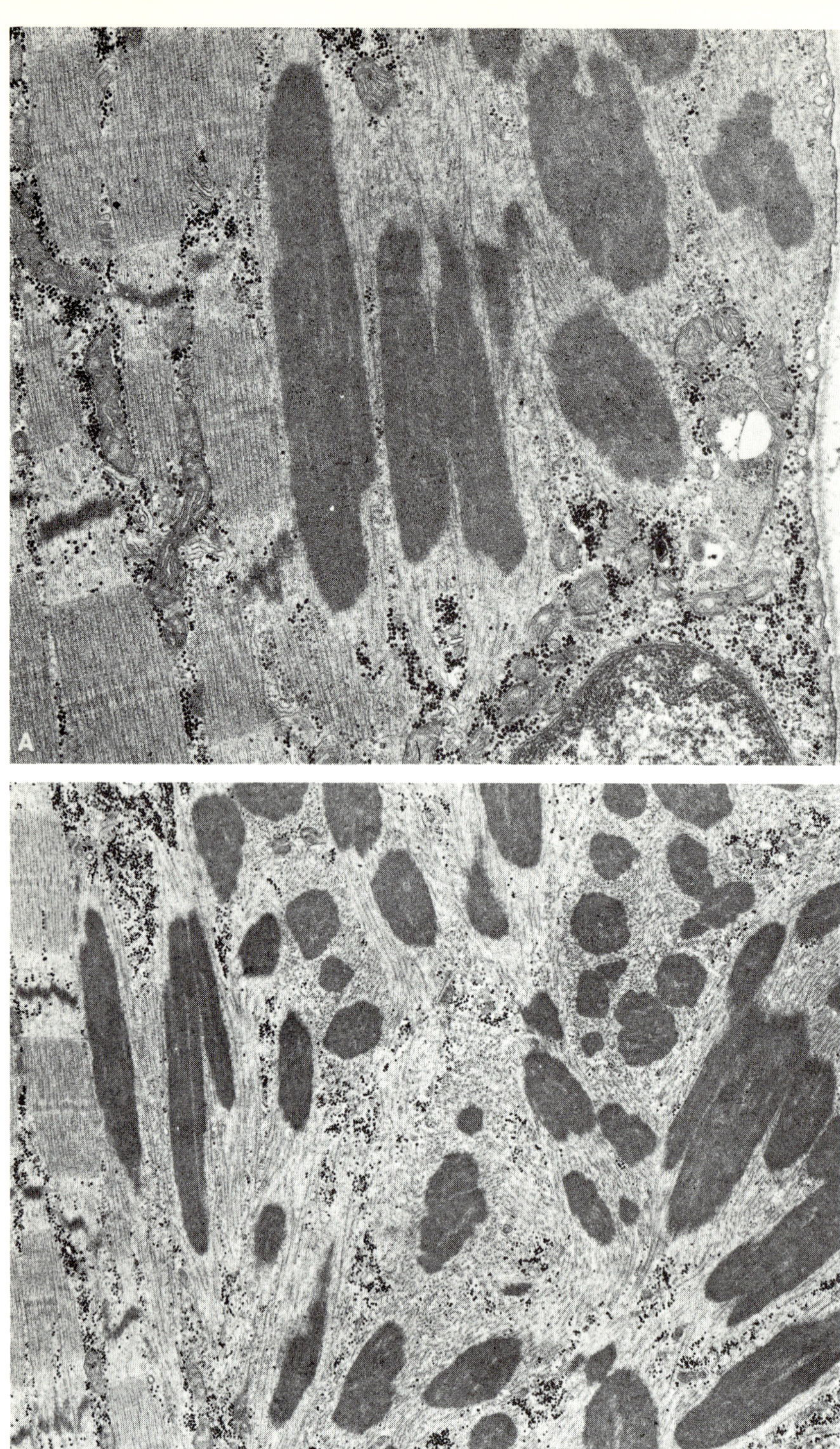

Fig. 5–8. Electron micrographs of same specimen in Figure 5–7. (A) From right to left, note the single dots of collagen in the endomysium, then the basement membrane, sarcolemma, glycogen particles (black dots), upper part of muscle nucleus, mitochondria, and myofilaments. The dark, irregularly shaped masses are the rod bodies (of about the same density as Z bands), from the ends of which are streaming the thin myofilaments. (B) An aggregate of rod bodies are sectioned in different planes; they are set in a matrix of filaments also oriented in many planes. Note spreading of Z lines in the three myofibrils (at left). (Courtesy of Dr. K. E. Astrom)

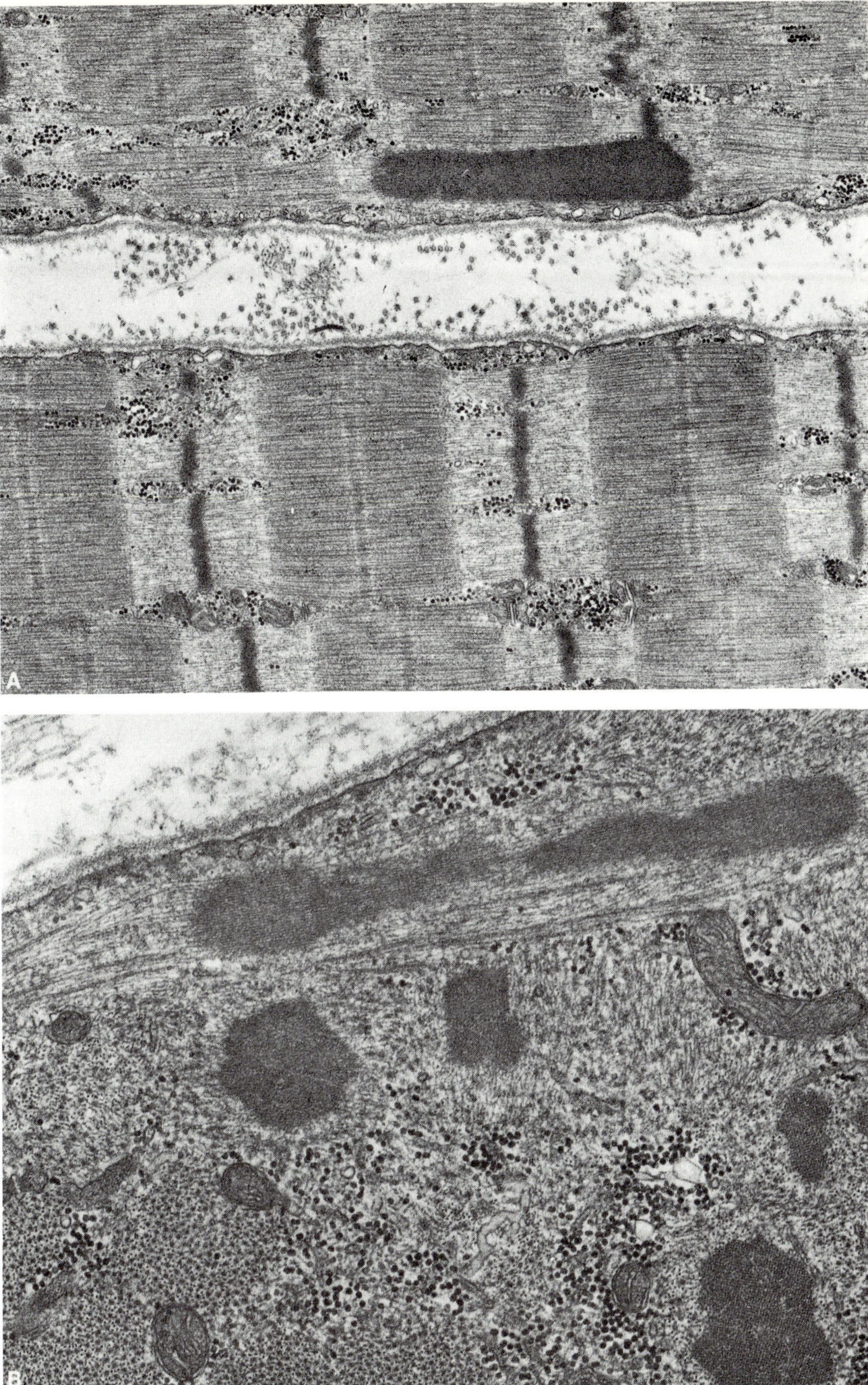

Fig. 5–9. Electron micrographs of same specimen in Figure 5–7. (A) A single rod body is seen near the sarcolemma of the upper fiber. Noted also are the sarcolemma and basement membrane, subsarcolemmal vesicles, probably part of transverse sarcoplasmic reticulum, distorted Z bands, and collagen of endomysium attached to the basement membrane. (B) Transverse plane at higher magnification. All the aforementioned structures are seen including the fine filaments attached to rod bodies and their lattice-like structure. In the region of the rod bodies thick and thin myofilaments are in disarray, in contrast to those of the properly organized myofibrils (lower left). (Courtesy of Dr. K. E. Astrom)

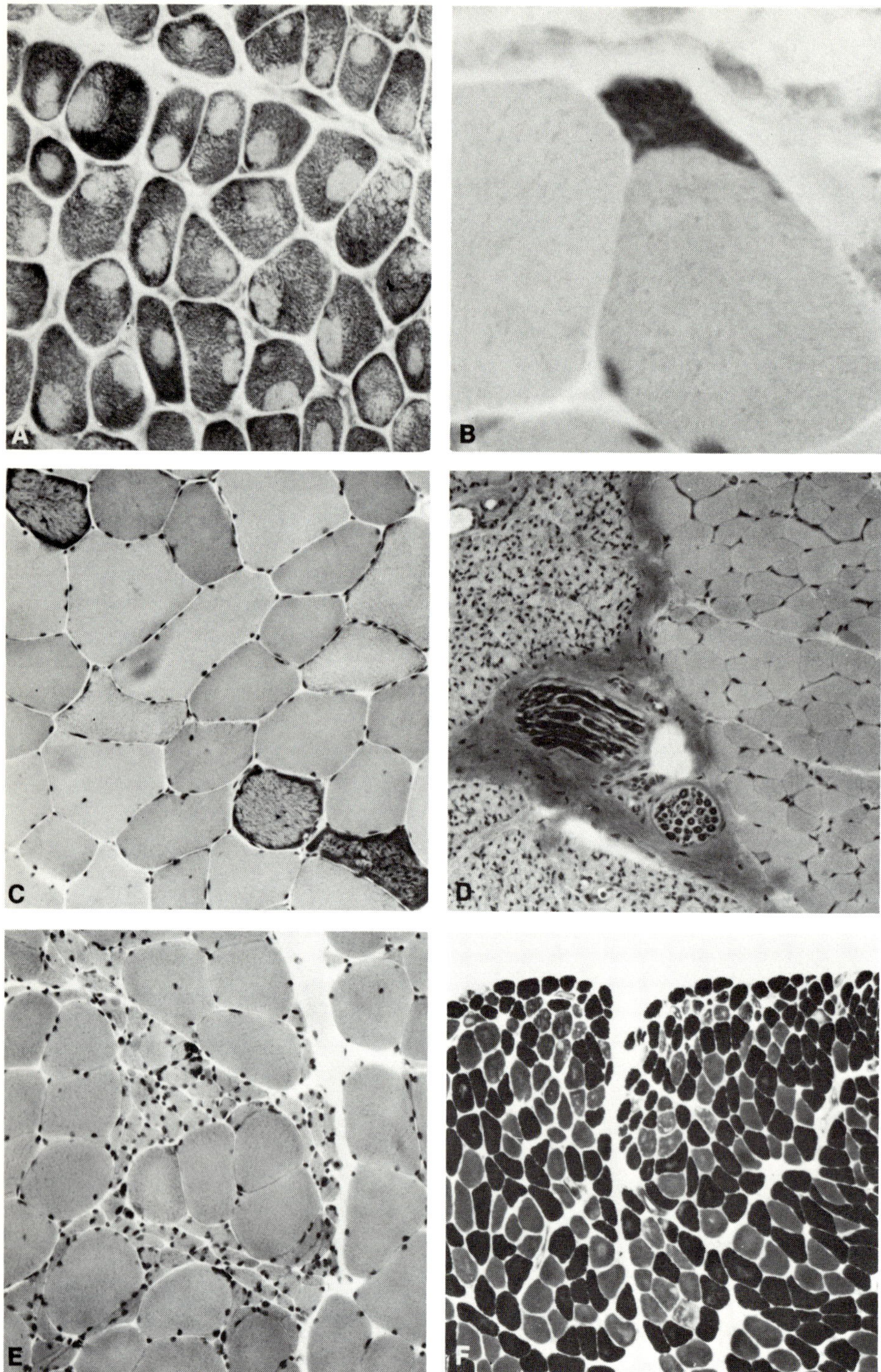

Fig. 5–10. (A) Central core disease appearing as cleared zones in fibers highly reactive in DPNH dehydrogenase preparation. (B) Cluster of rod bodies in muscle fiber from a case of polymyositis. (C) Three dark "ragged-rod" fibers from a patient with Kearns-Sayre ophthalmoplegia. There are abundant abnormal mitochondria in the periphery of these fibers as shown in an electron microscopic preparation from the same case (Fig. 15–11). (D) Group atrophy from denervation. On the left all muscle fibers are reduced to a uniformly small diameter; those on the right are more normal in size. Note nerve twigs (at center). (E) Group atrophy from denervation. (F) Perifascicular atrophy in dermatomyositis. (Courtesy of Dr. M. H. Brooke) (A. ×225; B, trichrome stain, ×500; C, D, trichrome stain, ×125; E, hematoxylin, eosin, ×125; F, ATPase, pH 9.4, ×50)

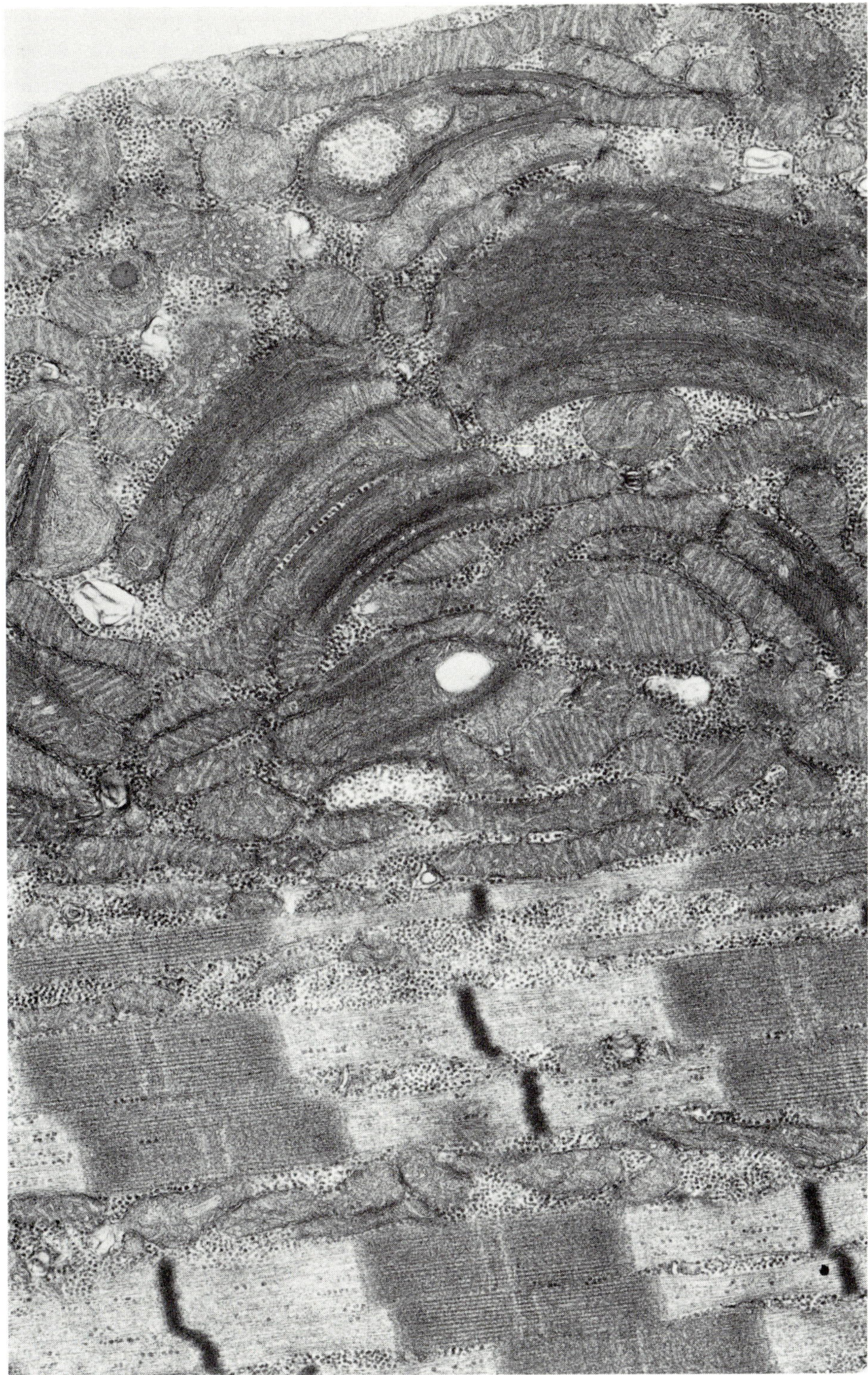

Fig. 5–11. Electron micrograph of fiber from a mitochondrial myopathy, Kearns-Sayre oph-thalmoplegia. The mass of abnormally formed mitochondria fills the upper two-thirds of the picture. (Photograph by Dr. H. E. Neville; Courtesy of Dr. M. H. Brooke)

early in life. In some instances weakness of particular groups of muscles are not recognized until the end of the first year of life or later. Rarely, in the mildest form of the disease the diagnosis does not become evident until adult years. Unlike the other congenital myopathies this one involves to some degree essentially all the striated skeletal muscles. Ptosis and ocular palsies are combined with weakness of facial, masticatory, lingual, pharyngeal, laryngeal, and cervical muscles in most of the patients. In the limbs distal weakness keeps pace with proximal weakness. The muscles remain thin and areflexic throughout life. Motor development is necessarily retarded, though improvement with maturation can occur; later, however, motor functions that have been acquired may be lost as the disease slowly advances. Several patients have shown signs of cerebral abnormality. Muscle enzymes in the serum have not been elevated. EMG shows myopathic potentials and fibrillations. The pattern of inheritance in the 20 or more reported cases is usually that of an autosomal recessive trait, though rare instances of a coincidence in parent and child suggests a mendelian dominant trait.

The outstanding feature of the disease is the smallness of muscles and of their constituent fibers and central nucleation. It is this latter attribute that led to the name myotubular myopathy; the fibers, according to Spiro et al.,[94] resembled the myotubular stage of embryonic muscle (Fig. 5–12). With growth, however, older subjects develop quite large fibers, but these too retain central nuclei. The muscle nuclei tend to be dark and seldom show heavy nucleolation or other signs of hyperplasia, and basophilia of the sarcoplasm is not seen. In the larger fibers both central and peripheral (subsarcolemmal) nucleation is observed, whereas the smaller ones have only central nuclei. A cleared zone is seen around most of the nuclei, and the number of nuclei per fiber is increased. In many microscopic fields every fiber has one or more central nuclei, and some degree of "rowing" is seen in longitudinal sections.

Histochemical stains demonstrate the usual proportion of type I and II fibers. More of the small fibers are of type I. Engel et al.[34] found that the preponderance of type I fibers and only

a small proportion of type II fibers had central nuclei. For that reason he referred to his cases as "type I fiber hypotrophy with central nuclei" despite the presence of similar changes in some of the type II fibers.

In the central parts of the fibers (in the cleared zones surrounding the nuclei) Campbell et al.,[13] Kinoshita and Cadman,[60] Vallat,[99] and Vital et al.[102] found, by electron microscopy, disintegration of myofilaments, scattered Z band material, autophagic vacuoles, granular debris, and lipofuscin. Such regions react positively in acid phosphatase preparations. The number of residual myofilaments is decreased. Such cleared zones differ from the sarcoplasmic masses of myotonic dystrophy, and disorientation of myofibrils (so frequent in myotonic dystrophy) is not found in centronuclear myopathy.

In Campbell's case large sections of the quadriceps showed a variation in the degree of involvement of different fascicles and a considerable loss of fibers as well as replacement fibrosis and fat cell infiltration. Spindles were not affected, nor were there changes in nerve fibers or spinal cord. Nerve endings on muscle are small and poorly ramified.

The obscure nature of the pathologic process invites speculation. The small central nucleated fibers do not really resemble typical myotubes, and as Campbell et al.[13] point out there is evidence of a central degenerative process leading in all probability to fiber loss. In the adult cases this cannot be demonstrated, possibly because of sampling difficulties. Such changes argue against a purely developmental abnormality even though there may have been an initial arrest of nuclear migration.

ELECTRON MICROSCOPIC PATHOLOGY

The critical reader who scrutinizes the clinical and morphologic characteristics of these progressive congenital myopathies naturally must wonder whether each constitutes a distinct and separate disease. Implied in the original descriptions of each is the notion that hypotonia and weakness, their principal clinical manifestations, are based on the ultrastructural

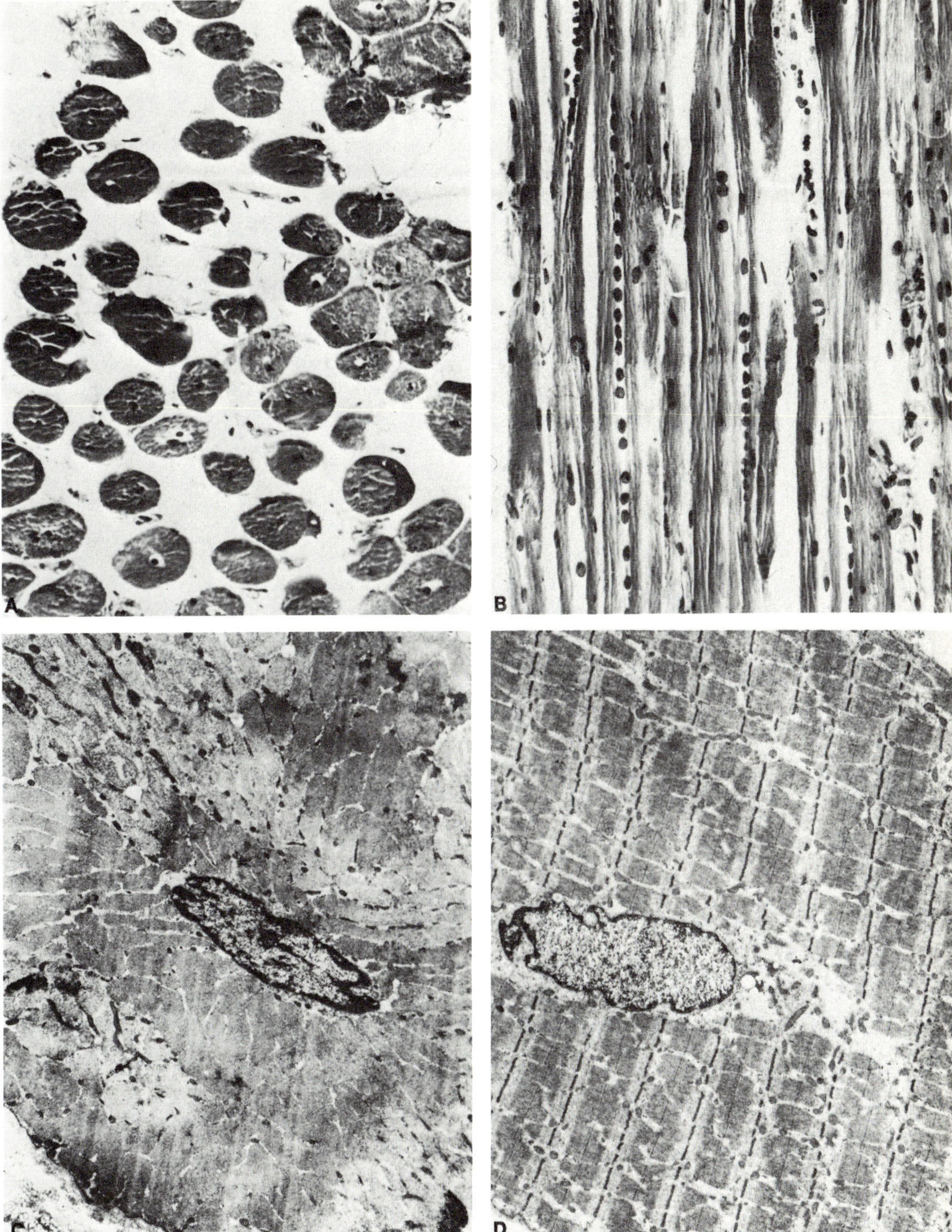

Fig. 5–12. Myotubular myopathy. (A) The central nucleus with perinuclear halo, marks each transversely sectioned fiber. (B) The central nuclei are rowed in the longitudinally sectioned fiber. (C) Electron micrograph of an obliquely cut fiber showing the central irregularly shaped fibers. (D) In this longitudinal section the complete integrity of other organelles, especially the myofilaments, is apparent. (Courtesy of Dr. Jean-Michel Vallat)

alterations of the fibers. Further, a certain specificity of the electron microscopic lesion in each disease is assumed. Both postulates are probably fallacious. There are certainly other more obvious changes in the muscles to explain the clinical findings, and it is notable that information concerning the extent of muscle hypoplasia or fiber loss is lacking because complete autopsies have not been performed on most of the patients. Furthermore, electron microscopy showed that each of the many subcellular changes lack specificity, being common to many diseases.

One of the major difficulties that has arisen pertains to the interpretation of minor alterations seen for the first time by electron microscopy. Standards of normality are uncertain or lacking altogether, and biopsy artifacts (nearly all the studies have been made on biopsy specimens) are as prominent and confusing as in light microscopy, only they are now magnified several hundred times. Moreover, there is the other obvious fact that most of the diseases under consideration have at present a recondite pathogenesis. Unfortunately few of the more common lesions of known pathogenesis have been examined with sufficient care under the electron microscope to serve as prototypes. The pathologist is uncertain as to the manner in which many of the alterations of organelles relate to one another even in diseases of definable pathogenesis.

Hudgson and Pearce,[56] Engel,[27, 29] Shafiq et al.,[86] Price,[78] and Mair and Tomé[68] presented critiques of the relative importance of the electron microscopic changes that characterize these diseases. (See Chapter 13 for a discussion of electron microscopic artifacts.)

AMYOTONIA CONGENITA, CONGENITAL HYPOPLASIA OF MUSCLE, BENIGN CONGENITAL HYPOTONIA

Many observations on arthrogryposis apply also to the floppy or hypotonic infant. There are many causes of hypotonia and delay in motor development in the neonate or infant. Hypotonia therefore is but a symptom of a variety of diseases, just as are contracture and pseudocontracture (arthrogryposis).

Here we confine ourselves specifically to neuromuscular diseases, setting aside those forms of hypotonia and inadequacy of motor development due to congenital diseases of the central nervous system (atonic diplegia of Foerster), the acquired lipid and glycogen storage diseases (Tay-Sachs and others), mongolism, cretinism, and achondrodysplasia. The neuromuscular diseases underlying congenital hypotonia can be subdivided into the following groups: (1) infantile muscular atrophy of Werdnig-Hoffman; (2) infantile muscular dystrophy; (3) myopathies (central core, nemeline, mitochondrial, myotubular); and (4) undifferentiated forms of hypotonia and muscular underdevelopment including congenital hypotonia of Walton and congenital hypoplasia of Krabbe. The amyotonia congenita of Oppenheim, once applied to all this latter group, has become obsolete.

Infantile muscular atrophy of Werdnig and Hoffmann is discussed in Chapter 10. Suffice it to say that it can usually be distinguished from congenital muscular dystrophy and the more benign or nonprogressive myopathies by the severity of generalized muscle weakness, fascicular twitchings, the characteristic findings in muscle biopsy (Fig. 5–13), and the progressive clinical course.

Congenital muscular dystrophy has already been mentioned, and further details on the entire subject appear in Chapter 6. The congenital nonprogressive myopathies have already been discussed. Only the undifferentiated group remains for further comment. Included here are cases with generalized thinness and weakness of muscle, often with retained reflexes, an inability to contract all or certain muscle groups strongly against resistance and gravity, the sparing of cranial muscles, the lack of EMG change, and the relatively normal appearance of all muscle fibers in a biopsy specimen. The only definite change is slight hypoplasia and this occurs without familial incidence. As shown by Walton[103] (Chapter 10), some of these children recover completely after a few months or years, whereas others slowly improve, being delayed in all motor accomplishments. They sit and stand late in the first

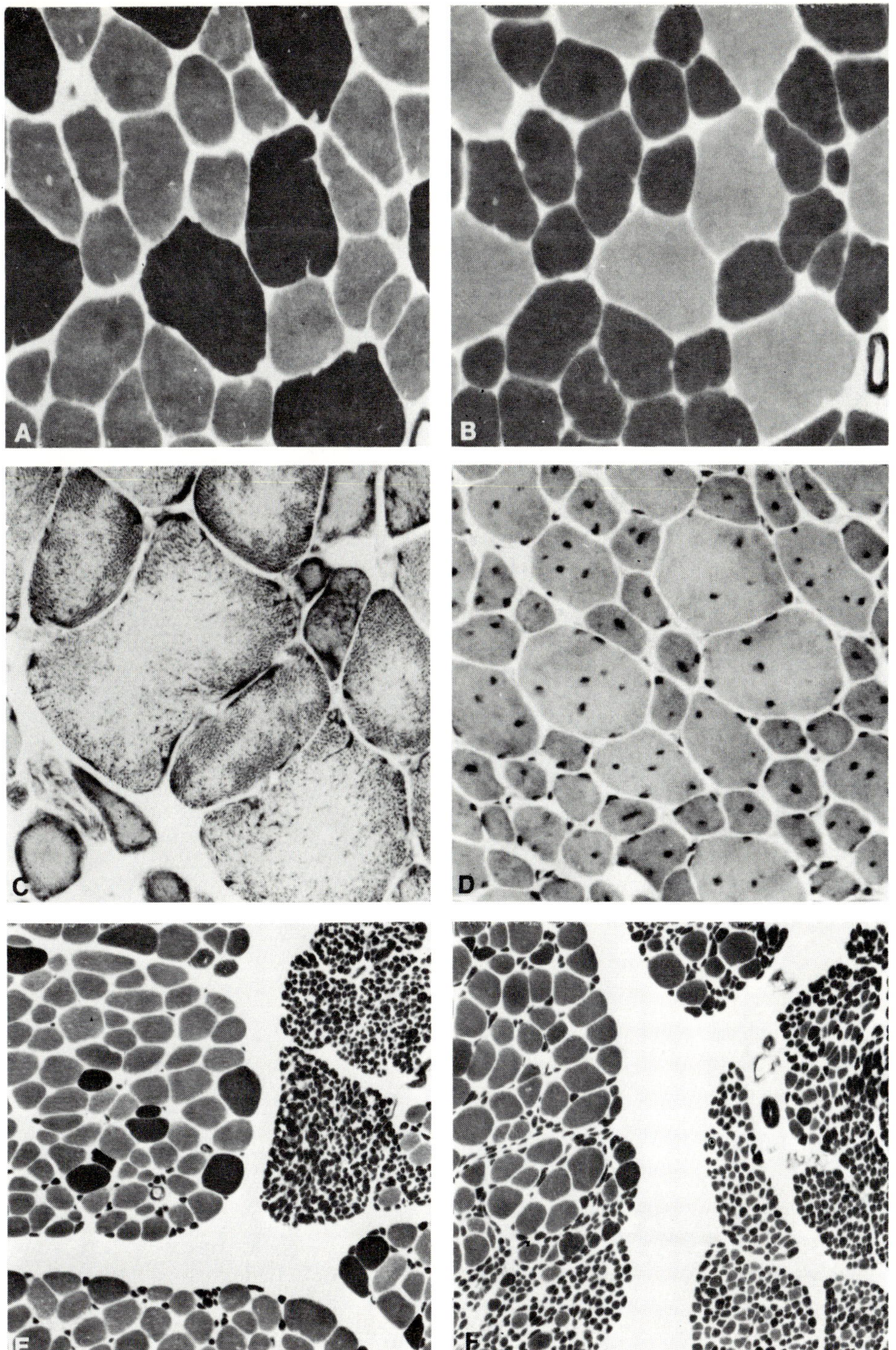

Fig. 5–13. (A and B) Pathology of infantile hypotonia. Nonprogressive myopathy in which the type I fibers are smaller than type II fibers. (C) Moth-eaten, whorled fibers and ring fibers from the same case in a DPNH dehydrogenase preparation. (D) Central nucleation of all fibers with smallness of type I fibers from a patient with a congenital nonprogressive myopathy (female, 22 years). (E and F) Transverse sections from a patient with Werdnig-Hoffmann disease. Most of the larger fibers are type I and the smaller ones type II, but with mixtures of the other types among both large and small fibers. (Courtesy of Dr. M. H. Brooke) (A, ATPase preparation, pH 9.4; B, ATPase, pH 9.6; A, B, ×175; C, ×310; D, hematoxylin, eosin, ×225; E,F, ATPase, pH 9.4, ×125)

year of life and walk late, sometimes not until the age of 3–5 years. Once they attain upright stance and locomotion they may continue to have difficulty in arising from a recumbent or sitting position and in running, climbing, and walking long distances. Arms are always affected to a lesser degree. The thinness of muscle becomes more evident with natural growth in weight and stature. Pathologic study reveals a uniform smallness of all muscle fibers, with no increase in connective tissue or fat and no evidence of degeneration. Hence the term given by Krabbe—universal hypoplasia.

In this latter group there are surely a number of diseases yet to be differentiated. A step in this direction was taken by Jerusalem *et al.*[59] when they reported their observations on the muscles of two brothers age 11 and 15 years who had been weak since birth, slow and clumsy in learning to walk, subject to frequent falls, and who were having difficulty in walking, arising from a squatting position, and climbing stairs. The muscles were somewhat thin but retained their tendon reflexes, and mental function was normal. The creatine phosphokinase (CPK) level in serum was normal in one and slightly elevated in the other, and the EMG showed a few myopathic potentials in one and was normal in the other. The condition had not progressed. Biopsy disclosed fine vacuolation between myofibrils, particularly in type II fibers, and in electron micrographs the fine vacuoles were membrane-bound and were either empty or contained amorphous debris. The limiting membranes were thought to be derived from sarcoplasmic reticulum—hence the name given to the condition was *sarcotubular* myopathy. Lack of glycogen, acid phosphatase, and fat in vacuoles and changes in other organelles excluded the other forms of congenital myopathy.

There are surely other types of hypotonia as well, some linked to certain medical diseases (celiac disease, mongolism) in which one cannot ascertain the morphologic basis. Such children tend to improve as they grow older though often with persistent deficiency of muscle power.

Brooke[9] recently reviewed his experiences with floppy children. With biopsy preparations he illustrates cases that do not conform to any of the above groups (see Figure 5–13, which shows several types of congenital hypotonia).

MUSCULOSKELETAL ABNORMALITIES WITH MALIGNANT HYPERTHERMIA

In some families disposed to malignant hyperthermia (Chapter 12) there is a particularly high incidence of ptosis, high-arched palate, dislocated patellas, and sometimes various types of abdominal hernias, dislocatable shoulders, and kyphoscoliosis.[8] Persistent elevation of CPK values in sera reflects a mild necrotizing myopathy that may be aggravated by Suxamethonium and halothane anesthesia, leading to a fatal hyperthermic accident.

CONCLUSION

Imperfections in embryogenesis seems to account for the large number and variety of anomalies of skeletal muscle. As judged by their strong familial incidence, many of these are determined by hereditary factors that interfere with the differentiation of individual muscles from the myotomes. Associated faults in other organs may occur and indicate that the maldevelopment must occur early. Some anomalies (e.g., the congenital absence of certain muscles) may be linked to inherited dystrophic disease of muscle whereas others appear as isolated defects. While it is certain that a congenital deficiency of skeletal muscle may be conditioned by a defect in the nervous system so that the trophic or maturation factors provided by innervation are lacking, all deformities due to an imbalance of muscle action cannot be explained in this way. As in clubfoot, a uniform smallness of muscle fibers with intact motor nerve supply can be demonstrated in some muscle groups. This finding requires elucidation by further observation and experiment, for it may prove fundamental to an understanding of several congenital deformities such as amyoplasia congenita, clubfoot, facial diplegia, and Sprengel's shoulder.

Skeletal abnormalities are prominent in congenital deficiencies of muscle. If movement of a limb is hindered during the growth period, its osseous development is stunted. Muscular imbalance in either a limb or the spine is almost invariably attended by bony deformity.

The recently differentiated group of congenital disfigurative myopathies fall between the developmental anomalies and the metabolic and dystrophic myopathies. At present their morphologic characteristics seem not to be sufficient for their complete identification.

REFERENCES

1. AFIFI AK, SMITH GW, ZELLWEGER H: Congenital nonprogressive myopathy central core disease and nemaline myopathy in one family. Neurology (Minneap) 16:371, 1966

2. BANKER BQ, VICTOR M, ADAMS RD: Arthrogryposis multiplex due to congenital muscular dystrophy. Brain 80:319–334, 1957

3. BECHTOL CO, MOSSMAN HW: Club-foot: an embryological study of associated muscle abnormalities. J Bone Joint Surg [Am] 32a:827–836, 1950

4. BETHLEM J, MEYJES FEP: Congenital non-progressive central core disease of Shy and Magee. Psychiatr Neurol Neurochir 63:246–251, 1960

5. BING R: Ueber angeborene Muskeldefecte. Virchows Arch [Pathol Anat] 170: 175–229, 1902

6. BONNEVIE K: Embryological analysis of gene manifestation in Little and Bagg's abnormal mouse tribe. J Exp Zool 67:443–520, 1934

7. BRANDT S: A case of arthrogryposis multiplex congenita. Acta Paediatr Scand 34:365–381, 1947

8. BRITT BA, KALOW W: Hyper-rigidity with hyperthermia associated with anesthesia. Ann NY Acad Sci 157:947–958, 1968

9. BROOKE MH: The pathologic interpretation of muscle histochemistry. The Striated Muscle. Edited by CM Pearson, FK Mostofi. Baltimore, Williams & Wilkins, 1973, pp 86–122

10. BROOKS B: Pathologic changes in muscle as a result of disturbances of circulation: an experimental study of Volkman's ischemic paralysis. Arch Surg 5:188–216, 1922

11. BROWN JB, MCDOWELL F: Syndactylism with absence of the pectoralis major. Surgery 7:599–601, 1940

12. BROWNE D: Congenital deformities of mechanical origin. Proc R Soc Med 29:1409–1431, 1936

13. CAMPBELL MJ, REBEIZ JJ, WALTON JN: Myotubular centronuclear or pericentronuclear myopathy. J Neurol Sci 8:425–443, 1969

14. CHRISTOPHER F: Congenital absence of the pectoral muscles. J Bone Joint Surg 10:350–351, 1928

15. CONEN PE, MURPHY EG, DONOHUE WL: Light and electron microscopic studies of the myogranules in a child with hypotonia and muscle weakness. Can Med Assoc J 84:983–986, 1963

16. COURTILLIER L: Contribution à l'etiologie et à la pathogenie de pied-bot congénital. Arch Gen Med 1:536–563, 1897

17. DANIS P: Les paralysies oculo-faciales congenitales. Ophthalmologica 10:113–137, 1945

18. DITTRICH RJ: Pathogenesis of congenital club-foot. J Bone Joint Surg 28:373–399, 1930

19. DUBOWITZ V, PEARSE AGE: Oxidative enzymes and phosphorylase in central-core disease of muscle. Lancet 2:23–24, 1960

20. DUBOWITZ V, PEARSE AGE: Enzymic activity of normal and dystrophic human muscle. J Pathol 81:365–378, 1961

21. DUBOWITZ V, PLATTS M: Central core disease with focal wasting. J Neurol Neurosurg Psychiatry 28: 432–437, 1965

22. DUBOWITZ V, ROY S: Central core disease of muscle: clinical, histochemical and EM studies of an affected mother and child. Brain 93:133–146, 1970

23. DURANTE G: Anatomie pathologique des muscles. Manuel d'Histologie Pathologique. Edited by V Cornil, L Ranvier. Paris, Felix Alcan, 1902, Vol II, pp 1–477

24. DURANTE G, COURTILLIER L: Atrophie muscularie par régression embryonnaire totale chez un enfant atteint de pied bot paralytique congenital. Bull Soc Obstet Paris 2:366–371, 1899

25. EALING MI: Amyoplasia congenita causing malpresentation of the foetus. J Obstet Gynaecol Br Emp 51:144–146, 1944

26. EDINGER L: Ruckenmark und Gehirn in einem Fall von Angeborene Mangel eines Vorderarms. Virchows Arch [Pathol Anat] 89:46, 1882

27. ENGEL AG: Ultrastructural reactions in muscle disease. Med Clin North Am 52:909–931, 1968

28. ENGEL AG, GOMEZ MR: Nemeline Z-disc myopathy: observations on the origin, structure and solubility properties of the nemeline structures. J Neuropathol Exp Neurol 26:601–619, 1967

29. ENGEL AG, MACDONALD RD: Ultrastructural reactions in muscle disease and their light microscope correlates. Muscle Disorders. Proceedings of the International Congress, Milan. Second Congress Series 199, 1969, pp 71–89

30. ENGEL WK: Critique of congenital myopathies and other disorders. Exploratory Concepts in Muscular Dystrophy and Related Disorders. Edited by AT Milhorat. International Congress Series 147. Amsterdam, Excerpta Medica, 1967, pp 27–40

31. ENGEL WK: Mitochondrial aggregates in muscle disease. J Histochem Cytochem 12:46–48, 1964

32. ENGEL WK: Selective and nonselective susceptibility of fiber types: a new approach to human neuromuscular diseases. Arch Neurol 22:97–117, 1970

33. ENGEL WK, FOSTER JB, HUGHES BP, et al: Central core disease—an investigation of a rare muscle cell abnormality. Brain 84:167–185, 1961

34. ENGEL WK, GOLD GN, KARPATI G: Type I fiber hypertrophy and central nuclei. Arch Neurol 18:435–444, 1968

35. ENGEL WK, REZNICK JS: Late onset rod-myopathy: a newly recognized, acquired and progressive disease. Neurology (Minneap) 16:308–309, 1966

36. ENGEL WK, WANKO T, FENICHEL GM: Nemaline myopathy: a second case. Arch Neurol 11:22–39, 1964

37. FAIRBANK HAT: Congenital elevation of the scapula: a series of eighteen cases, with a detailed description of a dissected specimen. Br J Surg 1:553–572, 1913–1914

38. FORD FR: Disease of the Nervous System in Infancy, Childhood and Adolescence. Second edition. Springfield, Ill, Charles C Thomas, 1944

39. GARROD AE, DAVIES LW: On a group of associated congenital malformations, including almost complete absence of the muscles of the abdominal wall, and abnormalities of the genito-urinary apparatus. Med Chir Trans 88:363–382, 1905

40. GATES RR: Human Genetics. New York, Macmillan, 1946, Vols I and II

41. GILLES DE LA TOURETTE: Pathogenie et traitment des pieds bots. Sem Med 16:517–522, 1897

42. GILMOUR JR: Amyoplasia congenita. J Pathol 58:675–685, 1946

43. GONATAS NK: The fine structure of rod-like bodies in nemaline myopathy and their relation to the (Z-disc). J Neuropathol Exp Neurol 25:409–421, 1966

44. GONATAS NK, PEREZ MC, SHY GM, et al: Central core disease of skeletal muscle: ultrastructural cytochemical observations in 2 cases. Am J Pathol 47:503–524, 1965

45. GONATAS NK, SHY GM, GODFREY EH: Nemaline myopathy: the origin of nemaline structure. N Engl J Med 274:3–35, 1966

46. GRAEFE A: Graefe-Saemisch Handbuch. Leipzig, 1880, 6, ii, p 60

47. GREENFIELD JG, CORNMAN T, SHY GM: Prognostic value of muscle biopsy in the floppy infant. Brain 81:461–465, 1959

48. GREIG DM: Congenital high scapula. Edinburgh Med J 31:22–44, 1924

49. GRUNER JE: Sur quelques anomalies mitochondriales observées an cours d'affections musculaires variées. C R Soc Biol (Paris) 157:181–182, 1963

50. GUBBAY SS, WALTON JN, PEARSE GW: Clinical and pathological study of a case of congenital muscular dystrophy. J Neurol Neurosurg Psychiatry 29:500–508, 1966

51. GUNN RM: Congenital ptosis with peculiar associated movements of the affected lid. Tr Ophthalmol Soc UK 3:283–287, 1882–1883

52. HEIJBROEK NI: Arthrogryposis multiplex congenita. Maandschr Kindergeneeskd 10:267–276, 1941

53. HENDERSON JL: The congenital facial diplegia syndrome; clinical features, pathology and aetiology. Brain 62:381–403, 1939

54. HENKEL HL: Muscles of congenital clubfoot. Z Orthop 108:604–632, 1971

55. HEUBNER O: Ueber angeborenen Kernmangel. Charite Ann 25:211–243, 1900

56. HUDGSON P, PEARSE GW: Ultramicroscopic studies of diseased muscle. Disorders of Voluntary Muscle Movement. Second edition. Edited by JN Walton. London, Churchill, 1969, pp 277–317

57. HULBERT KF: Congenital torticollis. J Bone Joint Surg [Br] 32:50–59, 1950

58. IRANI RN: Pathological anatomy of club-foot. Clin Orthop 84:14–20, 21–24, 1972

59. JERUSALEM F, ENGEL AG, GOMEZ MF: Sarcotubular myopathy. Neurology (Minneap) 23:897–906, 1973

60. KINOSHITA M, CADMAN TE: Myotubular myopathy. Arch Neurol 18:265–271, 1968

61. KLIPPEL M: Arrêt de développement du membre supérieur consécutif à un traumatisme datant de l'enfance: atrophie musculaire numérique. Presse Med 2:49, 1897

62. KRABBE K: Les lesions embryonnaires a la lumiére des defectuosités mammaire et pectorale de la syndactylie et da la microdactylie. Acta Psychiatr Neurol 24:539, 1949

63. LEADBETTER GW: Cited by White JW: South Med J 22:675–678, 1929

64. LE DOUBLE AF: Traité des Variations du Système musculaire de l'Homme. Paris, Schligher Frères, Editeurs, 1897

65. LICHTENSTEIN BW: Congenital absence of the abdominal musculature; associated changes in the genito-urinary tract and the spinal cord. Am J Dis Child 58:339–348, 1939

66. LIDGE RT, BECHTOL RC, LAMBERT CN: Congenital muscular torticollis; etiology and pathology. J Bone Joint Surg [Am] 39A:1165–1182, 1957

67. LITTLE CC, BAGG HJ: The occurrence of four inheritable morphological variations in mice and their possible relation to treatment with x-rays. J Exp Zool 41:45–91, 1924

68. MAIR WGP, TOMÉ FMS: Atlas of the Ultrastructure of Diseased Muscle. Edinburgh, Churchill Livingstone, 1972

69. MALL FP: On the frequency of localized anomalies in human embryos and infants at birth. Am J Anat 22:49–72, 1917

70. MCCLENDON SJ: Agenesis of the abdominal muscles. Arch Pediatr 51:673–676, 1934

71. MICHAUD J: Note sur la pathogénie du pied bot congénital à propos d'un exemple d'une déformité dé ce genre paraissant liée à une lesion congenitale de la moelle épiniere. Arch Physiol 3:586–598, 1870

72. MIDDLETON DS: The pathology of congenital torticollis. Br J Surg 18:188–204, 1930

73. MIDDLETON DS: Studies on prenatal lesions of striated muscle as a cause of congenital deformity. Edinburgh Med J 41:401–442, 1934

74. MOEBIUS PJ: Ueber angeborene doppelssitige Abducens-Facialis Lahmung. Munch Med Wochenschr 35:91, 108, 1888

75. MORLEY EB: Congenital defect of the pectoralis muscle. Lancet 1:1101–1102, 1923

76. NOVÉ-JOSSERAND R, VIANNAY C: Pathogénie du torticollis congénital (théorie ischémique). Rev Orthoped (2nd series) 7:397–425, 1906

77. OSLER W: Congenital absence of the abdominal muscles with distended hypertrophied urinary bladder. Bull Hopkins Hosp 12:331–333, 1901

78. PRICE HM: Ultrastructure of the skeletal muscle fiber. Disorders of Voluntary Muscle. Second edition. Edited by JN Walton. London, Churchill, 1969, pp 29–56

79. PRICE HM, GORDON CB, PEARSON CM, et al: New evidence for excessive accumulation of Z band material in nemaline myopathy. Proc Natl Acad Sci USA 54:1398–1406, 1965

80. RAINEY J, FOWLER JS: Congenital facial diplegia due to nuclear lesion. Rev Neurol Psychiatry 1:149–155, 1903

81. RESNICK E: Congenital unilateral absence of the pectoral muscles often associated with syndactylism. J Bone Joint Surg 24:925–928, 1942

82. RICHTER RB: Unilateral congenital hypoplasia of the facial nucleus. J Neuropathol Exp Neurol 19:33–41, 1960

83. ROBERTS JAF: The inheritance of a lethal muscle contracture in the sheep. J Genet 21:57069, 1929

84. ROCHER JL: Les raideurs articulaires congénitales multiples. J Med Bordeaux 43:772, 815, 1913

85. SEITELBERGER F, WARKO T, GAVIN MA: The muscle fiber in central core disease: histochemical and electron microscopic observations. Acta Neuropathol 1:223, 1961

86. SCHAFIQ SA, DUBOWITZ V, PETERSON J, et al: Nemaline myopathy: report of a fatal case with histochemical and electron microscopic studies. Brain 90:817–828, 1967

87. SHELDON W: Amyoplasia congenita. Arch Dis Child 7:117–136, 1932

88. SHY GM, MAGEE KR: A new non-progressive myopathy. Brain 79:610–621, 1956

89. SHY GM, ENGEL WK, SOMERS JE, et al: Nemaline myopathy: a new congenital myopathy. Brain 86:793–810, 1963

90. SHY GM, GONATAS NK, PEREZ M: Two childhood myopathies with abnormal mitochondria. I. Megaconial myopathy. II. Pleoconial myopathy. Brain 89:133–158, 1966

91. SILVERMAN FN, HUANG N: Congenital absence of the abdominal muscles; associated with malformation of the genito-urinary and alimentary tracts. Am J Dis Child 80:91–124, 1950

92. SPATZ H, ULLRICH O: Klinischer und anatomischer Beitrag zu den angeborenen Beweglichkeitsdefecten im Hirnnervenbereich. Z Kinderheilkd 51:579–597, 1931

93. SPIRO AG, KENNEDY C: Hereditary occurrence of nemaline myopathy. Arch Neurol 13:155, 1965

94. SPIRO AJ, SHY GM, GONATAS NK: Myotubular myopathy: persistence of fetal muscle in an adolescent boy. Arch Neurol 14:1–14, 1966

95. SPRENGEL H: Die angeborene Verschiebung des Schulterblattes nach oben. Arch Klin Chir 42; 545–549, 1891

96. STERN WG: Arthrogryposis multiplex congenita JAMA 81:1507–1510, 1928

97. THOMASEN E: Der angeborene Klumpfuss: Ueber die Mechanick der Deformität und ihre primäre Behandlung. Acta Orthop Scand 12:33–100, 1941

98. ULLRICH O: Angeborene Muskeldefekte und angeborene Beweglichkeitsstorungen in Gehrinnervenbereich. Handbuch de Neurologie. Edited by Springer, Bumke and Foerster. 1936, Vol 16, pp 139–181

99. VALLAT JM: Les hypotonies congenitales d'origine myogene ou myopathies congenitales. Thesis 86 presented 16 June, 1972. Bergeret edition. Bordeaux, France, 1972, pp 1–173

100. VANDER EECKEN H, PEARSON CM, ADAMS RD: Unpublished observations

101. VAN WEJNGAARDEN GM, FLEURY P, BETHLEM J, ET AL: Familial myotubular myopathy. Neurology (Minneap) 19:901–908, 1969

102. VITAL CL, VALLAT JM, MARTIN F, et al: Etude clinique et ultrastructurelle d'un cas de centronucléarie myotubulaire myopathy de l'adult. Rev Neurol 123:117–130, 1970

103. WALTON JN: Amyotonia congenita. Lancet 1:1023–1028, 1970

104. ZINTZ R: Dystrophische Veränderungen in ausseren in ausseren Augenmuskels und Schullermuskeln bei der sog. progressiven Graefeschen Ophthalmoplegie. Progressive Muskeldystrophie, Myotonie, Myasthenie. Edited by E. Kuhn. Berlin, Springer Verlag, 1966, pp. 109–114

chapter 6

MUSCULAR DYSTROPHIES

Progressive muscular dystrophy is a progressive, hereditary, degenerative disease of skeletal muscles. The innervation of the affected muscles, in contrast to that in the neural and spinal muscular atrophies, is sound. Indeed, the final proof that the origin of this disorder is in the muscle fibers themselves comes from the demonstration of intact spinal motor neurons, muscular nerves, and nerve endings in the presence of severe degenerative changes in the muscle fibers. The characteristic features of symmetrical distribution of muscular atrophy, the retention of faradic excitability in proportion to the remaining power of contraction, the intact sensibility and preservation of cutaneous reflexes, and the liability to heredofamilial incidence serve to set this group of diseases apart on clinical grounds alone.

Though some investigators[18] have applied the term muscular dystrophy to the degenerations of muscle resulting from deficiency in vitamin E and from the effect of certain viruses, we believe this practice to lead to terminologic confusion. The reaction of muscle fibers to deficiency in vitamin E and to the Coxsackie viruses presents a mixture of degenerative and regenerative changes. The intensity of cellular response not only distinguishes the histology of such conditions but also implies a fundamental difference in pathogenesis. We therefore reserve the term dystrophy for the classic types of purely degenerative muscular disease of genetic etiology. The relative inadequacy of regenerative activity in this disease in our opinion is a fundamental characteristic not hitherto appreciated. Also, a slow, deliberate type of onset and progression distinguishes it from many other myopathic affections. There is greater difficulty in classifying the benign, relatively

nonprogressive diseases such as central core, nemeline, mitochondrial, and myotubular diseases (Chapter 5). We prefer to call them myopathies (even though they are hereditary) because of their different course and the nature of the pathologic changes.

Although various myopathic disorders such as myoglobinuria and myotonia have long been known in mammals other than man, it is only in recent years that inherited affections of a dystrophic kind have been described in animals. The condition most closely resembling the muscular dystrophy of man is the autosomal recessive characteristic found in the Bar Harbor 129 strain of mice by Michelson *et al.*[108] in 1955. In some muscles in such animals at some phases of the disease, multiple foci of fiber destruction occur in a manner unusual in certain of the human dystrophies. This difference may reflect only the intensity of the disease process in relation to the life span of the mouse, for in all other respects it has close resemblance. Since the relationship still remains *sub judice,* we described the disease in mice in the chapter on experimental pathology. Similarly, the more intense pathologic changes and doubtful genetic relationship of diseases such as scrapie in sheep and certain myopathic syndromes in other animals led us to separate these diseases from the muscular dystrophies. A brief description of them appears in Chapter 3.

HISTORICAL NOTES

The differentiation of dystrophic diseases from those secondary to neuronal degeneration was at first greatly confused by ambiguity concerning the nature of progressive neural muscular atrophy, amyotrophic lateral sclerosis, and peripheral neuritis. The improvement of staining methods to demonstrate neural degeneration, particularly the Weigert method for myelin, and systematic studies of dystrophic muscle eventually made possible a clear distinction. The conception of a group of primary dystrophic diseases of muscle, and the chief histologic features of this class of disorder were not fully developed until the work of Erb[57] published in 1891. As pointed out by Growers,[71] isolated cases of muscular dystrophy had been commented upon earlier by Bell,[15] Partridge,[115] Aran,[4] and Little,[101] but the nature of the disease was not appreciated. Meryon[106, 107] observed families in which several members had developed progressive atrophy and weakness of muscles early in life. Postmortem examination of two of these cases showed the spinal cord and nerves to be healthy but the muscles to be the seat of "granular degeneration." Enlargement of muscles was mentioned only incidentally. Meryon[107] therefore first established in 1864 "an idiopathic disease of the muscles, dependent perhaps on defective nutrition." He added that "spontaneous twitchings are not apt to occur," but unfortunately still otherwise confused the condition with progressive neural muscular atrophy.

The French neurologist Duchenne, who had long pursued a study of abnormalities of muscular action, described a simple progressive muscular atrophy of childhood, often with familial history, in 1855,[49] but likewise considered the disease to be a juvenile form of progressive neural muscular atrophy. In the second edition of this famous monograph, in 1861, he described in addition and for the first time "hypertrophic paraplegia of infancy" as a distinct syndrome of unknown pathology, but with speculations as to the nature of the hypertrophy of muscles. Great interest in this condition was then aroused in Germany, and in 1866 von Eulenberg and Cohnheim[152] reported the findings at autopsy in a case of "hypertrophic paralysis." They remarked upon the absence of any changes in the central nervous system and the presence of fatty tissue, which they considered to be the degenerated contents of sarcolemmal tubes. In 1865 Griesinger[75] was

the first to make a surgical biopsy and confirm the presence of abundant adipose tissue in the large muscles. This procedure was in those days particularly hazardous, and the ethics of such scientific curiosity were strongly criticized.[49]

In 1865 Duchenne, unable to obtain pathologic confirmation in any of his cases, devised a trocar ("muscle harpoon") with a punch at the side that could be plunged into the muscle for biopsy without anesthesia. With this device he made microscopic examinations of the hypertrophic muscles at different stages of the disease. With the histologic evidence thus gained and increased experience with the course of the paralysis, he published his important memoir in 1868[48] on "pseudo-hypertrophic, or myosclerotic paralysis." He noted the increase of connective tissue and fat cells in the affected muscles and also the preservation of striation. He concluded that the disease was primarily one of the interstitial muscular tissue, an idea recently reemphasized by Bourne and Golarz.[24] Duchenne did not have an opportunity to conduct postmortem examinations until 1871, when with Charcot he confirmed the interstitial lesions in muscles and the absence of changes in the nervous system, and included these findings in a summary of the chief features of the disease in the third edition of his monograph in 1872.[49]

In 1872, as an appendix to this work, Duchenne amplified his earlier description of another syndrome, "progressive muscular atrophy of childhood"; from his clear account and illustrations of the familial incidence, myopathic facies, and halting course, it is now clear that the disease was the facioscapulohumeral type of dystrophy. Unfortunately he was unable to obtain pathologic material except from punch biopsy, which showed only fatty muscle fibers such as he had obtained in neural atrophy. In 1876 Leyden[99] drew attention to similarities between all familial progressive muscular atrophies; and Gowers[71] in 1879 already recognized that muscular hypertrophy and atrophy could occur in variable proportions in the same family, and even a pure muscular atrophy without hypertrophy. Erb[55] in 1882 drew attention to the absence of electrical reaction of degeneration in "progressive muscular atrophy" in young people. Not until 1884 did Landouzy and Déjerine obtain a necropsy on one of the cases of progressive muscular atrophy of childhood of Duchenne, upon which they published a brief note,[92] followed by their classic paper on "myopathie atrophique" (facioscapulohumeral atrophy) the following year.[93] In 1884 two months after Landouzy and Déjerine's first communication, Erb[56] published his full account of "the juvenile form of muscular atrophy," which differed only in later and less obvious involvement of facial muscles.

Since that time contributions to the knowledge of muscular dystrophy have been very numerous, particularly in relation to the classification of various types of the disease on the basis of distribution of the affected muscles, age of onset, and other clinical features. An interesting and valuable early American contribution was made by Sachs.[128] Erb's comprehensive monograph[57] in 1891 clearly described the chief histologic changes in the muscles, and he was the first to emphasize the general similarity of the disease process in all the different dystrophic syndromes. He suggested a classification into forms beginning in adult life and in childhood. The adult form was subdivided into two groups (hypertrophic and atrophic) and the childhood form into two types (with and without facial weakness). The cycle of changes in any single muscle fiber was thought to begin with enlargement followed by atrophy, and the splitting of fibers was prominent. An increase in interstitial tissue was a secondary feature. General reviews of the large literature have been written by Lorenz,[102] Batten,[10] Marinesco,[105] Spiller,[138] Bing,[19] Curschmann,[34] Hurwitz,[84] Shank *et al.,*[135] Tyler *et al.,*[145, 147, 148] and Walton and Nattrass.[158]

The possibility that isolated paralysis of the external ocular muscle could be a muscular dystrophy has seldom been given serious consideration owing to the extreme rarity of affection of these muscles in the course of the classic types of dystrophy. A group of cases, however, usually termed "chronic progressive nuclear ophthalmoplegia," has long stood apart from other types of ophthalmoplegia on account of the extreme chronicity of the paralysis, its symmetry, the commonly associated weakness of orbicularis oculi, and the fact that the mechanism of pupillary contraction and ac-

commodation is unaffected. Von Graefe,[153] although lacking pathologic material, in his original description of this disorder remarked on the resemblance between progressive ophthalmoplegia and progressive bulbar palsy (labioglossolaryngeal paralysis), and Gowers[72] later assumed a degeneration of the nuclei of the ocular nerves similar to that in progressive muscular atrophy. Moebius[112] held the same view of the essential pathology of the disease and named it progressive nuclear ophthalmoplegia.

In 1890 Fuchs,[67] on the basis of a muscle biopsy, was the first to propose that this disease represented a "myopathy" of the external ocular muscles. The arguments favoring the myopathic origin were summarized by Kiloh and Nevin[89] and are sufficiently convincing that most physicians have now accepted the notion that most if not all of these cases are a special form of muscular dystrophy. The following lines of evidence support this conclusion: (1) Although involvement of the ocular muscles is infrequent in progressive muscular dystrophy, unquestionable cases have been reported. (2) In a small proportion of cases of progressive ophthalmoplegia (10–25%) the facial, trunk, and limb musculature becomes involved in a characteristic myopathic pattern at some stage of the disease. (3) In a few fully autopsied cases[86, 95] the loss of nerve cells in the ocular nuclei was slight and was not accompanied by degeneration of the nerves to the ocular muscles, although there was almost complete ophthalmoplegia. (4) Biopsies of ocular muscles have in numerous cases demonstrated the typical changes of muscular dystrophy. Hutchinson[85] described the clinical features of the disease in detail, and Kiloh and Nevin[89] provided an analysis of cases recorded since his time. This disorder is the most chronic and most localized of the muscular dystrophies.

Taylor[140] was the first to record a late-life, dominantly inherited, paralytic disease affecting muscles innervated by oculomotor and vagal nerves. He assumed a nuclear or neural origin. The descendants of this same family, when studied by Victor *et al.,*[149] were shown by electromyography (EMG) and muscle biopsy to have a late-life dystrophy. Opportunity for complete autopsy examination was afforded later, and Moore *et al.*[113] confirmed the normality of the cranial nerves and their nuclei. The muscle lesions, more widespread than was realized from clinical examination alone, conformed to a proximal dystrophic pattern except that those of extraocular muscles could not be distinguished except in degree from senile muscle. These authors questioned whether this disease should be considered an inherited presenile myopathy.

The first unmistakable description of myotonic dystrophy was presented by Hoffmann[81] in 1896, and two further cases were described in 1900. He noted the distribution of the atrophy in the face, forearms, and sternomastoid muscles. Numerous reports of "Thomsen's disease with muscular atrophy" followed. Rossolimo[125] gave it the name myotonia atrofica, which has now fallen into disuse owing to confusion with amyotonia congenita. Batten and Gibb[11] and Steinert[139] recognized the peculiarity in the distribution of muscular atrophy in these cases and asserted that it was a special form of Thomsen's disease. Steinert, besides describing the characteristic distribution of muscular atrophy, drew attention to the testicular atrophy with reduced potency, baldness, and acrocyanosis, and indicated the importance of these as evidence of a widespread dystrophic process. Curschmann[33] also concluded that dystrophy with myotonia (dystrophia myotonica) and congenital myotonica (Thomsen's disease) are different diseases, both on clinical and genetic evidence; others[103] asserted that they are identical. Greenfield[74] was the first to arouse interest in the frequent association of familial cataract. In a later study Adie and Greenfield[2] gave a full description of the muscular histology. The most recent and complete reviews of this subject are those of Thomasen,[141] based on a study of 101 living patients, and de Jong,[37] involving almost the same number of subjects. Schroeder and Adams[133] provided the first complete electron microscopic study.

CLINICAL FORMS OF MUSCULAR DYSTROPHY

Although it is possible to distinguish a large number of separate types of this disease on clinical evidence, it is doubtful if any diagnostic, genetic, or prognostic purpose is served by further subdivision of the six main groups of the muscular dystrophies: severe generalized familial muscular dystrophy, mild restricted muscular dystrophy, progressive dystrophic ophthalmoplegia, oculopharyngeal dystrophy, dystrophia myotonica, and late distal muscular dystrophy.

SEVERE GENERALIZED FAMILIAL MUSCULAR DYSTROPHY (ERB'S CHILDHOOD HYPERTROPHIC AND ATROPHIC FORMS)

Severe generalized familial muscular dystrophy is a rapidly progressive type of primary muscle disease beginning usually in early childhood, with strong familial liability and occurring predominantly in males, with or without pseudohypertrophy. The most frequent and best known form is Duchenne's pseudohypertrophic muscular dystrophy. The disease commences in early life, before the 6th year in the majority of cases. Slight symptoms such as indisposition to walking or running at the usual age, a proneness to falling, and an enlargement of muscles may exist for some years before the disease is recognized. The two outstanding clinical characteristics are progressive alteration in the size of the muscles and weakness. The muscles may increase or decrease in size. Most frequently enlargement of the calf muscles first attracts attention, and the infraspinatus and deltoid muscles later become enlarged. Occasionally the triceps, quadriceps, and glutei are similarly affected. As a rare exception the initial hypertrophy may affect nearly every muscle in the body, as in one of Duchenne's cases in which even the facial musculature of the patient was hypertrophied (a "Farnese Hercules").[49] Hypertrophy of the thenar muscles was noted by Berger[17] and others but is more likely to be related to congenital myotonia.

The enlarged muscles have a firm, resilient consistency and are less strong than healthy muscles of the same size. The size of the muscles may increase under observation but eventually they become atrophied. It is of course important to distinguish the condition from true physiologic hypertrophy, which may precede pseudohypertrophy but is more often associated with a variety of conditions, few of which affect the very young child. In pseudohypertrophic dystrophy the muscles of the pelvis, lumbosacral spine, and shoulder girdle become wasted from the onset, and loss of muscular power in these groups is responsible for the initial defect in posture and movement. The weakness is usually first apparent in the musculature attached to the pelvis and lower spine, and the resulting instability of the pelvis may be manifest in a waddling type of gait long before atrophy of these muscles is clinically appreciable. Owing to an enfeeblement of the flexors and extensors of the hip and the extensors of the knee, equilibrium in the standing position becomes so precarious that the patient secures his balance by placing his feet far apart to increase his base of support. The waddling gait in which the body is inclined from side to side is, according to Duchenne, due to a weakness of the gluteus medius muscles. Weakness of the muscles of the lumbar spine, abdominal wall, and hip joints account for the lordosis upon standing and the opposite forward curvature when sitting. An unequal degree of weakness on the two sides may result in scoliosis, but usually the posture is symmetrical in the early stages.

Bilateral weakness of the extensors of the hip and knee joints always interferes with such activities as arising from the floor or from a chair and in ascending stairs. In order to rise from a seated or squatting position the patient habitually puts his hands first on his knees to keep them extended, and then pushes his trunk upward by working the hands up the thighs. This is called Gowers' sign. Arms are later affected especially at the shoulders and the pretibial muscles weaken. Ultimately all power over the hip, knee, shoulder, elbow, and ankle joints is lost, and the atrophy spreads to the periphery of the limbs. The hand, face, jaw, laryngeal, pharyngeal, and ocular muscles usually are relatively spared to the end. Weakness of the facial musculature, sternomastoid muscles, and diaphragm has been reported occasionally as a late development in the disease;[72, 105] and as an extreme rarity the extraocular muscles may be affected.[72]

The limbs are usually flaccid and loose. Shortening and contracture appear late, except for some degree of pes equinus. When the disability leads to confinement to chair or bed, some fixed flexion at the knees and elbows and kyphoscoliosis are likely to occur. The tendon reflexes are lost when the muscles concerned become involved. The bones of limbs are thin and undergo demineralization, and the appearance of ossification centers is delayed. In common with other muscular dystrophies, there is no disorder of smooth muscle, but hypertrophy of the heart has been reported by several writers.[167] Minor disorders of cardiac rhythm and a special liability to cardiac failure are frequent.

The disease is much more frequent in males than in females—in the proportion of 6:1 in Gowers' series. An onset after the age of 15 years is rare. It is usually transmitted by unaffected females. Progression of the disorder is very slow, but rapid worsening after a febrile illness such as pneumonia or typhoid fever is common. The essentially progressive nature of the disease is seen in the rarity of survival beyond the age of 20 years.

MILD RESTRICTED MUSCULAR DYSTROPHY (LANDOUZY-DÉJERINE FACIOSCAPULOHUMERAL DYSTROPHY; ERB'S ADULT ATROPHIC FORM WITH FACIAL INVOLVEMENT)

Mild restricted muscular dystrophy is a slowly progressive proximal myopathy involving primarily the musculature of the shoulder and often the face, with long remissions and often complete arrest, and familial incidence. A subvariety of this disease is a very slowly progressive form without facial weakness. Special types of early and of late onset are described but are unusual.

This type of disease, though less common than the first type mentioned, is not rare. There may be a strong familial incidence. Males and females are affected equally. The best known form is the facioscapulohumeral dystrophy of

Landouzy and Déjerine.[92-94] The onset of symptoms usually occurs between 6 and 20 years, though isolated cases with onset in early adult life are occasionally encountered. Usually the first symptom is a difficulty in raising the arms above the head, but in many cases inability to close the eyes completely owing to weakness of the facial musculature has already been noticed since early childhood. The physical signs are those which result from the weakness and atrophy of muscles. There is an inability to whistle or to purse the lips and a peculiar looseness and protrusion of the lips, likened by some to those of the tapir (the myopathic facies"). The eyes are not completely closed in sleep, and even when an effort is made to close them voluntarily they can be pulled open by the examiner with relative ease. The lower parts of the trapezius muscles and the sternal parts of the pectoral are almost invariably affected, leading to poor fixation of the scapulas. At this stage some compensatory hypertrophy of the deltoids, which tend to be less affected or spared, may be mistaken for pseudohypertrophy. Sternomastoid, serratus magnus, rhomboid, erector spinae, latissimus dorsi, deltoid, the spinatus muscles gradually become atrophic and weak as the disease advances. As a consequence, the bones of the shoulders and back are thrown into relief; the scapulas project and are elevated, and the collar bones are prominent. The anterior axillary folds slope down and out owing to wasting of the pectoral muscles. The biceps tend to suffer less than the triceps but both are eventually involved. As the triceps and biceps undergo atrophy the upper arm becomes smaller than the forearms. Slight lordosis and pelvic instability indicate that the muscles of the pelvic girdle and hip joint are also affected, but atrophy of these very rarely assumes the proportions seen in the severe Duchenne dystrophy. At this point the affection often is arrested and may never again progress.

It should be emphasized that our classification of this type of dystrophy as "mild" is with reference to its entire course. In many cases it leads to disability of mild degree compatible with a useful existence. In others the degree of disability produced by atrophy of the limb girdles may be very severe before the disease becomes arrested, yet the disease still lacks the inevitable momentum of the pseudohypertrophic type. One of our patients seen in 1935 was in exactly the same state in which he had been pictured in Déjerine's textbook in 1910. Another was a woman of 62 years, in the hospital for gallstones, who had noticed no increase in her scapulohumeral atrophy since the age of 22 years. Even longer survivals are reported by others. The disease may become arrested with the face alone affected. The atrophy may extend to the lumbosacral muscles but usually only in a late stage of the disease. Occasionally the proximal muscles of the arms are first involved, and this variety without facial weakness is called the "juvenile muscular dystrophy of Erb"[55] (also "limb girdle dystrophy" by Walton). Its onset, however, often after the age of 20 and sometimes as late as 45 years, and its very slow but continued progression, weak familial incidence, and frequent involvement of the facial musculature as a late event show its incomplete identity with the Landouzy-Déjerine type. Onset in the lumbosacral muscles is a rare variant (the femoral dystrophy of Leyden and Moebius).[99, 112] It should be pointed out that proximal or limb girdle involvement is not a pattern indicative only of muscular dystrophy, for chronic polymyositis and the Kugelberg-Welander form of progressive muscular atrophy have a similar distribution of muscle weakness.

A feature shared by this group of dystrophies is the occasional finding of congenital absence of muscles or of parts of muscles in patients who later become subject to the typical affection. Thus absence of one pectoral, brachioradialis, or biceps femoris muscle has caused a conspicuous deformity since birth. Similarly a congenital weakness of the face can exist until late in life before other signs develop. It is possible that cases of congenital absence of the sixth cranial nerves (congenital inability to abduct either eye) may belong also to this group. The external ocular muscles may certainly become affected as a late event in the classic type of the disease.[93] A slowly progressive dystrophy of the quadriceps beginning late in life may represent a variant of this type of dystrophy and in common with the other varieties reaches a stationary stage within 2–5 years. It is necessary to emphasize that prominence and firmness of the deltoid and gluteal musculature commonly occur as a compensatory physiologic hypertrophy in the early and mild stages of the Erb and Landouzy-Déjerine types, without in any way corresponding to the pseudohypertrophy of Duchenne's disease. Weakness of shoulder and hip fixation, however should provide adequate physiologic explanation for its presence.

Manifestations of cardiac disorder in mild types of muscular dystrophy are commonly overlooked. Tachycardia is frequently ventricular or auricular extrasystoles and abnormalities in size and voltage of any part of the electrocardiogram were present in about 7% of patients examined by Zatuchni and his associates,[167] who also reviewed the many observations by others. These investigators drew attention to the frequency of cardiomegaly in muscular dystrophy (10 of 26 cases usually in the Duchenne type). In addition to their own case they found records in the literature of 23 cases of sudden, or relatively sudden, death attributed to cardiac failure.

PROGRESSIVE DYSTROPHIC OPHTHALMOPLEGIA (OCULAR MYOPATHY OF VON GRAEFE-FUCHS)

Progressive dystrophic ophthalmoplegia is a very slowly progressive myopathy involving primarily and usually limited to the levators of the eyelids and the external ocular muscles. It results in ptosis and ophthalmoplegia, with familial incidence in approximately half of the cases.

This type of muscular dystrophy has an onset at any age from infancy to over 50 years, but most of the cases develop before the third to fourth decades of life. The onset is insidious and the course slowly progressive in most cases, though the disease may become arrested at any stage. The incidence in males and females is about equal, and there is some history of ptosis or ophthalmoplegia in the family in about half of the cases.

Ptosis of the eyelids is almost invariably the first symptom and is rarely preceded by weakness of other ocular muscles. The progressive weakness of lateral and vertical movement of the eyes develops gradually, so that the patient may be completely unaware of his difficulty. In the advanced stages of the disease all movements are paralyzed, and the eyes remain slightly divergent. Usually the eyes are affected simultaneously, but in rare instances one may be involved before the other, and diplopia may then become a prominent symptom. The smooth muscles of the iris, ciliary body, and orbit are spared. The facies of the patient, sometimes referred to as hutchinsonian, is characteristic, with the forehead constantly wrinkled and the head tilted backward in an effort to see beneath the drooping eyelids. The eyelids are abnormally thin because of atrophy of the levator muscles. Other muscles are affected in approximately 25% of cases, particularly the orbicularis oculi and other facial muscles. Thus in progressive ophthalmoplegia, as in myasthenia gravis and dystrophia myotonica, there may be a unique combination of weakness in closing *and* opening of the eyes. Paralysis of facial musculature may prevent wrinkling of the forehead. The muscles of mastication, sternocleidomastoids, and others of neck and shoulders are weak and wasted in some cases.

Myotonia, cataract, and endocrine disorders have not been observed, and their absence distinguishes this class of disease from dystrophia myotonica, with which it might be confused because of the ptosis of eyelids. The more extensive forms of the disease may resemble the mild rstricted dystrophy, and indeed Landouzy and Déjerine[93] described ocular palsy in one of their cases. The characteristic feature of progressive dystrophic ophthalmoplegia is that ptosis and ocular paralysis precede involvement of other muscles by many years. In view of this distinctive distribution and tendency to spontaneous arrest we set it apart from the other types of muscular dystrophy.

Opinions diverge as to whether all cases of progressive external ophthalmoplegia have a myopathic basis. Certainly it is difficult to decide on the basis of the appearance of the biopsied eye muscle, for in such chronic diseases dystrophy and partial denervation atrophy are difficult to distinguish. Some of the patients exhibit signs of a mild polyneuropathy and have an elevated cerebrospinal fluid (CSF) protein. Drachman[44] discussed this problem under "ophthalmoplegia plus."

A special variety of ocular paralysis in childhood is associated with retinitis pigmentosa, cardiac disorders, and dwarfism—the Kearns-Sayre Syndrome. It has been discussed with the congenital myopathies (Chapter 5).

OCULOPHARYNGEAL DYSTROPHY

Oculopharyngeal dystrophy is a disease dominantly inherited, affecting both sexes equally, and beginning usually after the 50th year of life. Ptosis of eyelids with relatively full eye movement and dysphagia are the two leading clinical manifestations. This genetic disease of cranial muscles appears to follow a course different from the ophthalmoplegic dystrophy. Studies by Barbeau[8] traced Taylor's original patients back to an early French immigrant from whom several hundred descendants with this disease have issued. Other families have since been identified.

Complaints referable to the musculature are not registered until the sixth decade, as a rule. Drooping of eyelids is the first observable manifestation and soon thereafter increasing difficulty in swallowing and change in voice become manifest. Progression is slow over a period of many years. Limitation of food intake leads to increasing cachexia. The face becomes gaunt and thin, which with the ptosis makes these patients readily identifiable in family albums. Inability to obtain adequate nutrition hastens the end of life unless prevented by a "feeding gastrostomy" or nasogastric intubation. Physical weakness from malnutrition may have prevented the detection of mild involvement of facial, lingual, neck, and shoulder muscles, which we observed to be slightly affected in the one available autopsied case.

The EMG is typically myopathic with a full interference pattern of small, brief motor unit potentials. There is neither fibrillation nor myotonia. The serum enzymes have been normal.

DYSTROPHIA MYOTONICA (MYOTONIC DYSTROPHY, STEINERT'S DISEASE)

Myotonic dystrophy is steadily progressive, familial, distal myopathy with associated weakness of the muscles of the face, jaw, neck, and levators of the eyelids. There is a tendency to myotonic persistence of contraction in the affected parts and testicular atrophy. Its distinguishing features, besides the distal distribution of muscular atrophy and the myotonia, are dystrophic features in nonmuscular tissues (such as cataracts), testicular atrophy, and other disorders of the endocrine glands. It is probable that many cases of muscular dystrophy beginning in the hands, forearms, and pretibial muscles are aberrant examples of myotonic dystrophy in which the myotonia was so mild as to escape notice or had not yet developed. Cases without myotonia have been found in families in which other members were typical.[60, 141] Certainly the distal myopathy of Gowers,[73] exemplified by a youth of 18 years with weakened and wasted anterior tibial and forearm muscles, atrophy of the sternomastoid muscles, and paresis of the orbicularis and frontalis muscles, presents some of its essential features. The late distal myopathy described by Ley and Titeca[98] in two brothers, with gradual onset of weakness in the hands at the age of 35 and 43, associated eventually with ptosis, cataract, dysphonia, heart block, and wasting of sternomastoids, certainly belongs to the group of dystrophia myotonica, though myotonia remained absent in the following 10 years.

The muscular wasting in typical dystrophia myotonica most commonly is not noticed until the third decade though we have seen a number of children with the typical facies. In the second generation the onset is earlier, not uncommonly during childhood. The small muscle of the hands are usually the first to show atrophy, though some loss of substance of the extensor muscles in the forearms is often evident very early in the course of the affection. In other cases ptosis of the eyelids and some thinness and lack of expression in the facial musculature has been present for many years, perhaps from earliest childhood. The masseters and sternomastoids are almost invariably wasted, associated with a general thinness and an exaggerated forward curvature of the neck (swan neck). In some families atrophy of the anterior tibial muscles is an early sign and at this stage differs from peroneal muscular atrophy only in the sparing of the peroneal muscles. The tendon reflexes may be lost or are much reduced. The disease progresses very slowly so that weakness and atrophy of other proximal limb and trunk muscles are present only many years later at an advanced stage of the disease. Contracture is rarely seen, and the thin, flattened hands are consequently soft and pliable.

Atrophy of the masseters leads to a narrowing of the lower half of the face, with a slender mandible. This appearance together with the ptosis and general weakness of the facial musculature constitutes the "myopathic facies" that is characteristic, yet different from the protruding "tapir" lip and transverse smile of facioscapulohumeral dystrophy. Recurrent dislocation of the jaw is common. Weakness of pharyngeal and laryngeal muscles soon becomes manifest in the weak, monotonous, nasal voice, and later dysphagia. The uterus of the pregnant myotonic woman may lack normal contractility, interfering with parturition.

The most striking characteristic is the myotonia, which is obvious both in the prolonged contraction of certain muscles following brief electrical or mechanical stimulation (percussion myotonia) and in the delay in relaxation after a strong voluntary contraction. This is most evident in the hands, face, and tongue, and occasionally in the limb musculature. Relaxation of an initial strong movement may be hampered by myotonia, but relaxation becomes more rapid with repetition of the movement. Gentle movements usually do not elicit myotonia, so that blinking of eyes, movements of facial expression, and the like are not impeded, whereas strong closure of the eyes or clenching of the fist is followed by a long delay in relaxation. The myotonia has been found to precede muscle atrophy by 2–3 or more years. The patient may be so accustomed to it, however, that he does not notice it. In children weakness precedes myotonia; the latter has not been observed in subjects less than 2–3 years of age.

The phenomenon of myotonia has the same EMG and pharmacologic characteristics as that of Thomsen's disease, and these aspects are discussed with that disease. In dystrophia myotonica, however, it is rare for the myotonia to be elicited from every part of the body musculature as it is in Thomsen's disease, and in many patients it is evident only in the hands, tongue, and facial musculature. Its relationship to the dystrophy of the muscles is not direct. For example, the tongue very commonly shows myotonia yet seldom becomes atrophic. The forearm flexors show strong myotonia after voluntary grasp of the hand, but the extensors atrophy first and usually show percussion myotonia better than the flexors. That myotonia can be lacking entirely in a family showing the typical distribution of muscular atrophy and cataracts was confirmed by Fearnsides.[60, 141] In other words there is no parallelism between the dystrophic and myotonic processes.

Myotonic dystrophy differs from the other forms of myopathy in several additional features, particularly in its association with obvious abnormality of endocrine glands. Testicular atrophy and either reduced libido or impotence have been noted in a considerable proportion of reported cases and are now an accepted part of the syndrome. Childless marriages are frequent. Hypothyroidism has also been described but with less frequency; the coldness of extremities, slow pulse, and loss of hair are often attributed to this. Feeblemindedness and onset of the dystrophy during childhood tend to occur in the second generation. Mild functional changes in cardiac activity, chiefly of the nature of bradycardia with lengthened P-R interval, have been described by a number of observers and are reviewed by Thomasen,[141] deWind and Jones,[43] and Fisch.[62]

Greenfield[74] was the first to call attention to the high incidence of cataract in myotonic families; this association was closely studied by Fleischer[64, 65] and Vogt[151] in a long series of reports. The lenticular opacities are best seen by a slit lamp and have the following characteristics. Small, regular opacities are found in the anterior and posterior cortex of the lens just beneath the capsule; they are yellow, blue, and blue green and are highly refractile. They are found in a large proportion of young myotonic patients. A stellate cataract appears later in the posterior cortex and may be small and symptomless, taking years to develop. The usual kind of senile cataract is found in elderly patients. Cataract, at first senile and later presenile, may occur in the generations preceding the dystrophic generation. Some patients with myotonic dystrophy have no lenticular opacities.

Although we adhere to the view that congenital myotonia (Thomsen's disease) is a separate condition (Chapter 12) with other disorders of muscular excitability, myotonia can exist alone for many years before dystrophic symptoms appear in some members of a family suffering from dystrophia myotonica. In such a case only the family pattern can clearly differentiate these conditions, except that the myotonia is usually more restricted in the dystrophic case.

The recent excellent descriptions of families suffering from paramyotonia congenita by Drager et al.[45] and French and Kilpatrick[66] have done much to clarify the relationships of myotonia to dystrophy. These authors describe families suffering from the weakness and slowness of movement induced by cold that are typical of the condition first described by von Eulenberg under the same title. Myotonia is then evident in the face and tongue, but not usually in the limbs. The mode of inheritance differs from that of dystrophic myotonia, and dystrophic features do not occur. Some members of each family suffered periodic attacks of flaccid weakness that resembled those of periodic paralysis. Paramyotonia is, then, the most limited form of myotonia and Thomsen's disease the most generalized; and in each the inheritance is different from that of dystrophia myotonica (Chapter 12).

Peripheral vascular changes are common, the affected extremities becoming blue and cold. Such changes are also seen in the other types of dystrophy when the distal muscles are involved, so that we attribute them to the effect of paralysis of dependent parts rather than to any specific feature of the disease.

Dysphagia is attributed to weakness and atrophy of the striated part of the esophagus. It is dilated in X-rays taken after a barium swallow.

LATE DISTAL MUSCULAR DYSTROPHY (WELANDER'S DISTAL DYSTROPHY)

Late distal muscular dystrophy is a very slowly progressive myopathy, its onset occurring during middle life, with atrophy in the hands, forearms, and pretibial muscles. There is considerable evidence establishing the existence of a distinct subvariety that first affects the distal muscles of the limbs—named myopathia distalis tarda hereditaria by Welander,[160] who patiently collected and analyzed 249 cases. The distal myopathy of Gowers[73] was described at a time when dystrophia myotonica had not been clearly differentiated. Many reports were subsequently made by others, without histologic examination and with imperfect differentiation from peroneal muscular atrophy and familial forms of progressive spinal muscular atrophy. Milhorat and Wolff[110] reported in 1943 a family of 12 members affected by a "progressive muscular dystrophy of atrophic distal type," with one autopsy. The onset was between the ages of 26 and 43, and the patients were incapacitated by spread of the atrophic process to proximal muscles within 5–15 years. In Welander's large series the age at onset varied 20–77 years, with a mean of 47 years. The mode of inheritance was dominant. Purely myopathic changes were demonstrated in 3 autopsies and 22 biopsies. The characteristic onset involved weakness and wasting of the extensors and small muscles of the distal parts of the upper and lower limbs. This most commonly began in the hands. There were no fasciculations, pain, or sensory disorders. Myotonia was absent, and the EMG in 71 cases showed only the small unit potentials seen in dystrophy. Three of 244 subjects, at ages 77, 76, and 72, had cataracts that were considered of senile type. The remaining patients had no cataracts and no significant endocrine disorders. The course of the dystrophy was extremely slow, and though some wasting of proximal muscles was observed in some patients within 10 years of onset it was seldom severe. In others the atrophy did not spread to the proximal muscles even 15–44 years after its onset.

OTHER VARIETIES

From time to time one sees cases of muscular dystrophy difficult to classify under any of the types mentioned. One of these is infantile muscular dystrophy described above in relation to undifferentiated forms of infantile hypotonia. Many such patients prove to be suffering from Werdnig-Hoffmann's disease, which is a spinal atrophy and not a muscular dystrophy, though earlier writers debated its relationship.[138] The remainder are a heterogeneous group, of which a small number present all the characteristics of a muscular dystrophy.[27, 154] This type of muscular dystrophy, manifest at or soon after birth, often rapidly progresses to death within 2–6 years. Rarely a family is reported with only very slow progression compatible with survival to middle life[143] or even to old age.[136] In another family studied by us[7] arthrogryposis was present in one child at birth, and a simple dystrophy in a second child; both children had great disability from a dystrophy that had clearly been present in utero. Such families have been reported so seldomly that no generalization regarding prognosis or inheritance seems possible at present.

It should also be remarked that a pure limb girdle dystrophy is occasionally observed in young girls at an age when only the male sex-linked Duchenne dystrophy is usually expected; or an otherwise typical Duchenne dystrophy with great enlargement of calves that is obviously pursuing an extremely slow course may be encountered during adolescence or early adulthood. This has been called the Becker type of pseudohypertrophic muscular dystrophy. Probably this is not the same disease as Duchenne's dystrophy because, unlike the latter, it is frequently associated with color blindness. This suggests a different locus of the abnormal gene on the X chromosome.

Marked hypertrophy of quadriceps may stand as the first sign of a uniquely restrictive muscular dystrophy with atrophy and weakness following some years later. An atrophic form of dystrophy may occasionally affect one arm or leg only for many years, but such asymmetries are more typical of progressive muscular atrophy.

Difficulty is also experienced in the classification of certain families like the one reported by Turner[143, 144] in which six members presented the syndrome of amyotonia in infancy, with very slow wasting of neck and shoulder muscles and weakness of the hands; there was no affection of the face, ptosis, or mention of cataract later in life. In an autopsy of one of these cases rows of central nuclei in normal muscle fibers were seen, as well as atrophic muscle fibers in random distribution. This last feature, the familial but not hereditary incidence, and the loss of tendon reflexes long before detectable atrophy, all suggest the probability of an infantile form of dystrophia myotonica. Milder forms of restricted scapuloperoneal dystrophy [77, 134] are also difficult to classify, though they have been described and a part of them surely belong to the group or relatively benign limb girdle dystrophies.

Universal dystrophy, in which all the skeletal muscles are affected, including those of the eyes, is another rare variety. In such a family, composed of three members of two successive generations (two males and one female), recently under the observation of the author, every skeletal muscle had become weak and more or less atrophic over a period of 10–15 years, beginning in adult life. There was no myotonia, and no evidence of cataracts or endocrine abnormality.

INHERITANCE

There have been numerous studies of the inheritance of the muscular dystrophies.[68] One of the most complete is by Bell,[16] whose report covers 1228 cases taken from the literature and the files of the National Hospital, Queen Square. Tyler and Stephens[147] reported on a Mormon family of 1249 members, of whom 159 were known to have facioscapulohumeral dystrophy. Three types of inheritance are manifest in these pedigrees: dominant, recessive, and sex-linked. The pseudohypertrophic muscular dystrophy is usually sex-linked, whereas the other forms tend to be either dominant or recessive.

Considerable attention has been given to identifying female carriers of Duchenne dystrophy. The creatine phosphokinase (CPK) levels are elevated in two-thirds of the carriers, and an even larger number are identifiable by the abnormally brief duration of action potentials in the EMG. Exercise increases the number of carriers revealed by the CPK values. Carriers of the Becker variant are not recognizable by these methods.

The recent work of Walton and Nattrass[158] showed that the true, halting, Landouzy-Déjerine type follows the pattern of a simple dominant gene whereas dystrophy following the pattern of an autosomal recessive inheritance usually takes the slowly progressive form described by Erb (juvenile muscular dystrophy). This excellent study also shows the frequency of mutation—spontaneous onset of dystrophy without family history—especially in the pseudohypertrophic type.

The family reported by Minkowski and Sidler[111] is a striking example of the operation of recessive factors in the production of myopathy in a small restricted community in Switzerland. These investigators found 13 cases of severe dystrophy in one generation of a family without previous history of the disease, except that the parents were both related to two known myopathic families 10 generations earlier. Rotach[126] reported on the next generation of this family, which presented two cases of moderately severe type (Leyden-Moebius), four of more slowly progressive, and one of mild stationary form of the disease. The last dilution of the gene by nonmyopathic families had therefore resulted in lessening of the severity and change in type of the affection. Statistical studies indicate that the sex-linked forms of the disease have a bad prognosis, and the dominant or recessive forms a more favorable one.[16] The mild late and distal forms usually have a dominant inheritance.

Several pedigrees of families with progressive ophthalmoplegia are on record. In two of them, the family described by Bradburne[26] in which there were 16 cases in five successive generations and the family observed by Faulkner[59] in which there were five cases in two generations, the inheritance was dominant. Males and females were affected alike. However, it is likely that the disease may also appear as a recessive trait, as it did in the exceptional family described by Gates[68] in which miosis and ophthalmoplegia were combined. In general the hereditary features of familial ptosis and dystrophic ophthalmoplegia are weak and uncertain.

Myotonic dystrophy, which affects both males and females, is also strongly familial. There have been numerous studies of the mode of inheritance, one of the most revealing of which is that of Fleischer[64, 65] which was later extended by Henke and Seeger.[79] The inheritance tends to be dominant with direct transmission from parent to child, but the dominance may be variable, for in some cases both parents are normal. The onset of the disease in the children is at an earlier age than in the parent (i.e., progressive inheritance or anticipation) and the severity increases in succeeding generations (i.e., potentiation).[2, 119] Fleischer, [64] Adie and Greenfield,[2] and Ravin and Waring[119] interpreted the sequence of events in any one family as being due to a dominant factor at first manifested by very slight signs such as senile cataract. In succeeding generations the cataract occurs earlier, and finally in one generation the full syndrome of myotonic dystrophy becomes evident. This is transmitted to the next generation, and it appears at an earlier age until it begins before the onset of maturity, causes sterility, and a termination of the disease in that family.

Congenital myotonia (Thomsen's disease) is usually present in several members of a family and can be traced through several successive generations, but from the fine study of Thomasen[141] it is clear that dystrophic features do not develop. We accordingly cannot classify congenital myotonia with the muscular dystrophies. The position of acquired myotonia is different. Most authorities regard it as a mild type of myotonic dystrophy, with all the potentialities of that disease (Chapter 12). The late onset of myotonic symptoms in an isolated member of a family, however, presents a difficulty that can be resolved only by the continued absence of cataract and of muscular dystrophy.

ETIOLOGY

Muscular dystrophy is conceded by all to be a group of hereditary diseases whose pathogenic mechanisms are unknown. The hereditary patterns indicate that both the severity of the diseases and the distribution of affected muscles in the individual patient are primarily determined by differences in the linkage of a faulty gene or genes. The meaning of such variation in linkage is not understood. Little is known of the way in which a genetic disturbance can produce atrophy and degeneration of muscle fibers at a certain period of life. One credible hypothesis postulates that through an inborn metabolic fault certain skeletal muscles, particularly those which develop first during fetal life (the axial and trunk musculature), cannot be sustained in a state of health.

BIOCHEMICAL AND METABOLIC CONSIDERATIONS

A concerted effort has been made within recent years to discover a fundamental biochemical defect or defects in the dystrophic muscle. Thus far, although several abnormalities in the chemistry of the muscle fiber have been described, their specific relationship to the dystrophic process has not been established.

CREATINE AND CREATININE METABOLISM

Historically the first biochemical defect to be noted in dystrophy was a decrease in urinary creatinine excretion. This change was observed by Rosenthal[124] in 1870 and later was corroborated by several others.[109] Subsequently in

1909 Levine and Kresteller[97] reported not only a diminished creatinine output but an increase in urinary excretion of creatine in this disease. The significance of this observation derives from the fact that creatine, an amino acid, is a prominent constituent of all striated muscular tissue. It may be ingested (exogenous creatine) but more commonly is synthesized in the liver from portions of three amino acids: glycine, arginine, and methionine.[109] It is delivered as preformed creatine to the skeletal muscles, which contain the largest amount of this compound of any organ or tissue (about 450 mg/100 g fresh muscle tissue). Creatinine is the anhydride of creatine (formed by the loss of one molecule of water) and is a normal degradation product of it. It is second in importance only to urea as an endproduct of amino acid catabolism. The creatinine content of muscles is low (about 5 mg%) since this compound is readly diffusible from the muscle fiber and eventually is excreted in the urine. The serum of a normal adult male contains creatine in amounts of 0.2–0.6 mg%; the levels are higher in the female (0.4–0.9 mg%). Creatinine serum levels range from 0.8 to 1.4 mg% and are increased only in serious renal disease. The renal threshold for creatine is about 0.5 mg%; hence none is excreted at lower serum levels. The average adult 24-hour urine excretion of creatine varies from 60 to 150 mg in normal males and is almost twice as high in females.[163] The daily excretion of creatinine is remarkably constant.[42] From one day to the next in an individual and among individuals it varies from 1.0 to 1.6 g.

In the muscle cell creatine is phosphorylated by the action of creatine kinase, with the aid of adenosine triphosphate (ATP), to form the high-energy compound creatine phosphate (phosphocreatine, PC), which is of vital importance in the contractile processes of the muscle fiber. The quantity of creatinine excreted daily in the urine is roughly proportional to the total body muscle mass because there is a spontaneous and predictable rate of degradation of creatine to creatinine, which is then completely disposed of by the kidney. Muscle contributes about 95% to this total. In actual practice this proportion is remarkably accurate in normal subjects as well as in those with one or other of the many wasting diseases.

With reference to creatine metabolism, the most characteristic features in progressive muscular dystrophy are (1) a decrease in creatinine content in the urine, (2) an increase in urinary creatine, (3) often a mild hypercreatinemia[109, 110] and (4) a low creatine phosphate and creatine content of dystrophic muscle fibers.[122] The first three alterations are *not* specific for muscular dystrophy but may occur in any situation in which there is a reduction of total body muscle mass such as in neurogenic atrophy, dermatomyositis, and amyotonia congenita. Hypercreatinuria may also occur in some diseases that do not primarily affect the muscles, e.g., hyperthyroidism, Addison's disease, and eunuchoidism in the male. From the studies of Shank *et al.*[135] it is evident that the decreased urinary output of creatinine is a more reliable indication of a diminished muscular bulk than are the urine and serum values for creatine since hypercreatinemia and hypercreatinuria may occur for any of four reasons:[100] (1) accelerated synthesis of creatine, so that creatine is introduced into the circulation from the liver at a rate in excess of its removal from blood by other tissues, especially muscle and kidney (methyltestosterone administration may stimulate such a sequence); (2) ingestion of large amounts of creatine; (3) wholesale destruction of muscle fibers or leak from them, as may occur in myoglobinuria, acute myositis, and perhaps rapid progression of muscular dystrophy; and (4) reduced muscle mass with inadequate space for disposition of creatine synthesized at the normal rate. The muscles of a normal subject are not saturated with creatine and if a test dose (1–3 g) is given orally, most is accepted by muscle, and little or no creatinuria results. In an individual with muscular dystrophy or another disease in which the muscle mass is lowered, there is impaired creatine tolerance because the remaining musculature is not capable of absorbing all the creatine and some spills out into the urine. Hence it can be concluded that the creatine tolerance test[110] is nonspecific and when positive indicates either a reduction in functional muscle mass or a saturation of muscle with creatine.

OTHER METABOLIC ABNORMALITIES IN MUSCULAR DYSTROPHY

Nearly all the general metabolic studies conducted in the dystrophic patient have yielded results that may be explained by the diminished bulk of the musculature (atrophy) or a net loss of muscle substance (i.e., destruction of muscle fibers). Metabolic balance determinations (dietary intake compared to excretory losses) have shown a negative balance (overall loss) of nitrogen, phosphorus, and potassium.[100] There is also an increased urinary excretion of amino acids[3] other than creatine. In muscular dystrophy an abnormal glucose tolerance curve (either high or flat) has been reported. However, this is not consistent and it was found to be normal by Tyler and Perkoff[146] and Danwski *et al.*[36] The latter workers noted that insulin and glucose, when given in combination, produce hyperglycemia in muscular dystrophy and hypoglycemia in control subjects. The inadequate disposal of glucose is compatible with the presence of a diminished muscle mass, although other explanations are also possible.

An abnormal quantity of the pentose sugar called ribose was recently reported as a feature of the dystrophic patient, but the ribosuria was not confirmed in another careful study.[123] Likewise, a low metabolic rate has been described in these patients, but when corrections were applied for the loss of muscle mass the metabolic rates approached normal. Danowski *et al.*[35] noted a wide range of results in their determinations of basal metabolic rate. Values for serum protein-bound iodine were in general higher in children with muscular dystrophy than in controls, and a few were even above the euthyroid range. The significance of this finding is not clear, but several explanations apart from the presence of muscular disease could be advanced for it.

Two recent studies of potassium and sodium concentrations in dystrophic patients have indicated that there is (1) a net decrease in the total body potassium and no change in the sodium (using the radioisotope dilution method[22]), and (2) a diminution in intracellular potassium and an elevation in sodium[164] when small-muscle biopsy samples were analyzed by the neutron activation technique using as a reference base the amount of noncollagenous nitrogen in the tissue. The significance of a lowered intrafiber potassium concentration is as yet uncertain but is likely to be the result of a secondary leakage of potassium from the diseased fibers, much the same as the intracellular enzyme leak from fibers (see below). It is probable that some potassium is linked to protein within muscle, and when these protein components are lost as part of the dystrophic process some potassium is displaced from the fiber. It is noteworthy that the serum levels of potassium are normal in muscular dystrophy.

SERUM ENZYME LEVELS

During the past several years considerable attention has been focused upon the increased activity of serum enzymes in various pathologic states. More than 20 reports have described an elevation of serum levels of one or several normally occurring intracellular enzymes in muscular dystrophy and certain other myopathies. In Table 6–1 are enumerated the various enzymes that have been measured up to the present and the trend in serum values usually found. As a general rule, the enzyme levels tend to be higher in early, active dystrophy during childhood (especially with pseudohypertrophy), whereas levels are often lower or in the normal range in the more slowly progressive adult variety or in a far-advanced childhood form in which there is a great deal of muscular wasting.[31, 46, 115, 162] It is remarkable that serial enzyme determinations made in several patients over a 4-year period by Pearson showed only slight variations from one test to another, with a gradual tendency to decline as muscular wasting progressed.

An occasional sibling of an affected child has a very high level of serum transaminase (and perhaps other enzymes) when clinical evidence of dystrophy is barely detectable. Such a case was described by Pearson[116] who has now seen four such children, all of whom subsequently developed unequivocal evidence of myopathy. The clinically unaffected relatives of dystrophic children have normal enzyme values,[115, 162] which eliminates the possibility that there is a genetically transmitted defect that accounts for the elevated serum enzyme activity.

In diagnostic problem cases with muscular weakness the determination of one or several serum enzyme levels may be of considerable value (Table 6–2). Unfortunately these changes are not specific for muscular dystrophy and may also be found in active cases of polymyositis or in other situations in which considerable muscle necrosis is occurring, i.e., acute myoglobinuria, extensive muscle trauma. Normal levels are usually present in muscular atrophy of spinal or peripheral nerve origin, even though the muscle wasting is proceeding

TABLE 6–1. Serum Enzymes in Muscular Dystrophy

Enzyme	Serum value	References
Creatine phosphokinase	Usually elevated	Ebashi et al.;[51] Pearson[116]
Aldolase	Usually elevated	Aronson & Volk;[5] Rowland & Ross;[127] Schapira et al.;[131] Thomopson & Vignos;[142] White & Hess;[161] White[162]
Phosphohexoisomerase	Often elevated	Pearson;[116] Schapira et al.;[132] White[162]
Lactic dehydrogenase	Usually elevated	Pearson;[116] Schapira & Dreyfus;[130] White[162]
Isocitric dehydrogenase	Rarely elevated	White[162]
Glutamic oxalacetic transaminase	Usually elevated	Dreyfus et al.;[47] Pearson;[115] White[162]
Glutamic pyruvic transaminase	Usually elevated	Pearson[116]
Acid phosphatase	Not elevated	White & Hess[161]
Alkaline phosphatase	Not elevated	White & Hess[161]

TABLE 6–2. Muscle Enzyme Activity in Duchenne Muscular Dystrophy

Serum

Increased	Normal
Aldolase	Acid phosphatase
Phosphohexose isomerase	Alkaline phosphatase
Lactic dehydrogenase	Isocitric dehydrogenase
Transaminases	DNAase
Malic dehydrogenase	Dipeptidases
Creatine phosphokinase	ATPase
	5-Nucleotidase

Muscle

Decreased	Normal
Creatine phosphokinase	Lactic dehydrogenase
Aldolase	Hexokinase
Phosphorylase	Cytochrone oxidase
Phosphoglucomutase	Succinoxidase
Phosphohexose isomerase	Succinic dehydrogenase
Malic dehydrogenase	Aconitase
Myosin	ATPase
Adenylic deaminase	Glycerophosphatase
Phosphodiesterase	Transaminase
	Phosphohexokinase
	5-Nucleotidase

rapidly, as well as in myotonia dystrophica, myasthenia gravis, and peroneal muscular atrophy.[47, 115, 127] Exceptions have been reported particularly in certain cases of progressive spinal and neuropathic atrophy.

The explanation of the serum elevations of these intracellular enzymes is not clear at the present time. In muscular dystrophy there appears to be a correlation with frank necrosis of muscle fibers. But in hypothyroid myopathy, where the CPK and aldolase are often elevated, there is no necrosis. Furthermore, in acutely denervated muscle (e.g., poliomyelitis) the rate of loss of muscle bulk is infinitely more rapid and yet the serum enzyme levels remain normal.[115] Perhaps there is an increase in permeability of the muscle membrane that permits a steady leakage into the serum. If this is the case, there must be a high turnover rate of many intracellular enzymes (and perhaps of other elements also) in dystrophic muscle because tissue enzyme analyses[47] have shown a

reduction in intracellular enzyme values that is only moderate and not sufficient to account for the total cumulative loss in the serum.

Thus one should not conclude from this discussion that an elevation in serum enzymes is specific for human muscular dystrophy or that their loss from the muscle fiber causes the weakness and wasting. Rather their appearance in the serum most likely results from a more fundamental defect in the dystrophic fibers that is yet to be specified. In confirmation, serum enzyme elevations have now been reported in several myopathies of animals, including the mouse,[159] chicken,[32] sheep, and calf.[23]

GLYCOLYSIS IN DYSTROPHIC MUSCLE AND OTHER CONSTITUENTS

Preliminary studies in three laboratories have disclosed that there is a quantitative deficiency of several intrafiber enzymes in muscular dystrophy (Table 6–2). In all of these studies the reference base for determining the quantity of muscle tissue in the specimen has been the amount of noncollagen protein present. Dreyfus et al.[46] demonstrated a true decrease in the glycolytic enzymes phosphorylase (both phosphorylase A and B) and aldolase, the activity of each of which was only one-third that of controls. The overall glycolytic (or glycogenolytic) activity of the dystrophic tissue was likewise reduced to about 30% of normal. These authors point out that a reduction in lactic acid production, usually an indication of decreased glycolytic activity, may also occur in denervation atrophy after several weeks.

Ronzoni et al.[121] analyzed biopsy material from 12 dystrophic patients and found several abnormalities in the dystrophic material that were not known to occur in atrophic muscle. They showed that the failure of glycolysis was not due to any lack of substrates or of the coenzymes ATP and DPN. Hexokinase, the primary enzyme in the utilization of glucose by muscle, was present in normal concentration. In contrast, phosphorylase was greatly reduced, as was aldolase. Both of these enzymes are known to be high in the blood serum of dystrophic patients. Aldolase, although present in only small amounts, was sufficient to allow later steps in the glycolytic cycle to continue. Lactic dehydrogenase, the final enzyme in the cycle, was present in normal or higher than normal amounts. Ronzoni *et al.* observed a difference between dystrophic and atrophic muscle; in atrophy all of these protein enzymes were reduced in proportion to the level of water-soluble protein, whereas in dystrophy only certain specific ones were diminished. The loss in phosphorylase and aldolase may be due to changes in the permeability of the membrane of the muscle cell. These workers also found in dystrophic muscle a disproportionately low creatine content and reduced lactic acid production (less than would be expected from the water-soluble protein content). There was no fault in creatine production but rather a failure in PC formation due to a lack of its formative enzyme, creatine kinase. There was no lack of ATP, which is the essential energy source for the synthesis of PC.

In another study Vignos and Lefkowitz[150] confirmed a markedly reduced glycolytic activity in abdominal muscles in 11 cases of juvenile dystrophic muscle and in 7 of neurogenic atrophy. Surprisingly, however, glycolysis in five or six muscles from cases of adult dystrophy was normal. They too obtained low levels of creatine kinase. The myosin content was disproportionately low in muscular dystrophy. In contrast, succinic dehydrogenase, a representative enzyme of aerobic respiration, was not reduced. The results of this important study are summarized in Table 6–3. These authors further observed that the fat content of the rectus abdominis muscle in dystrophy

and denervation atrophy was similar in amount and only slightly increased. This result, of course, does not correlate with the significant increase in fat that is seen grossly and histologically in many cases of dystrophy. The explanation for this discrepancy, in our opinion, lies in the fact that the rectus abdominis is rarely involved to any extent in the pseudohypertrophic process. It is likely that if a limb muscle in pseudohypertrophic dystrophy had been sampled, a significant increase in fat content and possibly an even more striking alteration of enzyme activity would have been found.

Horvath and Proctor[82] measured a number of tissue components in biopsies of 14 patients with progressive muscular dystrophy and compared their findings with those obtained in normal human and mouse muscle. They found an increase in total solids, fat, collagen, sodium, and chloride and a decrease in muscle proteins in the dystrophic tissue, consistent with the degree of muscle tissue replacement by fat and fibrous connective tissue. The potassium concentration of the residual muscle was lower than would have been expected on the basis of the calculated amounts of noncollagen protein or myosin. The protein composition of the remaining muscle was found to differ from normal muscle in that there was a disproportionate increase in water-soluble protein and the myosin- and alkali-soluble fractions showed great variations (sometimes increased and again decreased) without evident relationship to the residual muscle mass. In this respect the dystrophic muscle was found to differ from atrophic muscle.[78]

TABLE 6–3. Summary of Biochemical Determinations on Muscle from Dystrophy and Neurogenic Atrophy

Constituents or activity measured	Juvenile dystrophy	Adult dystrophy	Neurogenic atrophy
Creatine kinase	Marked reduction	Minimal reduction	Marked reduction
ATP	Moderate decrease	Normal	Marked decrease
Succinic dehydrogenase	Normal	Normal	Minimal reduction (?)
Glycolytic activity	Marked rduction	Normal	Marked reduction
Myosin	Minimal reduction	Moderate reduction	Normal (?)
Collagen nitrogen	Marked increase	Moderate increase	Marked increase
Water (% of wet weight)	Normal	Normal	Normal

From Vignos and Lefkowitz.[150]

Other studies have revealed a decrease in glycolytic enzymes in dystrophic muscle and impaired production of lactic acid during ischemic exercise. This is believed to be secondary, as is the late decrease in oxidative enzymes. The mitochondria of the dystrophic fiber are normal, but some of the tubular fragments of endoplasmic reticulum are morphologically altered.[117]

It is obvious that relatively few significant biochemical data have been secured and that it is impossible at this time to denominate the specific metabolic defects in muscular dystrophy.

HISTOCHEMICAL OBSERVATIONS

A series of histochemical studies of normal and diseased muscle have been conducted in recent years in Bourne's laboratory. In dystrophic muscle fibers, fat, glycogen, disulfide and sulfhydryl groups, and amino groups are not significantly altered in content or distribution (see references by Beckett and Bourne in Chapter 1). The cholinesterase activity of motor end-plates is well preserved in dystrophic muscle,[12] and in fact remnants of muscle fibers embedded in dense connective tissue or fat often give a strong cholinesterase reaction, suggesting to the investigators that the presence of this enzyme might retard degeneration of that segment of the muscle fiber. Two oxidative enzymes, succinic dehydrogenase and cytochrome oxidase, are not significantly altered,[12] except that there is a progressive reduction of both enzymes as advanced atrophy and fibrosis develop. Alkaline and acid phosphatase are not present in normal or dystrophic muscle to any extent, with the exception that the invading connective tissue and reactive cells show a high level of the acid enzyme.

A phosphorolytic enzyme, 5-nucleotidase, was found in high concentration in the interstitial connective tissue, reactive cellular elements, and capillary walls in dystrophic muscle, whereas it is present only in capillary walls in control muscle.[13] Since this enzyme has the ability to dephosphorylate adenosine monophosphate (adenylic acid), which is the starting point in the resynthesis of ATP, its presence may be of considerable significance. In fibers undergoing degenerative atrophy the enzyme had diffused throughout the atrophic portion.

In another publication Bourne and Golarz[24] extended their observations on 5-nucleotidase to the connective tissue of dystrophic muscle. When a large group of high-energy phosphate compounds (ADP, ATP, PC, and the like) were used individually as substrates on muscle sections, it was found that all except PC were actively dephosphorylated by the enzyme. Control muscles showed no such ability. Intermediate glycolytic compounds such as glucose-6-phosphate were not affected. Of the several coenzymes tested, three (DPN, TPN, and thiamine pyrophosphate) were dephosphorylated. The authors speculated upon the possibility that the fundamental defect in muscular dystrophy might not be in the muscle fiber but in the connective tissue which supports it. A more likely explanation is that it merely reflects the proliferative activity of the connective tissue, believed by most pathologists to be a secondary event.

Regrettably, systematic application of some of histochemical methods for fiber typing (Chapter 1) have proved to be relatively uninformative. In Duchenne dystrophy, although the ATPase stain at pH 9.4 may no longer clearly distinguish types I and II A and B fibers, all are agreed that the disease attacks all fiber types indiscriminately. The same is true of the limb girdle and facioscapulohumeral dystrophies, but strangely in myotonic dystrophy the type I fibers undergo atrophy, at least in the early stages of the disease. In facioscapulohumeral dystrophy where huge hypertrophied fibers are frequent they are usually of the type II group.

To summarize, the following biochemical abnormalities have been demonstrated in dystrophic human muscle: decreased glycolytic enzymes and relative normality of mitochondrial oxidative enzymes; increased cathepsins (lysosomal enzymes); decrease in isoenzyme V of lactic dehydrogenase; decreased muscle CPK; decreased myoglobin; smallness of polyribosomes; decreased total body and muscle potassium; increased serum CPK, aldolase, LDH, SGOT, SGPT; mild aminoaciduria; decreased protein synthesis; creatinuria and short half-life of labeled creatine; increased serum α_2-globulin and polysaccharides; and pentosuria. There is no agreement as to which if any of these are of primary importance in indicating the basic biochemical defect that leads to muscle necrosis or premature degeneration of its constituent fibers under conditions of natural activity.

PATHOLOGIC HISTOLOGY

In spite of overt differences in the course and distribution of the six "classic" types of muscular dystrophy, the essential changes in the muscles are similar in all, except that in myotonic dystrophy central nuclear proliferation and myofibrillar disorganization are more prominent. The histopathology has been described in detail and discussed fully by Erb,[57] Pick,[118] Durante,[50] Marinesco,[105] Jendrassik,[87] and more recently by Wohlfart and Wohlfart,[166] Slauck,[137] Curschmann,[34] Hassin,[76] Black and Ravin,[21] and Walton and Gardner–Medwin.[157]

In the cases of classic dystrophy that we examined at autopsy, the appearance of the muscles was characteristic and corresponded to that seen by others. In pseudohypertrophic muscular dystrophy the enlarged gastrocnemii look rather pale, reflecting their large content of fat. (In true hypertrophy at an early stage the muscles retain their dark red color.) Other muscles are unnaturally small and vary in color from yellowish to pinkish gray. Their pale translucent appearance has been likened to "fish flesh" by some observers. This alteration of color depends on the relative amounts of fat and fibrous tissue that replace the muscle fibers. Their high fat content causes them to float in water.

SIZE OF MUSCLE FIBERS

Probably the single most important feature of muscle dystrophy is the reduction in the number of muscle fibers. In a severely affected muscle no recognizable fibers may remain—or at most only a few of them—in a vast field of connective tissue and fat cells. This loss of muscle fibers appears to result from repeated segmental necrosis and exhaustion of the regenerative powers of fibers.

The remaining fibers in any given muscle specimen reveal variation in size; and there is a large amount of connective tissue and fat cells which led to the names myosclerotic atrophy and lipomatous myoatrophy in early medical writings. Both very large and very small fibers are scattered in haphazard arrangement throughout the muscle at all stages of the disease (Fig. 6–1). Examination of muscles that are only slightly affected and the occasional observation that enlargement and firmness of muscles may later give way to atrophy suggest that hypertrophy is the initial change in any particular muscle fiber and precedes degeneration or atrophy by months or years. The largest fibers may attain a diameter of up to 230 μ. In cross section the enlarged fibers have a rounded rather than the usual polygonal contour. Other fibers are smaller than normal and usually tortuous. Fibers as narrow as 10 μ may appear to have otherwise normal structure. Thus the reduced number of fibers in each motor unit under the light microscope accounts for the relatively small and more brief action-potential spike in the EMG. In some large fibers the earliest change appears to be homogenization or hyalinization of the fiber with some obscuration of the striation; many of the thin fibers also have this appearance. It is an alteration of uncertain significance, however, and may be due to a peculiarity of fixation. The staining reactions of such fibers resemble normal muscle in other respects. In some fibers the myofibrillar bundles are unusually dense and separated by clefts, an appearance which also results from mild degrees of physical damage during preparation of the sections. Actual splitting of muscle fibers into daughter fibers, each with distinct sarcolemma but lying within the same endomysial tube, is a common feature in all dystrophies; and in places five or more fibers may be seen bound together with few or no endomysial partitions (Fig. 6–2). The enlargement of fibers explains true hypertrophy in the early phase of dystrophy. It is

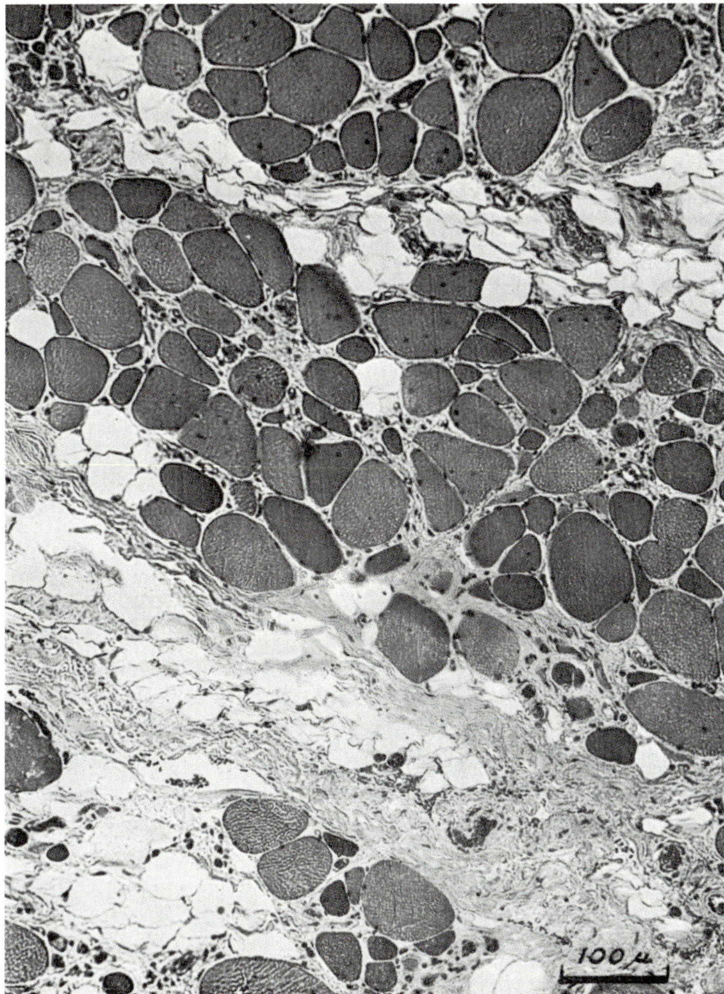

Fig. 6–1. Pseudohypertrophic muscular dystrophy. Section taken from enlarged calf muscle of 11-year-old boy whose brother was similarly affected. (iron hematoxylin, Van Gieson)

not the cause of the great augmentation in muscle bulk in pseudohypertrophic dystrophy, for it is offset by the smallness of many other fibers. Instead, the increase in endomysial fat cells is responsible for the abnormal size (Figs. 6–3 and 6–4).

NUCLEI

In all types of dystrophy the sarcolemmal nuclei of both the swollen and atrophied fibers are increased in number and are larger and more varied in shape than normal. The difference from normal fibers is slight, however, though the nucleoli are often more prominent. The large vesicular oval nucleus characteristic of intense regeneration is infrequent except in some cases of rapidly advancing childhood dystrophy. In many of the smallest fibers some nuclei stain darkly and are shrunken. In some fibers rows of sarcolemmal nuclei may assume a central position in the fiber. This is more characteristic of myotonic dystrophy than the other varieties,[2] and long chains of central nuclei are prominent in many otherwise unaltered fibers in that disease. If such chains are of any length, the nuclei tend to be shrunken and stain irregularly, appearing inactive and degenerate; under these circumstances there is no increase in sarcoplasm. The completely

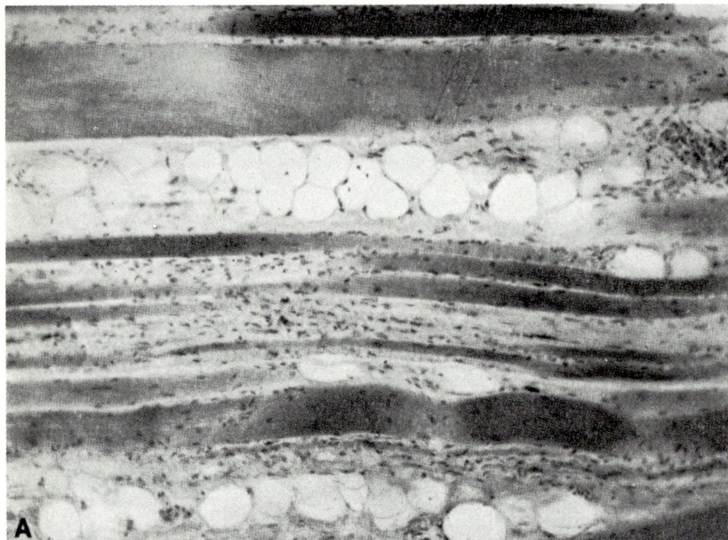

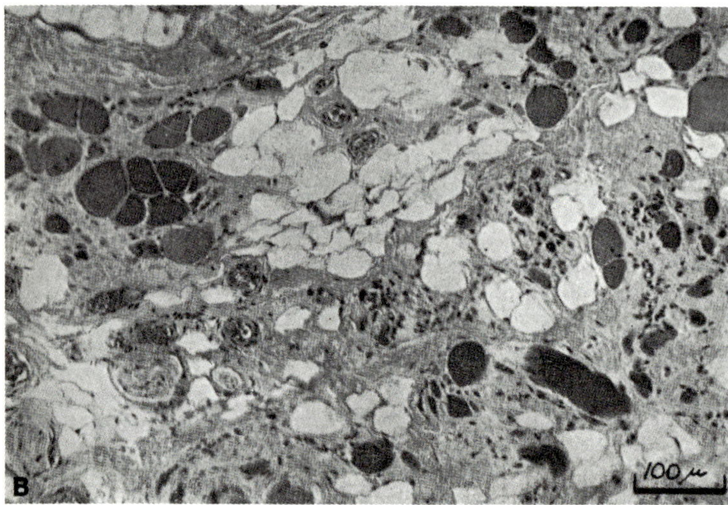

Fig. 6–2. Pseudohypertrophic muscular dystrophy. (A) Various terminal stages. A thin fiber (just below center) has broken into chains of dark small nuclei, some with and some without cytoplasm. (B) Transverse section of late stage showing fiber splitting, very small fibers, and nuclear residues. (hematoxylin, Van Gieson)

atrophic fibers are represented by clumps and chains of darkly staining nuclei (Figs. 6–2, 6–3, 6–5, and 6–6A), which, as after any other type of muscular atrophy, ultimately come to lie in rows in the connective tissue with no visible cell membrane. In places groups of dark pyknotic nuclei lying next to such a row or clump of residual muscle nuclei appear to be drawn into the wall of a fat cell (Fig. 6–19B). We have no evidence that such muscle nuclei are phagocytic at any stage, as has been maintained by Pick[118] and Hassin.[76] The fate of the nuclei appears to us to be the same as that in neural muscular atrophy.

DEGENERATION OF FIBERS

As the myopathic process advances, more and more of the muscle fibers atrophy and disappear. How this happens is a question of fundamental importance. Hyaline or granular degeneration is found in single fibers or groups of them and is followed by a regenerative response. Ultimately, however, this fiber is completely destroyed or replaced by rows of fat cells that come to lie between the muscle fibers, breaking up the homogeneity of the fasciculus. Fat is not deposited in the interstitial tissue by the degenerating fibers. The smallest fibers ap-

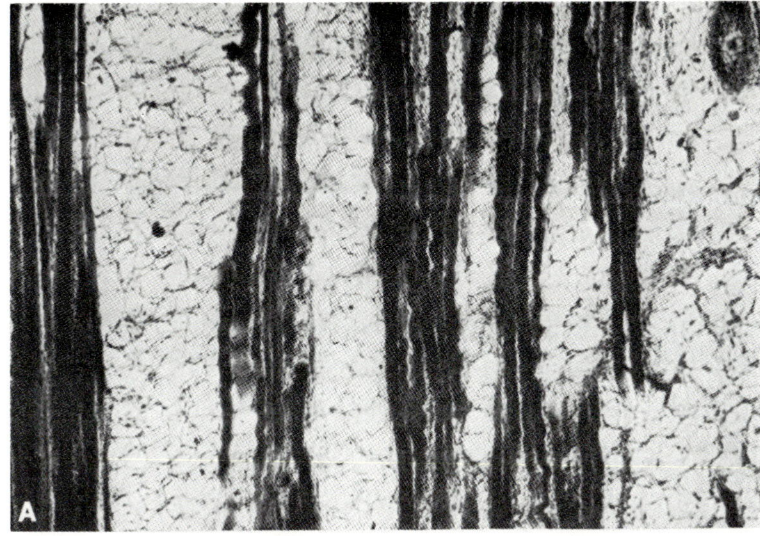

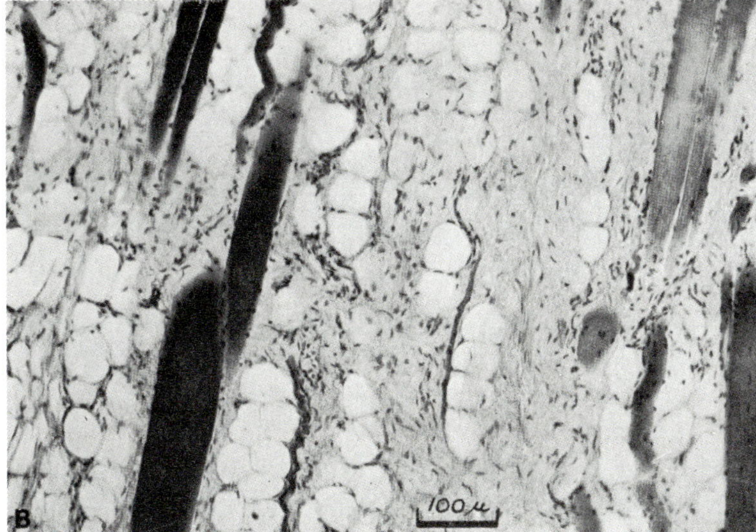

Fig. 6–3. Pseudohypertrophic muscular dystrophy. (A) Longitudinal section from large calf muscle showing small fasciculi and large accumulation of fat. Though most muscle fibers are small, atrophy is much less advanced than in Figure 6–2. (B) Section of muscle showing advanced changes. There are both very large and very small fibers, the latter distorted by fat cells. Nuclear remnants are seen in places. (A, silver impregnation, hematoxylin counterstain; B, hematoxylin, Van Gieson)

pear destined to undergo the type of fragmentation not unlike that seen in atrophy following denervation. Shrinkage and irregular, dark staining of nuclei appears to precede fragmentation of the fiber. Myoblastic spindle cells, each with a single nucleus, are occasionally found where a small fiber is disintegrating (Fig. 6–2A), and these must be regarded as regressive changes. Such fragmentation into degenerating myoblasts is a remarkable characteristic of mouse dystrophy (Chapter 3). In tissue culture of human dystrophic muscle, Geiger and Garven[69] found that the myoblasts

derived from biopsy specimens of patients suffering from muscular dystrophy become large and fail to proliferate. Others[20] dispute this claim and find no difference between normal and dystrophic muscle growing in artificial media. In infantile muscular dystrophy all the affected fibers appear to remain for a long period in their fetal form with oval plump central nuclei.

Occasionally fragments of slightly atrophic fibers are found, each surrounded by a sarcolemma (Fig. 6–6A). In this event the absence of the phenomena of regeneration is

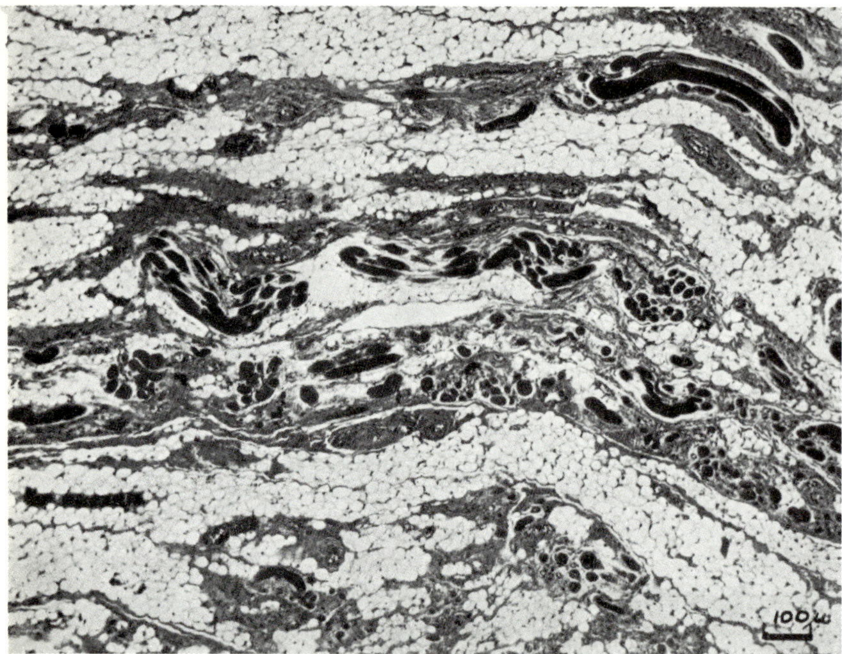

Fig. 6–4. Pseudohypertrophic muscular dystrophy. Section of biopsy specimen from calf muscle in which shrinkage during fixation has led to tortuosity of muscle fibers. (hematoxylin, eosin)

striking.[40] Necrosis of muscle fibers occurs but rarely in the more benign, chronic dystrophies, but in the Duchenne form it is readily demonstrated. Such fibers have flocculated eosinophilic contents and seem to be invaded by histiocytes. In this circumstance the relative paucity of regenerative changes is notable and may be viewed as a characteristic of the dystrophic process.[39] The lack of cellularity that is so characteristic of dystrophic muscle and distinguishes it sharply from polymyositis is also in part due to the inadequacy of regenerative changes.

Denny-Brown[39] found in two cases that if the site of a muscle biopsy is examined by a second biopsy 10–14 days later the cut ends of the muscle fibers exhibit degenerative changes with loss of myofibrils, without evidence of regeneration of muscle buds or new myofibrils (Fig. 6–6A). This phenomenon is particularly evident in the muscular dystrophy of mice (Chapter 3). There is no general agreement as to the meaning of this poverty of regeneration.

Denny-Brown regards it as a fundamental feature of the dystrophic process; but the author has the impression that regenerative changes, as seen in microscopic sections (proliferation of sarcolemmal nuclei, basophilia of sarcoplasm, increase in polyribosomes, poor alignment of myofilaments, sparse mitochondria, and branching of fibers), are always present in proportion to the degree of degeneration, and that the dystrophic muscle fiber cannot be distinguished from the normal fiber in respect to its reaction to injury. This opinion is based on a study of biopsy material obtained 10 days after deliberate chemical injury from seven patients with muscular dystrophy who showed such changes in all except the most atrophic fibers.[156] The observations of Baloh *et al.*[6] corroborate this impression.

Bands of connective tissue and fat cells separate the atrophic muscle fibers in the later stages of all forms of myopathy. It is not certain on the basis of microscopic sections whether this is a replacement fibrosis by the

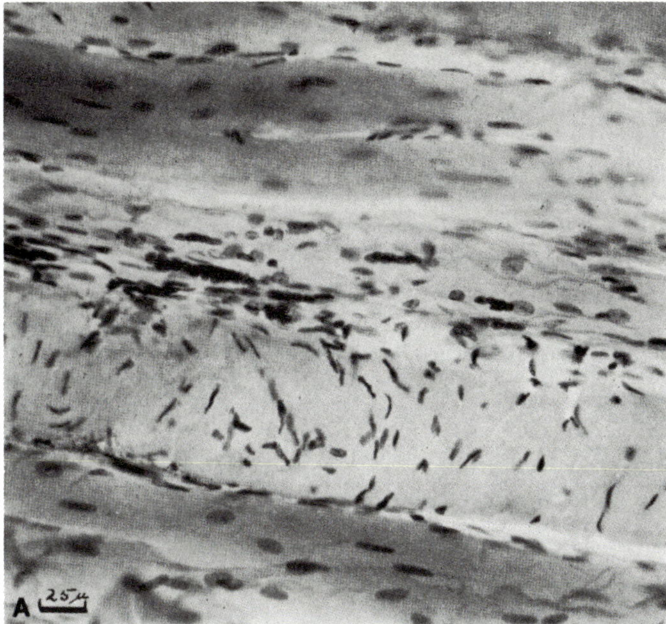

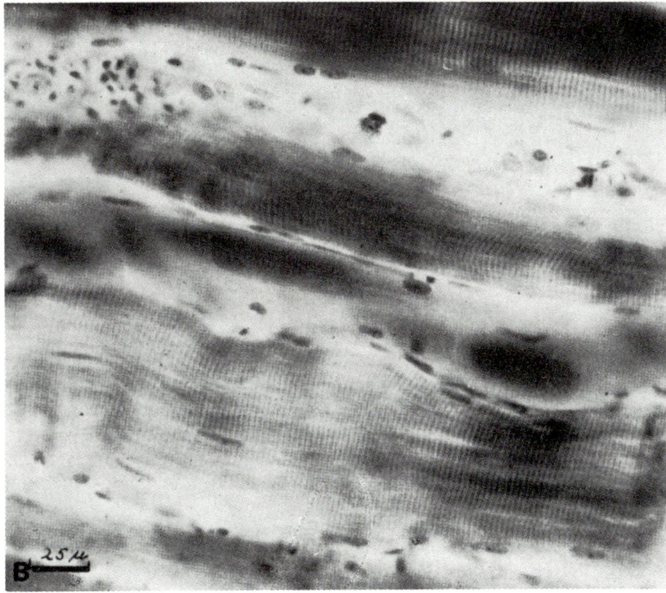

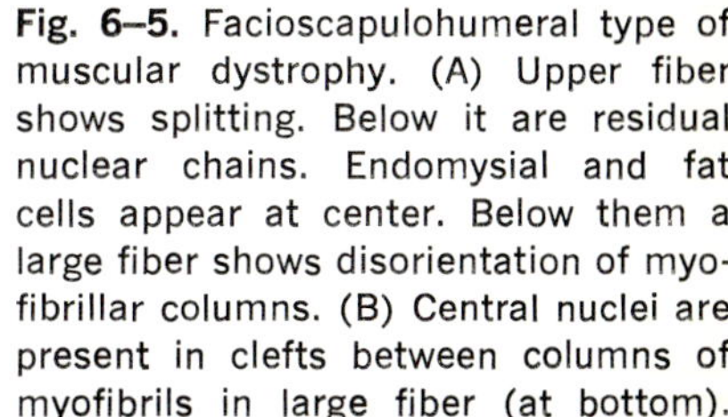

Fig. 6–5. Facioscapulohumeral type of muscular dystrophy. (A) Upper fiber shows splitting. Below it are residual nuclear chains. Endomysial and fat cells appear at center. Below them a large fiber shows disorientation of myofibrillar columns. (B) Central nuclei are present in clefts between columns of myofibrils in large fiber (at bottom).

endomysial connective tissue, as has been claimed by some workers, or merely a condensation of endo- and perimysium. Metaplasia of degenerating muscle fibers to form connective tissue, while suggested by some, has not been convincingly corroborated. Active fibroblastic reaction is not observed at any stage of the disease, though it is evident that the fibrosis is an active process. Inflammatory cells are not prominent at any time, but occasionally a small collection of mononuclear cells may be found around a venule (Fig. 6–19A).

FATTY TISSUE

In most of the involved muscles hyperplasia of adipose tissue in the form of fat cells (lipoma-

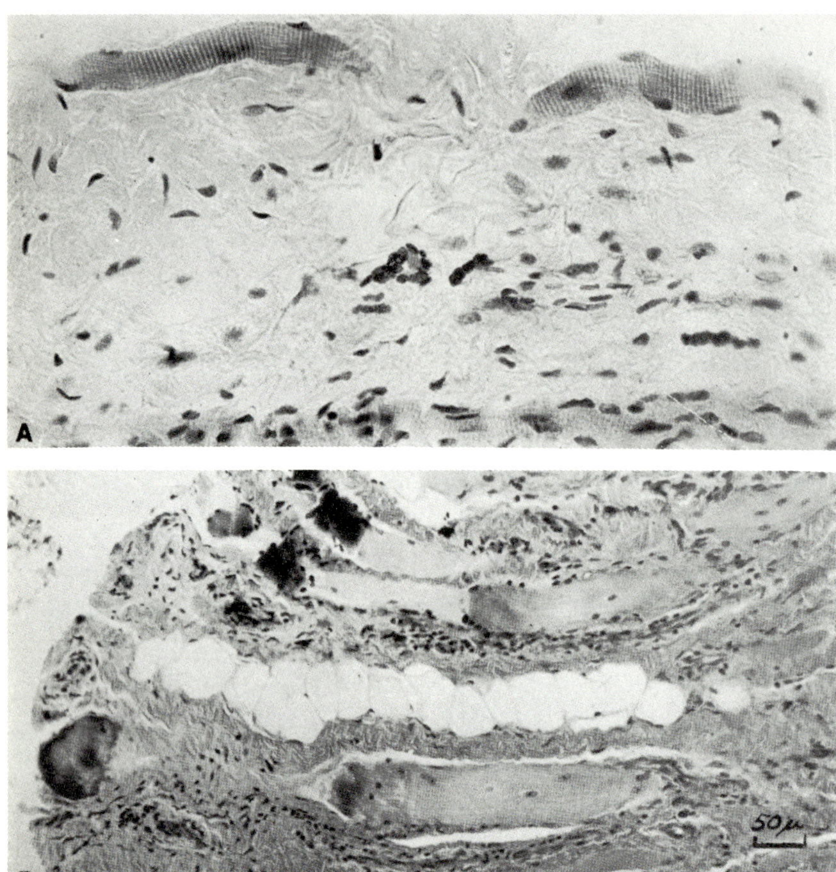

Fig. 6–6. (A) Pseudohypertrophic muscular dystrophy, showing details of muscle fibers in various stages of degeneration and fragmentation. (B) Mild arrested muscular dystrophy (Landouzy-Déjerine type). Biopsy section of biceps muscle taken 11 days after a previous biopsy showing muscular reaction at site of previous resection. Note retraction caps and regressive nuclear changes without evidence of regeneration. Sections stained with phosphotungstic hematoxylin showed loss of myofibrils. (A, hematoxylin, Van Gieson; B, eosin, methylene blue)

tosis) is in evidence. These cells may appear discretely or in rows. In the late stages of the disease there are only scattered atrophic muscle fibers in a mass of fatty tissue. Finally all the fibers disappear, leaving strands of connective tissue alone representing the former fasciculi (Figs. 6–20B and 6–23A). The origin of the fatty tissue is not known, but there is certainly more adipose tissue in dystrophic muscles in early and moderate stages of atrophy than in neural or spinal muscular atrophy of similar degree. The difference may be related only to the long duration of the process.

BLOOD VESSELS

Although thickened and with narrowed lumens, the intramuscular arteries in dystrophic muscle do not differ from those in denervative atrophy. Demos and Escoffier[38] and more recently Engel[54] who favor a vascular pathogenesis, offer controversial evidence of a circulatory disorder; the latter shows pictures of occluded intramuscular arteries, which we believed to be artifactual. Recent studies show no reduction in the circulation of the dystrophic limbs of patients and the small vessels are

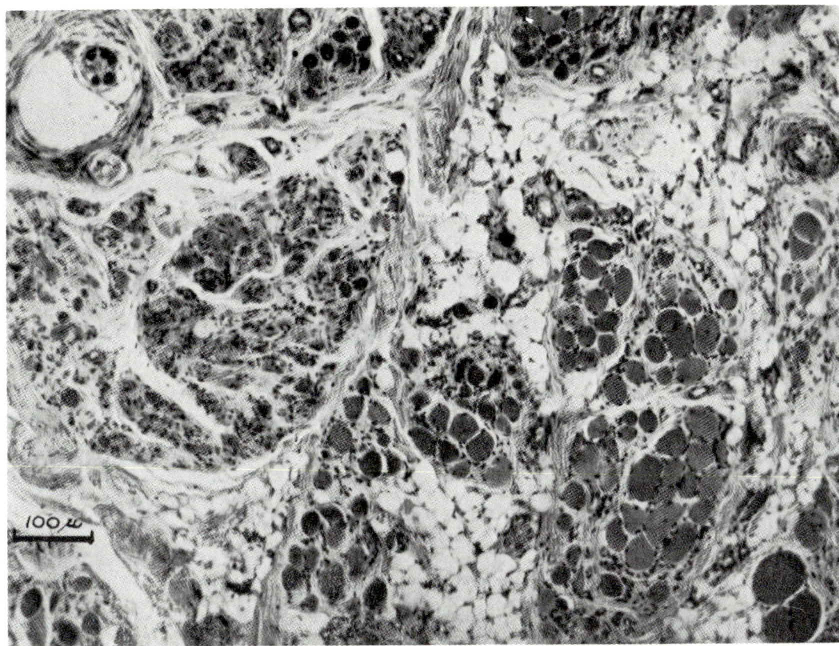

Fig. 6–7. Facioscapulohumeral (Landouzy-Déjerine) type of muscular dystrophy. Sections taken from sternomastoid muscle of 18-year-old patient who succumbed to infectious hepatitis. Note great variation in size of muscle fiber, interstitial fibrosis, and accumulation of fat. An intact muscle spindle is seen in upper left corner and two split muscle fibers to the right. (hematoxylin, Van Gieson)

Fig. 6–8. (A) Pseudohypertrophic muscular dystrophy. Preparation taken from same muscle as in Figure 6–1. Note motor endings on fibers to left and right. Nerve fiber to small atrophic muscle fibers presents a side branch just before reaching fiber. This branch also ends on muscle fiber, just out of focus. (B) Facioscapulohumeral type of muscular dystrophy. Nerve ending in coarse knobs on dystrophic muscle fibers and side branches wandering in connective tissue are seen. (A, Cajal method for nerve endings; B, Gros-Bielschowsky, hematoxylin)

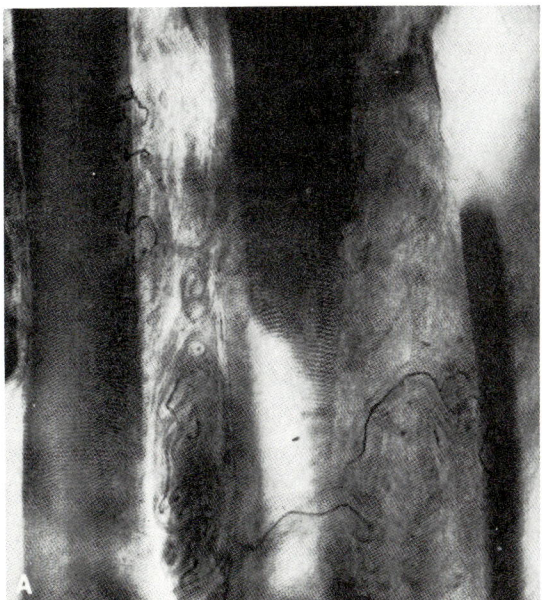

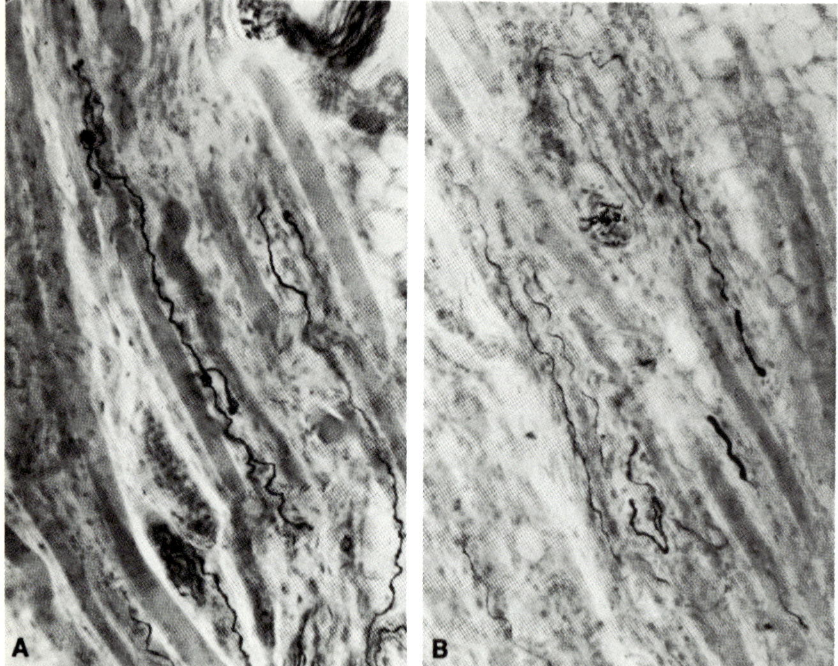

Fig. 6–9. Facioscapulohumeral type of muscular dystrophy. Same muscle as in Figure 6–10B, showing terminal knobs of motor nerve fibers ending blindly in connective tissue, and intact bundles of nerve fibers.

of normal appearance under the electron microscope.

NERVE FIBERS

The motor and sensory nerve fibers are not damaged. However, when the muscle fibers have reached an advanced state of degeneration the ramification of the motor end-plate does in fact show some atrophic changes consisting of shrinkage of the arborization within the sole-plate. Just before the motor axon enters the sole-plate, long filaments of it may bud into the endomysium (Fig. 6–8) as if seeking a new muscle fiber as the old one disappears.[25] When all or nearly all the muscle fibers have disappeared, only a few motor axons with clubbed terminations can be found in the connective tissue (Fig. 6–9). Muscle spindles are preserved (Fig. 6–7) even after all muscle fibers have disappeared,[9] and the same is true of sensory nerve fibers and other end-organs.

CENTRAL NERVOUS SYSTEM

There are no significant changes within the central nervous system. Nerve fibers within the muscle, the peripheral nerves, spinal ganglia, and spinal anterior and lateral horn cells are not affected to any significant degree. No numerical reduction in the anterior horn cells occurs even after the muscles have undergone virtually total degeneration. Similarly the lateral horn cells of the thoracic segments of the spinal cord are not affected at all, contrary to the report of Kuré and Okinaka.[91]

CARDIOVASCULAR SYSTEM

The intima and media of blood vessels within the muscles increase in cellularity and thickness, but to no greater degree than in other forms of muscular atrophy except possibly in myotonic dystrophy. Smooth muscle is not otherwise affected. The cardiac changes are less

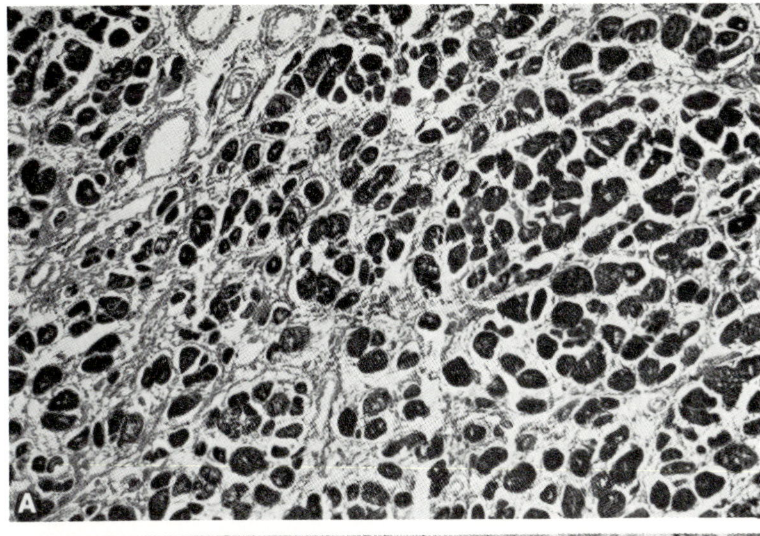

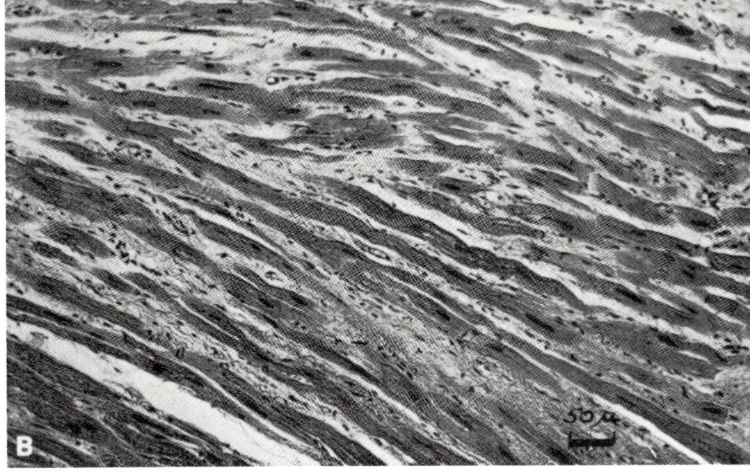

Fig. 6–10. Mild arrested muscular dystrophy (Erb type) with cardiomegaly. Sections of heart muscle showing myocardial fibrosis (A, phosphotungstic hematoxylin; B, phloxine, methylene blue)

clear-cut than those in skeletal muscle but are nonetheless definite. Cardiac dysrhythmias due to lesions of the conductive system are common.[28,43] Zatuchni *et al.*[167] described a case and summarized the reports of others.

The weight of the heart has varied from 140 to 600 g, and true hypertrophy is rare. The epicardial fat is increased, but fatty replacement of cardiac fibers is doubtful. Grayish flecks and small fibrotic areas are frequently present in the cardiac muscle. The endocardium rarely appears thickened, the valves are not involved, and the coronary vessels are unchanged except when there is coexistent, associated disease. Microscopically a fibrosis of the myocardium is the most striking

change, varying from finely diffuse sclerosis to large areas of scarring (Fig. 6–10). The muscle fibers vary in size and may show scattered degeneration with fragmentation and loss of striation. The nuclei in such areas show degenerative changes. No evidence of specific inflammatory reaction is present, but a patchy histiocytic cellular infiltration may be prominent. The condition of cardiac fibrosis thus described by Globus,[70] Cohen,[31] and Zatuchni *et al.*[167] presents some of the features characteristic of muscular dystrophy. The branched structure of the cardiac muscle fiber, however, obscures the great variation in fiber diameter that is so characteristic of the disease in skeletal muscle. Degeneration of Purkinje fibers was observed in one of our patients with heart block.

SPECIAL CHARACTERISTICS

Erb,[57] after reviewing these various histologic changes, concluded that the essential process was hypertrophy followed by atrophy and that it was associated with vacuolation, splitting of the muscle fiber, nuclear proliferation, increased endomysium, and transformation into fat cells (lipomatosis). Differences between the various types of dystrophy were related to accentuations of various aspects of this process. However, we already gave reasons for rejecting the idea of transformation of the highly differentiated muscle cells into fat cells.

The atrophic fibers often have granules that stain with fat stains, but these appear not to differ from those in some fibers in normal muscle or in atrophy following nerve section. Finkelnberg[61] described an interesting case in a child dying at 21 months of age after an onset of progressive dystrophic weakness at 11 months. Large muscle fibers in scattered groups of six to eight fibers were interspersed with uniformly small fibers of 3–10 μ in size. The large fibers were larger and more numerous than the B fibers of Wohlfart. The increase in endomysium and fatty transformation were already far advanced. Complete degeneration into chains of muscle nuclei had not occurred. The endomysial and fatty changes in this case strongly suggest that lipomatosis is a fundamental disorder of the endomysium. The affected muscle fibers had simply regressed to the fetal type. Krösing[90] is frequently cited as having established the possibility of transformation of muscle cells into fat cells, but reference to his data shows that he was confusing the fatty macrophages in acute necrosis with muscle cells.

We have not been impressed by vacuolation of muscle fibers as a feature of muscular dystrophy at any stage. Our experiments with mouse dystrophy (Chapter 3) incline us to the view that the large fibers represent a physiologic hypertrophy of relatively unaffected fibers. The

essential process that continues in dystrophic mouse muscle after it has been denervated is a progressive decrease in size associated with ultimate degeneration and fragmentation. Evidently fragmentation can occur in earlier stages when some event has accelerated the disease process; large or medium-sized fibers are then fragmenting, but very rarely with necrosis and phagocytosis. It is in this way that we imagine fragments of muscle fiber are left to undergo independent slow degeneration. It is therefore not surprising that occasionally (in our experience, rarely) the electromyographer finds a few isolated *fibrillating foci* in dystrophic muscle. Such foci should be discrete in true muscular dystrophy. Rarely such foci are so irritable that the insertion or movement of the EMG electrode evokes a whole series of small fiber potentials, but never so prolonged that confusion with true myotonia is possible. Reduction in voltage and duration of unit potentials remains the only regular EMG sign of muscular dystrophy and is obvious only when a great many motor units have lost some of their constituent muscle fibers.

The use of modern histologic techniques such as those involving the ultraviolet microscope has shed little additional light on the problem of the fundamental defect in the muscle fiber. Hoagland *et al.*[80] found that under ultraviolet light the sarcolemma, which is seen with greater clarity than by other techniques, exhibited an irregular absorption, giving it a frayed or eroded appearance. Myofibrillar disruption and areas of homogenization were seen in some fibers in muscles that had not yet been involved by the dystrophic process and which were said to appear normal in visible transmitted light. Excessive fraying of the endomysial connective tissue was also reported. The illustrations of some of these early changes are not altogether convincing, and we agree with Hoagland *et al.*[80] that a new technique of this kind must be applied to more control cases before artifacts can be excluded.

Such changes, particularly a vitreous homogenization, are claimed by Pearson[116] and Walton[155] to be the earliest ones seen with the light microscope. In infants who inherited Duchenne type muscular dystrophy, as many as half of the intact fibers underwent hyaliniza-tion even before degeneration began. We are unwilling to accept this change as valid be-cause it is so often found in biopsy material of all kinds, and such fibers cannot be followed into a later stage of phagocytosis.

The only myofibrillar change regularly found in muscular dystrophy has been poverty of staining in the terminal stage of fragmentation of the fiber. The ringbinden common in dys-trophia myotonica occur also in other forms of muscular dystrophy and are discussed with the other special characteristics of Steinert's disease.

Electron microscopic examination of dys-trophic muscle was undertaken by many in-vestigators with the expectation that the ultra-structure of the fibers might yield some clue as to the nature of the disease process. The results have been disappointing. Early in the course of disease when most of the fibers appear normal by light microscopy no definite changes are visible under the electron microscope. If any alterations are evident at this stage of the disease they consist of uninterpretable focal degeneration of myofilaments in a few of the sarcomeres of a myofibril. The thin myofila-ments of the I band appear to be most sus-ceptible to this change, and there may also be disorganization and streaming of Z bands. In more advanced stages of the disease the large muscle fibers are intact while atrophic ones show increased numbers of mitochondria and glyco-gen, Z bands that are imperfectly aligned, and accumulations of lipid droplets. Also some mitochondria may rupture or disintegrate or form peculiar crystalline inclusions, and there are many filamentous and myeloid bodies and other structures of uncertain origin. The plasma membranes of capillaries are thickened and of increased electron density; desmosomes, lipid inclusions, and membranous and lami-nated bodies appear in the cytoplasm of endo-thelial cells. The borders of muscle nuclei tend to be irregular, and cytoplasmic indentations give rise to intranuclear inclusions. Frankly necrotic fibers are reduced to amorphous masses in which phagocytes may be seen. Satellite cells are usually numerous where there is vigorous regenerative activity. No one type of alteration seems to antedate frank degeneration of the muscle fiber, and no ultra-structural lesion that corresponds to the early vitreous appearance described as pathologic by Pearson[116] has been identified. These several changes—which were documented by Hudgson and Pearce,[83] Fardeau,[58] and Mair and Tome[104] —are not in any way specific for muscular dystrophy.

SEVERE GENERALIZED MUSCULAR DYSTROPHY WITH PSEUDO-HYPERTROPHY (DUCHENNE)

In pseudohypertrophic muscular dystrophy the most prominent histologic feature is the oc-currence of groups of degenerating muscle fibers. Four or five fibers in a cluster disinte-grate over a longitudinal extent of a few sar-comeres, and the necrotic sarcoplasm is soon invaded by phagocytes. The process excites active regeneration as revealed by the forma-tion of satellite cells, myoblasts with mitotic figures, RNA-rich sarcoplasm, heavily nucleo-lated muscle nuclei, and protofibrils. Forking or branching occurs at these sites, and the newly formed shoots are thin and nucleated. The intervening connective tissue and blood vessels are not affected.

Another feature at a later stage is disruption of the muscle architecture by columns of fat cells. Fasciculi of muscle fibers may contain only 5–20 fibers, widely separated from the next fasciculus by fatty tissue. In places this striking pattern may be present when very few of the muscle fibers show severe atrophy (Fig. 6–3A). There is a slight increase in endomysial connective tissue. The blood vessels show no striking change. At this stage there may also be present an astonishing number of very small, but otherwise perfectly formed fibers exhibiting well developed striation. Most of these fibers run from one end to the other of the longi-tudinally oriented section; others are derived from the splitting of a single larger fiber, both or several branches remaining within a single endomysial sheath. These small fibers lie closely apposed to large hypertrophic fibers. Intensification of longitudinal striation owing to separation of the columns of myofibrils is a feature of some of the larger fibers. The increase of muscle nuclei is not great, and those centrally placed while frequent are less prominent than in myotonic dystrophy; the same is true for "rowing" of nuclei in longitudinal sections.

A few of the smaller atrophic fibers show poor transverse striation and have a moderate increase in the number of muscle nuclei. Vacuolation and hyaline and granular changes are a rarity. Only in the late stages are the characteristic rows of nuclei of extreme atrophy found (Fig. 6–6A), with rows of degenerating spindle cells. The motor nerve endings in one case examined by the author showed an unusually intricate arborization of the axis cylinder, but they were otherwise normal, even on very small fibers (Fig. 6–8A). A few knobs of axis cylinder ended freely in the connective tissue. Even at a late stage the amount of connective tissue was slight in relation to the large amount of fat. Splitting of the muscle fibers is sufficiently infrequent to require a careful search for its presence.

The effect of nerve section has not been reported, but Tyler[145] noted a case in which symmetrical hypertrophy of the calf muscles was first observed at the age of 3 years, and an attack of poliomyelitis involving only the right lower limb occurred at the age of 6 years. One year later the whole right limb was very atrophic, including the calf muscles, as compared with the left. Some bulk still remained in the calf muscle, however, and in the absence of any histologic study it must be assumed that the bulk of the muscle is made up of normally innervated large muscle fibers. It was these latter that had undergone denervation atrophy.

MILD RESTRICTED DYSTROPHY

In the mild restricted form of dystrophy degenerating fibers are rare and fibrosis is a notable feature, probably accounting for the liability to contracture noticed by Landouzy and Déjerine.[94] Such contracture, however, was present in only three of their seven cases. The increase in collagenous tissue is particularly evident in transverse section (Fig. 6–7, 6–11 and 6–12) in which the great thickening of the endomysium is at once evident. In parts of the muscles there is also an abnormal amount of fatty tissue but never to the degree observed in pseudohypertrophic dystrophy. The muscle fibers in the Landouzy-Déjerine and the Erb juvenile types also show a greater range of changes in each fasciculus. Huge fibers sometimes showing normal striation and nuclei, but more often pale and with central nuclei, lie mingled with atrophic fibers and nuclear remnants. Splitting of muscle fibers is moderately frequent (Figs. 6–5A and 6–11A) and is particularly obvious in transverse section, in which it is seen as two or three fibers within one endomysial ring. Vacuolation is occasionally seen, and pronounced longitudinal striation with unraveling of the myofibrils is frequent (Fig. 6–5A). This appearance is evidently the result of intrasarcolemmal rupture, for we have observed earlier stages of its development in myotonic dystrophy. All degrees of simple atrophy, with loss of striation, increasing cellularity of sarcoplasm, and spindle cells (Fig. 6–11A), may be found. The most atrophic areas of the muscles show strands of fatty tissue with dense bands of fibrous tissue containing residual chains and clumps of muscle nuclei, and here and there an intact muscle fiber. The motor nerve endings were found by Denny-Brown[39] to show early thickening of the terminal arborization, with side branches just before the axis cylinder reaches the motor plate (Figs. 6–8B and 6–9). These side branches wander irregularly in connective tissue, ending in bulbs, some of which become large and encapsulated. Such changes are present in the axis cylinders of hypertrophied muscle fibers. In more atrophic muscle fibers the termination of the axis cylinder in the end-plate and the abortive preterminal branches both become clubbed (Fig. 6–8B), and finally the axis cylinder ends in a large bulb in the connective tissue, no longer connected with the remnants of the muscle fiber (Fig. 6–9). As in pseudohypertrophic dystrophy, the nerve plexuses show no degeneration, and the muscle spindles, free endings, and autonomic nerve fibers remain intact.

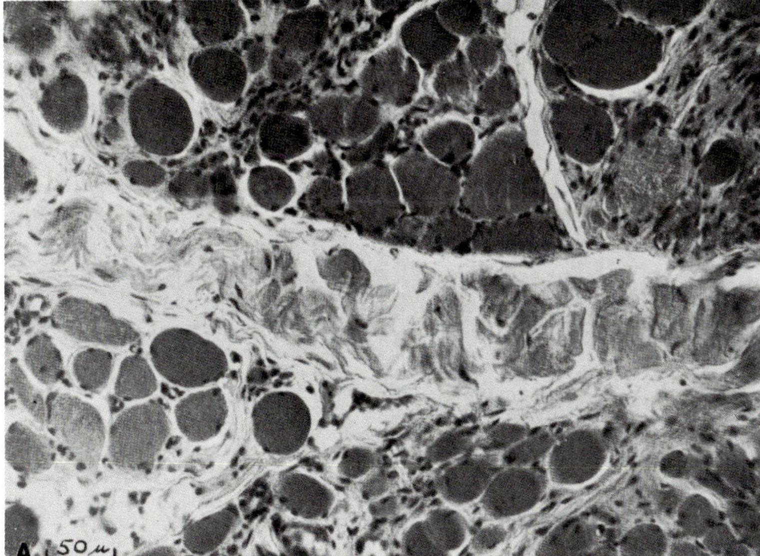

Fig. 6–11. Facioscapulohumeral type of muscular dystrophy, as seen in pectoralis major. (hematoxylin, Van Gieson)

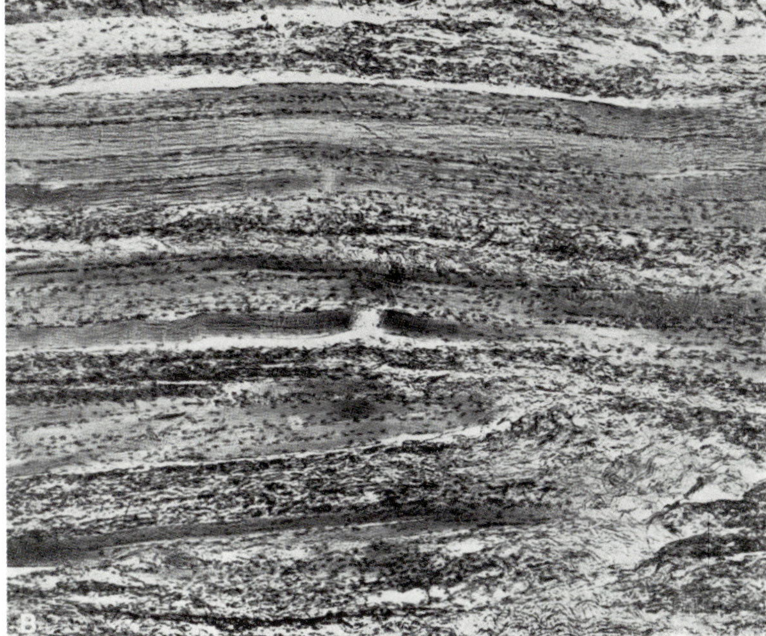

PROGRESSIVE DYSTROPHY OF EXTERNAL OCULAR MUSCLES

It has been particularly difficult to interpret the muscle changes in progressive dystrophy of the external ocular muscles. Atrophy of groups of fibers supplied by a single motor unit cannot be depended upon as a criterion of muscle atrophy because some motor units innervate only four or five muscle fibers in the eye muscles. Usually only fragments of ocular muscle, biopsied in the course of operation to correct strabismus, have been available for examination. The orbicularis muscle has shown the same changes as the external ocular muscles. In the reported cases the most convincing abnormalities have been the great variation in size of the muscle fibers and the presence of hypertrophied fibers,

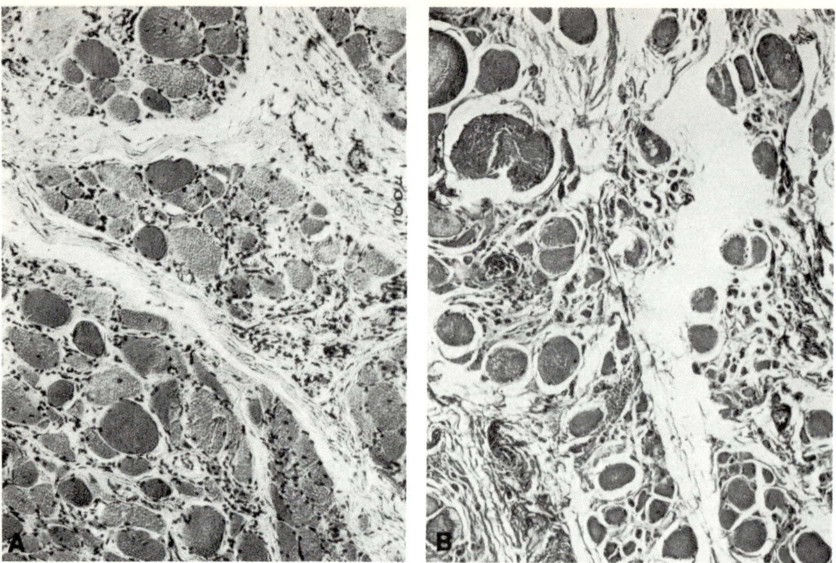

Fig. 6–12. Facioscapulohumeral type of muscular dystrophy. (A) Section of pectoralis major. (B) Juvenile (Erb) type of muscular dystrophy. Section of quadriceps muscle. Note pairs of muscle fibers within same endomysial tube. (A, hematoxylin, Van Gieson; B, hematoxylin, eosin)

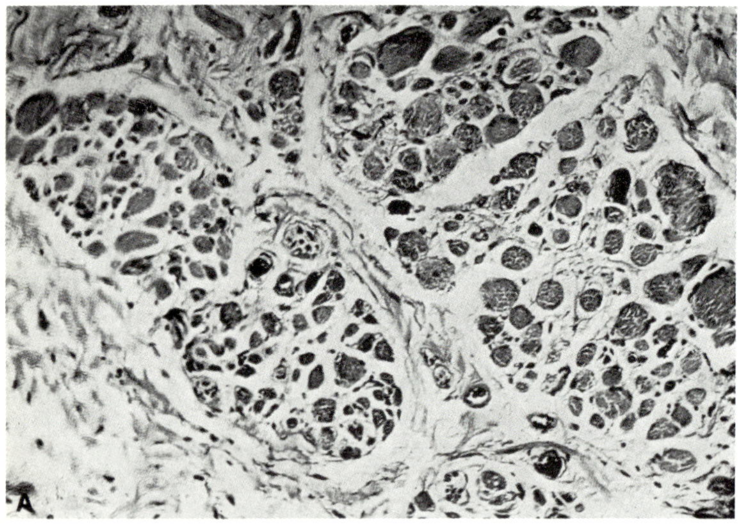

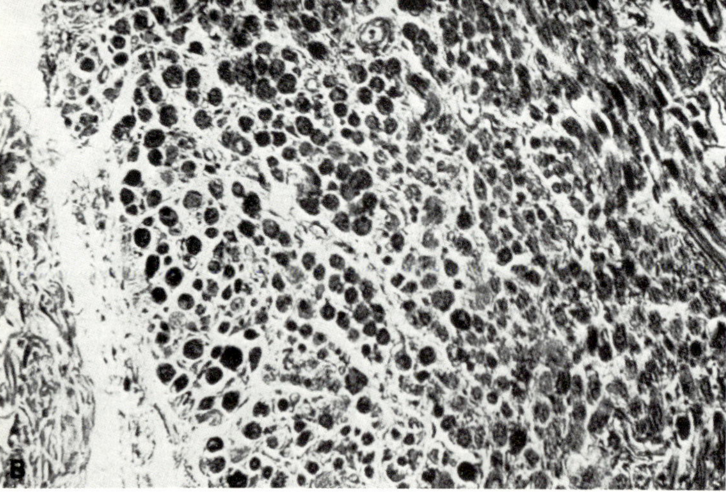

Fig. 6–13. (A) Section of extraocular muscle from a patient with progressive dystrophic ophthalmoplegia. (B) Section from normal extraocular muscle at same magnification. (A, From a section kindly supplied by Dr. S. Nevin)

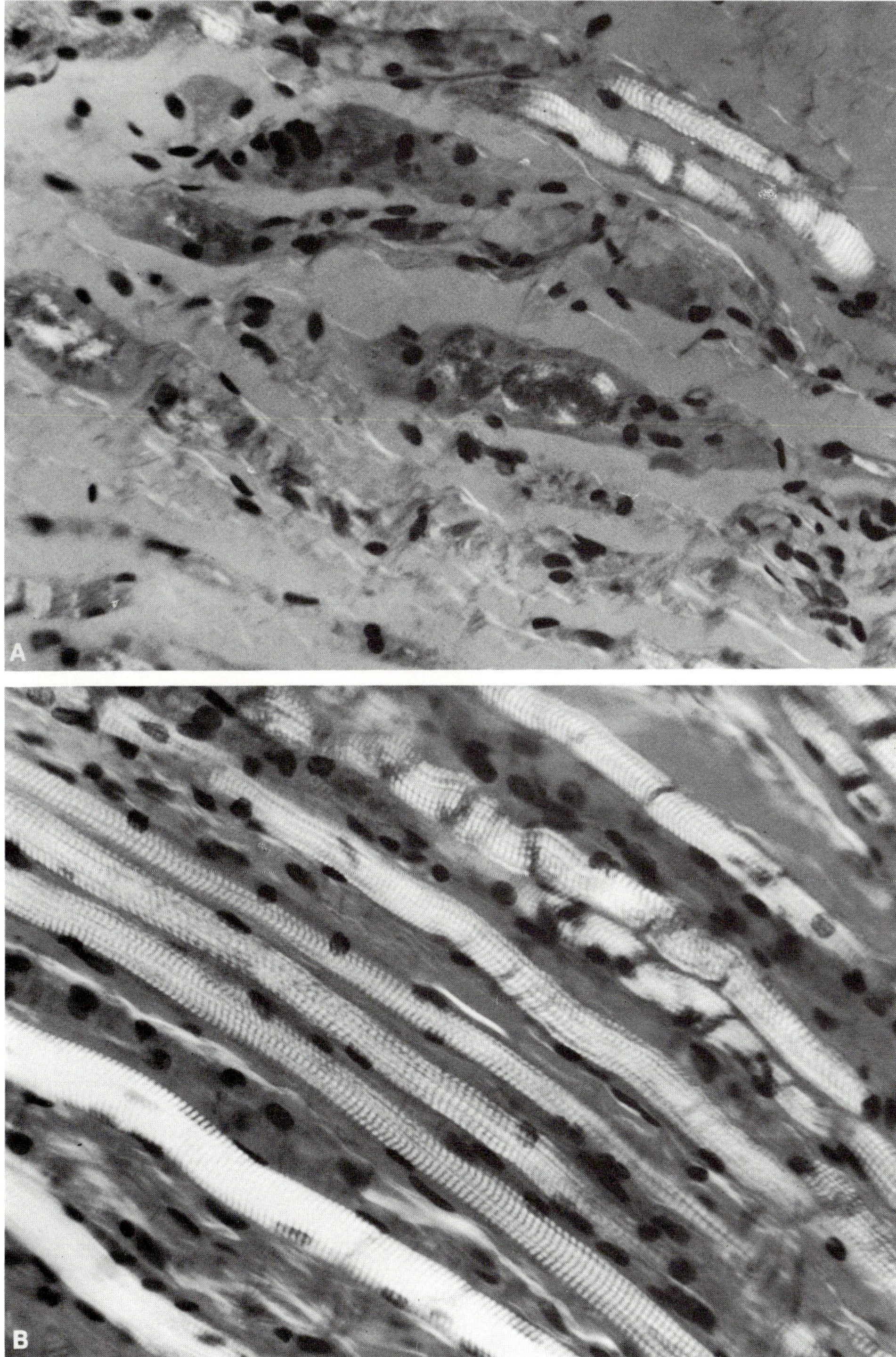

Fig. 6–14. (A) Longitudinal section of superior rectus muscle of eye viewed under crossed polaroids. Only a few muscle fibers remain, and the intervening connective tissue is greatly increased. (B) Normal superior rectus eye muscle from an individual of approximately the same age for comparison. (hematoxylin, eosin)

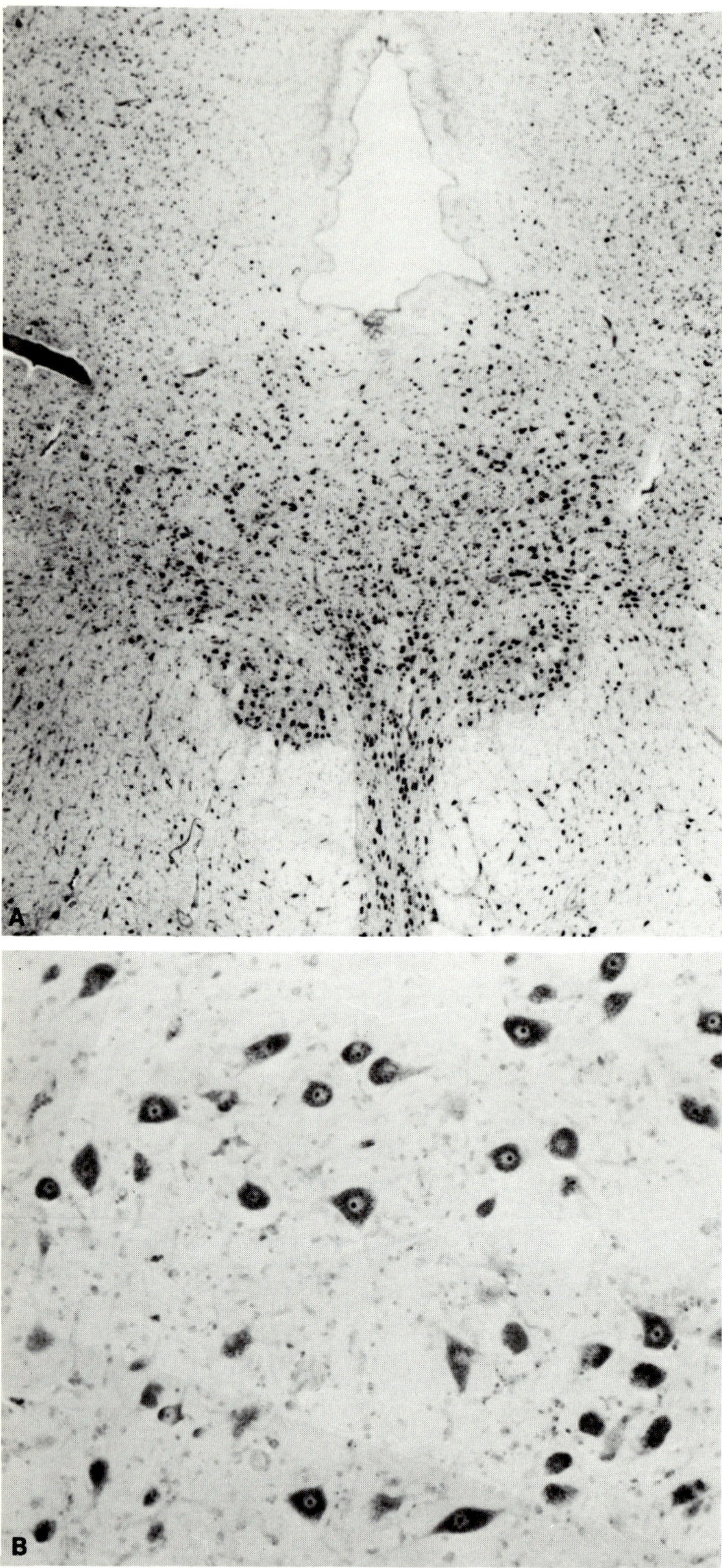

Fig. 6–15. (A) Low power magnification of oculomotor nuclei from same patient as in Figure 6–14, demonstrating the integrity of the neurons. (B) Higher magnification, showing normal appearance of the neurons in these nuclei. (Nissl stain)

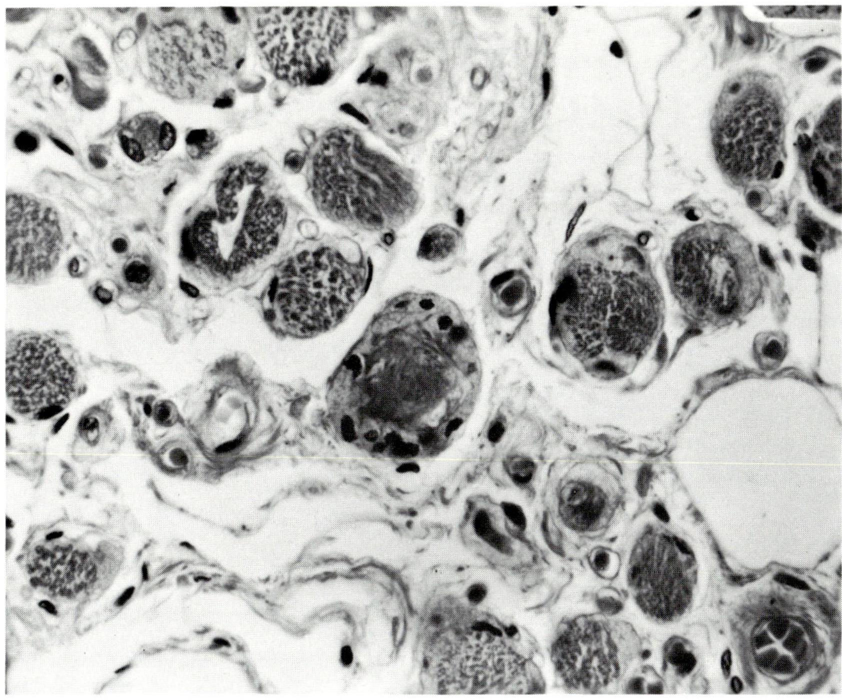

Fig. 6–16. Oculopharyngeal dystrophy. Cross section of residual fibers of superior rectus muscle. Aside from a diminution in the numbers of fibers there is considerable variation in size, peripheral sarcoplasmic masses, and ringbinden, as well as an increase in connective tissue. (hematoxylin, eosin, ×610)

some two or three times the size of those in a normal ocular muscle. Other fibers are reduced to very small dimensions. Normal, hypertrophic, and atrophic fibers are found side by side in haphazard arrangement, separated by variable amounts of connective tissue and fat cells.[89] Degeneration of isolated muscle fibers has been demonstrated (Figs. 5–10 and 5–11), and the accumulation of large numbers of abnormal mitochondria have led to the classification of such cases with the mitochondrial myopathies. Nerve bundles traversing the muscles are said to be normal. A further excellent description was given by Beckett and Netsky.[14] There have been no cases of progressive ophthalmoplegia in our pathologic material, but through the courtesy of Dr. S. Nevin we examined a biopsy specimen from one of his recently published cases and consider the muscle changes (Fig. 6–13) as typically myopathic.

A more detailed analysis of the ocular muscles and brain stem in a case of this disease can be found in the report of Cogan *et al.*[30] The majority of the muscle fibers had disappeared, leaving fat cells and condensed connective tissue sheaths (Fig. 6–14). Residual fibers displayed a wide variation in sizes, and in some the central parts of the fiber were impoverished of myofibrils and converted to eosinophilic masses surrounded by frothy appearing cytoplasm. The quantity of lipofuscin was increased. The nerve cells of the oculomotor nuclei were normal in number and form (Fig. 6–15). Such cases corroborate the theory of Fuchs[67] and Kiloh and Nevin[89] that this disease is dystrophic in nature. Other muscles are often affected to a lesser degree, and some patients also have a chronic polyneuropathy with increased protein in the CSF. However, it is not certain that all cases of ophthalmoplegia are the same.

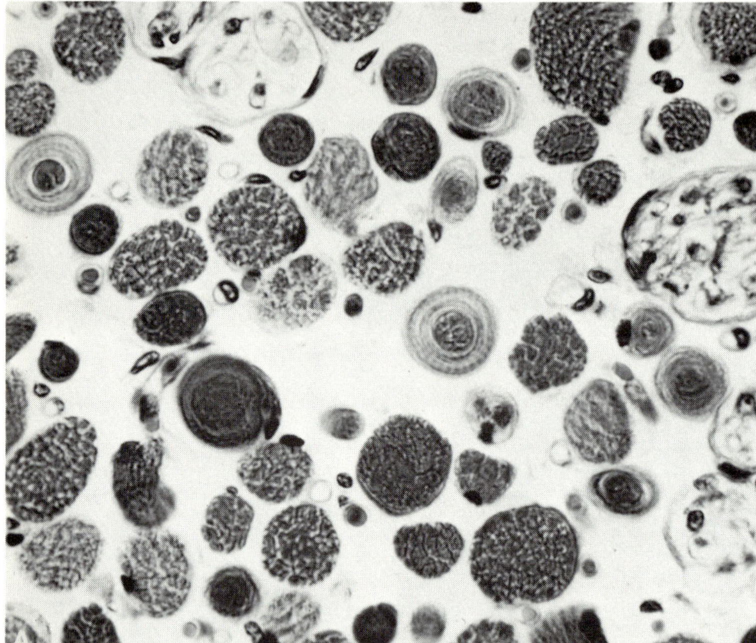

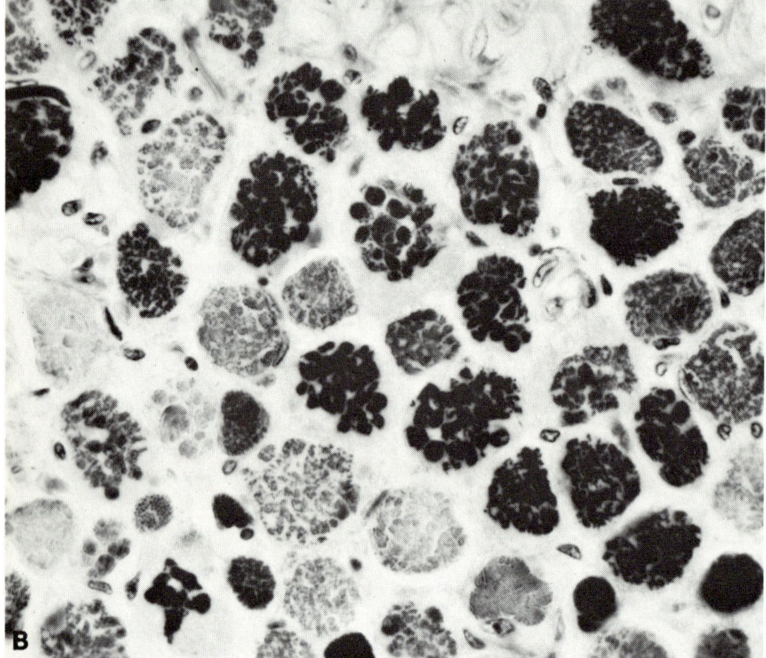

Fig. 6–17. Normal superior rectus muscle of 74-year-old man, showing light and dark fibers, clustering of myofibrils ringbinden, and nerve twigs. (phosphotungstic hematoxylin)

OCULOPHARYNGEAL DYSTROPHY

Biopsy examination by the author and colleagues[120] affirmed the myopathic nature of this disease; and the first and only complete postmortem study conducted in our laboratory with Rebeiz *et al.*[120] corroborated this impression, for the ambiguous and oculomotor nuclei retained their full complement of nerve cells. The oculomotor, trochlear, and abducens nerves including the intramuscular nerve twigs were also intact. The upper esophagus and

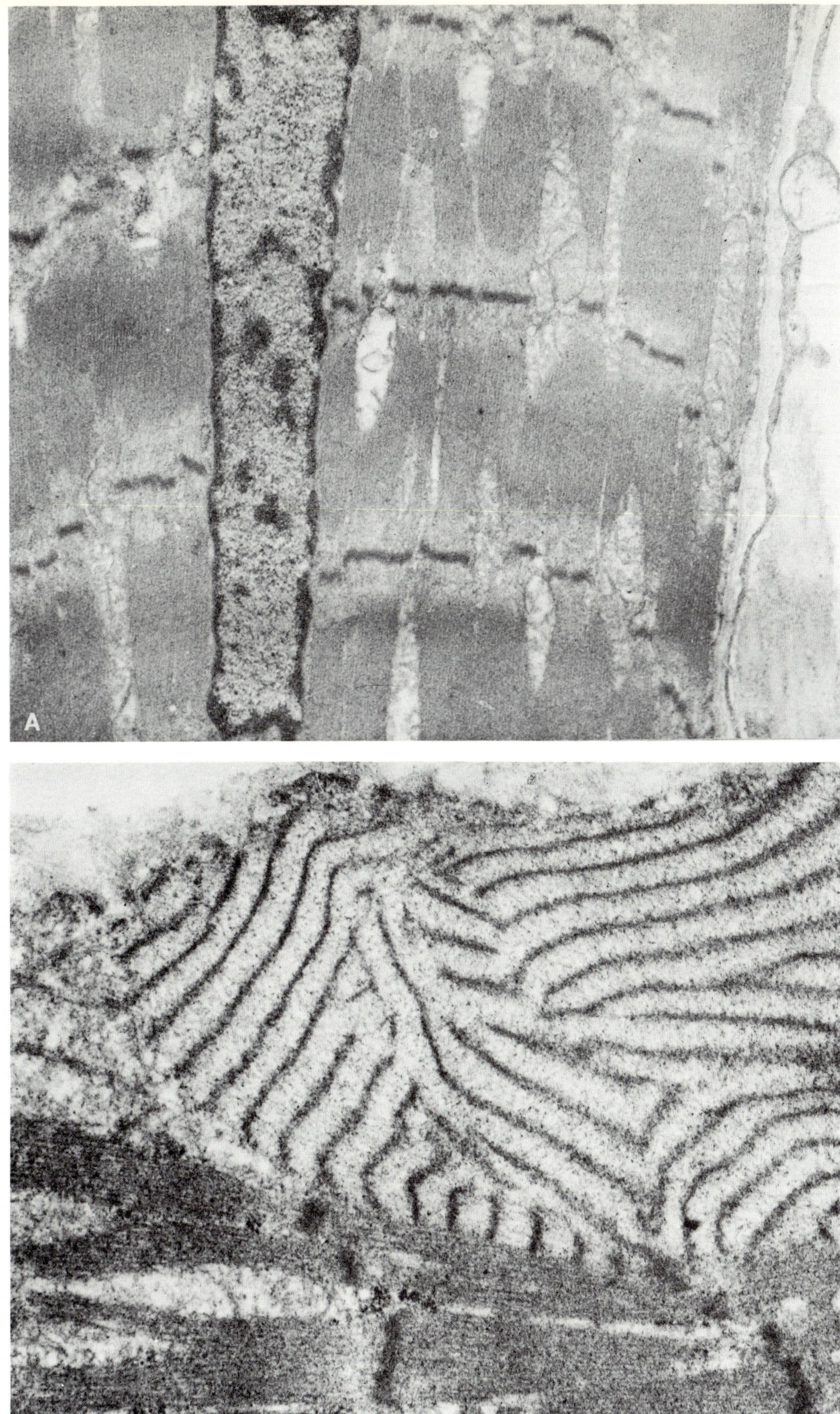

Fig. 6–18. Electron micrograph of same case as in Figure 6–16. Tissue is not well preserved (12 hours postmortem). (A) Muscle fiber with central nucleus, intact sarcolemma, and swollen mitochondria. (B) Array of electron-dense material of fingerprint appearance beneath sarcolemma. (A, B, ×30,000)

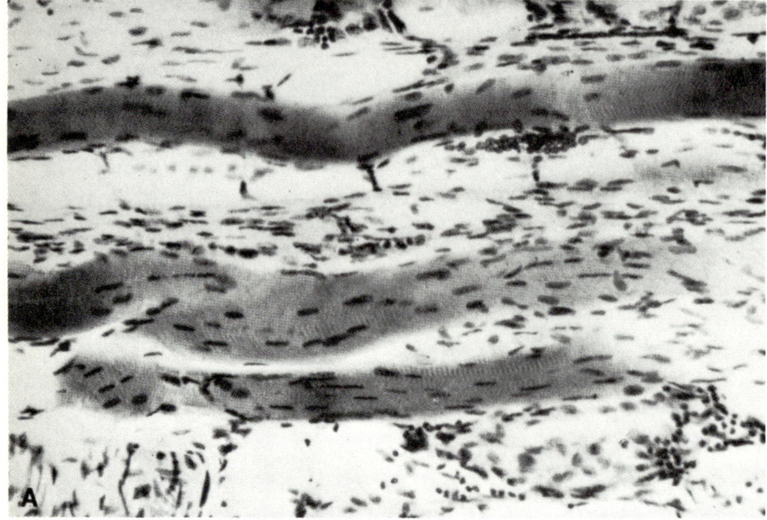

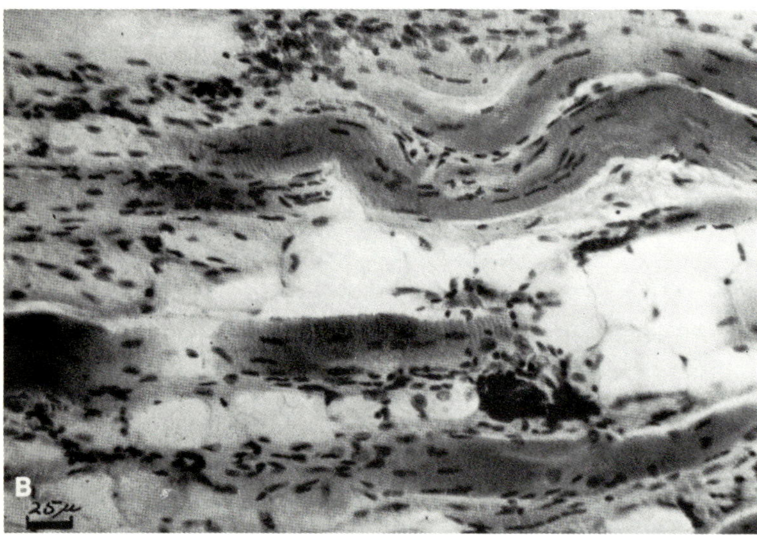

Fig. 6–19. Myotonic dystrophy. (A) Section from an extremely atrophic muscle showing three surviving muscle fibers separated by strands of connective tissue and fat, in which clumps and chains of residual muscle nuclei remain. Fibroblastic nuclei are of two kinds: pale-staining oval nuclei and darkly staining, narrow, often sinuous nuclei. Venule below has a clump of mononuclear cells near it. (B) Some activation of fat cells is proceeding next to a dense clump of pyknotic muscle nuclei.

pharyngeal constrictors were paper-thin and pale. In the pharyngeal, extraocular, and lingual muscles there was a marked reduction in the number of muscle fibers, amounting in the pharynx and upper esophagus to 80%. Indeed in some sections only two or three muscle fibers remained. Residual fibers varied in size and shape (Fig. 6–16). In the extraocular muscles they were smaller in diameter than those of normal control muscles (Fig. 6–17), and the nuclei in many occupied a central position (Fig. 6–18). Some fibers lost all or a part of their myofibrils, especially in the peripheral parts. The myofibril free sarcoplasm appeared eosinophilic and faintly fibrillar. Some fibers contained an excess of lipofuscin granules. Deep encircling or penetrating myofibrils (ringbinden) were discernible in the sarcoplasm of many of the fibers. Fat cells had increased in number (lipomatosis).

Comparison of two control cases of the same age disclosed only quantitative differences. We noted that the normal eye muscle at this age often contains fibers with loss of peripheral myofibrils and the formation of sarcoplasmic masses but to a lesser degree than in our dystrophic case (Fig. 6–17A,B). Moreover, connective tissue, lipofuscinosis, and central nucleation were much less prominent in the control muscles. Fiber number may be some-

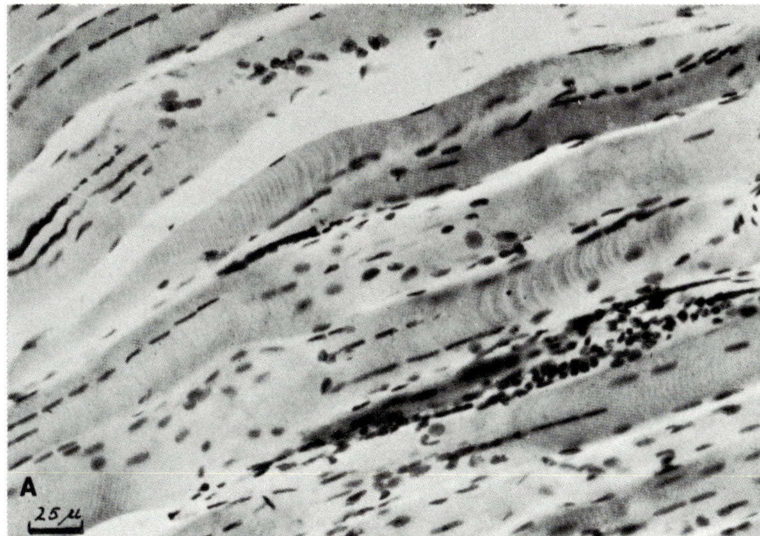

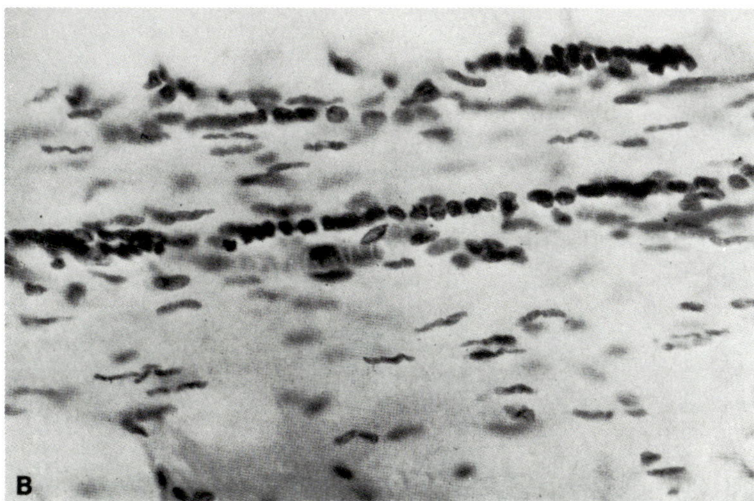

Fig. 6–20. Myotonic dystrophy. Sections from a severely atrophied muscle of the hand. (A) A small muscle fiber is undergoing degeneration, with densely staining cytoplasm, nuclear proliferation, and some evidence of phagocytosis. (B) A strand of connective tissue contains no muscle fibers—only chains of pyknotic muscle nuclei with no visible sarcolemma or myofibrils. Sinuous, narrow, fibroblastic nuclei are characteristic of such residual connective tissue. Some endothelial nuclei (at center) belong to a capillary containing erythrocytes.

what depleted at this age, and the range of fiber sizes widened.

In our autopsied dystrophic case the tongue, sternomastoid, deltoid, biceps, and sartorius were impoverished of fibers but to a lesser degree than the levator palpebral and pharyngeal muscles. Again the span of fiber sizes had increased. Under the electron microscope many of the residual fibers of the superior rectus and levator palpebral muscle were entirely normal, conforming to the images described by Cheng and Breinin.[29] Their plasma membranes had maintained a natural appearance, and their cytoplasm was filled with perfectly striated longitudinally disposed myofilaments and normally arranged transverse and longitudinal sarcoplasmic reticulum. It was not difficult to recognize both dark and light fibers in the dystrophic and the normal control cases. In all specimens with respect to the dark type I fibers the M band was missing but mitochondria were normally abundant. In the masses of relatively clear sarcoplasm, some myofilaments were present but in full disarray. Both actin and myosin filaments could be identified, as could Z bands, but the latter were disorganized. Some of the mitochondria were swollen but otherwise not unusual; granules of lipofuscin were numerous whereas glycogen granules were virtually absent. Odd structures having fingerprint pat-

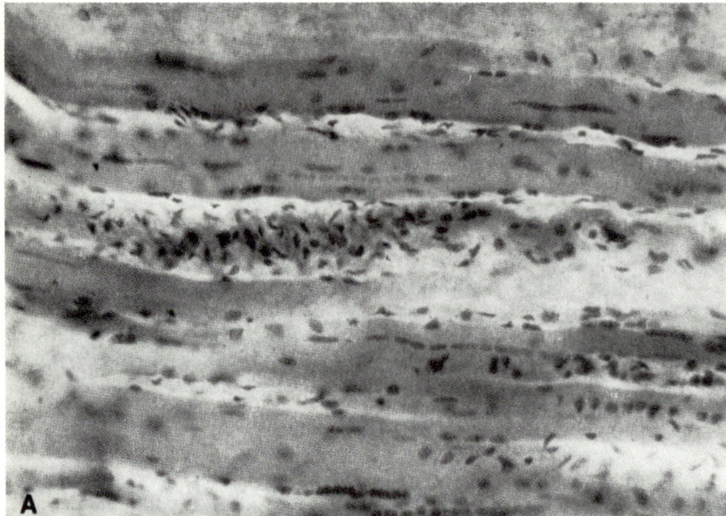

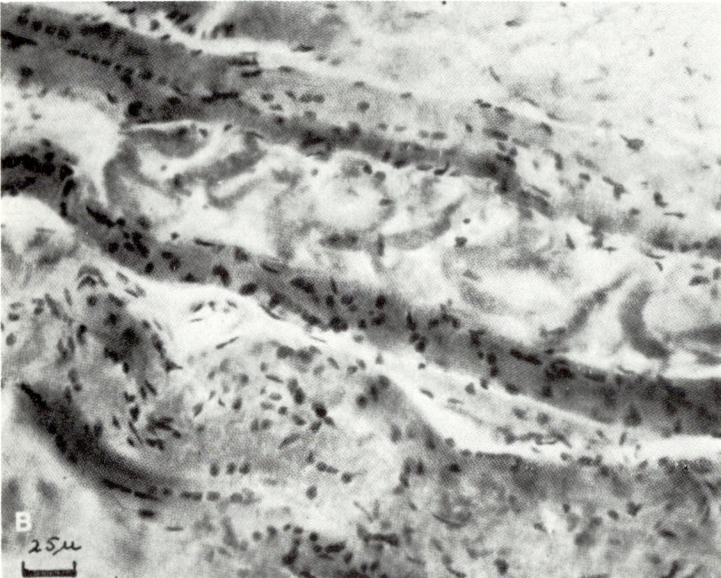

Fig. 6–21. Myotonic dystrophy. These nuclear and fibrillary changes are commonly observed in such biopsy specimens. (hematoxylin, Van Gieson)

terns were seen here and there in the subsarcolemmal region, related probably to Z bands. The latter were not unique for they have been seen by Cogan *et al.*[30] in extraocular muscles 1–2 months after enucleation of an eye. Fibroblasts and collagen were increased.

Since many qualitative differences were found between the dystrophic and normal senile eye muscles, we were attracted to the idea that the basic abnormality in oculopharyngeal dystrophy is of the nature of a presenile myopathy (muscular abiotrophy) rather than a true dystrophy.

MYOTONIC DYSTROPHY

Myotonic dystrophy is distinguished pathologically from other muscular dystrophies by the selective involvement of certain muscles such as eye, facial, sternomastoid, forearm, and leg; by the involvement of other tissues (testes, lenses); and by the presence in muscle fibers of rows of central sarcolemmal nuclei, sarcoplasmic masses, and myofibrillar disorganization. Enlargement of scattered muscle fibers without other change except for the presence of centrally placed muscle nuclei in long rows

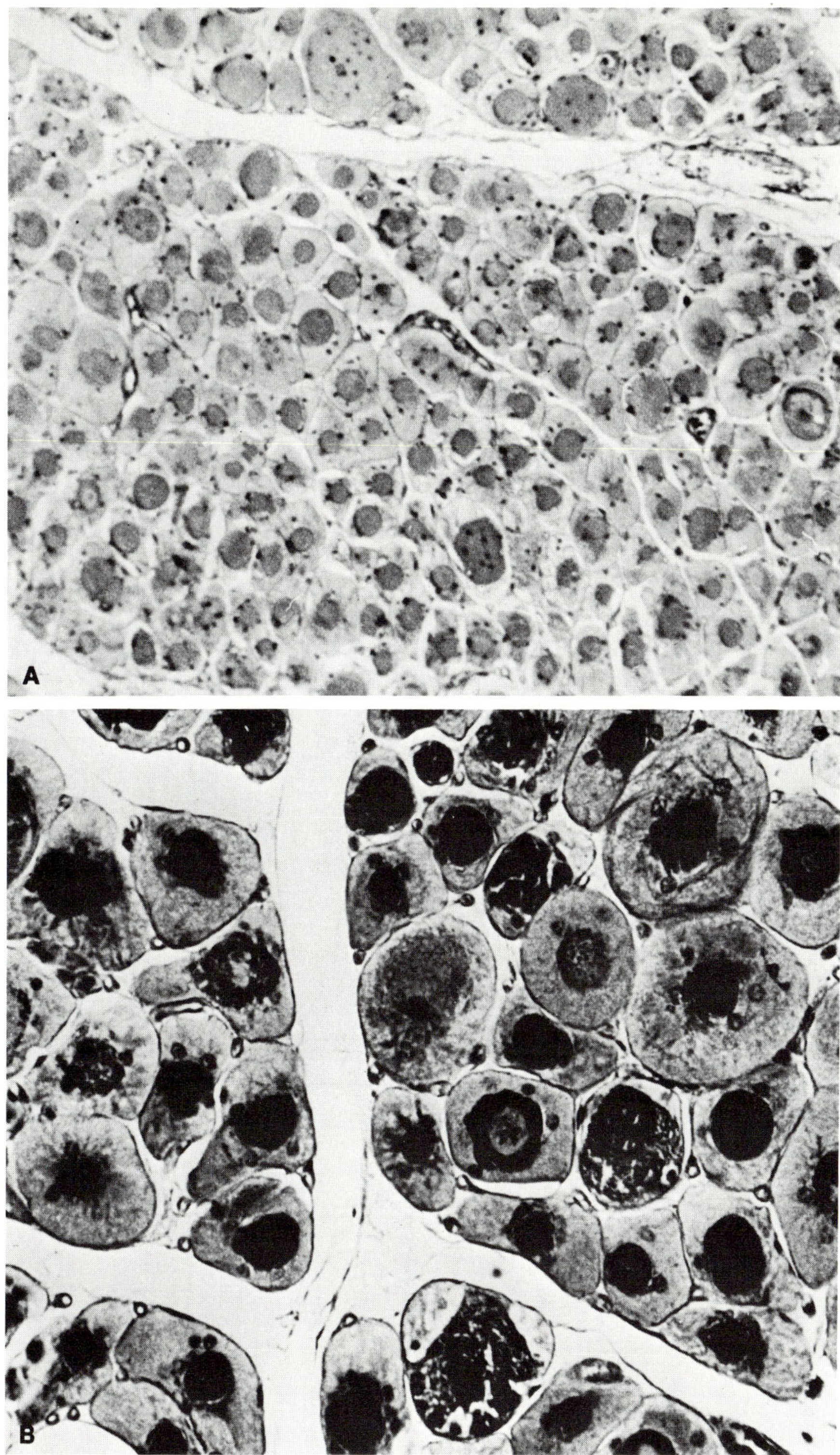

Fig. 6–22. Myotonic dystrophy. (A) Transverse section of anterior tibial muscle, showing peripheral, well nucleated sarcoplasmic masses and central core of preserved myofibrils. (hematoxylin, eosin ×350)

may be a very prominent feature (Figs. 6–19 through 6–21). It may be seen in muscles that do not manifest any sign of atrophy. Indeed, the presence of long chains of such nuclei in an otherwise intact muscle fiber is diagnostic of dystrophia myotonica. True hypertrophy—i.e., very large fibers filled with myofibrils and with peripheral flattened sarcolemmal nuclei—is always evident. In view of the high frequency of neuronal discharge (as well as the tendency to peripheral spasm) in this disease, true hypertrophy appears to be a natural consequence of myotonia.[41] We have found no splitting of muscle fibers. The large fibers are particularly liable to show contraction bands, clots, and segmentation, especially in biopsy specimens. The presence of small fragments and the absence of tissue reaction indicate that these changes are frequently a fixation artifact. The extraordinary excitability of a biopsy specimen is obvious to everyone who has handled the living muscle in this disease, and such fixation artifacts are difficult to avoid except by allowing the muscle to die before fixation. In addition, however, there is almost always some evidence that rupture of some of the large fibers has occurred during life, for segments of such fibers may be found embedded in connective tissue, with intact sarcolemma at the ends and evenly dispersed, small, pyknotic nuclei.

Isolated fibers may show recent severe degenerative changes such as nuclear proliferation, intense basophilic staining of cytoplasm, and even phagocytosis (Fig. 20A). Longer segments of a fiber containing a twisted skein of fibrillar material, no longer striated, with scattered distorted nuclei in their mesh probably also arise from antemortem rupture of the contents of the fiber, for the same appearance of a twisted skein may be seen in a recent rupture (Fig. 6–21), with nuclear distortion. All of these effects may be attributed to the abnormal excitability of the muscle and clearly may affect the abnormal muscle fiber in vivo as well as during biopsy or fixation. It is remarkable that reconstitution of myofibrils under such circumstances is entirely absent. Degeneration of muscle fibers was noted by many of the earliest investigators, and some of the distortions we discussed above were described in

detail by Marinesco,[105] who considered them to be the result of a special process which he called myotaxie.

Other changes regarded as characteristic by Wohlfart[165] are sarcoplasmic masses and myofibrillar disorganization (a kind of deep ringbinden). These occur mainly in the periphery of the fiber, as seen in transverse sections, as a ring of homogeneous sarcoplasm around a central fascicle of myofibrils (ringbinden). This appearance, described in normal muscle by Schäffer[129] and shown by him to be due to transverse section of a "contraction band" appears to us to be prominent in myotonic muscle. However, within the sarcoplasmic masses there are a faintly filamentous structure, an abundance of enzymes in histochemical stains, and large, active-appearing nuclei. Such disoriented packets of myofilaments constitute a special type of ringbinden, unlike those occurring in a normal fiber, with all the abnormal myofibrils being beneath the sarcolemma. The number of fibers showing sarcoplasmic masses and deep ringbinden varies from one specimen to another; every fiber is affected in some (Fig. 6–22), while in others one must search for a single example. In some biopsies such disfigurations are completely absent. Fiber loss is another feature, and many of the residual fibers are atrophic. Engel[52–54] and others have called attention to an atrophy of type I fibers early in the course of the illness.

Ultrastructural morphologic study of the muscle fiber in this form of disease has shown that the peripheral masses of finely filamentous sarcoplasm are indeed highly reactive cellular zones wherein myofilaments (thin and thick) occur in complete disarray. Only in some places are packets of them organized, but then their orientation is disturbed. This is the basis of the deeply seated ringbinden seen by light microscopy. Small clearings in sarcoplasm, especially in atrophic fibers, and vacuolation are not unusual. Tubular structures, accumulations of lysosomes and lipofuscin, complex arrangements of membranes near the infolded fiber surface, discontinuities in the T system, circular cytoplasmic inclusions, and myelin-like structures are also identifiable. This change might be interpreted as the expression of a

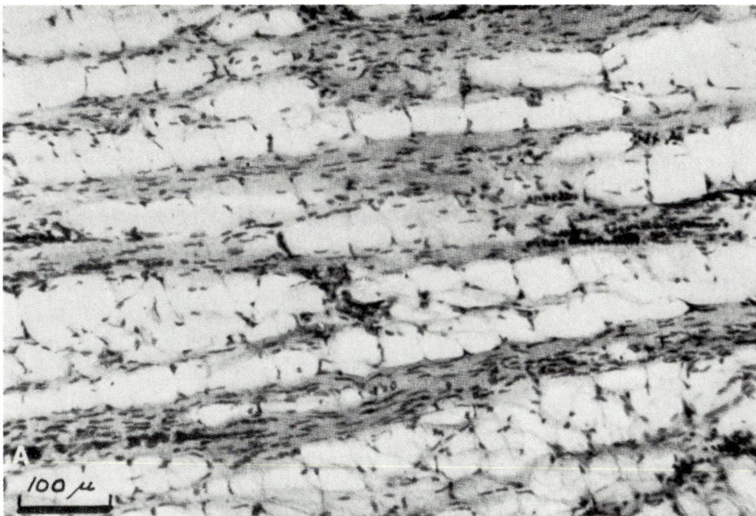

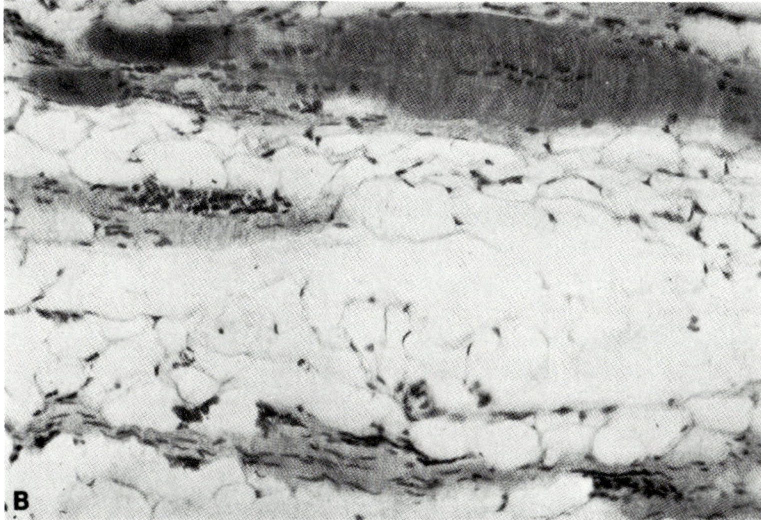

Fig. 6–23. Myotonic dystrophy. Sections from severely atrophic muscle. (A) Complete absence of muscle fibers. (B) One large hypertrophied muscle fiber, with fat and strands of endomysium. (hematoxylin, Van Gieson)

basic defect in myofilament formation and integration, but how it leads to degeneration of the fiber is unclear. Although sarcoplasmic masses and degenerating myofilaments have been seen in other of the muscular dystrophies by Lapresle and Fardeau[96] and Fisher *et al.,*[63] they occur with such regularity and prominence in myotonic dystrophy as to stamp it as a special disease.

Besides changes which probably reflect only great mechanical irritability, the atrophic muscles in dystrophia myotonica show precisely the same changes that are seen in other types of myopathy. Splitting of muscle fibers is very uncommon. There is much less connective tissue than in the muscles of the Landouzy-Déjerine type of this disease, and contracture is a correspondingly rare clinical feature. Accumulation of the fat cells is rare in mildly affected fasciculi, but in the last stage of atrophy (Fig. 6–23) it is indistinguishable from that of other types. Rows and clusters of darkly staining nuclear remnants are found lying in the connective tissue and between fat cells.

Testicular atrophy, with only a few seminiferous tubules remaining and either an increase or a decrease in interstitial cells of Leydig, is

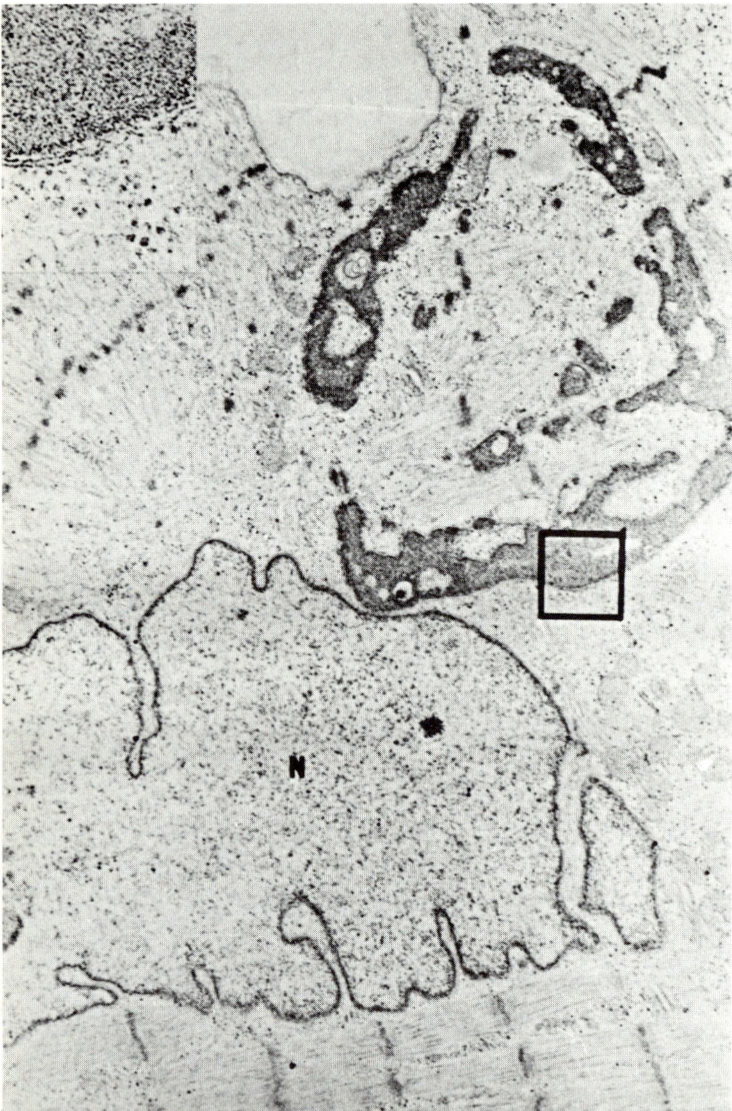

Fig. 6–24. Myotonic dystrophy. Electron micrograph of biopsy specimen. At bottom are the longitudinally oriented myofilaments; above these is a crenated muscle nucleus. Disoriented myofilaments with Z particles and an irregular mass of electron-dense material fill the sarcoplasmic mass.

the most constant extramuscular histologic finding. Changes in the pituitary, thymus, thyroid, and adrenal glands have been reported in some cases but often are so nonspecific as to make it unlikely that they are significant.[2, 21, 88] Cardiac muscle in the few autopsied cases has not shown any of the changes of the type seen in skeletal muscle.[21]

The author has had little experience with the lens lesion. Presumably the opacities which are high in lipid content are deposited under the most recently formed fibers. Again it may represent a membrane disorder, for drugs such as 20–25 diazocholesterol which produce myotonia also cause cataracts.

CONCLUSION

These detailed descriptions of the muscle tissue in the several hereditary muscle syndromes suggest that the unitary theory of Erb, which assumes that all the different forms of muscular dystrophy share one pathogenic mechanism, should be questioned. The enzymatically and

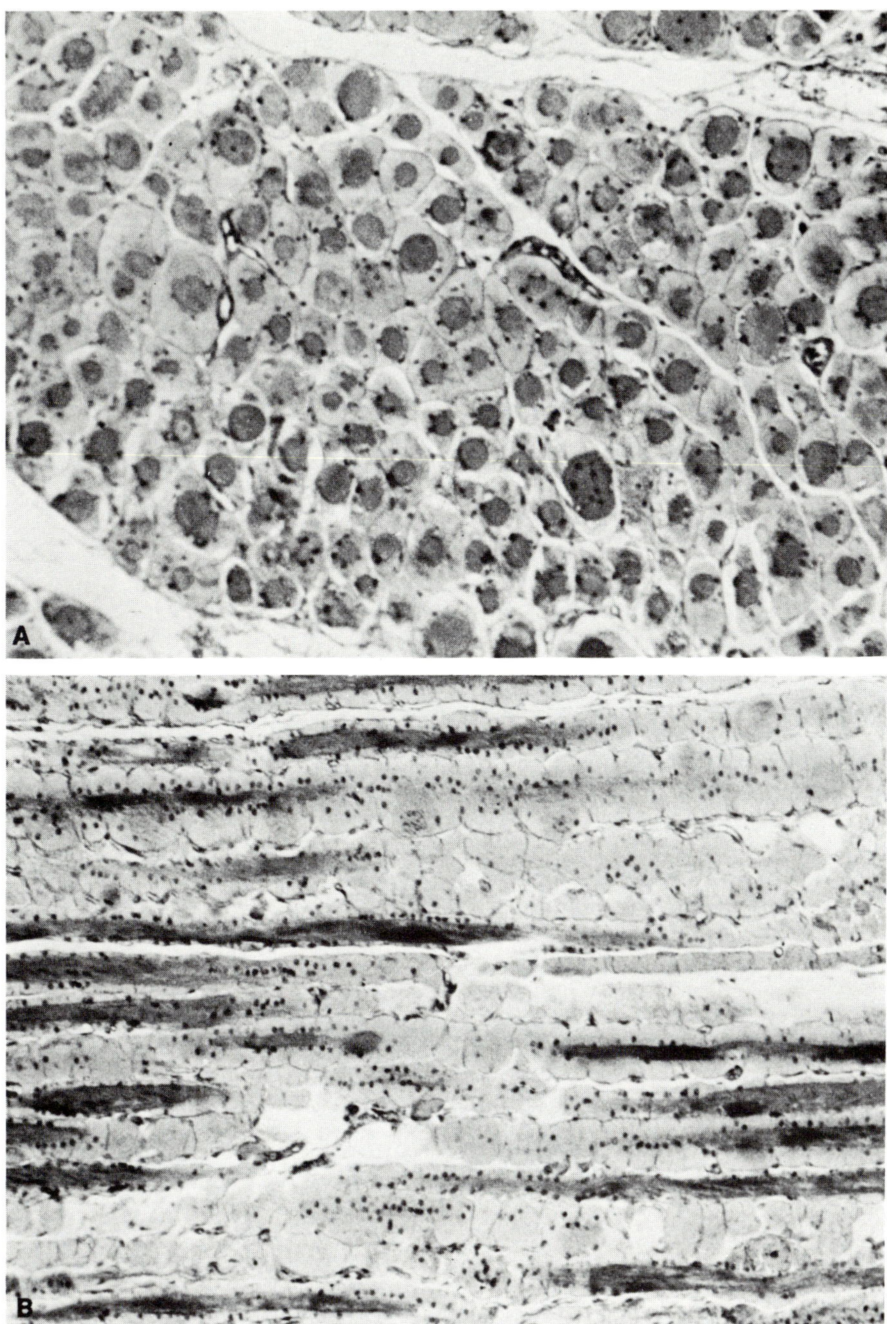

Fig. 6–25. Myotonic dystrophy. (A) Transverse section, showing highly typical pattern in which all muscle fibers have lost their peripheral myofibrils, only the central ones remaining as a core. Nucleation is prominent in zones of clear sarcoplasm. (B) Longitudinal section, cut imperfectly so that central cores of preserved myofibrils pass in and out of the plane of the section. Sarcolemma is festooned (?artifact). Note nucleation of sarcoplasmic masses. Peroneal muscle. (hematoxylin, eosin)

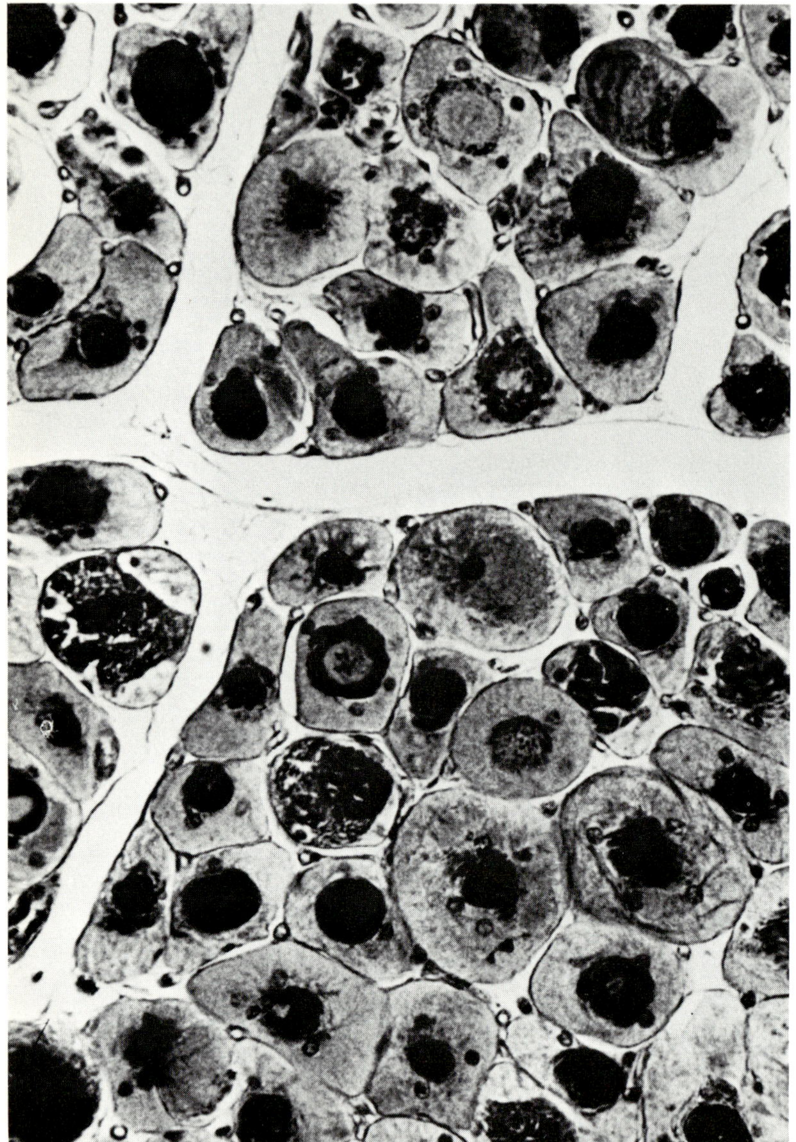

Fig. 6–26. Myotonic dystrophy. Note fibrillar texture of sarcoplasmic masses and deep ringbinden. (phosphotungstic hematoxylin)

cellularly active sarcoplasmic masses in myotonic dystrophy appear to separate this form of dystrophy from the others, and we are inclined to hold the same opinion about oculopharyngeal dystrophy. It would seem preferable therefore to speak of muscular dystrophy in the plural.

The most tantalizing problem, the solution of which has eluded all pathologists to date, is specification of the primary muscular ab-normality in any one of the dystrophies. The closer inspection afforded by the electron microscope has only magnified the acknowledged lesion without shedding light on its pathogenesis. Only deformities of mitochondria and sarcolemma, evidence of an unhealthy state of the muscle fiber, are seen in the myofilamentous degeneration. Could it be that the initial change is hypertrophy followed by atrophy or segmental necrosis with abortive

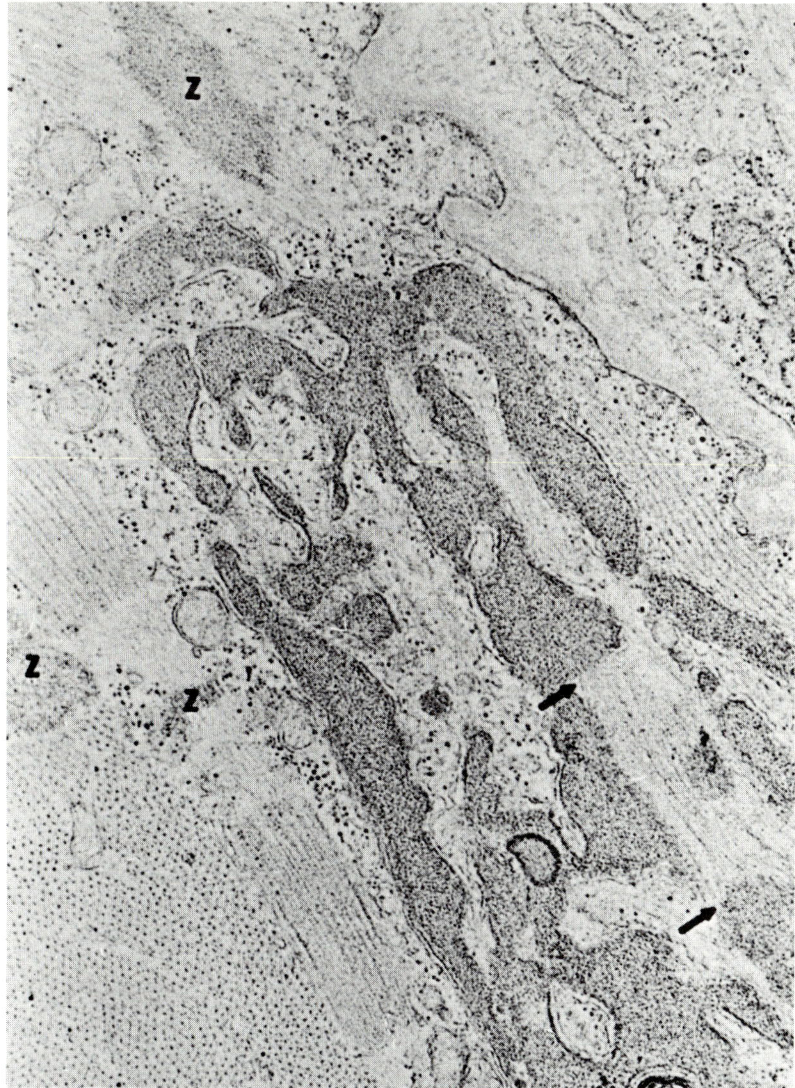

Fig. 6–27. Electron micrograph. Section of ringbinden and sarcoplasmic mass. Disorganized dark Z material is streaked through center of section. At lower left are transversely sectioned myofilaments near which are encircling myofilaments. To right of center are circular myofilaments, and some of the thin ones (between arrows) are attached to Z bands.

regeneration? Segmental necrosis and regenerative branching of reforming fibers could, it must be conceded, lead to hypertrophy of the parent fiber from which the branches are derived;[1] and of course with loss of some muscle fibers the residual ones may enlarge from increased work. Moreover, the newly formed branches are smaller than normal. Thus hypertrophy might be viewed as secondary effects of primary segmental necrosis. Why, though, if necrosis is regularly followed by regeneration, do muscle fibers eventually disappear? Is it a matter of the reparative processes being inadequate, or is necrosis so frequent as to ultimately exhaust the restorative powers of the muscle fibers? What happens to the atrophic muscle fibers; do they suffer the same fate as those fibers deprived of their innervation?

The author's bias is toward segmental necrosis as the primary lesion, attributing it to a genetically determined metabolic fault that prevents the fiber from surviving under conditions of even moderate functional activity. We find evidence of a neural or vascular factor in pathogenesis to be inconsistent with observed facts. Obviously the dystrophic process does not interfere with the natural development of muscle fibers nor prevent their immediate response to injury. Yet a series of necrotizing processes obviously terminates in death of the entire fiber. Of course once this happens, since the muscle cannot replenish its constituency of fibers, depletion in number (numerical atrophy) results in permanent failure of function. Paralysis in dystrophy seems equated essentially with fiber loss. Equally puzzling is the phenomenon of myotonia for which no morphologic basis, either in sarcolemma or endoplasmic reticulum, has been established.

The dystrophic process appears not to attack any one organelle of the muscle fiber. The myofibrils, for example, preserve their structural integrity until late in the disease. To imply that dystrophy involves a loss of support and conservation of contractile structure helps little in understanding the disease, for too little is known of its normal maintenance. The relatively normal quantities of contractile elements suggest that the dystrophic lesion is independent of the contractile process. The natural life history of a myofibril and its constituent myofilaments has yet to be written, though it is certain that muscle training can increase and disuse decrease their number. Again, such changes seem to occur in muscle quite independently of muscular dystrophy.

On the other hand, the notion that *segmental necrosis of fibers* should provide a specific characterization of muscular dystrophy is clearly untenable, for this change is observed in polymyositis and certain metabolic myopathies such as those due to glycolytic defects. In dystrophy it is the repeated necrosis of muscle fibers over a period of years that seems to set it apart. The only disease that corresponds even remotely to muscular dystrophy in its temporal profile is recurrent polymyositis where, interestingly, many of the cardinal histologic features of dystrophy eventually appear.

Regeneration, fiber loss, forking and branching, group atrophy and enlargement of single fibers, lipomatosis and fibrosis, which are common to both dystrophy and myositis, must therefore be regarded as nonspecific secondary changes. Again it is only the intensity of the necrotizing process in relation to infiltrates of inflammatory cells, a brief span of activity, and more recently the demonstration of virus particles in some cells that distinguishes chronic polymyositis from the muscular dystrophies.

REFERENCES

1. ADAMS RD: The giant muscle fiber: its place in myopathology. Modern Neurology. Edited by S Locke. Boston, Little, Brown, 1969, pp 225–240

2. ADIE WJ, GREENFIELD JG: Dystrophia myotonica (myotonia atrophica). Brain 46:73–127, 1923

3. AMES SR, RISLEY HA: Aminoaciduria in progressive muscular dystrophy. Proc Soc Exp Biol Med 68: 131–135, 1948

4. ARAN FA: Réchèrches sur une maladie non encore décrite du système musculaire (atrophic musculaire progressive). Arch Gen Med 24:5, 1850

5. ARONSON SM, VOLK BW: Studies on serum aldolase activity in neuromuscular disorders. I. Clinical applications. Am J Med 22:414–421, 1957

6. BALOH R, CANCILLA PA, KALYANARAMAN K, *et al:* Regeneration of human muscle: a morphological and histochemical study of normal and dystrophic muscle after injury. Lab Invest 26:319–328, 1972

7. BANKER BQ, VICTOR M, ADAMS RD: Arthrogryposis multiplex due to congenital muscular dystrophy. Brain 80:319–334, 1957

8. BARBEAU A: Oculopharyngeal muscular dystrophy in French Canada. Progress in Neuro-ophthalmology. International Congress Series 176. Amsterdam, Excerpta Medica, 1969

9. BATTEN FE: The muscle-spindle under pathological conditions. Brain 20:138–179, 1897

10. BATTEN FE: Myopathy, or muscular dystrophy. System of Medicine. Edited by TC Allbutt, HD Rolleston. New York, Macmillan, 1910, Vol 7, p 31

11. BATTEN FE, GIBB HP: Myotonia atrophica. Brain 32:187–205, 1909

12. BECKETT EB, BOURNE GH: Some histochemical observations on human dystrophic muscle. Science 126:357–358, 1957

13. BECKETT EB, BOURNE GH: 5-Nucleotidase in normal and diseased human skeletal muscle. J Neuropathol Exp Neurol 199–204, 1958

14. BECKETT RS, NETSKY MG: Familial ocular myopathy and external ophthalmoplegia. Arch Neurol Psychiatry 69:61–72, 1953

15. BELL C: The Nervous System of the Human Body. Second edition. London, Longman, 1830

16. BELL J: On pseudohypertrophic and allied types of progressive muscular dystrophy. The Treasury of Human Inheritance. Edited by E Fischer. London, Cambridge University Press, 1943, Vol IV, pp 283–342

17. BERGER O: Ueber Pseudohypertrophie der Muskeln. Arch Psychiatr Nervenheilk 14:625–642, 1883

18. BICKNELL F, PRESCOTT F: The Vitamins in Medicine. Second edition. New York, Grune, 1946

19. BING R: Kongenitale, heredofamiliare und neuromuskulare Erkrankungen, Handbuch d. inn. Med. Edited by G von Bergmann, R Staehln. Berlin, Springer, 1926

20. BISHOP A, GALLUP B, SKEATE Y, et al: Morphological studies of normal and diseased human muscle in culture. J Neurol Sci 13:333–350, 1971

21. BLACK WC, RAVIN A: Studies in dystrophia myotonica. Arch Neurol Psychiatry 44:176–191, 1947

22. BLAHD WH, BAUER FK, LIBBY RI, ET AL: Radioisotope studies in neuromuscular disease. 2. Studies in muscular dystrophy and myotonia dystrophica with sodium and potassium. Neurology (Minneap) 5:201–207, 1955

23. BLINGOE C, DYE WB: Serum transaminase in white muscle diseases. J Animal Sci 17:224–226, 1958

24. BOURNE GH, GOLARZ MN: Human muscular dystrophy as an aberration of the connective tissue. Nature (Lond) 183:1741–1743, 1959

25. BOWDEN REM, GUTMANN E: Observations in a case of muscular dystrophy, with reference to diagnostic significance. Arch Neurol Phychiatry 56:1–19, 1946

26. BRADBURNE AA: Hereditary ophthalmoplegia in five generations. Trans Ophthalmol Soc UK 32:142–153, 1912

27. BRANDT S: Werdnig-Hoffmann's Infantile Progressive Muscular Atrophy. Copenhagen, Munksgaard, 1950

28. BULLOCK RT, DAVIS JL, HARA M: Dystrophia myotonica with heart block. Arch Pathol 84:130–140, 1967

29. CHENG K, BREININ G: A comparison of fine structure of extraocular and interosseous muscle in the monkey. Invest Ophthalmol 5:535, 1966

30. COGAN DG, KUWABARA T, RICHARDSON EP: Pathology of abiotrophic ophthalmoplegia externa. Bull Johns Hopkins Hosp 3:42, 1962

31. COHEN S: Myocardial fibrosis in progressive muscular dystrophy. J Med (Cincinnati) 17:26–28, 1936

32. CORNELIUS CE, LAW GRL, JULIAN LM, et al: Plasma aldolase and glutamic-oxaloacetic transaminase activities in inherited muscular dystrophy of domestic chicken. Proc Soc Exp Biol Med 101:41–44, 1959

33. CURSCHMANN H: Ueber familiäre atrophische Myotonie. Dtsch Z Nervenheilk 45:161–202, 1912

34. CURSCHMANN H: Dystrophia musculorum progressiva (Erb). Handbuch der Neurologie. Edited by O Bumke, O Foerster. Vol 16, pp 431–497, 1936

35. DANOWSKI TS, BASTIANI RM, MCWILLIAMS FD, et al: Muscular dystrophy. IV. Endocrine studies. J Dis Child 91:356–364, 1956

36. DANOWSKI TS, GILLESPIE HK, EGAN TJ, et al: Muscular dystrophy. V. Blood sugar and serum electrolytes following insulin and dextrose, alone or in combination. J Dis Child 91:429–435, 1956

37. DE JONG JGY: Dystrophia Myotonica, Paramyotonia and Myotonia Congenita. Assen, Netherlands, Royal Van Gorcum Ltd, 1955

38. DEMOS J, ESCOFFIER J: Troubles circulatoires au cours de la myopathie: études arteriographiques. Rev Franc Etud Clin Biol 2:489–494, 1951

39. DENNY-BROWN D: The nature of muscular diseases. Can Med Assoc J 67:1–6, 1952

40. DENNY-BROWN D: The nature of polymyositis and related muscular diseases. Trans Stud Coll Physicians Phila 28:14–29, 1960

41. DENNY-BROWN D, NEVIN S: The phenomenon of myotonia. Brain 64:1–18, 1941

42. DEUEL HJ, SANFORD I, SANFORD K, et al: A study of nitrogen metabolism: effect of 63 days of protein-free diet on the nitrogen partition products in the urine and on the heat production. J Biol Chem 76:391–406, 1928

43. DEWIND LT, JONES RJ: Cardiovascular observations in dystrophia myotonica. JAMA 144:299–303, 1950

44. DRACHMAN D: Ophthalmoplegia plus: the neurodegenerative disorders associated with progressive external ophthalmoplegia. Arch Neurol 18:654–674, 1968

45. DRAGER GA, HAMMILL JF, SHY GM: Paramyotonia congenita. Arch Neurol Psychiatry 80:1–9, 1958

46. DREYFUS JC, SCHAPIRA G, SCHAPIRA F: Biochemical study of muscle in progressive muscular dystrophy. J Clin Invest 33:794–797, 1954

47. DREYFUS JC, SCHAPIRA G, SCHAPIRA F: Serum enzymes in the pathophysiology of muscle. Ann NY Acad Sci 75:235–249, 1958

48. DUCHENNE GB: Réchèrches sur le paralysie musculaire pseudohypertrophique ou paralysie myosclerotique. Arch Gen Med 6 (series 11):5, 179, 305, 421, 552, 1868

49. DUCHENNE GB: De l'Electrisation Localisee et son

Application a la Pathologie et a la Therapeutique. Third edition. Paris, Bailliere, 1872

50. DURANTE G: Anatomie pathologique des muscles. Manuel d'Histologie Pathologique. Edited by V Cornil, L Ranvier. Paris, Felix Alcan, 1902, pp 1–477

51. EBASHI S, TOGOKURA Y, MOMOI H, *et al:* High creatine phosphokinase activity of sera of progressive muscular dystrophy patients. J Biochem (Tokyo) 46:103–141, 1959

52. ENGEL WK: The essentiality of histo- and cytochemical studies of skeletal muscle in the investigation of neuromuscular disease. Neurology (Minneap) 12:778–790, 1962

53. ENGEL WK: Selective and nonselective susceptibility of fiber types: a new approach to human neuromuscular diseases. Arch Neurol 22:97–117, 1970

54. ENGEL WK: Duchenne muscular dystrophy: a histologically based ischemia hypothesis and comparison with experimental ischemic myopathy. The Striated Muscle. Edited by CM Pearson, FK Mostofi. Baltimore, Williams & Wilkins, 1973, pp 453–472

55. ERB WH: Handbuch der Elecktrotherapie. Leipzig, Vogel, 1882, p 389

56. ERB WH: Ueber die "juvenile Form" der progressiven Muskelatrophie ihre Beziehungen zur sogenannten Pseudohypertrophie der Muskeln. Dtsch Arch Klin Med 34:467–519, 1884

57. ERB WH: Dystrophia muscularis progressiva: klinische und pathologieanatomische Studien. Dtsch Z Nervenheilk 1:13–94, 173–261, 1891

58. FARDEAU M: Ultrastructural lesions in progressive muscular dystrophies: a critical study of their specificity. Proceedings, International Congress on Muscle Diseases, Milan, 1969. Edited by JN Walton, N Conal, G Scarlato. International Congress Series 199. Amsterdam, Excerpta Medicia, 1969, pp 98–108

59. FAULKNER SH: Familial ptosis with ophthalmoplegia externa starting in adult life. Br Med J 2:854, 1969

60. FEARNSIDES EG: A case of myotonia atrophica with a family history of cataracts, but no history of familial myopathy, and no myotonic manifestations. Rev Neurol Psychiatry 13:311–316, 1915

61. FINKELNBERG R: Anatomischer Befund bei progressive Muskeldystrophie in den ersten Lebensjahren. Dtsch Z Nervenheilk 35:453–460, 1908

62. FISCH C: The heart in dystrophia myotonica. Am Heart J 41:525–538, 1951

63. FISHER ER, COHN RE, DANOWSKI FS: Ultrastructural observations of skeletal muscle in myopathy and neuropathy with special reference to muscular dystrophy. Lab Invest 15:778–793, 1966

64. FLEISCHER B: Ueber myotonische Dystrophie mit Katarakt: Eine hereditäre familiäre Degeneration. Arch Ophthalmol 96:91–133, 1918

65. FLEISCHER B: Untersuchung von sechs Generationen eines Geschlechtes auf des Vorkommen von myotonischer Dystrophie und anderer degenerativer Merkmale. Arch Rassen Ges Biol 14:13–39, 1922

66. FRENCH EB, KILPATRICK R: A variety of paramyotonia congenita. J Neurol Neurosurg Psychiatry 20:40–46, 1957

67. FUCHS E: Ueber isolieren doppelseitge Ptosia. Arch Ophthalmol 36:234–259, 1890

68. GATES RR: Human Genetics. New York, Macmillan, 1946, Vol II, pp 955–1009

69. GEIGER RS, GARVIN JS: Pattern of regeneration of muscle from progressive muscular dystrophy patients cultivated in vitro as compared to normal human skeletal muscle. J Neuropathol Exp Neurol 16:532–543, 1957

70. GLOBUS JH: The pathologic findings in the heart muscle in progressive muscular dystrophy. Arch Neurol Psychiatry 9:59–72, 1923

71. GOWERS WR: Pseudo-Hypertrophic Muscular Paralysis: A Clinical Lecture. London, Churchill, 1879

72. GOWERS WR: A Manual of Diseases of the Nervous System. Philadelphia, Blakiston, 1888, Vol I, p 587

73. GOWERS WR: A lecture on myopathy, a distal form. Br Med J 2:89–92, 1902

74. GREENFIELD JG: Notes on a family of "myotonia atrophica" and early cataract, with a report of an additional case of "myotonia atrophica." Rev Neurol Psychiatry 9:169–181, 1911

75. GRIESINGER W: Ueber Muskelhypertrophie. Arch Heilkunde 6:1–13, 1865

76. HASSIN GB: The histopathology of progressive muscular dystrophy. J Neuropathol Exp Neurol 2:315–325, 1943

77. HAUSMANOWA-PETRUSEWICZ I, ZIELINSKA S: Zur neurologischen stellung des scapuloperonealen syndroms. Dtsch Z Nervenheilk 183:377–382, 1962

78. HELANDER E: On quantitative muscle protein determination. Acta Physiol Scand [Suppl 141] 41:1, 1957

79. HENKE K, SEEGER S: Ueber die Vererbund der myotonischen Dystrophie: Genetischer Beitrag zum Problem der Degeneration. Z Ges Anat 13:371–415, 1927

80. HOAGLAND CL, SHANK RE, LAVIN GI: The histopathology of progressive muscular dystrophy revealed by ultraviolet photomicrography. J Exp Med 80:9–18, 1944

81. HOFFMANN J: Ein Fall von Thomsen schwer Krankheit, compliciert durch Neuronitis multiplex. Dtsch Z Nervenheilk 9:272–278, 1896

82. HORVATH B, PROCTOR JB: Muscular dystrophy: quantitative studies on the composition of dystrophic muscle. Res Publ Assoc Nerv Ment Dis 38:740–766, 1960

83. HUDGSON P, PEARCE GW: Ultramicroscopic studies of diseased muscle. Disorders of Voluntary Muscle. Edited by JH Walton. London, Churchill, 1969, pp 277–317

84. HURWITZ S: Primary myopathies: report of 36 cases and review of the literature. Arch Neurol Psychiatry 36:1294–1316, 1936

85. HUTCHINSON J: On ophthalmoplegia externa or symmetrical immobility (partial) of the eye, with ptosis. Med Chir Trans 62:307–329, 1879

86. JEDLOWSKI P: Rev Otoneuroophthalmol 20:203, 1943; quoted by Kiloh and Nevin[90]

87. JENDRASSIK E: Hereditäre Krankheiten. Handbuch der Neurologie. Edited by M Lewandowsky. Berlin, Springer, 1911, Vol 2

88. KESCHNER M, DAVISON C: Dystrophia myotonica: a clinico-pathologic study. Arch Neurol Psychiatry 30:1259–1275, 1933

89. KILOH LG, NEVIN S: Progressive dystrophy of external ocular muscles (ocular myopathy). Brain 74:115–143, 1951

90. KRÖSING R: Ueber die Rückbildung und Entwickelung der quergestreiften Muskelfasern. Virchows Arch [Pathol Anat] 128:445–484, 1892

91. KURÉ K, OKINAKA S: Behandlung der Dystrophia musculorum progressiva durch kombinierte Injectionen von Adrenalin und Pilocarpine. Klin Wochenschr 9:1168–1170, 1930

92. LANDOUZY L, DÉJERINE J: De la myopathie atrophique progressive (myopathie hereditaire) debutant, dans l'enfance, par la face, sans alteration du systeme nerveux. C R Acad Sci (Paris) 98:53–55, 1884

93. LANDOUZY L, DÉJERINE J: De la myopathie atrophique progressive; myopathie sans neuropathie, debutant d'ordinaire dans l'enfance, par la face. Rev Med 5:81–117, 253–366, 1885

94. LANDOUZY L, DÉJERINE J: Nouvelles recherches cliniques et anatomopathologiques sur la myopathie atrophique progressive a propos de six observations nouvelles, dont une avec autopsie. Rev Med 6:977–1027, 1886

95. LANGDEN HM, CADWALADER WB: Chronic progressive external ophthalmoplegia. Trans Am Ophthalmol Soc 26:247–260, 1928

96. LAPRESLE J, FARDEAU M, MILHAUD M: Etude des ultrastructures dans les dystrophies musculaires progressives. Proceedings of the 5th International Congress of Neuropathology, Zurich, 1965. Edited by F Luthy, A Bischoff. New York, Excerpta Medica, 1966

97. LEVINE PA, KRISTELLER L: Factors regulating the creatinine output in man. Am J Physiol 24:45–65, 1909

98. LEY J, TITECA J: Étude physiopathologique de deux cas familaux de myopathic distale tardive. J Belge Neurol Psychiatry 33:231–253, 1933

99. LEYDEN E: Klinik der Rückenmarks-Krankheiten. Berlin, Hirschwald, 1876, Vol 2

100. LILIENTHAL JL, ZIERLER KL: Diseases of muscle. Biochemical Disorders in Human Disease. Edited by RHS Thompson, EJ King. New York, Academic Press, 1957, pp 445–493

101. LITTLE WJ: On the Nature and Treatment of the Deformities of the Human Frame. London, Longman, 1853

102. LORENZ H: Die Muskele Krankungen, Spec Pathologie und Therapie. Edited by H Nothnagel. Wien, Holder, 1898, Vol 2

103. MAAS O, PATERSON AS: The identity of myotonia congenita (Thomsen's disease) dystrophia myotonica (myotonia atrophica) and paramyotonia. Brain 2:198–212, 1939; 73:318–336, 1950

104. MAIR WGP, TOME FMS: Atlas of the Ultrastructure of Diseased Muscle. Edinburgh, Churchill Livingstone, 1972

105. MARINESCO G: Maladies des muscles. Nouveau Traite de Medecine. Edited by PCH Brouardel, A Gilbert, L Thoinot. Paris, Bailliere, 1910

106. MERYON E: On granular and fatty degeneration of voluntary muscles. Med Chir Trans 35:73–84, 1852

107. MERYON E: Practical and Pathological Researches on the Various Forms of Paralysis. London, Churchill, 1864, pp 200–215

108. MICHELSON AM, RUSSELL ES, HARMON PJ: Dystrophia muscularis: a hereditary primary myopathy in the house mouse. Proc Natl Acad Sci USA 41:1079–1084, 1955

109. MILHORAT AT: Creatine and creatinine metabolism. Res Publ Assoc Res Nerv Ment Dis 32:400–421, 1953

110. MILHORAT AT, WOLFF HG: Studies in diseases of muscle. XIII. Progressive muscular dystrophy of atrophic distal type; report on a family; report of autopsy. Arch Neurol Psychiatry 49:655–664, 1943

111. MINKOWSKI M, SIDLER A: Zur Kenntnis der Dystrophia musculorum progressiva und ihrer Vererbung. Schweiz Med Wochenschr 9:1005–1009, 1928

112. MOEBIUS PJ: Periodische Oculomotoriuslähmung. Dtsch Z Nervenheilk 17:294–305, 497–499, 1900

113. MOORE MJ, REBEIZ JJ, HOLDEN M, et al: Biometric analyses of normal skeletal muscle. Acta Neuropathol (Berl) 19:51–69, 1971

114. PARTRIDGE: Fatty degeneration of voluntary muscles. Trans Pathol Soc Lond 1:334, 1846–1848

115. PEARSON CM: Serum enzymes in muscular dystrophy and certain other muscular and neuromuscular diseases. I. Serum glutamic oxalacetic transaminase. N Engl J Med 256:1069–1075, 1957

116. PEARSON CM: Histopathological features of muscle in the preclinical stages of muscular dystrophy. Brain 85:109–120, 1960

117. PENNINGTON RFT: Biochemical Aspects of Muscle Disease, Disorders of Voluntary Muscle. Edited by JN Walton. London, Churchill, 1969, pp 385–410

118. PICK F: Zur Kenntnis der progressiven Muskelatrophie. Dtsch Z Nervenheilk 17:1–56, 1900

119. RAVIN A, WARING JJ: Studies in dystrophia myotonica. I. Hereditary aspects. Am J Med Sci 197:593–609, 1939

120. REBEIZ JJ, CAULFIELD JB, ADAMS RD: Oculopharyngeal dystrophy—a presenescent myopathy: a clinico-pathologic study. Progress in Neuroophthalmology. Proceedings of the International Congress of Neuro-Genetics and Neuro-Ophthalmology, Montreal, 1967. Edited by JR Brunette, A Barbeau. Amsterdam, Excerpta Medica, 1969, Vol II, pp 12–31

121. RONZONI E, BERG L, LANDAU W: Enzyme studies in progressive muscular dystrophy. Res Publ Assoc Res Nerv Ment Dis 38:721–729, 1960

122. RONZONI E, WALD S, BERG L, et al: Distribution of high energy phosphate in normal and dystrophic muscle. Neurology (Minneap) 8:359–368, 1958

123. RONZONI E, WALD SM, LAM RL, et al: Ribosuria in muscular dystrophy. Neurology (Minneap) 5:412–418, 1955

124. ROSENTHAL M: Handbuch der Diagnostik und Therapie der Nervenkrankheiten. Erlangen F. Enke, 1870, p 220

125. ROSSOLIMO G: Atrophische Form der Thomsen'-schen Krankheit. Neurol Centralbl 21:135–136, 1902

126. ROTACH F: Veiträg zum Verlauf und zur Verebund der Dystrophia musculorum progressiva in einem Inzurchgebeit. Schweiz Arch Neurol Psychiatry 80:157–161, 1958

127. ROWLAND LP, ROSS G: Serum aldolase in muscular dystrophies, neuromuscular disorders and wasting of skeletal muscle. Arch Neurol Psychiat 80:157–161, 1958

128. SACHS B: Progressive muscular dystrophies: the relation of the primary forms to one another and to typical progressive muscular atrophy. J Nerv Ment Dis 13:726–755, 1888

129. SCHÄFFER J: Beiträge zur Histologie und Histogenese der quergestreiften Muskelfasern des Menschen und eininger Wirbelthiere. Sitzubgsg Kaiserl Akad Wiss Wien Math Nat Classe 102:1–142 (Abth 3), 1893

130. SCHAPIRA G, DREYFUS JC: Lacticodéhydrase plasmatique an cours des myopathies. C R Soc Biol (Paris) 151:22–23, 1957

131. SCHAPIRA G, DREYFUS JC, SCHAPIRA F: L'elevation du taux de l'aldolase sérique, test biochemique des myopathies. Sem Hop 29:1917–1920, 1953

132. SCHAPIRA G, DREYFUS JC, SCHAPIRA F, ET AL: Glycogenolytic enzymes in human progressive muscular dystrophy. Am J Phys Med 35:313–319, 1955

133. SCHROEDER JM, ADAMS RD: The ultrastructural morphology of the muscle fiber in myotonic dystrophy. Acta Neuropathol 10:218–241, 1968

134. SEITZ D: Zur nosologischen stellung des sogenannten scapuloperonealen syndroms. Dtsch Z Nervenheilk 175:547–559, 1957

135. SHANK RE, GILER H, HOAGLAND L: Studies on diseases of muscle. I. Progressive muscular dystrophy; a clinical review of 40 cases. Arch Neurol Psychiatry 52:431–442, 1944

136. SHY GM, MAGEE KR: A new congenital non-progressive myopathy. Brain 79:610–621, 1956

137. SLAUCK A: Pathologische Anatomie der Myopathien. Handbuch der Neurologie. Edited by O Bumke, O Foerster. Berlin, Springer, 1936, Vol 16, pp 412–431

138. SPILLER WG: The relation of the myopathies. Brain 36:75–114, 1913

139. STEINERT H: Myopathologische Beitrage. I. Ueber das klinische und anatomische Bild des Muskelschwunds der Myotoniker. Dtsch Z Nervenheilk 37:58–104, 1909

140. TAYLOR EW: Progressive vagus-glossopharyngeal paralysis with ptosis: contribution to the group of family diseases. J Nerv Ment Dis 42:129, 1915

141. THOMASEN E: Myotonia, Thomsen's Disease, Paramyotonia and Dystrophia Myotonica. A Clinical and Heredobiologic Investigation. Denmark, Aarhus, Universitetsforlageti, 1948, pp 1–251

142. THOMPSON RA, VIGNOS PJ: Serum aldolase in muscle disease. Arch Intern Med 103:551–564, 1959

143. TURNER JWA: The relationship between amyotonia congenita and cogenital myopathy. Brain 63:163–177, 1940

144. TURNER JWA: On amyotonia congenita. Brain 72:25–34, 1949

145. TYLER FH: Studies in disorders of muscle. III. "Pseudohypertrophy" of muscle in progressive muscular dystrophy and other neuromuscular diseases. Arch Neurol Psychiatry 63:425–432, 1950

146. TYLER FH, PERKOFF GT: Studies in disorders of

muscle. VI. Is progressive muscular dystrophy an endocrine or metabolic disorder? Arch Intern Med 88:175–190, 1951

147. TYLER FH, STEVEN FE: Studies in disorders of muscle. II. Clinical manifestations and inheritance of facioscapulohumeral dystrophy in a large family. Ann Intern Med 32:640–660, 1950

148. TYLER FH, WINTROBE MM: Studies in disorders of muscle. I. The problem of progressive muscular dystrophy. Ann Intern Med 32:72–79, 1950

149. VICTOR M, HAYES R, ADAMS RD: Oculopharyngeal muscular dystrophy: a familial disease of late life characterized by dysphagia and progressive ptosis of the eyelids. N Engl J Med 267:1267, 1962

150. VIGNOS PJ, LEFKOWITZ M: A biochemical study of certain skeletal muscle constituents in human progressive muscular dystrophy. J Clin Invest 38:873–881, 1959

151. VOGT A: Die Kataract bei myotonischer Dystrophie. Schweiz Med Wochenschr 2:669, 1921

152. VON EULENBERG A, COHNHEIM R: Ergebnisse der anatomischen Untersuchung eines Falles von sogennanter Muskelhypertrophie. Verh Berl Med Ges 1:191–205, 1866

153. VON GRAEFE A: Verhandlungen ärztlicher Gesellschaften. Berl Klin Wochenschr 5:125–127, 1868

154. WALTON JN: The limp child. J Neurol Neurosurg Psychiatry 20:144–154, 1957

155. WALTON JN: Progressive muscular dystrophy: structural alterations in various stages and in carriers of Duchenne dystrophy. The Striated Muscle. Edited by CM Pearson, FK Mostofi. Baltimore, Williams & Wilkins, 1973, pp 263–291

156. WALTON JN, ADAMS RD: The response of the normal, the denervated and the dystrophic muscle cell to injury. J Pathol 72:273–298, 1956

157. WALTON JN, GARDNER-MEDWIN D: Progressive muscular dystrophy and the myotonic disorders. Disorders of Voluntary Muscle. Second edition. Edited by JN Walton. London, Churchill, 1969, pp 455–499

158. WALTON JN, NATTRASS FJ: On the classification, natural history, and treatment of the myopathies. Brain 77:169–231, 1954

159. WEINSTOCK IM, EPSTEIN S, MILHORAT AT: Enzyme studies in muscular dystrophy. III. In hereditary muscular dystrophy in mice. Proc Soc Exp Biol Med 99:272–276, 1958

160. WELANDER L: Myopathia distalis tarda hereditaria. Acta Med Scand [Suppl] 265:141–175, 1951

161. WHITE AA, HESS WC: Some alterations in serum enzymes in progressive muscular dystrophy. Proc Soc Exp Biol Med 94:541–544, 1957

162. WHITE LP: Serum enzymes: variations of activity in disease of muscle. Calif Med 90:1–8, 1959

163. WILDER VM, MARGULIS S: Creatinuria in normal males. Arch Biochem Biophys 42:69–71, 1953

164. WILLIAMS JD, ANSELL BM, REIFFEL L, *et al:* Electrolyte levels in normal and dystrophic muscle determined by neutron activation. Lancet 2:464–466, 1957

165. WOHLFART G: Dystrophia myotonica and myotonia congenita: histopathological studies with special reference to changes in muscles. J Neuropathol Exp Neurol 10:109–124, 1951

166. WOHLFART S, WOHLFART G: Mikroskopische Untersuchungen an progressiven Muskelatrophien. Acta Med Scand [Suppl] 63:1–137, 1935

167. ZATUCHNI L, AEGERTER EE, MOLTHAN L, ET AL: The heart in progressive muscular dystrophy. Circulation 3:846–853, 1951

chapter 7

INFLAMMATORY DISEASES (MYOSITIS)

The term myositis denotes an inflammatory reaction in muscle of a type usually related to the presence of infection. It may be acute, subacute, or chronic, and it may involve one or many muscles. There are two general classes of myositis. In one, a virus, parasite, pyogenic bacterium, or spirochete can be identified as the causative agent; in the other no cause can be assigned, but the lesion is presumed to be inflammatory because of the nature of the histopathologic changes. The latter class has been subdivided into polymyositis, dermatomyositis, and neuromyositis, depending on the association of skin and peripheral nerve lesions, and is closely related to the rheumatic or connective tissue diseases (rheumatic fever, scleroderma, lupus erythematosus, and polyarteritis nodosa), whose pathogenesis and essential pathology have not been determined.

True myositis is a morbid condition characterized by exudation of plasma, infiltration of the supporting tissues by neutrophilic leukocytes or other inflammatory cells, and damage to the parenchymal and interstitial elements of muscle. Activation of histiocytes, hyperplasia of surviving sarcolemmal nuclei of the type regularly promoted by destruction of muscle fibers, and proliferation of fibroblastic connective tissue constitute the usual reaction to injury. The implication of infectivity derives from the presence of inflammatory cells if a bacterial agent, virus, or parasite cannot be isolated. However, some caution must be exercised, for a mild infiltration of inflammatory cells may occur as a response to purely physical injury such as heat coagulation (Chapter 3). When the noxious agent is unknown, there always remains some doubt as to the nature of the lesion. This accounts for the uncertainty in the classification of several diseases in which degeneration of the muscle fibers and infiltrations of the connective tissues by inflammatory cells are the dominant pathologic findings. For such diseases the term myopathy with some qualifying adjective would be more accurate but the term myositis is retained because of its common acceptance and usage.

A distinction should also be drawn between true myositis, with its characteristic focal or diffuse infiltrations of inflammatory cells in the supporting tissues and prominent reaction of connective tissue elements, and Zenker's hyaline degeneration, with its typical destruction of muscle fibers in the absence of inflammatory cell infiltrates or connective tissue reaction. Both may occur in the course of a systemic infectious disease, but one is determined by localization of the infective agent in the muscle and the other by some obscure toxic or metabolic disorder of the muscle fibers induced by the infection. Zenker's hyaline degeneration is therefore set apart from myositis and will be discussed in Chapter 11.

There are, however, conditions in which slowly progressive muscular degeneration in chiefly proximal muscles is associated with primary changes in muscle fibers, with little or no reaction in the connective tissue; nevertheless, these conditions differ from Zenker's degeneration and classic muscular dystrophy. In the clinic such cases may closely resemble chronic progressive types of polymyositis, and we therefore discuss them under this heading, though a more precise designation is chronic myopathy of probable metabolic origin. Necrotizing, vacuolar, cachectic, and carcinomatous myopathies come within this category.

VIRAL MYOSITIS

Despite the existence of several types of viral myositis in animals (Chapter 3) the occurrence of this type of infective inflammation of muscle in man has had an uncertain status. Recently, however, an influenzal myositis has been identified, but its pathologic anatomy is incompletely demonstrated. Lundberg,[86] Middleton et al.,[100] and Mejlszenkier et al.[98] have all written articles on the subject. In 2 of 74 cases of an epidemic of influenzal type of illness in which there was a severe myalgia (Lundberg termed it myalgia cruris epidemica) in the calves that lasted for more than a week, degenerative changes of nonspecific type were found in biopsy preparations. Middleton observed a similar syndrome in 21 of 26 cases of influenza, and in the majority of these there was a rise in serum creatine phosphokinase (CPK). Mejlszenkier's patient was a child with proved influenza A infection whose calf muscles became swollen, painful, and tender; the CPK level was elevated 20 times above normal. The biopsy showed extensive muscle necrosis, and infiltration of neutrophilic leukocytes, lymphocytes, and plasma cells in the endomysium. The blood vessels and skin were normal. No inclusion bodies were seen. There was no way of deciding whether the influenza A virus (a myxovirus) that was isolated from the bronchial secretions had attacked the muscle cells or had induced necrosis by the formation of virus-antibody complexes on the surface of muscle cells.

EPIDEMIC PLEURODYNIA (BORNHOLM DISEASE, DEVIL'S GRIP)

Epidemics of this disease were first described by Finsen[46] in Iceland in 1856 and 1863 and reported by him in 1874. Since then sporadic outbreaks and epidemics have been observed in Finland, Sweden, Norway, Denmark, England, Germany, and the United States. The epidemics usually occur in the summer and early fall at about the same time as those of poliomyelitis. Children between the ages of 5 and 15 are most susceptible, though adults are frequently infected. The spread is probably by human contact; transmission by water or insects has also been postulated. A full exposition of the clinical and epidemiologic aspects of this disease appears in the monograph by Sylvest[155] and more recently in Scadding's article,[134] as well as in those of Finn et al.,[45] Waldenström,[167] and Gordon et al.[51]

Coxsackie group B virus has now been established as the causative agent of epidemic pleurodynia. The viruses of this group have a characteristic tendency to induce severe visceral lesions in suckling mice, especially diffuse myositis (Chapter 3), patchy myocarditis, and focal lesions in the central nervous system and fat pads. Group A strains produce a generalized myositis in suckling mice.

Epidemics of pleurodynia are usually associated with group B type 5 virus,[51] but other members of this group are also implicated on occasion. Group B viruses also cause in humans aseptic meningitis, acute benign pericarditis, and in the newborn myocarditis. Laboratory diagnosis is made by isolating the virus from the stools and by a rise or a high stationary titer of complement-fixing antibody to Coxsackie virus. Many clinically asymptomatic persons have virus present in the stools during an outbreak of pleurodynia. Dalldorf[28] wrote an excellent review of the clinical and virologic aspects of this group of pathogenic agents.

An acute onset of severe pain in the chest—chiefly at the costal margins and in the substernal region—and less often in back, shoulders, abdomen, or hips is usually the first symptom. The affected muscles are tender to pressure, and pain is induced by voluntary or reflex contraction. Pain on inspiration may be intense, and the respirations are therefore rapid and shallow. A pleural friction rub may be present. The temperature is slightly elevated. There is no skin eruption. The acute pain and fever last several days, and after a remission one or more relapses are not uncommon. In

Finn's series the average duration of pain in the chest was 8 days. Pleocytosis and protein elevation of the cerebrospinal fluid occur in some cases.

No fatal cases have been reported, and therefore the pathologic changes are not known. Welborn[172] obtained a portion of an affected muscle, the latissimus dorsi, in an acute case by biopsy and was unable to detect either gross or microscopic abnormality. Wolfart[175] studied muscle biopsies in several cases; all were negative. Lepine[82] and others, however, claim to have observed inflammatory infiltrates in biopsies of human muscle.

BACTERIAL MYOSITIS

ACUTE SUPPURATIVE MYOSITIS

Acute myositis with pus formation may occur with bloodstream infections or may be associated with infectious arthritis, pleuritis, or decubitus, from which it arises by direct extension; or it may follow crush or infected bullet wounds. Infection within a muscle such as the deltoid or gluteus may also be induced by a therapeutic injection. The commonest causative organisms are staphylcocci and streptococci. The rarity of abscess formation in muscle, even in the face of overwhelming septicemia or septic embolization of many viscera, indicates that muscle tissue does not provide a suitable medium for the growth of many bacteria. Not a single case of focal embolic polymyositis has come under our observation, though isolated reports of such cases can be found in the medical literature. The onset of systemic symptoms in localized suppurative myositis is usually acute with headache, fever, chills, and sweating, the principal manifestations. These symptoms are followed shortly by local pain and edema of the affected muscle or muscles. Fluctuation of the infected tissue occurs within 5–10 days and indicates the formation of an abscess that requires evacuation by surgical drainage.

In the early stages of suppurative myositis there is edema and cellular inflammation. The muscle fibers themselves remain intact or show changes such as swelling, obscuration of the cross striation, and increased granularity or vacuolation. As the process extends, abscesses are formed or a phlegmonous (diffuse interstitial) myositis develops. All semblance of muscle architecture is obliterated, though the outlines of necrotic muscle fibers may persist amid an edematous inflammatory mass (Fig. 7–1). There is a great increase in cellularity within a few days of the onset of the infection. Neutrophilic leukocytes, lymphocytes, and plasma cells become mixed with macrophages, proliferating fibroblasts, and the adventitial cells of blood vessels. Muscle abscesses become encapsulated by fibroblastic connective tissue (Fig. 7–2). In severe lesions in which there is much destruction of muscle, extensive fibrosis results and may lead eventually to contracture. *Streptococcus pyogenes* (Lancefield group A) may produce local pain, gas formation, and general toxemia, a syndrome closely resembling clostridial myositis.[92]

CLOSTRIDIAL MYOSITIS

Gas gangrene is an important wound infection encountered in both civilian and military practice. Healthy muscle is quite resistant to clostridial organisms, but trauma which induces necrosis renders it susceptible to infection. Saprophytic bacteria, the most important of which are *Clostridium welchii,* cause rapid decomposition of the tissue with the formation of gas bubbles (carbon dioxide from muscle glycogen) and edema.

Gas gangrene usually develops in deep puncture or lacerated wounds and is especially common in association with compound fractures. The disease begins symptomatically with an abrupt rise in temperature and pulse. There is general malaise, prostration, restlessness, and apprehension. The wound becomes more painful, swollen, and red. The surrounding skin assumes a red, then dusky, and finally yellowish brown or bronze coloration. A serosanguineous exudate can be expressed from the wound. Gentle palpation of the swollen tissues reveals crepitus, and the wound has a characteristic

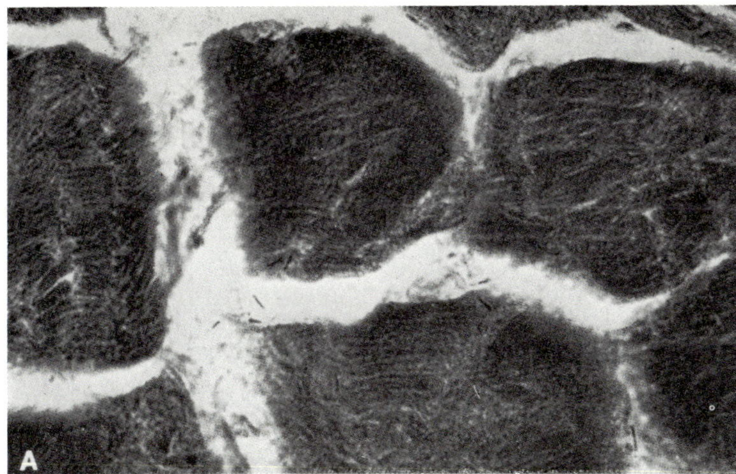

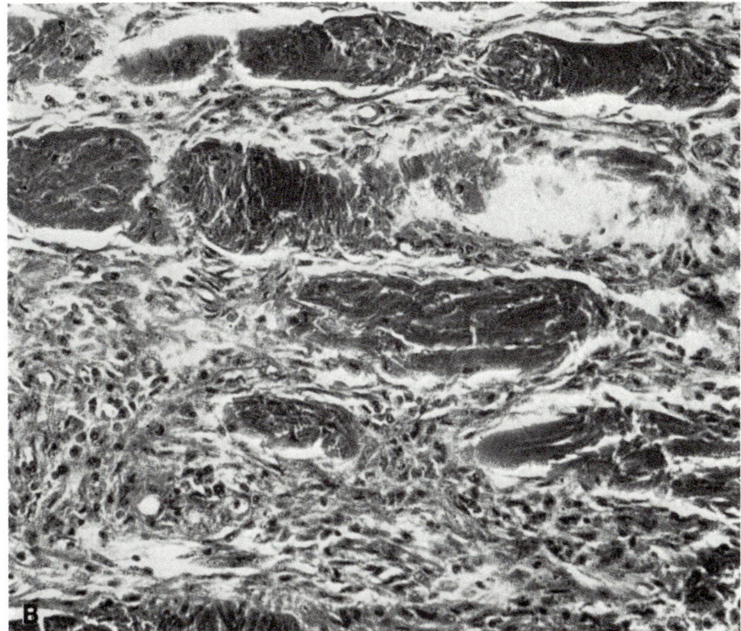

Fig. 7–1. Clostridial myositis. (A) Central necrotic zone with eosinophilic fibers and gas bacilli. (B) Necrotic muscle fibers showing floccular degeneration and phagocytosis. (H&E)

sweetish, acrid odor. The symptoms are due to the exogenous toxin produced by the clostridial organisms, to septicemia, and rarely to fat embolism from the solution of fat by the toxin.

The pathology of this form of myositis was described in detail by Govan[53] and Macfarlane and MacLennan.[91] More recent publications have added little to these earlier accounts. Rapid disintegration of the muscle is a striking feature. Grossly, in the early stages it may appear normal or semitranslucent but fails to respond to stimuli and does not bleed when cut. Later the muscle becomes pink gray and friable

and eventually a dirty blackish green. An interesting observation is that even in the early hours the involved muscle loses its normal toughness and elasticity as a result of the action of an enzyme, collagenase, which dissolves the supporting stroma. Such a muscle can be readily crushed between the fingers or smeared out on a slide. Microscopically, in the immediate vicinity of the wound there is necrosis involving all tissue elements. The muscle fibers show a coagulation necrosis with loss of striations and pale staining or disappearance of nuclei (Fig. 7–1). These dead fibers persist for

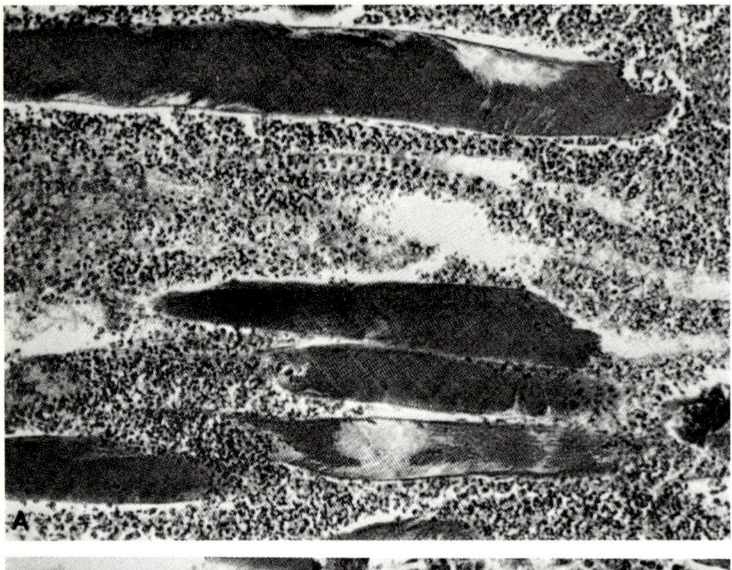

Fig. 7–2. Clostridial myositis. (A) Margin of fresh lesion showing necrotic fibers and exudation of fibrin and neutrophilic leukocytes. (B) Reparative process with proliferation of sarcolemmal nuclei and fibroblasts. (H&E)

some time but eventually become thin and disappear. In some places the fibers tend to fuse and later to liquefy. Vascuolation of fibers is conspicuous. Large numbers of gram-positive bacilli are visible both in muscle fibers and in connective tissues (Fig. 7–1A). The endomysial connective tissues and sarcolemma and all vascular and neural structures are destroyed. Exudation of fibrin is profuse.

Immediately beyond this zone of necrosis the changes vary somewhat depending on the severity of the infection. If mild, there is an intense infiltration of neutrophilic leukocytes and congestion of blood vessels, with isolated necrotic fibers (Fig. 7–2A). In severe cases the muscle fibers for some distance from the necrotic zone appear to be separated from one another by edema. The striations are preserved in some but not in others. Many of the sarcolemmal nuclei are pyknotic and either dark or pale in hematoxylin stains. Cells of the interstitial connective tissue are similarly affected. The connective tissue loses its argentophilic property and becomes granular. Only a few gram-positive bacilli are seen in this zone, and they are limited to the connective tissue. At the margin of the lesion, which is seldom well defined, hemorrhage and vascular congestion are pronounced. Some of the small blood vessels are occluded by fresh fibrin thrombi. Leukocytic infiltration is absent except at the border of the lesion. Frequently the clostridial infection spreads to surrounding tissues. Subcutaneous adipose tissue, for example, is often attacked and undergoes necrosis with nuclear pyknosis and karyolysis similar to that of muscle.

Several investigators have attempted to clarify the nature of the processes responsible for the rapid local and systemic effects of this infection.[48, 90] The bacillus elaborates large quantities of a toxin (theta toxin) and at least four enzymes (lecithinase, collagenase, hyaluronidase, and desoxyribonuclease). All these enzymes have potent chemical actions on tissue. Either alone or perhaps in combination with the toxin and breakdown products of necrotic muscle, they probably account for most of the local and systemic symptoms.

Treatment consists of wide debridement of necrotic muscle, the prompt use of an antitoxin locally and parenterally, and antibiotic therapy, especially penicillin. If the infection is controlled, the necrotic muscle tissue is eventually removed. Muscle regeneration is sporadic and inadequate (Fig. 7–2B) and a muscle defect and fibrosis is the usual result.

TUBERCULOUS MYOSITIS

Striated muscle is never involved by the initial or primary tuberculous infection except in rare cases of accidental inoculation by an infected needle. In the strict sense of the word there is no true primary tuberculous myositis. If muscle is implicated in a tuberculous process, it is usually by direct extension from a neighboring joint or cold abscess. Hematogenous spread to muscle, observed in very few cases,[124] is altogether exceptional. The high content of lactic acid is said to account for the rarity of generalized tuberculosis of muscle.

Tuberculous infection in muscle may assume three forms: (1) local extension from a neighboring tuberculous focus, (2) solitary metastatic tuberculous abscess or nodular tuberculosis, and (3) tuberculous polymyositis. The

first two are not in any way peculiar to muscle. As to the latter, we have observed only a few cases of nodular polymyositis of a tuberculous nature.

LOCAL EXTENSION FROM A NEIGHBORING TUBERCULOUS FOCUS

Some involvement of muscle in the region of a cold abscess is not at all uncommon. The infected material erodes the epimysium and spreads along the sheaths of the muscle. Muscle fibers and supporting connective tissue are destroyed by the caseating necrosis so characteristic of this disease. The extent of the necrosis and the degree of limiting fibroblastic reaction are believed to depend on factors of sensitivity and immunity that are beyond the scope of this monograph. The common sites of this local muscular involvement are the psoas, iliacus, and anterior abdominal muscles in extension from the common psoas abscess; the intercostal and other thoracic and paravertebral muscles in tuberculous empyema; and the quadriceps, femoral biceps, and gastrocnemius muscle in tuberculosis of the knee.

GENERALIZED OF MILIARY TUBERCULOSIS OF MUSCLE

Hematogenous spread to muscle was first reported by Reverdin;[126] and Delormé,[34] Lanz and de Quervain,[81] Kaiser,[69] and Plummer et al.[124] presented cases. If the lesion is small there may be no symptoms; a large focus is manifested by swelling and pain. The process, usually limited to one muscle, rarely involves several in succession. Either frank abscess formation with the formation of a sinus tract, or nodular sclerosis and finally calcification may result. The muscular infection is assumed to occur during the tuberculous bacteremia or septicemia that is responsible for dissemination of an active tuberculous focus in the lungs or, less often, another organ. The bacteria are said to lodge in the connective tissue outside intramuscular vessels and to involve the muscle fibers secondarily, as in other types of infection of muscle. Necrosis of tissue, activation of histiocytes to form epithelioid and giant cells, infiltrations of lymphocytes and plasma cells, and a variable fibroblastic proliferation result in the formation of typical tubercles. Tubercle bacilli can be seen in these tubercles but not in the muscle fibers. Muscle fibers adjacent to solitary or confluent tubercles are destroyed, and those at some distance undergo simple pressure atrophy. Some of the degenerating fibers lose their striations and show a granular or fatty change; others are hyalinized. Damaged muscle fibers may regenerate by budding, with the formation of multinucleated sprouts or clubs or muscle giant cells that are sometimes mistaken for Langhan's giant cells.

In miliary tuberculosis typical lesions may be found in muscles but are extremely rare. We have on a few occasions found such a lesion in a muscle biopsy from a child with obscure fever. The lesion consisted of a solitary Langhan's giant cell set among a cluster of lymphocytes and epithelioid cells. In several of the surrounding muscle fibers the sarcolemmal nuclei had increased in number and size, and the sarcoplasm tended to be basophilic. Rich and McCordock,[128] who studied the distribution of lesions in human cases of miliary tuberculosis and in animals that had received intravenous injections of tubercle bacilli, commented on the rarity of localization in organs other than the lungs, liver, spleen, and kidneys. The muscle lesions are asymptomatic.

NODULAR TUBERCULOUS POLYMYOSITIS

Nodular tuberculous polymyositis is a rare form of the disease; in some of the reported cases no attempt was made to distinguish it from Boeck's sarcoid. The distinction may in fact be very difficult. The clinical manifestations are those of tender nodular masses in muscles along with progressive weakness, atrophy, and reflex loss. Pseudocontractures may follow, the patient becoming completely incapacitated. Histologically the lesions are characteristic of tuberculosis with foci of epithelioid cells and lymphocytes and Langhan's giant cells in the perimysium (Fig. 7–3). In places muscle fibers are completely replaced by fibroblastic connective tissue. Some fibers

are in the process of granular or fatty degeneration; others, isolated in connective tissue, show regressive changes. Simple atrophy of muscle fibers adjacent to the lesions is a prominent finding (Fig. 7–3).

SYPHILITIC MYOSITIS

Whereas atrophy of muscle secondary to syphilitic disease of the spinal cord and the motor roots (syphilitic amyotrophy) was observed formerly not infrequently in the neurologic clinic of any large general hospital, a primary syphilitic inflammation of muscle (syphilitic myositis) is extremely rare. We have never observed such a case in either the clinic or the pathology laboratory. The medical literature contains numerous references to such a condition, but it is clear from a review of these cases that most were reported during the nineteenth century when the syphilitic infection was much more malignant than it is today and when diagnosis, without the aid of serologic tests, was much less accurate. It has been difficult therefore for the modern syphilologist to decide how many of the reported cases are authentic.

As far as can be determined, the syphilitic infection never occurs in muscle during the primary stages of the disease. Likewise, in secondary syphilis the myalgia, lassitude, and general enfeeblement, which formerly were quite frequent, do not appear to depend on a localization of *Treponema pallidum* in muscle or on any other recognizable affection of muscle. The few reported cases in which there was a syphilitic inflammation of muscle have occurred during the tertiary stages of the disease and have taken either of two forms: gumma or diffuse fibrosis of muscle.

MUSCLE GUMMA

Gummas are the most frequent tertiary manifestation of syphilitic muscle disease and were first described by Bouisson[11] and later Virchow.[162] Little has since been added to their original account of the condition, and the very

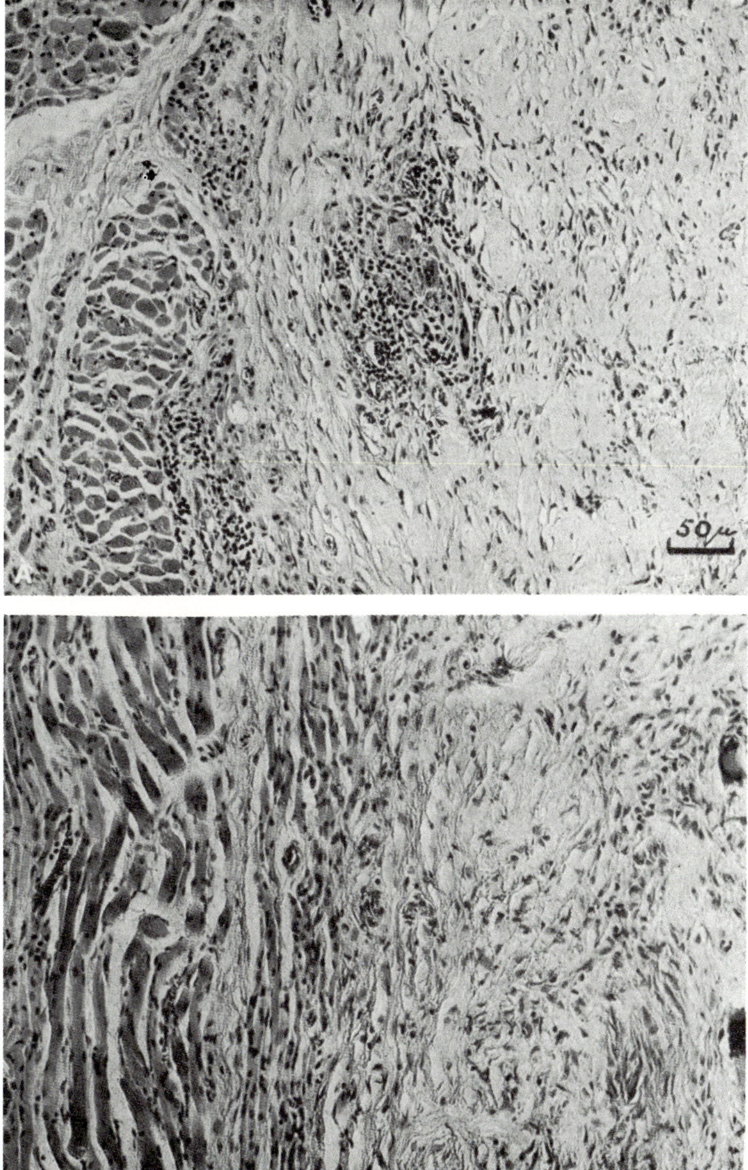

Fig. 7–3. Tuberculous myositis. Chronic generalized polymyositis with contractures in a 12-year-old boy with healed apical pulmonary tuberculosis. Note giant cells, fibrous tissue proliferation, and atrophy of neighboring muscle fibers. Caseous necrosis occurred in an adjacent part of the lesion. Tubercle bacilli were not found. (H&E)

few references that have appeared in recent years are given by Kierland and Underwood[73] in their case report of a solitary gumma of the sternocleidomastoid muscle. The lesion is solitary in most cases but may be multiple and rarely symmetrical on the two sides of the body. The sternocleidomastoid and biceps are involved most often, but isolated instances of gummas of the gluteal, femoral biceps, quadriceps, pectoral, trapezius, triceps, gastrocne-mius, forearm muscles, tongue, pharynx, larynx, eye muscles, diaphragm, and heart muscle are on record.

The clinical manifestations may be slight, and seldom is there more than a firm circumscribed swelling in the substance of a muscle. On macroscopic examination there is a well demarcated focus of yellowish gray tissue occupying one portion of the muscle. Microscopically the wall of the gumma is composed of dense fibrous connective tissue. Relatively homogeneous zones of necrosis in which all semblance of cell detail is lost, leaving only an argentophilic fibrous framework, are the most characteristic features of the lesion. Transitional forms of histiocytes, lymphocytes, and plasma cells infiltrate the connective tissue around such lesions. The hyperplastic connective tissue usually ceases abruptly at the edge of the lesion. Many of the adjacent muscle fibers have disappeared; others, in the process of degenerating, exhibit the same type of granular and fatty degeneration as that seen near a tuberculoma. Simple atrophy and regressive changes can also be observed. Proliferation of sarcolemmal nuclei and abortive attempts at regeneration can usually be found. Ultimately the lesion is transformed into a dense collagenous scar.

DIFFUSE SYPHILITIC MYOSITIS

Bouisson[11] presented one of the early accounts of diffuse interstitial myositis due to syphilis, and Notta[110] contributed additional cases. One or several muscles are affected, either simultaneously or in succession. Lesions of this type have been reported in the masseter, flexors of the fingers, pharynx, deltoid, biceps, sternomastoid, gastrocnemius, and external sphincter of the anus. The onset is gradual, with weakness, loss of tendon reflex, and pain if the periosteum of bone or contiguous nerves are involved. The muscle acquires an unusual firmness to palpation and becomes atrophied. Contractures are frequent. Once the process begins in any one muscle, it may progress to involve all of it and spread to adjacent muscles.

Histologically there is a diffuse and widespread loss of muscle fibers, and those which survive are separated by large amounts of fibrous connective tissue. Most writers have not commented upon cellular infiltrations or gummatous formations but have stressed the thickening of small arteries and the sparse perivascular lymphocytes, plasma cells, and mononuclear leukocytes. Some of the remaining muscle fibers have lost their striations or "have undergone fatty degeneration." Longi-

tudinal division of the fibers has been noted by Durante.[39] Relatively little evidence of regenerative activity can be seen. Osseous and cartilaginous transformation may be the end result of the disease process.

FUNGOUS MYOSITIS

Actinomycotic infection of muscle is the best known example of mycotic muscle disease. Nearly always there is a direct extension from a neighboring focus, e.g., from actinomycosis of the pleura or skin. Granulation tissue is abundant and has a tendency to fatty change, producing a grossly visible yellow color. The formation of abscesses and fistulas are the rule, and the characteristic yellow granules composed of colonies of the fungus *Actinomyces* can be seen in the draining material.

PARASITIC MYOSITIS

TRICHINOSIS

The most frequent and best known example of parasitic infection of muscle is trichinosis. As a disease it has plagued man and certain animals for several thousand years and may in fact have been partially responsible for the old Jewish law against eating the flesh of swine.[163] The earliest recorded observations of trichinae in human muscle were made by James Paget in 1835 (quoted by Gould[52]). The pathologic changes were studied extensively during the latter half of the nineteenth century, and the important monographs on muscle pathology of that era[39, 85] carried detailed descriptions of them which have not been improved upon in recent years. The most complete review of the subject was written by Gould[52] in 1945.

The causal agent, *Trichinella spiralis,* is a nematode that frequently infects the hog, rat, and (as a result of the ingestion of raw or poorly cooked pork) man. The encysted trichinae escape in the intestinal tract of man as the pork muscle containing them is digested.

These larvae develop and differentiate into adult male and female worms. The female measures about 3.0 mm and the male 1.5 mm in length. After fertilization the female burrows her way through the mucosa of the duodenum and jejunum and deposits several successive batches of embryos. The later gain entry into the lymphatics and are carried to regional lymph nodes, thence to the thoracic duct, and into the bloodstream. As many as 100 embryos have been demonstrated in 1 ml of blood. They are distributed to all parts of the body by the bloodstream and emigrate from the blood vessels to penetrate various tissues. They have been found in the liver, pancreas, heart, and brain, but only in striated muscle are conditions favorable for their growth and survival. All steps of the development and spread of trichinae have been followed in animals fed with infected pork, and most of the stages have been traced in man.

The symptomatology of trichinosis in humans is variable and depends to a large extent on the number of viable parasites ingested. If relatively few larvae enter the body there are no symptoms. When a large number are ingested —more than 100 per gram of muscle, according to Hall and Collins[56]—moderate to severe symptoms develop.

Three clinical stages in trichinosis are recognized. (1) The intestinal stage which is variable and when present appears from the second to the seventh day after ingestion of the infected meat. The symptoms are usually those of a mild gastroenteritis. (2) The stage of muscular invasion which begins at about the end of the first week and may last 5 weeks or longer. The accompanying symptoms are due to the effect on the tissues of the embryo trichinae or of hypothetical toxin elaborated by them. Chilliness and irregular, low-grade fever, pains in the trunk and extremities, muscular tenderness of variable degree, edema of the conjunctivae and orbital tissues, fatigue, and rarely prostration comprise the clinical picture. Muscle weakness may appear at this time, and when pronounced is sometimes associated with loss of tendon reflexes. This weakness may be generalized or limited to certain groups of muscles, e.g., ocular muscles with strabismus and diplopia, as in one of our

recent cases, or dysarthria, which was present in another. Unusually heavy infestations may provoke disorders of the central nervous system such as delirium, somnolence, coma, hemiplegia, or aphasia. Several such cases under observation at the Boston City Hospital were reported by Merritt and Rosenbaum[99] and Skinner.[142] In some there was a skin eruption. Evidence of myocardial involvement manifested by tachycardia and electrocardiographic changes is frequent. An eosinophilic leukocytosis commonly appears during this stage. This laboratory finding, a sensitivity to the trichina antigen when injected intradermally, and the appearance of antibodies aid in the diagnosis of obscure cases. (3) The stage of convalescence which is evidenced by the subsidence of symptoms and usually occurs during the second month of the infection. Persistence of rheumatic pains and stiffness, cachexia, and cardiac weakness have been reported but are exceptional. In our experience only in a minority of cases may all three stages be observed.

Trichinosis is seldom fatal even with quite heavy infestations. Myocarditis, with or without cerebral embolism, is the usual cause of death. The pathology of the disease in humans has been determined chiefly from muscle biopsies taken during the invasive stages of the illness and from chance autopsies of cases in which trichinosis was an incidental finding. In the United States this is often possible because of the high incidence of trichinosis—between 10 and 22% of the adult population in certain parts of the country.[56, 103]

PATHOLOGY

The most important pathologic changes are in the striated muscles. If the infestation is heavy the gross appearance of the muscle during the acute stage may suggest the diagnosis, for the muscle tissue is rather pale, soft, and granular, often termed "mealy." In the later stages, after the fifth week, small, elongated, grayish streaks can be detected, and the texture of the muscles may be more firm and resistant because of connective tissue proliferation.[23] From 6 to 18 months after infection, whitish specks of calci-

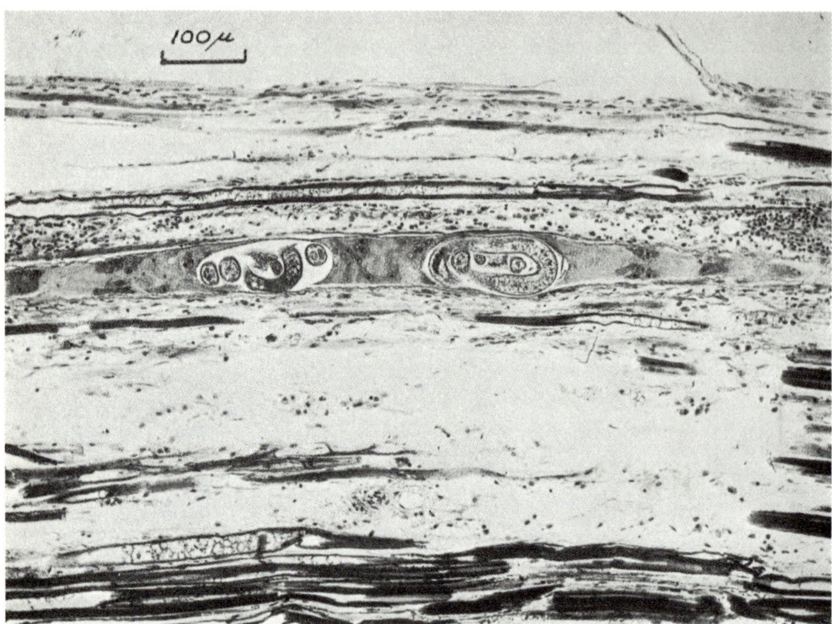

Fig. 7–4. Trichinosis. Two parasites are seen within same muscle fiber. (phosphotungstic hematoxylin)

fication are left, which grate upon the section knife. These are slightly elongated and ovoid, measuring 0.3–0.8 mm in length and 0.2–0.4 mm in width.

The young trichina as it first appears in the muscle fiber is a slightly curved worm 15–20 μ wide and approximately 250 μ long. As the parasite enlarges within the confines of the muscle fiber, it gradually curls up into the characteristic spiral form (Figs. 7–4 and 7–5) about the tenth day after invasion of the muscle. As it grows the adjacent sarcoplasm becomes more dense and homogeneous, as though compressed by the parasite. The embryo trichina causes degeneration of the part of the muscle fiber it invaded. Under the microscope the first change is eosinophilia and granular alteration of the sarcoplasm with loss of cross striations. This occurs during the first day or two of the invasion in parts of the muscle fiber adjacent to the larva.[108] This segment of the fiber then becomes edematous and fragmented, but its sarcolemmal sheath usually remains intact. Other changes indicative of damage to muscle substance, such as Zenker's hyaline or waxy degeneration, fragmentation,

or vacuolation, also occur. Several of these changes may be seen in one field (Figs. 7–6 and 7–7). The sarcolemmal nuclei rapidly enlarge and multiply [42, 55] and may be difficult to distinguish from invading phagocytic cells (Fig. 7–6).

The sarcoplasm of intact portions of the fiber becomes basophilic, and the nuclei migrate centrally from their subsarcolemmal position and also congregate at either end and along the sides of the parasite (Fig. 7–4). At this time an intense inflammatory reaction develops in the surrounding connective tissue (Fig. 7–5A). The cellular infiltrations consist predominantly of polymorphonuclear leukocytes, both eosinophilic and neutrophilic, but lymphocytes, plasma cells, and other mononuclear cells are usually present in smaller numbers. In severe infestations the cellular infiltrates may be diffuse (Fig. 7–5), but in milder ones they are localized around blood vessels and single muscle fibers (Fig. 7–6). Some of the parasitized muscle fibers are invaded by inflammatory cells. Neighboring muscle fibers may occasionally undergo hyaline degeneration and fragmentation, even though not invaded by a

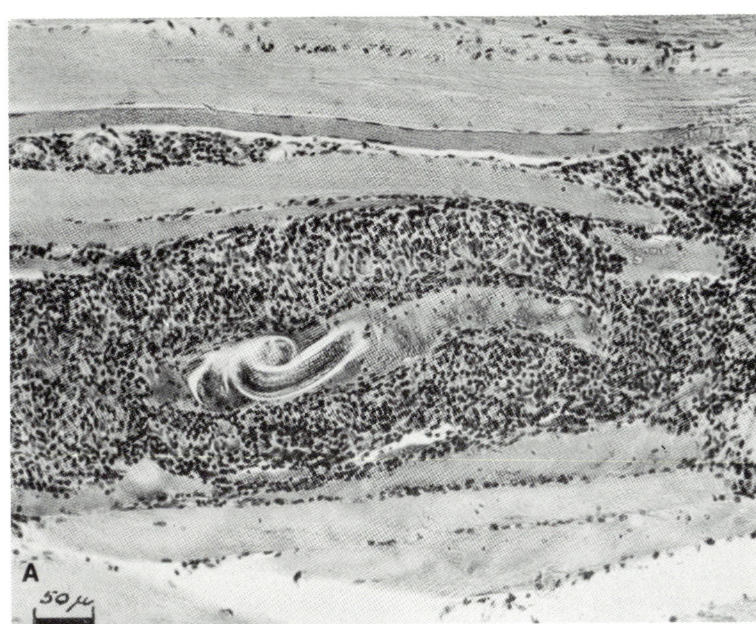

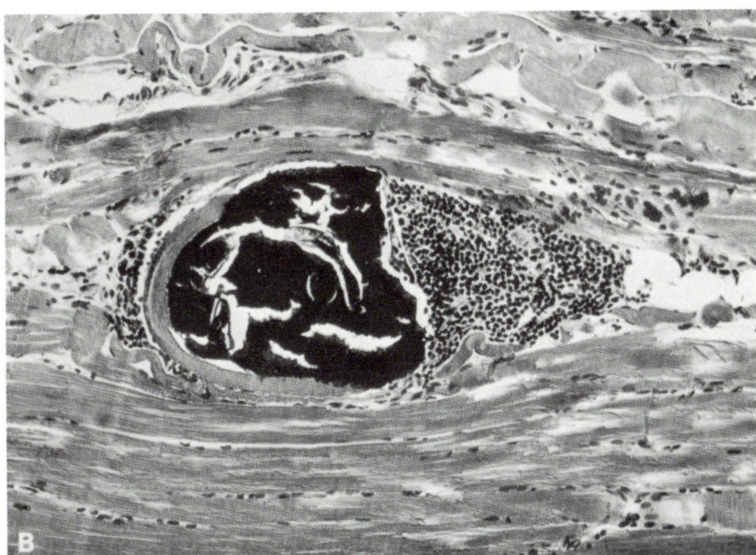

Fig. 7–5. Trichinosis. (A) A young parasite in a necrotic muscle fiber is surrounded by a large aggregation of inflammatory cells, mostly neutrophilic leukocytes. Sarcolemmal nuclei of parasitized muscle fiber have proliferated. (B) The capsule, seen well only on the left side of the calcified parasite, is thick and surrounded by lymphocytes. (H&E; ×200)

parasite, an effect attributed to a hypothetical toxin. If the fiber or part of the fiber which harbors the parasite becomes necrotic, as shown by the shrinkage and pyknosis of sarcolemmal nuclei and invasion by inflammatory cells, the parasite may die. The destroyed muscle is removed by the action of histiocytes, which may rarely form foreign-body giant cells around the parasite. After the trichina reaches a certain size, it apparently cannot invade another muscle fiber.

In the heart, liver, brain, and other organs the trichinae give rise to small inflammatory foci but soon die. Small foci of fibrosis may remain after the inflammatory reaction subsides.

When any part of the invaded fiber is destroyed, nuclear proliferation occurs in the remaining portion. The reaction is most active in the immediate vicinity of the parasite or in those parts of the fiber in which some crude change in the sarcous contents has occurred.

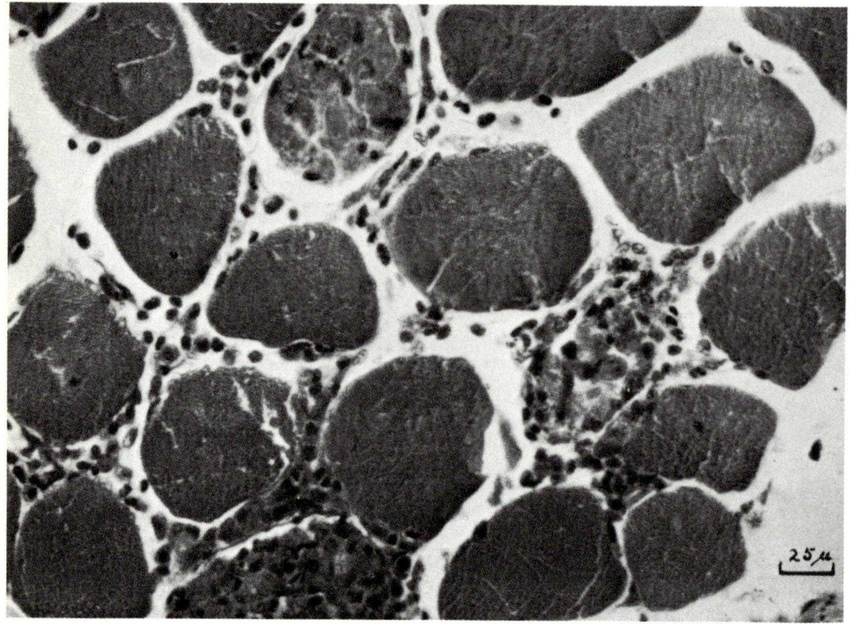

Fig. 7–6. Trichinosis. Section of a muscle biopsy specimen showing degeneration of isolated muscle fibers, which are replaced by macrophages and eosinophilic leukocytes. Eosinophilis, neutrophils, and lymphocytes are also seen in endomysial connective tissue.

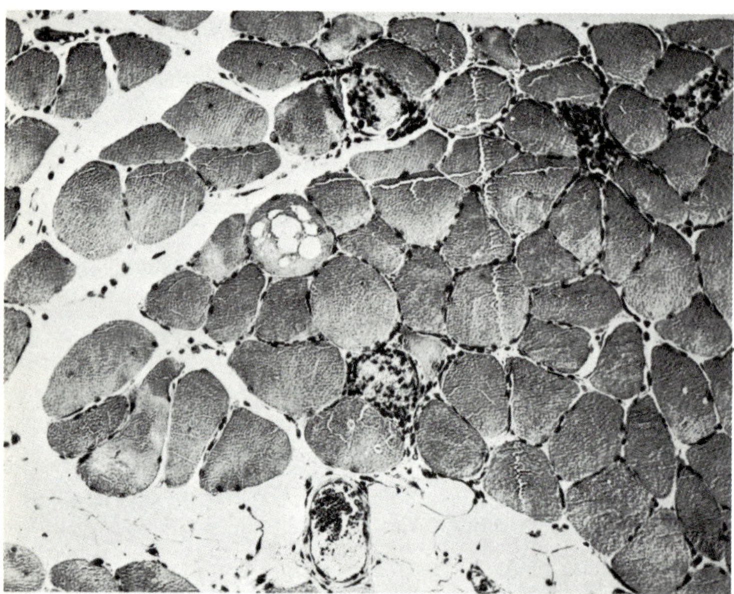

Fig. 7–7. Trichinosis. Section of muscle biopsy specimen showing various types of muscle fiber reaction. Parasites were difficult to demonstrate in this specimen, and only the presence of eosinophils in the cellular reaction led to their identification after further search.

The basophilia of the sarcoplasm and hyperplasia of sarcolemmal nuclei indicate regeneration. Neighboring muscle fibers become thin and atrophic. In biopsies at this stage the disease is difficult to differentiate from other types of polymyositis.

About the fifth or sixth week after invasion of the muscle fiber the parasites gradually become encapsulated. Many theories about formation of the capsule have been advanced. The explanation first offered by Virchow[163, 164] has much merit. He postulated that the inner part of the cyst wall was constructed from the compressed basophilic degenerated muscle sarcoplasm, and that an outer hyaline portion was contributed by thickened sarcolemma and adjacent connective tissue. Another interesting theory is that the capsule is formed of protein materials as a result of an antigen–antibody interaction in which the antigen is supplied by the parasite and the antibody from the circulation or from the invading inflammatory cells. The permanent capsule is completed at about the third month, by which time the surrounding inflammatory reaction has largely subsided. By now only small numbers of lymphocytes and histiocytes remain, and these disappear a few months later. The capsule consists of a thick oval ring with a small central parasite and later becomes a solid oval mass of hyaline connective tissue. Calcification occurs 6 months to 2 years after the initial infection and may include a portion or all of the capsule of the parasite (Fig. 7–5B). Within these capsules some of the parasites may remain viable for long periods of time—as long as 31 years,[81] but a progressively greater proportion die each year after the first. In the later stages encysted larvae, focal loss of muscle fibers, and fibrosis are the only residual abnormalities.

The muscles most susceptible to invasion by trichinae are the diaphragm, extraocular, tongue, laryngeal, jaw, intercostal, neck, back, abdominal, and limb muscles, in that order. This distribution of muscle damage is somewhat similar to that which follows the experimental administration of plasmocid (Chapter 3) and suggests that some common biochemical factors may determine localization of the disease in both instances.

The proclivity of the parasite to invade the muscle fiber is one of its remarkable features and is not understood. Surely the parasite possesses some particular metabolic requirement that can be satisfied only in a sarcous habitat. It is not clear whether the mechanism of muscle fiber destruction is invasion by the parasite or results from action of a toxin elaborated by the parasite. Muscle weakness can be accounted for by the effect of the interstitial inflammatory reaction; its transient character attests to the functional recovery of indirectly affected fibers. We have no entirely satisfactory explanation for the apparent absence of parasites in some biopsy specimens in which muscle fiber degeneration and infiltration of inflammatory cells are pronounced. It may be that in some cases the parasites are destroyed early in the course of the disease. The orbital and facial edema are presumably related to inflammation of the ocular and facial muscles. The derangement of the central nervous system has been attributed to embolization of the smaller cerebral arteries by trichinae (Hassin and Diamond)[58] or by blood clots from the heart secondary to myocarditis. It is remarkable that although the parasite has been demonstrated only rarely in the heart or brain, these organs nevertheless commonly show multiple small areas of infarct necrosis, each related to arteritis. It is possible that such lesions represent a diffuse toxic vascular reaction.[47] Such a reaction may also account for widespread areas of myositic atrophy in muscles in which parasites occupy relatively few of the fibers.

CYSTICERCOSIS

Cysticercosis is quite common in Eastern Europe and India[43, 88] but rare in the United States. It is caused by ingestion of the eggs of the pork tapeworm *Taenia solium*. The larval parasites, known as *Cysticercus cellulosae,* invade the intestine, enter the bloodstream, and are distributed to all parts of the body. The stage of invasion, like that of trichinosis, may be associated with muscular tenderness, fever, and eosinophilia. Nodules may be palpated in the tongue and other muscles, and there may

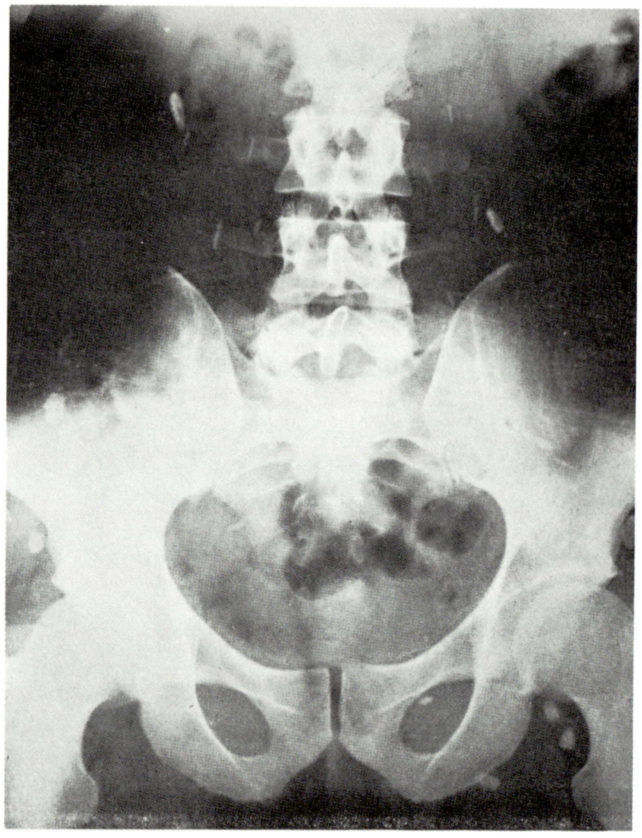

Fig. 7–8. Cysticercosis. Roentgenogram of pelvis and lumbar spine showing calcified cysts of various sizes in lumbar, iliac, and pelvic muscles, as well as in the abdominal wall.

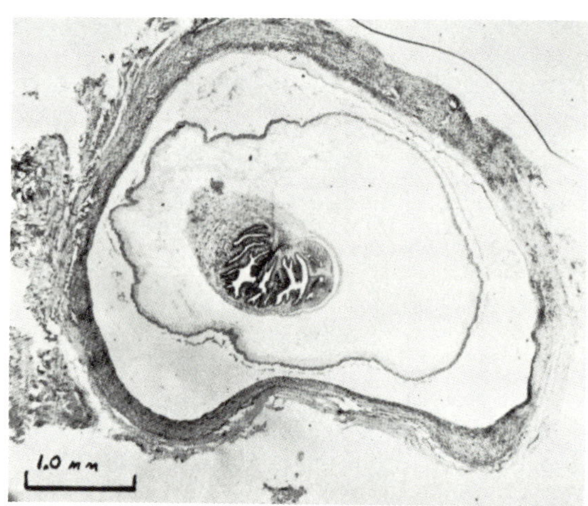

Fig. 7–9. Cysticercosis. Cyst has been teased away from muscle tissue and sectioned separately.

be enlargement and weakness of the calf muscles, as was reported by McGill[95] in a patient the author was privileged to examine. Usually the invasive stage is asymptomatic, and the disease is discovered later, either by the x-ray appearance of muscles (Fig. 7–8) or brain or both, or by the epilepsy which results from chronic brain lesions.

The muscular lesions consist of many small vesicles containing the parasites and a mild interstitial myositis with the presence of neutrophilic and eosinophilic leukocytes, lymphocytes, and histocytes. The appearance of a typical cyst, dissected free from a biopsy of muscle and sectioned separately, is shown in Figure 7–9. The parasites are situated in the interstitial connective tissue, in which they assume a spindle shape because of pressure from the surrounding muscle tissues. In time most of the inflammatory cells disappear, leaving only a few lymphocytes and mononuclear cells. The parasites ultimately become calcified. This does not occur until 1–5 years after the infestation and may be delayed even longer in the brain. The size of the calcified nodules varies greatly (Fig. 7–8), and in an early stage the appearance may be that of small rods, commonly thicker at one end. Later the appearance is that of thick, blunt, short rods. The diaphragm, proximal limb muscles, and tongue are preferred sites, and the diagnosis is commonly made by the characteristic shadows in roentgenograms.

ECHINOCOCCOSIS

Human echinococcosis of muscles is extremely rare, even in countries in which hydatid disease is common (South America, New Zealand, Australia, Iceland). Faust[43] states that the liver is most frequently involved, and on the basis of statistics compiled by various authors the muscles are invaded in only 0.7–9.1% of the cases. The embryo parasites of the tapeworm *Echinococcus granulosus* reach the muscle via the blood vessels and lodge in small capillaries, where they immediately become surrounded by neutrophilic and eosinophilic leukocytes, lymphocytes, and histiocytes. The embryos that survive form small cysts, which slowly enlarge and after about 5 months reach a diameter of 1 cm. In muscle the cysts are solitary, large, and very rarely multiple. About the cyst wall are epithelioid cells, giant cells, and eosinophils, all of which are surrounded by an outer layer of fibroblasts infiltrated with eosinophils. The surrounding muscle fibers are compressed, some are atrophic, and a few show granular and vacuolar degenerative changes. Usually no clinical evidence of muscle invasion is present unless the individual cysts rupture, forcing out

the scolices and causing a large number of daughter cysts to form. Sites of predilection in the musculature are the posterior parts of the trunk, inner side of the thigh, neck, and upper arm.

SARCOSPORIDIOSIS

Apart from the malarial parasites, only two types of sporozoa, Sarcosporidia and Coccidia, are known to infect humans. *Toxoplasma,* the causative agent of toxoplasmosis, probably belongs to this group, but there is some uncertainty as to its classification.

The Sarcosporidia are parasites occurring usually in the striped muscle of vertebrates, especially mammals. Although discovered by Miescher[101] in 1843, little has been learned concerning their life history. They are found in most domestic animals; in various wild animals, birds, and fish; and occasionally in man. The muscles most frequently parasitized are those of the heart, esophagus, larynx, diaphragm, and abdomen, but in some animals, especially mice, most of the skeletal muscles may be invaded. The Sarcosporidia are found in the striated muscle fibers and connective tissue and occasionally in smooth musculature as well. The exact manner whereby an animal becomes infested is unknown, but it is likely that the ripe spores are ingested in the food and gain entry to the musculature through the intestinal wall and the bloodstream. *Sarcocystis muris,* the sarcosporidium occurring in the skeletal muscles of the common mouse, has been transmitted[144] by feeding infected muscles to normal animals. The organisms, after being carried to the muscle, multiply and develop elongated, spindle-shaped, membranous capsules having rounded ends (Fig. 7–10) but no internal septa. The cysts are larger than those of *Toxoplasma,* commonly being 1 mm in width and sometimes as long as 5 mm; they are filled with myriads of sickle-shaped spores or sarcocysts (sporozoites), each with a distinct nucleus but no kinetoplast or parabasal body. The small sarcosporidial tubes are sometimes mistaken for trichinae.

A subgenus, *Balbianidae,* grow in the endo-

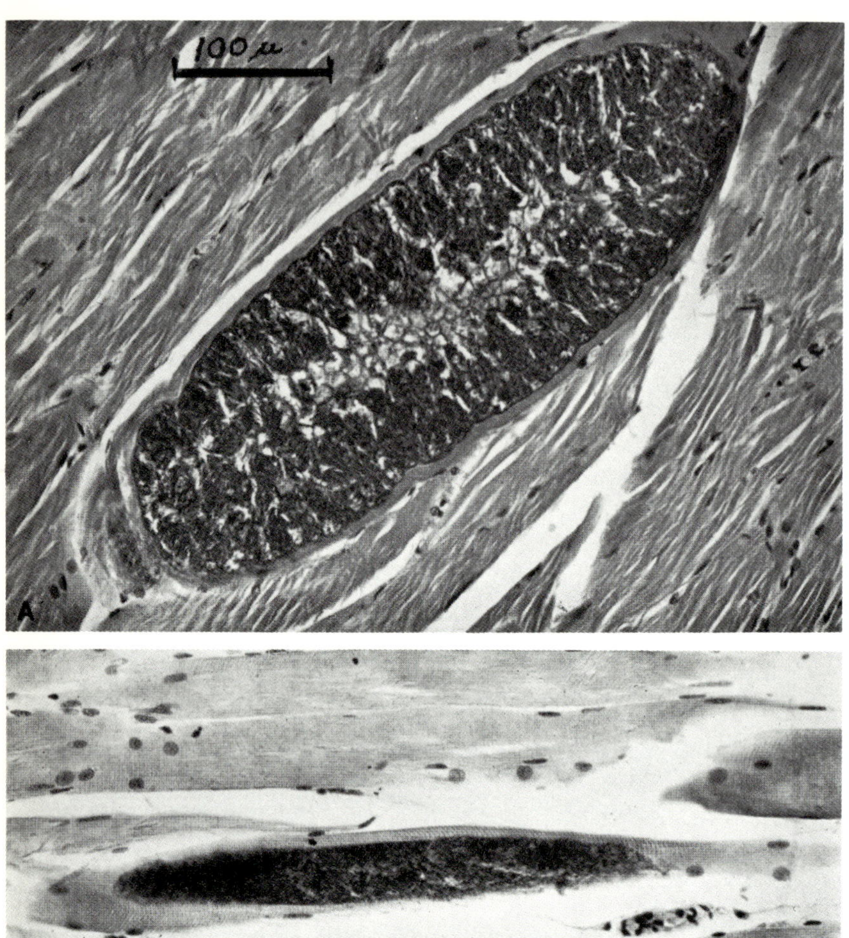

Fig. 7–10. Sarcosporidiosis in man. Note that in the early stage the cyst begins in a muscle fiber, as shown in the lower figure. (H&E)

mysium instead of in the muscle fibers. This variety has not been reported in man. Lorenz,[85] who reviewed this subject, describes microsporoids within the fibers of a patient having dermatomyositis of 5 months' duration that ended in recovery. Such a condition must have been an extreme rarity, for Lorenz never saw another case; but it emphasizes the wide etiologic possibilities in this class of muscle disorder.

The name *Sarcocystis lindermanni* is usually applied to the sarcosporidium in man, although there is no definite basis for the differentiation of organisms of this type into separate species. Relatively few human cases have been reported.[30, 44] Usually, discovery of the organisms has been quite accidental, occurring in cases in which there was no history of muscle disease. The two cases reported by Dastur and Iyer[31] are exceptions in that weakness, aching, and impaired tendon reflexes had been found. McGill and Goodbody[96] reported a case in which generalized periarteritis was possibly related to the infection, though the cysts in their case—one of which is shown in Figure 7–10—had typically excited no local reaction. Cysts have been found in heart muscle.[50] The relationship between human *Sarcocystis* and *Toxoplasma* has not been clarified.[135] Cathie and Cecil[17] found the positive dye test specific for *Toxoplasma* and pointed out that sheep sarcospores cannot be used in the antigenic test for *Toxoplasma* because of this specificity and also because many human sera contain a substance capable of modifying the staining reactions of the spores of *Sarcocystis tenella* from sheep.

Despite this lack of clarity, there is reason to believe that many of the reported cases of infection with sarcocystis are examples of toxoplasmosis.

TOXOPLASMOSIS

Toxoplasma, a protozoan parasite of uncertain classification, was first described by Nicolle and Manceaux[108] in North African rodents and by Splendore[148] in Brazilian rabbits. Spontaneous infections have since been reported in many other species of mammals, reptiles, and birds in many parts of the world. The first instance of human toxoplasmosis, that of an infant with encephalomyelitis and chorioretinitis, was published by Janku[67] and a second by Torres.[158] *Toxoplasma* was described but not identified in the first of these cases and was mistaken for *Encephalitozoon* in the latter. Wolf *et al.*[176, 180] in a series of publications presented other cases with full clinical and pathologic notes and correctly identified the organism. They deserve the credit for bringing this serious infection to the attention of the medical profession, for delineating the clinical syndrome in infants and children (mental retardation, hydrocephalus, convulsive seizures, cerebral calcification, and chorioretinitis), and for describing the pathologic changes (the granulomatous meningoencephalitis and chorioretinitis). Pinkerton and his associates[122, 123] reported adult cases with

fever, generalized maculopapular rash, and delirium—a clinical picture resembling that of typhus fever. Similar cases have been published by Sabin[132] and Callahan *et al.*[16] The latter paper includes a review of the literature with a complete bibliography as of that time. Acute myocarditis has been reported in adults,[125] as has acute chorioretinitis.[66] It is said that humans are probably infected from animals, mainly sheep, whose flesh is handled and eaten. There have been several reports of illness in a pet dog in the family of patients suffering from the disease, but the large number of possible animal vectors makes identification difficult.[125] Infants acquire the infection in utero from mothers who have been recently infected but have no symptoms of the disease. Their period of infectivity lasts only a few months.

Toxoplasma has been confused with other protozoan parasites, particularly with *Sarcocystis, Encephalitozoon, Leishmania,* and avian malaria. Some of the early cases of sarcosporidiosis, especially those in which the cysts are very small, are probably examples of toxoplasmosis. Parasitologists have not succeeded in classifying all the numerous strains of *Sarcocystis.* The *Toxoplasma* is a small organism (2–3 μ by 4–7 μ), is crescentic or curved, and occurs in pseudocysts.

Specific symptoms of skeletal muscle disease have not been reported, though *Toxoplasma* has been found in the skeletal muscle in 4 of 14 cases in which this tissue was examined, and in the myocardium in 6 of 16 cases.[16] The value of the muscle biopsy in clinical diagnosis was demonstrated in the case of an elderly woman admitted to one of the medical wards of the Boston City Hospital with irregular fever, delirium, and a purpuric skin eruption over the trunk, extremities, and face. There was laboratory evidence of liver disease, and slight changes in the electrocardiogram were present. *Toxoplasma* were found in a muscle biopsy. These parasites were isolated by intraperitoneal injection of a suspension of crushed muscle into mice. The patient developed a high titer of antibodies, $1:18,000$. The disease proved to be fatal, and the diagnosis was confirmed by postmortem examination.[4] In addition, there was a focal disseminated encephalomyelitis.

The pathologic change in the muscle may be

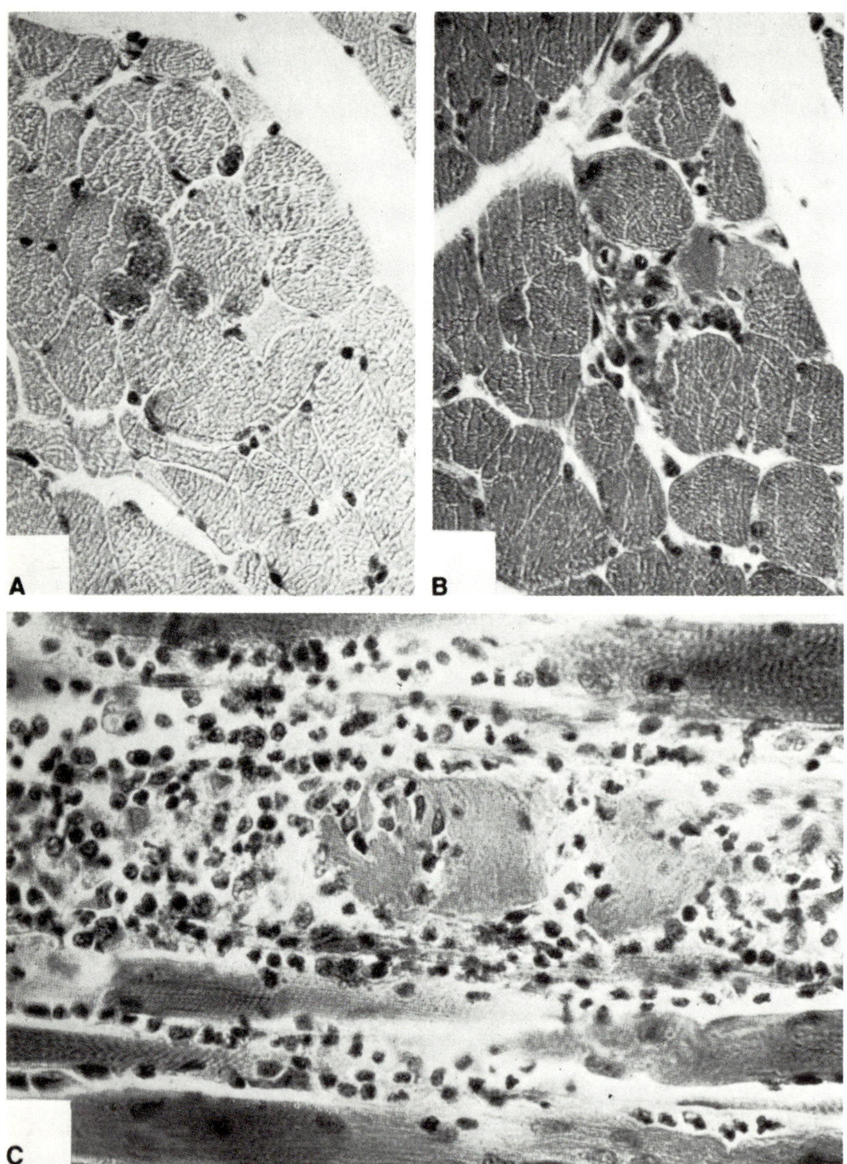

Fig. 7–11. Toxoplasmosis in muscle biopsy from an elderly woman suffering from myositis and otherwise unidentified spotted fever. (A) Cross section of gastrocnemius muscle showing pseudocyst within muscle fibers. (B) Degeneration of fibers with clusters of macrophages and hyperplastic sarcolemmal nuclei. (C) Collection of inflammatory cells: plasma cells, neutrophilic leukocytes, and lymphocytes. (phloxine, methylene blue, ×400)

characterized as a focal disseminated myositis. Parts of muscle fibers are destroyed in all the muscles. In each necrotic focus, which measures 50–100 μ in diameter, the damaged fibers are eosinophilic, hyalinized, and invaded by a few neutrophilic leukocytes (Fig. 7–11). In older lesions macrophages actively phagocytize the damaged fibers, and sarcolemmal nuclei proliferate from nearby healthy parts of the fiber (Fig. 7-11B). Lymphocytes, mononuclear leukocytes, and a few plasma cells infiltrate the adjacent endomysial connective tissue (Fig. 7–11C). Among some of these foci of inflammatory cells a few *Toxoplasma* can be indentified in Giemsa-stained sections. Giant cells are not usually found. There is evidence of regenerative activity in many of the surviving muscle fibers. Sarcolemmal nuclei are increased in number and are enlarged, and the sarcoplasm is basophilic. Many of the regenerating fibers are thin. In places pseudocysts are found; these are round or oval and range from 20–60 μ in diameter (Figs. 7–12 and 7–13). Some of these, although situated within the muscle fiber just beneath the sarcolemma, do not cause even the slightest inflammatory reaction (Fig. 7–11A and 7–12), and the parasitized fibers appear to be otherwise quite healthy. When there is evidence that the cyst has ruptured with liberation of the organisms (Fig. 7–13), the muscle fiber undergoes focal necrosis with resultant infiltration of inflammatory cells. This probably accounts for the difficulty in demonstrating pseudocysts in the necrotic and inflammatory foci.

TRYPANOSOMIASIS (CHAGAS' DISEASE)

Chagas' disease, caused by *Trypanosoma cruzi* and indigenous to South America, has occasionally been reported in Arizona, California, New Mexico, and Texas. Reduviid (kissing) bugs and rodents are the common hosts, and part of the parasite's cycle of development is passed in them. It is believed that the infection is transmitted to humans by the dejecta rather than by the bite of these bugs. *Trypanosoma cruzi* differs from other species pathogenic to man in that the causative organisms appear in the blood as a trypanosome and in the tissues in leishmanial form. In the blood it can be differentiated from *Trypanosoma gambiense* and *Trypanosoma rhodesiense* by its narrow undulating membrane, the extremely large kinetoplast, and the shape (usually in the form of a C or U). Multiplication takes place in endothelial cells and in the parenchymal cells of the viscera, the heart, and skeletal muscle fibers. In the cells the parasite loses its flagellum and undulating membrane.

The clinical manifestations of the disease are variable. Some individuals harbor the organisms but exhibit no signs of infection. The disease often takes an acute form in infants and children with fever; enlargement of liver, spleen, and lymph nodes; and edema of the face and eyelids. Symptoms of involvement of the nervous system (e.g., headache, drowsiness, or continuous sleep with periods of excitement or convulsions) predominate in the fatal cases. An ill-defined chronic form of the disease occurs in adults.

The principal gross pathologic changes, reviewed by Wolf and his associates,[181] are found in the heart and brain. The pericardial and pleural cavities contain a cloudy fluid, and the surfaces are covered with fibrinous exudate. The heart is enlarged, and the myocardium is pale and sometimes flecked with petechial hemorrhages or yellowish spots. The brain is swollen and congested, and occasionally there are scattered petechial hemorrhages in both the gray and white matter. Microscopic examination reveals a disseminated focal polymyositis, myocarditis, and encephalomyelitis. Parasites are localized within the fibers of skeletal muscles (Fig. 7–14). They appear as small, thin-walled cysts loaded with trypanosomes. In this form the small bodies within the pseudocysts are difficult to distinguish from *Toxoplasma* and Leishman-Donovan bodies. Parasitized muscle fibers undergo hyalinization, vacuolation, and fragmentation. Necrotic portions of the fibers are at first invaded by a few neutrophilic lymphocytes and later are phagocytized by macrophages. Proliferation of sarcolemmal nuclei in undamaged portions of the muscle fiber is prominent. Infiltrations of inflammatory cells, lymphocytes, plasma cells, and large mononuclears occur in the endomysial and perimysial connective tissues. A similar parasitic invasion and inflammatory reaction is found in the myocardium. There are fewer parasites in the brain, and these organisms usually occur in relation to small nodules composed of transitional forms of microglial cells and histiocytes, lymphocytes, and plasma cells, which in many respects resemble the nodules of typhus.

The diagnosis can be made by finding the parasites in the bloodstream. On the basis of

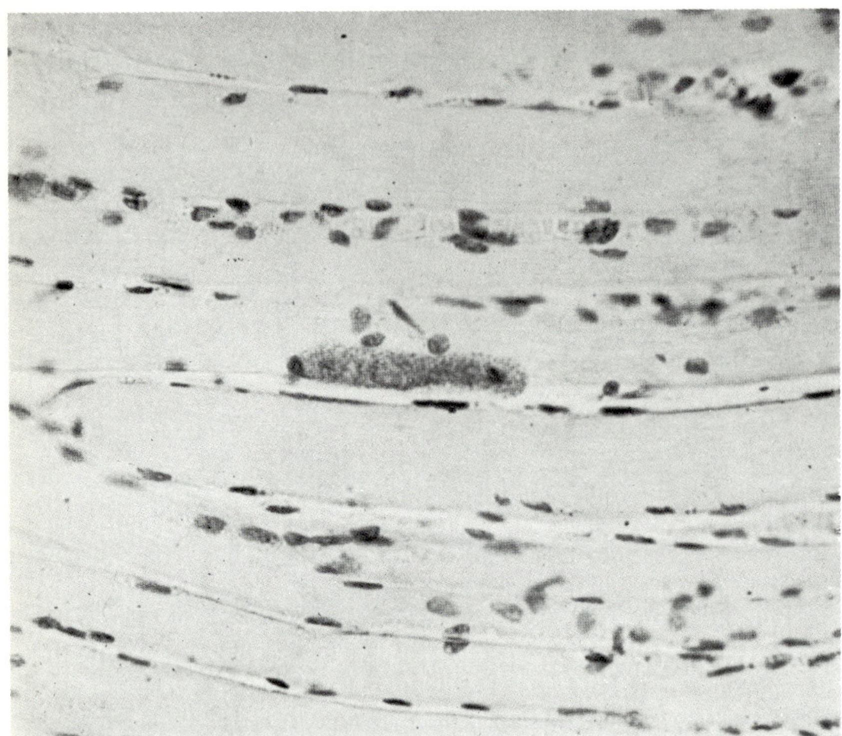

Fig. 7–12. Toxoplasmosis. Longitudinal section of cyst lying within a muscle fiber. (H&E)

Fig. 7–13. Toxoplasmosis. Ruptured pseudocyst in a fiber from triceps muscle of fatal human case. Crescentic shape, nucleus, and kinetoplasts of several liberated Toxoplasma are easily seen.

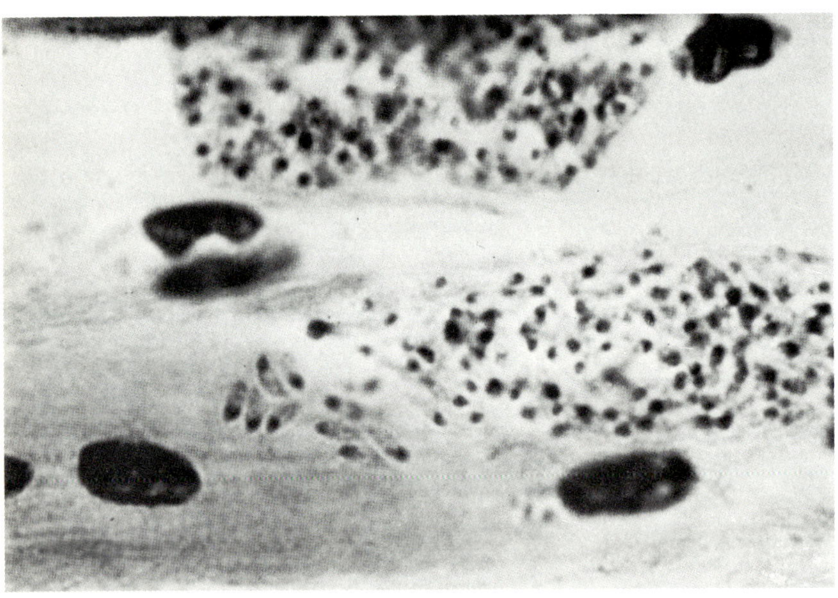

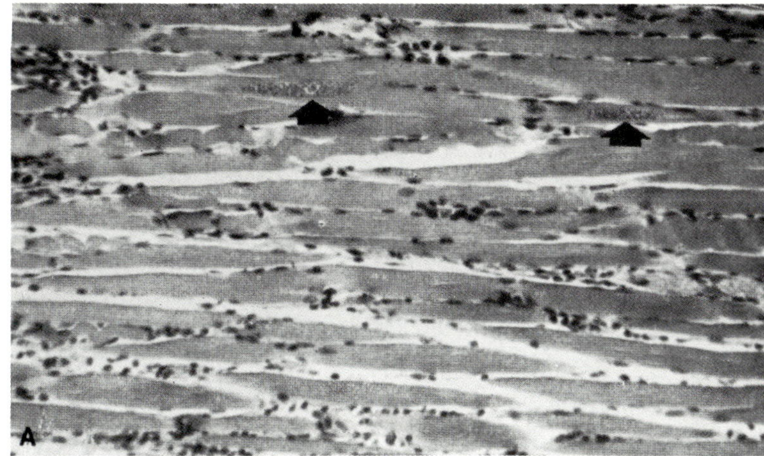

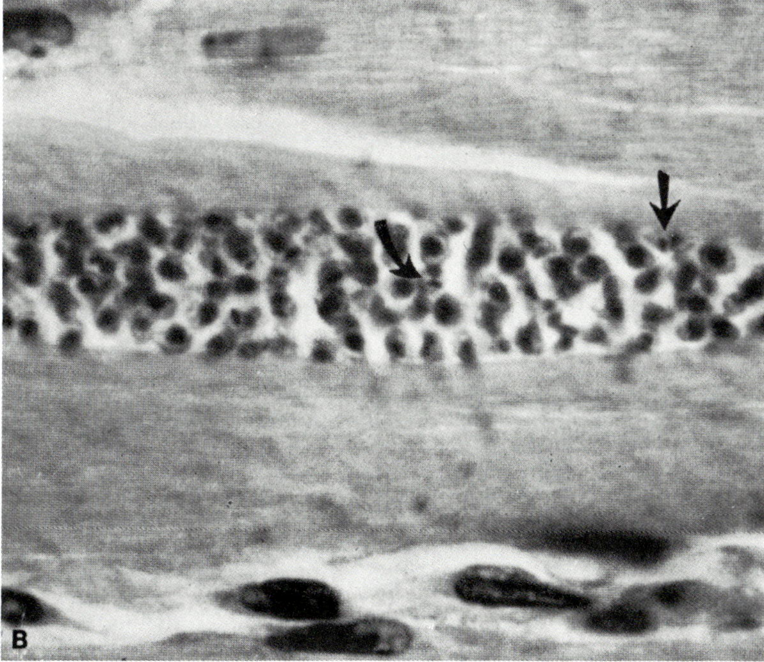

Fig. 7–14. (A) *Trypanosoma cruzi* myositis in mouse muscle. Note diffuse nature of inflammatory infiltrate and two parasite-laden cysts (arrows). (B) Leishmaniform stages of *T. cruzi* in a muscle fiber showing diagnostic kinetoplast, adjacent nucleus, and typical oval body (arrows). Delicate flagellum, arising from region of kinetoplast, is not seen. (H&E; ×2400)

the frequency of lesions in skeletal muscle, biopsy of muscle should be of assistance in diagnosis.

In other common diseases caused by protozoan parasites (e.g., malaria and leishmaniasis) lesions in skeletal muscles have not been reported. However, in reviewing the descriptions of the pathologic changes in these diseases, one gains the impression that the muscle tissue has not been examined systematically. In view of the tendency of other protozoan infections to be localized in skeletal muscle and for the organisms to lie dormant within muscle fibers,

a thorough study of muscle in malaria and leishmaniasis may be rewarding.

INFLAMMATORY REACTIONS OF UNKNOWN ETIOLOGY

POLYMYOSITIS, DERMATOMYOSITIS, AND NEUROMYOSITIS

The terms polymyositis, dermatomyositis, and neuromyositis serve no purpose other than to designate various combinations of inflammatory reactions in muscle, skin, and nerve, i.e., dermatomyositis refers to a combination of muscle and skin lesions and neuromyositis to nerve and muscle lesions. If only the muscles are involved, the condition is called polymyositis. In all these diseases the outstanding morbid change in muscle tissue is essentially the same: an acute or subacute degeneration of muscle fibers with a variable degree of inflammatory reaction.

The first observation of primary idiopathic myositis was made by Wagner[166] in 1863. However, the significance of this discovery was not appreciated until 1887, when other cases were simultaneously reported by Unverricht,[160, 161] Hepp,[63, 64] and Jackson.[65] Strümpell[152] summarized the literature up to 1891. Steiner[151] made a thorough study of two cases of this disease and defined it as "an acute, subacute or chronic disease of unknown origin characterized by a gradual onset with vague and indefinite prodromata followed by edema, dermatitis, and multiple muscle inflammation." This definition has not been improved upon during the past 50 years. Although many hundreds of cases have been reported, these surely represent but a small fraction of the total number, for we have encountered more than 20 cases per year at the Massachusetts General Hospital for a long time.

The subject was reviewed in a monograph by Antognetti *et al.*[5] and in the summary of a Belgian-Swiss symposium on muscle pathology.[115] The monograph by Walton and Adams[170] contains a complete bibliography up to 1958. Discussions of the basic lesion and pathogenesis are found in reference 116 and in separate chapters by Adams[1] and Kakulas[71] in the recent monograph on striated muscle.

CLINICAL MANIFESTATIONS

There have been many attempts to classify cases of idiopathic polymyositis, probably the most satisfactory being that of Eaton.[40] He placed in one category all cases in which the disease process was confined to the muscles and termed it polymyositis; those in which skin and muscles were involved were segregated, and for them he reserved the conventional term dermatomyositis. When an associated disease coexisted, he further qualified the diagnosis, e.g., polymyositis or dermatomyositis with lupus erythematosus, rheumatoid arthritis, or carcinoma of lung. The words acute, subacute, or chronic were added to convey some idea of the course. While relatively simple, this classification is also comprehensive and embraces all the different varieties we have observed clinically and pathologically with the possible exception of that which affects young children and has an unusually high mortality.[9] Walton and Adams[170] adopted this classification but regrouped their cases in order to stress the relationships between them and the connective tissue diseases or carcinomatosis.

Some idea of the frequency of this type of muscle disease and its clinical setting can be obtained from the series of 40 cases from the files of the Massachusetts General Hospital and Royal Victoria Hospital, Newcastle-on-Tyne, that constituted the substance of the report by Walton and Adams[170] (Table 7–1). In all of

TABLE 7–1. Classification of Idiopathic Polymyositis

Group I (14 cases)
  Polymyositis, acute, with or without myoglobinuria
  Polymyositis (subacute or chronic)
    Childhood
    Early adult life
    Middle life or late adult life (menopausal muscular dystrophy)
Group II (12 cases)
  Polymyositis with muscular weakness the dominant feature but with evidence of associated connective tissue disease; or dermatomyositis with severe muscular disability and often minimal or transient skin changes
Group III (8 cases)
  Severe connective tissue disease (rheumatoid arthritis, lupus erythematosus, scleroderma, rheumatic fever, or a combination thereof) with relatively slight muscle disability (polymyositis); or dermatomyositis with florid skin changes and muscular disability of secondary importance
Group IV (6 cases)
  Polymyositis with carcinoma (carcinomatous myopathy) or dermatomyositis in association with malignant disease

---

There were no cases of myasthenia gravis with polymyositis.

---

the cases mentioned in Table 7–1 the muscle disability was of sufficient degree to be recognized clinically; in none was the diagnosis of polymyositis based solely on finding the lesions of focal interstitial myositis in a muscle biopsy.

The age range was 4–20 years in group I, 7–54 years in group II, 25–58 years in group III, and 44–65 years in group IV. Females were more frequently affected by polymyositis with connective tissue disease, whereas the incidence of "pure" and carcinomatous polymyositis in males and females was approximately equal. There were three acute cases leading to fatal termination within a few weeks; these occurred in patients 7, 24, and (in a recent case not included in this series) 64 years old. The overall prognosis, excluding those with associated carcinoma, while not uniformly bad, was far from good. With follow-up as long as 10 years, it was found that 24 of the 34 cases (excluding those with carcinoma) were still alive. A few had recovered completely, but the majority of the survivors retained some degree of disability. Patients whose courses ranged from subacute to chronic were particularly likely to remain disabled, and not a few of these had relapsed one or more times after improvement. Oppenheim[114] recorded 2 deaths and 5 complete recoveries in 10 cases, a remission rate higher than that in our material. A number of milder cases are not admitted to hospital[36] and are not included in reports of verified cases.

Symmetrical weakness, atrophy, rarely tenderness, and induration of muscles, with a liability to fibrous contracture, appear in all types. There is some reduction in the electrical excitability, but a response to faradic stimulation remains as long as sufficient contractile substance persists, In Table 7–2 are shown some of the clinical manifestations observed in 100 cases of all ages abstracted from three reported series.

The more acute phases of polymyositis or dermatomyositis, perhaps more commonly seen in children[130, 136] may or may not be accompanied by fever and polymorphonuclear leukocytosis.[171] The muscles of the shoulder girdle are often the first to be affected, but soon the proximal muscles of all extremities become weak. Occasionally the subcutaneous tissue over the affected muscle is edematous and the overlying skin somewhat reddened. A diffuse erythema over the face and neck, spreading rapidly to the trunk and limbs, may precede or accompany the muscle affection, and a variety of other types of rash have been described. Desquamation is common. Pains in the muscles may be severe but more often are completely absent. In the course of days or weeks other muscles of the limbs are gradually involved, becoming weak or powerless, and the tendon reflexes disappear. The edema and induration of each muscle slowly subside, leaving it greatly reduced in bulk and shortened by fibrous tissue.

TABLE 7–2. Analysis of Some Clinical Signs and Symptoms in 100 Cases of Polymyositis and Dermatomyositis

Manifestations	Eaton[40] (41 cases)	Walton & Adams[170] (40 cases)	Pearson & Rose[119] (19 cases)	No. affected (of 100 cases)
Cutaneous features	22	23	16	61
Raynaud's phenomenon	10	10	2	22
Arthritic or rheumatic features	10	10	10	30
Muscular pain or tenderness	18	16	14	48
Muscular atrophy	19	?	15	34*
Fibrous contractures	11	8	14	33
Muscular weakness				
Proximal muscles	38	40	19	97
Distal muscles	24	14	6	44
Neck muscles	14	33	15	62
Dysphagia	22	21	14	57
Facial muscles	5	5	1	11

* From 60 cases. Atrophy was usually slight in degree.

The affected skin frequently remains indurated and sometimes pigmented. Pharyngitis and stomatitis are so common in some series of cases that Oppenheim proposed the name "dermatomucomyositis." Involvement of the muscles of the jaws and pharynx may cause a formidable problem in feeding, and the tenderness of the muscles and liability to flexor contractures make it difficult to provide nursing care to the most severely affected patients. Recovery is possible after a long stationary period, but involvement of the respiratory muscles commonly leads to a fatal termination. Residual contractures are sometimes a source of disability. Myocarditis has been reported in some cases.

The subacute and chronic varieties of the disease are much more insidious in onset and may resemble dystrophy in their course and distribution of muscle involvement. They are more common in adults of either sex than in children. After some vague prodromal symptoms such as malaise and fatigability, a weakness of the lower limbs gradually develops. The muscles of the thighs become weak and feel tight to the patient. Climbing stairs and rising from a chair are difficult. The feet and legs become weak and later very edematous. After many weeks the strength of the hand and forearm muscles is gradually lost. Pain on movement of the affected muscles is usually slight or absent except when rheumatoid arthritis or other connective tissue disease coexists. The cranial muscles are often spared until late, but eventually flexors and extensors of the neck, the laryngeal, pharygeal, and rarely tongue, muscles may be involved. Affection of respiratory muscles may be fatal. Muscle atrophy and contractures are pronounced. This chronic stage corresponds to Blau's[10] primary, generalized myositis fibrosa. Polymyositis is often associated with scleroderma or one of the other collagen diseases. In other cases skin lesions are absent or limited to redness with mild induration. Calcium deposits in the involved muscles and connective tissues (calcinosis) occur in advanced cases of poly- or dermatomyositis and scleroderma,[171] and simulate myositis ossificans.

Although the tendon reflexes are lost when the muscles are severely affected, there is usually no disturbance of cutaneous reflexes or of sensation. In some cases vague paresthesias of the extremities have been reported, and in rare cases there is even a symmetrical sensory disturbance in an area corresponding to the distribution of a peripheral nerve. Senator[137] gave the term neuromyositis to such cases. In view of the liability of exposed nerves to pressure lesions in chronically ill and wasted patients, it is doubtful if this represents a neuritis, i.e., neuromyositis. On the other hand, in some cases of acute infectious and toxic polyneuritis, in addition to serious involvement of the peripheral nerves, mild and clinically unsuspected muscular lesions which resemble in some ways those of polymyositis may coexist. Exceptionally there is pleocytosis in the cerebrospinal fluid. The combination of polyneuritis and polymyositis (neuromyositis of Senator 137) occurs more frequently in cases of occult carcinoma than in the other groups in Table 7–1. Apart from these exceptions, polymyositis and polyneuritis are distinct, unrelated clinical entities.

As becomes clear in the description of the pathologic changes, there is no essential difference between the most chronic changes seen in some muscles in chronic forms of polymyositis and those seen in the more rapid forms of muscular dystrophy. We mentioned that late and arrested forms of the facioscapulohumeral type of muscular dystrophy may appear to affect only the proximal muscles of the upper limb and then come to a virtual standstill. A similar and not uncommon affection after middle life is a gradual weakness of extension of the knees associated with symmetrical atrophy of the vastus internus muscles on both sides, without pain or tenderness. The knee jerks are lost early in the disorder, but there is no disturbance of other reflexes or of sensation. Spinal and shoulder muscles may be involved in some cases. After weeks to months, when difficulty in rising from a chair and instability of the knees have led to great uncertainty in walking, the affection may cease to progress. In other cases, however, the process becomes generalized, with involvement of laryngeal, pharyngeal, and finally respiratory muscles.

Also cardiac lesions have been reported. Some cases end fatally. The relationship of this disorder to polymyositis can be established only by muscle biopsy.[35, 107] It is subclassified by Denny-Brown[36] as necrotizing myopathy. We regard most cases of such late life muscular dystrophy, and the condition which Shy and McEachern[141] described as menopausal muscular dystrophy, as forms of chronic polymyositis, occurring during the climacterium or late years of life.

A number of laboratory findings are of value, both in confirming the diagnosis and in providing information concerning the nature of the pathologic process. Extensive inflammatory disease of muscle, whether acute or chronic, results in a large daily excretion of creatine and reduced urinary output of creatinine similar to that of progressive muscular dystrophy.[20] Serum levels of transaminase are greatly elevated, to as much as 10 times normal in some cases,[117] and creatine phosphokinase and aldolase levels are also raised. In denervation atrophy, by contrast, the serum levels of these two enzymes are usually normal, and in progressive muscular dystrophy they are increased, but to a lesser degree than in polymyositis and acute myoglobinuric myopathy.[117] The electromyogram reveals a typical "myopathic pattern," i.e., large numbers of exceptionally brief, low-voltage action potentials and in a surprising number of cases fibrillation potentials.[111] Presumably the altered action potentials, as in muscular dystrophy, may be ascribed to the depopulation of motor units, viz., loss of muscle fibers; the fibrillation is due either to a concomitant denervation of muscle fibers from inflammatory lesions in intramuscle twigs or to the survival of segments of muscle fibers that have been isolated from their end-plate region. A myasthenic type of reaction may be present.[41] The blood sedimentation rate is abnormally rapid in some of the acute and a few of the chronic cases of polymyositis. Positive latex fixation reactions, or other tests for a rheumatoid factor in the serum, occur in a significant percentage of cases of polymyositis and dermatomyositis,[118] probably signifying a pathogenetic common denominator between these disorders and rheumatoid arthritis.

PATHOLOGIC CHANGES

The skeletal muscles in all types of myositis, acute and chronic, usually appear abnormal to the naked eye. The tissue appears pale red, grayish red, or yellowish and is soft and friable or rubbery and firm according to the chronicity and duration of the disease. Microscopically the most striking change is a widespread degeneration of muscle fibers. In a late stage all the fibers in some muscles may have disappeared leaving a cellular fibrous tissue stroma. Most often many of the muscle fibers are still present and exhibit all stages of degeneration and regeneration (Fig. 7–15). Some are swollen and rounded in cross-sectional outline, and the myofibrils have disappeared leaving a waxy or hyalinized sarcoplasm. Vacuolation, granular degeneration, and fragmentation can be seen in many fibers. Phagocytosis of disintegrating fibers is common, and in fact sometimes empty sarcolemmal tubes are filled with macrophages. The sarcolemmal nuclei of some of the altered fibers are increased in number and size. Fibers of normal appearance may be found among the degenerating ones. Mild to marked perivascular and diffuse infiltrations of lymphocytes, plasma cells, histiocytes, and sometimes neutrophilic leukocytes occur in the endomysium and perimysium (Fig. 7–16). In some cases there is a slight hyperplasia of connective tissue in the intima and adventitia of small arteries and aterioles. In later stages, when all or nearly all of the muscle fibers have vanished, and fibrous connective tissue is thickened and in places only remnants of muscle fibers remain (Fig. 7–17). Multinucleated buds and branching fibers are seen here and there and represent abortive attempts at regeneration.

Acute Polymyositis. There are certain pathologic variations in the different types of myositis. The changes in the muscle fibers are studied to best advantage in the more acute forms of the disease, for many fibers show fragmentation and active phagocytosis of their contents (Fig. 7–15), and globular remnants containing clumps of pyknotic sarcolemmal nuclei are prominent. Always, however, when the

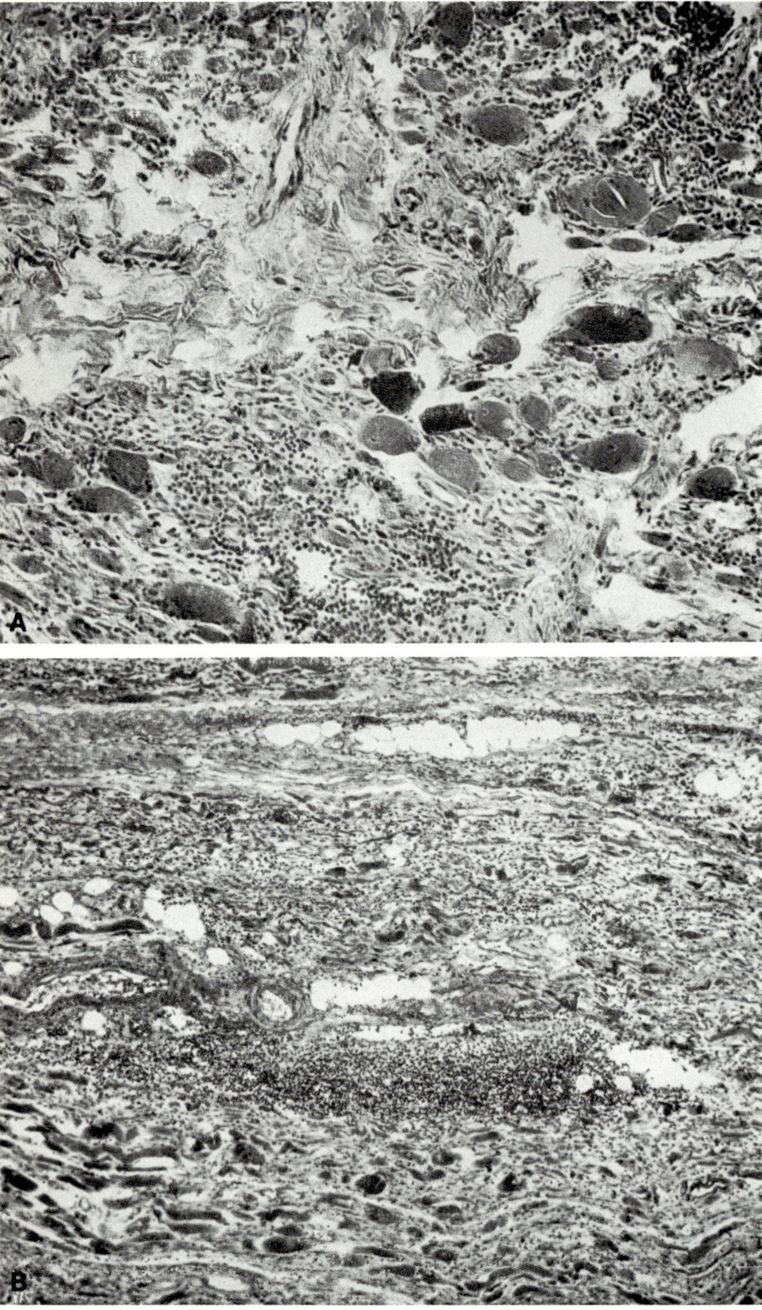

Fig. 7–15. Acute polymyositis (dermatomyositis). (A) Advanced destruction of muscle fibers with extensive infiltration of inflammatory cells. (B) Lower magnification. (phloxine, methylene blue; B, ×72)

stereology of the lesion is ascertained by serial sections, perivenous infiltration by lymphocytes and mononuclear cells is associated with muscle fiber necrosis. Large macrophages may be ingesting fragments of muscle tubes. The cellular reaction in the connective tissue is more intense. Hemorrhages have not been seen in our material, but in a few of the acute cases reported they are sufficiently numerous to justify the term hemorrhagic myositis.[157] In the less affected muscles or in a biopsy from an early stage of the disease, fragmentation of isolated groups of fibers, or even single muscle fibers, with phagocytosis may be found scattered in otherwise healthy muscle. The contents of such fibers stain darkly, and the sarcolemmal nuclei,

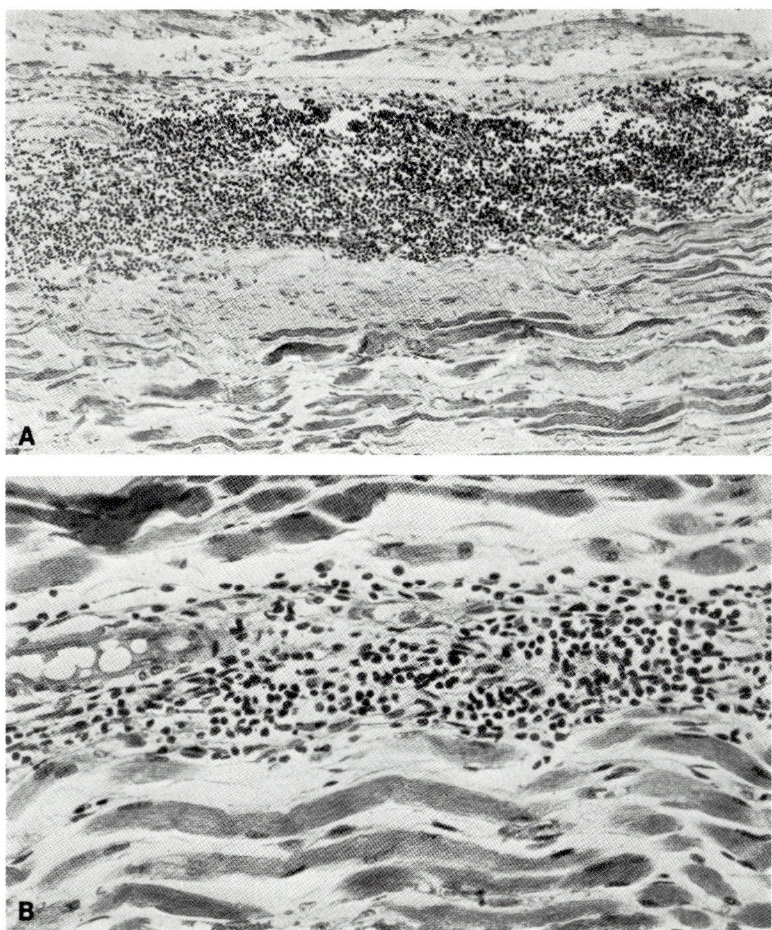

Fig. 7–16. Dermatomyositis, late stage. (A) Psoas muscle showing large perimysial infiltration of lymphocytes and a few histiocytes, with loss of adjacent muscle fibers and fibrosis. (B) Cellular infiltrate of lymphocytes, plasma cells, and histiocytes in lupus erythematosus. (H&E)

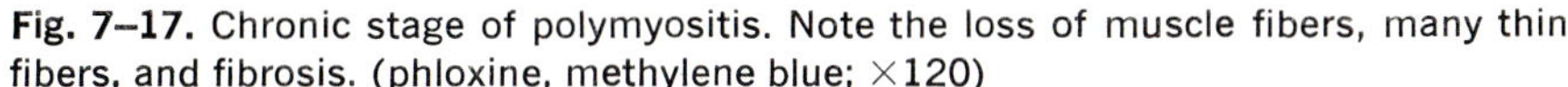

Fig. 7–17. Chronic stage of polymyositis. Note the loss of muscle fibers, many thin fibers, and fibrosis. (phloxine, methylene blue; ×120)

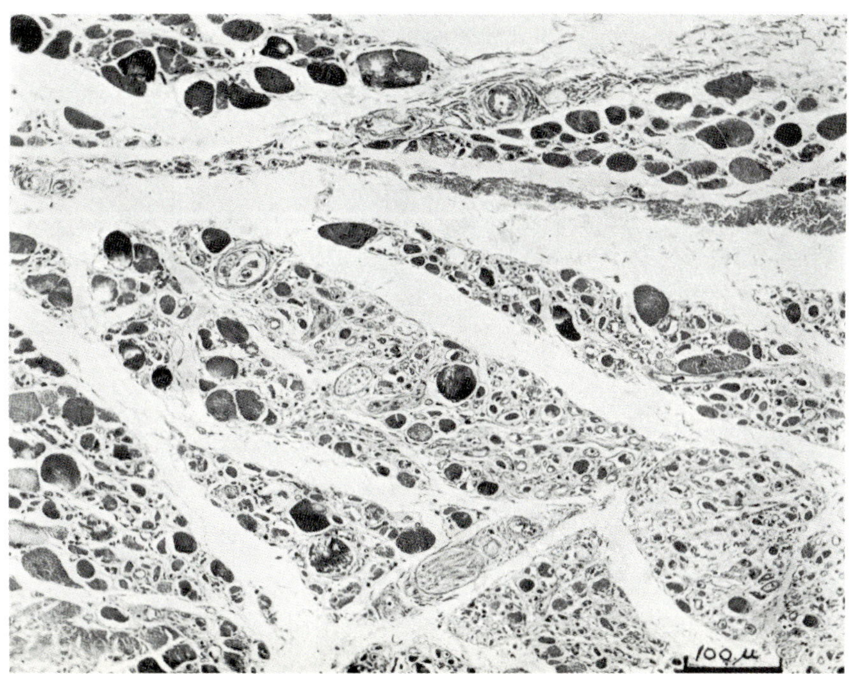

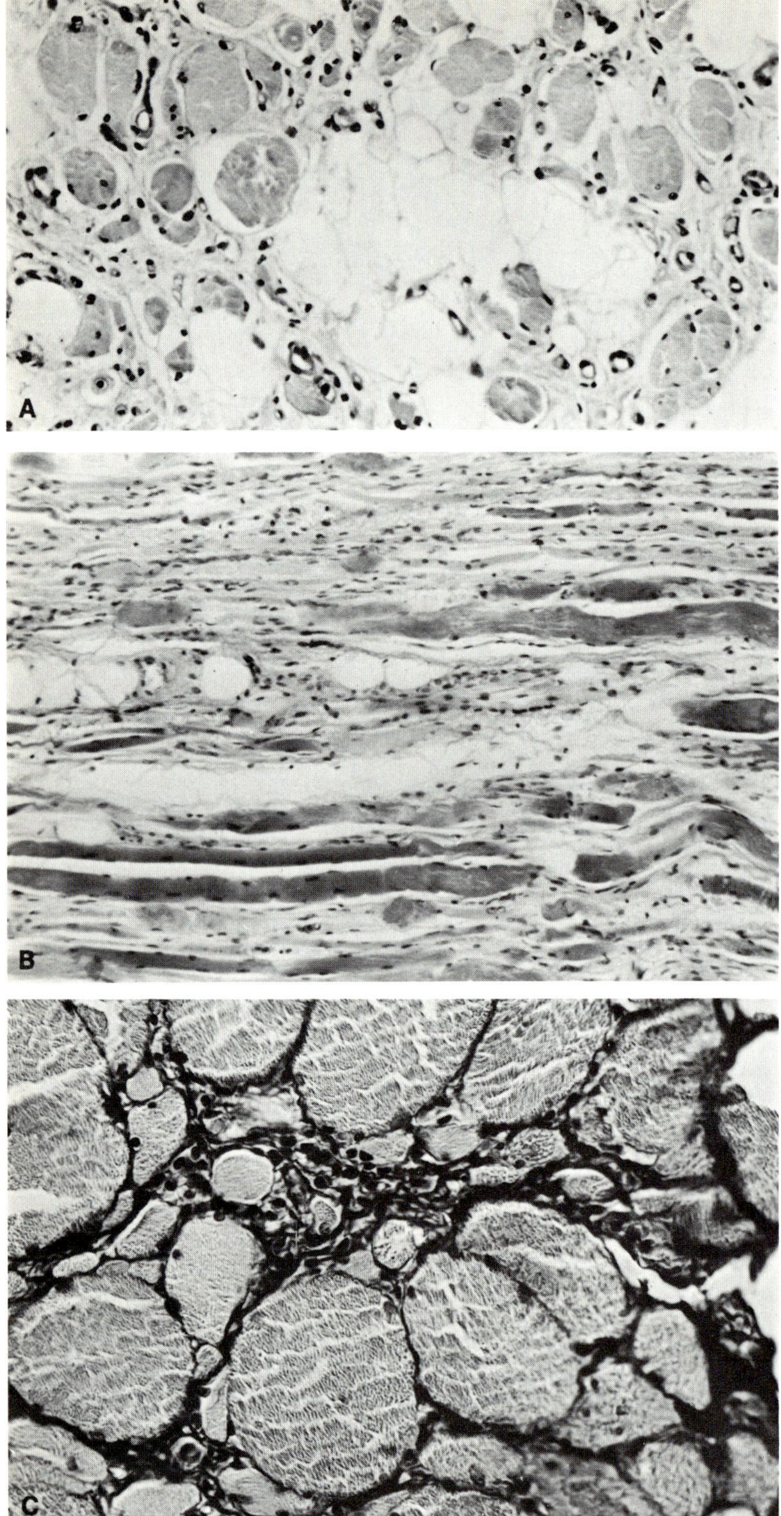

Fig. 7–18. Chronic polymyositis. (A) Transverse section, showing loss of muscle fibers, variation in size, branched fibers in tight clusters, fat cells, and fibrosis. (B) Longitudinal section: fiber loss, atrophy increase in endomysial fibroblasts, rows of lipocytes. (C) Transverse section showing fiber atrophy, hypertrophy, and focal fibrosis. (A,B, H&E; C, trichrome stain)

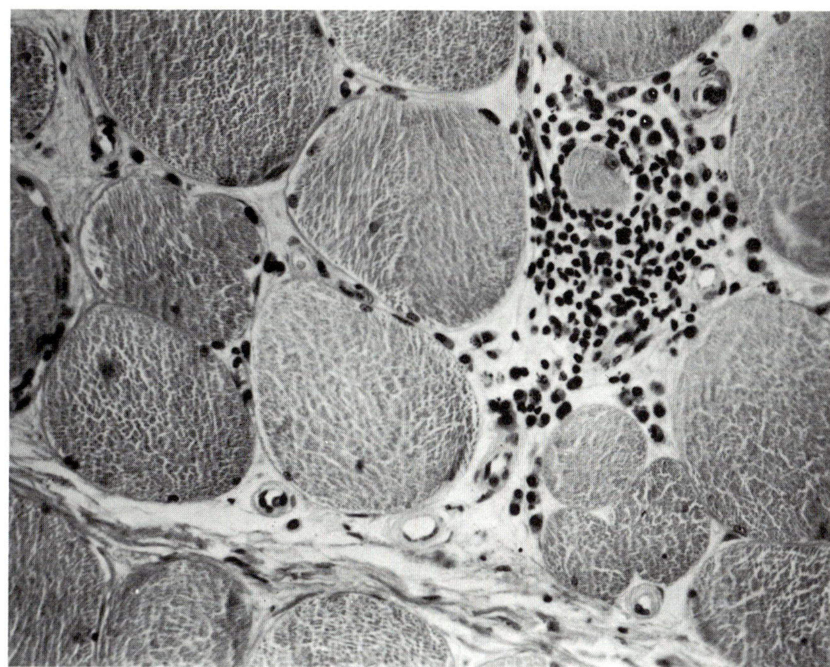

Fig. 7–19. Chronic polymyositis. Transverse section, showing an infiltrate of lymphocytes and plasma cells around an atrophic muscle fiber. The three small fibers are branches of a previously injured larger fiber. No other fibers are enlarged (>150 μ). (H&E)

though increased in number and often lying centrally in fissures among the masses of myofibrils or clumped between segmented hyaline masses, show no other regenerative activity. Large numbers of phagocytes lie clearly apposed to the sarcolemma, and some have penetrated the contents of the fiber. Such an appearance is common in an area of severe degeneration, and the unusual feature of the early stage of the disease is the isolation of such units in otherwise healthy muscular tissue.

In some cases of late-life myopathy an almost identical appearance is found. Muscle fibers still are in the process of disintegration, and active phagocytosis has been excited. In places basophilia of the hyaline contents of the fiber and proliferation of sarcolemmal nuclei are indications of regenerative activity. Of course there are now the additional effects of earlier destruction and repair, i.e., loss of fibers, fibrosis, and variation in size of residual fibers (Fig. 7–18). In reporting two such cases Nevin[107] emphasized the resemblance of this

process to the early stage of acute polymyositis on the one hand, and to that of dystrophic degeneration on the other; and Nattras[106] called attention to cases of chronic polymyositis misdiagnosed as progressive muscular dystrophy. We affirm the similarity in many of the pathologic changes. Fiber loss, variation in size of residual fibers, atrophy of the peripheral fibers in fascicles (Fig. 5–12), fiber splitting with groups of closely apposed small fibers (Fig. 7–18), giant fibers, fibrosis, and fat cell replacement were prominent in six of our autopsied cases in which death occurred 5 years or more after the onset of the polymyositis (Fig. 7–19). It must be assumed that all these changes common to two such dissimilar diseases must relate to some common factors (e.g., repeated segmental necrosis) and not to the essential causative factor in progressive muscular dystrophy (Chapter 6). In our opinion it is the nature of the inflammatory necrosis of the fiber and indeed the extent of infiltrates of inflammatory cells that set apart acute and subacute polymyositis from the purely degenerative conditions we know as muscular dystrophy.

The skin lesions vary from case to case. There may be a diffuse or patchy erythema and edema of the skin, erythema nodosum, erythema multiforme, or an eczematoid lesion. In the more chronic cases the erythema is supplanted by a brownish, tight, shiny, atrophic skin indistinguishable from that of scleroderma. In microscopic sections the dermis is edematous and the collagen fibers swollen and thickened. The papillae are flattened, and some of the basal cells are vacuolated. There is usually thinning of the cornified epidermis. The small arteries have thickened walls owing to an increase in connective tissues in the intima and adventitia. Infiltrations of lymphocytes, plasma cells, and large histiocytes occur beneath the epidermis and around blood vessels and sweat glands. Cellular infiltrations in the subcutaneous fatty tissue may be seen in some cases. When such changes in the interstitial tissues are prominent and muscle degeneration minimal, the prognosis seems to be somewhat better than in the cases with predominantly muscular changes.

Involvement of the peripheral and central nervous system is quite infrequent. In one of our cases small collections of lymphocytes and plasma cells were found in the endoneurium and perineurium of several of the peripheral nerves. An occasional myelinated fiber near these cell accumulations was fragmented, and there were numerous macrophages. In this same case there were 30–40 petechial hemorrhages in the white matter of the cerebral hemispheres, brain stem, and cerebellum.

Chronic Polymyositis (Chronic Myopathy). In the broad spectrum of polymyositis a wide variation of degrees of cellular reaction in the endomysium may be encountered. When extensive fibrosis had occurred (Fig. 7–19) or there is collagenization in sclerodermatous types of the disease, the typical eosinophilic floccular appearance of necrotic muscle fibers is seldom encountered. The muscle fibers instead are variable in size; many are quite small (Fig. 7–16) with an increased number of dark, flattened subsarcolemmal nuclei. Regeneration,

when it occurs, seems to be not at all prominent.

In one type of chronic polymyositis necrosis of isolated muscle fibers is the prominent feature throughout the illness. The degree of cellular infiltration of the connective tissue then varies with the intensity of muscle fiber degeneration and consists predominantly of histiocytes and macrophages. The muscular degeneration may be subacute and generalized, with emphasis on the proximal muscles, or more generalized and very slowly progressive over a period of several years. This is the type of disorder called necrotizing myopathy by Denny-Brown,[36] and when presenting as a chronic myopathy of quadriceps[35] during late life, is the late-life muscular dystrophy of Nevin[107] and others. It can cause slowly progressive atrophy of proximal muscles in earlier life, even early childhood, and is rarely familial, then closely simulating muscular dystrophy. Figure 7–20 shows the characteristic necrosis of isolated fibers seen in the deltoid muscle of a 22-year-old woman who had had weakness of the proximal muscles of her lower and upper limbs for 2 years. During the following 3 years a slight increase in weakness gradually occurred. Another view of this process is shown in Figure 7–21; the section was taken from a woman aged 36 years who had a 7-year history of progressive weakness of proximal muscles. Turner and Heathfield[159] described a patient suffering from this type of degeneration who had been reported by Denny-Brown[35] 22 years earlier. The muscle atrophy had progressed slowly, though the increase in disability was relatively slight.

A similar picture was found in the chronic atrophy of muscle that developed in Schmid and Mahler's case of the McArdle syndrome (Chapter 12), though the number of muscle fibers showing degeneration after an attack of muscle cramps was then much greater. Nevertheless, when necrosis of isolated muscle fibers is prominent and the connective tissue response minimal, there is always difficulty in differentiating polymyositis from dystrophy or a primary metabolic disorder of muscle. This type of change is also related to the Zenker type of degeneration associated with various infections. We have seen it in otherwise very

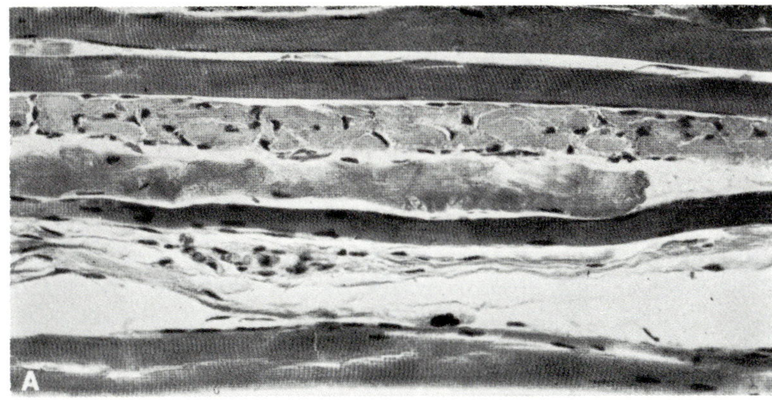

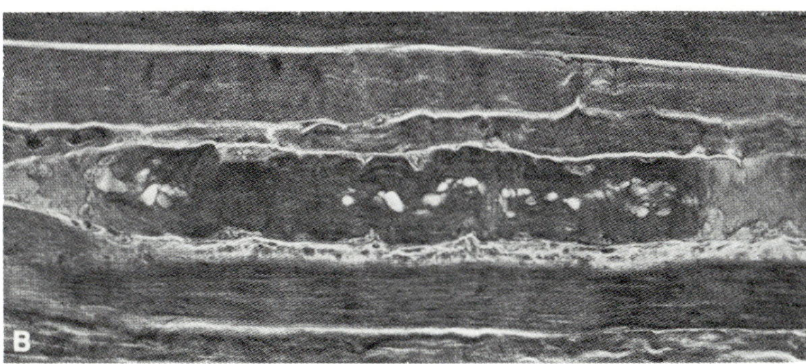

Fig. 7–20. Progressive type myopathy with necrosis of isolated fibers and slight interstitial change (necrotizing myopathy). (A) Eosinophilic necrosis of a fiber and invasion by histiocytes. (B) Early stage of degeneration with swelling and vacuolation. (C) Terminal stage with degenerate fragments and myoblasts. (A, H&E; B, phosphotungistic hematoxylin; C; luxol fast blue-cresyl violet)

mildly dystrophic muscle in a patient suffering from Landouzy-Déjerine muscular dystrophy who recently passed through an episode of toxemia of pregnancy. It was also observed by us following general reactions to penicillin and other antibiotics in patients sensitive to these substances; in all these conditions it is likely to persist for long periods of time. It can occur too as a reaction to thyrotoxicosis and to hyperinsulinism.

The converse may also be true—that the finding of a few foci of lymphocytes, monocytes, and even plasma cells are detected in a muscle biopsy from a typical dystrophic patient. Surely the inflammatory reaction is coincidental, just as it is an occasional finding in otherwise normal muscle.

Histochemical studies have been relatively uninformative in chronic polymyositis. All fiber types are affected but there tends to be an

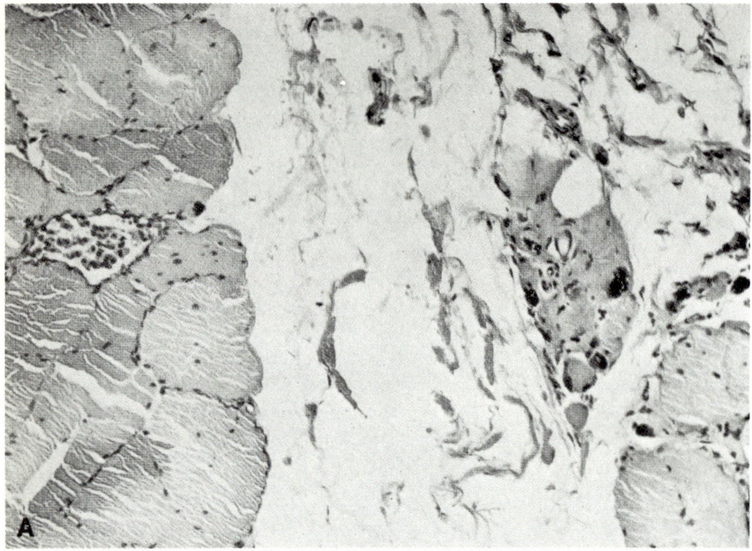

Fig. 7–21. (A) Necrotizing myopathy, showing necrosis of isolated muscle fiber with invading histiocytes. Remnants of other muscle fibers are seen at right. (B) Recent eosinophilic necrosis of isolated fibers in longitudinal section. (H&E)

atrophy of type II fibers, as in any disease which reduces the activity of muscle. (See recent monograph by Dubowitz, V. and Brooke, M. Muscle Biopsy, Saunders and Co., New York, 1974.) The atrophy tends to be perifascicular.

This small group of cases of chronic progressive polymyositis may differ from the above in respect to the presence of even more generalized but chronic degeneration of muscle fibers. Dermal changes are absent. The weakness was slowly progressive after an insidious onset in all muscles of the limbs in one of our patients, a 12-year-old boy. Death occurred after 7 months of progressive weakness. Biopsies of such lesions in a number of our middle-aged patients had a histologic appearance typified by that shown in Figure 7–22. At the periphery of each fasciculus the muscle fibers were reduced to shrunken tubes containing large numbers of dark, shrunken, sarcolemmal nuclei (Figs. 7–22 through 7–24). Less affected fibers showed a remarkable vacuolation, and for this reason we call it vacuolar myopathy.[36] As the

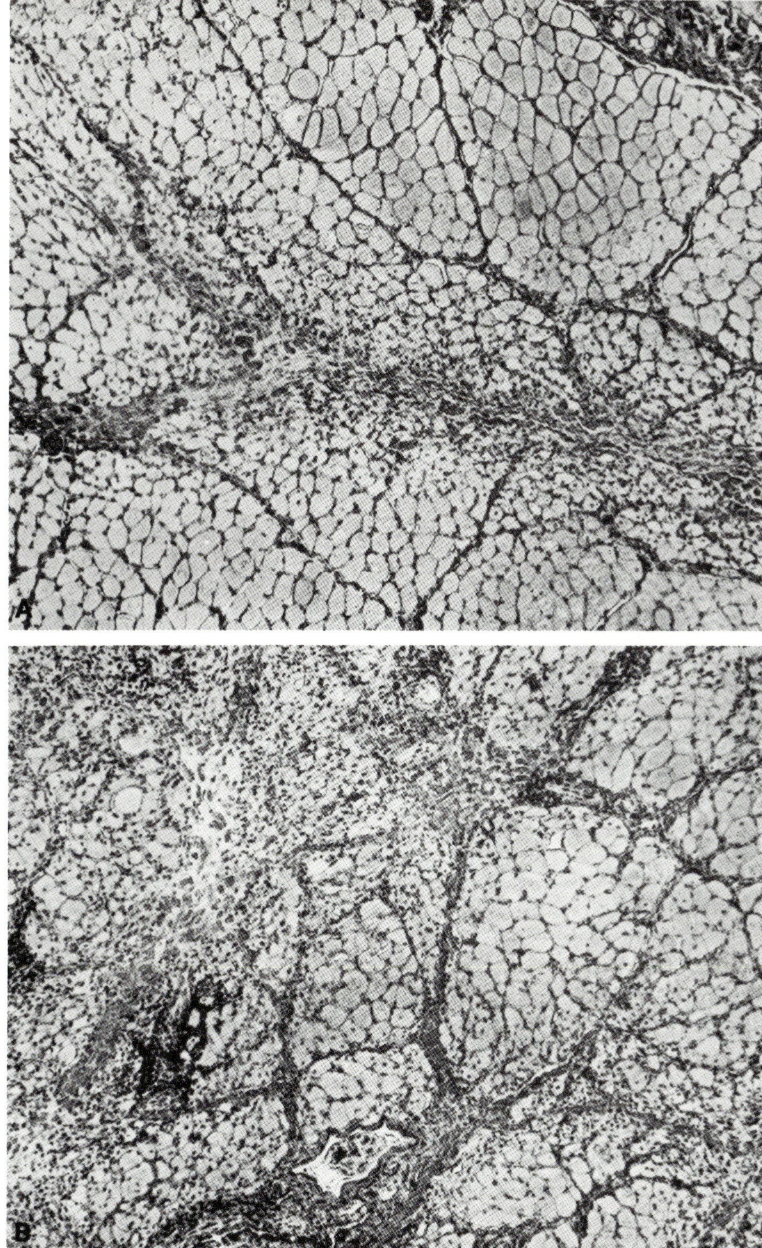

Fig. 7–22. Chronic progressive polymyositis. General view showing damage to peripheral fibers in several fascicles, and patchy accumulations of round cells. (hematoxylin, Van Gieson; ×84)

disease progresses the endomysial collagen is greatly increased in amount, and scattered lymphocytes and phagocytic cells are found in it. The muscle fibers nearer the center of the fasciculus are affected to a lesser degree and exhibit only slight vacuolation and hyaline change with loss of striations in some parts (Fig. 7–23). In longitudinal section the contents of the muscle fibers are coagulated in a segmental manner, with pyknotic muscle nuclei, separated by large vacuoles containing clumps of vesicular sarcolemmal nuclei (Fig. 7–24). Regenerative activity may be completely absent, as was noted by Univerricht, but this is exceptional in our experience.[160, 161] There is no endomysial reaction in direct relation to the degenerating fibers. At intervals, in cross or longitudinal section, small veins within the

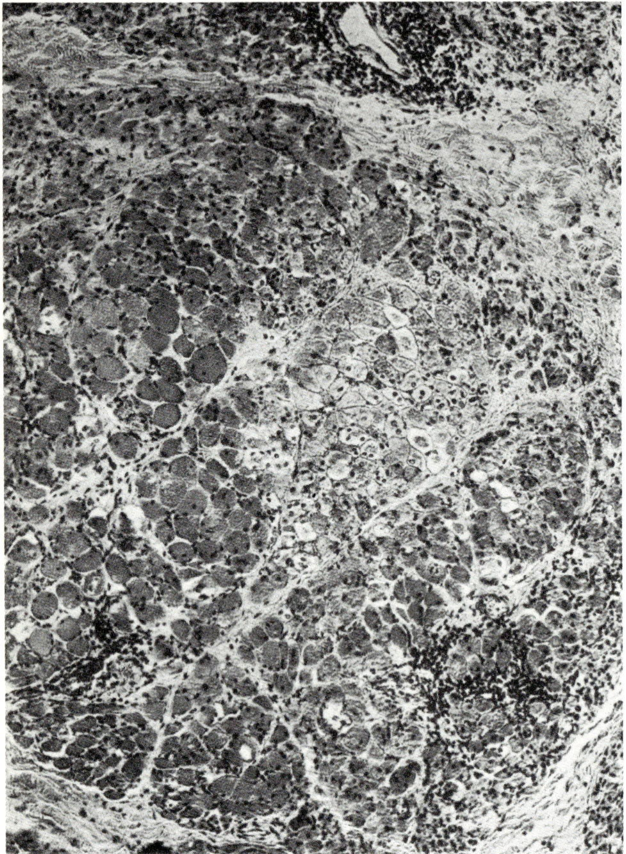

Fig. 7–23. Chronic polymyositis. Fasciculus from patient with slowly progressive ("vacuolar") type of the disease. (hematoxylin, Van Gieson; ×166)

muscle are surrounded by aggregations of lymphocytes ("lymphorrhages"), plasma cells, histiocytes, and a few mast cells (Fig. 7–23). Such accumulations are not related to any focal damage of the muscle fibers but appear to be regions of venous drainage. The nerve endings of the vacuolated atrophic fibers become thin and attenuated (Fig. 7–25), and the parent nerve fibers thicken, with bulbous enlargements just before they reach the muscle fibers, wandering aimlessly in the fibrous tissue, and ending in free bulbs (Figs. 7–25 and 7–26). The muscle spindles are resistant until a late stage, but the sensory endings then undergo similar changes. also leaving nerve fibers ending in terminal balls. Such changes in nerve endings are found in all types of polymyositis and are shown here only because they are universal in this form of the disease. These progressive forms of chronic polymyositis indeed share

many features with true muscular dystrophy. Myoglobinuria is likely to occur only in acute, severe attacks of these types of primary destruction of muscle fiber.[36]

An atrophy of muscle may occur in disseminated lupus erythematosus and also follow treatment with triamcinolone and related steroids. On the basis of a number of clinical reports it appeared to be reversible in early stages (Chapter 4), but in some patients a steadily progressive destructive myopathy ensues. Figure 7–27 shows the vacuolar degeneration associated with progressive nuclear changes and absence of regeneration that characterizes this condition. The tissue section is taken from a biopsy specimen obtained from a woman with rheumatoid arthritis who was treated with gold and cortisone and who experienced a progressive muscular degeneration that continued in the year following cessation

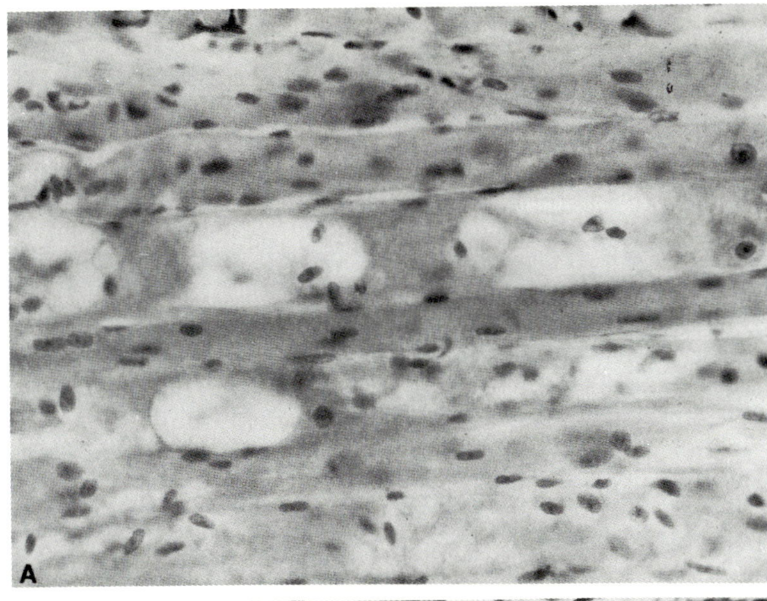

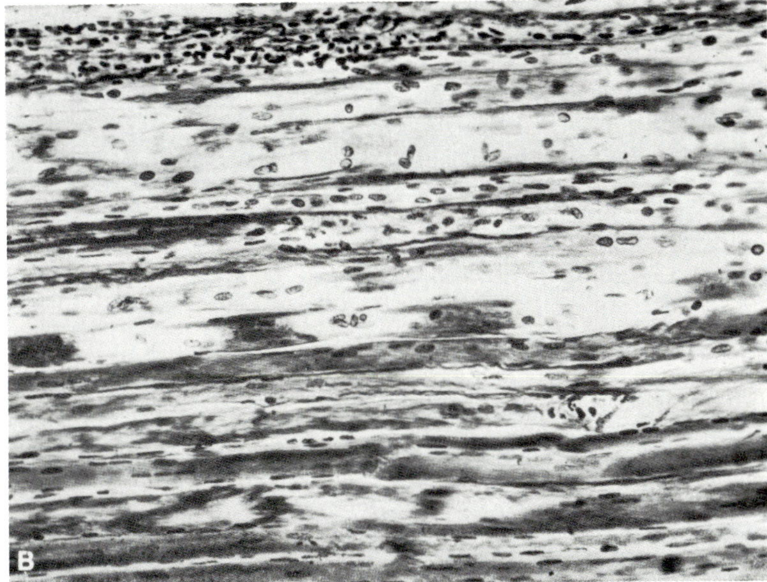

Fig. 7–24. Chronic progressive polymyositis. Vacuolar degeneration of individual muscle fibers. (A, hematoxylin, Van Gieson, ×360; B, Gros-Bielschowsky silver impregnation, ×130)

of treatment. Garcin *et al.*[49] found lesions similar to those of polymyositis in Behçet's disease.

All types of primary segmental degeneration of the muscle fiber should probably be classified as independent myopathies. Until more is known to their causation and specificity, however, only necrotizing and vacuolar changes are described here, for they are usualy encountered by the pathologist in biopsy tissue taken from

patients thought to have polymyositis. These are destructive conditions having poor prognoses.

Recent electron microscopic studies of acute and chronic idiopathic polymyositis and dermatomyositis have added little to our understanding of the early lesions, except for the demonstration of two types of viral particles (see below, *Etiology*). Rose *et al.*[131] after examining biopsy specimens from five patients

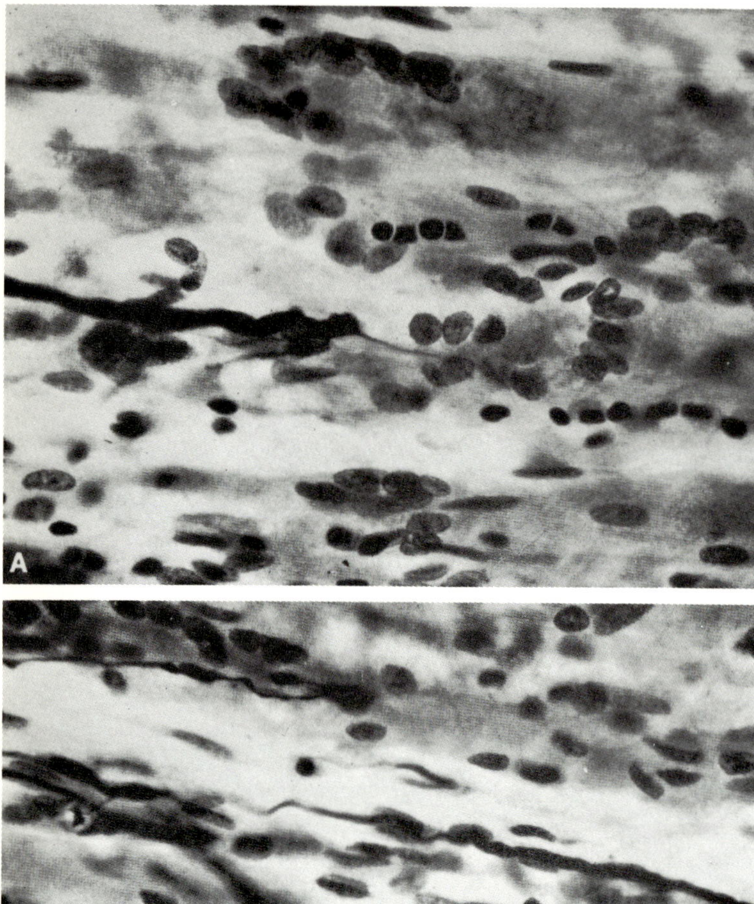

Fig. 7–25. Chronic polymyositis. (A) Nerve ending of severely affected muscle fiber showing atrophy of telodendria. Note clumping of endplate nuclei. Degenerated endplate is seen at top. (B) Late stage of myositis showing end-bulbs on nerve fibers no longer connected to endplates. (Gros-Bielschowsky, hematoxylin)

report dilatation of sarcoplasmic reticulum, degeneration beginning in the I band of thin myofilaments, and in the zones of complete segmental necrosis a degeneration of all organelles. These are the types of change characteristic of all degenerative lesions. The mitochondria are increased in number and some are distorted near such zones, and the triads are dislocated. Mair and Tome[93] call attention to an increased cellularity of capillaries with pericytes between the basement membrane and endothelium. Where regeneration is active, satellite cells are increased in number, myoblasts are present. and myotubes may be forming. Again, there is no feature of this change peculiar to polymyositis.[115, 131, 138]

Histochemical studies also have contributed little. Brooke and Kaplan,[13] who surveyed a

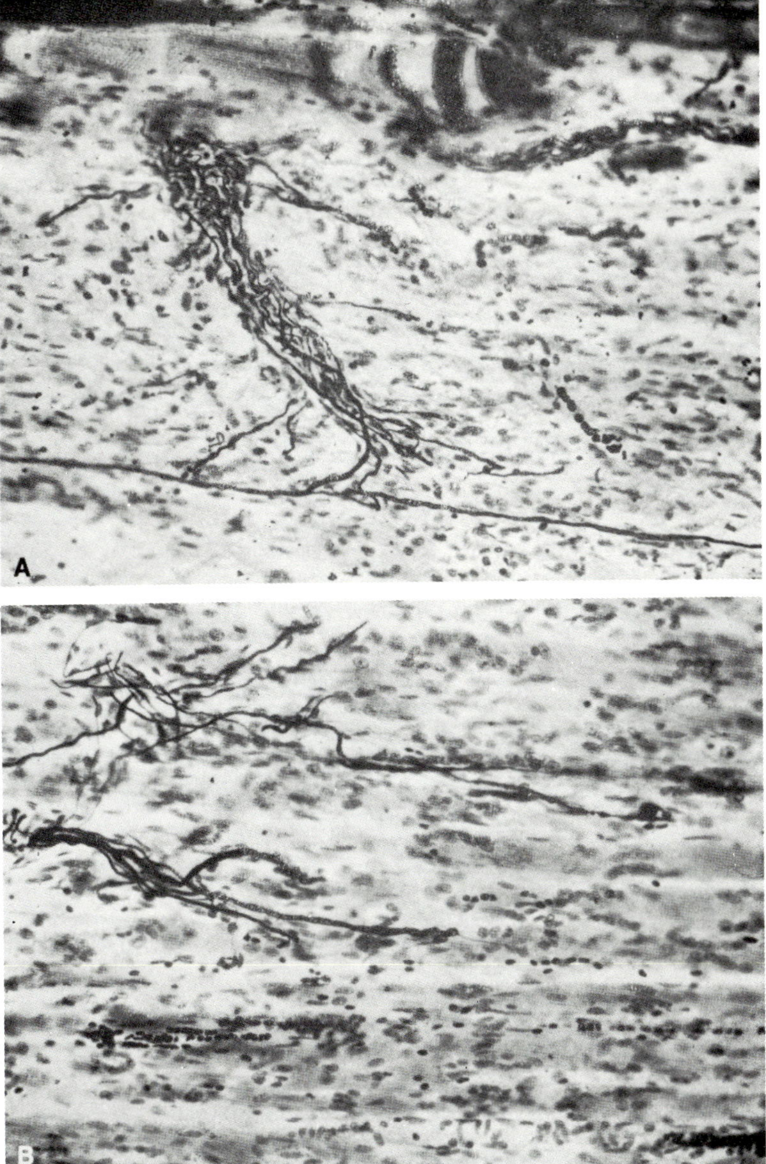

Fig. 7–26. (A) Chronic polymyositis. Thickened nerve fibers end in knobs or long leashes of fine filaments in severely affected fasciculus. (B) Degenerating end-plates. (Gros-Bielschowsky, hematoxylin)

number of biopsies from cases of polymyositis as well as polymyalgia rheumatica and rheumatoid arthritis, observed only an atrophy of type I fibers and cytoplasmic bodies, which they assume to be a degenerative product of Z disks.

In view of the extreme variability of the muscle lesions it may prove difficult to confirm the clinical diagnosis by a small, inadequate biopsy. In some instances the muscle may appear normal, so the lesion is not sampled. Again only its necrotic feature is visible, lead-

ing to a mistaken impression of some form of necrotizing or metabolic myopathy. The chronic stage can be virtually indistinguishable from an advanced stage of progressive muscular dystrophy.

ETIOLOGY

The etiology of these diseases has never been established. Because of the fever, pain,

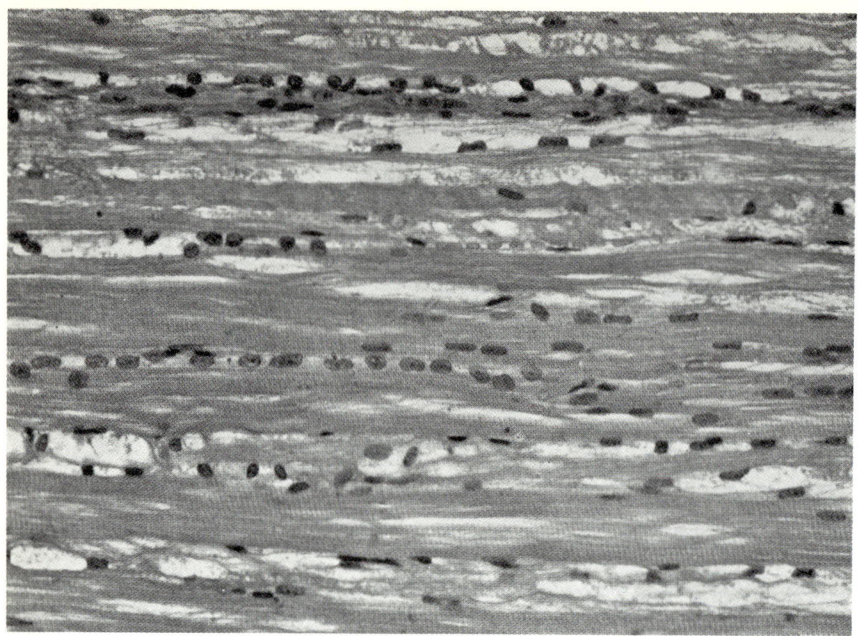

Fig. 7–27. Progressive vacuolar myopathy following treatment of rheumatoid arthritis with cortisone. Note dark shrunken degenerated muscle nuclei in large vacuoles. Thin basophilic fiber with dark pyknotic nuclei in upper part of field represents a terminal stage of lesion. (From biopsy section kindly provided by Dr. Sabina Stricht) (H&E)

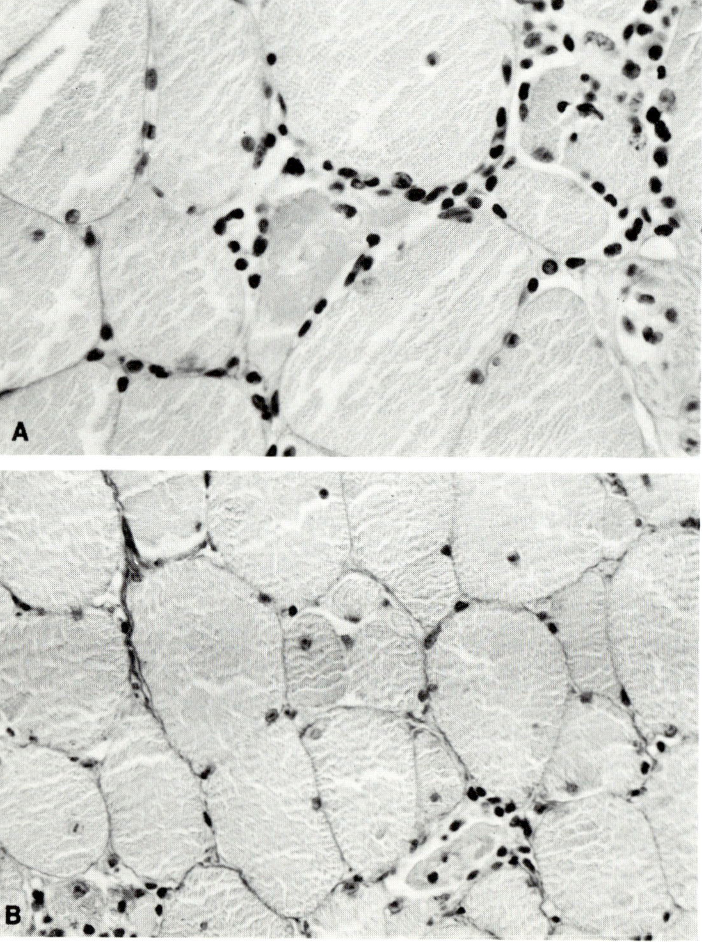

Fig. 7–28. Transverse section of early lesion of idiopathic polymyositis; deltoid muscle. (A) Lymphocytes and mononuclear cells separate the muscle fibers. The fiber at center has become eosinophilic and flocculated, and there is a zone of necrosis in the center of the fiber on the left side. Sarcolemmal nuclei are pale and nucleolated. (B) fiber at upper right has been injured and its sarcolemmal nucleus shows regeneration. The vein in upper part at left of center is surrounded by lymphocytes and the apex of the fiber to its left is showing early damage. Central nucleation prominent. (H&E)

and splenic enlargement in some cases, and the pathologic finding of inflammatory cells in the muscles, an infective cause is held to be responsible by most workers. The histologic changes in the more acute form of polymyositis include many features resembling those of experimental virus myositis and vitamin E deficiency, but in discussing these experimental studies we already noted the lack of specificity of these reactions and their production by toxic agents such as plasmocid and steroids. Indeed it may be that generalized polymyositis in man is not a single disease entity. Bacteriologic studies have not yielded convincing results. The report by Dalldorf and Sickles[29] of recovery of Coxsackie virus from two human cases of probably localized myositis, diagnosed clinically as acute poliomyelitis, raised the possibility of a filtrable virus as the etiologic agent. There is no proof, however, that the generalized clinical condition called polymyositis can be caused by the Coxsackie virus.

Shy and McEachern[141] obtained a favorable response to cortisone. O'Leary and Waisman[112] studied 40 patients who exhibited pronounced changes in the skin and mucous membranes. In addition to the usual types of skin lesion, they included cases of alopecia, stomatitis, and calcification of the subcutaneous tissues (Thiberge-Wissenbach syndrome). Such isolated cases of concurrent scleroderma or diffuse visceral involvement of the type seen in disseminated lupus erythematosus, and our own finding of associated rheumatic disease in more than half the cases, suggests that polymyositis must be causally related to the group of inflammatory connective tissue diseases that respond to corticosteroids.

Of considerable interest are a number of recent electron microscopic studies such as that of Chou and Gutmann[19] who demonstrated crystalline arrays in muscle cells from a case of polymyositis, which they interpret as picornavirus. Mastaglia and Walton[94] noted another type of cytoplasmic inclusion, believed to be a myxovirus. The patient Kakulas and Adams[71] presented as having an inclusion body polymyopathy proved on later study to have virus-like particles in sarcoplasm. The significance of such findings is difficult to ascertain. Virus particles have been observed at autopsy in the eye muscles of a patient who had no symptoms or signs or light microscopic evidence of muscle disease.[18] Transfer of such infections to another organism or proof of its infectivity by rising antibody titers has yet to be obtained.

The theory of causation, not incompatible necessarily with an infective etiology, which has received the most support is that of a cell-mediated autoimmune mechanism.[26, 32, 70, 133] The relationship of infiltrating cells to muscle fiber necrosis is the same as that demonstrated in experimental allergic polyneuritis, and led to the postulation of an autoallergic process in which transformed lymphocytes are attracted to the muscle fibers. Dawkins succeeded in producing polymyositis in the experimental animal by injecting sterile sarcoplasm with Freund's adjuvants. Kakulas[70] showed further that the transformed circulating lymphocytes were capable of destroying muscle fibers in tissue culture. Currie[26] and Saunders et al.[133] demonstrated destruction of tissue cultures of muscle cells by lymphocytes from human cases of polymositis. The experimental disease has been transferred to normal animals by the injection of sensitized cells. In serial sections of early lesions of human polymyositis Kakulas and Adams[72] noted that the segmental necrosis appeared always to occur adjacent to perivenous collection of lymphocytes and mononuclear cells. In other words, the earliest and smallest lesion is the perivenous infiltrate without which diagnosis must always remain in doubt. The segmental necrosis appears to be initiated by some product elaborated by transformed lymphocytes, probably an immunoglobulin, IgA or IgG, and is seen under the electron microscope to begin as a shredding of myofibrils, fragmentation of Z bands, and vacuolation of sarcoplasmic reticulum. Circulating antibodies to myofibrillar proteins have not been consistently demonstrated in human polymyositis. The necrotic reaction in turn is the excitant of the histiocytic reaction and at the same time induces the regenerative response including branching of the intact parts of the fiber. Fiber loss is poorly understood. Conceivably it represents an exhaustion of the recuperative powers of the fiber after a series of necroses. The connective tissue increases between fibers and fills the places formerly occupied by fully degenerated fibers.

The occasional direct relationship between injection of foreign protein (vaccine, antibiotic) and dermatomyositis also strongly suggests an allergic mechanism, though in most cases no such antigen can be identified. In view of the frequency of dermatomyositis in childhood, it is remarkable how rarely adult patients can give any history of attacks earlier in life. Nevertheless, types of polymyositis in which an inflammatory reaction is prominent without evidence of an infective agent may have an allergic mechanism, and it is in such cases that the value of steroid and immunosuppressant therapy seems most securely established.[112, 170]

ASSOCIATION WITH CARCINOMATOSIS

The association between dermatomyositis and malignant disease has been noted by many writers (Stertz, Bezecny, O'Leary and Waisman, Dostrousky and Sagher, McCombs and MacMahon, and others, reviewed by Walton and Adams[170]). Altogether, there are several hundred cases of this type in the medical literature, an incidence of approximately 15% of all reported cases of dermatomyositis. Males and females are about equally affected. The neoplasms described most frequently are carcinoma of stomach, breast, lung, and ovary, but similar tumors in gallbladder, bowel, kidney, uterus, larynx, vagina, esophagus, and parotid gland are also recorded. Lymphoma, leukemia, multiple myeloma, and sarcomas of soft tissues have also been conjoined with this muscle and skin disease. In some cases the illness took the form of acute dermatomyositis, in others a more chronic form. A chronic polymyositis with myasthenic features in association with carcinoma of the lung was recorded in a number of cases.[3, 41] Henson et al.[62] referred to this condition as carcinomatous myopathy. Denny-Brown, describing a unique form of progressive sensory neuropathy associated with bronchogenic carcinoma, noted the concurrence of a type of progressive muscular degeneration in this association. The type of polymyositis more commonly found in association with carcinomatosis is fairly rapid in onset and less evidently progressive, often showing remissions and relapses independently of the course or removal of the associated carcinoma.[159]

The type of histologic changes usually found in the muscles differs in no way from those encountered in the absence of malignancy. There is the same widespread muscle fiber degeneration, infiltrates of inflammatory cells, regeneration from surviving fibers, and fibrosis. No tumor cells are found in the muscles. The skin may or may not be involved. The relation of the myositis to the tumor has not been ascertained. It is not a chance coincidence, for the occurrence of malignancy in cases of polymyositis or dermatomyositis is five times greater than in the general population.[37] It is of interest that the polymyositis may regress as the tumor continues to grow. Operative removal, irradiation, or chemotherapy of the tumor, on the other hand, is sometimes accompanied by remission of the muscle disease. One may surmise that the tumor growth sets in operation some type of susceptibility to an unusual infective agent (virus?) or a hypersensitivity state which involves connective tissue and muscles. Such a hypothesis would explain the similarity of the lesion to that of other forms of polymyositis, its inconstant relationship to malignancy, and its conjunction with other diseases in distant tissues (nerve, skin, central nervous system). In other cases, the pathologic changes are first those of granular and pigmentary atrophy and then are more possibly related to a metabolic disorder.[36] Sheard,[140] in reporting three cases of dermatomyositis in association with visceral carcinoma, drew attention to the necessity of a search for malignant tumor, focal infection, and endocrine disturbance in every case. The author concurs.

DERMATOMYOSITIS IN CHILDREN

It is worth noting that epidemiologic studies of polymyositis by Medsger et al.,[97] which reveal an overall incidence of 0.5 cases per 100,000 population, show there are two age peaks: one during childhood, the other in late adult life. Black females were more often affected than black males, and white males more than white females. In two series of children with polymyositis[25, 136] there were twice as many females as males.

Cook *et al.*[25] and Banker *et al.*[9] call attention to the strikingly high incidence of a rapidly fatal form of necrotizing polymyositis (15 of 28 cases) in children below the age of 4 years. In seven autopsied cases the most striking changes were arteritis and phlebitis with necrosis of muscles, nerves, fat tissue, skin, and intestinal tract. The changes in the muscle were of the nature of infarct necrosis and denervation, akin to those described below under polyarteritis.

While there is no doubt that many of these children had fairly typical dermatomyositis at the start of the illness with erythematous lesions of skin in a butterfly distribution over face and extensor surfaces of joints and severe weakness in the limbs, the final picture clinically and pathologically resembles more that of a diffuse angiitis. We interpret these data as indicating in early childhood a more frequent conjunction of two inflammatory diseases of connective tissue: polyarteritis and dermatomyositis. On the other hand many cases of entirely typical dermatomyositis during either childhood or adolescence have been observed and these run a course no different from that of the adult. Polymyositis or dermatomyositis is seldom associated with malignancy at this age (one of Cook's patients had leukemia).

INTERSTITIAL (FOCAL) POLYMYOSITIS IN RHEUMATOID ARTHRITIS, RHEUMATIC FEVER, SCLERODERMA, LUPUS ERYTHEMATOSUS DISSEMINATUS

The first clear description of the inflammatory lesions of rheumatic fever was that of Aschoff,[7] and Curtis and Pollard[27] presented the first account of interstitial myositis in rheumatoid arthritis. Our conception of the pathologic changes in this and closely related diseases was expanded by the writings of MacCallum,[89] von Glahn and Pappenheimer,[165] Graeff,[54] Swift and McEwen,[154] Rich,[126] Steiner *et al.*,[149, 150] Clawson *et al.*,[21] Morrison *et al.*,[102] and Klemperer.[74] It has now been amply demonstrated

that acute rheumatic fever, lupus erythematosus disseminatus, scleroderma, and rheumatoid arthritis have one attribute in common—they are all accompanied by miliary inflammatory foci in skeletal muscle. These muscular lesions are often referred to as focal or interstitial myositis, a purely pathologic state not to be confused with the large, palpable tender foci in muscle that occur in sarcoidosis, tuberculosis, and fibromyositis.

The essential lesion of interstitial polymyositis is focal infiltration of inflammatory cells in the endomysial and perimysial connective tissues. The cellular aggregations are small (ranging from 50 to 300 μ in diameter), irregularly shaped, and multiple. Often as many as four or five such cellular infiltrations can be seen within one low-power microscope field (3–4 mm in width) as in Figure 7–29. They are composed predominantly of lymphocytes with a few plasma cells, histiocytes, and mast cells. Neutrophilic leukocytes are numerous in some cases and rare in others; eosinophilic leukocytes are seldom seen. In some of the larger infiltrations (Fig. 7–30) there are central collections of epithelioid cells and rarely a multinucleated giant cell.

The position of these cellular infiltrations within the muscles is quite variable. The most frequent localization is around small blood vessels, i.e., small veins and less often small arteries, arterioles, and capillaries in the perimysium. They may also be located in the epimysial and endomysial connective tissues, and their relationship to blood vessels is revealed best in serial sections. Similar infiltrations occur in the aponeuroses, tendons, and perineurium of intramuscular nerve trunks. The cellular infiltrates may rarely extend into the adventitia of intramuscular arteries in association with a slight proliferation of endothelial cells, and occasionally all the coats of the vessel wall may be involved.[147] Infrequently the entire wall of a small artery is infiltrated with inflammatory cells. Usually the blood vessels are not affected.

The muscle fibers, especially those immediately adjacent to an inflammatory focus, exhibit a variety of changes. The fibers may be hyalinized and eosinophilic or vacuolated. Often a short segment of a fiber, involving not

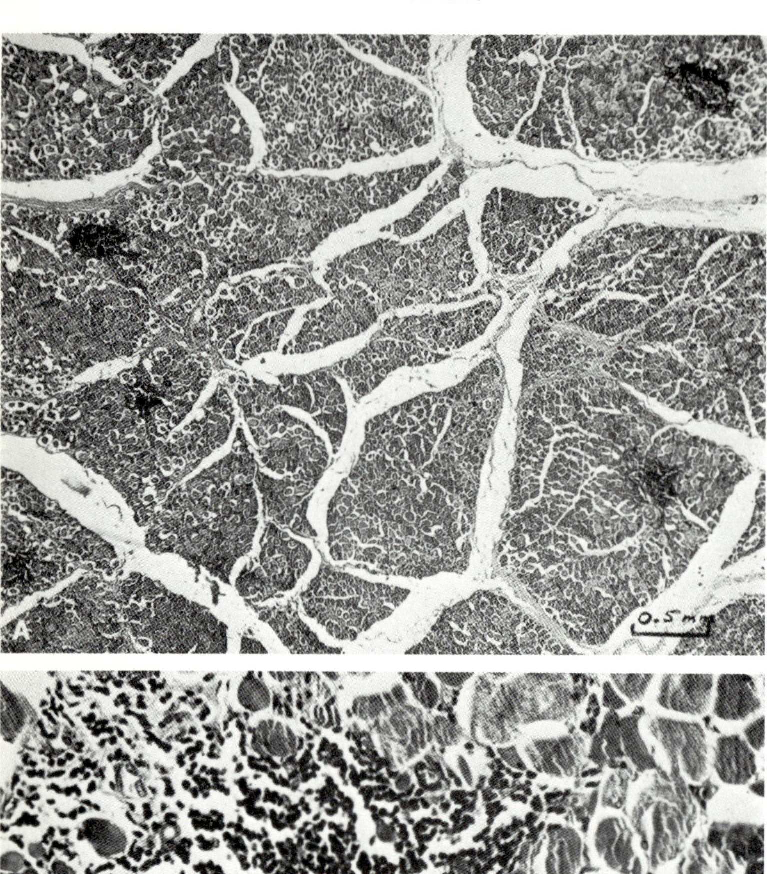

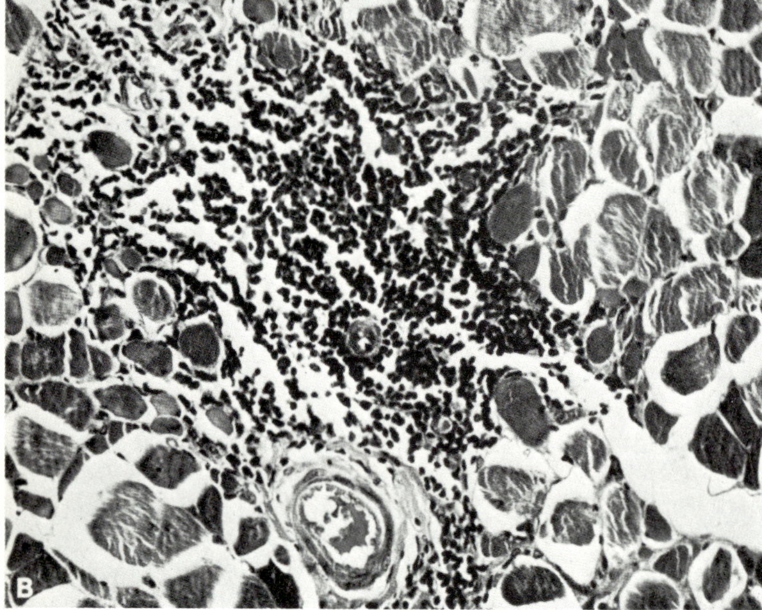

Fig. 7–29. Scleroderma. (A) Five discrete cellular infiltrations in endomysium are visible. (B) Higher magnification of one of these shows cells to be predominantly lymphocytes.

more than 10 to 15 sacromeres, degenerates. As the sarcoplasm disintegrates, macrophages appear in the endomysium next to the sarcolemma or within the necrotic segments (Fig. 7–31). Some of the sarcolemmal nuclei become shrunken and pyknotic and finally disappear, but others survive. These later enlarge, acquire a large nucleolus, and seem to multiply by simple fission. A hyaline sarcoplasm accumulates around them and gives rise to a thin basophilic fiber with prominent sarcolemmal nuclei. Muscle giant cells or muscle buds may form at the healthy ends of intact fibers. In some places in which parts of several contiguous muscle fibers have been destroyed the regeneration of fibers appears to be ineffective, and there is a

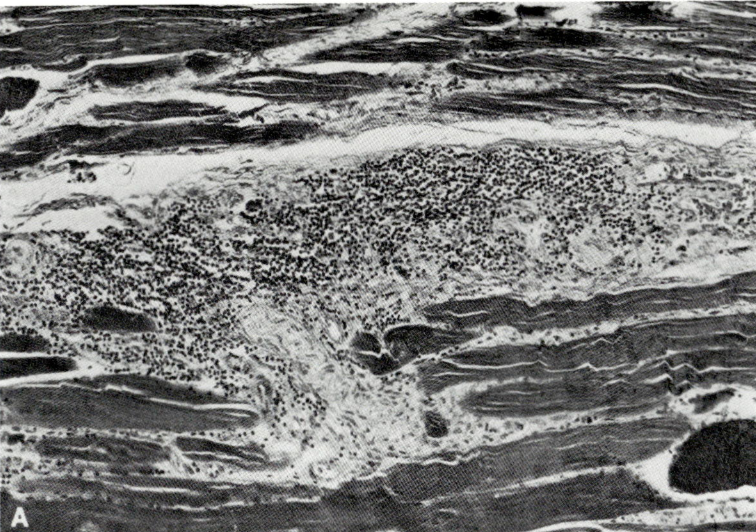

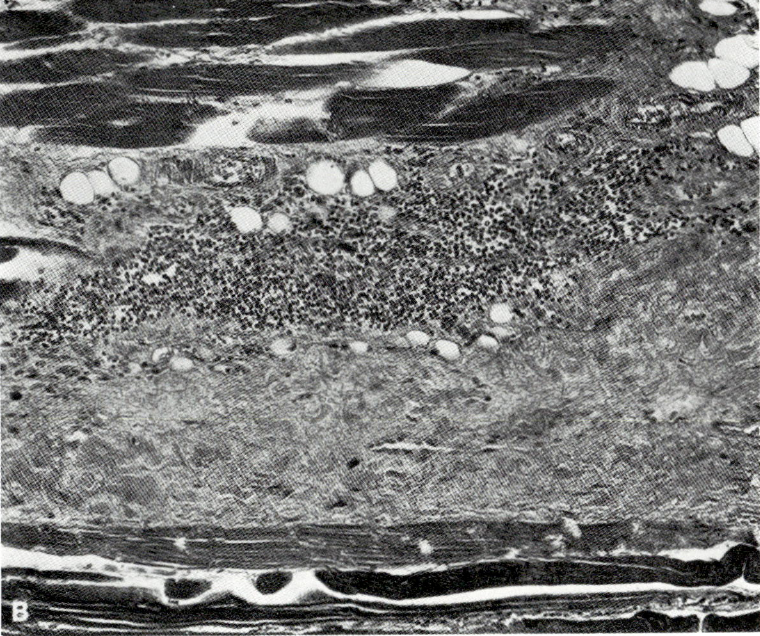

Fig. 7–30. Rheumatoid arthritis. Interstitial myositis. (A) Right gastrocnemius. (B) Triceps brachii. In addition to cellular infiltrates in perimysium—as well as around a muscular nerve in (a)—there is an increase in connective tissue. (H&E; ×155)

replacement by fibroblastic connective tissue (Fig. 7–16A). It is not uncommon for fibers that appear to lie some distance from such inflammatory foci to show similar changes. Atrophy of muscle fibers particularly type II ones with a relative increase in sarcolemmal nuclei is another common finding (Fig. 7–32). The average diameter of the fibers is less, and there is a greater variation in size of fiber than normal. In addition, quite large foci of infarct necrosis with fibrosis, due possibly to occlusion

of blood vessels either by thrombus or embolism, occasionally occur (Fig. 7–33B). In any one specimen there may be a mixture of inflammatory foci with acute degeneration of muscle fibers, healed foci with no inflammatory cells, fibrous scars, and a general atrophy of muscle with a slight increase in the number of fat cells. However, always the degeneration of muscle fibers is slight and insignificant in comparison to polymyositis.

These interstitial lesions have been found

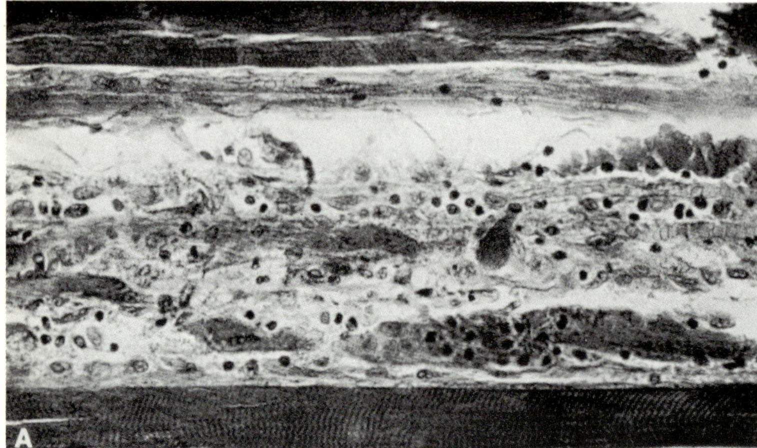

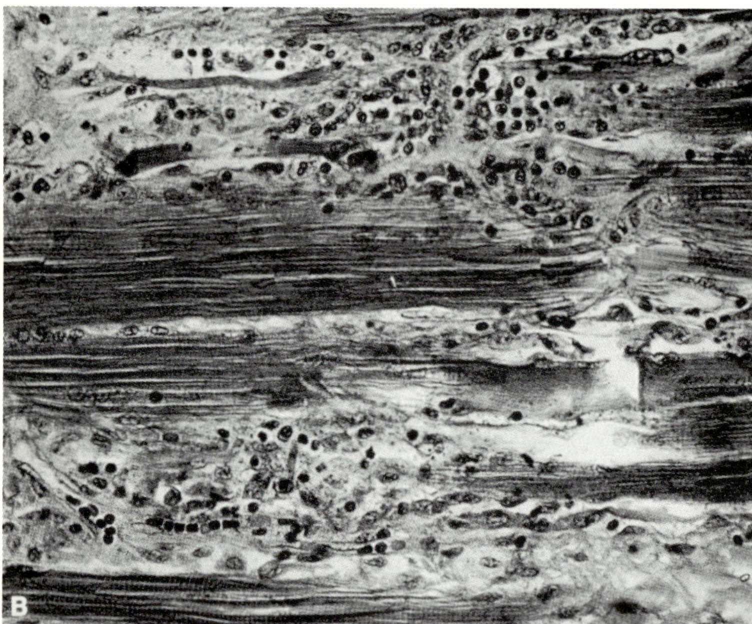

Fig. 7–31. Triceps muscle from patient with rheumatoid arthritis. Muscle fibers are seen in various stages of disintregration. Infiltrating cells are macrophages, lymphocytes, and plasma cells. (H&E; ×342)

in many different muscles, i.e., limb, trunk, ocular, pharyngeal, and diaphragmatic. Their occurrence cannot be foretold by any particular clinical symptom or sign. Indeed, in the majority of cases the clinician notes only moderate disuse atrophy without significant reduction in muscular power, reflex loss, or local tenderness. The pathologic changes are not limited to muscles or parts of muscles adjacent to inflamed joints or subcutaneous inflammatory nodules. Infarct necrosis is found when there has been a history of sudden aching pain in a muscle with swelling and discoloration of the overlying skin that had been present for some days.

RHEUMATOID ARTHRITIS

Interstitial polymyositis has been found in a high proportion of cases of rheumatoid arthritis—in all nine of the first cases examined by Steiner et al.;[149] in three or four cases with rheumatoid arthritis, splenomegaly, and leukopenia (Felty's

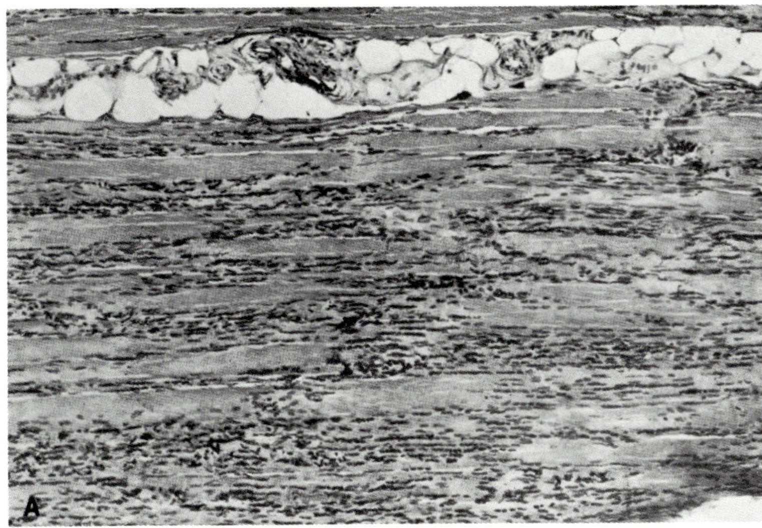

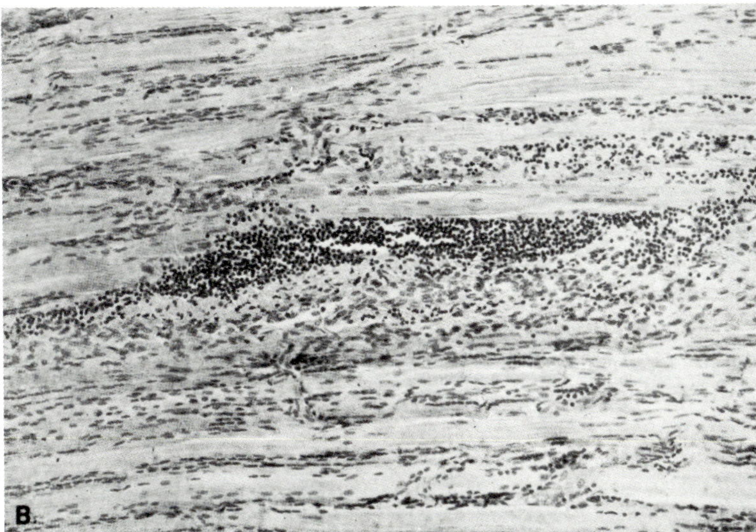

Fig. 7–32. Triceps muscle from a patient with rheumatoid arthritis. (A) Thin atrophic fibers with relative increase in sarcolemmal nuclei. (B) Perimysial infiltration of lymphocytes and plasma cells. (H&E; ×155)

syndrome) reported by Curtis and Pollard;[27] and in 13 of 16 cases recorded by DeForest *et al.*[33] Sokoloff *et al.*[147] noted an adventitial cellular infiltration in 5 of 57 rheumatoid arthritis cases. In addition to the interstitial myositis (Figs. 7–20, 7–24, 7–25, and 7–30) there may be other muscle changes in this disease. The general smallness of muscle fiber and the greater variability in size, probably indicative of disuse atrophy and cachexia, were already mentioned. In addition, extreme atrophy of fibers, with only rows of sarcolemmal nuclei and threads of poorly striated sarcoplasm remaining, may be found in some specimens (Fig 7–30A). This change probably indicates a neural muscular atrophy which is usually related to pressure palsy of exposed nerves.

RHEUMATIC FEVER

There is relatively little known about the state of the muscles in rheumatic fever. We were unable to find a single reference in the literature mentioning Aschoff bodies in the substance of skeletal muscle. Aschoff[8] evidently did not find the characteristic lesions in muscle or at least

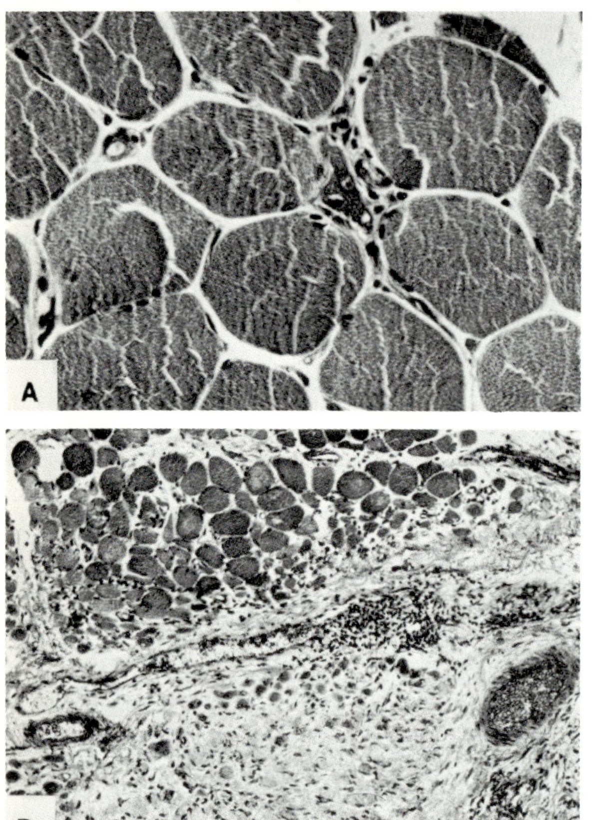

Fig. 7–33. Lupus erythematosus. (A) One shrunken basophilic fiber with enlarged sarcolemmal nuclei is surrounded by inflammatory cells in endomysium. (B) Most of muscle fibers in lower half of field have been replaced by connective tissue. Slight infiltration of lymphocytes and mononuclear cells is seen on one side of vein. (H&E; ×56)

did not mention them in his textbook. In several authoritative writings on the clinical and pathologic aspects of rheumatism, rheumatic myositis is mentioned, though a satisfactory description of the lesions cannot be found. "Growing pains" are often attributed to involvement of skeletal muscle, but without pathologic basis as far as we can learn. Wilson[174] in fact doubts this on clinical grounds alone for, as she points out, it is possible by palpation to demonstrate that the inflammation is localized in tendinous and peritendinous tissues that are tense, swollen, and tender. Poynton (cited by Wilson) presented a case of rheumatic fever in which there was a painful torticollis, but this case also lacked verification of actual pathologic muscle involvement. Lichtwitz[84] states that "muscle lesions in rheumatic fever, particularly in acute rheumatic polyarthritis, are common" and observes that they may be as widespread as the joint lesions, both in the

neighborhood of the joints and in muscles remote from affected joints. He quotes Risse[129] in support of this and refers also to the work of Klinge,[75–77] who produced an "allergic myositis" in the muscles of experimental animals and reported granulomatous lesions in the diaphragm in a patient who died of rheumatic fever and rheumatic endocarditis.[78] Graeff[54] observed granulomatous lesions in the muscles of patients dying of rheumatic fever, but it is not clear whether he was describing interstitial polymyositis or typical Aschoff bodies. Clawson et al.[21] reported a "nodular myositis" in biopsies of six successive cases of rheumatic fever, and Sokoloff et al.[147] found similar lesions in 7 of 21 cases; Aschoff bodies were not seen. Shaw[139] found Aschoff bodies, acute exudative foci containing neutrophilic leukocytes, and healed perivascular lesions in the diaphragm and the superior constrictor muscle of the pharynx. The muscles of the tongue and soft palate were normal. The rheumatic lesions were in the enveloping epi- and perimysial sheaths of the muscle and the aponeuroses and tendon bundles near their junction with the fibers, rather than in the substance of the muscle. The lesions in the pharyngeal muscles appeared to extend into the sheaths from the fibrous capsule of the faucial tonsils and in this respect were analogous to the cardiac muscle involvement adjacent to the annulus fibrosis and chordae tendineae.

A more complete survey of the skeletal muscles in rheumatic fever must be undertaken before any definitive statement on this subject can be made. This is not likely to be accomplished because of the diminishing frequency of the disease. All the more recent reports affirm the presence of interstitial polymyositis. If Aschoff bodes occur in muscle, they must be rare. They have not been observed in our rather limited material.

LUPUS ERYTHEMATOSUS

Muscle lesions are not uncommon in disseminated lupus erythematosus. Several different types of change have been observed. There may be a typical interstitial myositis indistinguishable from that occurring in rheumatoid arthritis and acute rheumatic fever, as shown in Figures 7–16B and 7–33B. Another frequent finding has been the acute hyaline necrosis or waxy degeneration of parts of single muscle fibers. The sarcoplasm appears homogeneous with loss of striation and has bright refractile eosinophilia. Parts of the fiber may fragment transversely (discoid degeneration) and are often invaded by macrophages. Occasionally a small fiber with basophilic cytoplasm and prominent sarcolemmal nuclei is seen. This probably represents a regenerating fiber (Fig. 7–33A). The specificity of such solitary fiber alterations for a disorder such as lupus erythematosus is quite debatable since in a recent survey of muscle biopsy material from routine autopsies similar regenerative changes were found in a number of biopsies from noncollagenous disease states. In two or three specimens we discovered lesions several millimeters in diameter in which all the muscle fibers had been destroyed and were replaced by fairly cellular fibrous tissue (Fig. 7–33B). Small numbers of macrophages containing blood pigment were also present. Lesions of this type are suggestive of infarct necrosis and may be due to embolism from an associated endocarditis (Libman-Sacks) or possibly to arteritis, but no occluded blood vessels were found in our cases. We have not observed the hyaline thickening or fibrinoid degeneration of the interstitial tissue of muscles or blood vessel walls which some pathologists believe to be the primary change in this disease. In those cases beginning as lupus erythematosus and terminating with dermatomyositis, the muscle lesions are exactly like those described under the heading polymyositis.

We encountered a remarkable vacuolar myopathy in a case of systemic lupus erythematosus.[120] A very high serum transaminase value had called attention to the possibility of a disease of the skeletal musculature in a 15-year-old boy who had some of the clinical findings of lupus erythematosus; a subsequent biopsy of a sternocleidomastoid muscle showed widespread vacuolation of segments of many muscle fibers. The "hydropic" vacuoles were either clear or contained a fine eosinophilic precipitate. Some vacuoles were loculated and were traversed by fine longitudinal septums composed of groups of myofibrils (Fig. 7–34) Myofibrils were absent within the confines of the vacuoles, and elsewhere were often displaced to one side, as though the contents of the vacuoles were under pressure. The sarcolemmal nuclei were large and occupied a central position in many places, often in association with the vacuoles. This change is not dissimilar to that seen in some of the muscle fibers in a vacuolar type of chronic progressive polymyositis described above (Fig. 7–24).

SCLERODERMA

The diffuse form of scleroderma may be accompanied by pathologic changes in the muscles. The most frequent muscle lesion, usually demonstrated by biopsy, is an interstitial myositis of the type already described (Fig. 7–29). When stiffness, swelling, and pain in the joints are present the skeletal muscles undergo severe

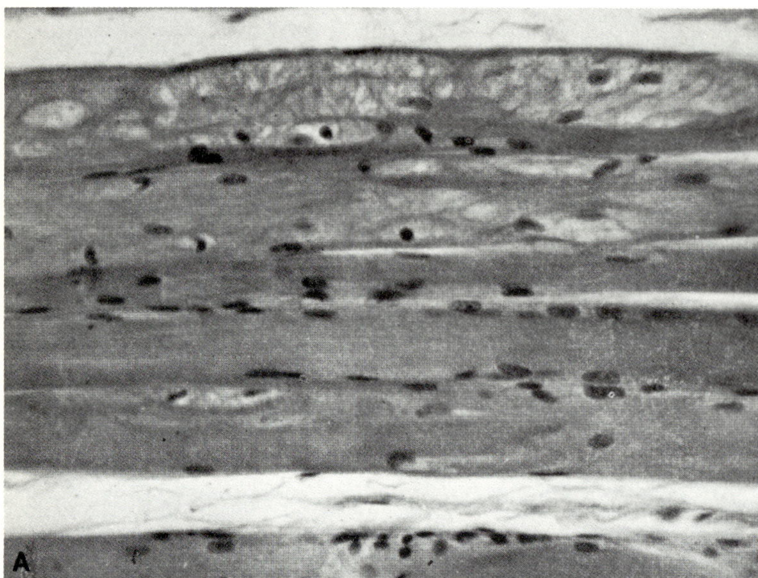

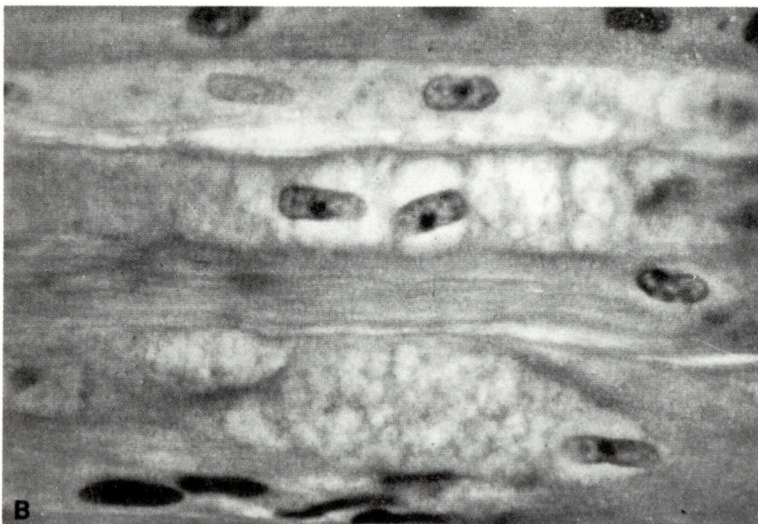

Fig. 7–34. Vacuolar myopathy in lupus erythematosus. (A) Varying degrees of solitary and multiloculated vacuolar myopathy. Note increase in sarcolemmal nuclei and their presence within vacuoles. (B) Latter finding is especially well shown here. Finely dispersed granular material within vacuoles is well visualized, as is displacement of myofibrils. (H&E; A, ×350; B, ×600)

atrophy and contractures may develop. In such cases there is pronounced atrophy of many muscle fibers, and others can be found in the process of degenerating. The sarcolemmal nuclei of these and other healthy appearing fibers are increased in number. The supporting tissues may or may not be infiltrated with inflammatory cells. In a few cases striated muscles, particularly those of the tongue, soft palate, and larynx, may exhibit the same extensive degeneration of muscle fiber and cellular infiltration of connective tissue as that which occurs in dermatomyositis. In our limited experience with this disease we have not been impressed with the thickening of endo- and perimysial connective tissues or with thickening of the walls of small intramuscular blood vessels, a prominent feature of the skin lesions.

SIGNIFICANCE OF INTERSTITIAL FOCAL POLYMYOSITIS

The specificity of interstitial polymyositis is open to question. Focal infiltrations of lymphocytes have been observed in many different diseases, some of which are surely not inflammatory. Buzzard,[15] in a clinicopathologic study of five cases of myasthenia gravis, found multiple foci of lymphocytes and other inflammatory cells in the muscles of all of them. These cellular infiltrates, which he termed lymphorrhages, occurred not only in skeletal muscles but also in other organs such as the heart, adrenal glands, liver, thyroid gland, and in one instance around vessels in the substance of several posterior root ganglia. Since he failed to find lymphorrhages in other fatal diseases, he suggested that they were the most characteristic change in myasthenia gravis. In nearly every one of Buzzard's cases some of the muscles showed an abnormal appearance of a small proportion of their fibers. Some were atrophic and others showed an acute hyaline segmental degenerative change. In reading these case records the diagnosis of myasthenia gravis could not be doubted, and there were no clinical signs of rheumatoid arthritis or related diseases. We have also seen collections of lymphocytes, though very rarely, in myasthenia gravis (Chapter 12). It seems to us, therefore, that Buzzard's lymphorrhages are indistinguishable from the cellular infiltrates of interstitial polymyositis.

Morrison *et al.*,[102] Clawson *et al.*,[21] and Sokoloff *et al.*[146] also expressed doubts as to the specificity of interstitial cell infiltrates in muscle. Clawson and his associates found them in 11 of 43 presumably healthy individuals dying of trauma, in 18 of 46 tumor cases, in 5 of 9 cases of cirrhosis of the liver, in 7 of 24 cases of hypertension, and in 8 of 31 cases of coronary arteriosclerosis.

We have noted a definite correlation between interstitial myositis and rheumatoid arthritis, lupus erythematosus, and scleroderma, and it is common in the less affected muscles of dermatomyositis (Fig. 7–16A). Such lesions are seldom found in random sections of skeletal muscle taken at autopsy. Therefore, while agreeing with Clawson and his associates that some degree of simple lymphocytic infiltration may be found in a number of quite unrelated diseases, multiple perivascular and adventitial nodules of inflammatory cells with destruction of contiguous muscle fibers occur with remarkable frequency in the so-called connective tissue diseases and rarely in other diseases. Moreover, the condition we chose to designate focal interstitial nodular polymyositis merges imperceptibly with the characteristic muscle lesion of polymyositis. Indeed, the latter may in some instances be only an advanced clinically recognizable form of interstitial nodular myositis. We have been unable to confirm the recent statement by Steiner and Chason[149] that the cellular infiltrations differ in composition and location in each type of these collagen group diseases. Probably in muscle, as in other tissues, no specificity can be assigned to slight lymphocytic infiltrations, but if there are foci of mixed inflammatory cells (i.e., lymphocytes, plasma cells, histiocytes, and neutrophilic leukocytes) and in addition definite degenerative changes in the muscle fibers and vascular lesions, the whole process acquires added significance. The former should be termed inflammatory cell infiltration of muscle, and the term interstitial myositis should be reserved for the latter. There is some evidence relating lymphocytosis to immunologic processes, and this may possibly apply to lymphocytosis of muscle. We have seen centrally placed reticulum cells (Fig. 7–35) in a few nodular infiltrates composed chiefly of lymphocytes. Such foci have the general aspect of lymph follicles not unlike those which may ultimately develop in any chronic lymphocytic inflammatory reaction. Probably lymphocytosis of muscles is an index of antibody formation.

FIBROMYOSITIS AND MYOGELOSIS

By definition fibromyositis is an inflammation of the fibrous tissues, especially those of the muscle sheaths, fascia, and aponeuroses, and probably of nerves as well. In theory the cause is any agent, microbic or toxic, that induces an inflammatory reaction. Cold, dampness, and trauma are predisposing factors. The generic syndrome consists of pain and stiffness, often of acute onset, in and around a group of muscles. The neck and shoulder are the most common sites and the lower thoracic and lumbar back muscles the next commonest. Firm, tender

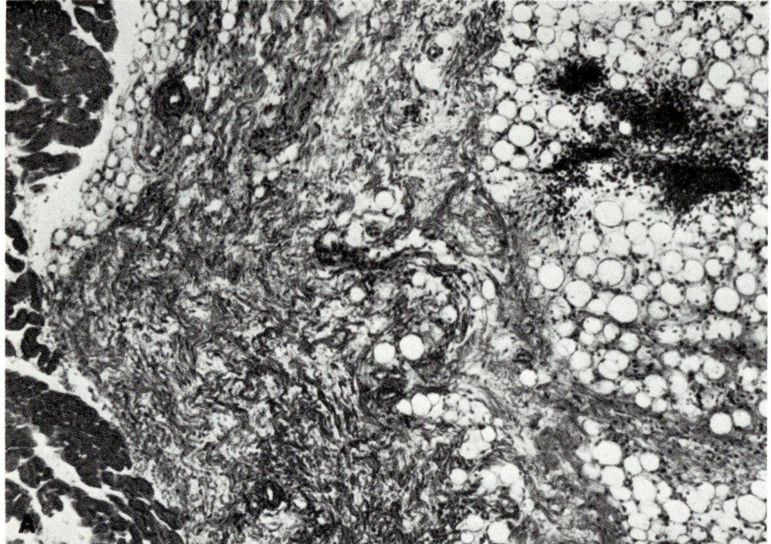

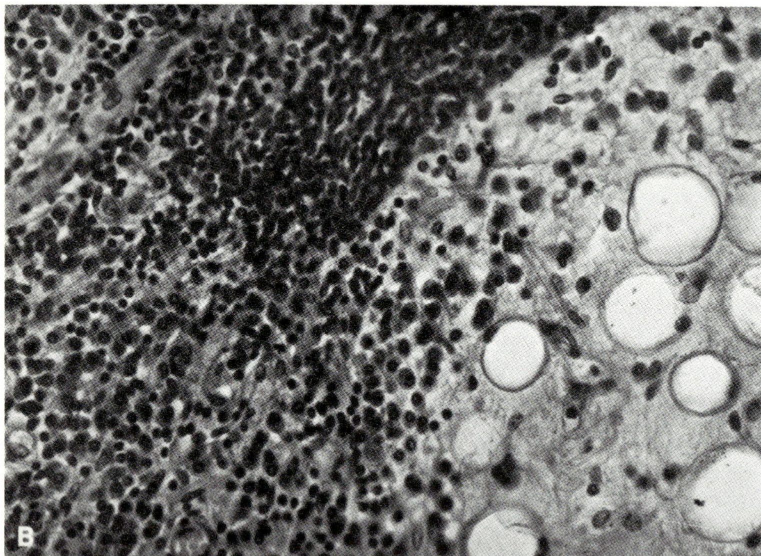

Fig. 7–35. Generalized polymyositis with scleroderma. (A) Muscle fibers are at left; also seen are the epimysium with fibrosis and a focus of lymphocytic activity and lymph follicle formation in epimysial fat (upper right). (B) Higher magnification of border of part of a lymph follicle just outside field of (A), showing border of follicle (at top) and portion of entering lymphatic vessel, with dense accumulation of various types of lymphocytes and endothelial cells.

nodules, sometimes several centimeters in diameter, are found in some cases. Active and passive movement of the affected parts and firm pressure on them induce pain. Low-grade fever may occur at the onset of the illness. Manifestations of involvement of a related peripheral nerve in the arm or leg may accompany the condition, which is then designated as a brachial or sciatic "neuritis." The disease is self-limiting, and fatalities are unknown. Occasionally it is the forerunner of one of the connective tissue diseases, especially rheumatoid arthritis; or it precedes by weeks or months clinical evidence of malignancy, such as lymphoma.

Myogelosis is another term applied to a painful state of one or several muscles in which a diffuse induration or discrete nodularity can be palpated. It was introduced, according to Jordan,[68] by Lange and Schade in 1921 and was popular especially among physiotherapists for the next decade. The chief clinical criteria were local pain and tenderness and a change in the texture of muscle, for the measurement of which special instruments were devised, i.e., the sclerometer and elastometer. The nodules in muscle were said to persist during sleep, anesthesia, and even for several hours post mortem. The examination of excised pieces of muscle consistently failed to reveal gross or microscopic abnormality, and it was therefore postulated that the local induration represented a jelling of muscle colloids. Other workers quite properly objected to this hypothesis and pointed out that myogelosis cannot be distinguished from local muscular spasm. In a biopsy of a muscle that contained a painful nodule of some weeks' duration, the only change we observed was a state of extreme contraction of the muscle fibers. We were unable to decide whether this was due to an increased local irritability of the muscle fibers or to a violent contraction induced by the fixative. Interest in the myogeloses has waned because of uncertainty as to the nature of the condition. The literature on this subject was reviewed in the monograph of Lange.[80] More recently Seyffarth, Clemmesen, and others reinvestigated the electromyographic aspects of this condition, which has also been called occupational myalgia, or painful myosis.[87] Poor relaxation of the muscles due to faulty posture is thought to result in cramps similar to those induced by fatigue.

Other reporters have been equally unsuccessful in defining the pathologic process. Hench *et al.*[60, 61] were unable to detect any definite abnormalities in specimens of "fibrositic" nodules, and other observers[14, 24, 143] reported only minor changes of doubtful significance, such as slight fibroblastic proliferation, serous or serofibrinous exodate, and thickening of the wall of blood vessels or of nerve sheaths. The most that one can say is that the pathologic basis of this large group of diseases is unknown.

LOCALIZED MYOSITIS

Over the years many generous colleagues have sent to the author biopsy specimens of thickened or nodular painful muscles. Tenderness, enlargement, and induration of all or part of the muscle were mentioned as the clinical findings. Biopsies have disclosed a striking lesion consisting of large numbers of inflammatory cells infiltrating a mass of interdigitating regenerated muscle fibers and strands of connective tissue. The mass may have reached such proportions as to have suggested, before biopsy, a muscle tumor. We have observed a number of similar pseudotumors that followed athletic injuries or overly energetic muscle stretching by zealous physiotherapists.

We have interpreted this reaction (in reality a pseudotumor) as an inflammation (infective?) with injury of both muscle fibers and connective tissue sheaths. It is the regeneration of both elements but particularly the latter which imparts the desmoid or pseudosarcomatous character to the lesion. Szaks[156] wrote about this lesion, referring to it as a proliferative myositis. No evidence of an identifiable microbe was obtained in these several cases. Eventually the process subsides leaving a pseudocontracture (fibrous contracture) of the muscle.

POLYARTERITIS NODOSA

A relatively uncommon disease, polyarteritis nodosa probably involves the peripheral nerves and muscles in 50–80% of cases. Arkin[6] estimated that the nerves are affected in 20% of cases and the muscles in 30%, but these figures appear to relate to predominant symptomatology. Muscular lesions in fact occur so often that muscle biopsy has become a standard diagnostic procedure.

The clinical neuromuscular manifestations include pains, paresthesias, sensory loss, muscular weakness or paralysis, and loss of tendon reflexes. Tender nodules along the arteries, for which the disease was named, and tenderness in the muscles are infrequent. The onset of symptoms is often acute or subacute, and progression is fairly steady, death occurring within a few months. Chronic cases are known, and recovery occurs in rare instances. The neurologic symptoms are often asymmetrical. They are usually accompanied by fever, tachycardia, hypertension, abdominal pain, melena, hematuria, asthma, and other systemic and visceral disorders characteristic of the disease.

Affection of the kidneys is especially frequent. It should be emphasized that lesions can often be demonstrated in skeletal muscles that are not weakened, atrophied, or tender.

The affected muscles in some instances appear normal grossly, whereas in others it is possible to observe numerous small hemorrhages on the surface and within the muscle tissue. In or near some of these lesions a thrombosed artery may be seen. Pronounced muscle atrophy is evident in cases in which the muscle nerves were damaged some weeks before death.

The primary change appears to be necrosis of part or all of the arterial wall with an intense inflammatory reaction. The media or muscularis is most often the site of involvement, though the process may extend to the adventitia or intima. The inflammatory reaction varies with the age of the lesion, at first consisting predominantly of neutrophilic and eosinophilic leukocytes and later of lymphocytes, plasma cells, and other mononuclear cells. As a consequence of the necrosis, several other vascular changes my develop. The vessel may be weakened at one point, permitting aneurysmal dilation and sometimes rupture with serious, even fatal hemorrhage. We saw a huge hematoma develop in one deltoid muscle. The vessel may become thickened and nodular (hence the name nodosa) owing to the formation of aneurysms or of periarterial nodules of inflammatory tissue. With intimal damage thrombi may form within the lumen. Following occlusion of one or several vessels there may be infarct necrosis of tissue within the territory supplied by these vessels. Since the necrosis is focal, a single section may not include the essential lesion, so that the only visible abnormality would be inflammatory cells surounding the vessel or infiltrating some portion of its wall.

The diverse character of the arterial lesions within the nerves and muscles, and at times even the spinal roots, accounts for the variety of changes that ocur in skeletal muscles. These are of the following types.

ARTERIAL LESIONS

The larger blood vessels in the epi-, peri-, and endomysium usually exhibit focal necrosis,

aneurysm formation, thrombosis, or merely an infiltration of inflammatory cells (Fig. 7–36). All these changes are not necessarily found in a single section or in one muscle. Also the lesions are of differing age and severity in the different muscles or parts of one muscle. It should be pointed out that polyarteritis nodosa involves the larger arteries, that the vascular lesions are necrotic, that the effects on muscle and nerve are by vascular occlusion and hemorrhage, and that infiltrations of eosinophils are common. Lymphocytosis of muscle and interstitial myositis may occur but are of course nonspecific.

FOCAL ISCHEMIC NECROSIS AND HEMORRHAGE WITHIN THE MUSCLE

Occlusion of one large or several small vessels within one part of a muscle may result in infarct necrosis of the muscle fibers and supporting connective tissue. The appearance of the necrotic muscle varies with the age of the lesion, and the changes are uniform within limits of a sharply circumscribed area. Early, the fibers are swollen, hyalinized, and eosinophilic (Fig. 7–37B); the sarcolemmal nuclei and endomysial fibroblasts are shrunken and pyknotic. Later the muscle fibers disintegrate and are removed by macrophages derived from the adventitial cells of blood vessels or by histiocytes. Finally there is fibroblastic proliferation and connective tissue replacement.

NEURAL MUSCULAR ATROPHY

As a consequence of ischemic necrosis of motor nerve fibers due to lesions in the nerve trunks, groups of muscle fibers undergo typical neural muscular atrophy (Fig. 7–37A). The general characteristics of this condition are described later.

SARCOIDOSIS (BOECK-BESNIER-SCHAUMANN DISEASE)

Boeck's sarcoid, sometimes called benign miliary lupoid and benign lymphogranu-

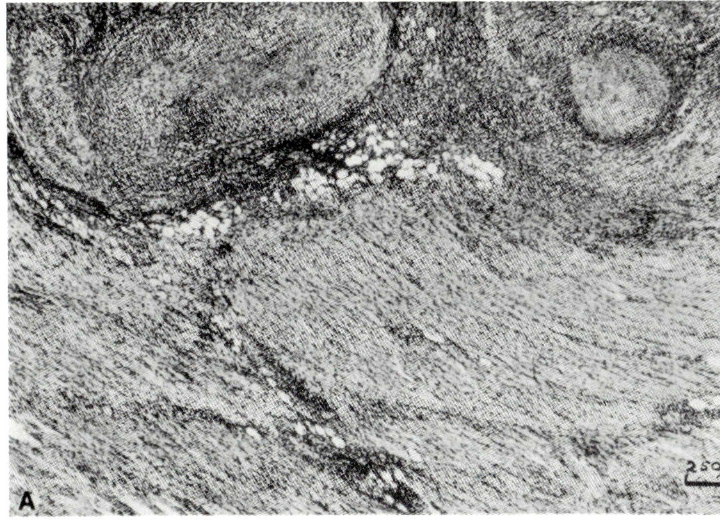

Fig. **7–36.** Polyarteritis nodosa. Section of anterior tibial muscle showing arterial lesions in muscle. (A) Artery (upper right) is thrombosed. (H&E)

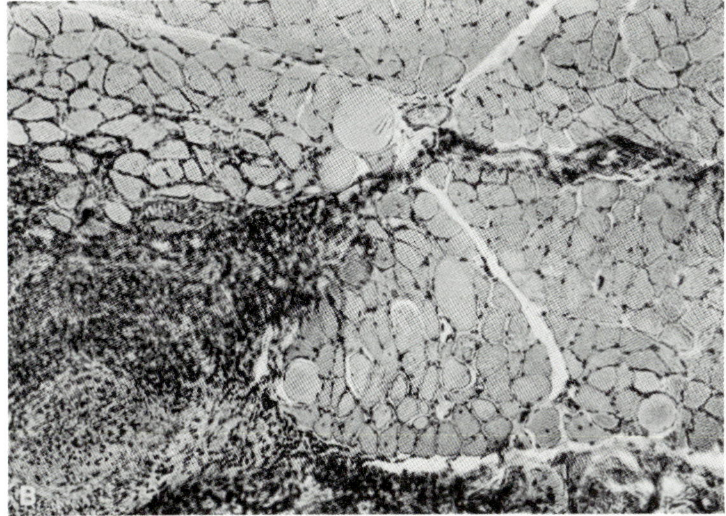

lomatosis, is a disorder characterized by formation of nodular and plaque-like lesions in the skin and subcutaneous tissues. The bones, lungs, tonsils, lymph nodes, and other organs may be involved. The disease is benign, in contrast to tuberculosis, which it closely resembles. Localization in the peripheral or central nervous system has been reported by several authors.[167] The facial nerve is often affected in one form of the disease, uveoparotid fever,[59] and other peripheral nerves or spinal roots have been involved in some cases.

Sarcoidosis of striated muscles was first described by Licharew[83] in 1908. In his case there was a palpable nodularity in the sternomastoid, biceps, and gastrocnemius muscles. Sundelin[153] gave a detailed account of multiple disseminated tumors in muscles that resembled tuberculosis histopathologically yet differed sufficiently to cause him to doubt this etiologic agent. His illustrations of the microscopic structure of these nodules showed them to be characteristic of Boeck's sarcoid. Snorrason's case[145] was of interest because the disease presented as progressive fibrous myositis. A number of papers appeared within recent years[57, 79, 104, 121, 168] on sarcoidosis of skeletal muscle, and each included comments on the value of muscle biopsy as a diagnostic pro-

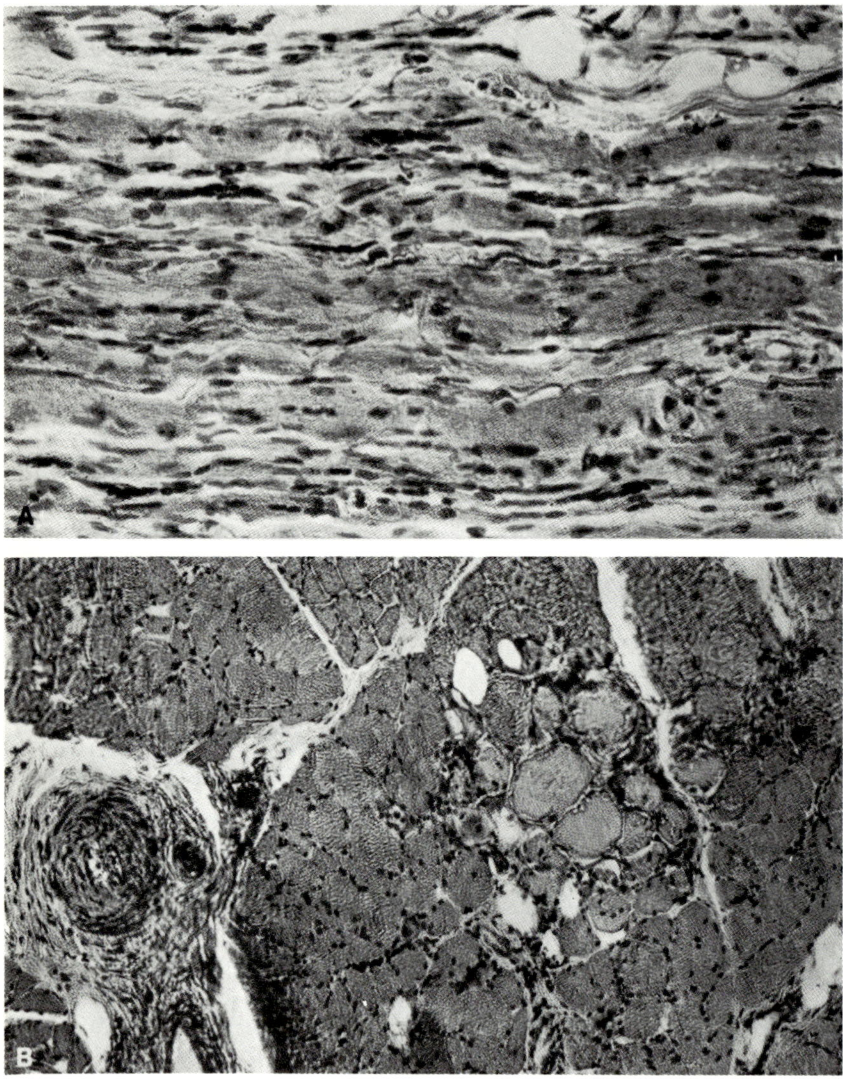

Fig. 7–37. Polyarteritis nodosa. (A) Typical picture of a fairly advanced neuro-muscular atrophy. (B) Small infarct with swollen hyalinized fibers. There is some thickening in wall of small artery in which a characteristic lesion is a few milli-meters proximal to this point. (H&E; A, ×320; B, ×210)

cedure. Several stressed the desirability of making routine histologic sections at multiple levels in the biopsy block in order to offer a better chance of finding the typical sarcoid granulomas. Random muscle biopsy from clinically unaffected muscle has yielded a surprisingly high incidence of positive results. Phillips and Phillips[121] found sarcoid lesions in biopsies from 3 of 4 patients, and Wallace et al.[168] observed them in 23 of 42 patients.

The etiology of sarcoidosis remains in doubt, and there are some physicians who still believe that it is a particularly benign form of tuber-culosis. Certainly from a review of the reported cases of sarcoidosis of muscle, it appears that efforts to culture or stain tubercle bacilli in the lesions of sarcoid were not always conducted with sufficient care to exclude tuberculosis.

In addition to the mild symptoms that ac-company sarcoidosis, the muscle involvement may present as a localized weakness and pain with either a palpable nodularity of one or several muscles or a progressive atrophy of muscles. In some cases there may be muscle involvement without specific symptoms.[2, 57, 104] The skin and mucous membranes may or may not be affected. If they are, the combination of skin and muscle lesions is properly designated as a form of chronic dermatomyositis but must be distinguished from that described above. In several reported examples one or several nerves has been indirectly affected so that localized sensory impairment, loss of tendon reflexes, and neural muscular atrophy may occur. Several patients observed on one of the medical wards of the Massachusetts General Hospital presented with the clinical picture of a sensori-motor polyneuritis.

The typical lesion is made up of a group of sharply circumscribed nodules separated by connective tissue septa. These nodules are composed largely of histiocytes or epithelioid cells; the latter are large and have lightly stippled nuclei and a prominent nucleolus. Lymphocytic infiltration, though present, is slight. In some places a few neutrophilic leukocytes, plasma cells, and mast cells are seen. Giant cells of Langhans' type are frequent (Fig. 7–38), yet necrosis of the caseous type (so common in tuberculosis) and tubercle bacilli cannot be demonstrated. These granulomas are situated within the connective tissue sheaths of muscle. They replace the muscle fibers, and those fibers which survive in the adjacent tissue are apt to be thin. Evidence of recent destruction of muscle fibers is found occasionally. With extensive involvement of the muscle, a large proportion of the fibers are replaced by connective tissue in which there are scattered granulomatous nodules. Thus the total bulk of the muscle is reduced and its contractile power weakened. Harvey[57] recently described three cases of a clinically evident myopathy due to extensive invasion and replacement of many skeletal muscles by the granulomas of sarcoid. Weakness was present in the first case for 20 years, in the second case intermittently for 5 years, and in the third for 15 years. Proximal muscles were usually weaker than the distal ones, and the clinical picture resembled to a remarkable degree that

of certain cases of chronic polymyositis of unknown etiology. Furthermore, corticosteroids improved strength and function in Harvey's third case and in the one described by Ammitzboll.[2]

When muscular nerves are destroyed by sarcoidosis, a typical localized muscular atrophy results. Hence the muscle disorder in Boeck's sarcoid may be of two types, one a neural atrophy and the other a localized inflammatory destruction of the muscle fibers. These muscle lesions are easily distinguished histologically from the interstitial (nodular) myositis described earlier in which a single giant cell may sometimes be seen, and from idiopathic dermatomyositis and polymyositis.

Granulomatous lesions similar to those of sarcoid are occasionally found in muscle biopsies from patients with any one of the several connective tissue diseases and from those with miliary or generalized tuberculosis. Also a form of giant cell myositis has been observed in patients with thymomas and myasthenia gravis, thyroiditis and myocarditis. [Cf Namba *et al.*[105] for recent review of literature]. In our opinion the tendency to classify every granulomatous lesion with a few Langhans' giant cells as sarcoid has resulted in the inclusion of many patients with other diseases, including typical polymyositis.

CHRONIC ORBITAL CELLULITIS WITH MYOSITIS (INFLAMMATORY PSEUDOTUMOR OF THE ORBIT, CHRONIC INFLAMMATION OF THE ORBIT)

Among patients who seek medical advice for unilateral proptosis and ophthalmoplegia there are a considerable number, variously estimated at 8–50%, in whom the orbital disease is due to a nonspecific inflammation. This fact has long been appreciated by ophthalmologists, who in the past have cautiously termed the condition inflammatory pseudotumor of the orbit.[173] More recent pathologic studies disclosed an inflammatory reaction in the extraocular muscles, and emphasis on this feature of the pathologic process has led to its designation

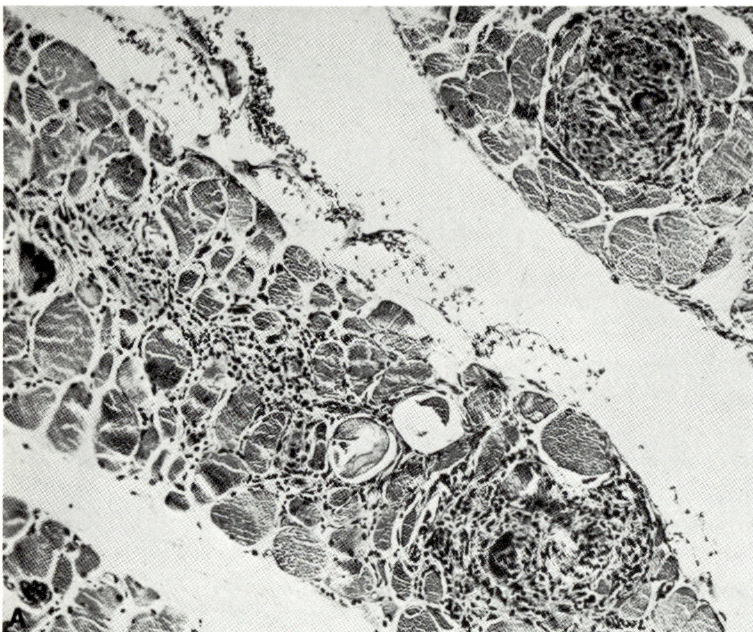

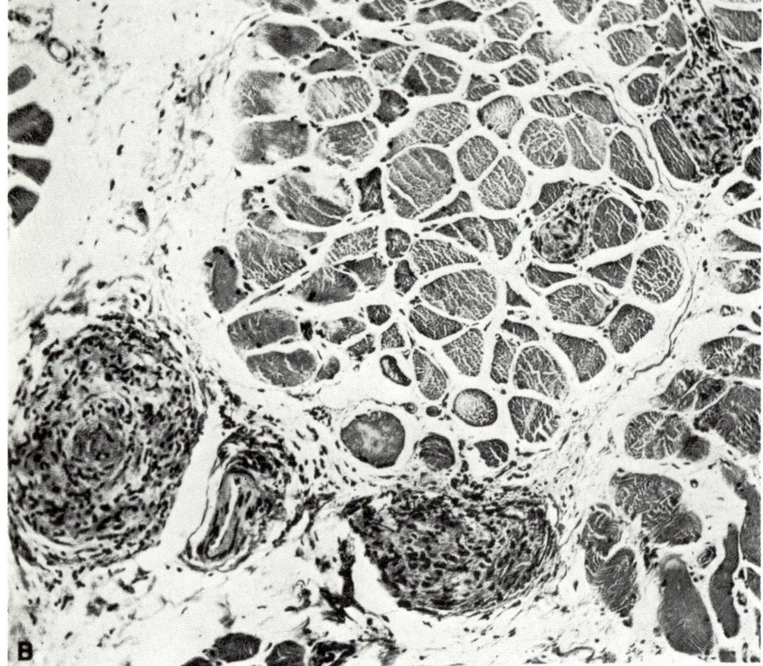

Fig. 7–38. Sarcoidosis. Section from involved muscle showing characteristic granulomatous formations with giant cells. In central fasciculus in (A) there has been destruction of muscle fibers and atrophy of other scattered fibers. (H&E)

as a chronic orbital myositis.[38] We are not in agreement with the latter view, for an analysis of the published cases and an examination of several biopsy specimens of orbital fat and ocular muscles from the Howe Laboratory of Ophthalmology through the courtesy of Dr. David Cogan make it clear that the inflammatory disease is rarely restricted to the muscles and usually involves all the orbital tissues, including fat, muscle sheaths, extraocular and optic nerves, and at times even the sclera and uveal tract of the eye. For this reason the terms orbital cellulitis or inflammation of the orbit are preferred.

A full account of the symptomatology is given in the above references and in Walsh's *Clinical Neuro-ophthalmology*.[169] The disease is nearly always unilateral, though in a few cases the other eye has been involved, either concurrently or at a later date, sometimes years later. Slight prominence of the eye or conjunctival irritation with lachrimation, photophobia, redness of the conjunctiva, and edema of the eyelid may be presenting symptoms. Diplopia due to obvious paresis of extraocular muscles occurs in the majority of cases. Any one or all of the extrinsic ocular muscles may be affected, and the intrinsic ones always escape. The proptosis progresses, but seldom exceeds 4–5 mm; as a rule, the globe is pushed forward, though it may be displaced in any direction. Orbital pain is a prominent symptom in some cases and is completely absent in others. Vision is disturbed rarely. The preauricular lymph node on the side of the involved eye is enlarged in all cases.[22] There is no evidence of disease in other organs. Under the mistaken diagnosis of tumor, the orbit is often explored, and either a discrete inflammatory mass or only a slight edema of the orbital tissues is found. In many cases after operation, or spontaneously, there is a remission of symptoms and complete recovery. Potassium iodide has produced dramatic cure in many cases, with the consequence that some ophthalmologists regard the response to this drug as a diagnostic test. The total duration of the illness is several weeks to months.

Prior to operation there is often confusion as to diagnosis. Syphilis, tuberculosis, fibrous dysplasia, eosinophilic granuloma, or Hand-Schüller-Christian disease can usually be excluded by clinical and laboratory tests. Malignant exophthalmos or hyperthyroidism, the uveoparotid syndrome of Heerfordt, and orbital tumor are more difficult to differentiate. The strict unilaterality of the disease, ptosis, preauricular lymphadenopathy, and absence of signs of thyroid disease should distinguish it from malignant exophthalmos, and the lack of parotid gland enlargement or signs of disease of skin and viscera from sarcoidosis.

The morbid anatomy of this disease is imperfectly known, though a few satisfactory pathologic specimens have been provided by enucleation of the eye for a suspected orbital tumor. Ophthalmic surgeons have often remarked on the edema of the orbital tissues, the matting together of orbital structures, and the pale appearance of the ocular muscles. When an inflammatory mass is found, it may be so discrete as to simulate a tumor. Usually, however, the orbital fat, eye muscles, and nerves are implicated in a mass of granulation tissue, consisting of fibroblasts and collagen, blood vessels, and numerous infiltrations of lymphocytes, mononuclear leukocytes, plasma cells, and eosinophilic leukocytes. Giant cells are seen occasionally, but well formed tubercles cannot be found. The sclera of the eye, the sheaths of the extraocular nerves, and the muscles are invaded by inflammatory cells. Fibers of the ocular muscles may be completely replaced by fibrous tissue, or single fibers may be widely separated by connective tissue. Evidence of muscle fiber degeneration and abortive regeneration of sarcolemmal nuclei are observed in some cases. Infiltrations of inflammatory cells in the walls of small arteries and veins and particularly in the perivenous spaces have led some ophthalmologists to conclude that endarteritis or phlebitis was a primary disorder, but this has not been more pronounced in cases we studied than would be expected in any chronic inflammation.

The etiologic agent in this disease is still disputed. Cultures for bacteria have been consistently negative; the Wassermann and tuberculin skin tests also have yielded negative results. Braley and Alexander[12] isolated a filtrable virus, called the Brooks virus after the patient from whom it was obtained. This work has not been confirmed.

CONCLUSION

A study of the morbid anatomy of inflammatory diseases of muscle affords a basis for the delineation of two major pathologic syndromes, i.e., the microbic and parasitic forms and idiopathic dermato- and polymyositis. Existing knowledge of the microbic and parasitic varieties is reasonably complete. There is an unmistakable inflammatory reaction with destruction of muscle fibers and changes in the supporting tissues; and the causative agent can be seen or isolated from the infected muscles. Occasional cases of acute disseminated polymyositis due to toxoplasmosis, trichinosis, miliary tuberculosis, or septicemia in which there is little or no destruction of connective tissue may be difficult to distinguish from the mildest forms of idiopathic polymyositis, herein classified as interstitial polymyositis, unless the causative organism is seen.

Our knowledge of the second category of inflammatory diseases of muscle is limited, and there is not even universal agreement that idiopathic dermato- and polymyositis represent an inflammatory disease.[141] Nor is the relationship of this disease to interstitial polymyositis settled. In our opinion the two important changes in muscle in the idiopathic form of polymyositis—i.e., degeneration of muscle fibers and infiltration of the connective tissue (usually near veins) by inflammatory cells—appear to be more or less independent and to vary in intensity in any given case. Prominence of degenerative change in many muscles determines to a large degree the classification of any given case as polymyositis, but it is the infiltration of inflammatory cells which stamps the whole process as one of myositis rather than myopathy or dystrophy. While it is conceded that one or two infiltrations of inflammatory cells may occur under a wide variety of conditions, the intensity of this change in the group of diseases under consideration indicates that it is an integral part of the pathologic process and not secondary to the disintegration of muscle fibers. The opposite condition of prominent infiltrations of inflammatory cells with little or no degeneration of muscle fibers usually decides the issue as to whether the case falls in the category of connective tissue disease, the condition then being classed as interstitial polymyositis. The latter, in its mildest forms, lacks specificity in pathologic diagnosis except when it stands in close relationship to clinical and other laboratory evidence of one of the so-called connective tissue diseases. We observe, between these two extremes (idiopathic dermato- or polymyositis and interstitial polymyositis) a wide spectrum of pathologic

change, not always revealed by the limited method of muscle biopsy.

In idiopathic dermato- and polymyositis and interstitial polymyositis the pathologic data suggest to us the operation of two separate but closely related factors, one an infiltration of inflammatory cells, the other degeneration of the muscle fiber. One might term these the inflammatory connective tissue and the myopathic factors. It can only be assumed at the moment that of these two factors the connective tissue factor is common to all varieties of so-called connective tissue disease, whether the lesions are principally localized in joints, skin, heart, nerve, or muscle. The myopathic factor is based on an interaction between the inflammatory cell and muscle. Moreover, an absence of proved viral, microbial, or parasitic agents suggests that the inflammatory reaction is of an allergic nature, possibly an autoallergy.

We agree with those who maintain that interstitial polymyositis is a part of the pathology of rheumatoid arthritis, rheumatic fever, lupus erythematosus, and scleroderma, while admitting that an occasional perivenous infiltration or lymphocytosis of muscle may occur in many diseases and even in individuals who were presumably well until the moment of death. In our own pathologic material it has not been possible to distinguish between rheumatoid arthritis, rheumatic fever, lupus erythematosus, and scleroderma on the basis of muscle biopsy alone. Distribution of the disease in other tissues and laboratory data are more important in separating the subvarieties of this form of myositis than are the pathologic changes in a small part of one muscle. Also our experience has been that a microscopic section of muscle does not enable the pathologist to distinguish between idiopathic dermato- and polymyositis and the form of the disease that accompanies carcinomatosis. In conditions in which interstitial inflammation is minimal or nonexistent and primary degeneration of muscle fibers is prominent, particularly with necrosis, vacuolation, granular change, or hyalinization of the core of myofibrils, a general metabolic fault must be suspected.

This whole group of idiopathic degenerative and inflammatory diseases of skeletal muscle deserves careful study, for they may provide clues to the nature of other obscure myopathic diseases.

REFERENCES

1. ADAMS RD: The pathologic substratum of polymyositis. The Striated Muscle. Edited by CM Pearson, FK Mastofi. Baltimore, Williams & Wilkins, 1973, Chap 15, pp 292–300

2. AMMITZBOLL FA: A case of Boeck's sarcoid with isolated localization in the musculature. Acta Rheumatol Scand 2:1–10, 1956

3. ANDERSON HJ, CHURCHILL-DAVIDSON HG, RICHARDSON AT: Bronchial neoplasm with myasthenia: prolonged apnea after administration of succinylcholine. Lancet 2:1291, 1953

4. ANDRUS S, KASS EH, ADAMS RD, ET AL: Toxoplasmosis in the adult. Arch Intern Med 89:759–782, 1952

5. ANTOGNETTI L, SCOPINARO D, ZINOLLI L: Aspetti di Patologia Musculare. Rome, Ditta Luigi Editore, 1957

6. ARKIN A: A clinical and pathological study of periarteritis nodosa. Am J Pathol 6:401–426, 1930

7. ASCHOFF L: Zur myocarditisfrage. Verh Dtsch Ges Pathol 8:46–53, 1904

8. ASCHOFF L: Pathologische Anatomie. Jena, Fischer, 1936

9. BANKER BQ, VICTOR M, ADAMS RD: Arthrogryposis multiplex due to congenital muscular dystrophy. Brain 80:319–334, 1957

10. BLAU A: Primary generalized myositis fibrosa: report of 2 cases with histopathology. Mt Sinai J Med NY 5:432–443, 1938

11. BOUISSON F: Mémoire sur les tumeurs syphilitiques des muscles et de leurs annexes. Gaz Med Paris 3:542, 563, 583, 594, 1846

12. BRALEY AE, ALEXANDER RC: Virus studies in lymphomatoid diseases of the ocular adnexa. Am J Ophthalmol 30:1369–1380, 1947

13. BROOKE M, KAPLAN H: Muscle pathology in rheumatoid arthritis, polymyalgia rheumatica and polymyositis. Arch Pathol 94:101–118, 1972

14. BUCKLEY CW: Fibrositis, lumbago and sciatica. Practitioner 134:129–134, 1935

15. BUZZARD EF: The clinical history and postmortem examination of five cases of myasthenia gravis. Brain 28:438–483, 1905

16. CALLAHAN WP, RUSSELL WO, SMITH MG: Human toxoplasmosis: a clinicopathologic study with presentation of 5 cases and review of the literature. Medicine (Baltimore) 25:343–397, 1946

17. CATHIE IAB, CECIL GW: Sarcospores and toxoplasma serology: an investigation of their alleged cross-reaction. Lancet 1:816–818, 1957

18. CAULFIELD JB, REBEIZ J, ADAMS RD: Viral involvement of human muscle. J Pathol 96:232–234, 1968

19. CHOU CM, GUTMANN L: Picovirus-like crystals in subacute polymyositis. Neurology (Minneap) 20:205–213, 1970

20. CHRISTIANSON HB, O'LEARY PA, POWER MH: Urinary excretion of creatine and creatinine in dermatomyositis. J Invest Dermatol 27:431–441, 1956

21. CLAWSON BJ, NOBLE JF, LUFKIN NH: Nodular inflammatory and degenerative lesions of muscles from 450 autopsies. Arch Pathol 43:579–590, 1947

22. COGAN DG: Personal communication, 1951

23. COHNHEIM J: Zur pathologischen anatomie der trichinen-krankheit. Virchows Arch [Pathol Anat] 36:161–186, 1866

24. COLLINS DH: Fibrositis and infection. Ann Rheum Dis 2:114–126, 1940

25. COOK CD, ROSEN FS, BANKER BQ: Dermatomyositis and focal scleroderma. Pediatr Clin North Am 979–1016, 1963

26. CURRIE S: Destruction of muscle cultures by lymphocytes from cases of polymyositis. Acta Neuropathol (Berl) 15:11–15, 1970

27. CURTIS AC, POLLARD HM: Felty's syndrome: its several features, including tissue changes, compared with other forms of rheumatoid arthritis. Ann Intern Med 13:2265–2284, 1940

28. DALLDORF G: Coxsackie viruses. Ann Rev Microbiol 9:277–296, 1955

29. DALLDORF G, SICKLES GM: An unidentified, filtrable agent isolated from the feces of children with paralysis. Science 108:61–62, 1948

30. DARLING ST: Sarcosporidiosis in an East Indian. J Parasitol 6:98–101, 1919

31. DASTUR DK, IYER CGS: Sarcocystis of human muscle. Bull Neurol Soc India 2:25–27, 1955

32. DAWKINS RD: Experimental myositis associated with hypersensitivity to muscle. J Pathol 90:619–625, 1965

33. DEFOREST GK, BUNTING H, KENNEY WE: Rheumatoid arthritis: the diagnostic significance of focal cellular accumulations in skeletal muscles. Am J Med 2:40–44, 1947

34. DELORMÉ: De la myosite tuberculeuse. Cong Franc Chir 5:555–559, 1891

35. DENNY-BROWN D: Myopathic weakness of quadriceps. Proc R Soc Med 32:867–869, 1939

36. DENNY-BROWN D: The nature of polymyositis and related muscular diseases. Trans Stud Coll Physicians Phila 28:14–29, 1960

37. DOWLING GB: Scleroderma and dermatomyositis. Br J Dermatol 67:275–290, 1955

38. DUNNINGTON JH, BERKE RN: Exophthalmos due to chronic orbital myositis. Arch Ophthalmol 30: 446–466, 1943

39. DURANTE G: Anatomie pathologique des muscles. Manuel d'Histologie Pathologique. Third edition. Edited by V Cornil, L Ranvier. Paris, Félix Alcan, 1902, Vol III

40. EATON LM: The perspective of neurology in regard to polymyositis. Neurology (Minneap) 4:245–263, 1954

41. EATON LM, LAMBERT EH: Electromyography and electrical stimulation of nerves in diseases of motor unit: observations on myasthenia syndrome associated with malignant tumors. JAMA 63: 1117, 1957

42. EHRHARDT O: Zur kenntniss der muskelveränderungen bei der trichinose des kaninchers. Beitr Anat Allg Pathol 20:1–42; (des Menschen) 43–50, 1896

43. FAUST EC: Human Helminthology. Third edition. Philadelphia, Lea, 1949

44. FENG LC: Sarcosporidiosis in man: report of a case in a Chinese. Chinese Med J 46:976–981, 1932

45. FINN JJ, WELLER TH, MORGAN HR: Epidemic pleurodynia: clinical and etiologic studies based on 114 cases. Arch Intern Med 83:305–321, 1949

46. FINSEN J: Iagttagelser angaaende sygdomsforholdene i Island Copenhagen (1874). Cited in Epidemic myalgia, or pleurodynia. JAMA 102: 460–461, 1934

47. FOLEY JM: The pathological changes in a case of encephalopathy due to trichinosis. J Neuropathol Exp Neurol 13:393, 1954

48. GANLEY OH, MERCHANT DJ, BOHR EF: The relationship of toxic fraction of a filtrate of Clostridium perfringens type A to the pathogenesis of clostridial myonecrosis. J Exp Med 101:605–615, 1955

49. GARCIN R, LAPRESLE J, HEWITT J: Déterminations musculaires au cours d'une maladie de Behçet (ou grande aftose de Touraine). Rev Neurol (Paris) 117:345–361, 1967

50. GILMORE HR JR, KEAN BH, POSEY FM JR: Case of sarcosporidiosis with parasites found in heart. Am J Trop Med 22:121–125, 1942

51. GORDON RB, ZENNETTER EH, SANDROCK RS: The varied clinical manifestations of Coxsackie virus infections. Arch Intern Med 103:63–75, 1959

52. GOULD SE: Trichinosis. Springfield, Ill, Charles C Thomas, 1945

53. GOVAN ADT: An account of the pathology of some cases of Cl. welchii infection. J Pathol 58:423–430, 1946

54. GRAEFF S: Rheumatismus und rheumatische Erkrankungen. Berlin, Urban und Schwarzenberg, 1936

55. GRAHAM JY: Beitrage zur naturgeschickte der trichina spiralis. Arch Mikr Anat 50:219–275, 1897

56. HALL MC, COLLINS BJ: Studies on trichinosis. II. Some correlations and implications in connection with the incidence of trichinae found in 300 diaphragms. Pub Health Rep 52:512–527, 1937

57. HARVEY JC: A myopathy of Boeck's sarcoid. Am J Med 26:356–363, 1959

58. HASSIN GB, DIAMOND IB: Trichinosis encephalitis: a pathologic study. Arch Neurol Psychiatry 15: 35–47, 1926

59. HEERFORDT CG: Ueber eine febris uveo-parotidea subchronica und der glandula parotis und der uvea des anges lokalisiert und häufig mit peresen cerebrospinalen nerven kompliciert. Arch Ophthalmol 70:254–273, 1909

60. HENCH PS, BAUER W, FLETCHER AA, ET AL: Rheumatism and arthritis: review of American and English literature. Ann Intern Med 10:754, 1936

61. HENCH PS, BOLAND E, CRAIN DC, ET AL: Rheumatism and arthritis: review of American and English literature, part II. Ann Intern Med 28:309–451, 1948

62. HENSON RA, RUSSELL DS, WILKINSON M: Carcinomatous neuropathy and myopathy: a clinical and pathological study. Brain 77:82–121, 1954

63. HEPP P: Ueber einen fall von acuter parenchymatoser myositis, welche geschwulste bildete und fluctuation, vortäuschte. Berl Klin Wochenschr 24:389–391, 1887

64. HEPP P: Ueber pseudotrichinose, cine besondere form von acuter parenchymatoser polymyositis. Berl Klin Wochenschr 24:297–299, 322–326, 1887

65. JACKSON H: Myositis universalis acuta infectiosa, with a case. Boston Med Surg J 116:498–500, 1887

66. JACOBS L, NAQUIN H, HOOVER R, ET AL: Comparison of toxoplasmin skin tests, Sabin-Feldman dye tests, and complement fixation tests for toxoplasmosis in various forms of uveitis. Bull Hopkins Hosp 99:1–15, 1956

67. JANKU (1923): Quoted by Callahan et al.[16]

68. JORDAN HH: Myogeloses: the significance of pathologic conditions of musculature in disorders of posture and locomotion. Arch Phys Ther 23:36–41, 1942

69. KAISER F: Zur kenntniss der primären muskeltuberculose. Arch Klin Chir 77:1033–1072, 1905

70. KAKULAS BA: Destruction of differentiated muscle cultures by sensitized lymphoid cells. J Pathol 91: 495–503, 1966

71. KAKULAS BA: Observations on the etiology of polymyositis. The Striated Muscle. Edited by CM Pearson, FK Mastofi. Baltimore, Williams & Wilkins, 1973, pp 485–497

72. KAKULAS BA, ADAMS RD: Principles of myopathol-

ogy as illustrated in the nutritional myopathy of the rottnest quokka. Ann NY Acad Sci 138:90–101, 1966

73. KIERLAND RR, UNDERWOOD LJ: Gumma of skeletal muscle: report of a case. Am J Syph Gonorrh Vener Dis 32:491–493, 1948

74. KLEMPERER P: The pathogenesis of lupus erythematosus and allied conditions. Ann Intern Med 28:1–11, 1948

75. KLINGE F: Die eiweissüberempfindlichkeit (gewebsanaphylaxie) der gelenke: experimentelle pathologisch-anatomische studie zur pathogenese des gelenkrheumatismus. Beitr Anat Allg Pathol 83:185–216, 1929

76. KLINGE F: Das gewebsbild des fierberhaften rheumatismus. I. Das rheumatische frühinfiltrat. Virchows Arch [Pathol Anat] 278:438–461, 1930

77. KLINGE F: Das gewebsbild des fierberhaften rheumatismus. II. Das subakutchronische stadium des zellknötchens. Virchows Arch [Pathol Anat] 279:1–15, 1931

78. KLINGE F: Das gewebsbild des fierberhaften rheumatismus. XII. Zusammenfassande kritische betrachtung zur frage der geweblischen sonderstellung des rheumatischen gewebeschadens. Virchows Arch [Pathol Anat] 286:344–388, 1932

79. KRABBE KH: Muscular localization of benign lymphogranulomatosis. Acta Med Scand [Suppl 234] 136:193–198, 1949

80. LANGE M: Die Muskelhaerten (Myogelosen). Munich, Lehmann, 1931

81. LANZ O, DE QUERVAIN F: Ueber hämatogens muskeltuberculose. Arch Klin Chir 46:97–138, 1893

82. LEPINE P: Personal communication

83. LICHAREW: Demonstration. Dermatol Centralbl 11:253–254, 1908

84. LICHTWITZ L: Pathology and Therapy of Rheumatic Fever. New York, Grune, 1944

85. LORENZ H: Die muskelerkrankungen. Handbuch Spec Path Therapie. Edited by H Nothnagel. Vienna, Holder, 1904, Vol XI, Part III, pp 1–727

86. LUNDBERG A: Myalgia crusis epidemica. Acta Paediatr Scand 46:18–31, 1957

87. LUNDERVOLD A: Occupational myalgia: electromyographic investigation. Acta Psychiatry Neurol 26:359–369, 1951

88. MACARTHUR WP: Cysticercosis as seen in the British Army, with special reference to the production of epilepsy. Trans R Soc Trop Med Hyg 27:343–363, 1937

89. MACCALLUM WG: Rheumatism: the Harrington lecture. JAMA 84:1545–1551, 1925

90. MACFARLANE MG: On biochemical mechanisms of action of gas gangrene toxins. Mechanisms of Microbial Pathogenicity. Cambridge, Cambridge University Press, 1955, p 72

91. MACFARLANE RG, MACLENNAN JD: The toxaemia of gas-gangrene. Lancet 2:328–331, 1945

92. MACLENNAN JD: Anaerobic infections in war wounds of the Middle East. Lancet 2:63–66, 1943

93. MAIR WGP, TOMÉ FMS: Diseased Human Muscle. Baltimore, Williams & Wilkins, 1972

94. MASTAGLIA FL, WALTON JN: Coxsackie virus-like particles in skeletal muscle from a case of polymyositis. J Neurol Sci 11:5493–599, 1970

95. MCGILL RP: Cysticercosis resembling myopathy. Lancet 2:728–730, 1948

96. MCGILL RJ, GOODBODY RA: Sarcosporidiosis in man with periarteritis nodosa. Br Med J 2:333–334, 1957

97. MEDSGER TA, DAWSON WN, MASI AT: The epidemiology of polymyositis. Am J Med 47:715–723, 1970

98. MEJLSZENKIER JD, SATRAN AP, HEALY JJ, ET AL: The myositis of influenza. Arch Neurol 29:441–443, 1973

99. MERRITT H, ROSENBAUM H: Involvement of the nervous system in trichiniasis. JAMA 106:1646–1649, 1936

100. MIDDLETON PH, ALEXANDER RM, SZYMANSKI MT: Severe myositis during recovery from influenza. Lancet 2:533–535, 1970

101. MIESCHER F: Ueber eigenthümliche schlauche in den muskeln einer hausmaus. Berl Verhandl Naturf Ges Basel 5:183–190, 1943

102. MORRISON LR, SHORT CL, LUDWIG AO, ET AL: The neuromuscular system in rheumatoid arthritis. Am J Med Sci 214:33–49, 1947

103. MOST H, HELPERN M: The incidence of trichinosis in New York City. Am J Med Sci 202:251–257, 1941

104. MYERS GB, GOTTLIEB AM, MATTMAN PE, ET AL: Joint and skeletal muscle manifestations in sarcoidosis. Am J Med 12:161–169, 1952

105. NAMBA T, BRUNNER NG, GROB D: Idiopathic giant-cell polymyositis. Arch Neurol 31:27–31, 1974

106. NATTRAS FJ: Recovery from muscular dystrophy. Brain 77:549–570, 1954

107. NEVIN S: Two cases of muscular degeneration occurring in late adult life, with a review of the recorded cases of late progressive muscular dystrophy (late progressive myopathy). Q J Med 5:51–68, 1936

108. NEVINNY H: Ueber die veranderungen der skeletmuskulatur bei trichinose. Virchows Arch [Pathol Anat] 266:185–238, 1927

109. NICOLLE C, MANCEAUX L: Sur une infection a corps de Leishman (ou organismes voisins) du gondi. C R Acad Sci 147:763–765, 1908

110. NOTTA A: Memoire sur la rétraction musculaire syphilitique. Arch Gen Med 24:413–440, 1850

111. O'LEARY PA, LAMBERT EH, SAYRE GT: Muscle

studies in cutaneous diseases. J Invest Dermatol 24:301–310, 1955

112. O'LEARY PA, WAISMAN M: Dermatomyositis: a study of forty cases. Arch Dermatol Syph 41: 1001–1019, 1940

113. OPPEL TW, COKER C, MILHORAT AT: The effect of pituitary adrenocorticotropin (ACTH) in dermatomyositis. Ann Intern Med 32:318–323, 1950

114. OPPENHEIM H: Ueber die polymyositis. Berl Klin Wochenschr 40:381–385, 416–419, 1903

115. PALMEIRO J, BEHREND R, WECHSLER W: Electronmikroskopische befunde an der Skeletmuskulatur bei Polymyositis. Acta Neuropathol (Berl) 7: 26–43, 1966

116. PATHOLOGIE MUSCULAIRE: etudies presentées à la première réunion neurologique Belgo-Suisse. Acta Neurol Belg, 1954

117. PEARSON CM: Serum enzymes in muscular dystrophy and certain other muscular and neuromuscular diseases. I. Serum glutamic oxalacetic transaminase. N Engl J Med 256:1069–1075, 1957

118. PEARSON CM: Rheumatic manifestations of polymyositis and dermatomyositis. Arthritis Rheum 2:127–143, 1959

119. PEARSON CM, ROSE AS: Myositis: the inflammatory disorders of muscle. Res Publ Assoc Res Nerv Ment Dis 38:422–478, 1960

120. PEARSON CM, YAMAZAKI JN: Vacuolar myopathy in systemic lupus erythematosus. Am J Clin Pathol 29:455–463, 1958

121. PHILLIPS RW, PHILLIPS AM: The diagnosis of Boeck's sarcoid by skeletal muscle biopsy. Arch Intern Med 98:732–736, 1956

122. PINKERTON H, HENDERSON RG: Adult toxoplasmosis: a previously unrecognized disease entity simulating the typhus-spotted fever group. JAMA 116:807–814, 1941

123. PINKERTON H, WEINMAN D: Toxoplasma infection in man. Arch Pathol 30:374–392, 1940

124. PLUMMER WW, SANES S, SMITH WS: Hematogenous tuberculosis of skeletal muscle: report of a case with involvement of the gastrocnemius muscle. J Bone Joint Surg 16:631–639, 1934

125. POTTS RE, WILLIAMS AA: Acute myocardial toxoplasmosis. Lancet 1:483–484, 1956

126. REVERDIN JL: Note dans un cas de tuberculose musculaire primitive. Cong Franc Chir 5:560–568, 1891

127. RICH AR: Hypersensitivity in disease. Harvey Lect 42:106–147, 1946–1947

128. RICH AR, MCCORDOCK H: Inquiry concerning the role of allergy, immunity and other factors of importance in the pathogenesis of human tuberculosis. Bull Hopkins Hosp 44:273–422, 1929

129. RISSE H: Polymyositis acuta und acuter gelenkrheumatismus. Dtsch Med Wochenschr 23:232–234, 1897

130. ROBERTS HM, BRUNSTING LA: Dermatomyositis in childhood: a summary of 40 cases. Postgrad Med 16:396–404, 1954

131. ROSE AL, WALTON JN, PEARCE GW: Polymyositis: an ultramicroscopic study of muscle biopsy material. J Neurol Sci 5:457, 1967

132. SABIN AB: Toxoplasmic encephalitis in children. JAMA 116:801–807, 1941

133. SAUNDERS M, KNOWLES M, CURRIE S: Lymphocyte stimulation with muscle homogenate in polymyositis and other muscle-wasting disorders. J Neurol Neurosurg Psychiatry 32:569–575, 1969

134. SCADDING JG: Acute benign dry pleurisy in the Middle East. Lancet 1:763–766, 1946

135. SCOTT JW: Sarcosporidia—a critical review. J Parasitol 16:111–130, 1930

136. SELANDER P: Dermatomyositis in early childhood. Acta Med Scand [Suppl 246] 138:187–203, 1950

137. SENATOR H: Ueber acute polymyositis und neuromyositis. Dtsch Med Wochenschr 19:933–936, 1893

138. SHAFIQ SA, MILHOVAL AT, GORYCKI MA: An electron microscopic study of muscle degeneration and vascular changes in polymyositis. J Pathol 94:139–147, 1957

139. SHAW AFB: Topography and pathogenesis of lesions in rheumatic fever. Arch Dis Child 4: 155–164, 1929

140. SHEARD C: Dermatomyositis. Arch Intern Med 88:640–658, 1951

141. SHY GM, MCEACHERN D: The clinical features and response to cortisone of menopausal muscular dystrophy. J Neurol Neurosurg Psychiatry 14: 101–107, 1951

142. SKINNER JC: Neurologic complications of trichinosis: report of 2 cases. N Engl J Med 238: 317–319, 1948

143. SLOCUMB CH: Fibrositis. Clinics 2:169–178, 1943

144. SMITH T: The production of sarcosporidiosis in the mouse by feeding infected muscular tissue. J Exp Med 6:1–21, 1901

145. SNORRASON E: Myositis fibrosa progressive-lymphogranulomatosis benigna Boeck. Nord Med 36:2424–2426, 1947

146. SOKOLOFF D, WILENS SL, BUNIM JJ, ET AL: Diagnostic value of histologic lesions of striated muscle in rheumatoid arthritis. Am J Med Sci 219:174–182, 1950

147. SOKOLOFF L, WILENS SL, BUNIM JJ: Arteritis of striated muscles in rheumatoid arthritis. Am J Pathol 27:157–173, 1951

148. SPLENDORE A: Sur un nouveau protozoaire para-

site du lapin: deuxieme note preliminaire. Bull Soc Pathol Exot 2:462–465, 1909

149. STEINER G, CHASON JL: Differential diagnosis of rheumatoid arthritis by biopsy of muscle. Am J Clin Pathol 18:931–939, 1948

150. STEINER G, FREUND HA, LEICHTENTRITT B, ET AL: Lesions of skeletal muscles in rheumatoid arthritis (nodular polymyositis). Am J Pathol 22:103–145, 1946

151. STEINER WR: Dermatomyositis, with report of a case which presented a rare muscle anomaly but once described in man. J Exp Med 6:407–442, 1903

152. STRÜMPELL A: Zur kenntniss der primaren acuten polymyositis. Dtsch Z Nervenheilk 1:479–505, 1891

153. SUNDELIN F: Tumeurs multiples disséminées dans les muscles des extremités et rappelant la tuberculose par leur structure histologique. Acta Med Scand 62:442–460, 1925

154. SWIFT HR, MCEWEN C: Rheumatic fever, Oxford System of Medicine. Edited by H Christian. New York, Oxford, 1938

155. SYLVEST E: Epidemic Myalgia: Bornholm Disease (H Anderson, tr) New York, Oxford, 1934

156. SZAKS L: Tumorous pseudosarcomatous lesions of striated muscle proliferative myositis. Zentralbl Allg Pathol 114:239–244, 1971

157. THAYER WS: Notes on a case of acute hemorrhagic polymyositis. Boston Med Surg J 147:313–317, 1902

158. TORRES CM: Sur une nouvelle maladie de l'homme caracterisée par la presence d'un parasite intracellulaire très proche du toxoplasme et de l'encephalitozoon, dans le tissue musculaire cardiaque, les muscles du squelette, le tissue cellulaire souscutané et le tissue nerveux. C R Soc Biol 97:1778–1796, 1797–1804, 1927

159. TURNER JWA, HEATHFIELD KWG: Quadriceps myopathy occurring in middle age. J Neurol Neurosurg Psychiatry 24:18–21, 1961

160. UNVERRICHT H: Polymyositis acuta progressiva. Z Klin Med 12:533–549, 1887

161. UNVERRICHT H: Dermatomyositis acuta. Dtsch Med Wochenschr 17:41–44, 1891

162. VIRCHOW R: Ueber die natur der constitutionell-syphilitischen affectionen. Virchows Arch [Pathol Anat] 15:217–336, 1858

163. VIRCHOW R: Oarstellung der Lehre von den Trichinen. Berlin, Reimer, 1864

164. VIRCHOW R: Ueber trichina spiralis. Virchows Arch [Pathol Anat] 18:330–346, 1860

165. VON GLAHN WC, PAPPENHEIMER AM: Specific lesions of peripheral blood vessels in rheumatism. Am J Pathol 2:235–249, 1926

166. WAGNER E: Fall einer seltnen muskelkrankheit. Arch Heilkunde 4:282, 1863

167. WALDENSTRÖM J: Some observations on uveoparotitis and allied conditions with special references to the symptoms from the nervous system. Acta Med Scand 91:53–68, 1937

168. WALLACE SL, LATTER R, MALIA JP, ET AL: Muscle involvement in Boeck's sarcoid. Ann Intern Med 48:497, 1958

169. WALSH FB: Clinical Neuro-ophthalmology. Baltimore, Williams & Wilkins, 1947, pp 1112–1114

170. WALTON JN, ADAMS RD: Polymyositis. Edinburgh, Livingstone, 1958

171. WEDGWOOD RJT, COOK CB, COHEN J: Dermatomyositis: report of 26 cases in children with discussion of endocrine therapy. Pediatrics 12:447–466, 1953

172. WELBORN MB: Epidemic pleurodynia ("devil's grip") in Cincinnati. Am J Med Sci 191:673–678, 1936

173. WILLIAMSON-NOBLE FA: Inflammatory pseudotumor of the orbit. Br J Ophthalmol 10:65–78, 1926

174. WILSON MG: Rheumatic Fever. New York, Commonwealth Fund, 1940

175. WOHLFART G: Personal communication, 1950

176. WOLF A, COWEN D: Granulomatous encephalomyelitis due to an encephalitozoon (encephalotoxic encephalomyelitis), a new protozoan disease of man. Bull Neurol Inst NY 6:306–371, 1937

177. WOLF A, COWEN D: Granulomatous encephalomyelitis due to a protozoan (toxoplasma or encephalitozoon): identification of cases from the literature. Bull Neurol Inst NY 7:266–290, 1938

178. WOLF A, COWEN D, PAIGE BH: Human toxoplasmosis: occurrence in infants as an encephalomyelitis: verification by transmission to animals. Science 89:226–227, 1939

179. WOLF A, COWEN D, PAIGE BH: Toxoplasmic encephalomyelitis: a new case of granulomatous encephalomyelitis due to a protozoan. Am J Pathol 15:657–694, 1939

180. WOLF A, COWEN D, PAIGE BH: Toxoplasmic encephalomyelitis: the experimental transmission of infection to animals from a human infant. J Exp Med 71:187–214, 1940

181. WOLF A, KABAT SA, BEZER AE: Effect of cortisone in activating trypanosomal encephalitis and myocarditis in rhesus monkeys. J Neuropathol Exp Neurol 11:84–85, 1952

chapter 8

TUMORS

Muscle tumors can be divided into two general categories, those that are primary and develop from a constituent element of muscle proper, and those that are secondary and arise either as distant metastases or by direct extension from an adjacent tumor. The principal morphologic features of these two groups of tumors are described on the following pages.

TUMORS COMPOSED OF MUSCLE CELLS

Tumors of mesenchymal or mesodermal origin that differentiate into striated muscle cells are exceedingly uncommon. According to the histogenetic classification in common usage among pathologists, there are two types of tumor derived from myoblastic tissue. These are the rhabdomyoma and the rhabdomyosarcoma. The latter is subdivided into four types, the embryonal cell sarcoma, the botryoides, the pleomorphic, and alveolar sarcoma. The granular cell myoblastoma is now proved to be a tumor of nervous derivation and omitted from further discussion.

RHABDOMYOMA

Rhabdomyoma is a curious tumor, extremely rare, and seldom found where it might be expected—in skeletal muscle. There are several forms of it, difficult to classify because of the diverse conditions under which they occur. Ewing[8] subdivides them into the following groups: (1) congenital rhabdomyoma, which appears to be the result of an embryogenic disturbance in the development of an organ, e.g., rhabdomyoma of the heart, and probably of the uterus, bladder, and prostate; (2) metaplastic rhabdomyoma, which may arise in organs that contain no striated muscle—hence its hypothetical derivation from smooth muscle cells, e.g., the rare rhabdomyoma of the gastrointestinal tract and esophagus; (3) teratomatous rhabdomyoma, which takes its origin from the muscle cells of a teratoma, e.g., the rhabdomyoma of the sex glands; (4) teratoid rhabdomyoma, a tumor of possible teratomatous origin and exemplified by the peculiar adenomyosarcoma of the kidney, and the mixed sacral tumors containing muscle. Some of these tumors, such as the congenital rhabdomyoma of the heart, are invariably benign whereas the tendency of others such as those arising in teratomas is to change into the highly malignant rhabdomyosarcoma. Because of their common origin in teratomas and teratoid tumors and their highly malignant character, we prefer to class these last two varieties with the teratomas.

The rhabdomyoma may be solitary or multiple and may take a nodular, flattened, rounded, or polypoid form. At times well circumscribed, in other instances it may extend diffusely throughout an organ. The unfixed tissue on section is gray and soft; intervening strands of firm, white connective tissue traverse and at times encapsulate the tumor. The more cellular rhabdomyomas may be mottled reddish yellow.

Under the microscope the rhabdomyoma is found to be composed of large cells of diverse form and size and, in parts in which differentiation is more perfect, of bundles of parallel and intertwining striated muscle fibers sup-

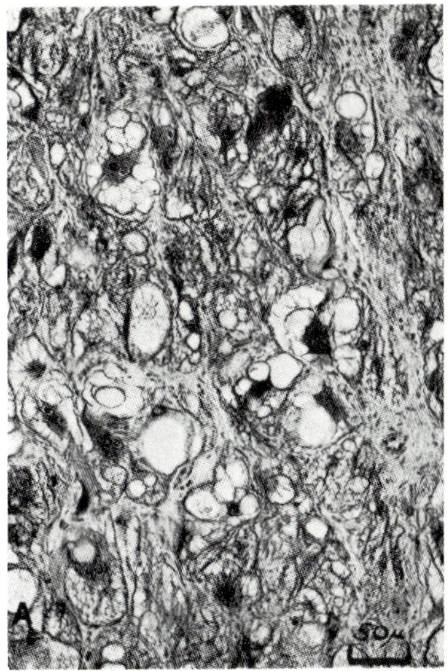 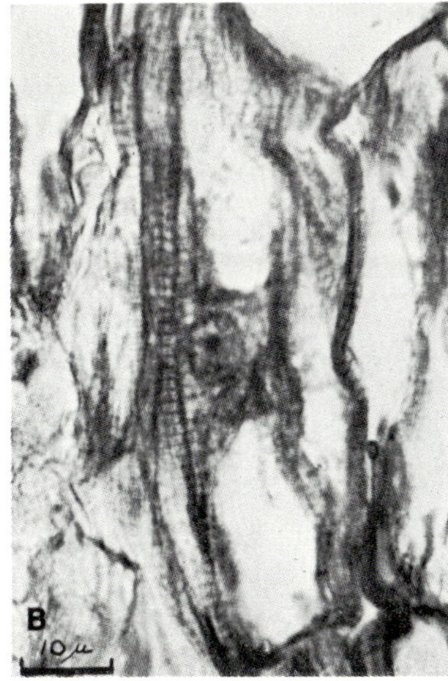

Fig. 8–1. (A) Rhabdomyoma of heart. Low-power view showing the pleomorphism of cells and their heterogeneous arrangement. Several large vacuolated "spider cells" can be seen. (B) High-power magnification of the same tumor. Central nucleus and traversing bundles of cross-striated myofibrils of a single cell are visible. (phosphotungstic hematoxylin) (A, ×650; B, ×1400)

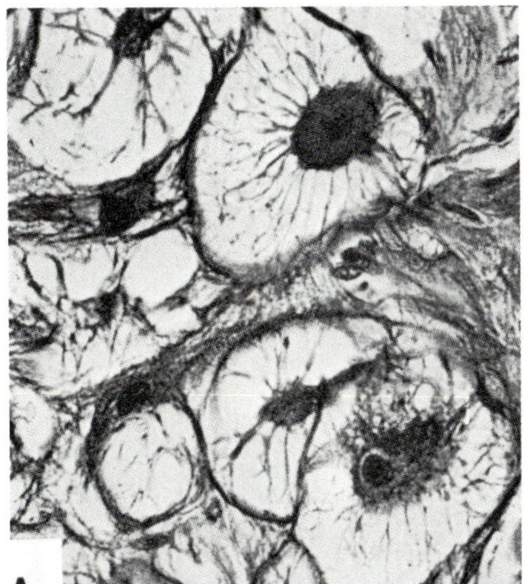 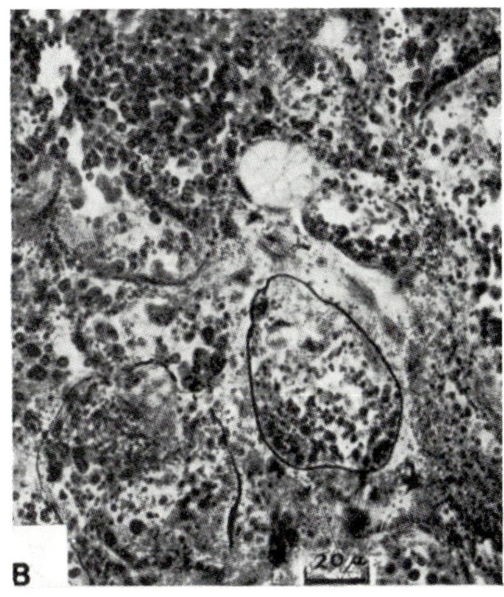

Fig. 8–2. (A) Several "spider cells" in one field in rhabdomyoma of heart. Note bundles of striated myofibrils between several of the large cells. Many of the delicate peripherally radiating intracellular strands are also cross-striated. (B) Rhabdomyoma showing many large and small intracellular glycogen droplets. Several cells have been outlined to show their margins. (A, phosphotungstic hematoxylin, ×800; B, Best's carmine glycogen stain)

ported by adult or embryonal connective tissue (Fig. 8–1). Completely differentiated striated muscle fibers are seldom present. The cell type that has attracted most attention in the past is one which is large and has abundant, coarsely vacuolated cytoplasm. Striated myofibrils in concentric arrangement can be seen near the nucleus and between the vacuoles in some of these cells, termed "spider cells" (Fig. 8–2A). Smaller cells, of spindle form, usually lack myofibrils. Multinucleated cells which are actually incomplete muscle fibers with granular sarcoplasm and peripheral rows of striated myofibrils are characteristic. Mitotic figures are absent. The stroma consists of loose connective tissue and is usually not abundant; it sometimes undergoes a myxomatous change.

There are two special varieties of rhabdomyoma: congenital rhabdomyoma of the heart and metaplastic rhabdomyoma.

CONGENITAL RHABDOMYOMA OF THE HEART

Congenital rhabdomyoma of the heart is a very distinctive lesion. Wolbach's report[27] in 1907 of 12 cases collected from his own material and from the literature is an authoritative reference on this subject. Batchelor and Maun[3] presented a comprehensive review of the literature up to 1945 and were able to find 63 cases. The tumors usually present as multiple, circumscribed masses within the wall of the heart and may project into the cardiac chambers or onto the external surface. A peculiar loose tissue in which there are many huge vacuolated cells constitutes the main structure of the tumor. The cells are large, rounded, and multinucleated; striations are visible between the vacuoles in many of them (Figs. 8–1 and 8–2A). The resemblance of these cells to Purkinje cells of the conducting system of the heart has impressed some pathologists; others have commented on their similarity to the cells of the fetal heart muscle and upon their abundant glycogen content (Fig. 8–2B). It is believed that this tumor is the result of an embryogenic disturbance in the development of the heart. Congenital rhabdomyoma of the heart is of particular interest to neurologists because of its frequent association with tuberous sclerosis.

METAPLASTIC RHABDOMYOMAS

Metaplastic rhabdomyomas are likely to occur wherever smooth muscle is found, i.e., in stomach, intestine, kidney, testis, prostate, and uterus. They are quite rare and often difficult to differentiate from the teratomatous rhabdomyomas (described later). Most frequently these growths are benign, but some may grow like sarcomas and metastasize. The chief histologic features are the islands of cross-striated muscle mixed with smooth muscle cells. Some pathologists hold that smooth muscle cells may acquire cross striation as a kind of metaplasia. Others object on the ground that smooth and striated muscle cells are of different derivation. The author favors the latter viewpoint for it seems unlikely that one well differentiated cell changes into another. In some organs such as the uterus both striated and smooth muscle may be seen during embryonic life, and the persistence of the former could account for the later development of a rhabdomyoma. A more credible theory is that which postulates the mixed origin of the tumor from cell rests or from pluripotential mesenchymal cells.

RHABDOMYOSARCOMA

PLEOMORPHIC TYPE

The pleomorphic rhabdomyosarcoma arises within the skeletal muscles of the body and, as the name implies, is of an extremely malignant nature. The original and now classic study was made by Stout[24] in 1946. This tumor makes up about 10–15% of sarcomas of the somatic soft tissues and ranks high among all the malignant, soft tissue tumors of childhood. The type described by Stout[24] is now called the pleomorphic rhabdomyosarcoma. Approximately two-thirds originate in the muscles of the extremities, usually the legs. Less than 5% occur in the tongue, in contrast to the high incidence of "myoblastomas" in that site. Table 8–1 is a composite of 247 cases compiled from several quite large series[12, 20, 25] of cases. Stout[24] emphasizes the extremely malignant character of this tumor. In his series of 121 cases, of which 108 received some form of treatment, only four symptom-free 5-year survivals were observed. Here, however, the method of therapy is of importance, for enucleation of the tumor was followed by a local recurrence in more than 60% of cases. Pack and Ariel,[20] using a carefully planned operation the essential feature of which was wide excision of the affected area, obtained 5-year survivals in 22 of 65 cases (33.8%).

The presence of a tumor is usually the only clinical symptom, and trauma does not appear to play a significant role in its development. The mass is usually deep-seated in the muscle

TABLE 8–1. Common Sites of Rhabdomyosarcoma

Site	Cases
Lower extremity	*116*
Buttock	9
Leg (unspecified)	29
Thigh	74
Lower leg	2
Knee	1
Foot	1
Upper extremity	*57*
Shoulder	6
Arm (unspecified)	42
Upper arm	1
Forearm	6
Hand	2
Trunk	*40*
Back	16
Chest wall	11
Abdominal wall	2
Groin	7
Diaphragm	3
Psoas region	1
Head and neck	*34*
Head	12
Tongue and pharynx	14
Neck	8
Total	*247*

and has a limited mobility. It may grow slowly or rapidly, more often the latter. The greatest incidence is during the fifth and sixth decades of life, about 50% occurring at these ages. Excluding several questionable cases of extremely long duration, Stout[24] found that the tumor had been present for an average of 10.7 months before treatment was instituted.

The gross characteristics of the tumor were most adequately described and illustrated by Pack and Ariel,[20] who noted the great variation in its appearance from one case to another. The neoplasm may be solitary, multilobulated, or even apparently multicentric in origin. It may present as a flattened nodule between muscle sheaths but more often takes the form of a poorly circumscribed intramuscular mass. Whatever its gross appearance, however, it is never truly encapsulated and always infiltrates the surrounding tissue. The consistency of the tumor also varies in accordance with the amount of connective tissue elements contained in it, but usually it is soft and often hemorrhagic or necrotic owing to the fact that

its blood supply has been compromised. Hemorrhagic necrosis may be so extensive in some cases as to resemble a large hematoma. The color of the tumor varies with the amount of congestion and hemorrhage—hence it may be pale yellow, pink, or dark red.

The tumor is very cellular, and the cells exhibit no special arrangement. Most are polymorphic, some being spherical and others ovoid, irregularly polyhedral, or fusiform. Cells which resemble normal striated muscle cell are rare. Many tumors are composed of such poorly differentiated cells that it is difficult to distinguish them from other sarcomas (Fig. 8–3). Stout[24] describes four types of tumor cells: (1) large, rounded cells with one or several central nuclei and a strongly acidophilic cytoplasm; (2) strap-shaped cells with two or more nuclei arranged in tandem and a cytoplasm which may show longitudinal fibrils; (3) racquet-shaped cells with a single nucleus at one expanded end and a tapering body; and (4) giant cells with peripherally arranged vacuoles separated by thin strands of cytoplasm, the "spider-web" cell. Some of these cell types are illustrated in Figure 8–4. According to Wolbach[28] and others, the distinguishing features of the tumor cells are the abundant acidophilic cytoplasm and the presence of deeply staining granules and delicate short fibrils in bizarre arrangement. The latter may be sheaflike or distributed in geometric patterns resembling at times a mitotic figure. Wolbach believes these granules, which are usually paired, to be centrioles because of their staining reaction, size and surrounding clear zone of centroplasm. The delicate fibrils form in pairs between the centrioles. The granules are 0.2 μ in diameter, and the fibrils vary in length from 3.0 to 5.0 μ. The granules multiply by simple division, and the fibrils split longitudinally, usually retaining connection with a single granule. Complex arrangements of centrioles and fibrils are found in many tumor cells. A more orderly arrangement of these cytoplasmic structures becomes evident in the differentiated cells. Fibrils with definite cross striation occur in cells that are presumed to be more mature (Fig. 8–4B). In areas the tumor may resemble fibrous tissue, an appearance which some pathologists, on doubtful grounds, attribute to a

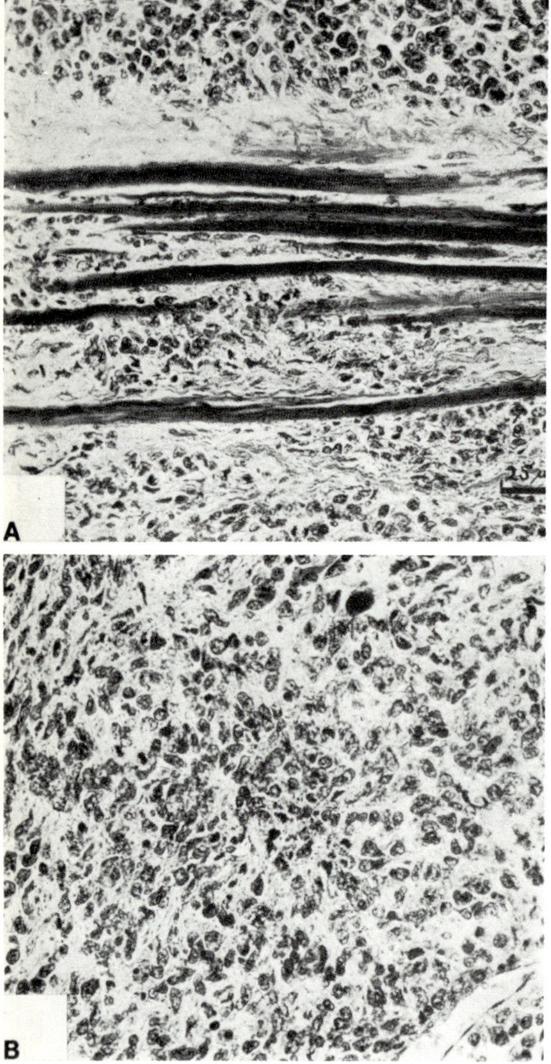

Fig. 8–3. (A) Example of pleomorphic rhabdomyosarcoma surrounding several fairly normal muscle fibers. (B) Poorly differentiated rhabdomyosarcoma. Note disorderly arrangement of cells, pleomorphism of cell types, and scanty supporting stroma. Myofibrils could not be found in this portion of the tumor. (phosphotungstic hematoxylin; ×400)

transformation of rhabdomyoblasts into fibroblasts.

This sequence of multiplication of centrioles, formation of centriole clusters, dispersion of centrioles,* development of fibrils connecting

* The centriole or centrosome is a minute body in the cytoplasm of a cell. It consists of two parts, the central granule and an enveloping sphere of differentiated protoplasm. Some histologists speak of the central granule as the centriole and the enveloping sphere as the attraction sphere. The reader is reminded that the centriole is not the same as a chondriosome. The latter term, which is synonymous with mitochondria, chondriomes, basal filaments, and vegetative filaments, is used to designate all filaments and rodlike structures in the cytoplasm of cells.

the centrioles, and the continued division of centrioles and fibrils, which Wolbach observed in both rhabdomyoma of the heart and rhabdomyosarcoma of a skeletal muscle, is the basis of his theory of the origin of myofibrils. He believes that the granule of the centriole gives rise to the dark Q band of the fiber and that a contraction of fibrillar material accounts for the Z band. The suggestion that the myofibrils originate from centrioles seems plausible when it is recalled that the motor apparatus of the spermatozoa and blepharoplasts of ciliated cells and flagellates are also believed to originate from centrioles.

The granular cytoplasm of the tumor cells,

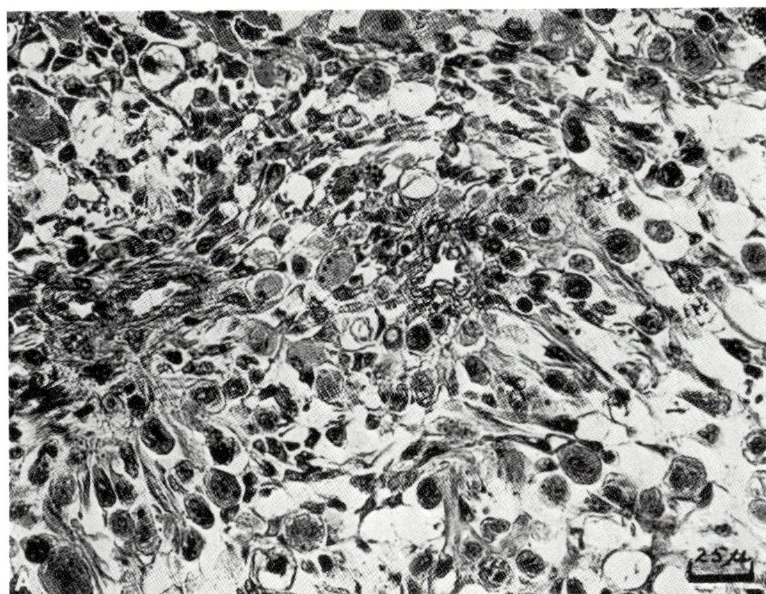

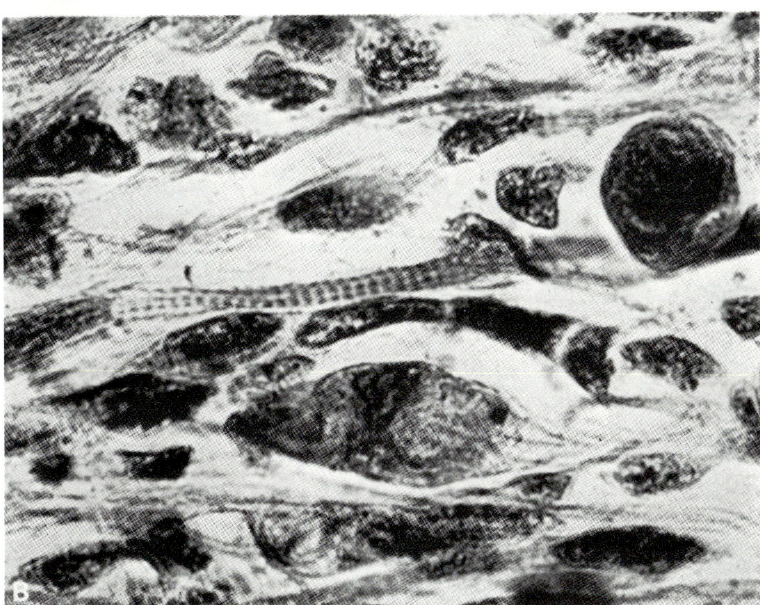

Fig. 8–4. (A) Cellular portion of rhabdomyosarcoma. Rounded and racquet-shaped cells described by Stout[24] are obvious. Clusters of centrioles and myofibrils are barely discernible in the cytoplasm of some cells. (B) High-power view of striated portion of strap-shaped cell surrounded by several giant and atypical cells. Presence of such cross-striated tumor cells established diagnosis of rhabdomyosarcoma. (A, ×650 phosphotungstic hematoxylin; B, ×1300 phosphotungstic hematoxylin)

which usually stains deeply with acidic dyes and contains glycogen, together with the faint or prominent cross striations, longitudinal myofibrils, or a vague linear arrangement of the cytoplasmic granules (as demonstrated in Mallory's phosphotungstic acid hematoxylin stain), usually permits accurate identification of the rhabdomyosarcoma. In some cases a prolonged search under high magnification is necessary, but once these features are demonstrated the diagnosis is certain. Stout found striations in 9 of his 14 cases. Occasionally cross striations may be demonstrated only in the metastases and not in the primary tumor. One must not mistake altered fibers of invaded muscle for tumor cells, for the normal muscle fibers may be surrounded by invading tumor tissue (Fig. 8–3A), and this may be a source of error in the diagnosis of an otherwise extremely undifferentiated tumor which may not have arisen from muscle tissue.

Because of the malignant character of rhabdomyosarcoma, extensive removal of the primary growth, including sacrifice of a good margin of healthy tissue, must be done as early as possible. Only rarely have patients had a lengthy survival after minimal surgery. The choice between radical excision and amputation depends upon a number of factors, including the location and the degree of malignancy. As mentioned earlier, Pack and Ariel[20] presented the most favorable 5-year cure rate (33.8%) for 65 cases treated by radical surgical procedures. This has not been the experience of others, perhaps because of a more conservative surgical approach. Radiation therapy has only slight effect and should not be used prior to surgery, since it often induces necrosis and hemorrhage into the body of the tumor. Furthermore, a course of irradiation should not delay definitive surgical treatment, which is imperative at the earliest moment. Postoperative radiation therapy has often been used, but its value following extensive excision or amputation is almost impossible to assess.

TUMORS OF STRIATED MUSCLE ARISING IN TERATOMAS; MIXED TUMORS

Teratomas frequently contain striated muscle fibers. Tumors of this class may arise in the neck in connection with the branchial clefts and thyroid gland, in the nasopharynx as polypoid growths, and in the cranial cavity, abdominal cavity, and sacrococcygeal region. In their gross appearance they may vary greatly in size—some reaching astonishing proportions —and present all of the varied formation of teratomas in general. There is a frequent tendency of one element to overgrow and suppress other elements, at times making difficult their identification, which depends on the finding of cells or special structures derived from more than one germ layer. Two of these tumors in which striped muscle cells play an important role are the rhabdomyoma and rhabdomyosarcoma of the genital organs and the adenomyosarcoma of Wilms. In the other teratomas muscle cells seldom contribute significantly to the total mass of the tumor.

Cappell's work[5] indicates that malignant rhabdomyomas of palate and vagina may properly be classed as rhabdomyosarcomas.

EMBRYONAL RHABDOMYOSARCOMA

The embryonal rhabdomyosarcoma along with the alveolar rhabdomyosarcoma is the most frequent soft tissue tumor in individuals less than 21 years of age. Occurrence in males and females is about equal, and the age range of 13 patients was from birth to 23 years (average 8 years). The common sites of localization are the tissues of the head, neck, orbit, and urogenital tract. The clinical behavior resembles that of the more common pleomorphic type of rhabdomyosarcoma (rates of growth, pattern of metastases, etc.).[11] A rapidly expanding painless mass in one of the aforementioned tissues stands as the usual mode of expression. The majority of the cells are small and undifferentiated, but some are spindle-shaped and may be distinguished from those of the fibrosarcoma only by the presence of longitudinal and sometimes transversely striated myofibrils. The latter are found in approximately one-third of cases. Its greater cellularity and arrangements of cells and stroma are also diagnostically useful criteria. The tumor is rapidly fatal (two-thirds of the patients die within 15 months), and no one mode of therapy (excision, radiation, chemothrepay) has a demonstrable advantage.

TERATOMATOUS RHABDOMYOMAS OF GENITAL ORGANS

Teratomatous rhabdomyomas of the genital organs bear close resemblance to the embryonal cell rhabdomyosarcoma from which cell type it is derived. The usual site is the uterus or vagina; these rare tumors may present themselves as polypoid sarcomatous masses in the vagina. Another name is sarcoma botryoides. In the male the counterpart of this tumor appears to be the malignant rhabdomyoma of the prostate. These tumors may form before birth or develop during childhood and occasionally arise in adult years. The principal clinical finding is the large polypoid mass which protrudes from the vagina with hemorrhage and fetid discharge. The prostatic tumor in the male causes obstruction of the neck of the bladder. Dysuria, fever, and cachexia are later developments. As the neoplasms enlarge, they extend into the adjacent pelvic tissues.

The histologic structure is remarkably complex, giving support to the theory that these tumors are a consequence of some disorder of embryonic development. The predominant cells are spindle-shaped, and many transitions between these and definite striated muscle cells have been traced. Islands of cartilage occur frequently, and myxomatous degeneration is not uncommon. Metastases are rare.

The rhabdomyosarcoma of the testis is another example of this type of tumor. It may take the form of a nodular mass of huge size, sometimes several thousand grams in weight. Spindle cells, muscle cells, cartilage, and epithelial cysts are the constituent elements.

TERATOID ADENOMYOSARCOMA (EMBRYOMA OR WILMS' TUMOR)

Teratoid adenomyosarcoma is a peculiar tumor of embryonal type which at times exhibits a marked degree of anaplasia. In the majority of cases the tumor occurs during the first 3 years of life, and it is rarely seen after the tenth year.[15] The tumor is one of the largest of all renal growths; some have weighed several thousand grams. Occasionally the tumors are bilateral. The rate of growth is rapid, and the course is nearly always fatal.

The tumor tissue lies within the capsule of the kidney, and although usually demarcated from the normal kidney tissue it may merge with it in places. Some parts of the tumor are soft and others firm and fibrous; the color is variable depending upon the amount of hemorrhage. The cellular composition of the tumor is diversified. Tubules of cylindrical or cuboidal cells are surrounded by broad fields of undifferentiated spindle cells. Smooth muscle, striated muscle, fat cells, cartilage, and bone often exist in some parts of it. Local extensions and metastases are exceptional; the liver and the lungs are the most frequent sites of metastasis. The

exact origin of these embryonal tumors is unknown.

ALVEOLAR CELL RHABDOMYOSARCOMA

The alveolar cell rhabdomyosarcoma was brought to the attention of pathologists in 1956 by Riopelle and Thériault.[23] Table 8–2 gives the localization of the cases reviewed by Enzinger and Shiraki.[7]

TABLE 8–2. Alveolar Rhabdomyosarcoma, 109 Cases

Site	No. of cases
Head and neck	20
Trunk	20
Upper extremities	29
Lower extremities	30

Those of the arms often arise in the hand or forearm muscles, especially the flexor ones. Sixteen in this series occurred in the perineal, perianal, or perirectal region and 12 in the orbits. The average age of the patients was 15 years and the sex incidence nearly equal.

Grossly the tumor, which presents as a deep-seated painless mass or swelling, resembles the pleomorphic form, being gray, yellow, or pinkish white and rather granular on the cut surface. Hemorrhage is present in a few.

The term alveolar applies to the tendency of the small, poorly differentiated cells to aggregate in little islands or pockets separated by bands of fibrous connective tissue. The small round, oval, or elongated cells may be difficult to distinguish from those of a soft tissue sarcoma, anaplastic carcinoma, or reticulum cell sarcoma. The cell size ranges from 10 to 15 μ, and the nuclei are oval and finely stippled, with or without conspicuous nucleoli. Mitoses are variably frequent. In approximately one-third of the cases, in the more pleomorphic parts bulbous, club-shaped, racket-shaped, or spindle-shaped cells with cross-striated myofibrils are visible in preparations stained by Mallory's PTH or the Masson technique. Multinucleated giant cells with central eosinophilic sarcoplasm and a wreath of nuclei are found in many of the specimens. PAS stains show the cells to be rich in glycogen.

This tumor, like the other types of rhabdomyosarcoma, is highly malignant. Seventy-five percent of patients die within the first year and over 90% within a few years regardless of the mode of therapy. Both lymphatic and blood-borne metastases develop—in the lymph nodes, lungs and pleura, bones, and pancreas, and less often in the heart, adrenals, testes, liver, and kidneys.

Transplantation of such tumors into tissue culture media results in the harvesting of both spindle and giant cells but without myofibrils.[17] Bassler and Voth[2] failed to see myofilaments with the electron microscopy.

TUMORS OF BLOOD VESSELS AND SUPPORTING TISSUES

Other tumors arising in skeletal muscle originate from nervous and vascular elements and from the supporting stroma of the muscle. These include lipomas, fibromas, neuromas, angiomas, synoviomas, desmoid tumors, and others. In general these growths when present in muscle do not differ from similar tumors in other organs. There are two of these tumors, however, which occur with such regularity in muscle as to warrant special consideration: angiomas and desmoid tumors.

ANGIOMA

Angiomatous or hemangiomatous tumors often arise in skeletal muscle. The tumors are similar to those found elsewhere in the body, especially the skin, and are usually classified according to the predominating vascular structure as cavernous, arterial, venous, or capillary, or at times in some combination such as cavernous-arterial angioma and the like.

Jenkins and Delaney[13] summarized certain biological characteristics of these tumors in muscles. Up to 1953 Jones[14] was able to find 358 reported cases in the literature. The pres-

ence of a tumor mass is the commonest symptom, being present in almost all cases. Pain occurs in about 58% and functional impairment or deformity in 25%. The age of the patient is of distinct importance. Seventy-nine percent occur before the age of 21 years and 94% within the first 30 years. There is no definite sex predominance, and hereditary influence does not appear to play a role. The site of origin of the tumor is in the lower extremity in 14%, the upper extremity in 25%, and in the head and trunk in the remaining 33%. The most commonly involved muscle is the rectus femoris, in 17% (Table 8–3).

TABLE 8–3. Hemangiomas of Muscle: Frequency of Tumors at Various Sites

Site	Cases
Lower extremity	107
Buttock	6
Thigh	59
Leg	39
Foot	3
Upper extremity	64
Shoulder	9
Arm	22
Forearm	26
Hand	7
Trunk	56
Chest	36
Back	8
Abdomen	22
Head and neck	28
Head	22
Neck	6
Not stated	1
	1
Total	256

From Jenkins and Delaney.[13]

The gross appearance of the tumors is characteristic only in so far as they are frequently blue or reddish. Some diffusely infiltrate the surrounding muscle, whereas others are more or less sharply circumscribed. Microscopically they are composed of vascular elements in a connective tissue stroma. Arteries, arterioles, veins, and capillaries may predominate in certain areas or in the entire tumor, but cavernous spaces are more frequent.

The supporting connective tissue stroma may be scanty or quite abundant, at times dense and scirrhous; and in the stroma remnants of striated muscle are present in various stages of atrophy and degeneration. In the central portions of the tumor the muscle fibers are usually completely destroyed, whereas at the periphery irregular extension of blood vessels between muscle fibers produces varying degrees of muscle fiber change. Some tumors are separated from the adjacent muscle by a fibrous connective tissue capsule.

Treatment in nearly all cases has been by local excision. Surgery should be conducted under tourniquet control of bleeding if this is feasible. In many cases in which complete control of bleeding has not been possible, extensive hemorrhage has occurred though death from operative intervention has not been reported. There is no mention in the literature of death due to an angiomatous tumor of muscle, nor has definite sarcomatous change been reported. Recovery has been complete in all cases in which excision was performed, except an occasional recurrence usually after incomplete removal of the tumor mass.

DESMOID TUMOR

Desmoid tumors are benign fibrous tissue neoplasms that originate in musculoaponeurotic structures throughout the body. They do not arise from muscle fibers per se but from the supporting connective tissue of the muscles and from the aponeuroses. However, these tumors are so intimately related to muscular tissue that any discussion of striated muscle would be incomplete without a few words about them.

Grossly and histologically desmoids resemble fibromas but differ in their peculiar property of local invasion of muscle tissue and their lack of encapsulation. Histologically the diagnosis depends on the demonstration of fibrous tissue engulfing and destroying striated muscle fibers. As Musgrove and McDonald[18] and Booher and Pack[4] pointed out, morphologically they have the local neoplastic property of invasion, but histologically they are nonmalignant.

The tumor may cause pain that is usually mild and of aching character; or, as often happens, the only evidence of growth is the presence of a nodule. Approximately two-thirds of desmoid tumors occur in the musculature of the anterior abdominal wall, and the remainder in many other muscles throughout the body. Table 8–4 shows the distribution of 77 tumors observed by Pearman and Mayo[21] at the Mayo Clinic.

TABLE 8–4. Sites of Desmoid Tumors

Site	Cases
Ant. abdominal wall (rectus abdominis, external, and internal oblique or transversalis)	55
Pectoralis major	5
Scapular muscles	4
Rectus femoris	3
Gluteal muscles	2
Sternomastoid	1
Posterior belly of digastric	1
Biceps brachii	1
Extensor carpi ulnaris	1
Hamstring muscles	1
Dorsum of foot	1
Masseter	1
Scar of radical operation on breast	1
Total	77

Desmoid tumors vary from 1 to 15 cm in diameter. They have a firm rubbery consistency and on cut section are whitish pink with interlacing bands of white fibrous tissue, much like a uterine fibroid. Central areas of myxomatous and cystic degeneration are common. Microscopically the tumor is composed of fibroblasts, varying in number in different areas. The cells are embedded in dense fibrillar tissue that resembles collagen (Fig. 8–5). Mitoses are quite rare. The incorporated muscle fibers show varying degrees of degeneration with loss of striation and hyalinization. Often multinuclear collections in a pale protoplasmic mass are the only remnants of these muscle fibers.

The pathogenesis of desmoid tumors has not been definitely determined. Of the theories advanced, only two remain tenable today. The first is that the tumor is related to trauma. Many of the abdominal growths arise after

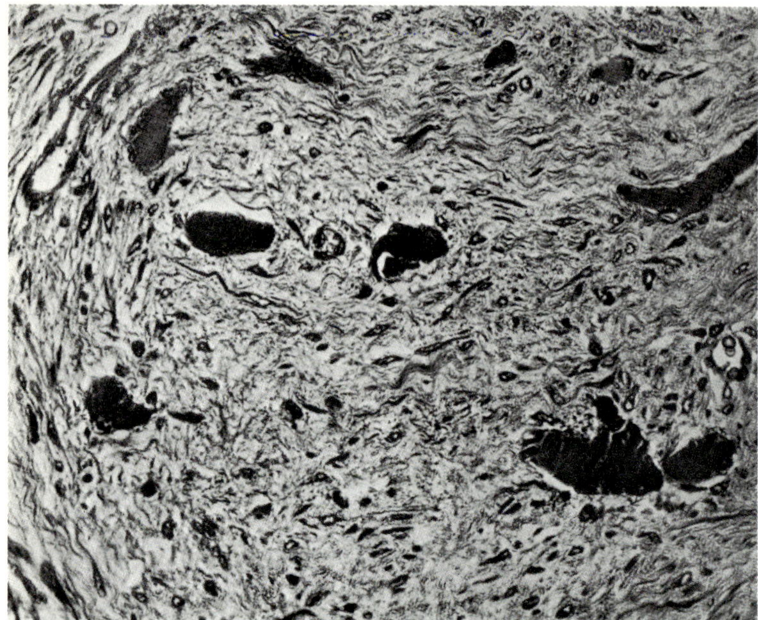

Fig. 8–5. Desmoid tumor of abdominal wall muscle. The few dark masses are muscle fiber remnants that have lost their striations. Stroma is composed of dense fibrillar, poorly cellular, fibrous connective tissue. (phosphotungstic hematoxylin; ×485)

pregnancy with a prolonged labor and violent muscular contractions. Elsewhere in the body Musgrove and McDonald found that in 14 of 34 tumors the patient definitely linked the onset of the tumor with trauma.[18] Desmoids also arise at the site of a previous surgical incision. Here they resemble a keloid, but of course in the keloid the fibrous tissue overgrowth is limited to the skin and subcutaneous tissue with no invasion of striated muscle.

The second theory is that their formation is related to an endocrinologic disturbance. The majority of these tumors occur in women (70% of Musgrove and McDonald's cases), and bioassays of the tumor tissue reveal a high concentration of estrogenic and anterior pituitary gonadotropic substances. Their similarity to uterine fibroids has been mentioned, and Lipschütz[16] succeeded in producing connective tissue tumors in the uterus, other abdominal organs, and the abdominal wall in guinea pigs in 100% of animals by injecting estrogenic substances. Moreover, it is possible to prevent the formation of these experimental fibroids by various other steroid hormones (testosterone, progesterone, desoxycorticosterone).

The differential diagnosis is not difficult if the characteristic striated muscle infiltration by a fibroma-type growth is present. Most difficulty may arise in differentiation from a low-grade fibrosarcoma.[4] The latter, however, is usually more cellular, and mitoses are far commoner. The desmoid must also be distinguished from the keloid, the neurofibroma, and the sclerosing hemangioma or hemangioendothelioma.

Although desmoid tumors do not metastasize, their malignant property of local invasion dictates that they be treated by radical excision. The penalty for conservatism is a very high rate of recurrence. Endocrine therapy in the prevention and treatment of these tumors has been suggested on the basis of experimental work, but a critical evaluation of the effectiveness of these measures is not possible as yet.

SECONDARY TUMORS

It is a noteworthy fact that the commoner malignant tumors of the human body seldom spread to skeletal muscle. In a survey of 500 consecutive cancer necropsies Willis[26] observed metastases to skeletal muscle in only four cases, two of epidermoid carcinoma of the head and neck and two of thyroid carcinoma. Sarcomas, tumors of blood-forming organs, and less common tumors also tend to grow in other tissues far more readily than in skeletal muscle.

When one considers the abundant vascular supply to muscle and the frequency with which blood-borne metastases occur in the lungs, liver, and bones, it seems clear that there must be some property of muscle which is inimical to the extensive growth of secondary tumors. As Willis points out, there is nothing peculiar in the arteriolar or capillary circulation in muscle to account for this relatively low incidence of metastatic gowths. It is not likely, as has been suggested by some pathologists, that the contractile activity of muscle compresses the tumor cells through the capillaries and into venous channels, since most tumor emboli average 50–150 μ in diameter, a size much larger than most capillaries. Other explanations must therefore be sought for the regression and destruction of metastatic tumor emboli in muscle. Variations in pH and the accumulation of lactic acid and other tissue metabolites are probably important in this respect.

Cases in which muscle seems capable of sustaining the growth of secondary tumor implants are uncommon. Such a case was reported by Muslow.[19] Pearson[22] examined a similar case in which tumor tissue was widely disseminated in all the skeletal muscles and heart. Called a carcinosarcoma, this was a large and necrotic tumor mass that arose from the left diaphragm and invaded and compressed the left lower lobe of the lung. Microscopically it was an anaplastic growth resembling a sarcoma. The predominant cell was spindle-shaped, large, and irregular in outline, with a faintly acidophilic cytoplasm and a large finely granular nucleus. Many atypical mitotic figures were present. Occasionally groups of these cells or smaller, rounder ones were clustered in a modified acinar pattern. There was relatively little stroma, and invasion of blood vessels by tumor cells was evident. Some parts of the tumor were extensively necrotic. The muscular tissue adjacent to metastatic nodules was compressed (Fig. 8–6B), and tumor cells infiltrated

the muscle for a short distance beyond the compact edge of the main tumor mass.

Pearson[22] made a more systematic survey of the metastatic spread of tumor to striated muscle and found it to be more frequent than was reported by Willis. Eight or nine muscles were selected (Chapter 4) and systematically examined at autopsy in 38 cases of malignancy. Metastatic tumor emboli or growths were found microscopically in a total of 14 muscles in six of these cases. The iliopsoas contained tumor four times, the diaphragm three times, and the other muscles at least once. Three of the tumors were lymphomatous (reticulum cell sarcoma, lymphoblastoma, and giant follicle lymphoma with sarcomatous change). The rest were carcinomas originating in the ampulla of Vater, the stomach, and the male breast. Each of the metastatic tumors retained in its muscle metastases the growth pattern of the primary tumor.

Carcinomas are more apt to involve muscle by direct extension from an adjacent tumor than by metastases from a distant tumor. As stated previously, the mode of secondary tumor growth in muscle varies with the type of primary tumor. Highly malignant infiltrative tumors may extend along fascial planes (Fig. 8–7A) and between muscle fibers (Fig. 8–6A). The reticulum cell sarcoma illustrated in Figure 8–9 behaved in this manner.

As noted by many early pathologists,[6] the neoplastic cells may invade the sarcolemmal sheaths in the same manner as some parasites and multiply within the striated substance of individual muscle fibers[9, 10] (Fig. 8–7B).

In order to obtain some idea of the behavior of secondary carcinoma or sarcoma in muscle it is necessary to examine sections from the main mass of tumor, from its advancing border, and from the muscle at some distance from the tumor. At the growing edge of the tumor there are usually many islands of infiltrating tumor cells. Sarcomatous growths tend to acquire a perivascular distribution, whereas carcinomas may penetrate the sarcolemmal sheaths; both follow fascial planes and connective tissue sheaths, which offer the least resistance to tumor growth. Carcinomatous cells enter the muscle fiber through a focal dissolution of the

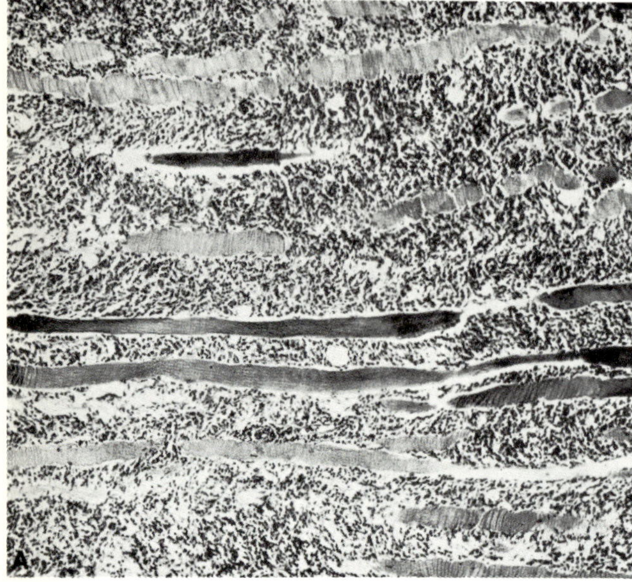

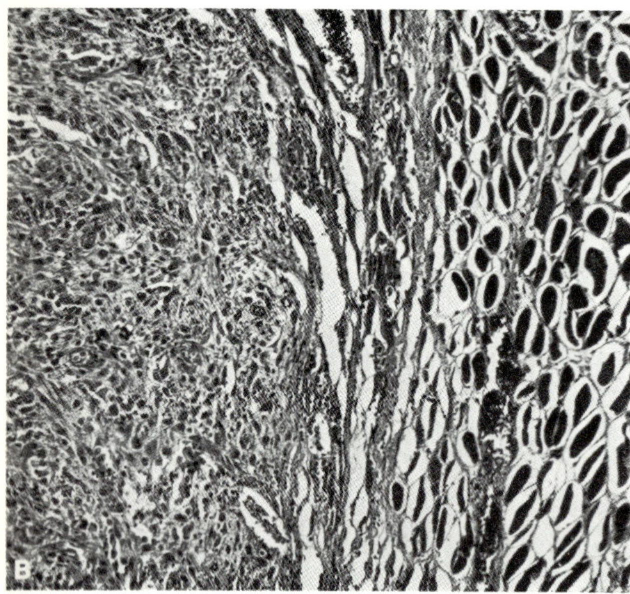

Fig. 8–6. (A) Extensive invasion of skeletal muscle by oat cell carcinoma of the lung. Tumor tissue has grown along endomysium and has separated muscle fibers. These fibers are undergoing cloudy swelling and hyaline degeneration. (B) "Carcinosarcoma" of the kidney invading the intercostal muscle. Tumor tissue at margin of nodule has compressed muscle fibers into thin strands. (A, phosphotungstic hematoxylin, ×92; B, H&E, ×100)

sarcolemma[25] and extend as cords of cells along the longitudinal axis of the fiber. The substance of the invaded muscle fibers becomes waxy or hyalinized (Fig. 8–7B). The sarcolemmal nuclei proliferate. The sarcolemma may persist as an investing sheath for the tumor cells.

Surviving muscle fibers demonstrate a variety of modifications, from simple atrophy to pigmentary and other types of degeneration (Fig. 8–8). Vacuolar degeneration is especially frequent. Granular change is relatively rare, and hyaline degeneration is most commonly seen in muscle fibers invaded by tumor cells or extensively surrounded by them (Fig. 8–6A). The connective tissue is increased in amount and is rich in tumor cells, activated fibroblasts, and histiocytes. Scattered lymphocytes and sarcolemmal cells may be present.

At a greater distance from the tumor in fascicles in which there are no tumor cells there is a slight increase in the interfascicular con-

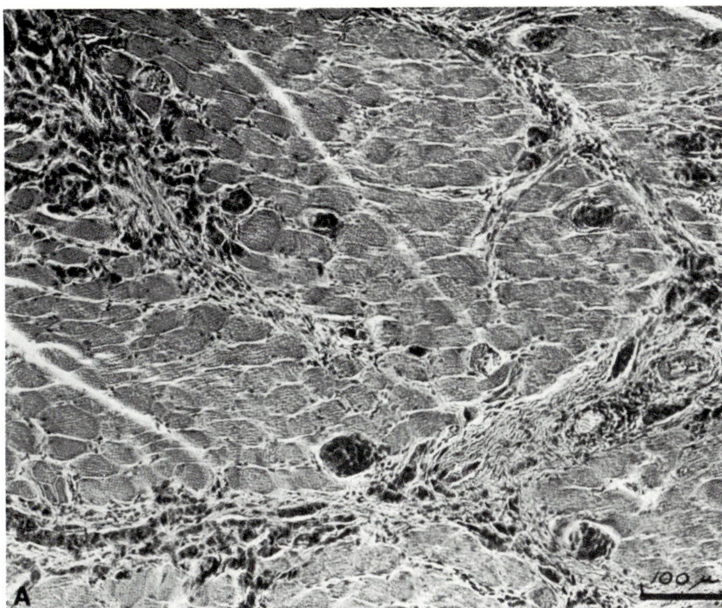

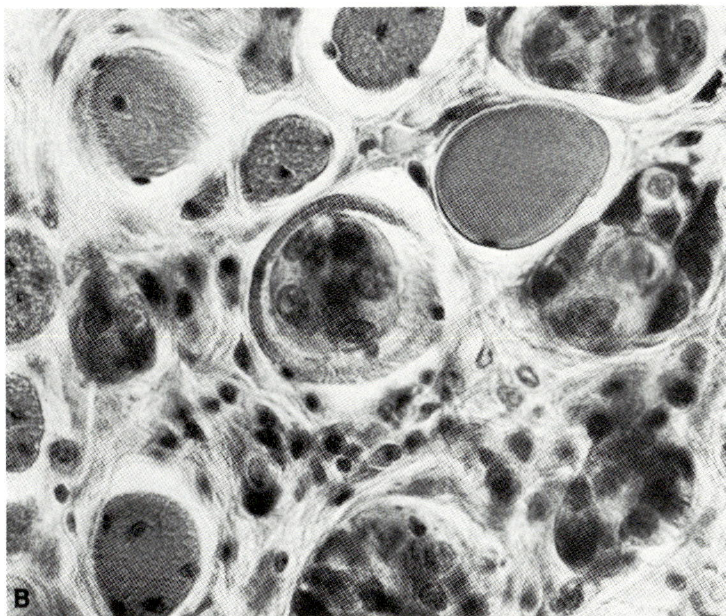

Fig. 8–7. (A) Lymphatic and perimysial spread of breast carcinoma into adjacent pectoral muscle. (B) Section of same tumor shown in (A), demonstrating carcinomatous invasion of muscle fibers. Some fibers are completely replaced by tumor cells. Hyaline degeneration can also be seen. (H&E, ×590)

nective tissue. The muscle fibers are of average size, but the sarcolemmal nuclei are increased in number and size. These are disposed in long rows or in clusters. Some have lost their subsarcolemmal position and are seen at all depths within the fiber Scattered atrophic fibers and others showing vacuolation or cloudy swelling are also present.

LOCAL EFFECTS OF TUMOR GROWTH (COMPRESSION ATROPHY)

Other aspects of this subject of considerable interest to pathologists are the local effects of the tumor growth on muscle tissue[1] and the general or systemic effects (e.g., cachexia) of the tumor as manifested in all the body tissues including muscle. The former can best be appreciated by examining sections taken from the advancing edge of the tumor. These mechanical effects differ, as might be expected, according to the rate of tumor growth and manner of invasion. When the tumor originates in the substance of the muscle and remains encapsulated it displaces the muscle fibers and causes atrophy by compression. When a malignant tumor infiltrates the muscle, there is a clear zone of invasion in which tumor cells are mixed with degenerating and regenerating muscle fibers. The neighboring fibers also suffer a mechanical compression (Fig. 8–6B), which is quite similar in all types of tumors and requires separate consideration. The effects of the cachexia associated with malignant tumors of the nonmuscular tissue is discussed in Chapter 11.

Atrophy of muscle from compression is found in the region adjacent to any benign or malignant tumor and near inflammatory masses such as abscess or tuberculoma, encysted parasites, and exostoses. This type of change is a simple atrophy of fibers which may or may not be accompanied by degenerative changes.[6] If the compression has lasted for any time most of the muscle fibers in parts of the muscle nearest a tumor have disappeared. Amid richly cellular connective tissue only a few atrophic fibers occurring singly or in small bundles can be identified. These surviving fibers have lost their cylindrical shape and appear as peculiarly formed masses of sarcoplasm (Fig. 8–8). Some are fusiform, whereas others are moniliform or have diverse shapes. They cannot be traced for any distance in longitudinal sections and may terminate in cones or bifurcated ends. Many of these fusiform or pleomorphic masses of sarcoplasm still retain visible striation. Some have many nuclei in their most swollen parts; others are acellular and are joined by strands or bridges to the nucleated masses of sarcoplasm. Histiocytes and often macrophages can be identified in the connective tissue. Lymphocytes and other inflammatory cells are seldom numerous except when the compression atrophy is due to granulomatous or inflammatory masses.

The fate of these syncytial sarcoplasmic bodies is difficult to determine. The sarcolemmal nuclei in some become small and darkly stained, and with further nuclear regression of the muscle fiber completely disappears. At a certain phase of dissolution one or several droplets of fat may form within such a mass— true fatty degeneration. Other fibers lose their striation and their coloring (myoglobin) and come to resemble connective tissue so closely that they can be distinguished only by the character of their nuclei (Fig. 8–8). This transformation into fibrous tissue was emphasized by Durante, who believes that in this way the muscle fibers contribute to formation of the tumor's capsule.

At a greater distance from the neoplasm simple atrophy dominates the pathologic picture. The atrophic fibers are either isolated or occur in small groups intermingled with more or less healthy fibers. The thin fibers are often of uneven caliber and follow a tortuous course. Their striation may be retained but is obscured by a fine granularity similar to that seen in cloudy swelling. Some of the fibers are partially decolorized, as if they have lost their myoglobin. The nuclei of such fibers have proliferated, often to a marked degree. These nuclei first enlarge, then acquire a more prominent nucleolus and divide by simple fission. The daughter nuclei, which are smaller and more deeply stained, may be quite numerous, as many as 30–40 being found in some short segments of muscle fibers. They are arranged in long rows or in clusters, and some nuclei no longer retain their subsarcolemmal position. This atrophy is not uniform through the whole length of the muscle but is most pronounced at points at which the compressive effect is greatest. Alongside atrophic fibers there may be scattered hypertrophic ones attaining a transvese diameter of 80–90 μ. The striation of these fibers is often rather indistinct, and the sarcolemmal nuclei are increased in number and size.

All degrees of degeneration can be seen in muscle fibers, particularly those undergoing simple atrophy. In some there is the classic hyaline or waxy degeneration and discoid segmentation. Granular change is less common than fatty degeneration. There is longitudinal fissuration of fibers, resulting, so it is believed,

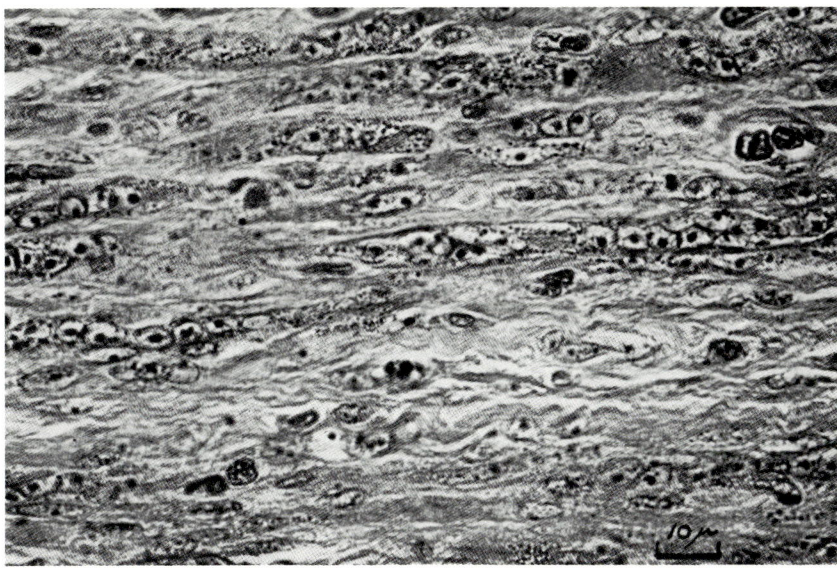

Fig. 8–8. Compression atrophy of muscle fibers adjacent to expanding nodule of carcinoma. Note rows of sarcolemmal nuclei and scanty remnants of sarcoplasm which contain clusters of pigment granules. (phosphotungstic hematoxylin)

Fig. 8–9. Longitudinal section of muscle invaded by reticulum cell sarcoma. (×220)

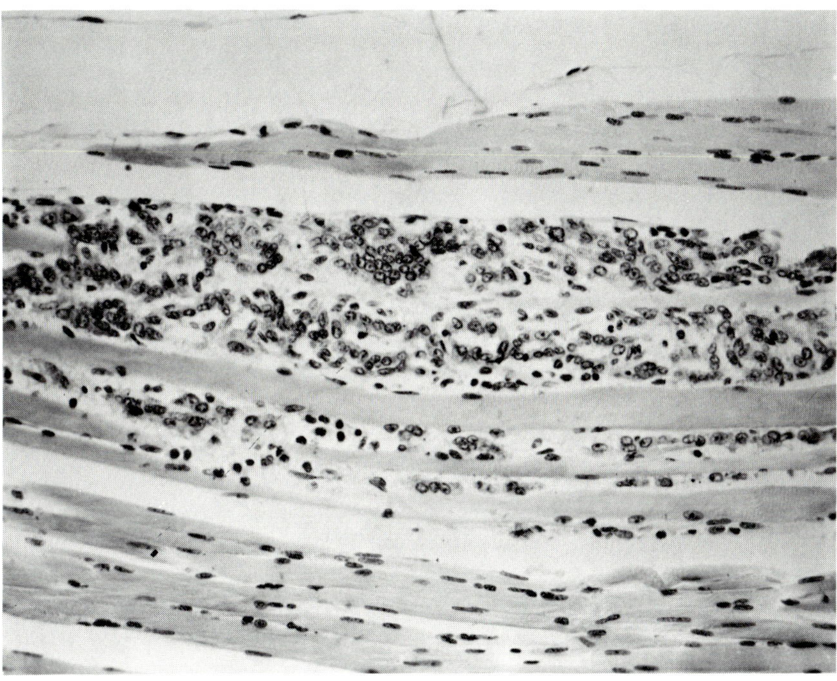

from an augmentation of sarcoplasm. Vacuoles are not unusual in atrophic and degenerating fibers. These develop within the sarcoplasm, either in the center or near the surface of the fiber. Appearing first as small opaque splotches on the fiber, they grow and extend along the fiber to form first little canals and then larger vacuoles. Pigmentary atrophy can be detected in many fibers, in some of which striation is conserved. The first sign of this change is the accumulation of grains of yellowish pigment at each end of the sarcolemmal nuclei (Fig. 8–8). The striation of some of these fibers is indistinct and the protoplasm pale and granular. As atrophy proceeds the muscle fiber is reduced to a small aggregate of nuclei surrounded by granules of lipofuscin and a thin sheet of structureless sarcoplasm.

At a still greater distance from the tumor the surviving muscle fibers are larger and more numerous. Their cross striation may be indistinct, while the longitudinal striation is exaggerated. The myofibrils appear to be separated by an excess of sarcoplasm, and in transverse sections Cohnheim's fields are unusually prominent. Abortive regeneration may lead to splitting or branching of muscle fibers and may form a veritable muscle plexus that resembles cardiac muscle. Blood vessels and sheets of collagenous and fibrillar connective tissue extended between the fibers.

In summary, then, some of the fibers near a muscle tumor disappear as a consequence of simple atrophy, whereas others undergo hyaline, fatty, or granular degeneration. Regressive changes exceed regenerative activity.

CONCLUSION

Tumors arising from elements of the striated muscles of the body like those of the highly differentiated neurone are uncommon. The highly malignant rhabdomyosarcoma is fortunately among the rarest of these tumors and may be difficult to recognize, even after the most detailed histologic study unless the characteristic striated tumor cells are found. The origin of the so-called granular cell myoblastoma is still in doubt. In the light of recent work, indications are that it may arise from the elements of nervous tissue, though derivation from muscle is by no means excluded. The isolated reports of heteromorphic or metastasizing myoblastomas probably represent unusual examples of anaplastic rhabdomyosarcomas. Tumors arising from the supporting tissues of the muscles are less rare but are usually benign and only of importance because of the effects of local swelling and compression.

Despite the relatively large vascular supply of the muscular tissue, metastatic tumor growths in muscle are uncommon. The reasons for this relative immunity are not readily understood, but it is likely that they are concerned with the metabolic and chemical environments of muscle fibers.

Carcinomas are more apt to invade muscle by direct extension from a primary growth in another tissue or organ (i.e., breast). Such growths may expand and compress the adjacent fibers, thereby producing a simple compression atrophy and some spotty degenerative change as well. More malignant tumors infiltrate along fascial planes and destroy the muscle fibers by direct invasion or by severe compression and ischmia.

REFERENCES

1. ANZINGER FP: The changes occurring in striped muscle in the neighborhood of malignant tumors. Am J Med Sci 123:268–284, 1902

2. BASSLER P, VOTH D: Pathologie und submikroskopische Morphologie des Sogenennten Sarcoma Botryoides der grossen Gallenganze. Krebsforschung 65:44–45, 1962

3. BATCHELOR TM, MAUN ME: Congenital glycogenic tumors of the heart. Arch Pathol 39:67–73, 1945

4. BOOHER RJ, PACK GT: Desmomas of the abdominal wall in children. Cancer 4:1052–1065, 1951

5. CAPPELL DF: Tumors of striated muscle. Ann R Coll Surg 2:80–92, 1948

6. DURANTE G: Anatomie pathologique des muscles. Manuel d'Histologie Pathologique. Edited by V Cornil, L Ranvier. Paris, Félix Alcan, 1902, Vol II, pp 1–477

7. ENZINGER FM, SHIRAKI M: Alveolar rhabdomyosarcoma—an analysis of 110 cases. Cancer 24:18–31, 1969

8. EWING J: Neoplastic Diseases: a Treatise on

Tumors. Philadelphia, Saunders, 1940, pp 240–248

9. HARTZ PH, VAN DER SAR A: Chloroleukemia: report of a case with special reference to its neoplastic nature. Am J Pathol 18:715–727, 1942

10. HASSIN GB: Carcinoma of muscle tissue as a cause of laryngeal paralysis. J Neuropathol Exp Neurol 6:358–368, 1947

11. HORN RC JR, ENTERLINE HTA: A clinicopathological study and classification of 39 cases of rhabdomyosarcoma. Cancer 11:181–199, 1958

12. HURLEY JV: Rhabdomyosarcoma of skeletal muscle. Aust NZ J Surg 24:45–55, 1954

13. JENKINS HP, DELANEY PA: Benign angiomatous tumors of skeletal muscles. Surg Gynecol Obstet 55:464–480, 1932

14. JONES KG: Cavernous hemangioma of striated muscle. J Bone Joint Surg [Am] 35A:717–728, 1953

15. LADD WE: Embryoma of the kidney (Wilm's tumor). Ann Surg 108:885–902, 1938

16. LIPSCHÜTZ A: Induction and prevention of abdominal fibroids by steroid hormones and their bearing on growth and development. Cold Spring Harbor Symp Quant Biol 10:79–90, 1942

17. MCALLISTER RM, MELNYK J, FINKELSTEIN JZ, ET AL: Cultivation in vitro of cells derived from human rhabdomyosarcoma. Cancer 24:520–526, 1969

18. MUSGROVE JE, MCDONALD JR: Extra-abdominal desmoid tumors; their differential diagnosis and treatment. Arch Pathol 45:513–540, 1948

19. MUSLOW FW: Metastatic carcinoma of skeletal muscles. Arch Pathol 35:112–114, 1943

20. PACK GT, ARIEL IM: Tumors of the Soft Somatic Tissues; a Clinical Treatise. New York, Hoeber, 1958

21. PEARMAN RO, MAYO CW: Desmoid tumors. Ann Surg 115:114–125, 1942

22. PEARSON CM: The incidence and type of pathologic alterations observed in muscles in a routine autopsy survey. Neurology (Minneap) 9:757–766, 1959

23. RIOPELLE JL, THÉRIAULT JP: Sur une forme méconnue de sarcoma des parties molles, le rhabdomyosarcome alvéolaire. Ann Anat Pathol (Paris) 1:88–111, 1956

24. STOUT AP: Rhabdomyosarcoma of the skeletal muscles. Ann Surg 123:447–472, 1946

25. VOLKMANN R: Zur histologic des Muskelkrebses. Virchows Arch Pathol Anat 50:543–549, 1870

26. WILLIS RA: Pathology of Tumors. St. Louis, Mosby, 1948

27. WOLBACH SB: Congenital rhabdomyoma of the heart: report of a case associated with multiple nests of neuroglia tissue of the meninges of the spinal cord. J Med Res 16:495–519, 1907

28. WOLBACH SB: A malignant rhabdomyoma of skeletal muscle. Arch Pathol 5:775–786, 1928

chapter 9

TRAUMATIC AND CIRCULATORY DISEASES

TRAUMATIC DISEASES

Muscle can be traumatized either by externally applied force or by tearing incident to violent exertion, the latter occurring especially when opposing muscle groups are brought into action simultaneously. Direct injury to muscle causes the muscle fibers and their connective tissue sheaths to shatter. The extent of the necrosis depends upon the amount of force applied, the presence or absence of fracture of adjacent bones, and in addition on the degree of hemorrhage, edema, and injury to blood vessels. Phalen[40] discussed this problem in relation to trauma of the anterior tibial muscles and stressed the importance of ischemia. Blomfield[6] related the more extensive posttraumatic necrosis of some muscles, such as the biceps or anterior tibial muscles, to a solitary source of blood supply (Chapter 2). Repair of these traumatic lesions is variable. Trauma to large areas of muscle results in extensive parenchymal destruction and subsequent scar tissue replacement or even calcification and bone formation. After minor injuries (such as stab or bullet wounds or those following surgical incision) regeneration of muscle can be almost complete, with relatively little scar formation.

MUSCLE RUPTURE AND FASCIAL TEARS

The commonest types of muscle trauma are those which follow prolonged exercise in an untrained subject or violent exertion and contraction of certain muscles leading to muscle herniation, rupture, hemorrhage, and edema. Muscle hernias are produced by protrusion of a muscle belly through a rent in the overlying fascia and epimysium. The hernia usually does not attain maximal size for several days. A localized pain and swelling appear during exertion; with further use of the muscle, a soft elastic tumor unattached to the skin appears. Upon contraction of the affected muscle there is often pain and the tumor hardens—a reliable point of differentiation from subfascial lipomas and other similarly placed tumors. In addition, the hernia may be reduced by digital manipulation when the muscle is relaxed. The muscles usually involved are the belly of the biceps and less frequently the rectus femoris and gastrocnemius. Histologically a mild inflammatory infiltrate can occur about the margins of the protruding muscle. Long-existing hernias may show a mild fibrosis. Incarceration of the herniated muscle segment may result in necrosis, but this is rare.

Another condition to be distinguished from a fascial tear is an actual rupture of muscle tissue during violent exertion. The rupture may occur in the muscle substance, at the junction of muscle and tendon, or at the origin of insertion of tendon into bone. However, the majority of ruptures (66% according to Gilcreest[16]) are in the fleshy belly of the muscle, and the tendinous ends of muscle remain intact. This subject was reviewed extensively by Maydl,[32] Hackenbruch,[18] Lorenz,[30] von Meyenburg,[51] and in the American literature by Gilcreest.[16] These authors report that muscle rupture occurs largely in individuals engaged in certain occupations, and such injury can therefore be called an occupational hazard. Gilcreest observed frequent rupture of the neck muscles in loaders and packers, back muscles in stevedores, biceps and triceps in baseball pitchers and weight lifters, adductors of the thigh in horseback riders, and calf muscles in boxers, runners, and mountain climbers. Rupture may occur also in severe tetanus, generalized convulsions, or in electric shock injuries, which induce violent contractions in antagonistic muscles concurrently, as well as in individuals in poor condition who make an unusual movement. The pain and swelling may then be mistaken for thrombophlebitis.[43]

The usual sequence of events is sudden overloading of a muscle when it is already maximally contracted, producing an audible snap, moderate to severe pain, swelling, and a local bulge in the involved muscle. If the rupture occurs at the lower end of a muscle, the local bulge or swelling appears in the upper end; conversely, if the tear is in the upper portion, the swelling appears in the lower end. The rupture, which may become complete only after 24 hours, does not cause complete paralysis unless all the tendinous attachments are torn. Undoubtedly many cases of partial muscle rupture are overlooked or diagnosed as sprain, "charley-horse," or rheumatism.

In certain pathologic conditions muscle rupture may happen with even mild exertion. For example, McMaster[33] observed cases of rupture in trichinosis, typhoid fever, tuberculosis, and some general infections. It is likely that some of these ruptures are incidental to Zenker's degeneration of localized groups of muscle fibers. A hematoma may form at the site of the injury. Mercer[34] enumerated the muscles most frequently subject to partial or complete tearing and rupture. They are the short extensors of the toes, the gastrocnemius, plantaris, tibialis posterior, rectus femoris, biceps brachii, deltoid, trapezius, sternomastoid, and abdominal muscles.

HISTOLOGIC CHANGES

In muscles subjected to trauma, either by direct external force or indirectly through violent contraction, there is rupture of many muscle fibers and tearing of their sarcolemmal sheaths, as well as a variable degree of damage to the endomysial sheath of connective tissue. In adjacent parts which are less damaged there is a severe internal derangement of the sarcoplasm without destruction of the sarcolemmal sheaths. This derangement may take the form of only a variation in staining reaction or may consist of granular, vacuolar, or waxy degeneration depending on the severity of the trauma. These changes can be seen shortly after injury. Interstitial edema and extravasation of blood, often in the form of large hematomas, probably play an important role in increasing the muscle destruction by pressure and ischemia. Cellular infiltrates of neutrophilic leukocytes and later of lymphocytes and mononuclears, phagocytosis of necrotic muscle, and variable degrees of regeneration appear later. Associated ischemic damage of the muscle tissue due to vascular injury is frequent (Fig. 9–1). When the parallel arrangement of endomysial tubes is destroyed by the injury, fibrous tissue proliferation is extensive and prevents effective regeneration. The gap in the muscle is then eventually replaced by a large mass of scar tissue.

MUSCLE HEMORRHAGE

Hemorrhage into the substance of a muscle may be caused by trauma, by thrombosis of intramuscular veins in the course of a severe infection such as typhoid fever, and in hemorrhagic states such as during anticoagulant therapy, scurvy, and thrombopenic purpura. Ralis[41] reports it to be a not infrequent finding in the paravertebral and hip muscles of infants delivered by breach.

The most frequent site of muscle hemorrhage is probably the rectus abdominis. Here it is often associated with Zenker's waxy degeneration, which renders the muscle excessively fragile and liable to injury from any excessive movement such as a strain, a convulsion, or even rough palpation of the abdomen. The dimensions of these hemorrhages vary from ecchymoses to hematomas several centimeters in diameter. The extravasated blood may ap-

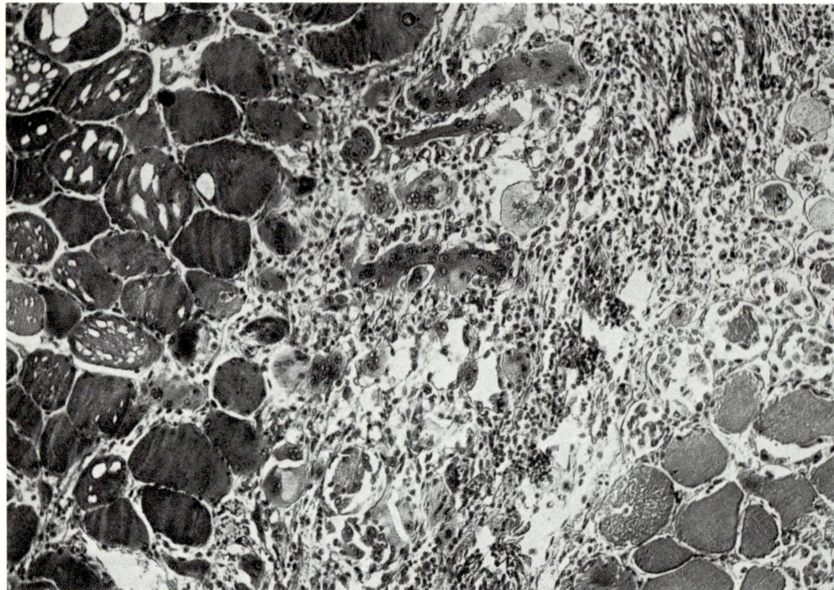

Fig. 9–1. Muscular injury accompanying fracture of bone, showing transition from complete necrosis of muscle fibers on right of figure to partial damage and vacuolation on left. Regenerative activity is evidenced by budding and giant cells at center. (phosphotungstic hematoxylin)

Fig. 9–2. Zenker's degeneration and hemorrhage into rectus abdominis muscle of 68-year-old patient dying of lobar pneumonia.

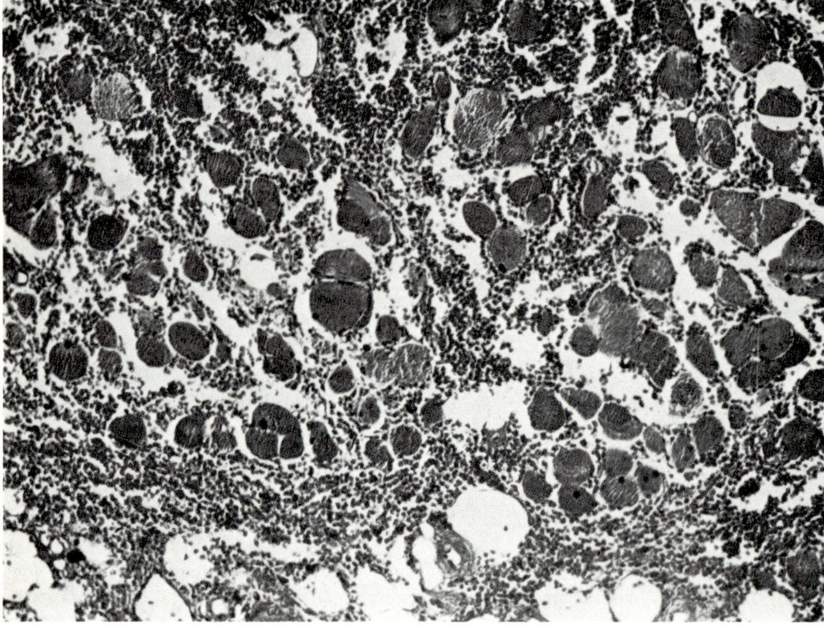

pear as streaks between the fascicles of muscle fibers or as discrete hematomas. The latter may become encapsulated, and cartilage or bone may form in their place. The limb muscles, being more often exposed to injury, are the commonest sites of traumatic hemorrhage. In certain hemorrhagic diseases such as hemophilia bleeding into the iliopsoas muscles is noteworthy because of its frequency and the likelihood of damage to the lumbar plexus and femoral nerves.

HISTOLOGIC CHANGES

Microscopic examination reveals a dissociation of the muscle fibers by red blood corpuscles (Fig. 9–2). The intervening muscle fibers may appear healthy, or they may manifest hyaline degeneration. The reactions to the effused blood are similar to those in other tissues. If the hemorrhage is small, the blood is removed with facility, and if large it may form a clot which encapsulates and persists for months or years as a recognizable mass of red corpuscles or of blood pigment. Several pathologists have commented on the ease with which blood is removed from muscle, and it has been suggested that active movement hastens resorption through lymph channels. At the margins of the clot, red corpuscles are phagocytized by macrophages, and orange hemosiderin pigment can be detected within these cells as early as the sixth or seventh day after injury. Vascular connective tissue forms around the clot. Fibroblasts increase in number, elongate, and produce a network of reticular fibrils. Collagenous connective tissue is laid down later. Organization of the clot by connective tissue may be complete or incomplete. If complete, a white cicatrix with scattered hemosiderin-containing macrophages in and around it eventually forms. This may calcify and form an osteoma. If incomplete, the periphery of the hematoma becomes organized and is composed of firm, white connective tissues, and the center is converted into a cyst the content of which may be brick-red, dark brown, or black. Around such hematomas there is evidence of cellular repair of damaged fibers and regeneration, usually by budding from surviving fibers.

TRAUMATIC MYOSITIS OSSIFICANS

As the name implies, traumatic myositis ossificans is a condition in which there is formation of bone in muscle as the result of injury to the muscle substance. It has also been termed traumatic ossifying myositis, ossifying hematoma, calcified hematoma, and traumatic parosteal bone formation.

In general, two types of the disease have been outlined, according to the form of the initiating trauma. The first group includes ossifications which appear in muscle following a single severe traumatic blow to or tear of a muscle; in the second group are those which arise following repeated minor trauma to a muscle or muscle group. The latter has been called myositis ossificans circumscripta.

According to Fay[12] this condition was first reported by Copping in 1740 and Freke in 1741. Hasse[19] in 1832 published a clear and concise description of ossifications in the pectoralis major and deltoid muscles in Prussian infantry recruits:

"A few days after the start of exercise those predisposed to the disease perceive a small red painful swelling on the part of the left shoulder where the musket presses. If this is neglected, a number of hard, movable gland-like tumors are formed in the muscle. These soon change into large masses of solid cartilaginous consistency, and then in 4 to 7 weeks into a solid mass of bone, which, according to its extent, more or less impedes the motion of the arm."

Ossification of solitary muscles is associated with certain occupations or avocations. The usual sites are in those muscles exposed to repeated injury. No age group is immune, but the condition occurs most often in active young men. Any athletic event which requires a high degree of exercise, exertion, and physical contact may traumatize muscle. Horseback riders and jockeys develop "rider's bone" in the adductors of the thigh; cavalrymen develop an osseous plate on the outer side of the thigh in places repeatedly traumatized by the sabre; infantrymen form a "drill bone" or "exercise bone" in the deltoid or pectoral muscles from the impact of a discharging rifle; in fencers and baseball pitchers portions of the biceps and brachialis anticus ossify due to the peculiar and sudden overextension of the forearm required in these activities.

Statistical surveys reveal that the muscles which most frequently undergo ossification are the quadriceps femoris and the brachialis anticus. Schultz (quoted by Makins[31]) reported 233 cases from the German army. All but three were in the above-mentioned muscles. Binnie[2] surveyed the early literature on this subject; Table 9–1 was compiled from his findings.

It is obvious that ossification may occur in almost any muscle but is seen most frequently in anterior thigh muscles and brachialis anticus. Also it is probable that the incidence is much higher than has been reported, for a considerable number of asymptomatic cases have been discovered by routine roentgenograms taken for other reasons. Many of these ossifications reportedly disappear spontaneously after variable periods of time.

TABLE 9–1. Distribution of Traumatic Myositis Ossificans in 65 Cases Reported in the Literature

Site	Cases
Shoulder	*4*
Unspecified	1
Deltoid	1
Pectoralis major	2
Upper arm	*18*
Unspecified	1
Biceps brachii	2
Brachialis anticus	13
Triceps brachii	2
Forearm, flexors	*1*
Hip	*3*
Gluteus maximus	1
Iliopsoas	2
Thigh	*31*
Anterior thigh (muscle unspecified)	15
Rectus femoris	3
Vasti (lateral, medial, intermediate)	9
Sartorius	1
Adductor magnus	3
Lower leg, gastrocnemius	*4*
Digastric	*1*
Diaphragm	*1*
Rectus abdominis	*1*
Cremasteric	*1*
Total	*65*

Adapted from Binnie.[2]

In ossification following a single injury, the usual history is that of a severe traumatic episode such as the kick of a horse, a fall, a blow, or rupture of a muscle from sudden exertion or strain. Indeed, it is probable that the expression "charley-horse," commonly applied by atheletes to any form of traumatic localized tenderness in muscle, originated from the old gymnastic exercise of jumping onto a padded wooden "horse," sustaining a sudden impact upon the stretched adductor muscles of the thighs. Bleeding into the muscle and hematoma formation is a common antecedent event, although this does not seem to be essential in all cases. Within a period of 1–4 weeks after the initial trauma, a firm and often painful mass appears in the affected muscle. It is readily palpable and may progressively limit motion, depending upon its position and extent. The ossification which follows repeated small traumas, on the other hand, may be entirely asymptomatic or may cause a slow, progessive limitation of motion and mild discomfort.

On the basis of x-ray studies, two general patterns of calcification have been described:[24] a feathery pattern in a muscle due to ossification of a hematoma that has dissected along muscle fasciculi, and ossification of a solitary hematoma, producing an irregular calcified structure of variable density. The calcifications are first detected in 10–20 days as faint shadows, which thereafter gradually increase in density.

The initial injury is to the muscle and adjacent soft tissue and is associated with hemorrhage of variable degree. Then follows a low-grade aseptic inflammation. Injury to adjacent bone and periosteum may be present. There is extensive proliferation of the intramuscular connective tissue, and soon islands of bone and cartilage arise in these thickened connective tissue septa (Fig. 9–3). The muscle fibers are not primarily involved in the process of the ossification but are incorporated and compressed in the fibrosing and calcifying tissue. Some destruction of muscle may be secondary to the effects of pressure and displacement. In some well advanced ossified lesions, muscle fibers of healthy appearance may persist amid dense fibrous tissue and spicules of bone (Fig. 9–4).

The bone may be formed directly from ele-

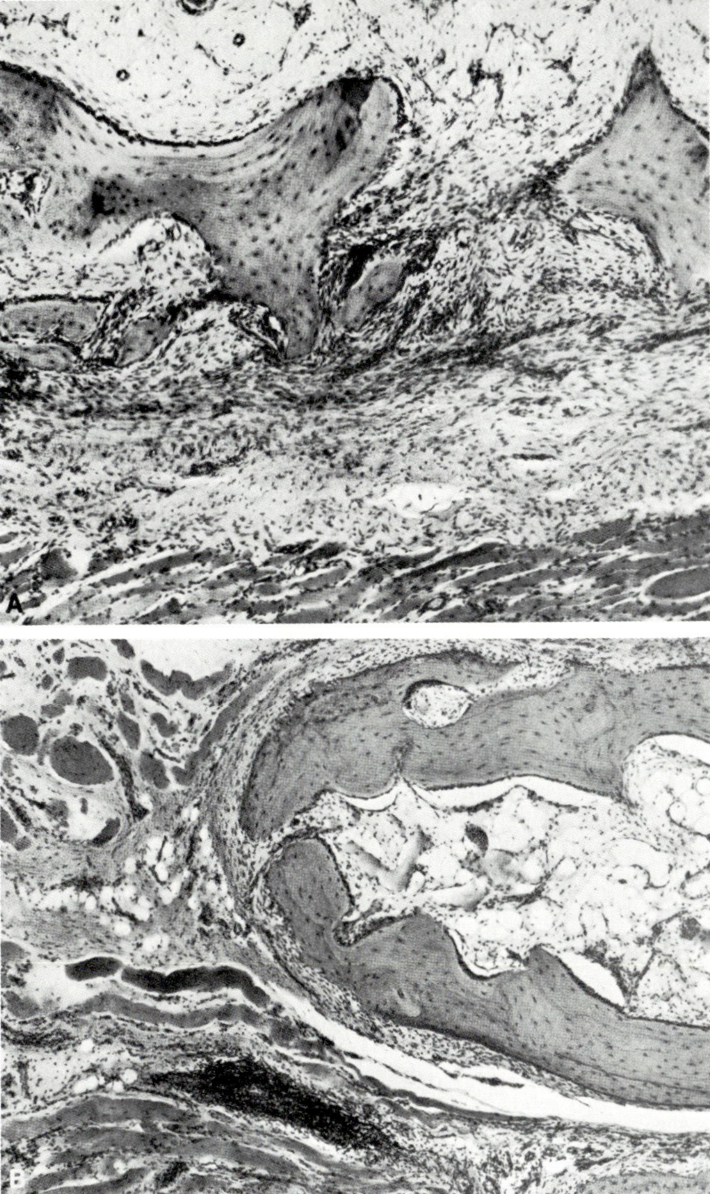

Fig. 9–3. Traumatic myositis ossificians. Biopsy of biceps taken 4 months after supracondylar fracture of right humerus. Obliquely oriented muscle fibers terminating in connective tissue are seen below. Osteoid tissue with rows of osteoblasts and myxomatous tissue appear above. (H&E)

ments of the fibrous tissue, or it may arise in islands of cartilage, i.e., endochondral bone formation (Fig. 11–18). The newly formed bone is generally of the cancellous type and resembles or is identical with the bone of callus formations in healing fractures. When viewed at the operating table, the bony lesion is often surrounded by a gelatinous substance that resembles bursal tissue. Beneath this material is a fibrous capsule from which irregular bands of fibrous connective tissue radiate inwardly.

The fibrous tissue seems to be the mother substance upon which the ossification occurs.

PATHOLOGIC CHANGES

Geschickter and Maseritz[15] have given a detailed histologic description of the pathologic changes in traumatic myositis ossificans. These authors described hemorrhage, degeneration of muscle, and hyperplasia of connective tissue

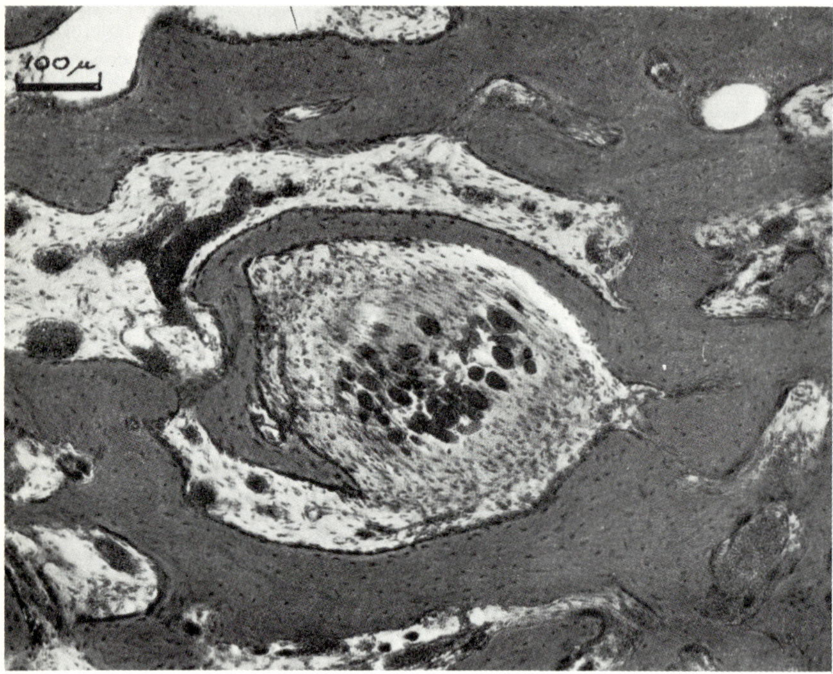

Fig. 9–4. Myositic ossifications. Incorporation of a small group of muscle fibers in a ring of osteoid tissue. Note also large dilated blood vessels between spicules of bone.

as the changes preceding ossification. Later, in any given area, organization of hemorrhage, cartilage formation, endochondral ossification, periosteal ossification, myxomatous tissue, angiomatous tissue, calcification en masse, and hyalinization and ossification within capillaries may occur. The inner portions of the connective tissue may calcify en masse, or islands of osteoid tissue with intervening marrow spaces may develop. The latter contain fat, collections of lymphocytes, and dilated capillaries filled with red blood cells (Fig. 9–3B). True bone marrow usually does not develop. Aggregations and small collections of myxomatous material resembling osteomucin may be seen (Fig. 9–3B). These collections appear to form a nucleus for bone formation. In other areas the osseous tissue arises directly from connective tissue elements or from recently formed cartilage. Occasional small atrophic or normal muscle fibers are present near and in the spaces between the bone trabeculae (Fig. 9–4). Each trabecula is enclosed in a thick tunic of cuboi-

dal osteoblasts similar to those present in any osteoblastic lesion. Geschickter and Maseritz found no essential difference between the ossifications following trauma and those of the progressive type.

PATHOGENESIS

Many theories have been advanced for the formation of bone in muscle and are reviewed by Carey.[9] The most tenable are the following: (1) traumatization of the periosteum of an adjacent bone with the displacement of osteoblasts into the muscle and the subsequent formation of bone; (2) activation by trauma or hemorrhage of periosteal implants already present in muscle; (3) metaplasia of the pluripotential intermuscular connective tissue into bone; and (4) metaplasia of fibrocartilage, a normal constituent of many muscle tendons, into bone.[22] It is clear that in addition to the above, other factors favoring bone formation must operate, for only a very small percentage of muscle strains or contusions eventually ossify. No abnormality of the calcium or phosphorus mechanism or of the phosphatase enzyme system has been demonstrated.

MALIGNANT DEGENERATION

Several cases of malignant degeneration of the ossified tissue with formation of osteogenic sarcoma have been reported. Pack and Braund[37] personally saw two such cases and located five similar ones in the literature. There is not complete agreement on this point, however, for some pathologists believe that osteogenic sarcomas may grow slowly and be mistakenly diagnosed as myositis ossificans in an early stage.

TRAUMATIC NECROSIS OF PRETIBIAL MUSCLES (ANTERIOR TIBIAL SYNDROME)

During the past few years several reports appeared in the literature concerning a peculiar syndrome in which the pretibial muscles of one or both legs became acutely swollen, painful, and paralyzed after a period of vigorous exercise. More than 20 cases have been reported to date, but the condition is more common than this indicates, for we have heard of several unpublished examples, some of which ended fatally. Moreover, there is reason to believe that the milder forms of the same condition constitute the troublesome pretibial stiffness ("shin splints") of athletes in training.

The first accounts of this condition were those of Severin[45] in 1943 and Sirbu et al.[46] in 1944. Severin's case occurred in a healthy 25-year-old man who developed "acute myositis" of the anterior tibial region soon after moderate exertion. During the subsequent year a slowly progressive fibrosis with contracture developed, with partial loss of contraction power. Microscopically after 1 year the muscle was largely replaced by dense, coarse scar tissue. The case of Sirbu et al.[46] was one of a series of cases illustrating the soft tissue com-

plications of leg fractures. The patient was a 38-year-old Filipino soldier who developed severe pain over the anterior aspect of both legs following a prolonged march. The anterior tibial muscles of both legs were swollen and tender and at operation bulged through the incised fascia as gray-white masses. Following removal of a portion of the crural fascia, the right leg recovered fairly well, but the muscles of the left were almost completely gangrenous and had to be excised.

Horn[23] described two further cases in which the same condition had developed in young healthy soldiers after prolonged marching. Phalen[40] related the details of four cases of ischemic necrosis of the anterior crural muscles incident to trauma. In three of these fracture of the tibia and fibula had been attended by necrosis of pretibial muscles, and in a fourth painful swelling and paralysis of the pretibial muscles of one leg had occurred in a young man who had walked a half mile. The case described by Pearson et al.[38] was that of a young Greek boy in whom the condition developed bilaterally after strenuous broad jumping on a concrete pavement. Hughes[25] reported three additional cases, and Carter et al.[10] contributed data on nine more. In all of the latter cases the pretibial syndrome had occurred in one leg; in six of these the condition followed exertions of a football game, in one case marching, and in two a transfusion of blood into the leg veins (internal saphenous specified in one case). Tillotson and Coventry[49] added one case, and many others have been reported in recent years.

CLINICAL FEATURES

The clinical features of the recorded cases have been remarkably uniform. Usually the patient is a young adult male who participated in vigorous running, jumping, or walking for long distances without previous training. Sometimes pain in the pretibial regions of one or both legs begins during the exercise or follows it within a few hours. Aching pain, firm swelling, and paralysis of the pretibial muscles comprise the chief symptoms. Erythema, glossy edema, and heat are noted in the overlying skin. Low grade fever, neutrophilic leukocytosis, and albuminuria occur in some cases. The tibialis anterior, extensor digitorum longus, and extensor hallucis longus are weakened, and attempts to flex or extend these parts passively cause discomfort. Carter *et al.*[10] noted paralysis of the extensor digitorum brevis and an area of sensory loss on the dorsum of the foot between the first and second toes in several of their cases. The latter is due to implication of the anterior tibial nerve. The peroneal, gastrocnemius, and intrinsic muscles of the foot are spared. The only paralysis is of dorsiflexion of the foot and toe. Foot-drop is less than would be expected in complete paralysis of these muscles from a nerve lesion.

Arterial pulsations of the foot are retained, and Carter *et al.*[10] could demonstrate no abnormality of the blood vessels by arteriography of the leg in five cases. The affected muscles do not respond to either faradic or galvanic stimulation, and no electrical activity is seen in the electromyogram.

The pain, dermal erythema, and systemic symptoms gradually subside in the course of a few days, but the paralysis may persist. In our case there was no return of muscle contraction at the end of a year.[38] We were told of two soldiers who went into circulatory collapse and died—apparently from renal failure—after attempts to manipulate the paralyzed limbs during the acute phases of the illness. Postmortem findings were not available. Myoglobinuria and collapse or lower nephron nephrosis may have been responsible for such fatalities. This sequence of events also may have operated in the two remarkable cases of Robinson[42] in which scattered muscle necroses were associated with posttraumatic uremia.

The morbid anatomy of the muscles of the pretibial syndrome has been studied at the operating table and in biopsies. When the anterior crural fascia is incised, the acutely swollen muscles bulge into the wound. The muscle tissue varies in appearance according to the age of the disease. During the first days the muscles are soft and friable, but soon some parts become reddish gray or grayish white. After some weeks the muscles are partially or completely replaced by white, fibrous tissue.

HISTOLOGIC CHANGES

Microscopic pathologic changes were portrayed by our case. Several small pieces of the anterior tibial muscle were examined on the 17th day of the illness. Some of these pieces contained apparently normal muscle fibers, whereas in others the normal architecture of the muscle was severely disrupted and there was marked fibrous tissue replacement (Fig. 9–5A). In some areas the muscle fibers were observed in all stages of degeneration. Some were swollen, intensely acidophilic, and devoid of nuclei and striations (Fig. 9–5); others were granular and vacuolated and their nuclei shrunken and pyknotic. Extravasated red corpuscles and hemosiderin-containing macrophages were conspicuous in several places. In one fasciculus, hemorrhage separated both necrotic and regenerating fibers. Some fibers were surrounded and invaded by macrophages. All that remained of many muscle fibers were the sarcolemmal sheaths. Many of these were partially collapsed and contained granular clumps of acidophilic material that was being phagocytized by macrophages.

At the borders of the lesions extensive regenerative activity was evident (Figs. 9–5 and 9–6). Here the muscle fibers were small and contained large numbers of well stained nuclei. Longitudinal splitting was commonly observed in new fibers, and the sarcoplasm had a well developed longitudinal striation. Abortive buds showing regressive changes were common (Fig. 9–6). The interstitial fibrous connective tissue was increased in amount and was present in the form of broad bands and slender strands that separated the remaining muscle fibers. The connective tissue was cellular; fibroblast nuclei were increased in number and size. There were, in addition to histiocytes, scattered lymphocytes and a few plasma cells in the endomysial connective tissue (Fig. 9–5B). Arterial walls were thickened, but no occluded vessels were seen.

The extensive necrosis of muscle fibers, hemorrhage, proliferation of fibrous tissue, and muscle regeneration indicate an acute destructive process. This condition can be distinguished from a poly- or dermatomyositis, trichinosis, acute suppurative myositis, and

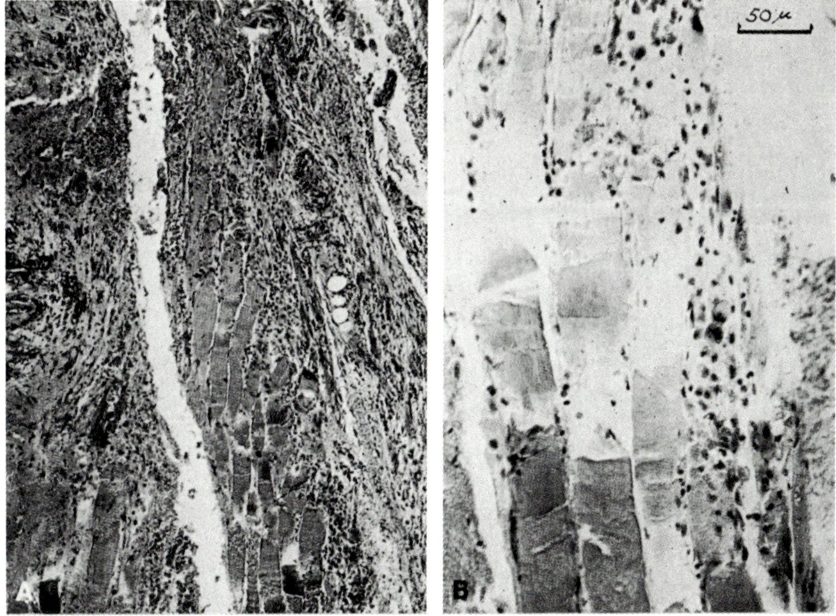

Fig. 9–5. Anterior tibial syndrome. (A) Center of a fasciculus is necrotic and hemorrhagic. At borders there is regenerative activity and fibrosis. (B) Higher magnification showing transition from hyaline degeneration and phagocytosis to regenerative activity. (phloxine, methylene blue)

Fig. 9–6. Anterior tibial syndrome. (A) Transverse section showing regenerating fibers separated by interstitial edema. (B) Longitudinal section of fibers in same state, ending in multinucleated buds exhibiting regressive changes having met hemorrhage and fibrosis. (phosphotungstic hematoxylin)

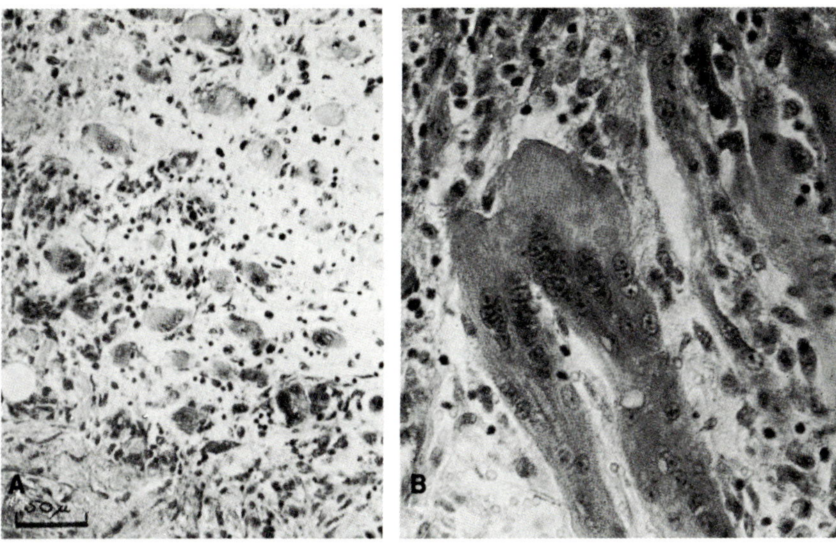

diffuse vascular disease (with which it can be confused clinically) by the extent of the damage to both parenchymal and interstitial elements and by the hemorrhage.

PATHOGENESIS

The pathogenesis of the pretibial syndrome is not known, but several interesting possibilities have been suggested. The most likely one is that strenuous exercise of an untrained muscle liberates excess quantities of metabolites, leading to temporary swelling of the muscle. The pretibial muscles—confined as they are within the rigid anterior tibial compartment formed by the interosseous membrane, the fibula and tibia behind, and the anterior tibial fascia in front— swell and increase the pressure within this compartment to such a degree that ischemic necrosis results. Hughes[25] pointed out that muscles may swell as much as 20% from prolonged activity. This swelling and hemorrhage in the tight pretibial chamber causes the circulation of the undamaged portions of the muscles to become embarrassed. This would explain why only pretibial muscles are affected in this way and why incision of the anterior crural fascia has proved beneficial. This explanation applies also to the two cases of Carter et al.[10] in which a transfusion into a leg vein with increased pressure due to extravasation of blood led to pretibial necrosis.

There are objections to this hypothesis on the grounds that all muscles in the pretibial compartment are not affected to the same degree, the extensor hallucis longus being most vulnerable and the extensor digitorum longus the least. We suggest that in milder degrees of the phenomenon those muscles nearest the venous exit from the compartment suffer least. However, these differing degrees of damage have not been demonstrated in most of the cases, and the full extent of the pathologic changes has not been ascertained. The suggestion of Horn[23] that there is traumatic thrombosis of the anterior tibial artery was refuted by the arteriographic studies of Carter et al.[10] Admittedly, vascular spasm or venous occlusion may play a part in some cases.

Mumenthaler et al.,[36] in a recent analysis of 103 cases, propose a broader view of pathogenesis of pretibial muscle neurons, since 40 of their cases were due not to simple exercise-induced swelling but to arterial ischemia, and 10 followed orthopedic surgery for lower leg fractures. This suggests that there is an overlap, especially in some of the fracture cases, with Volkmann's contracture (see below).

"SHIN SPLINTS"

The possible relation of the pretibial syndrome to the mild and relatively common affliction of muscles known as "shin splints" and to traumatic myalgia is of interest. Every athletic coach is well acquainted with the tenderness and pain of pretibial muscles that occur when runners, broad jumpers, and hikers begin to train. The affected muscles become firm and swollen, and any attempt at active or passive flexion and extension at the ankle is painful. The condition does not appear to differ from the milder stiffness and tenderness of any muscle following unaccustomed exercise. Although histologic studies of the overly taxed muscles have not been made, for obvious reasons one suspects necrosis of scattered fibers similar to that found in autopsy surveys of muscle, which would also explain the transient rises in serum creatine phosphokinase and adolase levels.

The cause of shin splints is assumed to be recurrent trauma of the unconditioned muscles. Rest and cold applications may be required in severe cases, but most athletes "walk off" their ailment.

DAMAGE TO MUSCLE AFTER EXPOSURE TO COLD

During World Wars I and II when large numbers of men were compelled to live and work under inclement conditions, the deleterious effects of lowered temperature on the extremities and the exposed parts of the body became quite manifest. Frostbite, trench foot, and immersion foot led to numerous casualties. The high incidence of injuries of this type provided new impetus to their investigation, and many clinical and experimental studies of these subjects were reported. These are summarized in several publications.[5, 13, 14, 50]

Frostbite is produced by the action of extreme cold on the exposed parts of the body, particularly the feet, hands, ears, and nose. The frozen parts become white and bloodless and later red, swollen, and very painful during the process of thawing. In mild frostbite only the superficial tissues are injured, with ulceration of epidermis, formation of blisters, and later desquamation. In severe degrees the part may become gangrenous. Degeneration of muscle fibers with waxy or hyaline changes (Zenker's type of degeneration) and fibrosis were reported by Smith et al.[47] in true frostbite.

Immersion foot or trench foot occurs usually after the feet have been immersed in cold water for a period of days. This was particularly common in men surviving calamities to ships or airplanes at sea, who were forced to sit for days in life rafts with their feet hanging in cold water. Coldness, numbness, and swelling of the feet were the usual symptoms, and edema, increased warmth, and cyanosis appeared soon after rescue. Bullae, hemorrhage, severe pain, and sensory impairment developed soon afterward. Gradual recovery usually took place. On the basis of a study of human biopsy material it appeared that the skeletal muscles were affected much less often than the epidermis and superficial tissues. There were two types of changes in muscles: (1) secondary atrophy of distal muscles due to degeneration of peripheral nerves (observed by Blackwood[3] and Friedman et al.[14]); and (2) severe degeneration of isolated muscle fibers or groups of fibers in a patchy distribution, manifesting the same alterations as those described in experimental cooling of a muscle (Chapter 3). Phagocytosis of dead muscle fibers, proliferation of sarcolemmal nuclei, and occasional muscle giant cells were seen. Fibrosis was a later reaction. Encapsulated infarcts sometimes occurred, resembling those of Volkmann's contracture.

Friedman et al.[14] attempted to reproduce this disorder in experimental animals. They exposed the hindlegs of rabbits to cold water (35–39°F) for 3–4 days, and the animals were sacrificed 4–135 days later. They found coagulation of the sarcoplasm of the muscle fibers with hyaline masses and clumps of granular material. There was interstitial edema and a moderate infiltration of neutrophilic leukocytes and lymphocytes. Proliferation of sarcolemmal nuclei was present by the fourth day after the experiment and became more pronounced with regeneration of muscle fibers in succeeding days. Within the space of 4 months, recovery was complete and the muscles appeared to be normal.

It is concluded that in immersion foot the essential change is a disturbance of circulation, probably due to slowing and stagnation of blood. Some of the changes within skeletal muscles, especially the patchy necrosis and encapsulated infarcts, must be on the basis of a partial ischemia. Yet there can be little doubt that cold may damage both nerve and muscle directly and that neural atrophy can also complicate the picture.

MUSCLE NECROSIS IN CARBON MONOXIDE ASPHYXIA

Muscle necrosis after carbon monoxide asphyxia is included here because we believe it to be due largely to the result of pressure ischemia and not to the specific effects of carbon monoxide. Hedinger[20] described two cases of necrosis of muscle secondary to severe carbon monoxide poisoning and collected 12 somewhat similar cases from the literature. Evidently muscular involvement, a rare complication of this type of poisoning, may appear at a variable interval after the stage of acute toxicity. Myocardial necrosis has been mentioned as another complication. The lesions are usually limited to the muscles of one or more extremities. Severe swelling and pain are noted in the paralyzed part. Ulcerations may develop in the overlying skin, and there is often evidence of a peripheral neuropathy. In one of the cases described by Hedinger there was mild edema of both forearms 20 hours after the poisoning, and by the fifth day this had progressed to a painful swelling of both arms. By the tenth day the edema had subsided but a slowly progressive fibrosis with contractures ensued. After 3.5 months scar tissue had formed within the affected muscles. Fibrous contractures resulted, and when these were freed surgically, a muscle biopsy showed "chronic myositis" with many muscle fibers in a state of waxy degeneration and enclosed in a mass of dense fibrous tissue. Peripheral nerves (sciatics, peroneals, etc.) may also suffer compressive injury during asphyxial coma and result in denervation atrophy of corresponding muscles.

HISTOLOGIC CHANGES

Such muscles, when examined early, have been described as yellow brown or grayish. This color is characteristic of necrotic muscle. There are a variety of degenerative microscopic changes (waxy, vacuolar, hyaline, and fatty) in the muscle fibers, with large or small hemorrhages, leukocytic infiltrations, active phagocytosis, proliferation of sarcolemmal nuclei, and the early appearance of calcification. The damaged muscle tissue later becomes enveloped in dense scar tissue which progressively shrinks, producing the fibrous pseudocontractures seen clinically. Calcific masses in the scar tissue may become evident on x-ray study.

It seems likely that this entire picture may be secondary to prolonged application of pressure. (One of the patients described by Hedinger was unconscious for several hours and apparently lay on her arms during this time.) The skin ulcerations and subsequent edema and contractures are similar to those often observed in Volkmann's ischemic necrosis. In these conditions muscle necrosis is the primary disorder and is responsible for the subsequent edema, cellular reaction, and replacement fibrosis. It is possible that the combination of carbon monoxide with the myohemoglobin interferes with oxidative enzymes of muscle tissue and renders it more susceptible to pressure and ischemia. Anoxemia, if sufficiently severe, may cause necrosis of muscle. We have observed lesions of this type in the heart and skeletal limb muscles of humans who survived for some 24–48 hours after being subjected to prolonged anoxemia or circulatory collapse. Similar lesions may be found in limb muscles in individuals who have been insensate for several days because of a major cerebral disease, especially if

they have been kept immobile under a hypothermic blanket. This is the subject of a report by Penn et al.[39] who noted myoglobinuria also. The alcoholic patient may injure his muscles in the same way.[28]

MUSCLE NECROSIS FROM ELECTRICAL INJURIES

High voltage current including lightning bolts may cause intense destructive lesions of many tissues including muscle. The lesions are at the site of and along the pathway of the current. The expected reactions to regional myonecrosis occur. Liberation of myoglobin may lead to fatal myoglobinuria and lower nephron nephrosis.

CIRCULATORY DISEASES

Owing to the rich collateral circulation of muscle tissue, infarction is relatively infrequent as a primary disease process. For example, the main artery to a limb may be ligated without critical ischemia of any part of that limb, as long as the circulatory system of the body is healthy. If, however, the heart is weak, the peripheral arteries are atherosclerotic, or thrombosis is so extensive as to interfere with collateral circulation, gangrene occurs. Also the final effect of arterial occlusion is modified by venous obstruction (Chapter 3).

The circumstances under which ischemic necrosis of healthy muscle usually occurs are as follows: (1) occlusion of the main artery to a limb by embolism, thrombosis, or other less well defined means such as the fracture and tight splinting leading to Volkmann's contracture; (2) thrombosis of multiple peripheral arteries and veins as in atherosclerosis; (3) thrombosis of multiple intramuscular vessels as in polyarteritis nodosa; (4) swelling and hemorrhage in certain muscle groups enclosed in rigid compartments such as the pretibial muscles, in which traumatism appears to result in extensive ischemic necrosis.

GANGRENE OF AN EXTREMITY

The most characteristic example of gangrene is that which results from slow and gradual occlusion of an artery to the foot and leg by atherosclerosis and thrombosis. The pathologic findings may not be different, however, from the more abrupt occlusion by an embolus. The tissues deprived of blood become dry and mummified. The skin is wrinkled and soon becomes discolored owing to hemolysis of red blood corpuscles, at first yellowish or greenish yellow, and finally black from the sulfide of the iron formed.

This phenomenon is accompanied by coldness of the part, neuralgic pains, numbness and paresthesias, and paralysis of movement. The muscle tissue is found in a state of massive necrosis. The necrotic tissue retains its natural color for a time; but soon it becomes hard and friable with patches of yellowish green discoloration, and at the margin where some circulation is preserved it is a mottled red and yellow. Some foci of rose color appear later. The yellow coloration represents destroyed muscle, and the rose color regenerating muscle. The tissues become granular and lose their staining reactions. If the gangrene is dry, little or no cellular reaction to the necrosis occurs. The muscle fibers appear to be congealed and may preserve a recognizable form as a sequestrum for a period of weeks to months.

HISTOLOGY

The early stages of ischemia in human muscles following damage to limb arteries was reviewed by Scully and Hughes.[44] In microscopic sections the fragile fibers tend to be fragmented. There is considerable variation in the staining reaction; in places the cross striation is clear, but elsewhere there is no evidence of fibrillary structure. The sarcolemmal nuclei are dark and shrunken, and after a few weeks they disappear completely. Discoidal degeneration is common. If the gangrene is moist, meaning that the circulation is at least partially restored, the typical reactions to necrosis appear (Fig. 9–7). Here, as at the margins of the zone of dry

Fig. **9–7.** Infarct necrosis of muscle. Muscle fibers are hyalinized, eosinophilic, and widely separated by edema. Neutrophilic leukocytes infiltrate endomysium. (phloxine, methylene blue)

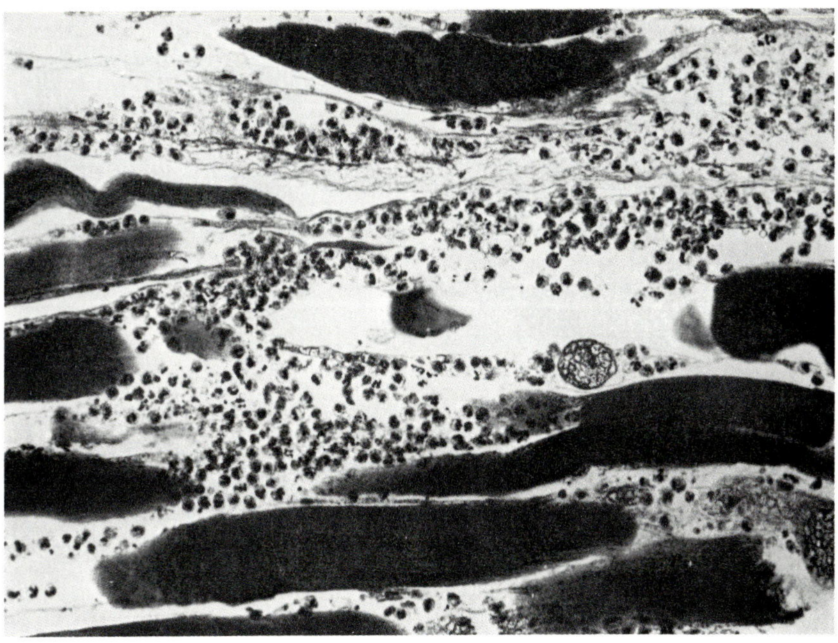

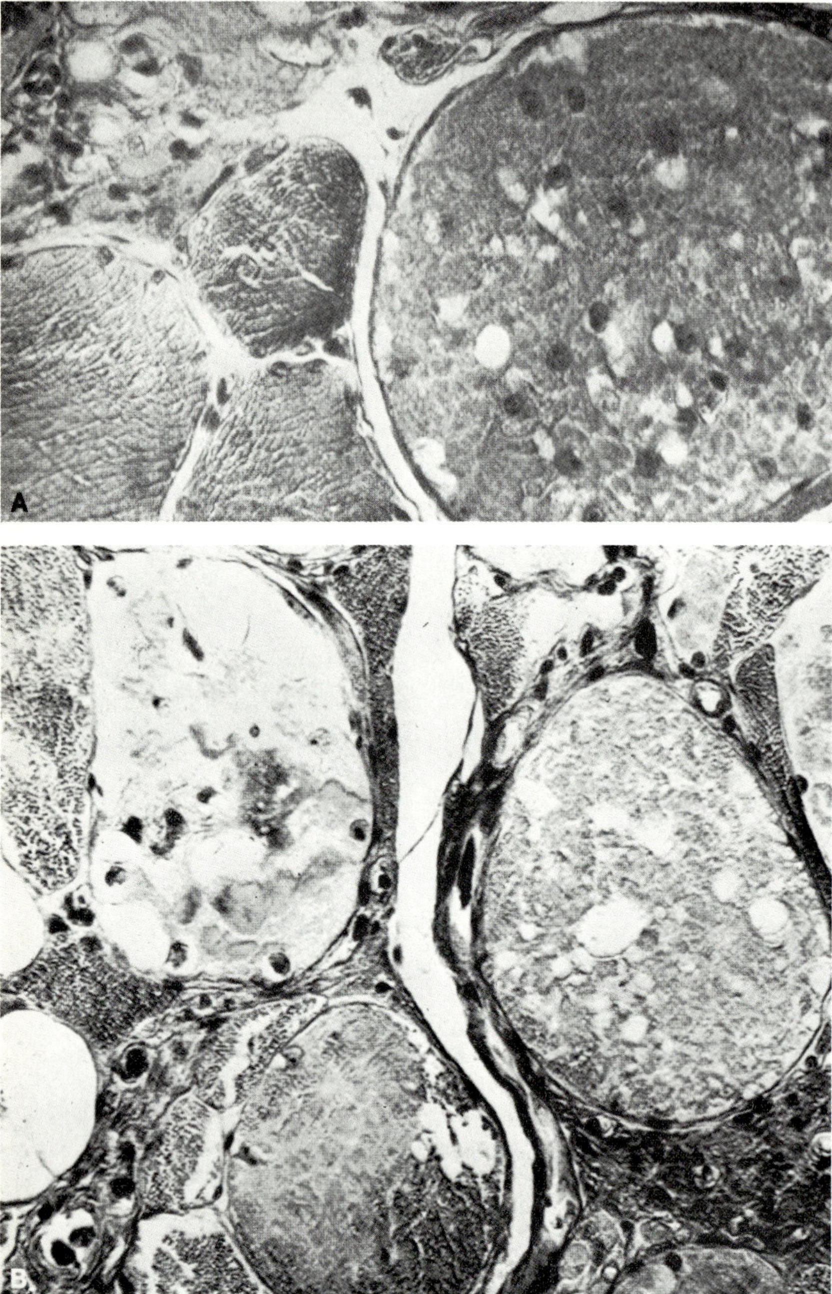

Fig. 9–8. Early infarction of gastrocnemius muscle in patient with thrombosis of femoral artery. Note swelling of fibers, flocculation of sarcoplasm, and invasion by macrophages. (toluidine blue, eosin)

gangrene, the striations of the muscle fibers disappear within a few days, and the fibers acquire a waxy or hyaline appearance. Sarcolemmal nuclei become shrunken and pyknotic, and finally disappear. Neutrophilic leukocytes invade the dead tissues, and within 48–72 hours macrophages and histiocytes invade and begin to remove the damaged fibers (Fig. 9–7). Some of the surviving fibers may have swollen enormously (i.e., become edematous) and exhibit a striking vacuolation. This is illustrated in Figure 9–8, which shows a section taken from the gastrocnemius muscle of an amputated leg (thigh amputation with femoral artery thrombosis). Fibroblasts, which are more resistant to ischemia than the muscle fibers, also proliferate and add to the cellularity around blood vessels and between any surviving muscle fibers.

The sarcolemmal nuclei of portions of muscle fibers which escape injury multiply and form regenerating buds. New fibers are formed at the edge of the infarct in this manner, or within the damaged muscle by an embryonal type of proliferation of surviving sarcolemmal nuclei (Fig. 9–9). The young fibers can be distinguished from the old ones by their small size and bluish coloration in hematoxylin and eosin stains.

There are irregularly shaped foci of muscle fiber degeneration in parts of the muscle which lie adjacent to a massive infarct. In these foci, which are usually small, the muscle fibers rapidly disappear and are replaced by vascularized connective tissue. Muscle fibers undergoing atrophy or degeneration can be found in or near the islands of connective tissue. In these lesions the small blood vessels have thickened walls due in part to hyperplasia of endothelial cells; it is impossible to determine whether this is related to a primary disorder of the vessel wall or to a secondary reaction of the vessels to necrosis and parenchymal atrophy.[4]

Intramuscular nerves are also affected.[7] Fragments of degenerating nerve fibers can be found in some nerve trunks, while others con-

Fig. 9–9. Extensive embryonic type of regeneration of muscle fibers, with some budding, following infarct necrosis. Regenerated muscle fibers are easily recognizable by their basophilia, which contrasts with remaining acidophilic hyalinized fibers. (phloxine, methylene blue; ×230)

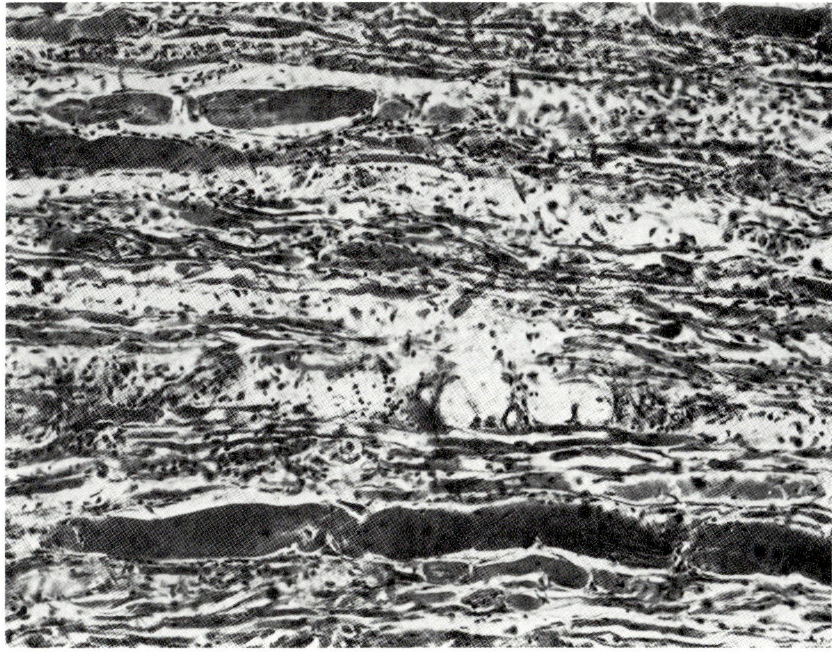

tain no identifiable fibers. In older specimens there are thin regenerating axons with end-bulbs either in the nerve trunks or wandering freely in the connective tissue sheaths, as if in search of muscle fibers.

In some muscles whose blood supply has been compromised by vascular occlusion massive necrosis does not occur. Instead there is either a diffuse interstitial fibrosis or multiple foci of muscle necrosis with replacement fibrosis. The muscle in these conditions is less conspicuously altered and to the naked eye either appears diffuse pale red gray or is flecked by foci of yellowish green discoloration. This state of diffuse interstitial fibrosis has a characteristic appearance under the microscope. Individual muscle fibers or fascicles of muscle fibers are separated by bands of connective tissue, the cellularity and vascularity of which decrease with the passage of time. The muscle fibers retain a clear striation and nuclei a healthy appearance. Hyalinized or vacuolated fibers may be observed. Occasional groups of muscle fibers all of a uniform small size suggest denervation atrophy. Extravasation of blood may or may not be in evidence. Scattered collections of lymphocytes and histiocytes infiltrate the perimysium and surround blood vessels. Although there is seldom opportunity to study the various stages in the development of this process, it may be supposed that initial edema and congestion of blood vessels cause both diffuse alteration of muscle and hyperplasia of connective tissue. The multiple foci of ischemic necrosis and fibrosis have the same general appearance .as those which lie near a massive infarct of muscle.

Both the diffuse interstitial fibrosis and multiple focal necrosis and fibrosis represent the effects of lesser degrees of ischemia. The same diffuse interstitial fibrosis has been observed after traumatic swelling and hemorrhage in muscles of the anterior tibial compartment (the pretibial syndrome), and after a traumatic occlusion of the femoral vein.[7] Multiple foci of ischemic necrosis and fibrosis have been found in cases with both arterial and venous occlusion of the leg vessels, and after trauma in which hemorrhage and swelling may have resulted in ischemia by compressing intramuscular blood vessels. In gangrene of an extremity muscle necrosis is much less extensive than is that of the skin and subcutaneous tissues.

ISCHEMIC NECROSIS OF ISOLATED MUSCLES

The arrangement of intramuscular blood vessels differs from one muscle to another. Some, such as the gastrocnemius, the long head of the biceps, and the anterior tibial muscles, depend on a single large nutrient artery for their blood supply. If in the event of injury this artery becomes thrombosed, a considerable portion of the muscle undergoes infarct necrosis. These peculiarities of vascular supply probably account for the massive necrosis of isolated muscles that sometimes follows injuries to the arm and leg.

The pathologic changes are similar to those described under *Gangrene of an Extremity,* above. However, the extent of necrosis is less, and regeneration of muscle and corresponding recovery of contractile power are more complete.

THROMBOSIS OF INTRAMUSCULAR VESSELS

The disease in which occlusion of multiple intramuscular blood vessels occurs most characteristically is polyarteritis nodosa (Chapter 7). Similar lesions are observed rarely in atherosclerosis and giant cell arteritis (temporal arteritis).

ARTERIOSCLEROSIS OF INTRAMUSCULAR VESSELS

Generalized arteriosclerosis and hypertension may affect the intramuscular arteries. In cases of longstanding severe hypertension, particularly of the malignant variety, the small arteries and arterioles usually exhibit some degree of hyperplastic change. However, necrotizing arteriosclerosis with "smudging" or fibrin impregnation of the vessel wall, or an infiltration of inflammatory cells, seldom occurs even in the most malignant phases of hypertension and nephrosclerosis. This aspect of the subject was explored by Kernohan *et al.*[26] and Barker *et al.*[1] Biopsies of the pectoralis major and deltoid muscles from 150 patients with hypertensive vascular disease were examined. In a large proportion of this series there was a thickening of the walls of the arterioles due chiefly to hypertrophy and hyperplasia of the muscularis coat and to a lesser extent of the intima and adventitia. The lumen was narrowed, and the ratio of the thickness of the vessel wall to the diameter of the lumen was increased. Proliferation of subendothelial fibroblasts was noted in some cases. Changes of this type may be found in any skeletal muscle. Occlusion of small arteries and arterioles has not been observed in our material, and there are no significant alterations of the muscle fibers. However, capillary structure is known to be altered in diabetes mellitus and aging. The main alteration is a thickening of the basement membrane.[11, 27]

Atherosclerosis seldom extends to the intramuscular arteries, and when it does only the larger arteries are involved. Wartman,[53] who attempted to determine the distribution of atherosclerosis in the different organs, observed atheromatous deposits within the arteries of skeletal muscles infrequently. This is explained by the fact that most of the intramuscular arteries are small and it is of course a common observation that atheromatous plaques are rarely seen in blood vessels less than 250 μ in diameter.

The use of muscle biopsy as a means of determining the nature of a diffuse vascular disease therefore has serious limitations. While affording in some instances conclusive evidence of polyarteritis nodosa and often indicating the presence of hyperplastic arteriolosclerosis, it is not likely to divulge the presence of the necrotizing arteriolitis of malignant hypertension or of atherosclerosis. In reviewing the reports of other pathologists we have been impressed with the frequency of erroneous interpretations of arterial narrowing when, in fact, the vessels were normal to our eye. Also faulty sectioning and redundancy of the endothelium may give an illusory appearance of thrombosis.

VOLKMANN'S ISCHEMIC CONTRACTURE

In 1872 Richard von Volkmann[52] published an account of a rapidly progressive posttraumatic muscle contracture. He was the first to distinguish between paralysis of muscles due to nerve injury and that due to traumatic lesion of the muscles. At first he regarded this syndrome as an "inflammatory myositis," but in later papers he stated that these contractures were invariably caused by a stoppage of arterial blood due to tight bandages and splints. In 1884 Leser[29] reported seven similar cases and presented a clear account of the clinical aspects of this condition. The term Volkmann's contracture was introduced by Hildebrand[21] in 1890. Since then many full descriptions and additional case reports have appeared in the medical literature; two of the most interesting that can be recommended to the student of pathology are those of Brooks,[8] an experimental study of muscle ischemia, and of Griffiths,[17] who made this the subject of his Hunterian Lecture to the Royal College of Surgeons in 1940.

CLINICAL ASPECTS

The clinical aspects of Volkmann's contracture are too well known to be detailed here. In brief, the first change, which usually begins a few hours to as late as several days after receipt of the injury, consists of a burning pain in the distal part of the extremity (hand or rarely foot) associated with swelling, cyanosis, and sometimes pallor. The pain is especially severe on the attempted passive extension of the fingers. The radial pulse is usually absent during the first days but may return later. The fingers are cool. Within a few days the fingers and sometimes the wrist begin to assume a flexed position due to firm shortening of forearm muscles. Moreover, these muscles are incapable of full voluntary, reflex, or electrical excitation. Once established, the condition is for the most part permanent. After 2–3 months there is fibrous transformation of the muscles, subcutaneous tissue, tendons, and synovial sheaths, all of which have become fused into a continuous mass of fibrous tissue. The pain, accompanied by the subjective sensations of burning, tingling, and formication, persists. Deep pressure, heat, and cold may be unusually painful, whereas cutaneous sensation may be impaired.

Volkmann's contracture is rare, having occurred in only 8 of 21,000 fractures reported to the Organized Fracture Clinics of England.[17] The condition nearly always follows a fracture but has been observed after injury to soft parts alone. Fracture of the lower end of the humerus is most often responsible for this condition; next in order of frequency are fractures of the forearm. The largest number of cases occurred between the ages of 2 and 18 years. Thomson and Mahoney[48] reported on 42 cases observed in a large children's hospital in Toronto during the years 1931–1947. The most interesting of these were nine cases of acute ischemia of the lower limb following simple fracture, usually of the femur, in which there was no evidence of damage to the femoral vessels.

HISTOLOGIC CHANGES

The morbid anatomy, as usually seen in the later stages of the disease, is fairly uniform. The muscle tissue no longer looks like muscle at all. It has been converted to a hard cylindrical mass of whitish yellow or greenish yellow tissue adherent by fibrous adhesions to adjacent structures. It cuts with difficulty. Usually small amounts of muscle tissue are still visible. Microscopically, by the time the subacute or chronic stage is reached, islands of necrotic fibers devoid of living nuclei are surrounded by a capsule of connective tissue. In the latter and surrounding these sequestra of muscle fibers there is usually a zone of cellular activity with proliferating fibroblasts and active macrophages. Syncytia of sarcolemmal nuclei (muscle giant cells) are present here. In still later stages the muscle tissue is largely or entirely replaced by collagenous connective tissue.

PATHOGENESIS

The pathogenesis of Volkmann's ischemic contracture is still being argued. Many workers in the past and some at the present time maintain that arterial occlusion is the primary factor. Nerve injury, which coexists in a certain percentage of cases, is not responsible. The experimental findings of Brooks[8] (Chapter 3), which were confirmed by Middleton,[35] led to the conclusion that venous obstruction was more important than arterial. Griffiths[17] criticized these experiments and rejected the notion of venous obstruction because the typical pathologic picture of ischemic contracture was not reproduced. He was of the opinion that arterial occlusion could be demonstrated in every case and reported a similar fibrous pseudocontracture after embolism of an extremity. We can conclude only that the pathogenesis of Volkmann's contracture has not been settled. Experimental venous occlusion in animals has a more damaging effect on muscle than does arterial occlusion (Chapter 3). However, there is no doubt that in most of the biopsies the lesions resemble those produced by experimental occlusion of nutrient arteries. Unfortunately the blood vessels in such lesions have not been studied with the care that a primary vascular lesion merits. Attempts to elucidate the clinical condition by studying experimental lesions in animals have obvious shortcomings. There is need of further investigation of this subject.

CONCLUSION

It is clear that there are many features in common between the lesions in muscle provoked by trauma and by vascular occlusion. In each instance the pathologic process is unselective as compared to that induced by toxic and metabolic disorders. Pannecrosis of whole regions of the muscle is frequent. Not only the muscle fibers but also connective tissue sheaths, blood vessels, nerves, and sensory organs all suffer to a degree proportionate to the intensity of the etiologic factor. Extravasation of blood, either in the form of large hemorrhages or petechiae, is a frequent accompaniment of each pathologic process. In the acute stages the histologic picture is notable for its pleomorphism. Tissue destruction of variable extent is associated with active phagocytosis of degenerate muscle fiber and red blood corpuscles. Migration of neutrophilic leukocytes, as a reaction to necrosis, adds to the complexity of the cellular picture. Multinucleated muscle buds and bands of sarcolemmal nuclei represent attempts at a regenerative process that are doomed to failure because of the magnitude of the lesion. In the more chronic stages the reparative process is remarkably heterogeneous. Broad fields and brands of connective tissue which replace the destroyed fibers have lost completely their orientation to the surviving muscle. This is the result of damage to the endomysial tubes (Chapter 3). In the course of time, especially in the fibrosis which follows hemorrhage, conditions become appropriate for the deposition of lime salts, and myositis ossificans is likely to occur.

The factor of physical disruption and ischemia are to some extent inseparable. In every case of severe traumatism there is some degree of vascular damage which leads to thrombosis and infarction or of collapse of small thin-walled blood vessels due to tissue swelling. Similarly, in vascular occlusion acute tissue swelling may compress and damage healthy fibers in areas in which the blood supply would be sufficient to maintain the viability of the tissue. Therefore on the basis of a small biopsy it may be impossible to distinguish between infarct necrosis and traumatic necrosis.

There is still uncertainty as to the comparative effects of venous and arterial occlusion on human skeletal muscle. Meager pathologic data suggest that venous occlusion is more likely to produce a diffuse interstitial fibrosis, whereas arterial occlusion may have no effect whatsoever, or it may result either in massive necrosis or multiple small foci of ischemic damage, depending on the angioarchitecture of the involved muscle. This is in keeping with the experimental findings of Brooks.[8] Since all these conditions lead to a considerable proliferation of fibrous tissue, with resulting contracture, the relative importance of the several factors involved varies from case to case, especially in the condition called Volkmann's contracture.

REFERENCES

1. BARKER NW, KEITH NM, KERNOHAN JW: Arterioles in hypertensive disease. Mayo Clin Proc 6:172–174, 1931

2. BINNIE JF: On myositis ossificans traumatica. Ann Surg 38:423–440, 1903

3. BLACKWOOD W: Studies in the pathology of human "immersion foot." Br J Surg 31:329–350, 1943–1944

4. BLACKWOOD W: A pathologist looks at ischemia. Edinburgh Med J 51:131–143, 1944

5. BLACKWOOD W, RUSSELL H: Experiments in the study of immersion foot. Edinburgh Med J 50:385–398, 1943

6. BLOMFIELD LB: Intramuscular vascular patterns in man. Proc Soc Med 38:617–618, 1945

7. BOWDEN REM, GUTMANN E: The fate of voluntary muscle after vascular injury in man. J Bone Joint Surg [Br] 31B:356–368, 1949

8. BROOKS B: Pathologic changes in muscle as a result of disturbances of circulation: an experimental study of Volkmann's ischemic paralysis. Arch Surg 5:188–216, 1922

9. CAREY EJ: Multiple bilateral traumatic periosteal bone and callus formations of the femurs and left innominate bone. Arch Surg 8:592–603, 1924

10. CARTER AB, RICHARDS RL, ZACHARY RB: The anterior tibial syndrome. Lancet 2:928–934, 1949

11. DANOWSKI TS, KHURANA RC, GONZALEZ AR, ET AL: Capillary basement membrane thickening and pseudodiabetic myopathy. Am J Med 51:757, 1971

12. FAY OJ: Traumatic periosteal bone and callus formation: the so-called traumatic ossifying myositis. Surg Gynecol Obst 19:174–190, 1914

13. FRIEDMAN NB: The pathology of trench foot. Am J Pathol 21:387–433, 1945

14. FRIEDMAN NB, LANGE H, WIENER D: Pathology of experimental immersion foot. Arch Pathol 49:21–26, 1950

15. GESCHICKTER CF, MASERITZ IH: Myositis ossificans. J Bone Joint Surg 20:661–674, 1938

16. GILCREEST EL: Rupture of muscles and tendons, particularly subcutaneous rupture of biceps flexor cubiti. JAMA 84:1819–1822, 1925

17. GRIFFITHS DL: Volkmann's ischemic contracture. Br J Surg 28:239–260, 1940

18. HACKENBRUCH P: Ueber interstitielle Myositis und deren Folgezustand, die sogennante rheumatische Muskelschwiele. Beitr Klin Chir 10:73–101, 1893

19. HASSE GR: Ossification of certain muscles (abstract). Am J Med Sci 11:501, 1832

20. HEDINGER C: Muskelveränderungen bei Kohlenmonoxydvergiftung: ihre Bezeihungen zum Ver-

schüttungssyndrom. Schweiz Med Wochenschr 78:145–151, 1948

21. HILDEBRAND O: Ein Fall von geheilter, auf Ischämie bewehender Muskelcontractur. Dtsch Z Chir 30:98–100, 1890

22. HIRSCH EF, MORGAN RH: Causal significance to traumatic ossification of the fibrocartilage in tendon insertions. Arch Surg 39:824–837, 1939

23. HORN CE: Acute ischemia of the anterior tibial muscle and the long extensor muscle of the toes. J Bone Joint Surg 27:615–622, 1945

24. HOWARD C: Traumatic ossifying myositis. US Naval Med Bull 46:724–730, 1946

25. HUGHES JR: Ischaemic necrosis of the anterior tibial muscles due to fatigue. J Bone Joint Surg [Br] 30B:581–594, 1948

26. KERNOHAN JW, ANDERSON EW, KEITH NM: The arterioles in cases of hypertension. Arch Intern Med 44:395–423, 1929

27. KILO C, VOGLER N, WILLIAMSON JR: Muscle capillary basement membranes related to aging and diabetes mellitus. Diabetes 21:881–905, 1972

28. LAFORCE FM: Crush syndrome after ethanol. N Engl J Med 284:1104, 1971

29. LESER E: Untersuchungen über ischaemische Muskellähmungen und Muskelcontracturen. Samml Klin Vort 249:2087–2114, 1884

30. LORENZ H: Die Muskelerkrankungen, Specielle Pathologie und Therapie. Edited by H Nothagel. Vienna, Hölder, 1904, Vol XI, pp 1–727

31. MAKINS GH: Traumatic myositis ossificans. Proc Soc Med 4:133–142, 1911

32. MAYDL K: Ueber subcutane Muskel und Sehnenzerreissungen, sowie Rissfracturen, mit Berücksichtingung der analogen, durch directe Gewalt entstandenen und offenen Verletzungen. Dtsch Z Chir 17:306–361, 513–547, 1882

33. MCMASTER PE: Tendon and muscle ruptures. J Bone Joint Surg 15:705–722, 1933

34. MERCER W: Orthopaedic Surgery. Sixth edition. Baltimore, Duthie, R.B. Arnold, 1964

35. MIDDLETON DS: The pathology of congenital torticollis. Br J Surg 18:188–204, 1930

36. MUMENTHALER M, MUMENTHALER E, MEDICI V: Das tibialis anterior—Syndrom nach operationen am Unterschenkel. Arch Orthop Unfallchir 66:201–219, 1969

37. PACK GT, BRAUND RR: The development of sarcoma in myositis ossificans. JAMA 199:776–779, 1942

38. PEARSON C, ADAMS RD, DENNY-BROWN D: Traumatic necrosis of pretibial muscles. N Engl J Med 239:213–217, 1948

39. PENN AS, ROWLAND LP, FRASER DW: Drugs, coma and myoglobinuria. Arch Neurol 26:336–343, 1972

40. PHALEN GS: Ischemic necrosis of the anterior crural muscles. Ann Surg 127:112–120, 1948

41. RALIS Z: Muscle hemorrhage in babies born by breech. Arch Dis Child 45:709, 1970

42. ROBINSON GL: Scattered muscle necroses associated with post-traumatic uraemia. Lancet 1:799–800, 1950

43. ROBINSON NA: Spontaneous rupture of gastrocnemius presenting as thrombophlebitis. Arch Surg 38:385–388, 1972

44. SCULLY RE, HUGHES CW: The pathology of ischemia of skeletal muscle in man. Am J Pathol 32:805–829, 1956

45. SEVERIN E: Umwandlung des Musculus tibialis anterior in Narbengewebe nach Überanstrengung. Acta Chir Scand 89:426–432, 1943

46. SIRBU AB, MURPHY MJ, WHITE AS: Soft tissue complications of fractures of the leg. Calif West Med 60:53–56, 1944

47. SMITH JL, RITCHIE J, DAWSON J: Clinical and experimental observations on the pathology of trench frost-bite. J Pathol 20:159–190, 1915–1916

48. THOMSON SA, MAHONEY LJ: Volkmann's ischemic contracture and its relationship to fracture of the femur. J Bone Joint Surg [Br] 33B:336–347, 1951

49. TILLOTSON JF, COVENTRY MB: Spontaneous ischemic necrosis of the anterior tibial muscle: report of case. Mayo Clin Proc 25:223–227, 1950

50. UNGLEY CC, BLACKWOOD W: Peripheral vasoneuropathy after chilling: immersion foot and immersion hand. Lancet 2:447–451, 1942

51. VON MEYENBURG H: Die quergestreifte Muskulatur. Handbuch der Speziellen Pathologischen Anatomie und Histologie, Edited by F Henke, O Lubarsch. Berlin, Springer, 1929, Vol 9, pp 299–509

52. VON VOLKMANN R: Krankenheiten der Bewegungsorgane. Handbuch der Chirurgie. Edited by FJ Pitha, T Billroth. 1872, Vol 2, p 846

53. WARTMAN WB: The incidence and severity of arteriosclerosis in organs from 500 autopsies. Am J Med Sci 186:27–35, 1933

chapter 10

SPINAL AND NEURAL MUSCULAR ATROPHIES

Muscular weakness, atrophy, and loss of tendon reflexes occur in a variety of dissimilar diseases. Those which are progressive and systematized can be classified into three main groups: muscular, neural, and spinal. Primary muscular atrophy, usually designated as myopathy or muscular dystrophy, was discussed in Chapter 6. In spinal muscular atrophy the fault is in the anterior horn cells of the spinal cord, and in neural muscular atrophy the lesion is in the peripheral nerves or motor nerve roots. In spinal and neural atrophy the muscle wasting is secondary to degeneration of nerve fibers.

DISEASES OF THE SPINAL CORD (SPINAL MUSCULAR ATROPHY)

Destruction of the motor cells in the anterior horns of the spinal cord or motor nuclei of the brain stem with resulting atrophic, areflexic muscle paralysis is common to a number of conditions. The best known and most frequent of these are acute anterior poliomyelitis and "motor system disease" (amyotrophic lateral sclerosis and related diseases). Syringomyelia, syphilitic amyotrophy, intramedullary tumors, and traumatic and other destructive lesions of the gray matter of the spinal cord produce similar effects.

ACUTE ANTERIOR POLIOMYELITIS

Acute anterior poliomyelitis is the best example of a disease which leads to acute atrophic paralysis. The usual clinical picture comprises flaccid paralysis with loss of tendon reflexes of acute onset, often preceded by general symptoms such as fever, vomiting, diarrhea, sore throat, and headache. Signs of meningeal irritation are almost invariably present. Usually the paralysis is asymmetrical. Paralysis of respiratory muscles and disturbance of the respiratory regulating center of the medulla are the usual causes of death.

The pathologic changes of the skeletal muscles in cases of poliomyelitis have received relatively little attention. Descriptive details can be found in the publications of Kopits,[71] Bastos Ansart,[6] d'Harcourt and Mazo,[40] Wohlfahrt and Wohlfart,[108] Hipps,[63] Carey et al.,[21] Dublin et al.,[42] Denst and Neubuerger;[39] but only the latter reports, and those of Mittelbach,[80] and Drachman et al.,[41] which stress clinicopathologic correlations, are accessible to most American students.

The paralysis and subsequent atrophy of muscle are secondary to disease of the motor neurons. The degree of paralysis and atrophy depend on the severity of the disease, i.e., on the number of anterior horn cells that are altered or destroyed. In the acute stages of the disease the muscle fibers appear to be quite normal even though their motor neurons have been completely destroyed. There may be a few scattered infiltrations of inflammatory cells in the muscles, as with any muscle disease or even normal muscle, but they are rare and do not correlate with the tenderness of muscles. The reduction in size of the paralyzed fibers becomes evident within a few weeks after degeneration of the motor nerve fibers and end-

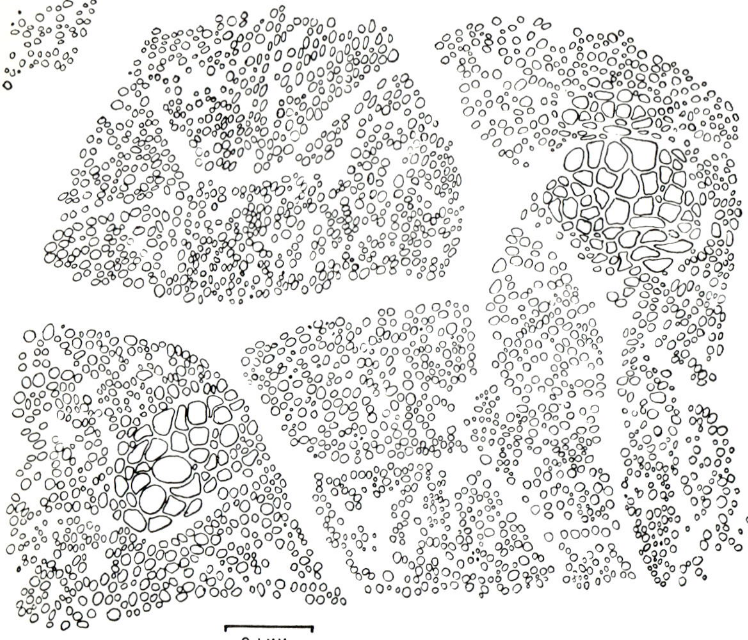

Fig. 10–1. Poliomyelitis, 6 weeks after acute phase. Camera lucida drawing from portion of a large section of pectoralis major muscle, showing normal muscle fibers only in two small groups (upper right and lower left). Muscle appeared to be completely paralyzed.

plates. The atrophy of the muscle fibers, as in all types of secondary muscle atrophy, is attended by an appreciable loss of myofilaments and a relative increase in sarcolemmal nuclei. The longitudinal and cross striations are retained. In the most severely wasted muscles all of the motor units may be atrophic, but commonly one or two isolated bundles of 20–100 intact muscle fibers indicate that some motor units have survived (Fig. 10–1). The muscle fibers retaining innervation may undergo hypertrophy (Fig. 10–3), and the number of fibers within intact motor units is increased owing to collateral regeneration of terminal nerve fibers.

Though all the paralyzed fibers become small, there is great variation in size. After 4 months the increase in sarcolemmal nuclei becomes more obvious, and irregular clumps of four or more muscle nuclei, darker than the rest, begin to form at fairly regular intervals along some of the muscle fibers (Fig. 10–2). After 6 months of paralysis short linear rows of shrunken dark nuclei may appear (Figs. 10–2B and 10–3A,C), with irregularly scattered rows of fat cells and occasional groups of small round cells (Fig. 10–3C). The rows of shrunken nuclei lie in very thin sarcolemmal tubes filled with unstriated sarcoplasm, but

after longer intervals are found lying free in the connective tissue. The process of fragmentation in the terminal stage of atrophy (Chapter 3) appears to occur in scattered fibers over large areas of the paralyzed muscles. Transverse sections present an even mixture of severely atrophic fibers, and nuclear remnants (Fig. 10–4). There is great variation in size of fiber, though all are small. In section the clumps of nuclei are very prominent (Fig. 10–4B), and the endomysial reticulum, though condensed, is not thickened. The replacement of fibers by fat cells appears to be progressive after 6 months of paralysis. Muscle fibers presenting only moderately advanced atrophy, with faintly striated myofibrils, continue to be found for many years, but their number progressively diminishes. With this exception, it is characteristic of poliomyelitis that all the muscle fibers are in much the same stage of atrophy, in contrast to the successive stages of atrophy present in many chronic types of multiple neuritis or progressive spinal muscular atrophy.

In most microscopic preparations of late poliomyelitic atrophy the larger muscle fibers (presumably normal) exhibit a variety of degenerative and regenerative myopathic changes. Mittelbach[80] illustrated "target fibers" among

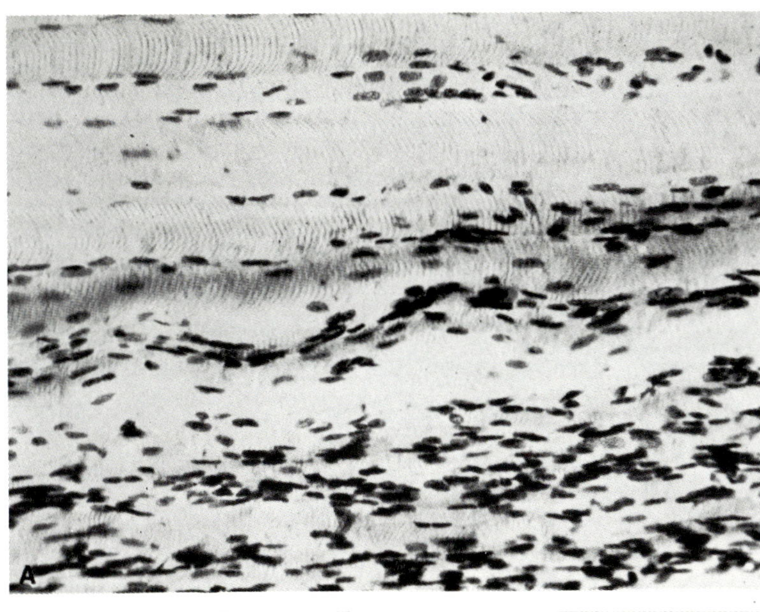

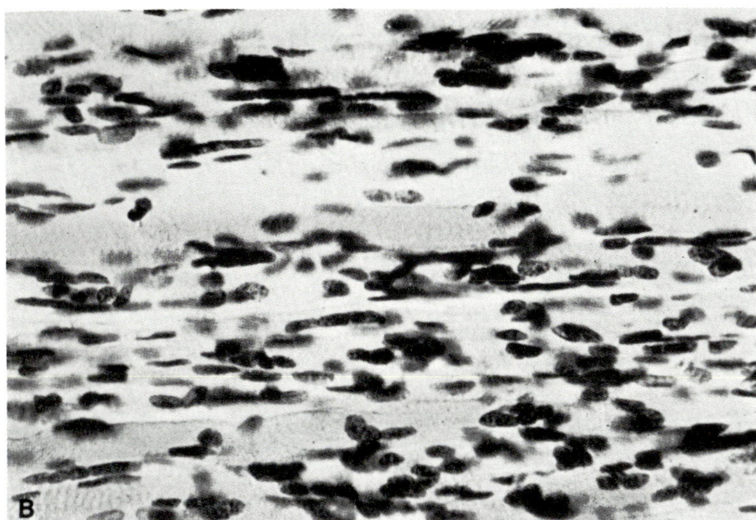

Fig. 10–2. Longitudinal section of muscle paralyzed by poliomyelitis for 4 months. In upper part of (A) there are several fibers of an intact motor unit. In lower part of (A) and in (B) muscle fibers, though striated, are of smaller caliber. (B) Note clumping of nuclei in some fibers, and short columns of shrunken nuclei in other very small fibers and in fragments of fibers. (hematoxylin, Van Gieson)

them and Drachman *et al.*[41] showed hyaline necrosis, phagocytosis, proliferation of sarcolemmal nuclei, basophilic sarcoplasm, and branching of regenerating fibers. This results in marked variation in the size of intact fibers; thus the composite picture comes to resemble progressive muscular dystrophy. In all probability these changes are related to trauma or overuse of weakened muscles for they do not occur in a disease such as Werdnig-Hoffmann's infantile muscular atrophy where degeneration of anterior horn cells occurs in a relatively immobile infant.

There is seldom much increase in fibrous connective tissue in relation to the shrunken muscle fibers (Fig. 10–4), and when fibrosis is prominent it may possibly be related to contracture or overstretching. Even then fibrosis affects the perimysium more than the endomysial sheaths of the atrophic fibers. The late degenerative changes in the small fibers, which occur after 6–12 months or longer, were described by Hipps,[63] who studied the microscopic pathologic changes in chronic cases by biopsy, and who believes that granular degeneration of muscle fibers is followed by fibro-

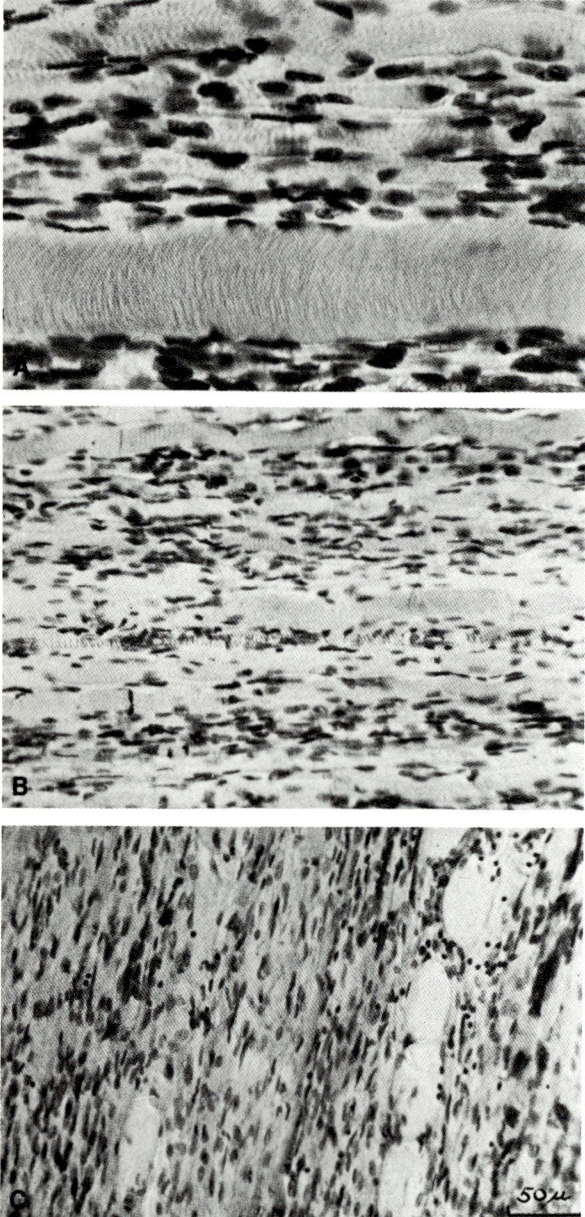

Fig. 10–3. (A and B) Longitudinal sections of muscles from patient with poliomyelitis after 4 months of paralysis. (3) Another patient after 6 months. (A) One large fiber remains among many atrophic ones. Long slender muscle fibers and chains of muscle nuclei are present. (C) Nearly all muscle fibers are very small, and in some degeneration is present. Fat cells are beginning to appear, and a few mononuclear cells and histiocytes are present. (hematoxylin, Van Gieson)

sis. In some of his sections fibrous tissue and fat cells had completely replaced the muscle fibers. He remarked also on transverse bands of connective tissue, which in his article were pictured as extending obliquely across muscle fibers. They were interpreted as representing a tear in muscle but are probably normal intramuscular aponeuroses.

The most severely atrophic muscles, biopsied after an interval of years, consist of loose areolar tissue in which occasional isolated muscle fibers and muscle spindles alone betray the original structure, as Kopits[71] showed. Regeneration of atrophic muscle fibers is a moot point. While observed by d'Harcourt and Mazo[40] and Kopits[71] in "abortive" form, they were not found by Hipps.[63] It appears that the final breaking up of slender muscle fibers in the process of degeneration, demonstrated by Hayem[62] in infantile paralysis long ago by teas-

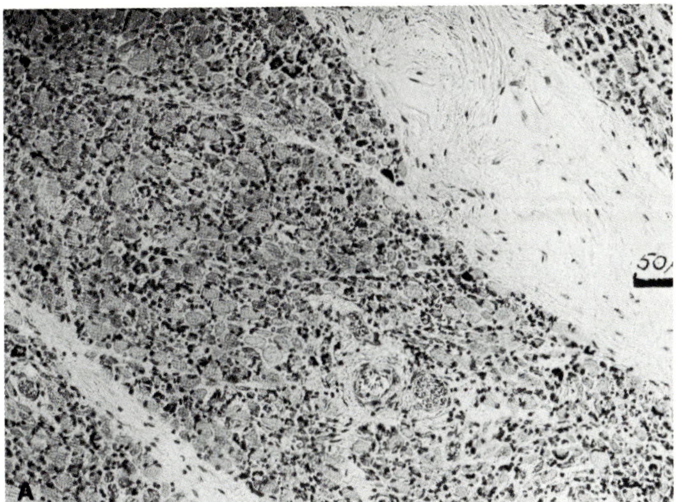

Fig. 10–4. Poliomyelitis of 6 months' duration. (A) Transverse section of gastrocnemius shows the great variation in fiber size, though all fibers are very small. Onset of degeneration is occurring in many fibers. (B) Same muscle at higher magnification. Semicircular nuclei are nuclear clumps in section. (hematoxylin, Van Gieson)

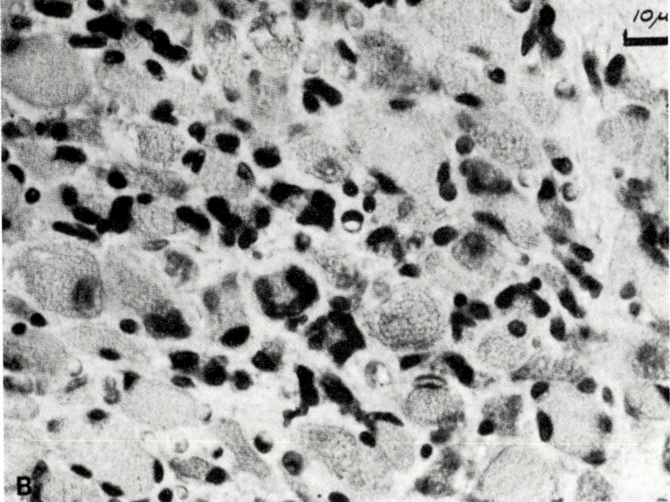

ing methods, has been misinterpreted as "regeneration" by the first three authors. The degenerative phase which ultimately follows poliomyelitis differs in no way from that of any other neural or spinal atrophy of sufficient duration. We have not observed regeneration in our own material except in a few of the large (hypertrophic), normally innervated fibers. Trauma to the delicate atrophic fibers during the first months of atrophy probably hastens the process of degeneration. The walls of the small blood vessels become thickened.

The changes in the peripheral nerve fibers and the motor end-plates were studied by Chor,[25] Horányi-Hechst,[66] and Carey et al.[21] Chor employed a modification of the Ranson technic for staining the fine terminal twigs of the motor nerve. His material included experimental animals in which a peripheral nerve had been sectioned or poliomyelitis induced, and humans dying of poliomyelitis. He found exactly the same changes in the motor nerve fiber and the motor end-plate after nerve section and after poliomyelitis. In the latter condition the changes were less uniform, a difference he attributed to the plurisegmental innervation of skeletal muscle.

Woolf and Till[110] took muscle biopsies from the "motor point" of the muscle—i.e., the site at which electrical excitability was maximal (entrance of nerve)—and studied the intramuscular nerves, their terminal expansions, and the subneural and subsarcolemmal apparatus. Degenerative changes were found as well as delicate collateral branches. Some muscle fibers had as many as three motor end-plates. These authors also interpreted the hyperneurotized fibers and extravagant plexuses as abortive regenerative changes. In the state they termed "chronic terminal motor neuropathy," it is implied that the earliest changes are in the terminal portions of the surviving axons.

Collateral regeneration from intact muscle nerves was observed in animals by Edds,[45] and Wohlfart[106] showed that it also occurs in man. Collateral nerve twigs invade the empty neurilemmal tubes of the degenerated nerve fibers and establish contact with the isolated muscle fibers at their original end-plate regions. Motor units in this way may increase greatly in size, accounting, in the opinion of Wohlfart, for the excessively large size of the action potentials of motor units in both poliomyelitis and in motor system disease and for the discrepancy between the degree of anterior horn cell degeneration and number of atrophic motor units. The resulting increase in the "terminal innervation ratio" has been quantitated by the biopsy method of Coërs and Woolf.[26]

Carey et al.,[21] using a special metallic impregnation technic (a Cajal modification of Ranvier's gold chloride) found that the earliest morphologic change in the motor nerve fibers in acute poliomyelitis was a ball-like retraction or an abnormal expansion of the terminal branches of the axis cylinders. This was followed within 36 hours by granular degeneration and dissolution of many of the end-plate fibers. In three patients dying of bulbar poliomyelitis Carey found almost complete degeneration of the fibers in respiratory muscles within 26 hours after the first neuromuscular symptoms. The terminal parts of the axon became moniliform and then fragmented. There were aggregates of cells which he had difficulty in identifying. The other morphologic changes were variable, being quite advanced in some fibers and not perceptible in others. Carey concluded that degeneration of the nerve fiber proceeded centripetally, quite unlike the degeneration which follows section of a peripheral nerve, in which he found that all parts of the entire

peripheral segment degenerated at the same time. He also observed swelling and loss of striation of some muscle fibers, proliferation of perivascular reticular connective tissues, an increase in mesenchymal and endothelial cells, passive hyperemia, and vascular proliferation in the form of endothelial sprouts, the formation of lipoblasts from cells in the reticular tissue, and an infiltration of lymphocytes.

We are skeptical of many of Carey's findings. He relied principally on a gold chloride stain—which we find to be unpredictable, especially on postmortem material. This stain is also not suitable for studying and identifying many of the cells in muscle. The morphologic changes in the finer nerve endings are probably due to a variation of the stain. This is borne out by the fact that he demonstrated the same changes in sensory nerve fibers and muscle spindles, which are known not to be affected by poliomyelitis. With silver impregnation methods in experimental poliomyelitis we found the degeneration of peripheral nerves and motor end-plates to be identical with that which follows section of anterior nerve roots. Absorption of the terminal ramification of the end-plates took 2–3 days, and fragments of axis cylinders in the peripheral nerve persisted for more than 1 week. Also, we doubt the existence of a significant degree of lymphocytic infiltration in muscles paralyzed by poliomyelitis. We have not observed them even when the virus was inoculated into muscle. Occasional foci of mononuclear cells (Fig. 10–3C) may be found in any atrophic muscle, and at times in normal muscle. Certainly the work of Bodian[9, 10] shows that poliomyelitis virus attacks the nerve cell directly, and not the motor axon or ending.

As Drachman et al.[41] emphasized, the finding of late degenerative changes in the muscle fibers of healthy, normally innervated motor units in acute poliomyelitis, a disease known to affect only motor neurons, discredits the theory that they are myopathic when observed in such diseases as peroneal muscular atrophy or primary muscular atrophy.

Shortening of the paralyzed muscles of poliomyelitis is presumably related to the effect of immobilization on the connective tissues, for it may still be present when no contractile elements remain. The ultimate deformity of trunk and limbs depend on this as well as on imbalance of residual muscles.

"MOTOR SYSTEM DISEASE" "THE AMYOTROPHIES," (PROGRESSIVE MUSCULAR ATROPHY OF ARAN-DUCHENNE, AMYOTROPHIC LATERAL SCLEROSIS OF CHARCOT, PROGRESSIVE BULBAR PALSY OF DUCHENNE)

These are all variations of the same disease, and since corticospinal as well as anterior horn cell degeneration is common to all of them, they are generally referred to as motor system disease. As a rule the disease begins during the fifth decade, but exceptionally it develops as early as the second or as late as the ninth decade. It is seldom familial, though Koerner[70] and later Kurland and Mulder[74] described a large number of such cases on the island of Guam. Also a hereditary form of spinal muscular atrophy impossible to distinguish from the sporadic type is well known. In a small group of patients an earlier poliomyelitis is followed after 20–30 or more years by progressive muscular atrophy. The outstanding clinical features, well described by Strümpell[96] in 1888, are muscular weakness and atrophy, with or without spasticity. Either the limb muscles or the bulbar muscles (tongue, larynx, and pharynx) are first affected; the ocular muscles are always spared. Fascicular twitchings are found in almost every case, and muscle cramps are common. At first limited to one side, the condition soon becomes symmetrical; the corticospinal tract becomes involved sooner or later in most cases, as shown by spasticity, hyperactive tendon reflexes, and extensor plantar reflexes. Sensation remains unimpaired, and with few exceptions the intellect remains sound. The abdominal reflexes are long preserved, and the bladder and rectal sphincters are affected late, if at all. "Pseudobulbar palsy" with dysphagia, dysarthria, hyperactive jaw jerk, and poor emotional control may occur. The onset of the disease is insidious, and the

course is intermittently but inevitably progressive to a fatal termination in 2–6 years or more. Dysphagia from the bulbar paralysis resulting in aspiration pneumonia is the usual cause of death.

Cases failing to develop hyperactive tendon reflexes or spasticity are called progressive muscular atrophy. The paralysis is then entirely flaccid, for corticospinal degeneration, if present at all, is masked by precocious atrophy. If bulbar musculature is involved initially and prominently, the name progressive bulbar palsy is given. The atrophy often begins in the hands and forearms asymmetrically, although the legs may also be affected first. In the former pattern spastic paralysis becomes evident in the lower limbs about the time that atrophic changes are becoming obvious in the upper limbs and bulbar musculature. Ultimately atrophy overtakes the spastic muscles as well. This combination, the most common variety of the disease, is called amyotrophic lateral sclerosis.

A clinical variant of motor system disease was described in two papers by Wohlfart *et al.*[105] and Kugelberg and Welander[73] as a heredofamilial spinal muscular atrophy, proximal in distribution of weakness, and often mistaken for the Landouzy-Déjerine or Erb limb girdle progressive muscular dystrophy. The disorder affects both sexes and appears usually at about 9 years of age (range 2–17 years). The muscles of the pelvic girdle and thighs are weakened first, followed by those of the shoulder girdle, upper arms, and anterior neck. The reflexes gradually diminish in intensity and are lost during the late stages of the disease. The distal musculature is involved later, usually to a mild degree. Fasciculations are common. The disease is slowly progressive; most subjects can still walk 20 years after the onset of symptoms. The electromyographic and muscle biopsy findings are typical of degeneration of the spinal motor neurons.

Since these early reports other syndromes have been described by many authors, and the extensive literature is summarized by Gardner-Medwin *et al.*[56] The age of onset in their large series ranges from 1–50 years or more; both sexes are affected, and the inheritance is autosomal recessive. There are no signs of disorder of the corticospinal system; intelligence is normal; and sensory impairment is absent. Muscles of the thigh and lower leg weaken and undergo atrophy long before those of the trunk and upper extremities, and the cranial muscles are seldom involved. Such cases span the gap between Werdnig-Hoffmann disease and the hereditary form of adult motor system disease. A special form called the scapuloperoneal muscular atrophy of Brassard[14] was studied recently by Kaeser[69] who traced the condition as a mendelian dominant through five generations of a Swiss family. He contrasts it with atypical cases of Charcot-Marie-Tooth peroneal muscular atrophy and the shoulder-leg form of muscular dystrophy studied by Seitz[87] and Hausmanowa-Petrusewicz and Zelinska.[61] On the basis of genetic data Emery[47] finds that there are at least five patterns of familial progressive muscular atrophy—proximal (sometimes only quadriceps), distal, scapuloperoneal, facioscapulohumeral, and bulbar; probably an ocular group should be added. Surprisingly in several of these so-called benign forms of benign spinal muscular atrophy the electromyogram shows both denervative and myopathic patterns, and serum levels of creatine phosphokinase and aldolase are occasionally elevated.

The pathologic changes in both the sporadic and hereditary forms of this disease consist of degeneration of the anterior horn cells in the spinal cord, of motor nuclei in the brain stem, and in some instances of Betz cells in the motor cortex, with secondary degeneration of the corticospinal and corticobulbar tracts. The anterior spinal roots become thin and gray, and the motor nerve fibers in the peripheral nerves degenerate. The muscular atrophy is secondary to disease of the motor cells of the spinal cord. Darkschewitsch[31] in 1904 was one of the first to point out that the muscle fibers atrophy in small groups. Slauck[89-93] confirmed this observation and correctly attributed it to the innervation of a group of muscle fibers by one nerve fiber. The muscular changes have been studied by Wohlfahrt and Wohlfart,[108] Hassin and Dublin,[42] Wohlfart[105] and Mastaglia and Walton.[79]

The muscle atrophy is nearly always pronounced at the time of death. The distribution of this atrophy varies; either the distal or proximal muscles of the limbs are most severely wasted. The wasted muscles tend to retain their dark color.

Certain microscopic findings are regarded as typical of this disease, especially if the atrophy is not too advanced. A transverse section discloses one or a few groups of atrophic muscle fibers in a compact group or scattered among normal ones in the primary muscle bundles. These atrophied fibers are not only smaller than their neighbors but display among themselves a wide variation in size, as if all of them within any one unit did not undergo atrophy at the same pace. This is now understandable from the studies of Edström and Kugelberg[46] (Chapter 3).

In moderately advanced stages of the disease the morphologic changes become more characteristic, taking a form that suggests successive disseminated degeneration of individual anterior horn cells over a considerable period of time. All parts of the muscle in a transverse microscopic section usually exhibit this picture. The normal hypertrophied and atrophic fibers each form discrete groups (Fig. 10–5). In any one group of atrophied fibers there is a certain uniformity of size, though much variation can be noted between the groups (Figs. 10–6 and 10–8B). Figure 10–7 shows a comparison of normal (hatched) and diseased fiber sizes in Werdnig-Hoffmann disease and amyotrophic lateral sclerosis.[1] Occasionally all the fibers in a primary muscle bundle comprising several motor units are markedly atrophic. Normal sized or enlarged groups of hypertrophied fibers are common in this stage of the disease. The atrophic groups are not sharply delineated, and it is difficult to determine how many fibers belong to any one group. When a group is sharply circumscribed the number of fibers varies from about 10–50.

As in other types of denervation atrophy, the loss of diameter of muscle fibers proceeds rapidly until the fiber reaches a size of 10–15 μ. Subsequent atrophy is slow. This is probably the reason that the variation in size of fibers is much less in the advanced stages of the disease, when the picture comes to resemble the diffuse homogeneous secondary muscle atrophy of any severe neural disease. The groups of muscle

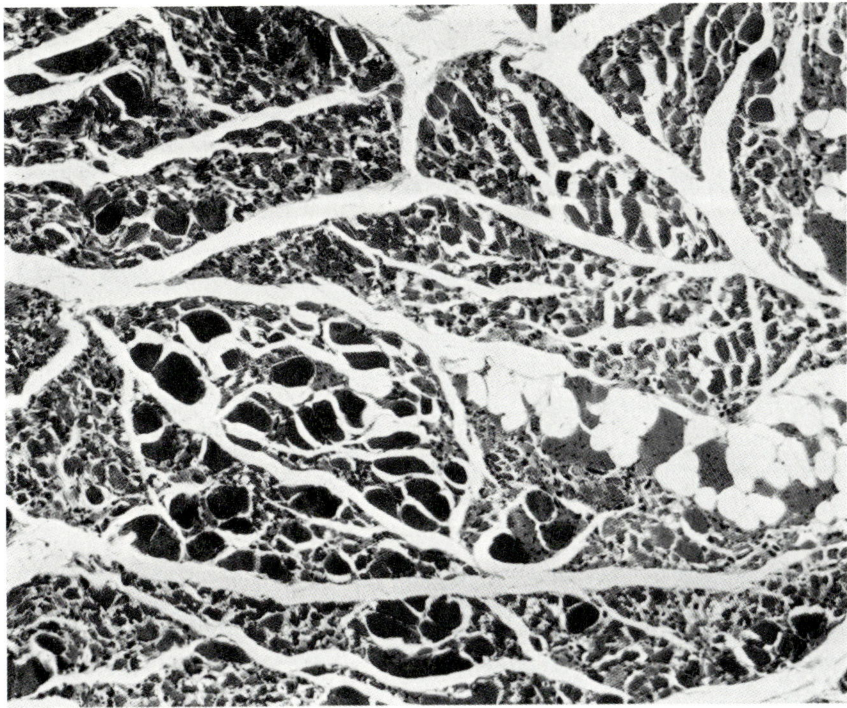

Fig. 10–5. Transverse section of muscle from patient with amyotrophic lateral sclerosis. Note variation in size of muscle fibers in different parts of field. At bottom is a motor unit that is only slightly atrophied. There are some fat cells present (at right). (H&E; ×100)

fibers which survive until the late stages of the disease are smaller than the atrophic ones in the early stages of the disease, ranging from 5 to 25 fibers each.[105] Of these, several may be of subnormal size, and groups of hypertrophied fibers measuring up to 150 μ or more may be found among them. Hyaline or vacuolar degeneration of large muscle fibers is rare, but here and there degenerative myophagia and regeneration can be seen and there is also granular degeneration and fragmentation of atrophic fibers (Fig. 10–7).

The polygonal contours of the muscle fibers, as seen in transverse sections, are often retained as the fiber undergoes atrophy. Nevertheless some become round and may assume flattened, oval, or pleomorphic shapes—due possibly to the process of fixation and sectioning. This pleomorphism is more obvious in paraffin-embedded than in celloidin-embedded sections.

Histochemical studies of moderately atro-phied muscle show that the small fibers contain little or no fat droplets and no glycogen, RNA, or acid phosphatase. Succinic dehydrogenase and Sudan black preparations fail to differentiate type I and II fibers among the atrophic groups. However with ATPase and phosphorylase methods the two types are easily identified in both the atrophic and hypertrophied groups. Another feature not evident in the usual histologic preparations is the occurrence of one histochemical type in large groups of fibers.

The sarcolemmal nuclei increase in size, and their nucleoli become more prominent. Sometimes rows of sarcolemmal nuclei persist after all visible sarcoplasm has disappeared. The uncertainty as to whether this increase is real or only apparent, due to the decreased size of the fibers, was already discussed in relation to neural atrophy in general. Wohlfart[105] expressed the opinion, based on quantitative histologic study of skeletal muscle, that the in-

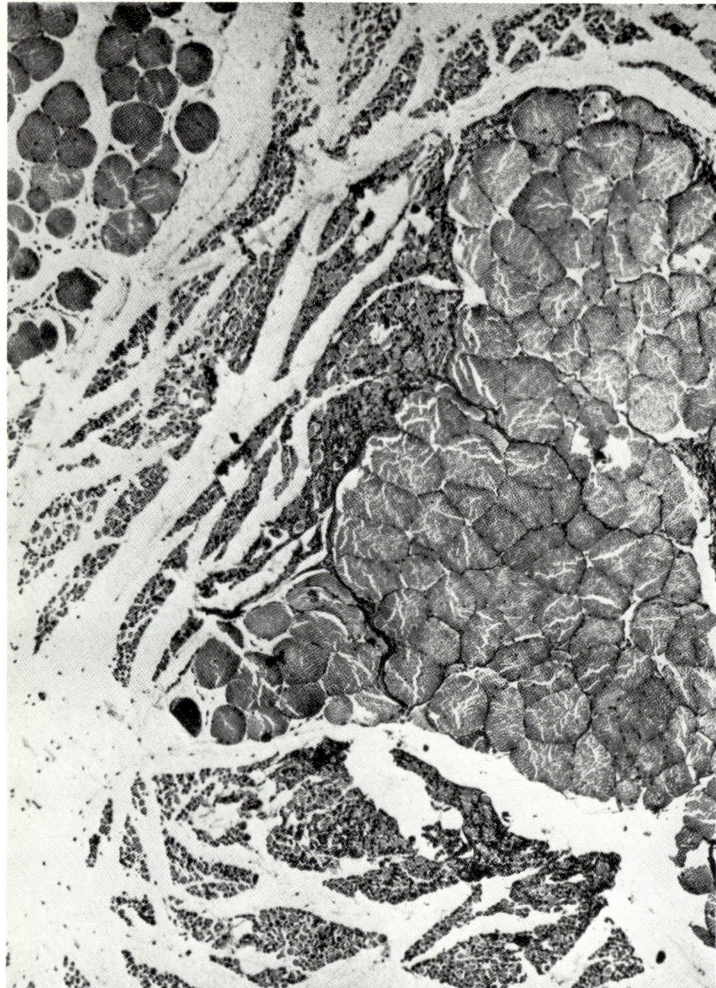

Fig. 10–6. Transverse section of atrophic very weak muscle (anterior tibial) from a patient with progressive muscular atrophy. Sound motor unit is surrounded by extremely atrophic cellular fibers. Part of motor unit in upper left corner is slightly atrophic. There is no increase in connective tissue or fat. (H&E; ×80)

Fig. 10–7. Longitudinal section of quadriceps muscle from a patient with progressive muscular atrophy. All fibers are atrophic. Some retain their striation; others are granular and have broken into oval-shaped masses. Interlacing bands in lower part of field are fibroblastic connective tissue. (phosphotungstic hematoxylin)

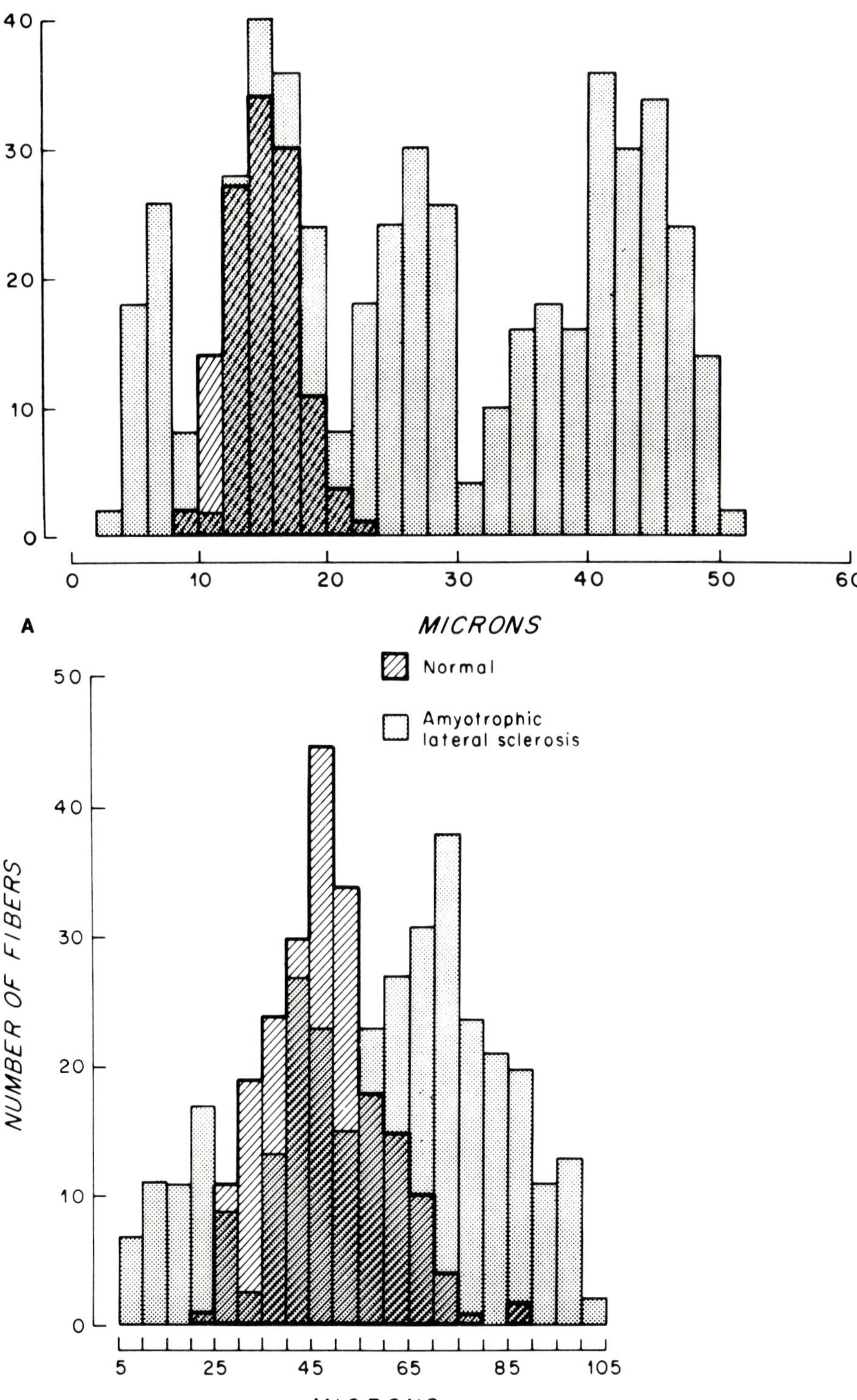

Fig. 10–8. (A) Muscle fiber size in 51-month-old patient with Werdnig-Hoffmann disease (dotted bars) compared to that in a 78-month-old normal subject (crossed bars). Fiber size ranged from 3 to 52 μ in the patient and from 9 to 22 μ in the normal subject. (B) Comparison of muscle from a patient with amyotrophic lateral sclerosis with normal muscle.

crease is only relative. The nuclei retain their hypolemmal position in the atrophic fibers but often become central in hypertrophied ones, not unlike those in certain forms of muscle dystrophy.

Under the electron microscope the atrophic fibers contain relatively more intermyofibrillar sarcoplasm, an increase in numbers of mitochondria and prominence of the sarcoplasmic reticulum and T-tubules. The nuclear membranes are infolded, often giving rise to cytoplasmic inclusions within the nucleus; and the plasma membrane shrinks while the basement membrane becomes redundant. Disorganization of the myofibrils, absence of mitochondria and depletion of SR characterizes the core of the targetoid fibers often seen in biopsies of denervated muscle.

The interstitial connective tissue increases with the progression of the atrophic process, and yet active fibroblastic proliferation is not observed. The blood vessel walls are thickened, with narrowing of the lumen (vascular atrophy). Cellular infiltrations in the interstitial tissue are no more frequent than in normal muscle. In the late stages there is moderate infiltration by fat cells, but seldom to the degree of that seen in progressive muscular dystrophy.

The intrafusal fibers and capsule of the muscle spindles were studied by Lapresle and Milhaud[75] and Cazzato and Walton.[22] In most of the cases of motor system disease the spindles remain unaltered. Exceptionally, the muscle fibers are atrophic.

The cause of the sporadic form of motor system disease is unknown. The old terms abiotrophy and degeneration are now deprecated because they imply an intrinsic lack of viability or a premature senescence for which there is no evidence. A slow viral destruction of motor neurons was recently proposed. Of course in the familial types the cause is genetic, but we have no idea of the mechanism that prevents the survival of certain groups of neurons for the lifetime of the individual.

The atrophic paralysis and the spasticity depend on destruction of the anterior horn cells and the corticospinal tracts, respectively. The mechanism of the fascicular twitchings (fasciculations) is less clear. These are coarse spontaneous twitchings of motor units occurring here and there through the weakened muscle (Chapter 2). They must be distinguished from contraction fasciculations, fibrillations, and myokymia.[37] Some of these fascicular twitches persist after complete Novocain block of motor nerve fibers and therefore appear to originate in the periphery of the motor axon. The conception of a primary disorder of excitability of motor neurons has much to recommend it.[35] The fasciculations are intensified by prostigmine and abolished by quinine and curare. They are particularly evident at stages in which the disease is progressing rapidly.

INFANTILE MUSCULAR ATROPHY (WERDNIG-HOFFMANN) AND AMYOTONIA CONGENITA (OPPENHEIM)

Amyotonia congenita, infantile muscular atrophy, the several congenital myopathies, and arthrogryposis or amyoplasia congenita are clinical entities in which the predominant features are muscular weakness, hypotonia, and reflex loss, presenting at birth or during the first months of life. They differ somewhat in their clinical manifestations, but the essential pathologic change in all is a failure of development or a degeneration of the motor neurons of the spinal cord. Of the three, amyoplasia is the least well known and the most uncertain of classification among the "nuclear amyotrophies."[16] In one of our cases the primary disorder in amyoplasia appeared to be a lack of development of spinal motor neurons. There is, however, lack of agreement between the published accounts of the pathologic changes in this disease; we chose therefore to describe it with other congenital defects of skeletal muscle (Chapter 5).

The first systematic study of infantile muscular atrophy was made by Werdnig,[104] who in 1891 reported two early infantile hereditary cases. Hoffmann[65] in 1893 wrote of a family of four children who all suffered from a similar disease. Thomsen and Bruce[98] also reported a case of progressive muscular atrophy in a child. More recently Walton[103] summarized the clinical findings in 67 cases followed for varying periods of time, up to 20 years in one case. It was pointed out that in about half the cases the disease was evident at birth or during the first 2 weeks of life; in the remainder the first symptoms appeared during the first year. The course was invariably progressive, and 75–80% died within the first 4 years of life. Exceptional patients lived on to adolescence or even adulthood, severely disabled with muscular atrophy, contractures of limbs, and spinal deformity, the disease seemingly having reached a stationary state. Weakness of trunk and limb muscles and extreme laxity of joints due to hypotonia were the earliest symptoms. Absence of tendon reflexes was almost invariable, and plantar reflexes were of the usual infantile type. The peculiar immobility with infrequent movements of small range contrasted with the alert facial expressions of these infants and their good nutritional status. Their characteristic posture was supine, with the head turned to one or other side and the arms abducted at the shoulder and flexed at elbows; the legs were flexed and externally rotated at hips and flexed at knees (frog posture). The chest was thin, the abdomen protuberant, and breathing largely diaphragmatic, often with thoracic retraction on inspiration. When lifted, the body hung limply and tended to slip through the examiner's hands. Fasciculation was visible only in the tongue. In late stages, dysphagia and pooling of pharyngeal secretions were noted. The eye muscles were not involved. As the months and years advanced, atrophy became more evident, and contractures and deformities developed. Earlier in life subcutaneous fat obscured the atrophy, which could be seen only if x rays of the limbs were taken. The sphincteric action was preserved, and sensation was intact. Death was usually due to enfeeblement of intercostal muscles and subsequent pneumonia.

In 1900 Oppenheim[83] identified another group of infantile muscular atrophies which he and subsequent writers believed to be separable on the basis of nonfamilial incidence, onset at the time of birth, and lack of progression. He called this disease myatonia congenita, but others after him introduced the term amyotonia congenita in order to avoid confusion with myo-tonia congenita (Thomsen's disease). Oppenheim emphasized the striking flaccidity of muscles and the hyperextensibility of joints, the generalized weakness without paralysis, and the absence of tendon reflexes. Subsequent case reports[12, 72, 100, 101] delineated the clinical syndrome with greater exactitude, often under different titles, i.e., benign congenital hypotonia, universal muscular hypoplasia, and congenital myopathy. Soon after birth these infants are observed to be limp and floppy; the diagnosis of amyotonia congenita is made because of the slack musculature and later the failure to sit, stand, and walk at the usual times. In contrast to cases of infantile muscular atrophy, they are not paralyzed even though their movements are relatively more feeble than normal. They usually retain their tendon reflexes, a point contrary to Oppenheim's statement; at no time is muscle atrophy pronounced, nor are fasciculations demonstrable by inspection (tongue) or electromyography. The ocular and other cranial muscles appear to function normally. When upright stance and locomotion are acquired at a later date, the first movements are clumsy, and falls are frequent. As the years pass some of these children (perhaps half) recover completely; the remainder continue to be somewhat weak, with hypotonia and unnaturally thin muscles. Electrical testing in some cases reveals no abnormality, and in others the pattern of numerous polyphasic potentials of short duration on volitional movement suggests a myopathic disorder. Muscle biopsy has disclosed only relatively thin muscle fibers.[58, 101] Motor unit atrophy and changes of infantile muscular dystrophy have not been observed.

The specificity of these diseases has been challenged on clinical as well as on pathologic grounds. Greenfield and Stern,[58] from a review of the literature and their own material, maintain that the diseases are clinical variants of the same pathologic process. Differences as to familial incidence, time of onset, and clinical course were said not to distinguish them. This view has been affirmed by a majority of clinical workers, who were impressed by the number of cases that combined the classic features of amyotonia congenita and infantile muscular atrophy. However, recent data prove this view to be incorrect, for the groups of floppy children on closer scrutiny prove to have a variety of diseases, some nonprogressive. It is possible that the weak hypotonic child with absent tendon reflexes who later gradually recovers suffers from infantile polyneuritis. No satisfactory pathologic studies of such cases are available, but the electromyographic and biopsy studies of Chambers and MacDermot[23] in two cases and of Byers and Taft[20] in several others provide evidence that at least some cases can be classified as polyneuritic. Furthermore several new congenital myopathies (central cord disease, nemeline, pleoconial and megaconial, and myotubular) have come to light (Chapter 5).

Our position on this controversial problem is in general agreement with that of Brandt[12]— that amyotonia is a *symptom complex* common to Werdnig-Hoffmann's disease, congenital muscular dystrophy, infantile polyneuritis, and the several congenital myopathies, all of which have not yet been differentiated. It is believed that the recent contributions of Walton,[103] Greenfield and his colleagues,[58] Schreier and Huperz,[86] Shy and Engel[88] and their colleagues, as well as unpublished observations of our own, indicate the existence of several categories of disease, now designated benign congenital hypotonia and universal muscular hypoplasia, which conforms in general to Oppenheim's original description of amyotonia congenita. This point will probably never be settled with finality because of the meagerness of the clinical description in Oppenheim's original report and the lack of pathologic data.

Infantile muscular dystrophies must be set apart in another group (Chapter 6). These too are characterized by atrophy, hypotonia, and loss of reflexes, indistinguishable at first from amyotonia congenita but later marked by progression and the appearance of atrophy, usually of proximal muscles. Turner[100, 101] described a family of 13 siblings, 6 of whom were affected. The pathologic findings were characteristic of muscular dystrophy. These observations support the suggestion made long ago by Batten[7] that some cases of infantile muscular atrophy are essentially myopathic, others neuropathic and myelopathic. Probably the family reported as having muscular infantilism by Gibson[57] ex-

emplifies the myopathic form of this disease, as does also the family described by Banker *et al.*[5] and by Levesque *et al.*[76]

ETIOLOGY AND PATHOGENESIS

The pathogenesis of infantile spinal muscular atrophy is unknown. Genetic factors were investigated recently by Brandt.[12, 13] In 70 families with 112 cases of infantile progressive muscular atrophy the pattern of inheritance indicated an autosomal recessive trait. Both sexes were equally affected, thus ruling out a sex-linked factor. The disease is self-limiting; it cannot of course be transmitted from one generation to the next. The evidence of inheritance is based only on the finding of multiple cases in one sibship and in pairs of identical twins. Variations in age of onset and rapidity of progression led Byers and Banker[19] to distinguish three subtypes of disease; the third and most benign appears to merge with the Wohlfart-Kugelberg-Welander syndrome.

PATHOLOGIC ANATOMY

The pathologic anatomy of many of the cases diagnosed as Oppenheim's disease and nearly all of Werdnig and Hoffmann's forms are the same. Many hundreds of cases have been examined post mortem and reported in the literature. Recently Lewey,[77] Burdick *et al.*,[17] Epstein,[51] Brandt,[12] Thieffry *et al.*,[97] and Greenfield and Stern[58] have written reviews of the subject. The lesions in the nervous system, about which there has been some disagreement, consist of a numerical reduction of the motor nerve cells in the anterior horns of the spinal cord and in motor nuclei of the medulla, pons, and midbrain, alteration (chromatolysis and cytoplasmic inclusions) of and supposed degenerative changes in many of the surviving cells, thinness, aggregation of neurotubules and poor myelination of the anterior roots of spinal cord, and a slenderness and poor myelination of the peripheral nerves. Most writers[3, 28, 29, 81, 90] expressed the opinion that degeneration of anterior horn cells is the principal morphologic change, whereas others,[8, 27, 51, 53, 82] incorrectly

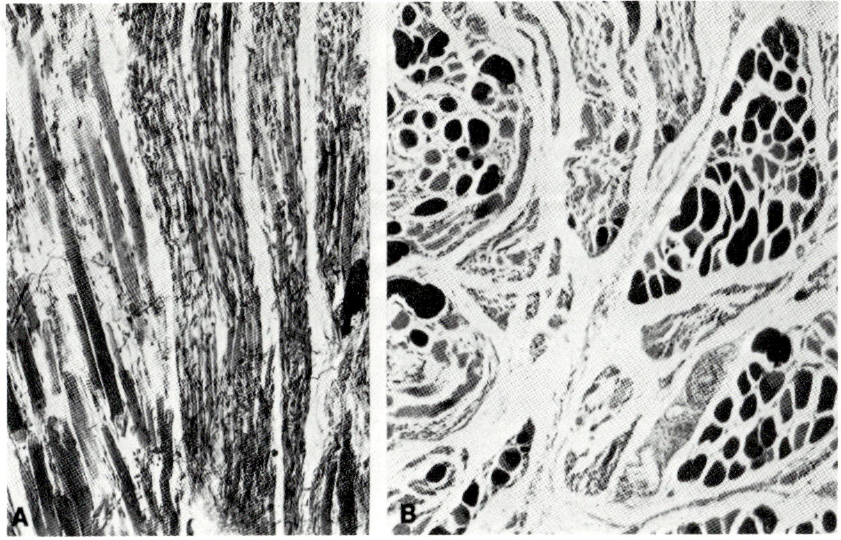

Fig. 10–9. (A) Longitudinal section of gastrocnemius muscle from patient with infantile muscular atrophy. In left part of field, muscle fibers are well developed but do not traverse field because of obliquity of section. Those on the right are thin and underdeveloped. There is some fraying artifact of connective tissue. (B) Cross section of muscle from a patient with amyotrophic lateral sclerosis showing varying degres of atrophy of motor units. (H&E)

in our view, regard the disease as a retardation in motor development.

HISTOLOGIC CHANGES

At autopsy the muscles of the limbs, shoulder girdles, and trunk are small, soft pink or light red, and translucent. Rigor mortis is usually absent. The internal architecture of the muscle is preserved, the relations of epi- and perimysium being undisturbed. A great variation in the size of muscle fibers is the outstanding finding. The majority of the fibers are very small; many are so narrow as to be little more than chains of sarcolemmal nuclei in thin, tortuous threads of sarcoplasm (Fig. 10–9A). These small fibers are called atrophic by some authors and embryonic or fetal by others. Longitudinal and transverse striations are retained. Other groups of muscle fibers appear to have developed normally, and many are very large, presumably hypertrophied, and measure up to 150 μ in width.[4] The atrophic fibers evidently persist for long periods but a few finally degenerate. Moderate amounts of fat and

fibrous connective tissue separate the bundles of small muscle fibers and eventually replace them. Muscle spindles are quite normal in every respect, and nothing remarkable is noted in connection with the blood vessels. Nerve bundles can be seen within the muscle, but there are no motor end-plates on the small fibers, a fact commented upon by a number of writers.[8, 51] The pathologic picture is typical of muscle atrophy secondary to disease of the spinal cord or peripheral nerves. The extreme thinness and embryonic appearance of the muscle fiber is doubtless due to the early age at which the motor neuron degeneration occurs. These changes in the muscles have been described in detail by Spiller,[94] Councilman and Dunn,[30] Greenfield and Stern,[58] Bielschowsky,[8] Forbus and Wolf,[54] Conel,[28] Hassin,[60] and Brandt.[12]

Recent studies of the histochemistry of the atrophic and large fibers reveal findings essentially in line with those of the benign forms of progressive muscular atrophy. Although Fenichel and Engel[52] found that both the extremely atrophic and hypertrophied fibers were type I (oxidative-rich) and all the normal fibers type

II, Dubowitz[43] could not confirm their observations. In his preparations the small fibers were of both types indicating an onset of disease after the 26th week of fetal life when fiber type differentiation occurs. The abnormally large or hypertrophied normal fibers tended to be of one type, usually type II (phosphorylase- and ATPase-rich). Presumably this is a manifestation of collateral regeneration of fibers by rapid-twitch motor neurons. From these facts Dubowitz argues that the thin muscle fibers have been denervated after a period of normal development.

In electron micrographs the shrunken fibers are of the same appearance as in adult motor system disease except that satellite cells are more numerous.

Of interest in this disease is the relative paucity of signs of degeneration and regeneration in the larger muscle fibers with intact innervation. The explanation of this may be that the patient is never able to use his muscles to any degree and cannot subject them to stress or injury.

The weakness of the voluntary muscles contrasts with the normality of involuntary ones. The diaphragm (in all except two cases cited by Burdick *et al.*[17]), the intestinal musculature, myocardium, bladder, and sphincters are not affected.

In general the fatal forms of this syndrome most closely resemble motor system disease. It is noteworthy that the relatively slight increase of fibrous tissue in the muscles is correlated with only a slight tendency to contracture. In this respect it may be contrasted with the condition of amyoplasia congenita; the greater liability to contracture in the latter may be related to the more pronounced connective tissue reaction (Chapter 5).

OTHER SPINAL CORD DISEASES CAUSING FLACCID ATROPHIC PARALYSIS

In syringomyelia, traumatic necrosis of the spinal cord, intramedullary tumor, necrotizing myelopathy (myelitis), and syphilitic amyotrophy the changes in the muscle fibers are similar to those outlined above. The distribution of atrophy varies according to the segmental or root innervation involved. Minor differences in the muscle atrophy are related to the extent and the chronicity rather than to the nature of the spinal disease.

DISEASES OF PERIPHERAL NERVES

The changes in muscle consequent to disease of peripheral nerves and spinal roots have received relatively less attention than the above, and many of the details are still unknown or disputed. The early literature on this subject was reviewed by Lorenz[78] in 1904 and later by von Meyenburg.[102] Recent reports of particular merit are those by Brodal and Refsum[15] and Wohlfart.[105] In these publications one senses that some of the disagreement arises from the fact that the writers were describing a number of different diseases of quite variable duration. Further, in affections of peripheral nerves, whether traumatic or inflammatory, it is common for part of the disturbance of motor function to result from an interference of neural conduction without destruction of the axis cylinder continuity. Such is the "periaxial segmental neuritis" of Gombault, which takes the form of demyelination of segments of the axis cylinder. This change can be found in a wide variety of polyneuritides[95] and is similar to the blockage of conduction commonly found in ischemic lesions of nerve (tourniquet paralysis, pressure palsy) when the damage falls short of actual necrosis.[36] Under these circumstances the muscular changes are only those of disuse atrophy, in spite of a muscular paralysis which may last many weeks. Such may be the case also in acute porphyria, if the neural damage is of mild degree.[38] Byers and Taft[20] described in detail several cases of a multiple peripheral neuropathy that occurred during childhood. Each case was initially diagnosed as muscular dystrophy or progressive muscular atrophy.

The pathologic anatomy of the striated muscles in polyneuritis conforms to the same general pattern as was outlined in spinal muscle atrophy.[85, 92] If the patient succumbs to his disease within 2–3 weeks of the onset of motor paralysis, the pathologic changes in the muscle fibers are unremarkable. Although the muscles are soft and somewhat wasted, atrophy of individual muscle fibers is difficult to demonstrate, and one cannot be certain usually of any consistent increase in sarcolemmal nuclei or other cellular constituents within the muscle during the first 2 weeks.

If the interval between paralysis and death is longer, definite changes take place and can be demonstrated by routine microscopic sections. The principal finding is atrophy of part or all of the muscle fibers. In some cases there is in a single muscle a mixture of groups of sound, often hypertrophic, fibers, moderately atrophied fibers, and severely atrophied fibers. This indicates that the disease in the peripheral nerve has not affected all nerve fibers at the same time. The most atrophic groups of muscle fibers belong to the nerve fibers involved early in the disease, and the groups of normal-sized or enlarged muscle fibers are innervated by fibers which have escaped and have responded to the demands of overwork. The pathologic changes in muscle in progressive neural atrophy are similar to those of progressive spinal muscular atrophy, except that the groups of atrophic muscle fibers are less likely to be disseminated. Usually all the muscle fibers are in approximately the same stage of atrophy, when it is possible to differentiate degrees of muscular atrophy after the first few weeks or months. A diffuse homogeneous neural muscular atrophy is therefore commonly found in the usual types of acute and subacute disease of the peripheral nerves, but in distinction to poliomyelitis the degree of atrophy may vary from one muscle to another.

It is often impossible to correlate the histologic changes in the peripheral nerves with the atrophy of striated muscles. As pointed out by Wohlfart,[105] there may be diffuse, homogeneous neural muscular atrophy when all the peripheral nerves are affected simultaneously— or the contrary, a picture of progressive neural atrophy when the nerve fibers are in various stages of degeneration. One source of error in such an attempted correlation between muscle fiber atrophy and severity of nerve fiber de-

generation is the impossibility of distinguishing between motor and sensory nerve fibers.

There is an apparent increase in the sarcolemmal nuclei in the atrophic fibers, just as there is in spinal muscular atrophy. In the most chronic forms of neural muscular atrophy these nuclei are slender and densely hyperchromatic and lie close together, forming chains. Sometimes they are so dense that they have the appearance of calcific deposits in a hematoxylin and eosin stain. The loss of transverse striations, which has been stressed by some writers, is not consistent. We have often observed the preservation of some degree of striation as long as a sarcolemmal sheath is visible. Connective tissue appears to increase slightly in amount, and yet there is seldom any evidence of fibroblastic activity. In late stages the prominence of clumps of dark, shrunken, nuclear remnants may be a striking feature.

The muscle spindles may be unusually conspicuous. The spindle capsule may be thick, the space between the capsule and intrafusal muscle fibers wide, and the fibers themselves thin. This we observed in both spinal and neural muscular atrophy. In all probability some atrophy of both the intra- and extrafusal portions of the spindle fibers occurs, but since they are naturally small to begin with, atrophy is less obvious than in the large muscle fibers. Wohlfart[105] remarked that the muscle spindles may degenerate in polyneuritis and suggested that changes in the sensory organs of muscle might afford a means of distinguishing, histologically, spinal and neural muscular atrophy. This opinion is not in accord with our observations, nor does it agree with those of earlier writers. Swelling and disintegration of muscle spindles (Fig. 10–10). are rare findings. Pure deafferentation of spindles results in hypernucleation of the equatorial zones of the muscle fibers.

In most of the subacute and chronic diseases of peripheral nerves there are abortive or successful attempts at regeneration. Microscopically the muscle, which may already have been heterogeneous, becomes even more so. Reinnervation of the muscle fibers within any one unit is not uniform, and some seem to enlarge more rapidly than others. The result is

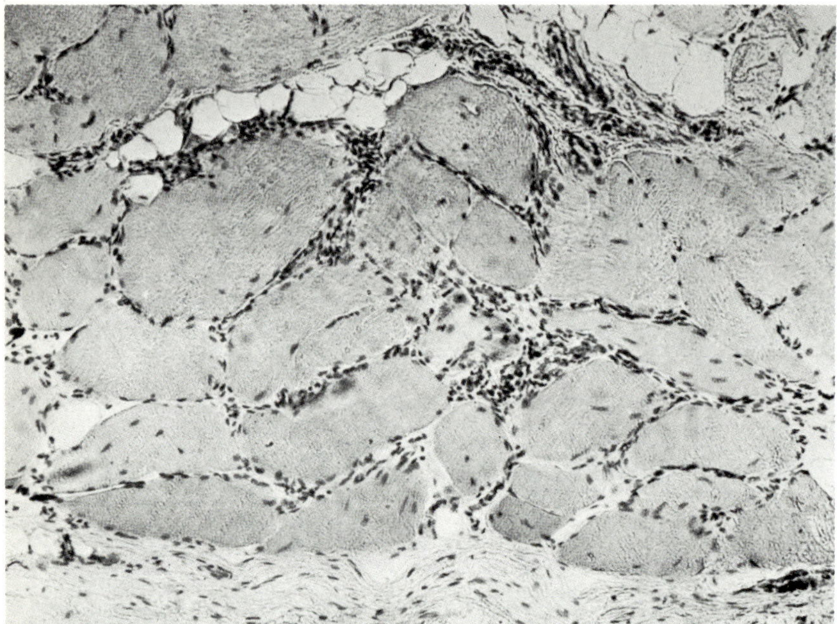

Fig. 10–10. Chronic polyneuritis. Two muscle spindles lying in atrophic muscle show great distention of capsular space with invasion by mononuclear cells. (hematoxylin, Van Gieson)

that few groups of fibers of a similar size can be found. Furthermore, in the course of time muscle fibers that have survived and those which have been reinnervated undergo hypertrophy. It is not uncommon to see single fibers with a cross-sectional diameter over 150 μ. This mixture of atrophic, normal, and hypertrophied fibers is a source of much confusion among pathologists and at times is mistaken as a mark of muscular dystrophy.

Histochemical stains show that both type I and II fibers are subject to denervation atrophy in most diseases. The large or hypertrophied ones, in contrast to the normal checkerboard mixture of type I and II fibers, occur in groups, all of uniform type. As in spinal muscular atrophy, this has been interpreted as due to collateral regeneration of nerve fibers from type I or II motor neurons.

CLASSIFICATION

In the clinical and pathologic differentiation of diseases of peripheral nerves there is great advantage in distinguishing localized forms from those that are generalized—of mononeuropathy from polyneuropathy. Mechanical, vascular, and neoplastic factors, and rarely localized infection are more likely to be the causal factors in mononeuropathies whether single or multiple (mononeuropathy multiplex) and metabolic disorders, and toxins and altered immune mechanisms in the polyneuropathies. Common usage has sanctioned the terms neuritis and polyneuritis for all these forms in spite of the absence of inflammatory reaction in all except a few of them such as leprosy. The terms neuropathy and polyneuropathy are preferable, particularly since the ultimate mechanism of many if not all these conditions is either metabolic or ischemic. It is doubtful if any classification based upon the elementary structure primarily damaged (i.e., whether parenchymatous or interstitial, demyelinating or axon destroying) is either useful or valid. Though leprosy, for example, leads to an intense interstitial reaction, the effect on the nerve fibers may be from ischemia. We therefore use here a classification based on the accepted etiology.

MONONEUROPATHY

TRAUMATIC LESIONS OF SINGLE NERVES AND PLEXUSES

Certain types of injury, particularly gunshot and other penetrating wounds, may sever peripheral nerves, whereas fractures and dislocations more often damage them without interrupting their continuity. After severance or severe damage to a single nerve, the degree of paralysis and atrophy depends on whether a muscle or group of muscles is entirely dependent on the nerve for innervation. With complete paralysis the affected muscles become perceptibly atrophied within 2–3 weeks, and the bulk of the muscle is reduced markedly by the second or third month. The ultimate fate of the muscles is determined to some extent by the care they are given and whether nerve regeneration occurs. If neglected, the denervated muscle may be shortened to a considerable extent owing to contraction by fibrous tissue. This can largely be prevented by avoiding injury, by maintaining adequate circulation, and by gentle, passive stretching.

Microscopic sections of the palsied muscles at the end of 2–3 weeks usually show no significant changes. Within 4–6 weeks the atrophy becomes evident microscopically. Most of the muscle fibers are moderately thin, but there is a greater than normal variability in the size of the muscle fibers. The sarcolemmal nuclei are increased in number and size, particularly in the more atrophied fibers. Small numbers of mononuclear histiocytes and a few lymphocytes collect around the blood vessels.

At the end of 6 months nearly all the fibers are reduced to thin ribbons with only enough sarcoplasm to separate the muscle nuclei. The latter are quite dark and form long chains. The longitudinal striations are usually more easily seen than the cross striations; neither are visible in some of the thinnest fibers. The endomysial connective tissue increases slightly in amount and is somewhat more cellular than usual. The later stages of muscle atrophy in human nerve injury are described by Bowden and Gutmann.[11] Thin muscle fibers may be found for years after nerve injury, but many degenerate into multinucleated clumps or masses with granular cytoplasm. It should be stressed that in the muscles denervated by a traumatic nerve lesion the muscle fibers in all the motor units are in approximately the same stage of degeneration.

INFLAMMATORY DISEASES

The facial muscles are frequently paralyzed as a result of lesions in the facial nerve. This nerve is usually affected by an inflammatory process within the fallopian canal, giving rise to Bell's palsy. There is a rapid development of paralysis of all facial muscles to the same degree. In many cases the paralysis is transient, and normal function is recovered within a few weeks. Only in exceptional cases is there a permanent flaccid paralysis. The nerve may be damaged in the same situation by fracture of the skull. Involvement of the nerve in its course through the parotid gland or elsewhere occurs in Boeck's sarcoid. The changes in the muscle are those of simple neural atrophy in all these conditions.

Sciatic and brachial neuritis give rise to muscle weakness and loss of tendon reflex but seldom cause very pronounced changes in the muscle. Pressure palsies often result in only a transient paralysis without marked atrophy, for the slighter degrees of pressure or ischemia may produce only a localized block in conduction without degeneration of axis cylinders. The destroyed nerve fibers usually regenerate quite effectively. Motor paralysis may occur in herpes zoster owing to spread of the inflammatory process from the sensory ganglion either to the adjacent motor root or to the anterior horn of the spinal cord. Complete muscle paralysis and atrophy are rare. Boeck's sarcoid may involve, in addition to the facial nerve, one or several peripheral nerves or roots, and the effect on the muscles depends on whether the nerves are completely interrupted.

Leprous neuritis is undoubtedly the most frequent and widespread form of polyneuritis in the world at large and is in fact the only true neuritis in the sense of being an inflammatory process within nerve caused by direct neural invasion of the infective agent. The characteristic lesion is granulomatous with lymphocytes, mononuclear and plasma cell infiltrates, giant cells, and a heavy overgrowth of peri-

and endoneural connective tissue that results in palpable thickening of the nerve.[33, 106] The chronic course with long delay between the time of invasion of skin and the first symptoms of neural involvement (2–6 years), the successive involvement of certain nerves (ulnar, external popliteal, medial cutaneous nerve of forearm, median, radial, anterior tibial, trigeminal, and occipital), and the chronic progressive course constitute the characteristic clinical picture. The atrophy, weakness, and anesthesia relate to the nerve lesions per se, whereas trophic changes in skin and nails and absorption of the distal parts are the result of repeated trauma to the anesthetic parts of the body.

Dastur,[32] who provided an excellent summary of the clinical and pathologic features of the disease, emphasizes the predominantly atrophic nature of the muscle lesions. Motor unit atrophy in weakened muscles was observed in the majority of cases along with degeneration of intramuscular nerves and motor endplates. In a few specimens intramuscular nerve twigs were the site of small granulomatous lesions that extended to adjacent muscle tissue, imparting a picture of denervation atrophy plus granulomatous myositis. Leprous bacilli were demonstrated within the granulomas often within the cytoplasm of Schwann cells.

VASCULAR

Infarct necrosis due to occlusion of nutrient arteries by arteritis, diabetic vascular disease and atherosclerosis also have been identified. Multiple vascular lesions in diabetic nerves, plexuses and roots give rise to a mononeuropathy multiplex. The muscles undergo denervation atrophy.

Other forms of localized inflammation due presumably to coxsackie and other viral infections also occur but little is known of their morbid anatomy.

POLYNEUROPATHY

ACUTE VARIETIES

Although there are many types of acute and subacute polyneuropathies (acute idiopathic polyneuritis of Landry-Guillain-Barré, polyneuritis of infectious mononucleosis, acute porphyria, diphtheritic, uremic, carcinomatous, and the like), their effect on muscle is essentially denervation. Although the peripheral nerve lesions differ in type and topography, such changes are seldom localized to intramuscular nerve twigs with sufficient clarity to enable the myopathologist to make the diagnosis from muscle specimens alone. Each case must be studied by surveying many sections of proximal and distal nerves of legs, trunk, arms, and cranium; spinal and autonomic ganglia; roots; spinal cord; and brain. Foreknowledge of the clinical data is essential in planning the strategy of the pathologic analysis.

When death occurs within a few days of the paralysis (usually the disease is an acute idiopathic polyneuritis) there has been too little time for muscle atrophy to develop even though paralysis was complete. Occasionally some of the intramuscular twigs are infiltrated with lymphocytes and mononuclear cells, and the myelin sheaths are fragmented (Fig. 10–11). After paralysis of several weeks, the changes of denervation become evident as in Figures 10–12 and 10–13 from a porphyric polyneuropathy. Correspondence between the degree and duration of paralysis and of denervation atrophy is widely discrepant during the first weeks or months of a polyneuropathy because in many of these acute forms the nerve lesions are of the Gombault segmental demyelinative type. Since the axis cylinders remain intact, the muscle is not denervated and function is soon restored by remyelination and not by regrowth of axons. Any thinning of muscle fibers then is more a matter of disuse than denervation. The diseases in which segmental demyelination may predominate are acute idiopathic polyneuritis and diphtheritic polyneuritis. Alcoholic beriberi, uremia, and carcinomatous, lead, arsenical, and porphyric disorders are more apt to destroy axis cylinders and lead ultimately to denervation atrophy of muscle. The topography of denervation paralysis also varies from one disease to another. In beriberi the distal muscles of the feet and legs are always the most affected. In acute idiopathic polyneuritis, the muscles of face, trunk, and limbs are all involved. Diabetes, polyarteritis nodosa, and sarcoidosis

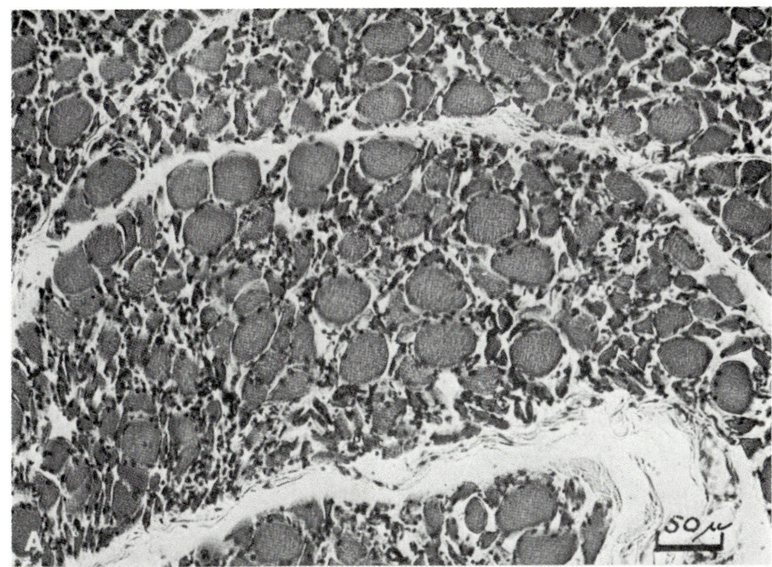

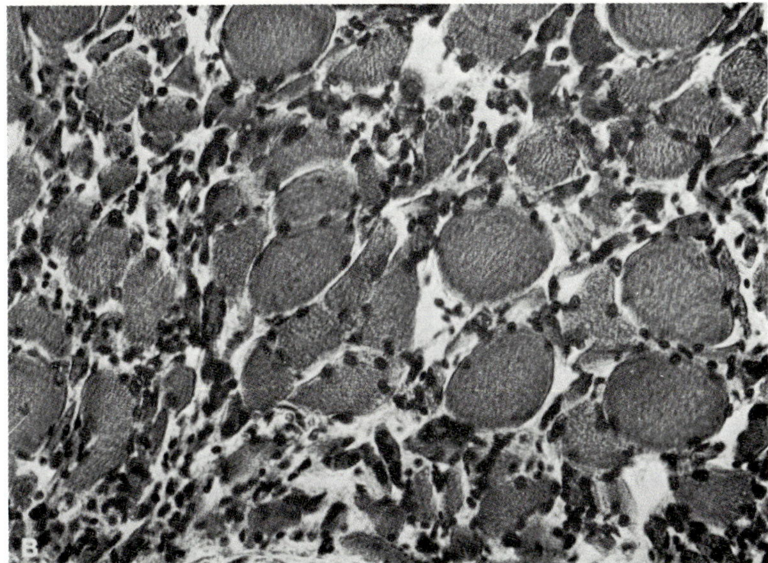

Fig. 10–11. Section from patient with acute infective polyneuritis who died on eighth day of illness. There is one hyalinized muscle fiber with hyperplastic sarcolemmal nuclei (at center). Note also slight infiltration of endomysium by lymphocytes and mononuclear cells. (hematoxylin, Van Gieson)

are more likely to cause a multiple mononeuropathy of variable distribution.

CHRONIC VARIETIES OF POLYNEUROPATHY

The more chronic types of peripheral nerve diseases cause a severe degree of muscular atrophy which may at times resemble the spinal muscular atrophies or the muscular dystrophies. The distal distribution of the muscle atrophy, sensory changes, and sometimes palpably en-larged nerves are usually sufficient to distinguish this class of disease.

Chronic Progressive and Relapsing Forms. In every medical clinic there are a few patients with chronic progressive or chronic relapsing polyneuritis. Males and females, usually of adult age, are equally disposed. The condition may, however, affect children, as pointed out in the report of Byers and Taft.[20] Weakness and atrophy of the muscles of the legs and arms, reflex loss, ataxia, and variable sensory

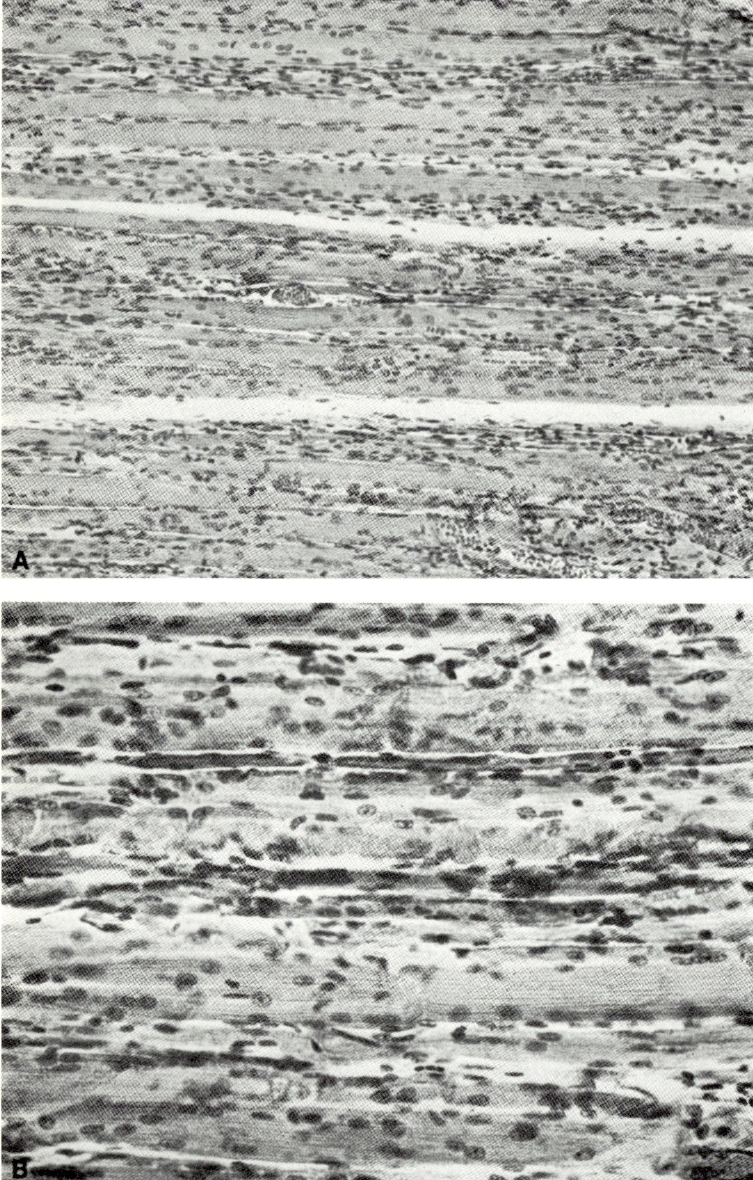

Fig. 10–12. Acute porphyria. Muscle (extensor carpi radialis) from autopsy some months after an unusually severe episode of damage to the peripheral nerves showing advanced neural atrophy of greater proportion of muscle fibers, with an increase in sarcolemmal nuclei. (A, ×200; B, ×400)

changes comprise the essential features of the clinical picture. Tinnitus, dizziness, deafness, and night blindness are sometimes associated. Usually there is no clue as to the etiology and pathogenesis of the disease.[67] Isolated autopsy reports have confused rather than clarified the clinical problems.

We examined the muscles in a few such cases but found only the changes of chronic neural atrophy. In this group the muscle spin-

dles occasionally show degenerative changes (Fig. 10–10). If the duration has been more than 6 months, clumps of residual pyknotic nuclei may be found lying free in the connective tissue without visible sarcolemma (Fig. 10–14), and a mixture of recent and advanced changes in the muscle fibers is evident. In some muscles the contrast between unaffected, recently affected, and degenerated motor units may be very marked in the corresponding groups of

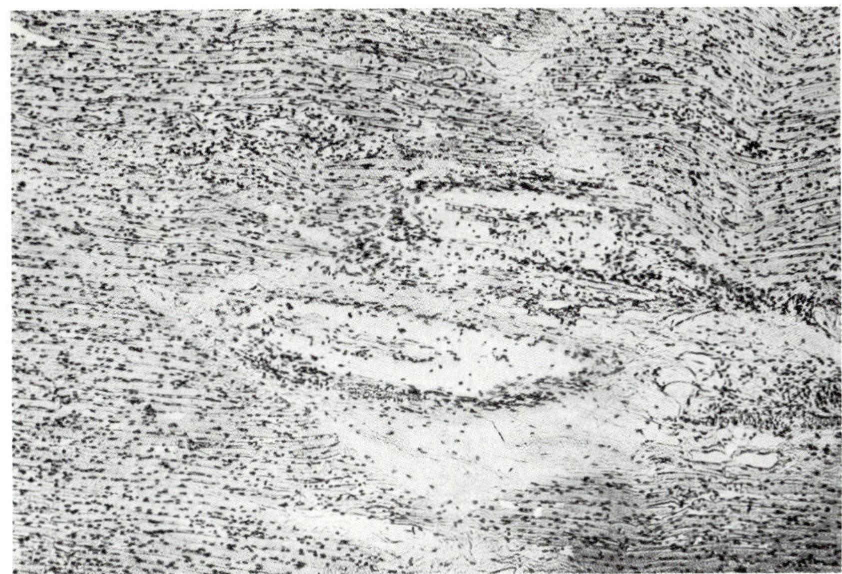

Fig. 10–13. Porphyria. Longitudinal sections of same muscle as in Figure 10–12, showing variation in size of fibers and degenerative changes in some. (H&E; A, ×160; B, ×320)

Fig. 10–14. Chronic progressive peripheral neuropathy of unknown etiology. Some muscle fibers are reduced to clumps of nuclear remnants, and others are in various intermediate stages of neural atrophy and degeneration. (hematoxylin, Van Gieson)

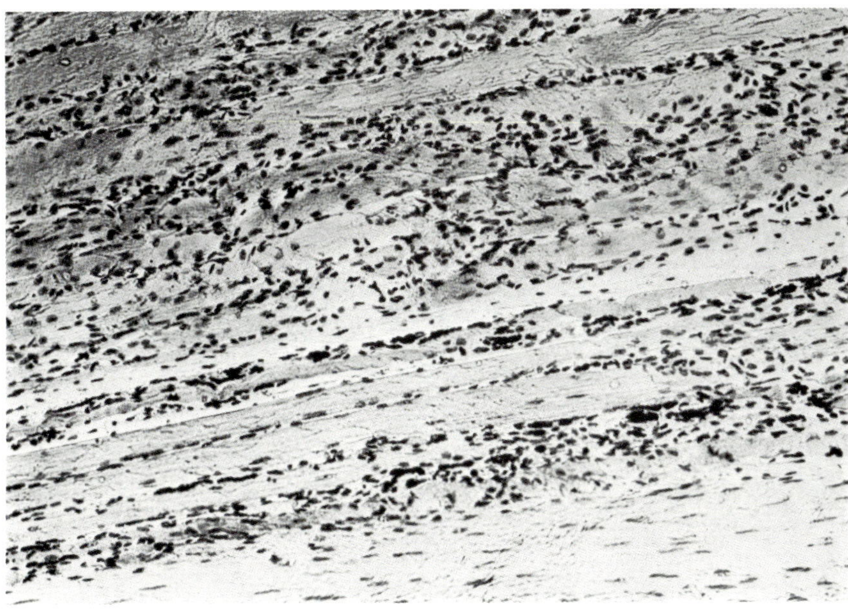

muscle fibers (Figs. 10–15 and 10–16). Bronchogenic carcinoma was present in two instances of progresssive sensory ataxia described by Denny-Brown.[34] In both cases the muscles showed degenerative changes (referred to under polymyositis, Chapter 7), in spite of the presence of intact motor nerve endings.

Peroneal Muscular Atrophy of Charcot-Marie-Tooth. Charcot-Marie-Tooth disease was described almost simultaneously by Charcot and Marie[24] and by Tooth[99] in 1886 and soon thereafter by Hoffmann[64] in 1889. Charcot and Marie referred to the disease as a progressive muscular atrophy beginning in the feet and legs and affecting the hands and forearms only after several years. Tooth, and later Hoffmann, emphasized its neuropathic features. England and Denny-Brown[50] recently showed how the pattern of sensory loss also points to a radicular pathology.

This type of muscular atrophy or amyotrophy is a heredofamilial disease having its onset during juvenile or adult years. The muscular weakness and atrophy first affect the extensor and abductor muscles of the feet. The symmetrical atrophy of these muscles produces first a mild pes equinovarus and later a pes cavus deformity of the feet. In time all muscles below the middle third of the thigh are affected, producing a "stork leg" or an "inverted champagne bottle" appearance of the legs. Eventually the intrinsic hand and the forearm muscles are involved, producing an inability to oppose the thumb and a mild clawing of the hands. The face, trunk, and proximal limb muscles escape the process entirely. Contractures are infrequent. Fascicular twitchings are often seen. Subjective sensory disorder, pains, paresthesias, and cramps are frequent and usually there is diminution of vibratory and position sense in the feet. Ataxia is rare. The ankle reflexes disappear early in the disease, but patellar reflexes are usually retained. The cerebrospinal fluid protein may be elevated, and the gold sol curve is abnormal at times. From a genetic point of view the modes of transmission are varied and often confusing. Allan[2] concluded that the disease is due to unit traits conditioned by a single defective gene. The pattern of inheritance may be dominant, sex-linked recessive, or simple recessive, but in each family it preserves a very distinctive pattern.[50] Jacobs and Carr[68] recently reviewed the orthopedic problems presented by the condition.

The pathologic changes consist of chronic degeneration of peripheral nerves and nerve roots. There may be loss of nerve cells in the spinal ganglia and secondary degeneration of posterior roots and posterior columns of the spinal cord. There is disagreement as to whether other spinal tracts are involved, but we believe that other involvement of tracts in the spinal cord is coincidental.

For a considerable number of years there was a dispute as to the nature of the muscle lesions in this disease. After the appearance of Slauck's papers,[89, 91, 92] however, it seemed certain that the muscular changes were secondary to disease of motor nerve fibers and were not of the primary dystrophic or myopathic type. So far relatively few additional studies have been made, and these are not in complete agreement. Wohlfahrt and Wohlfart,[105] in their extensive investigations of muscle disease, reported two cases of progressive neural muscular atrophy in which the muscle lesions were of the dystrophic type, and Gallinek[55] described a case in which there was a mixed picture of atrophy and dystrophy. Pette[84] doubted the clinical diagnosis in the cases of Wohlfahrt and Wohlfart and argued that in his material the muscle changes were typical of spinal or neural muscular atrophy, an opinion corroborated by the detailed study of the anterior tibial muscle in one case by Brodal and Refsum,[15] and by Wohlfart.[107] Haase and Shy,[59] in their report on biopsies of muscle in a series of cases of this disease, stressed the findings of central nucleation, degeneration, phagocytosis, and regeneration of large muscle fibers, and cited them in their argument that the disease is really a neuromyopathy.

Our material on chronic peroneal muscular atrophy is limited but supports the view expressed above that the muscle atrophy is seconday to disease of the motor fibers in the peripheral nerves. Cross sections of the peroneal or anterior tibial muscles show an advanced degree of muscle atrophy. The majority of the muscle fibers in some microscopic fields are extremely small, some reduced to 8–10 μ in width (Fig. 10–17). Normal and hypertrophied

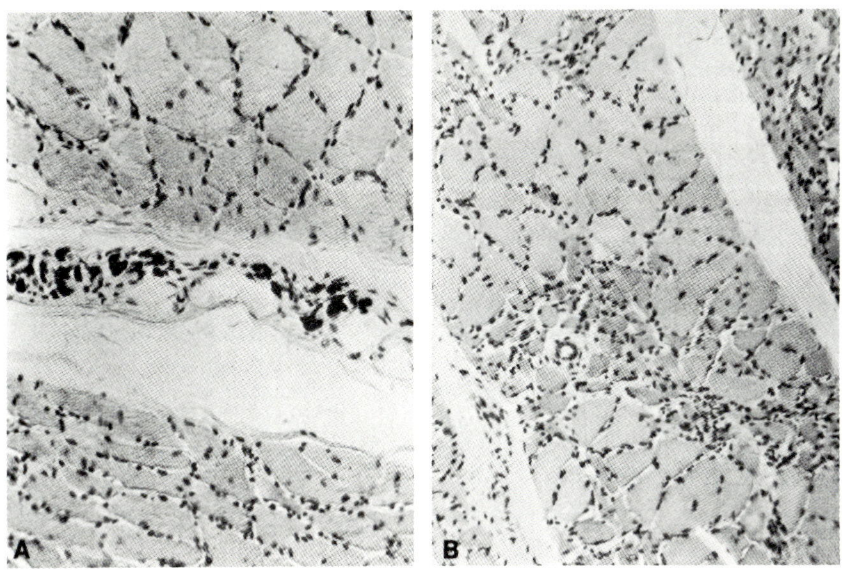

Fig. 10–15. Chronic polyneuropathy of unknown origin. (A) Pretibial muscle, showing normal muscle fibers (at top) and partially atrophic fibers (at bottom), separated by a strand of muscle fibers represented only by residual strands of nuclear remnants. (B) Partially degenerate group (motor unit) is sandwiched between two groups of unaffected fibers.

Fig. 10–16. Chronic polyneuropathy. Fibers of normal appearance, with others showing recent nuclear proliferation, lie next to an area of old atrophy, where nuclear residue alone represents muscle substance. (hematoxylin, Van Gieson)

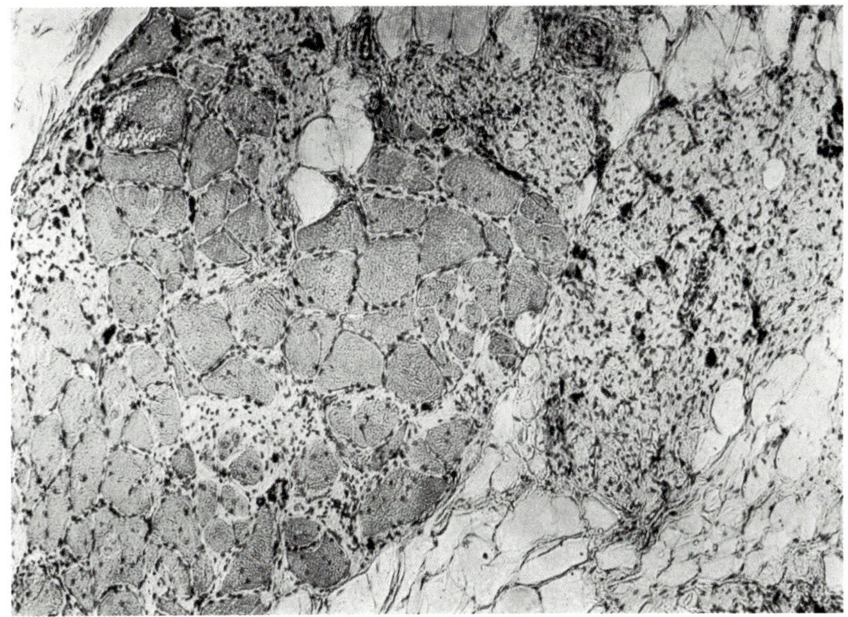

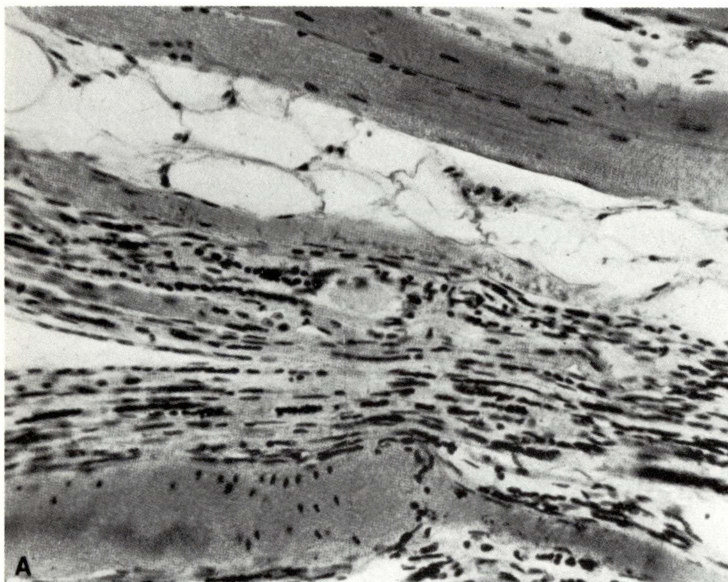

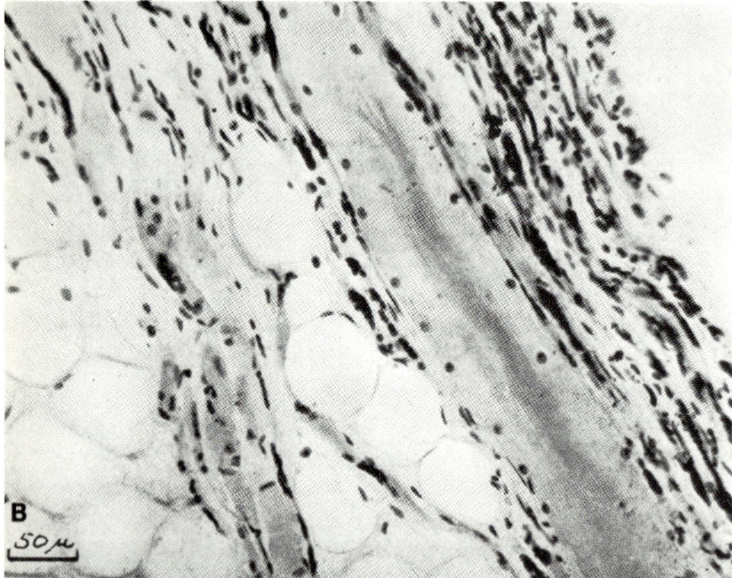

Fig. 10–17. Personal muscular atrophy. (A) Bundle of muscle fibers in late stages of neural atrophy contains fragments of degenerated fibers. At lower margin an atrophic fiber has evidently suffered antemortem rupture, with abortive nuclear proliferation followed by waxy degeneration of swollen traumatized segment (granular degeneration). (B) Large fiber is undergoing autolysis, interpreted as later stage of traumatic change seen in (A). Relationship of fat cells to nuclear remnants is also shown. (hematoxylin, Van Gieson)

fibers are present in other parts of the muscle and may be scattered among the trophic units (Figs. 10–17 through 10–19). In other words, while the atrophic fibers may be grouped in one part of the field and the normal and enlarged ones in another part, there is definite tendency in many places for these fiber types to overlap so that in any one area large, medium-sized, and small fibers are intermingled

Some of the atrophic fibers are so small as to be almost unrecognizable except at high magnification. They consist of slender tortuous strands of sarcoplasm, sometimes with visible striations but often hyalinized (Fig. 10–17). There are long rows of sarcolemmal nuclei in small muscle fibers. Fragmentation into spindle cells and other mononuclear forms is frequently observed. These nuclei are smaller and more darkly stained than the sarcolemmal nuclei of the less atrophied or normal fibers. Occasionally one or two muscle fibers present evidence of recent traumatic rupture (Fig. 10–17), but

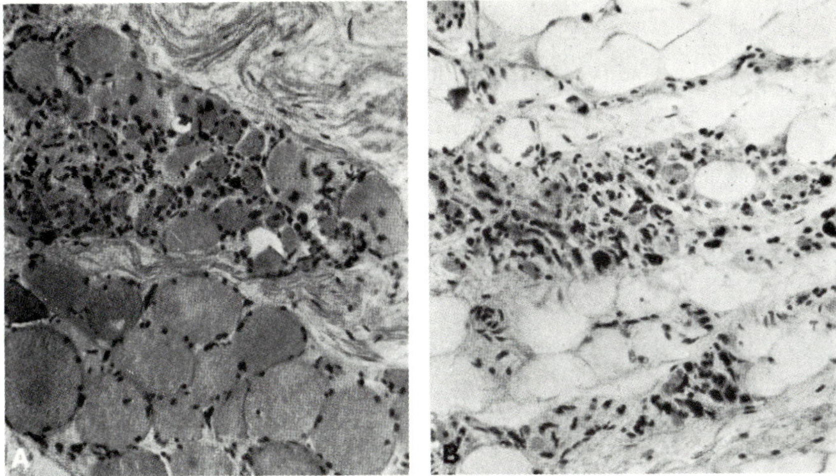

Fig. 10–18. Peroneal muscular atrophy in transverse section of severely affected muscle fibers. (A) Note various stages of atrophy in same fasciculus and hypertrophy of residual normal muscle fibers. (B) Degenerative changes are far advanced, with fatty replacement. (hematoxylin, Van Gieson)

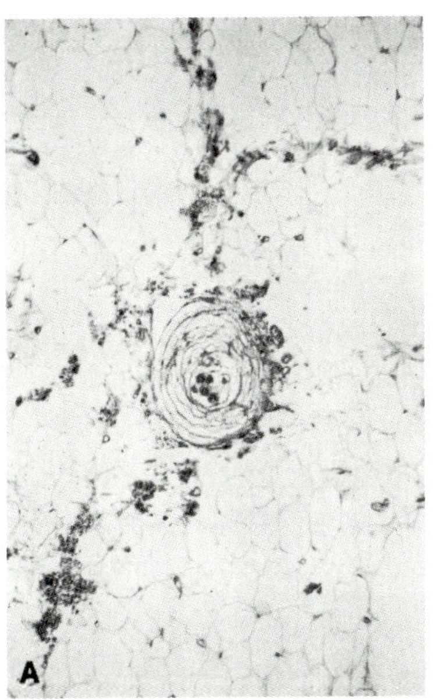

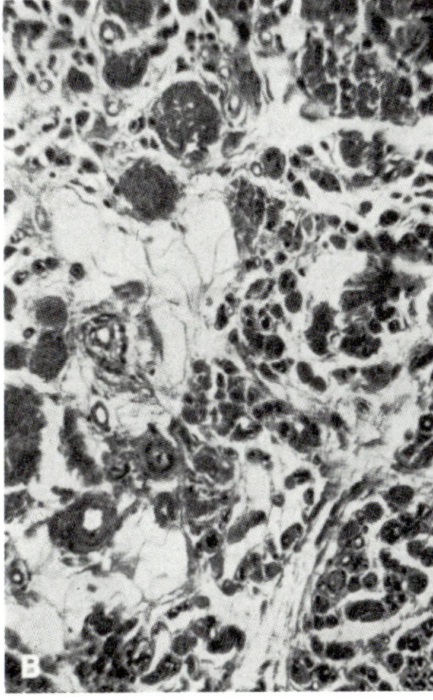

Fig. 10–19. Peroneal muscular atrophy. (A) Biopsy specimen shows a muscle spindle isolated in fatty tissue. Neighboring fibrous strands contain remnants of muscle fibers. (B) Very atrophic muscle in another part of same section. Numerous small circular structures are thick-walled capillaries. (H&E)

this is rare. Such fibers serve to emphasize the fragility of muscle undergoing atrophy. In the medium-sized fibers the nuclei retain their hypolemmal position and do not migrate centrally. There is little or no increase in connective tissue or fibroblasts, but fatty tissue is prominent. The muscle spindles may alone represent the muscular tissue (Fig. 10–19A), but the smallness of intrafusal fibers and lamination of the capsule are also manifestations of a chronic atrophic process. As in all very advanced atrophies the blood vessels, even the capillaries, appear to have thick walls (Fig. 10–19B).

In addition to these essentially atrophic changes, there is no question but that normal sized and hypertrophied fibers often show central condensation of sarcoplasm surrounded by a clearer zone (target fibers) as well as scattered degenerative and regenerative changes. It is this constellation of findings and the less clear grouping of atrophic and large fibers that has caused so much confusion as to the nature of the muscle disease. When judged by the distribution of the atrophied fibers, the rarity of muscle degeneration, the relative absence of fibroblastic proliferation, and infiltration of fat cells, this pathologic picture is undoubtedly that of a chronic neural muscle atrophy and not a dystrophy. This difficulty in distinguishing a denervation atrophy from a dystrophy in chronic neuropathies has been appreciated by all muscle pathologists for the past century. Durante[44] in his monograph calls attention to the problem and cautions against attributing the muscle lesion to whatever disease has affected the nerves or spinal cord. Drachman et al.[41] quite correctly point out that their occurrence during the late stages of poliomyelitic atrophy proves that some other factor was involved.

The target fiber with its central core of congealed sarcoplasm surrounded by a ring of loosely structured tissue forming a kind of ring was actually described in muscle biopsies of peronal muscular atrophy by Engel[48, 49] and has been confirmed by many authors. They are most frequent in the larger nonatrophied motor units. The election microscopic appearance is one of Z line smeared and dislocated along with actin and myosin filaments and sarcoplasmic reticulum. Glycogen and mitochondria are not visible in the core. Seen in 1921 by Buzzard and Greenfield[18] in amyotrophic multiple sclerosis and by others in poliomyelitis, the suggestion has been that target fibers are a sign of denervation. However, they have not been reproduced in denervation experiments in the experimental animal and are sometimes found in other human diseases of muscle. Engel[48, 49] in further studies relates them to "central core disease." Their significance therefore remains uncertain.

Chronic Hypertrophic Polyneuritis of Déjerine and Sottas. Chronic hypertrophic polyneuritis closely resembles peroneal muscular atrophy. It is sometimes hereditary, begins often during early childhood, and causes chronic progressive weakness and mild loss of sensation involving first the legs and later the hands. The chief point of difference is in the enlargement of the peripheral nerves (which are usually easily palpable owing to hyperplasia of the endoneurial connective tissue) and the frequent disorders of the sympathetic nervous system (miosis and the like). The muscle lesion in the few cases described was similar to that of any other chronic neural atrophy.

Diabetic Polyneuropathy. Diabetic "neuritis," a not infrequent complication of very chronic diabetes mellitus, is primarily a sensory neuritis and is the result of degenerative changes in the peripheral nerves. Pain is a prominent feature, and ataxia occurs late in the disease. Muscular weakness and atrophy in pronounced degree occur more frequently when one nerve is severely damaged, e.g., peroneal or femoral nerve. The muscle lesion then conforms to the general type of neural atrophy. We have not found any evidence of primary muscle change except in areas of ischemia due to vascular disease.

CONCLUSION

The morphologic changes in skeletal muscle in diseases which destroy either the motor neurons in the spinal cord or peripheral nerves are similar. The essential process is a progressive atrophy of the individual muscle fibers. The ultimate fate of the isolated, inactive muscle fibers appears to be fragmentation and degeneration. We find no evidence of transformation of muscle cells into fibroblastic or fatty tissue, though fat cells in time replace the muscle tissue if the process is sufficiently prolonged. The most advanced stage of atrophy is followed by clumping of darkly staining muscle nuclei within very thin sarcolemmal tubes. Rows of degenerate mononuclear myoblasts are sometimes seen. Residual clumps of pyknotic muscle nuclei without visible membrane or cytoplasm appear to be a final stage. Their long persistence in this form suggests that they contain a low metabolic residue and for a long time conserve viability.

In denervation of muscles of whatever cause, normal or hypertrophic muscle fibers undergo a variety of degenerative changes which bear no relation to the primary disease process. There are reasons for suspecting that they are due to trauma or overuse of the weakened muscle. Such changes are responsible for much of the difficulty in distinguishing chronic denervation from dystrophy in end-stage lesions. Fibrous tissue replacement and infiltration with fat cells seem to occur independently. Any differences between the various forms of spinal and neural atrophy depend on whether the disease process naturally affects all or part of the motor neurons simultaneously or in succession. Reinnervation modifies the pathologic picture.

In an earlier chapter it was proposed that the first stage of denervation atrophy is a reversion of muscle to a fetal state. This is most obvious in the nuclear amyotrophies of infancy, in which the muscle fibers in some motor units appear to have regained their fetal appearance owing to the fact that the motor nerve cells are destroyed so soon after innervation that full maturation and development of the fibers never occur.

REFERENCES

1. ADAMS RD: The giant muscle fiber: its place in myopathology, Modern Neurology. Edited by S Locke. Boston, Little Brown, 1969, Chap 16, pp 225–240

2. ALLAN W: Relation of hereditary pattern to clinical severity as illustrated by peroneal atrophy. Arch Intern Med 63:1123–1131, 1939

3. ANDERSON LR, REEVES DL: Amyotonia congenita (Oppenheim's disease): report of a case with autopsy findings. Bull Los Angeles Neurol Soc 5: 210–217, 1941

4. ASBURY AK, ARNASON BG, ADAMS RD: The inflammatory lesion in idiopathic polyneuritis: its role in pathogenesis. Medicine (Balt) 48:173–215, 1969

5. BANKER BQ, VICTOR M, ADAMS RD: Arthrogryposis multiplex due to congenital muscular dystrophy. Brain 80:319–334, 1957

6. BASTOS ANSART M: Tratimiento postural de la paralisis infantil. Arch Med Chir Espec 32:493–503, 1930

7. BATTEN FE: Progressive spinal muscular atrophy of infants and young children. Brain 33:433–463, 1911

8. BIELSCHOWSKY M: Ueber Myatonia congenita. J Psychol Neurol 38:199–233, 1929

9. BODIAN D: The virus, the nerve cell, and paralysis: a study of experimental poliomyelitis in the spinal cord. Bull Hopkins Hosp 83:1–107, 1948

10. BODIAN D: Histopathologic basis of clinical findings in poliomyelitis. Am J Med 6:563–578, 1949

11. BOWDEN REM, GUTMANN E: Denervation and reinnervation of human voluntary muscle. Brain 67: 273–313, 1944

12. BRANDT S: Hereditary factors in infantile muscular atrophy. Am J Dis Child 78:226–236, 1949

13. BRANDT S: Werdnig-Hoffmann's Infantile Progressive Muscular Atrophy. Copenhagen, Munksgaard, 1950

14. BRASSARD J: Etude clinique sur une forme héréditaire d'atrophie musculaire progressive débutant par les membres infériers (type fémoral avec griffe der orteils). Thesis, Paris, 1886

15. BRODAL A, REFSUM S: Progressive neural muscular atrophy. Acta Psychiatr Neurol 17:99–122, 1942

16. BUCHANAN D: Some disorders of the motor unit in infancy and childhood. Med Clin North Am 34: 147–164, 1950

17. BURDICK WF, WHIPPLE DV, FREEMAN W: Amyotonia congenita (Oppenheim): report of 5 cases with necropsy; discussion of the relationship between amyotonia congenita, Werdnig-Hoffmann disease, neonatal poliomyelitis and muscular dystrophy. Am J Dis Child 69:295–307, 1945

18. BUZZARD EF, GREENFIELD JG: Pathology of the Nervous System. London, Constable, 1921

19. BYERS RK, BANKER BQ: Infantile muscular atrophy. Arch Neurol 5:140–164, 1961

20. BYERS RK, TAFT LT: Chronic multiple peripheral neuropathy in childhood. Pediatrics 20:517–537, 1957

21. CAREY EJ, MASSOPUST LC, ZEIT W, ET AL: Anatomic changes in motor nerve endings in human muscles in early poliomyelitis. J Neuropathol Exp Neurol 3:121–130, 1944

22. CAZZATO G, WALTON JN: The pathology of the muscle spindle: a study of biopsy material in various muscular and neuromuscular diseases. J Neurol Sci 7:15–70, 1968

23. CHAMBERS R, MACDERMOT V: Polyneuritis as a cause of "amyotonia congenita." Lancet 1:397–401, 1957

24. CHARCOT JM, MARIE P: Sur une forme particuliere d'atrophie musculaire progressive souvent familiale debutante par les pieds et ies jambes et

atteignant plus tard les mains. Rev Med 6:97–138, 1886

25. CHOR H: Nerve degeneration in poliomyelitis; changes in the motor nerve endings. Arch Neurol Psychiatry 29:344–358, 1933

26. COËRS C, WOOLF AL: The Innervation of Muscle: A Biopsy Study. Springfield, Ill, Charles C Thomas, 1959

27. COLLIER J, HOLMES G: The pathological examination of 2 cases of amyotonia congenita with the clinical description. Brain 32:269–284, 1909

28. CONEL JL: Distribution of affected nerve cells in a case of amyotonia congenita. Arch Neurol Psychiatry 40:337–351, 1938

29. CONEL JL: Distribution of affected nerve cells in amyotonia congenita (second case). Arch Pathol 30:153–164, 1940

30. COUNCILMAN WT, DUNN CH: Myotonia congenita: report of a case with autopsy. Am J Dis Child 2:340–355, 1907

31. DARKSCHEWITCH L: Die pathologische anatomie der muskeln. Handbuch du path Anat d Nervensystems. Edited by E Flatau, L Jacobsohn, L Minor, Berlin, Karger, 1904, pp 1218–1270

32. DASTUR DK: The motor unit, in leprous neuritis: a clinico-pathological study. Neurol Bull Neurol Soc India 4:1–27, 1956

33. DEHIO K: On the lepra anesthesia and the pathogenetical relation of its disease-appearance. (Translated by HG Biswas) leprosy in India 24:78, 1952

34. DENNY-BROWN D: Primary sensory neuropathy with muscular changes associated with carcinoma. J Neurol Neurosurg Psychiatry 11:73–87, 1948

35. DENNY-BROWN D: Interpretation of the electromyogram. Arch Neurol Psychiatry 61:99–128, 1949

36. DENNY-BROWN D, BRENNER C: Paralysis of nerve induced by direct pressure and by tourniquet. Arch Neurol Psychiatry 51:1–26, 1944

37. DENNY-BROWN D, PENNYBACKER JB: Fibrillation and fasciculation in voluntary muscle. Brain 61:311–334, 1938

38. DENNY-BROWN D, SCIARRA D: Changes in the nervous system in acute porphyria. Brain 68:1–16, 1945

39. DENST J, NEUBUERGER KT: A histologic study of muscles and nerves in poliomyelitis. Am J Pathol 26:863–881, 1950

40. D'HARCOURT J, MAZO L: Contribución al estudio de la anatomia patológica de los músculos en la poliomielitis infantil: consecuencias clínicas y terapéuticas que se derivan de este estudio. Arch Neurobiol 13:333–342, 1933

41. DRACHMAN DB, MURPHY SR, NIGAM MP, ET AL: Myopathic changes in chronically denervated muscle. Arch Neurol 16:14–27, 1967

42. DUBLIN WB, BEDE BA, BROWN BA: Pathologic findings in nerve and muscle in poliomyelitis. Am J Clin Pathol 14:266–272, 1944

43. DUBOWITZ V: Enzyme histochemistry of skeletal muscle. 3. Neurogenic muscular atrophies. J Neurol Neurosurg Psychiatry 29:23–28, 1966

44. DURANTE G: Anatomie pathologique des muscles. Manual d'Histologie Pathologique. Third edition. Edited by V Cornil, L Ranvier. Paris, Félix Alcan, Vol 2, 1902

45. EDDS MACV: Collateral regeneration in partially denervated muscles of the rat. J Exp Zool 129:2, 1955

46. EDSTRÖM L, KUGELBERG E: Histochemical composition, distribution of fibers and fatigability of single motor units. J Neurol Neurosurg Psychiatry 31:424–433, 1969

47. EMERY AEH: The Nosology of the spinal muscular ahophies J. Med. Gen. 8:481–495, 1971

48. ENGEL WK: Muscle target fibers, a newly recognized sign of denervation. Nature (Lond) 191:389–390, 1961

49. ENGEL WK: A critique of congenital myopathies and other disorders. Exploratory Concepts in Muscular Dystrophy and Related Disorders. Edited by AT Milhorat. Amsterdam, Excerpta Medica, 1967, pp 27–40

50. ENGLAND AC, DENNY-BROWN D: Severe sensory changes and trophic disorder, in peroneal muscular atrophy (Charcot-Marie-Tooth type). Arch Neurol Psychiatry 67:1–22, 1952

51. EPSTEIN JA: Amyotonia congenita: a report of 3 cases with a review of the literature. J Mt Sinai Hosp 16:149–174, 1949

52. FENICHEL GM, ENGEL WK: Histochemistry of muscle in infantile spinal muscular atrophy. Neurology (Minneap) 13:1054–1066, 1963

53. FOOT NC: Report of a case of amyotonia congenita with autopsy. Am J Dis Child 5:359–373, 1913

54. FORBUS WD, WOLF FS: Amyotonia congenita (Oppenheim's disease) in identical twins. Bull Hopkins Hosp 47:309–322, 1930

55. GALLINEK A: Zum Wesen der neuralen Muskelatrophie. Dtsch Z Nervenheilk 114:74–94, 1930

56. GARDNER-MEDWIN D, HUDGSON P, WALTON JN: Benign spinal muscular atrophy arising in childhood and adolescence. J Neurol Sci 5:121–158, 1967

57. GIBSON A: Muscular infantilism. Arch Intern Med 27:338–350, 1921

58. GREENFIELD JG, STERN RO: The anatomical identity of the Werdnig-Hoffmann and Oppenheim forms of infantile muscular atrophy. Brain 50:652–686, 1927

59. HAASE GR, SHY GM: Pathological changes in muscle biopsies from patients with peroneal muscular atrophy. Brain 83:631–637, 1960

60. HASSIN GB: Amyotonia (myatonia) congenita: Oppenheim's disease: a clinical-pathological report of a case. J Neuropathol Exp Neurol 4:240–249, 1945

61. HAUSMANOWA-PETRUSEWICZ I, ZELINSKA S: Zur neurologischen stellung des scapuloperonealen syndroms. Dtsch Z Nervenheilk 183:377–382, 1962

62. HAYEM G: Recherches sur l'Anatomie Pathologiques des Atrophies Musculaires. Paris, Masson, 1877

63. HIPPS HE: The clinical significance of certain microscopical changes in muscles of anterior poliomyelitis. J Bone Joint Surg 24:68–80, 1942

64. HOFFMANN J: Ueber progressive neurotische Muskelatrophie. Arch Psychiatr Nervenheilk 20:661–713, 1889

65. HOFFMANN J: Ueber chronische spinale Muskelatrophie im Kindsalter, auf familiärer Basis. Dtsch Z Nervenheilk 3:427–470, 1893

66. HORÁNYI-HECHST B: Zur Histopathologie der menschllehen Poliomyelitis aenta anterior. Dtsch Z Nervenheilk 137:1–54, 1935

67. HYLAND NN, RUSSELL WR: Chronic progressive polyneuritis with report of a fatal case. Brain 53:278–289, 1930

68. JACOBS JE, CARR CR: Progressive muscular atrophy of the peroneal type (Charcot-Marie-Tooth disease): orthopaedic management and end-result study. J Bone Joint Surg [Am] 32A:27–38, 1950

69. KAESER HE: Scapuloperoneal muscular atrophy. Brain 88:407–418, 1965

70. KOERNER DR: Amyotrophic lateral sclerosis on Guam: a clinical study and review of the literature. Ann Intern Med 37:1204–1220, 1952

71. KOPITS I: Beiträge zur Muskelpathologie histologische Befunde an Muskeln, Nerven und Blutgefässen in Spät- und Endstadien peripheren Lähmungen, mit besonderer Berücksichtigung der Poliomyelitis Anterior Acuta. Arch Orthop Unfallchir 27:277–403, 1929

72. KRABBE KH: Congenit generalisetret muskelaplasi 338th meeting of Danish Neurol Soc, Feb. 27, 1946 (see Brandt[13])

73. KUGELBERG E, WELANDER L: Heredofamilial juvenile muscular atrophy simulating muscular dystrophy. Arch Neurol Psychiatry 75:500–509, 1956

74. KURLAND LT, MULDER DW: Epidemiologic investigations of amyotrophic lateral sclerosis. 1. Preliminary report of geographic distribution, with special reference to the Mariana islands, including clinical and pathologic observations. Neurology (Minneap) 4:355–378, 438–448, 1954

75. LAPRESLE J, MILHAUD M: Pathologie du fuseau neuro-musculaire. Rev Neurol (Paris) 110:97–122, 1964

76. LEVESQUE J, LEPAGE F, BOESWILLWALD M, ET AL: Congenital familial muscular dystrophy simulating Werdnig-Hoffmann-Oppenheim disease. Arch Fr Pediatr 13:202–207, 1956

77. LEWEY FH: Myatonia congenita (Oppenheim). Am J Dis Child 63:76–88, 1942

78. LORENZ H: Die muskelerkrankungen. Handbuch der Speziellen Pathologie und Therapie innerer Krankheiten. Edited by H Nothnagel. Vienna, Holder, 1904, Vol II, pp 1–727

79. MASTAGLIA FL, WALTON JN: Histological and histochemical changes in skeletal muscle from cases of chronic juvenile and early adult spinal muscular atrophy (the Kugelberg-Welander syndrome). J Neurol Sci 12:15–44, 1971

80. MITTELBACH F: Die Begleitmyopathie bei Neurogenen Atrophien. Monographien auf dem gesampt gebeit der Neurol u Psychiat, Heft 113. Berlin, Springer-Verlag, 1966

81. NEUMANN P: Zur patholgischen anatomie der myatonia congenita. Dtsch Z Nerveneilk 71:95–115, 1921

82. NIELSEN JM: Myatonia congenita of Oppenheim: report of a unique case. Am J Dis Child 35:82–86, 1928

83. OPPENHEIM H: Ueber allgemeine und localisierte Atonie der Muskulatur (Myatonia) im frühen Kindsalter. Monatsschr Psychiatr Neurol 8:232–233, 1900

84. PETTE H: Neurale muskelatrophie. Handbuch der Neurologie. Edited by O Bumke, O Foerster. Berlin, Springer, 1936, pp 497–524

85. ROSIN A: Beitrag zur lehre von der muskelatrophie. Beitr Pathol 65:487–534, 1919

86. SCHREIER K, HUPERZ R: Congenital generalized muscular hypoplasia. Ann Pediatr 186:241–248, 1956

87. SEITZ D: Zur nosologischen stellung des sogenannten scapulo-peronealen syndroms. Dtsch Z Nervenheilk 175:547–552, 1957

88. SHY GM, ENGEL WK: A critique of congenital myopathies and other disorders. Exploratory Concepts in Muscular Dystrophy. Edited by AT Milhorat. Amsterdam, Excerpta Medica, 1967, pp 27–40

89. SLAUCK A: Beiträge zur kenntniss der muskelpathologie. Z Neurol Psychiatr 71:352–356, 1921

90. SLAUCK A: Ueber myatonia congenita und infantile progressive spinale muskelatrophie. Dtsch Z Nervenheilk 67:1–28, 1921

91. SLAUCK A: Ueber progressive hypertrophische neuritis (Hoffmannsche krankheit). Z Gesamte Neurol Psychiatr 92:34–77, 1924

92. SLAUCK A: Histopathologische untersuchungen bei neuraler myopathie. Klin Wochenschr 2:2245–2247, 1928

93. SLAUCK A: Pathologische anatomie der myopathien. Handbuch der Neurologie. Edited by O

Bumke, O Foerster. Berlin, Springer, 1936, Vol IX, pp 412–431

94. SPILLER WG: Generalized or localized hypotonia in childhood: myatonia congenita. Univ Penn Med Bull 17:342–346, 1905

95. STRANSKY E: Ueber discontinuirliche zerfallsprozesse an der peripheren nervenfaser. J Physiol Neurol 1:169–199, 1902

96. STRÜMPELL A: Ueber spinale progressive muskelatrophie und amyotrophische seitenstrangklerose. Dtsch Arch Klin Med 42:230–260, 1888

97. THIEFFRY S, ARTHUS M, BARGETON E: Wernig-Hoffman: 40 cases with 11 autopsies. Rev Neurol (Paris) 93:621–644, 1955

98. THOMSEN J, BRUCE A: Progressive muscular atrophy in a child. Edinburgh Hosp Rep 1:361–383, 1893

99. TOOTH HH: The Peroneal Type of Progressive Muscular Atrophy. London, Lewis, 1886

100. TURNER JWA: The relationship between amyotonia congenita and congenital myopathy. Brain 63:163–177, 1940

101. TURNER JWA: On amyotonia congenita. Brain 72:25–34, 1949

102. VON MEYENBURG H: Die quergestreifte muskulatur. Handbuch der Speziellen Pathologischen Anatomie und Histologie. Edited by F Henke, O Lubarsch. Berlin, Springer, 1929, pp 299–507

103. WALTON J: Amyotonia congenita. Lancet 1:1023–1028, 1956

104. WERNIG G: Zwei fruhinfantile hereditäre fälle von progress der muskelatrophie unter dem bilde der dystrophie ober auf neurotischer grundlage. Arch Psych Nervenheilk 22:437–480, 1891

105. WOHLFART G: Muscular atrophy in diseases of the lower motor neuron; contribution to the anatomy of the motor units. Arch Neurol Psych 61:599–620, 1949

106. WOHLFART G: Collateral regeneration in partially denervated muscles. Neurology (Minneap) 8:175, 1958

107. WOHLFART G, FEX J, ELIASSON S: Hereditary proximal muscular atrophy—a clinical entity simulating progressive muscular dystrophy. Acta Psychiatr Neurol Scand 30:395–406, 1955

108. WOHLFAHRT S, WOHLFART G: Mikroskopische untersuchungen an progressiven muskelatrophien; unter besonderer rücksichtsnahme auf rückenmarks-muskel-befunde. Acta Med Scand [Suppl] 63:1–137, 1935

109. WOIT O: The spinal cord, peripheral nerve and skin patches of leprosy macules. (Translated by HG Biswas) Leprosy in India 24:133, 1955

110. WOOLF AL, TILL K: The pathology of the lower motor neurone in the light of new muscle biopsy techniques. Proc R Soc Med 48:189–198, 1955

chapter 11

MISCELLANEOUS DISEASES

*toxic, metabolic, and
endocrine diseases*

All diseases of skeletal muscle which at the moment cannot be classified under any of the preceding chapter headings are taken up here and in the following chapter. In a few, such as primary amyloidosis and myositis ossificans, there is a characteristic histologic process. In others, exemplified by myasthenia gravis, myotonia congenita, and familial periodic paralysis, a maximal disturbance of function subtly involves derangements of conduction of the motor nerve impulse and is attended by no significant histologic change in the muscle fiber itself. The latter group is considered in Chapter 12.

STRIATED MUSCLE CHANGES ASSOCIATED WITH ACUTE AND CHRONIC INFECTIOUS DISEASES

The effects of several physical agents on the sarcoplasmic constituents of muscle were reviewed in Chapter 3. Changes of similar character are known to accompany many infectious diseases and are assumed to be related to a toxic or metabolic disturbance incident to the systemic infection. Thus the disease process in the strict sense is a toxic or metabolic myopathy and not a myositis. The special pathologic aspects of this muscle disorder and the circumstances under which it occurs are considered in further detail here.

Louis is credited by Zenker[192, 193] for the original observation in 1830 of a gross morphologic change in skeletal muscle in a fatal case of typhoid fever. In 1844 Rokitansky (quoted by Zenker[192]) noted at postmortem examination a rupture of the rectus abdominis muscle due supposedly to the same type of lesion, and Virchow[181] presented the earliest description of the microscopic pathology in 1857. However, the first intensive study of this pathologic process was carried out by Zenker,[192] and his name later became attached to it. Zenker examined the skeletal muscles of 120 patients dying of typhoid fever in the epidemics in Dresden between the years 1859 and 1862. He referred to the lesion as a hyaline degeneration of muscle fibers and regarded it as a characteristic feature of that disease. Other pathologists subsequently found the same hyaline degeneration in a number of different infectious diseases and concluded that it was nonspecific. Further information on this subject can be obtained from the excellent paper of Forbus.[56]

Extensive surveys of skeletal muscle in all types of infectious disease were conducted by several pathologists (Hayem[69] in 1876, Durante[48] in 1902, Lorenz[100] in 1904, Marinesco[106] in 1910, Jewesbury and Topley[84] in 1912, von Meyenburg[183] in 1929 and others). In addition to Zenker's hyaline degeneration other diverse pathologic alterations have been encountered, such as edema, vacuolation, and atrophy of fibers. Pigmentary and fatty degeneration and amyloid deposits have also been described in infectious and other general diseases of chronic type.

These publications establish the existence of three types of skeletal muscle change in infectious disease: (1) Zenker's hyaline degeneration, related either to a microbial toxin or a metabolic disturbance and occurring in both acute and chronic infections, is the most frequent. (2) Atrophy of muscle fibers sometimes associated with vacuolation and edema accompanies chronic infections that lead to malnutrition and cachexia and may occur in the absence of infections either as a disorder of nutrition or as a consequence of aging. (3) Focal myositis due to metastatic infection, the least common, was dealt with in Chapter 7.

ZENKER'S HYALINE OR WAXY DEGENERATION

Hyaline or waxy degeneration may occur in any muscle but is especially common in the large muscles of trunk and pelvic, and shoulder girdles. It is most frequent in the rectus abdominis and other abdominal muscles, the abductor muscles of the thigh, the diaphragm, biceps brachii, pectoralis major, subscapularis, and gastrocnemius. Small muscles are affected less often, but lesions have been seen even, according to Zenker, in the laryngeal muscles and in the tensor tympani. An identical lesion in the heart is believed to account for tachycardia and at times unexpected death. Wells[186] observed hyaline degeneration of the muscle fibers of the diaphragm in fatal cases of lobar pneumonia, and he reasoned that some of the deaths from pneumonia may be due to progressive failure of the respiratory musculature.

There is no exact clinical counterpart of acute hyaline degeneration of muscle. It usually causes no definite symptoms and is a chance finding at autopsy (Chapter 4). When the muscle lesions are severe it is possible they may bear some relation to the general weakness, fatigue, and malaise that characterize the convalescence from prolonged infectious diseases such as typhoid fever, influenza, brucel-

losis, and infectious hepatitis. If hemorrhage complicates Zenker's degeneration there may be pain, local swelling, and subsequent discoloration of the subcutaneous tissues. Zenker's degeneration is found in a wide variety of serious diseases but is most marked in the patients who exhibit severe toxic manifestations. Therefore it is unlikely to be the cause of the muscular aches of lesser febrile states. It may well be the origin of muscular contractures which follow severe typhus, though in our experience these have tended to occur in the distal parts of the limbs.

On microscopic examination the first alteration, recognizable within a few hours of its inception, is an uneven swelling and tortuosity of the muscle fibers. The substance of the fiber is often changed; there may be a cloudy swelling with obscuration of striations in some places and their preservation in others. Fragmentation into horizontal segments (Bowman's discoid degeneration) is a frequent finding. The sarcolemmal nuclei are slightly swollen and may proliferate quite early, as noted by Waldeyer[185] in 1865.

At a more advanced stage, during the second and third week, a wide variety of changes is demonstrable. The most striking feature is the presence of cylindrical highly refractile, brightly eosinophilic masses within the muscle fibers (Fig. 11–1). In places faint transverse striations or a transverse fragmentation indicate their origin from the contents of the muscle fiber. Other fibers are finely granular. The sarcolemmal tube is still identifiable at this stage and can be seen not only to enclose the hyaline segments but to bridge the gaps between them. It is in these empty spaces, where the sarcous substance has broken, that phagocytic cells appear. A single fiber, if fortuitously cut in perfect longitudinal section, may display all these different changes; one portion of the sarcoplasm may appear granular, other parts may be hyalinized and fragmented, and the remainder may retain the natural appearance of a healthy fiber. It is for this reason that Hayem[69] in 1876 applied the term granulovitreous degeneration to this pathologic process. Other fibers, less numerous than the above, have intact striations but are vacuolated (Fig. 4–3). It should be pointed out that granular

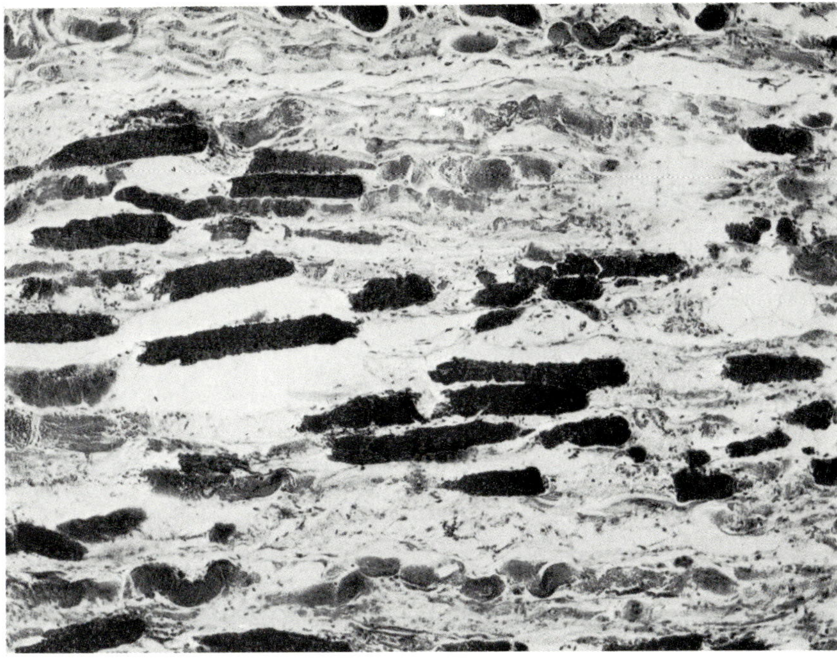

Fig. 11–1. Zenker's degeneration in rectus abdominis muscle from a patient who died of typhoid fever. Refractile and deeply eosinophilic qualities of some hyalinized muscle fibers accounts for their dark appearance. (H&E)

degeneration is less marked in this type of muscle disorder than it is in focal myositis and in severe neural atrophy. The lesion also varies in extent. Large areas of the muscle may have been transformed into hyaline segments, and the appearance is then roughly comparable to the effect of ischemia or the action of physical agents. More often, however, the waxy segments are interspersed with healthy fibers, and each row of them is the result of damage to a single muscle fiber.

During all this period the essential change is confined to the substance of the muscle fiber. There are no significant modifications of the connective tissue, aside from a generally loose reticulated appearance which, though usually attributed to edema, is difficult to judge in routine microscopic preparations. Infiltrations of inflammatory cells are not prominent, as Zenker pointed out. Scattered neutrophilic leukocytes are occasionally seen in or near the freshly necrotic parts of the muscle fiber. Later a few mononuclear cells infiltrate the empty or partially collapsed sarcolemma or collect around the blood vessels.

The necrotic fibers do not remain hyalinized for long. Soon they assume a granular appearance and become vacuolated. By appropriate stains it can be shown that some of these vacuoles contain fat. These remnants of sarcoplasm become surrounded by mononuclear leukocytes or histiocytes, which proceed to remove them by phagocytosis. Contrary to the opinion of Durante[48] and others, these phagocytes or myophages can be distinguished from the sarcolemmal nuclei (Fig. 4–2A) by their compact single or multiple nuclei, their rather heavy nuclear chromatin and lack of prominent nucleoli, their distinct cell membranes, and by the presence of particulate material in their cytoplasm. Forbus[57] proved their reticuloendothelial origin by demonstrating their capacity to assimilate vital dyes.

The earliest sign of regenerative activity may be seen in places in which an undamaged part of a muscle fiber adjoins a hyaline segment. As stated above, the proliferation of sarcolemmal nuclei occurs at an early stage, probably beginning within 24–48 hours. These nuclei become plump and hyperchromatic and acquire a large nucleolus. They appear to have undergone division by a simple mitotic process; however mitoses are rarely if ever seen. At first these cells are lined up in rows just beneath the sarcolemma, but later they fill the collapsed sarcolemmal tube, being arranged in compact cellular cords. Some pathologists attempted to distinguish two types of nuclei among them—a muscle cell nucleus and a satellite cell nucleus —but careful study in different stages of reaction convinces us that they are all of one type, differing only in relation to the sarcolemma, and mixed in places with the phagocytes mentioned above. Increasing amounts of sarcoplasm collect between and around these syncytial masses of hyperplastic nuclei, which replace the hyaline masses of necrotic muscle substance as the latter are removed by phagocytosis.

Volkmann[182] in 1893 remarked that one rarely sees regeneration by free budding from intact parts of the fiber. Instead the new fiber is formed by a blending of these rows of proliferating sarcolemmal nuclei (which are really a series of "seedlings"); this process was referred to by Durante[48] as regeneration of embryonic type (Fig. 4–2). Within a few days myofibrils begin to appear in the sarcoplasm between these cells. Young fibers can be distinguished from the old ones for a long time by their larger and more rounded nuclei, their basophilic staining reaction, and their delicate striation.

The type of phagocytic and regenerative activity depends ordinarily on the extent of the damaged areas. Usually the necrotic segments are separated from each other by healthy muscle fibers and perimysium. Each fragment then shows waxy changes and early and rapid phagocytosis throughout its entire length. Regenerative activity is exhibited not only at the ends but also throughout the damaged segment where there are surviving sarcolemmal cells. This regenerative process is possible of course only when a certain portion or all of the sarcolemmal nuclei survive, and in many respects it resembles the type of reparative process which follows experimental muscle damage by the agent plasmocid (Chapter 3). Survival of the interstitial tissue and the sarcolemma with

its nuclei is the reason for the perfect alignment of new fibers.

In other parts whole regions of the muscle may have become necrotic. The fibers in the center then retain a hyaline appearance for a longer period. The sarcolemmal nuclei are also necrotic, and even the connective tissue in the central parts of the lesion may be damaged to some degree. Phagocytosis and later regeneration then proceeds from the borders of the necrotic area by budding, as after experimental production of heat necrosis or ischemic necrosis. In both types of injury the formation of aberrant and abortive muscle buds is seldom seen unless hemorrhage or rupture of the muscle totally destroys the architecture of the endomysial tubes.

This pathologic picture may be modified by certain complications, the two most frequent of which are rupture of muscle and hemorrhage. Ruptures involve a few fibers or fascicles of fibers and may occur in the course of muscle activity. The rupture seems to be related to the friability of the damaged muscle tissue. The most frequent sites are the rectus abdominis and less often the muscles of the thorax, neck, and pelvic girdle. Reddish or mottled red and white discoloration can be seen in the vicinity of the rupture. The brick red or brownish coloration of extravasated blood may be noted if the rupture occurred days or weeks before death. Upon microscopic examination the muscle substance has disappeared, leaving large fields of only connective tissue, red corpuscles, and blood vessels. Vitreous changes may be detected in neighboring fibers in the early stage of such lesions; later the usual cellular reactions of Zenker's degeneration as well as phagocytosis of red corpuscles and the presence of blood pigment are in evidence. Muscle giant cells are present together with attempts at muscle regeneration by budding. These histiocytic and connective tissue reactions are similar to those which occur after traumatic rupture or hemorrhage.

Hemorrhage, the other frequent complication, may develop in the same muscles that are most liable to rupture. It is generally believed that hemorrhages in muscle during the first hours or days of an infectious disease are the result of focal vascular damage from bacterial emboli or their toxins and that hemorrhage later in the course of the illness is nearly always associated with Zenker's degeneration and rupture of muscle. The hemorrhage may be small, the size of an ecchymosis, or may form a large hematoma. Aside from the histologic evidence of hyaline degeneration, the gross and microscopic changes do not differ from those already outlined in Chapter 9 on muscle hemorrhage.

Localization of bacteria in the affected muscle may produce abscess formation with its characteristic morphology. This was discussed under bacterial myositis. "Muscle collapse" is another infrequent complication that we have never seen. It was described by Zenker as a condition in which all the muscle fibers become thin and pale, lose their striation, and assume irregular tortuous contours. It is not clear whether this condition affects a single muscle or many of the skeletal muscles, representing in the latter instance simply atrophy following prolonged malnutrition or some other process.

Zenker's waxy or hyaline degeneration was reported with greatest frequency in typhoid fever, but Durante remarked on its high incidence in smallpox. Similar lesions have also been found less often in scarlet fever, diphtheria, tuberculosis, staphylococcic and streptococcic septicemias, tetanus, cholera, paratyphoid A and B, infectious hepatitis, and yellow fever, as well as in a number of noninfectious diseases such as advanced malignancy and decompensated cirrhosis. The characteristic condition was also described by Wolbach and Frothingham[190] in epidemic influenza and by Forbus[56] in pneumonia. It seems that this type of toxic myopathy is liable to occur in any severe infection. At present, however, since typhoid fever and smallpox—the two diseases with the highest incidence of myopathy—are almost nonexistent, it is not a frequent pathologic finding. We are impressed with the fact that in the commoner acute infectious diseases of today, hyaline, granular, and fatty degenerations of muscle are slight or often entirely absent.

The distinctive feature of toxic myopathy is that it is commonly confined to the sarcous substance, which undergoes acute degeneration and spares to a considerable degree the sarcolemma, connective tissues, and blood vessels (Chapters 3 and 4). It illustrates convincingly the completely different susceptibility and diverse pathologic reactions of the muscle parenchyma and its stroma. When there is any significant degree of fibrous tissue reaction or infiltration of the stromatous tissue by inflammatory cells, the possibility of a true myositis must be considered even though an infective agent cannot be demonstrated. Much of the uncertainty as to the status of the muscle lesions which von Meyenburg,[183] Jeghers et al.,[82] and Sheldon[159] observed in Weil's disease arises from a failure on the part of pathologists to distinguish these parenchymal and interstitial reactions. In Weil's disease the calf muscles are mentioned specifically as being contracted postmortem and showing mottled discoloration, jaundice, and hemorrhagic spots or streaks 0.1–0.5 cm in diameter. Microscopically these lesions are focal in distribution and in the early stages consist of vacuolation, granularity or hyalinization, and fragmentation of the sarcoplasm. Some of the hyalinized segments may be yellowish green. A slight infiltration of neutrophilic leukocytes and mononuclear cells occurs in the connective tissue, occasionally in necrotic portions of the muscle fiber. Phagocytosis of degenerate muscle substance, proliferation of sarcolemmal nuclei, and slight proliferation of connective tissue complete this picture. In our material, which consists of only a few examples of this lesion, the connective tisssue reaction and cellular infiltration have been slight, and the pathologic change closely resembles Zenker's hyaline degeneration. It is notable, however, that the shredded appearance, characteristic of necrosis in the state of extreme contraction, is frequently present (Fig. 11–2A). In each case we attempted, without success, to demonstrate leptospiral organisms in the muscle lesions (Fig. 11–2B). In a recent report Sheldon[160] convincingly demonstrated small quantities of leptospiral antigen (presumably the organisms themselves) in the margins of focal necrotic muscular lesions by using a fluorescein-labeled antibody as the indicator. This evidence suggests that the muscle lesion in Weil's disease represents a true spirochetal myositis. However,

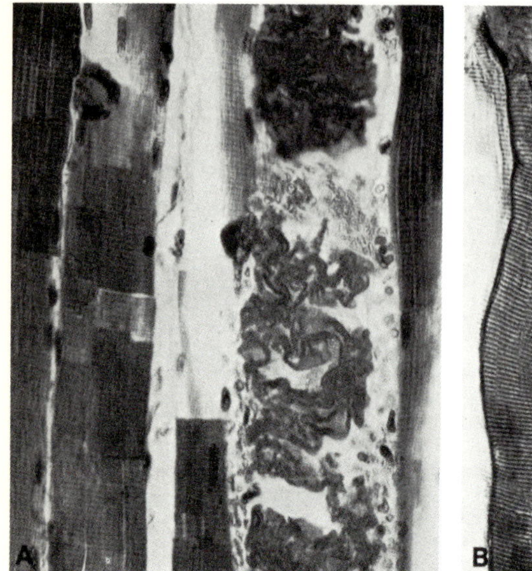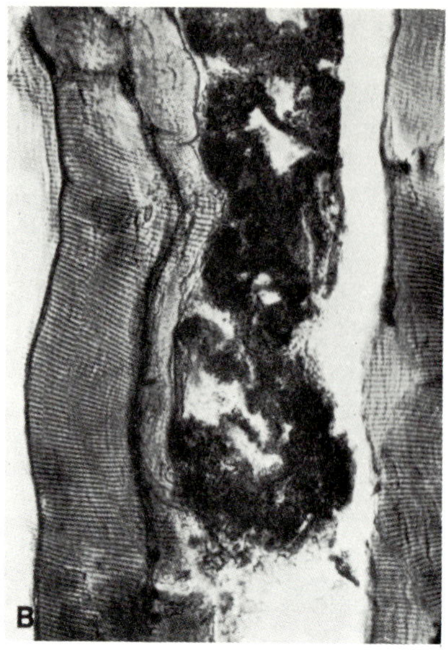

Fig. 11–2. Gastrocnemius muscle from a patient who died of leptospiral myopathy. Degenerated muscle fibers have frayed appearance in H&E stain (A), and granular appearance in Levaditi stain (B). Latter stain revealed large numbers of spirochetes in liver and kidneys, but none in the muscle.

we prefer to withhold final decision on this problem until the organisms themselves have been identified within the lesion.

The possibility of jaundice producing changes in the skeletal muscles was investigated by Houghton and Parker,[73] who examined the muscles in an unselected group of patients suffering from acute and chronic hepatic insufficiency with jaundice and found no definite abnormalities. Negative results were also obtained in a few cases of lower nephron nephrosis, which may accompany liver disease.

ATROPHY IN CHRONIC INFECTIOUS DISEASES

A variety of pathologic factors may operate in chronic infectious diseases. In addition to the toxic myopathy of either primary or secondary infection, malnutrition leading at times to advanced cachexia, polyneuropathy with neural muscular atrophy, and disuse atrophy from disease of joints may be combined in a complex pathologic picture. The actions of bacteria and their toxins are the same, whether they occur in an acute or a chronic infection, and were already described. The effects of wasting and cachexia are considered separately below.

EFFECTS OF EXOGENOUS TOXINS

There is very little available information on the effects of exogenous toxins, probably because muscle is not affected by most of the poisons to which man is exposed. Of the various forms of human intoxication phosphorus poisoning causes the most definite changes in skeletal muscle. Phosphorous is likely to be ingested accidentally or intentionally in the form of a rat poison. Exposure of workers engaged in the manufacture of matches occurred in former times. In acute intoxication there is hemorrhagic inflammation of the stomach followed some days later by acute necrosis of the liver and intense jaundice.

The outstanding change in the muscle fibers is at first a fine granularity and later the formation of fine droplets of fat stained readily by osmic acid. The granules enlarge and become fat droplets of much larger size than the interstitial fat of the normal muscle fiber. Degeneration of the muscle fiber proceeds and is attended by phagocytic removal of the degenerated muscle substance and proliferation of sarcolemmal nuclei if the patient survives for a sufficient length of time. These changes were studied by Fritz et al.[60] In 1863, Habershon[66] in 1867, Bergeat[13] in 1888, and others. In humans, lead, arsenic, and most other heavy metals have no direct effect upon the muscle fiber except through the medium of a peripheral neuropathy with the usual neural muscle atrophy.

EFFECTS OF EMACIATION AND CACHEXIA

A number of pathologists investigated the effects of chronic malnutrition on muscle[17, 20, 68 69, 130] and Durante[48] provided us with one of the most complete descriptions of the lesions in all types of wasting diseases. The effects of such diseases as cancer, tuberculosis, sprue, and other nutritional disorders are well known from the clinical standpoint. Along with the loss of subcutaneous fat all the muscles become thin and proportionately enfeebled. The trunk and limb muscles often exhibit unusual excitability to mechanical stimulus. A sharp blow delivered to the belly of the muscles by a percussion hammer causes immediate local contraction in the form of a visible ridge. The ridge persists for some seconds and may at times give origin to a small wave which slowly moves along the muscle. This reaction is known as myoedema and is contrasted to the brief idiomuscular contraction (Chapter 2). Since there is but little accompanying change in the electromyogram, this is ascribed to a propagated contraction wave in the contents of the fiber. As a rule, in the muscle atrophy of emaciation the strength of contraction is better than expected from the size of the atrophic muscles.

Upon macroscopic examination the muscles are thin, closely applied to the skeleton, and at times brownish red. Fat is usually lacking in the perimysial connective tissue. Histologically the fibers show atrophy and some degree of degeneration. The most striking aspect of nutritional muscle atrophy is probably the smallness and variation in the size of the fibers (Fig. 11–3A). There is an admixture of polygonal and rounded fibers of different sizes. Some of the small fibers are reduced to one-fourth or one-fifth the volume of the large ones (Fig. 2–27). In calculating the size of the fibers of biceps, rectus abdominis, and pectorals Jewesbury and Topley[84] found the diameters of normal fibers to be 12.5 μ in a case of carcinoma of breast and 20 μ in chronic pulmonary tuberculosis (normal 23.6 μ). There was in fact a significant reduction and greater variation in size of muscle fibers in all their cases of wasting disease. In sections stained by phosphotungstic acid hematoxylin the smallest fibers appear very dark. This darkness is due to the condensation of myofibrils, which are packed so closely it is impossible to count them accurately. It is clear, however, that the sarcoplasm is greatly reduced, and the mitochondria are very small and few in number, accounting for the pallor in hematoxylin and eosin staining.

In longitudinal sections the atrophic fibers are scattered haphazardly amid healthy ones, and there is usually no tendency for the atrophy to be grouped. The small angulated fibers occur in clusters of 2–5 or more if the process is severe. Often it is clearly in the pattern of motor units (Fig. 11–3B). The thin fibers are tortuous but of fairly even caliber throughout their length. They retain their striation except in places in which cloudy swelling or granulofatty degeneration is superimposed. The sarcolemmal nuclei at first enlarge, proliferate, and increase in number, but later they persist as rows of small dark nuclei. Some fibers are reduced to such a state of atrophy that chains of these nuclei are the only visible remains of the muscle fiber. Degenerating fibers become pale and of irregular width, lose their striations, and finally disintegrate. The granularity of such fibers in hematoxylin and eosin stains has coarse fat droplets as its counterpart in fat stains. This appearance of the degenerating fiber led to the designation of this process as granulofatty degeneration. The sarcolemmal nuclei multiply to form chains, but the activity of these nuclei appears to be reduced. Vacuoles occur in some fibers but whether they correspond to the autophagic vacuoles as seen in electron micrographs of chloraquin myopathy (Chapter 3) is not known. Phagocytosis of muscle fibers is much less prominent than in

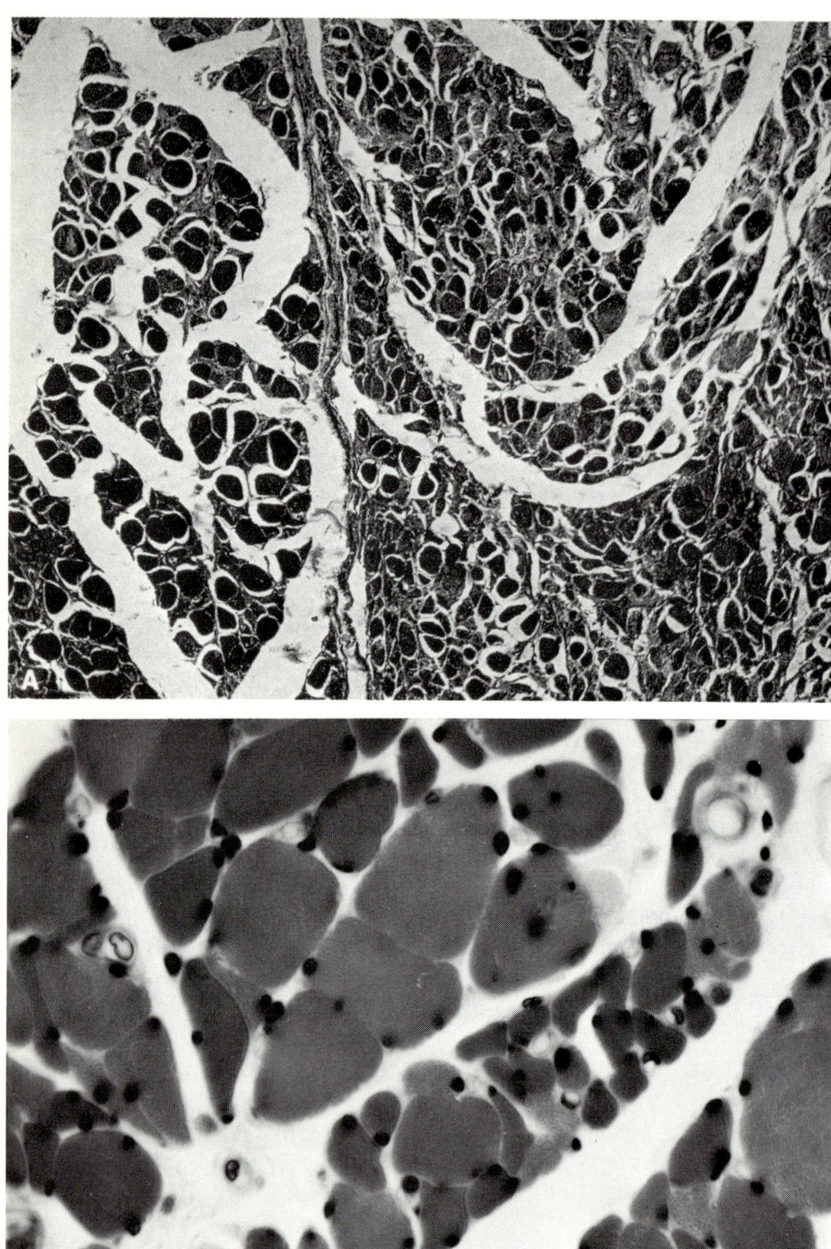

Fig. 11–3. (A) Transverse section of quadriceps muscle from an elderly man dying of chronic pulmonary tuberculosis. Smallness of muscle fibers and variability in size were noteworthy. (B) Transverse section of sartorius muscle of a man aged 84 years. Note variations in fiber sizes and the grouping of small and larger fibers. (H&E; B, ×610)

Zenker's hyaline degeneration. A few scattered hypertrophied fibers are usually seen.

Active proliferation of new fibers is not prominent, though occasionally multinucleated buds can be found. The connective tissue seems to be increased in amount, but when allowance is made for the collapsed and severely atrophied fibers it is clear that this is largely a condensation of the sheaths. The blood vessels are rather cellular, and activated histiocytes and macrophages appear around many of them. There is some reduction in the interstitial fat. Muscle glycogen is not changed.[84] Deposits of lipochrome pigment may appear but are not prominent except in elderly patients.

These manifestations of malnutrition and cachexia in musculature are diffuse but always more severe in leg than arm muscles and more in those of arm than trunk. There is some correlation between the degree of atrophy and the intensity of histologic change.[84, 87]

Some pathologists remarked upon differences in the distribution of muscular lesions in carcinomatosis, tuberculosis, and other diseases. Klippel states that in cancerous patients the leg muscles are more affected than other muscles, and Fraenkel[58] observes that in tuberculosis the thigh, diaphragm, leg, hand, neck, and thoracic muscles are involved, in decreasing order of intensity. These distinctions were not confirmed by our own observations or by those of others.[67, 105, 175] Acute inanition alone may be an important factor for, as Hayem[69] showed, the fibers may diminish 30% or more within a few weeks under these circumstances.

The "group atrophy" seen in Figure 11–3B has been observed in cachexia, and according to Engel[52] it affects type II phosphorylase-rich (dark ATPase) fibers more than others. The reason for this predilection is unclear. In the study of Tomlinson *et al.*[175] it did not correlate with age, being found often even in young patients lying in coma for many weeks. One possibility yet to be affirmed is that it is the consequence of a mild polyneuropathy affecting mainly the terminal branches of neurons. As yet physiologic and nerve biopsy studies have not been performed. There is also the possibility that total inactivity leaving postural mechanisms intact might cause a differential wasting of those motor units involved in phasic activ-

ities. Moreover, regeneration after segmented fiber necrosis with injury of sarcolemma may give an illusion of "group atrophy" when the new branches form compact groups 4–10 small fibers. The muscular lesions described above differ from those of Zenker's hyaline degeneration in that there is a conspicuous atrophy and granulofatty degeneration more often than hyaline degeneration, as well as a relative lack of regenerative activity.

The severity of the changes in some cases and their absence in others suggest that specific metabolic disorders or infections must also be operating in some cases of extreme debilitation due to diseases associated with cachexia. Thus in an unusual syndrome associated with bronchogenic carcinoma described by Denny-Brown[35] there were acute degenerative changes in the muscles; and dermatomyositis has been reported in association with malignant tumor (lung, breast, ovary, stomach) in many cases.[45, 188] In a recent survey of routine autopsy material, [134] we noted a significant number of acute degenerative changes in muscle fibers, especially in persons harboring a malignancy. In none of these cases had there been clinical evidence of myositis or of muscular weakness, other than that which could be attributed to the general effects of a chronic disease. It appears likely that the presence of tumor in these instances created a metabolic defect comparable to that seen in vitamin E deficiency. The presence of degenerating muscle buds in such muscles suggests the possibility of superimposed traumatic rupture of attenuated fibers.

SENILE MUSCULAR ATROPHY

Muscle wasting and general loss of muscular strength and agility during the senium are commonplace. The hands become thin and bony, and the interosseous spaces may be so prominent as to suggest progressive muscular atrophy. The arm and leg muscles tend to be thin and flabby. However, despite their reduced bulk the muscles are capable of exerting a surprising degree of power. Tremulousness, slowness, and incoordination of movement are also part of the clinical picture but are related to changes in the nervous system.

The age at which these changes in the muscles begin is highly variable. The muscle atrophy is merely a part of a more general atrophy of organs and tissues. Parenchymal elements shrink and slowly disappear. The capacity for cellular regeneration seems to diminish progressively with age. Fibrous tissue replacement (sclerosis) occurs, probably as a secondary reaction.

On gross examination the muscles are dark red, and sometimes with a brownish tinge but never as dark as is seen in brown atrophy of the heart. The fascicles stand out with unusual distinctness and are tough, as shown by their resistance to cutting.

Microscopically there is an overall diminution in bulk of the individual muscle fibers. This atrophy occurs in a more or less patchy irregular pattern. In many muscles isolated "knots" of scattered muscle nuclei are all that remains of a fiber. Atrophy of small groups of angulated fibers is frequent, particularly in the leg muscles, similar to that of cachexia. The muscle fibers are also diminished in number; some have disappeared leaving no trace. The atrophied fibers have slightly increased numbers of nuclei (relative) and many preserve their striations. On cross section this extreme variability is nearly as prominent as it is in the atrophy of chronic malnutrition (Fig. 11–3), suggesting the possibility of a common factor operating in these two conditions. Characteristic also is the occurrence of granules of yellow-brown pigment (lipochrome or lipofuscin) at each pole of the nucleus. This deposit of pigment in conjunction with simple atrophy or fatty degeneration is commonly referred to as pigmentary degeneration; however, this is not a reliable sign of degeneration but rather simple atrophy of old age.[84] Lacunae and vacuoles may be seen in some fibers, and others may show loss of striation, pallor, granularity, and an irregularity of caliber like that seen in fatty degeneration. There are scattered hypertrophic fibers mixed with the atrophic ones.

During middle age and beyond it has been said that some of the muscle fibers are transformed into peculiar structures with a homogeneous outer shell which in cross section appears as a ring beneath the sarcolemma the so-called *ringbinden*.[189] The significance of these structures is disputed. There is no doubt their number increases in the eye muscles, but for those found in most biopsy specimens where they are unattended by other change we favor the explanation of Erb,[53] who regards them as contraction band artifacts caused by acidic fixatives. The deeply situated ones in highly nucleated sarcoplasmic fibrillar masses are surely pathologic (i.e., the change in myotonic dystrophy). The amount of connective tissue is increased between individual fibers. It may be dense or loose and fibrillar. Also during aging muscle fat cells are sometimes prominent and in some cases are arranged in long rows, giving the muscle a honeycomb appearance. This state has been referred to as fatty infiltration of the aged. This pathologic change, according to Inglemark and Gustafsson,[80] is easily studied in the calf muscles of women, in whom there is a notable increase in the variability of the cross-sectional diameter of the fibers with advancing age and a progressive increase in the quantity of intramuscular fat cells. The fat cells in the earlier phases of this condition tend to have a perivascular localization and later occur interstitially and in the subfascial connective tissue as well. The nutritional status of the individual does not significantly alter the quantity or localization of this muscular fat; it appears to represent an almost obligatory replacement of the atrophic fibers in the aged.

In the older medical writings, particularly in the French literature, there are numerous references to myosclerosis of the aged. This term refers to a progressive atrophy of muscles and enfeeblement that occurs during late life. It results in contractures of limbs and shortening or retraction of the wasted muscles and their tendons. According to Hayem,[68] myosclerosis results in one form of senile paraglegia. It has been inferred that this is a myopathy of the senile, but the high liability to contracture when muscles are immobilized suggests that these muscle lesions are secondary to other factors, e.g., as in cerebral paraplegia in flexion (any degenerative disease of frontal lobes and basal ganglia).

Our opinion about these aging changes was modified by recent studies. The group atrophy of leg muscles, which we found in more than 80% of individuals over 60 years of age dying of systemic disease, is strongly suggestive of denervation (Fig. 11–4). Rebeiz and Adams[142] noted a corresponding loss of anterior horn cells in the lumbosacral segments in several patients. Doubtless other factors must be operative when similar changes occur in cachexia.

AMYLOIDOSIS

There are two forms of amyloidosis: primary and secondary. The primary form is rare and usually appears as solitary or multiple tumors in the larynx, tongue, and elsewhere. The skeletal muscles may be affected generally. In the secondary form, which is associated with chronic suppurative diseases, rheumatoid arthritis, multiple myeloma, and tuberculosis, amyloid is deposited in the viscera, particularly in the spleen, kidney, and liver. The relationship between the two forms of amyloidosis cannot be stated at present; probably one merges with the other. In secondary amyloidosis involvement of the peripheral nerves and muscles is rare. Therefore only the so-called primary form of the disease is considered here.

Primary systemic amyloidosis is of unknown etiology. Among the more constant clinical findings are a firm enlargement of the tongue and occasionally of other skeletal muscles due to amyloid infiltration, ophthalmoplegia, progressive weakness and fatigue of the musculature of the limbs, and rarely slowness of contraction and relaxation that may simulate myotonia. Symptoms of peripheral nerve involvement were present in several of our cases and in those of Kernohan and Woltman.[85] There is no evidence of chronic suppuration, and amyloidosis of the viscera does not occur to any significant degree.

The principal pathologic feature of the disease is the generalized deposition of amyloid in the interstitial tissue of the smooth and striated muscles. The extent of amyloid deposits in the skeletal muscles in this disease apparently varies from case to case and depends to some extent on the diligence with which one searches for them. Koletsky and Strecher[89] found reports of extensive muscle infiltration in 10 of the 24 cases (42%) in the literature up to the time of their publication. Dahlin[31] noted widespread involvement in each of seven cases in which such tissue was available for study.

Grossly, muscles containing considerable amounts of amyloid are unnaturally firm, and confluent nodular masses of translucent pearly gray lardaceous material are occasionally seen within them. Less severely involved muscles may show a slight to moderate increase in firmness with small focal gray deposits or a general paling of the muscle substance. Gross tests for amyloid by means of the iodine and sulfuric acid stain are strongly positive.

Microscopically a patchy or diffuse deposition of amyloid in the reticular or collagenous connective tissues separates the individual muscle fibers, and in some places nodular masses of amyloid replace the muscle fibers. Under the electron microscope the amyloid has the form of packets of thin fibrils 50–150 Å thick. Many of the individual muscle fibers appear normal even when surrounded and compressed by this substance, but varying degrees of fiber atrophy are usually found (Fig.11–5). Koletsky and Strecher[89] observed that bands of amyloid in the interstitial connective tissue seemed to invaginate and possibly invade the individual fibers, but this is illusory. Some of the muscle fibers appear to have undergone compression atrophy, and the effects of denervation are seen in others, viz., neural atrophy. In most cases amyloid deposition is more prominent in smooth muscle (media of the walls of the small arteries, arterioles, and small veins) than in the endomysial connective tissue (Fig. 11–5). Amyloid can be reliably demonstrated in the tissues by several staining methods: (1) crystal violet, which stains it red or violet; (2) Congo red, which stains it pale red and is also the basis of the intravenous test for amyloidosis; (3) Masson's trichrome and Mallory's aniline blue, in which amyloid stains like collagen; and (4) phosphotungstic acid hematoxylin, which provides an excellent contrast between the reddish orange of amyloid and the dark blue of the surrounding muscle tissue.

The relationship of primary amyloidosis to "amyloid neuritis" is not clear. Probably they

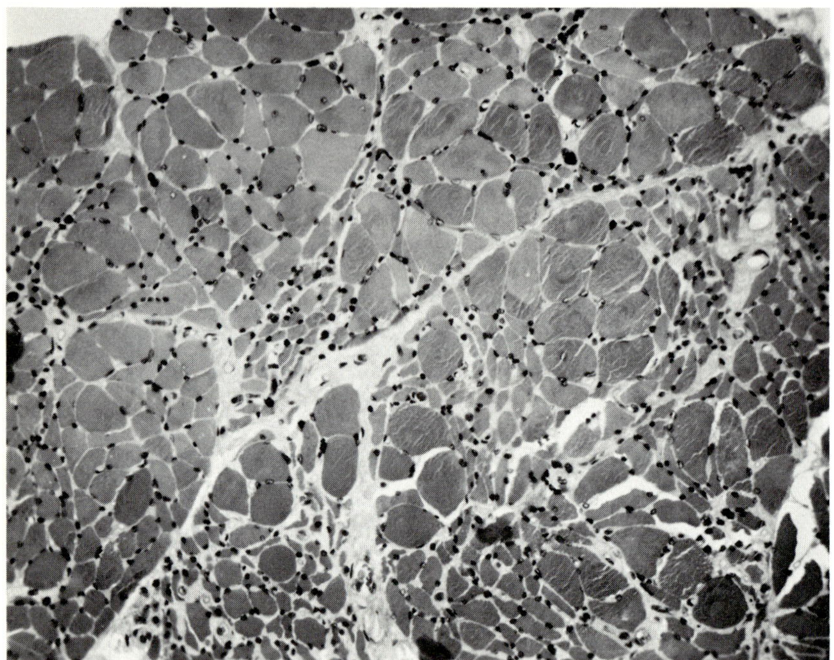

Fig. 11–4. Transverse section of gastrocnemius muscle of a 78-year-old woman. No neurologic disease was known to have occurred. The pattern (grouping) of small and large fibers is clearly indicative of denervation atrophy. (Note that counts of anterior horn cells in the lower lumbar segments showed them to have diminished by approximately 30%.) (H&E; ×220)

Fig. 11–5. Primary amyloidosis with involvement of skeletal muscle. There are deposits of homogeneous material which gave a positive test with iodine, a definite staining reaction with crystal violet, and by the periodic acid-Schiff method can be seen in the walls of small intramuscular blood vessels and perimysial connective tissue. (phosphotungstic hematoxylin)

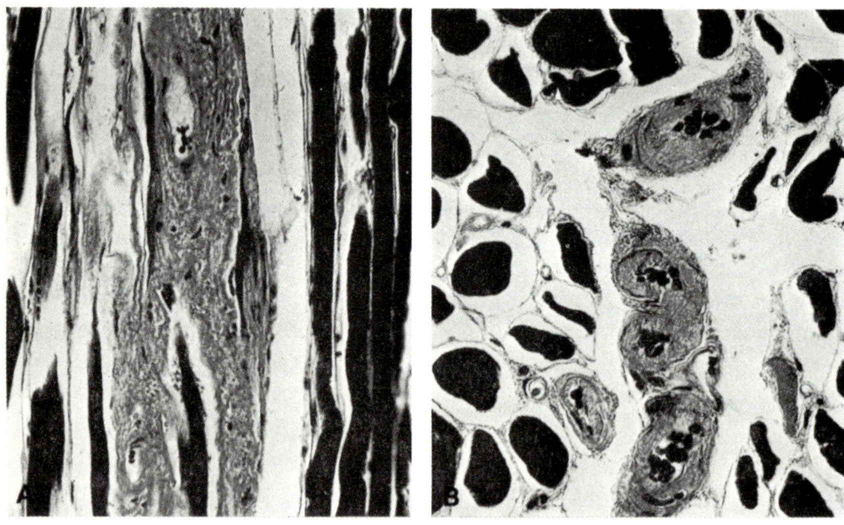

are the same disease. Kernohan and Woltman[85] reported several cases of this type, some of which were accompanied by diabetes mellitus. The clinical findings were essentially those of chronic progressive polyneuropathy of mixed sensory-motor type. Not more than 1% of the muscle tissue contained amyloid, except the diaphragm, which was more severely affected. Many of the large nerves were infiltrated with small amounts of homogeneous material that stained like amyloid. This material was deposited in the walls of and around the intramural blood vessels and elsewhere in the endoneurial connective tissue. Nerve fibers were degenerated and nerve cells in the spinal ganglia altered. Findley and Adams[55] reported a somewhat similar case in which cutaneous hypersensitivity was the principal evidence of peripheral nerve affection. One form of universal pain insensitivity and autonomic paralysis is associated with the Portuguese familial amyloidosis of Andrade.[3] Skin, gastrointestinal tract, liver, heart, and other organs were involved, and death was the result of cardiac failure. Denny-Brown[36] briefly reported a case of amyloidosis of the peripheral nerves and ganglia (related to a progressive sensory neuropathy) with deafness, and also found deposits of a substance resembling amyloid in the ganglia in a case of hereditary sensory neuropathy. Additional study of this group of cases is required before the pathology and symptomatology can be accurately defined.

GLYCOGEN STORAGE DISEASES

Our knowledge of the glycogen storage diseases was advanced considerably by the excellent contributions of Cori[30] who elucidated the cellular metabolism of glycogen and closely related carbohydrate substances. A new classification based on the specific biochemical defect has emerged. Von Gierke's disease, originally believed to be a single entity, now must be regarded as one of a family of closely related diseases, each due to a deficiency of a specific intracellular enzyme. In each variety of the disease, glycogen accumulates in particular tissues because of a "metabolic block."

At least seven separate entities, each conditioned by a defect in a particular intracellular enzyme, have been identified. Most have significant glycogen deposits in skeletal muscle fibers, often with accompanying muscular weakness.

In order to understand something of the nature of these disorders, a brief description of glycogen and related carbohydrate metabolism is necessary. The processes of glycogenesis and glycogenolysis (viz., the formation and breakdown of glycogen or animal starch) are central to the entire intracellular system of carbohydrate metabolism, as depicted in Figure 11–6. The starch is an extremely large branched molecule composed of many hundreds of glucose units in specific chemical linkages. The complete molecule resembles a tree with many secondary, tertiary, and further branchings. Two specific enzymes are necessary for the synthesis of glycogen. These are a phosphorylase and a "branching" enzyme.*

Degradation of this molecule, by means of which carbohydrate becomes available from this storehouse, requires the action of a "debranching" enzyme and the same phosphorylase already mentioned. Two of the rarer glycogen storage diseases (types III and IV) are due to deficiencies in either the branching or the debranching enzymes. Type I is caused by lack of glucose-6-phosphatase, which normally releases glucose into the circulation from the cell (Fig. 11–6). Since this enzyme is normally absent from skeletal and cardiac muscle, and these organs are not affected in type I disease, nothing further is said of this type. Only the cells of the liver and kidney, which do contain this enzyme, accumulate abnormal quantities of glycogen. The principal characteristics of the six other types of glycogen storage are outlined in Table 11–1, as modified from Cori[30] and Recant.[143]

In type II, Pompé's disease, also called cardiomuscular glycogen storage disease, the fundamental defect is one involving acid maltase. This proves to be the most frequent form of glycogen storage disease of muscle. More than 100 cases have been reported.[29, 40, 41, 59] It most

---

* Recent evidence implicates another pathway, the uridine triphosphate circuit, in the storage of glycogen (Chapter 2).

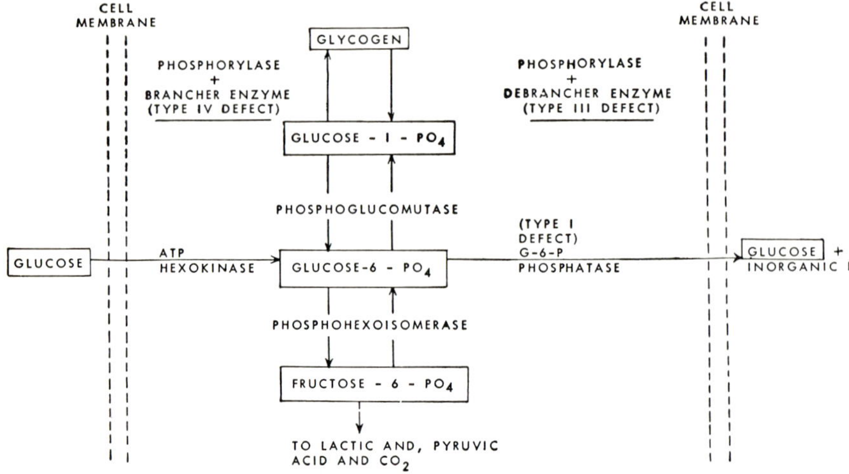

Fig. 11–6. Steps in the metabolism of muscle glycogen.

often begins during infancy and is manifested by a paucity of movement and enfeeblement of muscle action, hypotonia, enlargement of tongue, and areflexia. The effects of the disease are severe, and few of the patients survive for longer than 9–12 months. Feeding difficulties, intermittent cyanosis, dyspnea, and cardiac failure terminate life. The heart is usually massively enlarged. Friedman and Ash[59] called particular attention to the fact that the skeletal musculature may be affected to a much greater extent than the heart. This was true in five of

their cases in which the initial diagnosis was amyotonia congenita or Werdnig-Hoffmann's disease rather than some form of cardiac disease. The muscles were weak and movements sparse. The muscles felt unnaturally firm and rubbery, as in pseudohypertrophic muscular dystrophy. There was respiratory distress, and all the infants died before the ninth month. In some cases the tongue was enlarged, and in others the spleen. Muscle biopsy is often helpful in revealing the nature of the defect. There is striking cardiomegaly in many cases. The

TABLE 11–1. Summary of Glycogen Storage Diseases

Type	Other name	Enzyme deficiency	Organs affected	Glycogen structure
I	von Gierke's	Decreased or absent glucose-6-phosphatase	Liver and kidney	Normal
II	Pompé's disease (cardiomuscular glycogen disease)	α-1-4 glucosidase (acid maltase)	Generalized: extreme in skeletal and heart muscle	Normal
III	(Limit dextrinosis) Forbes disease	Decreased debrancher activity	Liver, heart, skeletal muscle	Abnormal: absent or very short outer branches
IV	Andersen's disease	Decreased brancher activity	Liver (cirrhosis), possibly other	Abnormal: very long inner and outer chains
V	McArdle's disease	Absent phosphorylase	Skeletal muscle only	Normal
VI	diSant'Agnese	Phosphoglucomutase	Skeletal muscle only	Normal
VII		Phosphofructose kinase deficiency	Skeletal muscle only	Normal

* diSant'Agnese *et al.*[42] and Thomson *et al.*[171]

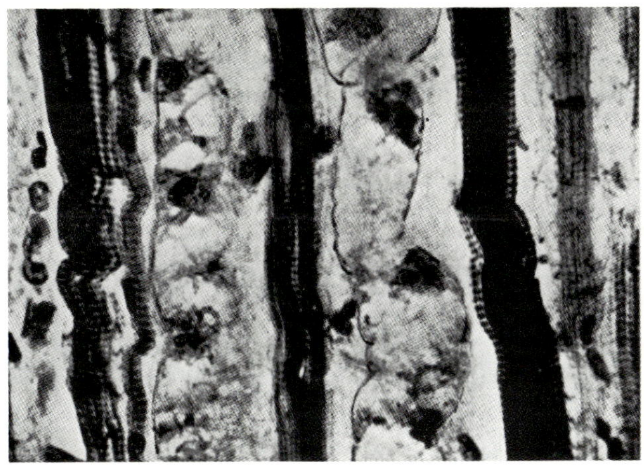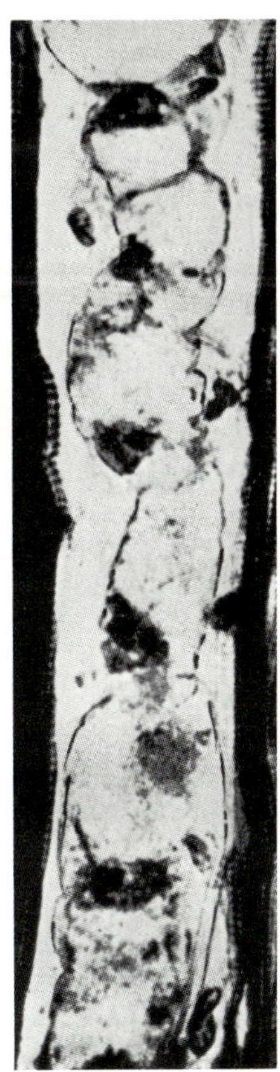

Fig. 11–7. Skeletal muscle from infant dying of Von Gierke's glycogen storage disease. Pale, unstained material that has completely replaced several muscle fibers was demonstrated to be glycogen by Best's carmine stain.

myocardial fibers are distended by large clear glycogen vacuoles, which lend a "lacework" pattern to them. Each fiber appears as a hollow cylinder surrounded and occasionally traversed by thin myofibril strands. Nuclei lie free within the vacuolated fiber. The tongue is consistently and markedly affected, the diaphragm less so; and the musculature of the neck, trunk, and limbs has shown variable quantities of stored glycogen. There has been no systematic pathologic study of the degree of involvement of the entire bodily muscle system.

When the muscle tissue stores a large excess of glycogen, a process best illustrated by type II or cardiomuscular glycogen storage disease, the principal histologic change is vacuolation and basophilic granular degeneration of sarcoplasm. The first observations of this phenomenon were made by Humphreys and Kato in 1934.[79] Subsequently a number of cases were studied by diSant'Agnese *et al.,*[40, 41] Clement and Godman,[29] and Friedman and Ash.[59] The

vacuolation may range from a few small, centrally placed, clear vacuoles to complete replacement of all the sarcous substance by large vacuoles in an expanded sarcolemmal sheath (Fig. 11–7). The vacuoles contain a material that gives a positive reaction for glycogen. The earliest signs of vacuolation consist of elongated cleared areas which separate the myofibrils in the center of the fiber. They are oriented parallel to the long axis of the fiber. Some of the adjacent myofibrils exhibit a basophilic change; and rounded basophilic refractile bodies and granules may appear in the vacuolated areas. As more vacuoles accumulate, the muscle fibers enlarge and the myofibril bundles become thinner and fewer. Eventually such fibers appear as ballooned tortuous sheaths containing many basophilic granules and refractile bodies in which intertwine a few thin myofibrils. Pyknotic nuclei, small amounts of granular sarcoplasm, and nuclear debris are compressed against the sarcolemma. Often such involved fibers lie next to normal appearing or only partially affected ones. In the lesser affected fibers, with central axial vacuoles, the peripherally placed nuclei appear to be normal. An increased diameter of the glycogen-laden fibers is shown best in cross section; some are more than twice the size of a normal fiber. Humphreys and Kato[79] observed fibers that exceeded 90 μ in diameter in the rectus abdominis of an infant less than 1 year of age. In transverse section the glycogen-distended fibers, the partially vacuolated ones, and the normal fibers form a striking microscopic picture that cannot be confused with any other disease. Glycogen is not found in the interstitial tissue although smooth muscle cells in the walls of arteries may be vacuolated and take a stain for glycogen.

Special fixation and staining methods are desirable for histologic demonstration of glycogen. Fluid fixatives such as absolute alcohol or Rossman's fluid preserve almost all the glycogen. Staining by Best's carmine or Bauer's periodic acid-Schiff methods are specific (Chapter 2). The latter can also be used after aqueous fixation, but with considerable loss of glycogen-positive material. Loss of stainable granules following digestion of the sections with diastase affords histochemical proof that the deposit is glycogen. In order to evaluate roughly the glycolytic capacity of the tissue under study, a fresh piece of muscle may be fixed in absolute alcohol and another left at 37°C for 4 hours before fixation. Both are then compared after staining by Best's method.

In type II disease most of the other tissue cells of the body also contain variable amounts of glycogen. Clement and Godman[29] pointed out that the motor neurons may be distended by glycogen and degenerate, as in lipid storage diseases. The neuronal lesions may contribute to the hypotonia and muscular flaccidity seen in many of these cases. Schnabel[156] found glycogen deposits in Schwann cells and large dorsal root ganglion cells in his case, as well as diffuse stainable glycogen in the astrocytes of the brain.

More benign forms of type II disease were reported by Hudgson *et al.*[76] and Engel.[51] These studies are of special interest because they provide excellent ultrastructural detail. The clinical picture, interestingly, is indistinguishable from that of a relatively mild limb girdle dystrophy or chronic polymyositis. There is a slowly advancing weakness of pelvic and pectoral girdle muscles, and the sternocleidomastoid and trunk muscles are less involved. In one case only the iliopsoas, glutei, and adductors were weakened and much later the pectoral, sternomastoids, and facial muscles. In another the weakness involved mainly the respiratory muscles (intercostals and diaphragm). The tendon reflexes are diminished but not lost. The serum levels of muscle enzymes are raised to 2–3 times normal, and the electromyogram (EMG) is distinguished by increased irritability shown as heightened insertion activity and brief myotonic potentials. Normal levels of serum glucose and lactic acid are obtained during epinephrine and glucagon injections. The rise in lactic acid in blood during ischemic exercise is normal. The electrocardiogram has been unremarkable in these adult cases.

The muscle biopsies divulge the diagnosis by demonstrating a type of change visible under the light and phase microscope (Fig. 11–8). The affected fibers (25–75%) are finely vacuolated. The fine vacuoles are 2–4 μ in diameter; larger ones range as high as 50–60 μ. Type I fibers are more affected than type II. The fiber sizes are more variable than normal, and many of the sarcolemmal nuclei have assumed a central position. Necrosis and phagocytosis of fibers is infrequent. Nonetheless, there are signs of regeneration and fiber splitting, as if some degree of sarcolemma permeability or fiber necrosis had occurred. The vacuolar material is positive for acid phosphatase, indicating that some of the vacuoles are autophagic, but far more important are the large number of sequestered membrane-bound packets of glycogen granules, as well as dispersed glycogen, which displace other organelles. The relative amounts of dispersed and membrane-bound glycogen varies in this and the other glycogen-storage diseases. The membranes of the packet appear to arise from the t-system (Fig. 11–9). Golgi cisternae appear also to fuse with these membranes. There are large electron-dense particles in the mitochondria. The glycogen storage leads to zones of myofibrillar degeneration. Hug *et al.*[77, 78] as-

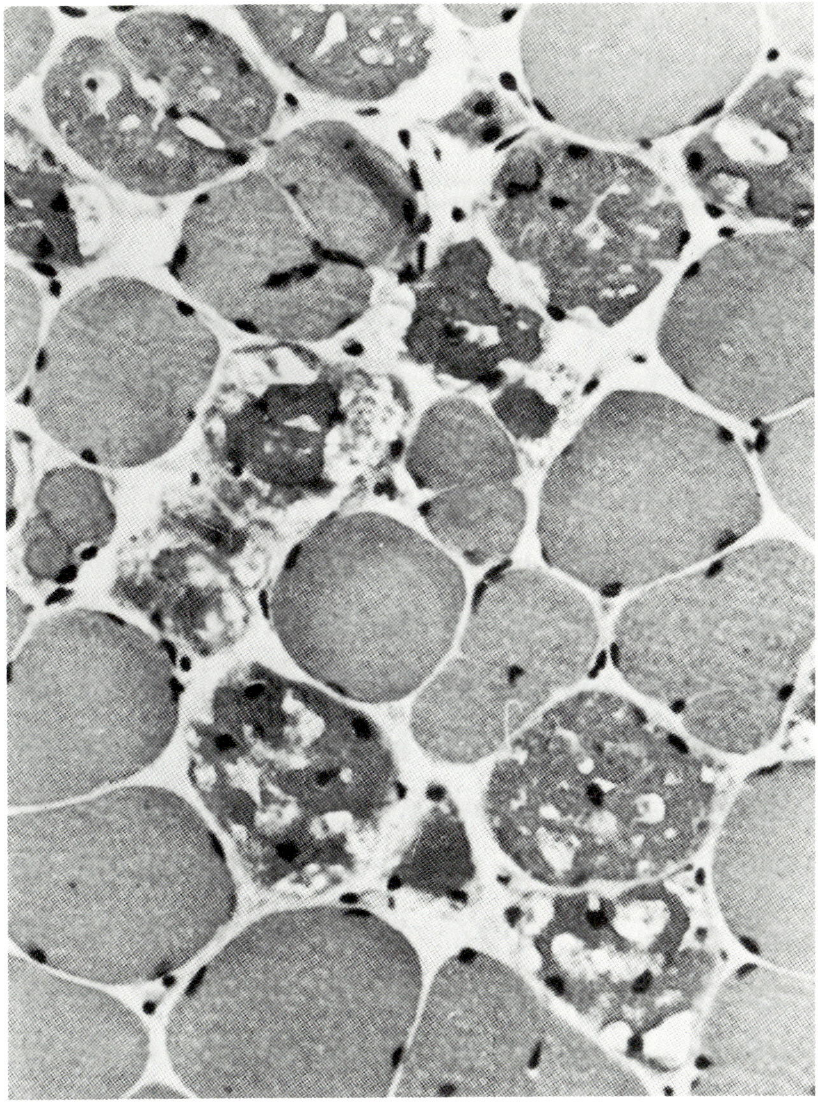

Fig. 11–8. Acid maltase deficiency (adult type). Approximately one-third of the fibers contain multiple vacuoles of varying size. Examples of fiber splitting are also visible. (Courtesy of Dr. A. G. Engel) (frozen section, H&E; ×175)

cribe the membrane-bound glycogen accumulation to an enzymatic defect in lyosomes. Electron microscopic studies of the infantile form by Bandhuin et al.,[7] Cardiff,[26] and Smith et al.[164] show similar lesions but with more devastating effects on the muscle fibers. This description of light and electron microscopy of Pompé's disease applies as well to the other types of glycogen storage disease of skeletal muscle.

Type III glycogenosis (Forbes disease or limit dextrinosis) is not ordinarily seen by a clinical myologist since skeletal muscles are seldom affected. However, Oliner et al.[129] reported a patient who presented at the age of 50 years with a chronic progressive myopathy associated with glycogen levels in skeletal muscles of 5.5% and in the liver of 11%. The glycogen had the short outer chains known to accompany the nearly complete absence of

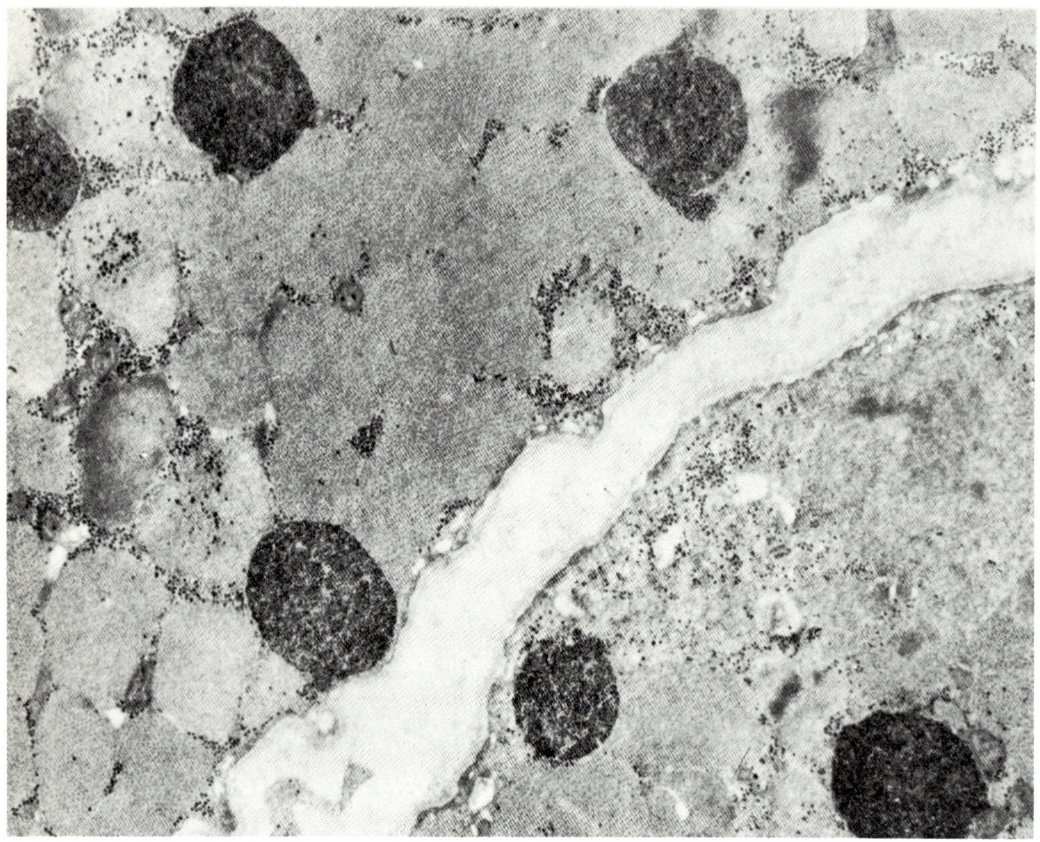

Fig. 11–9. Acid maltase deficiency (childhood type). The six dense accumulations of glycogen are surrounded by single or double membranes which, according to Engel, are derived from the transverse sarcoplasmic reticulum. (Courtesy of Dr. A. Engel) (electron micrograph; ×13,600)

amylo-1,6-glucosidase, the debrancher enzyme that degrades the arborizations of the glycogen molecule.

In type IV (Andersen's disease) there is primary liver involvement with childhood hepatic cirrhosis. Too few cases have been found to draw a clear clinical picture. Muscle was involved to a minor degree, both clinically and pathologically.

Type V is McArdle's disease. The discovery of this type of glycogenesis was made by McArdle[107] who observed an unusual clinical case in 1951 in which a disturbance of muscle function appeared to be related to a derangement of glycogen breakdown. The patient was a young man of 30 years whose complaints of weakness, stiffness, and aching pain in the muscles after exercise had been present throughout life. He was of small stature with slight thoracic scoliosis but was otherwise well proportioned, and the muscles were of average size and consistency. After light exercise these symptoms developed in the muscles that had been used most vigorously, e.g., in the calves and thighs after walking 100 yards, in the forearm muscles after flexing the fingers against resistance, or in the masseters after chewing. Immediately after exercise the affected muscles were abnormally firm and remained in a shortened state. The fingers, for example, remained partially flexed and could not be voluntarily extended. Localized painful swellings appeared in the weakened muscles. Passive extension encountered resistance and was painful. With rest these symptoms would subside within 5–10 min, but a longer time was required if the exercise was strenuous. The neurologic examination was otherwise entirely negative; muscle hypertrophy and myotonia were not present. Mood depression and intercurrent infections had an adverse effect on the symptoms. Contraction of muscles of the arm when the circulation was occluded provoked the symptoms within 1 min. The effect of cold was not mentioned. The family history was incomplete, but no other instance of the disease was known to have occurred.

A thorough investigation of this remarkable syndrome brought to light several facts, the most important of which was that the blood lactate and pyruvate did not rise during vigorous muscle contraction, as they do in a normal subject, but instead fell. Furthermore, the serum potassium rose from 14.8 to 18.6 mg/100 ml, whereas there was no change in normal control subjects. There was slight lowering of the inorganic phosphates. No defect in glycogen breakdown in the liver was demonstrated; a rise of blood sugar occurred after Adrenalin. The EMG revealed no electrical activity when the weak contractured muscles were at rest; during voluntary contraction normal action potentials were observed. No electrical activity could be obtained from the localized muscle swellings by means of a fine coaxial needle electrode. It was assumed, therefore, that the muscle was in a state of contracture in the physiologic sense. A muscle biopsy was not performed.

No further reports of this remarkable condition appeared during the next 6 years until a similar case was studied by Schmid and another by Pearson.[134] Schmid's patient was a man aged 52 years who first began to notice cramping pains and weakness in the limbs following exercise at the age of 20. The cramps could last 1–2 hours after moderate exercise followed by weakness lasting 1–2 weeks. In severe attacks the urine became dark for 12 hours or longer. In 1948 this patient's condition was reported as an example of spontaneous myoglobinuria by Kreutzer et al.[91] A fall in blood lactic acid during exercise was noted at that time. During the following years the musculature of the thighs and upper arms became progressively more atrophic as a result of re-

peated attacks. The cramps became less prominent and the weakness more severe and prolonged. Episodes of myoglobinuria became less frequent and could no longer be demonstrated at age 52. At this time the proximal muscular atrophy might have been mistaken for muscular dystrophy, except that there was no affection of the muscles of the trunk, and there were only two other affected members in a large sibship.[153] A failure in production of lactic acid with exercise was found by Schmid and Mahler[154] together with striking histologic changes in a muscle biopsy. The least affected muscle fibers in this patient, whom Denny-Brown studied, and to whom Shy[161] also referred, presented accumulations of granules, which stained dark red with Best's carmine and purple with periodic acid-Schiff reagent, aggregated in crescents under the sarcolemma (Fig. 11–10). These latter were of all sizes, and in some an invasion of histiocytes indicated a reactive process. Fibers in which the sarcolemma was raised in the entire perimeter had undergone necrosis, and many remnants of such necrotic fibers were found. In some segments of the affected muscles, severe destruction of muscle fibers with replacement by fatty tissue had occurred. There was no clinical evidence of myocardial damage or of disorder.

Chemical analysis of muscle biopsies by Schmid and Mahler[154] showed a fivefold increase of glycogen content, and yet incubation of a muscle homogenate produced only a fourth the quantity of lactate appearing in normal muscle. The defect could be corrected by the addition of crystalline phosphorylase, or by the use of glucose-1-phosphate as substrate. The anabolism of glucose in the muscle of UDPG-linked synthesis was unaffected.[155]

The glycogen in circulating leukocytes was within the normal range in this case, whereas it is increased in hepatic types of glycogen storage disease. The incubation of shed blood showed a normal rate of glycolysis. Administration of glucagon, epinephrine, and particularly glucose and insulin lowered the blood level of phosphorus and nonesterified fatty acids and increased the lactate concentration, indicating that utilization of glucose was not

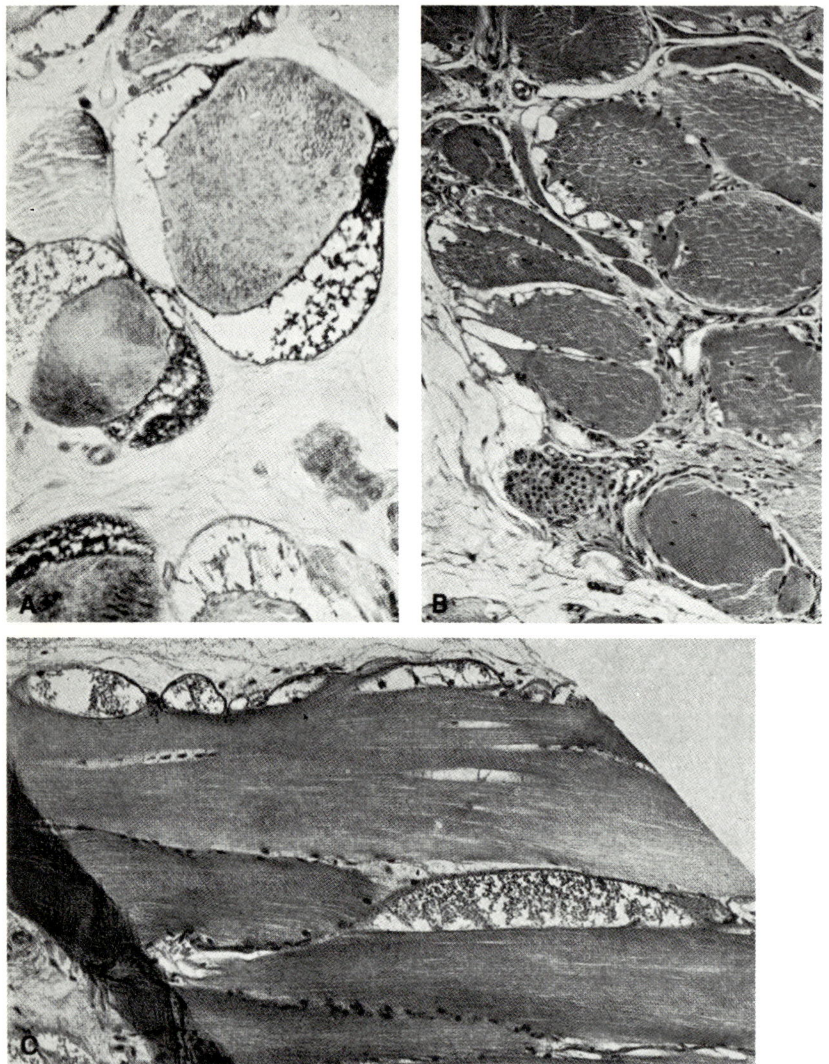

Fig. 11–10. Sections from biopsy of chronic McArdle syndrome. (A) Granular material is lying under sarcolemma. (B) Necrosis of isolated fibers and multiple blebs under sarcolemma of others. (C) Longitudinal section shows sarcolemmal blister containing bright red granules. Fiber above also shows histiocytes in questionable older accumulations, with chains of central nuclei. (A, PAS; B, methylene blue, eosin; C, Best's carmine)

impaired. These measures greatly increased the tolerance for muscular exercise, and later a report[152] showed that administration of glucagon constituted satisfactory treatment of the condition.

Independently Pearson[134] encountered a young man, 19 years old, who had a disorder similar to that described by McArdle.[107] He showed a marked limitation of exercise toler-

ance that had been present for as long as he could recall. He could walk slowly or at a moderate pace for long distances on a level surface, but climbing a slight incline rapidly induced fatigue and weakness in the exercised muscle groups, which if further exercised underwent severe cramping. Any muscle groups exercised sufficiently reacted in the same manner. Muscular atrophy was not present. Red

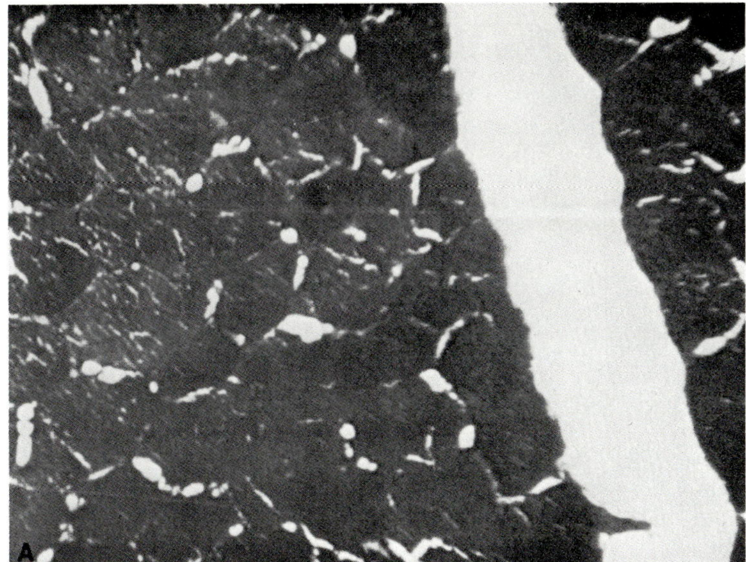

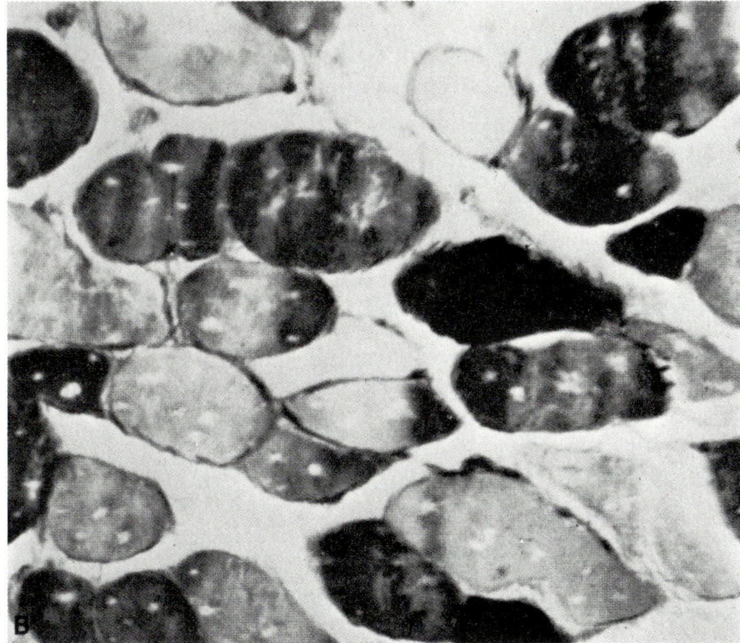

Fig. 11–11. Sections of rectus femoris muscle. (A) Control subject. (B) Young man with phosphorylase deficiency myopathy. Note significant increase in stainable glycogen in diseased muscle. (PAS; ×850)

urine (myoglobin) was passed on one occasion after intensive exercise. As in the cases described above, there was no rise in serum lactate or pyruvate after intensive or ischemic exercise. It was found that continuous infusion of several metabolites including sodium lactate, glucose, and fructose[136] improved exercise tolerance; of these, fructose was most effective since exercise was greatly improved at plasma fructose levels of 35 mg/100 ml, without change in glucose level. For glucose to be effective, a plasma glucose level above 160 mg/ml was necessary.

Biopsy of the rectus femoris muscle revealed a completely normal histologic picture. However, when sections were stained with Best's carmine or with the periodic acid-Schiff reagent, an intense, uniform coloration of all fibers was seen (Fig. 11–11B), in contrast to considerable variability in staining of control mus-

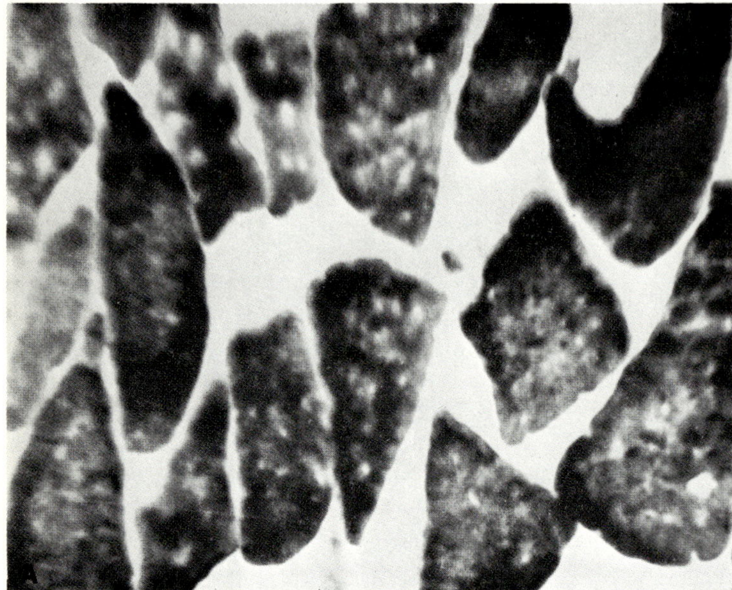

Fig. 11–12. Sections of rectus femoris muscle stained for phosphorylase and branching enzyme. (A) Control subject. (B) Patient's muscle. Normal muscle (A) contains both deep blue and reddish purple granules that signify the presence of these enzymes. Granulation is absent in the defective muscle (B). (Takeuchi method;[169] ×980)

cle. Furthermore, histochemical technics for the demonstration of phosphorylase and branching enzyme, utilizing the methods of Takeuchi,[167] disclosed a total absence of both enzymes in the patient's muscle (Fig. 11–12) and normal amounts of each in control muscle. Both the inactive phosphorylase b and the active phosphorylase a were deficient. Branching enzyme could not be demonstrated in the abnormal muscle because the detection of its presence depends also upon an intact phosphorylase system.

In further studies on this patient, Mommaerts, with Pearson and others,[119] found that homogenates of muscle showed no lactic acid formation from endogenous glycogen, whereas added hexosephosphates enabled glycolysis to proceed at a normal rate. Moreover, the glycogen content of the muscle was increased fourfold over that of control muscle, and phos-

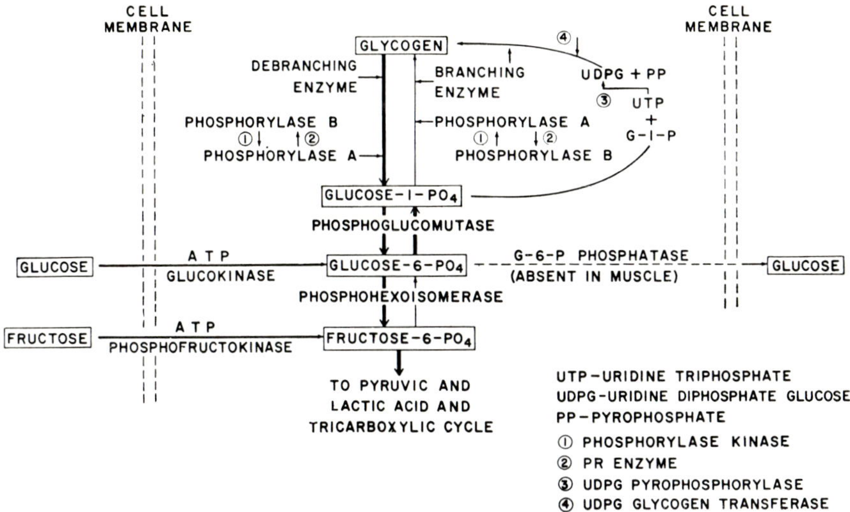

Fig. 11–13. Transformations of glucose and glycogen in muscle.

phorylase activity could not be demonstrated.[137] Other enzymes related directly or indirectly to glycogen metabolism (phosphoglucomutase, phosphorylase kinase) were present in the muscle of the patient, and the glycogen had a normal branched structure. The presence of the UDPG-glycogen transferase of Leloir and Cardini[98] was demonstrated and glycogen synthesis in the absence of phosphorylase was ascribed to the action of this enzyme. A full report of studies of this case appeared more recently.[137] di Mauro et al.[39] found that the transferase of Leloir (also called synthetase) diminishes as glycogen increases. The several enzymes and substrates involved in this segment of the glycolytic pathway are shown in Figure 11–13.

Schotland et al.[157] examined a muscle biopsy from a patient with McArdle's disease that revealed changes similar to those of the other glycogen storage diseases. There were large aggregates of glycogen particles in the intermyofibrillar space of the I band and under the sarcolemma. The myofilaments were disorganized at the level of the I band, presumably by the glycogen accumulation. They suggested that this may be an earlier stage of the random fiber degeneration seen with the light microscope.

The discovery of a defect in a single enzyme system in striated muscle adds to the growing list of similar defects observed in other tissues[196] (e.g., galactosemia, certain types of glycogen storage disease, alkaptonuria). The myopathy due to phosphorylase deficiency is the first such specific defect to be recognized in skeletal muscle, and it raises questions of great interest in the general understanding of metabolic diseases of muscle.

Although there are few known sufferers, Mahler has seen five other patients with similar clinical and laboratory findings, and the author has been called into consultation on several others. Cramps leading to muscle necrosis and myoglobinuria are unusual, but it is possible that lesser degrees of the disorder can masquerade as cases of chronic polymyositis. Conversely, it is possible that some of the cases described as spontaneous myoglobinuria are suffering from this condition, particularly those whose first attacks occurred in the age range 12–25 years (group I of Korein et al.[90]). Dreyfus and Alexandre[46] report a case of myoglobinuria with McArdle's disease in which the serum CPK value is in excess of 10,000, equivalent to the destruction of 100 g of muscle. Determining that the lactic acid content of blood fails to rise with exercise is a relatively simple test to perform. The meaning of the fundamental defect in phosphorylase is unclear.

It may be noted that moderate to severe exercise is necessary to produce muscle damage owing to the fact that muscle must utilize glycogen in addition to glucose and fats in all grades of fatiguing contraction. Thus Schmid's older patient could walk an unlimited distance on level ground but immediately brought on an attack of weakness by climbing a few steps. The other two patients were younger and had a somewhat greater exercise tolerance.

The mechanism by which contracture is produced in this disorder is not clear. Electromyograms reveal reduction of voltage and shortened duration of action potentials in severely affected muscles, comparable to the findings in other types of primary muscle disease. There is no fibrillation. During exertion the interference pattern tends to become complicated by high-frequency discharges, but finally as contracture sets in virtually all electrical activity ceases. Whether the onset of contracture is related to some abnormal shift of electrolytes or a shift in pH is unknown. The glycogen accumulations were found in muscles that were comparatively little affected and appeared hypertrophied (gastrocnemius) as well as those that were atrophic (quadriceps).

Although the cellular acculumation of glycogen in phosphorylase deficiency myopathy closely resembles that seen in glycogen storage disease, it is not as extensive. The incidence and course of the two diseases differ greatly, but it is of some interest that Zellweger et al.[191] noted muscle irritability in a more chronic type of hepatic storage disease.

Glycogenosis type VI is probably represented in the case of diSant'Agnese et al.[42] The patient was a 2.5-year-old girl whose retarded motor development and feeble musculature action were noted when she was 6 months old. The liver and heart were not affected. Thomsen's case[171] probably should also be placed in this category. The patient here was also a child with mild weakness and contracture of gastrocnemii. Possibly the cases of Lehocsky et al.[97] should be added: two young adults, brothers, who had had muscle cramps on exertion, a syndrome rather like that of McArdle's disease. The debrancher enzyme

was reduced in the red blood cells. There was vacuolar myopathy.

PHOSPHOFRUCTOKINASE DEFECT

Over the years the author has encountered a number of patients who were hampered throughout their lifetime by painful spasms of skeletal muscles during activity. Exercise of their muscles under ischemic conditions induced an electrically silent contraction (a true contracture), and the lactic acid level in venous blood from the active muscles did not rise. Thus they conformed to all the recognized criteria of McArdle's disease except that phosphorylase was present in their muscles. The glycogen content of the muscles was increased. The liver and heart appeared normal. The cases of Satayoshi et al.[150] and Layzer et al.[96] fall into this group; they showed the enzymatic defect to be a phosphofrucktokinase abnormality. The muscle lesions resemble those described above. di Mauro et al.[39] report both synthetase and phosphorylase activity to be present; actually the synthetase activity had increased in amount.

MUSCLE DISORDERS ASSOCIATED WITH DISEASES OF ENDOCRINE GLANDS

There are numerous reports of muscle disorders associated with a wide variety of diseases of the endocrine system. The clinical effects of the muscle disorders may have been quite evident, being manifested by muscle weakness and paralysis, or the opposite, spasm of the whole muscle or of isolated fasciculi. Morphologic changes were slight or absent in many of these cases. It must therefore be supposed that the muscle affection is related in some way to a derangement of the innervative or conductive mechanism or to some enzyme system necessary to muscle contraction; correction of the hormonal disorder usually results in complete restoration of muscular function.

MYOPATHY IN THYROID DISEASES

There is an intimate but poorly understood metabolic relationship between the thyroid gland and the skeletal musculature. During periods of hypo- or hyperfunction of the thyroid, associated clinical muscular syndromes are seen with a frequency far greater than would be expected to occur by chance alone. Various classifications of these muscular disorders have been offered; the one presented below is a slight modification of that suggested by Millikan and Haines,[116] who wrote one of the most authoritative treatises on this subject.

A. Thyrotoxicosis
  1. Thyrotoxic myopathy
     a. Chronic
     b. Acute
  2. Myasthenia gravis
  3. Periodic paralysis
  4. Exophthalmic ophthalmoplegia
B. Hypothyroidism and altered muscle function
  1. Hoffmann and Debré-Semelaigne syndrome

HYPERTHYROIDISM

Chronic Thyrotoxic Myopathy. Chronic thyrotoxic myopathy is the most frequent and undoubtedly the most striking form. The principal clinical manifestations are weakness, fatigability, muscular atrophy, and weight loss, all of which may be slowly progressive over a long period or may develop more rapidly within several weeks. The muscle weakness may be generalized or begin in certain muscle groups and later spread to others. It does not progress to complete paralysis, though the patient may become bedridden. The most severely affected muscle groups are those of the shoulder and pelvic girdles, and all patients complain of difficulty in climbing stairs. Atrophy is also prominent in these groups of muscle, but tendon

reflexes are retained. Coarse fascicular twitchings with occasional cramps were present in some reported cases, together with fatigue, which though less dramatic than in myasthenia gravis is liable to be mistaken for it.[111] Weakness of bulbar muscles occurs in some cases. The degree of atrophy, the lack of response to Prostigmin, and reversal by treatment of the hyperthyroidism distinguish this disease from myasthenia gravis or progressive muscular atrophy. Many cases of this thyrotoxic myopathy are reported in the literature, the most complete discussions being those of Thorn and Eder,[173] McEachern and Ross,[111] Morgan and Williams,[120] and Millikan and Haines.[116] The age incidence ranged from 22–66 years, with a somewhat greater frequency in the older age groups.[173] The sex ratio is in favor of males in a proportion of about 3:1. Complete restoration of power may be obtained by treating the underlying thyrotoxicosis.

Milder forms of this condition have for a long time been recognized in Graves' disease—particularly weakness of the iliopsoas and femoral muscles. Since most patients with thyrotoxicosis manifest some degree of muscle weakness, the diagnosis of chronic myopathy in thyrotoxicosis is customarily entertained only when the extent and severity of the clinical weakness become extreme. Probably the same biochemical disorders of muscle occur in all cases of hyperthyroidism, being more severe in chronic thyrotoxic myopathy.

A notable peculiarity of chronic thyrotoxic myopathy is that the muscle symptoms may appear before the symptoms and signs of thyrotoxicosis become evident, and the latter (exophthalmia, tachycardia, sweating, and the like) are often inconspicuous (masked hyperthyroidism). Thorn and Eder[173] suggested that the metabolic defect may be due to a failure to maintain adequate creatine synthesis in the face of abnormal metabolic demands or to an inadequate utilization and storage of creatine in the involved muscles. This is supported by some of the findings of Zierler.[195]

More recently Dennys and Hoffmann[38] performed in vitro studies on isolated muscle fibers of hyperthyroid animals; they demonstrated a brevity and enfeeblement of the contractile process related to interaction between thin and thick myofilaments and the resorption of calcium into vesicles of the sarcoplasmic reticulum. They conclude that these abnormalities are based on a depletion of high-energy phosphate stores.

Very little has been written about the pathologic changes of the muscles in this disorder since the observations of Askanazy[5] in 1898. He studied four cases in detail and stated that one of the most prominent features of the muscular disorder was a "generalized lipomatosis" in which rows of fat cells infiltrated between muscle bundles and at times between individual fibers. Such muscles were pale reddish. Microscopically, in addition to the fatty infiltration there was atrophy of muscle fibers, with an overall decrease in their diameters, progressing in some fibers to extreme wasting of sarcoplasm. Also reported were the formation of nuclear clumps resembling giant cells; the pale staining of fibers with granularity, homogenization, or fragmentation of the sarcoplasm; vacuolation and yellowish granular pigment collections in some degenerating fibers; and spotty or diffuse fatty degeneration of the fibers of some muscles. However, the muscle tissue in other patients with pronounced weakness has appeared normal histologically.

Asboe-Hansen et al.[4] studied muscle biopsy specimens from the quadriceps femoris or the brachial biceps in 10 patients with progressive exophthalmos and 21 patients with thyrotoxicosis. No mention was made of general or focal muscle weakness (exclusive of extraocular muscles) in these cases. After fixation with formalin and routine staining procedures, the muscles were essentially normal. However, in specimens fixed in basic lead acetate, subsarcolemmal semilunar structures (half-moons) were seen in cross section to occupy portions of many fibers in all the patients with exophthalmos and in 9–21 of those with uncomplicated Graves' disease. Spindle-shaped areas, elongated in the long axis of the fibers, were observed in longitudinal sections. Special staining procedures apparently identified the content of these structures to be an acid mucopolysaccharide resembling hyaluronidase. Twenty-five control cases failed to show similar structures. Their significance is still uncertain. Others have either failed to confirm the existence of these crescents or have identified them as PAS-positive accumulations of glycogen.

We examined the muscles in several cases of this disease and found only a minimal degree of atrophy and variation in the size of individual fibers. Other muscles appeared entirely normal. It is therefore probable that Askanazy's patients had some other muscle disease, e.g., Zenker's degeneration or polymyositis. We rarely observed the infiltration of fat cells and degenerative changes mentioned by Askanazy in muscles of limbs and trunk, though these changes were evident in the ocular muscles (Fig. 11–14); nor have we observed the semilunar structures described by Asboe-Hansen and co-workers in cases of moderately severe chronic thyrotoxic myopathy. The technics of fixation and staining outlined by these authors were used.

Several other investigations of thyrotoxic myopathy[63, 75] are in agreement with the above statements—that there are either no abnormalities or only varying degrees of atrophy and fat cell infiltration or of degeneration of small foci within fibers. Under the electron microscope Pearce[133] and Engel[50] found the sarcolemma to be thrown into sharp folds (projections) in which there were pinocytotic vesicles. In many of the mitochondria the cristae were replaced by amorphous material, scattered dense granules, and membranous fragments. The latter were sometimes seen outside mitochondria as though liberated through mitochondrial degeneration. Some of the mitochondria were extremely long, measuring up to 4 μ. There were also dilatations of the transverse tubular system. A few fibers contained deposits of subsarcolemmal glycogen. Type I fibers were more affected than type II. When these changes became advanced there was some degree of myofilament degeneration. The mitochondrial lesions are of special interest because of the demonstrated effects of thyroxine on mitochondria. Hoch[72] speaks of thyrotoxicosis as a disease of mitochondria. The sarcolemmal folding could be due to muscle fiber atrophy since a similar change is noted in denervation atrophy (Chapter 10). The altered transverse sarcoplasmic reticulum might interfere with transverse transmission of the action potentials, though there is more evidence that the parts of

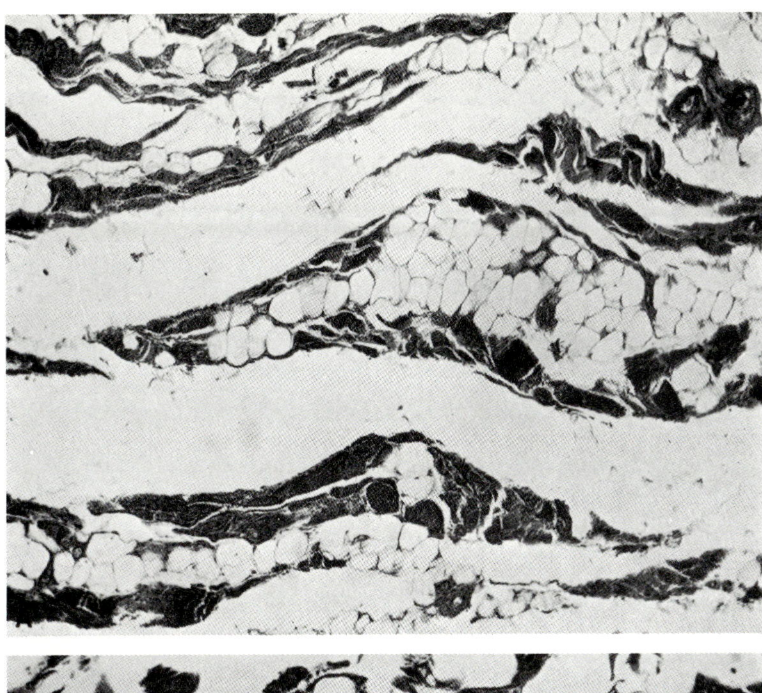

Fig. 11–14. Lipomatosis of a biceps muscle from patient who died of hyperthryoidism. (H&E)

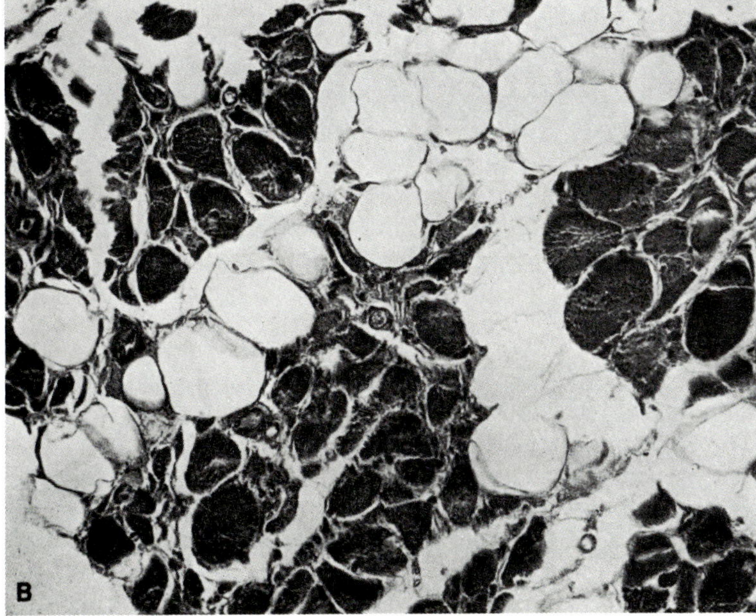

the contractile process involved in the sustained interaction of the myofilaments and their release are defective.

Acute Thyrotoxic Myopathy. A condition called acute thyrotoxic myopathy or "acute thyrotoxic encephalomyopathy" was described by Laurent,[95] Waldenström,[184] and others and appears to involve bulbar paralysis, generalized weakness, and tremors, rapidly progressing to delirium and coma. Strong[165] described a case of subacute type, intermediate between the acute and chronic forms. All types of thyrotoxic myopathy are relieved to some degree by treating the associated hyperthyroidism, and most patients derive no benefit from Prostigmin. Indeed, Strong's patient was greatly worsened by Prostigmin, developing weakness in the limbs and fascicular twitchings. The cases of McEachern and Ross[111] are exceptional in that the weakness was more obviously myasthenic in character, coarse fasciculations were present, and relief was obtained both from Prostigmin and treatment of the thyrotoxicosis. Muscle wasting was not present in one of their cases and was "most evident in small muscles of hands and feet" in another. We are inclined to consider most of these cases as a combination of thyrotoxicosis and myasthenia gravis (see below); the fasciculation, not usually evident in thyrotoxic myopathy, is unexplained. Indeed, from another point of view it can be stated that the more acute the process, the more prominent are the signs of bulbar and extraocular paralysis and less certain is the muscular atrophy. It is in such cases that the action of Prostigmin is most evident. From this point of view the acute cases of Laurent[95] seem to be a fulminating type of myasthenia gravis. Millikan and Haines[116] in their treatise on the neuromuscular disorders in thyroid disease denied the existence of the acute form of thyrotoxic myopathy. They concluded that most or all such patients have had a combination of myasthenia gravis and thyrotoxicosis.

Thyrotoxicosis and Myasthenia Gravis. The combination of thyrotoxicosis and myasthenia gravis has been reported in many cases.[19, 110, 116, 163, 173] Millikan and Haines apparently had the largest experience with these entities (25 cases). They pointed out that 80% of patients with this combination have been women in the age range 19–67 years. In almost all cases the myasthenia gravis was typical, and the muscle weakness responded well to Prostigmin. Diagnosis of the two diseases offered no difficulty since hyperthyroidism is not seen in uncomplicated myasthenia gravis, and the same is true of Prostigmin-reactive myasthenia in hyperthyroidism. Ophthalmoplegia may occur in both conditions, so it is important to know which disease is affecting the extraocular muscles. Under such circumstances it was found that Prostigmin lessens and curare worsens the paralysis of myasthenia gravis, whereas the ophthalmoplegia of Graves' disease was unaffected by these drugs.

The sequence of development and relationship of myasthenia gravis and hyperthyroidism have been matters of considerable debate in the literature. Millikan and Haines remarked that the hyperthyroidism preceded the myasthenia gravis in 48% of their cases; the reverse was true in 32%, and in 20% the two diseases appeared simultaneously. While a see-saw relationship was said to exist between the two conditions[110]—the myasthenia worsening as the hyperthyroidism responded to therapy and the reverse—this was not found in Millikan and Haines' cases.

The appearance of hyperthyroidism before, during, or after detectable myasthenia gravis has been seen in about 5% of cases,[116] indicating a statistically significant relationship. The reverse relationship is less distinct, however, since only a fraction of 1% of patients with hyperthyroidism ever have myasthenia gravis.

It can be concluded from the available evidence that in this combination we observe two separate diseases in a single patient who perhaps has a tendency to develop autoimmune diseases. The special weakness of hyperthyroidism is merely added to that of myasthenia gravis and does not respond to Prostigmin. The myasthenia gravis is not worsened by the hyperthyroidism. On the other hand, the hypothyroidism which may follow the treatment of thyrotoxicosis was observed by Eaton and others to have an adverse affect on myasthenia gravis, in that the Prostigmin requirement rises. Each disease should be treated as a separate entity. The hazards of this combination are greater than the mortality rates of each disease taken individually. Twenty-five per cent of Millikan's and Haines' patients died, usually from the myasthenia gravis, and all three patients reported by Silver and Osserman[163] succumbed. This combination must be differentiated from thyrotoxic muscular weakness and from polymyositis, in which the basal metabolic rate may be increased. We have not found any description of pathologic changes in the literature and have not personally examined the muscles in a case of myasthenia gravis and hyperthyroidism. The pathology of myasthenia gravis is discussed in Chapter 12.

Thyrotoxic Periodic Paralysis. A condition closely resembling familial periodic paralysis, associated with thyrotoxicosis in more than 30 cases,[168] is called thyrotoxic periodic paralysis. It is believed that bouts of periodic paralysis are greatly increased in frequency and severity by thyrotoxicosis or by thyroid feeding. In support of this point McEachern[109] noted the high frequency of hyperthyroidism and goiter in the series of periodic paralysis reported by Shinosaki and Tsuji. This condition may be the same as Basedow's paraplegia, an obscure disorder described in the earlier medical literature. It seems reasonable to conclude that the two are separate diseases occurring in a single individual. In a case of Hildebrand and Kepler,[71] a familial history of periodic paralysis was obtained, and the patient himself experienced mild and infrequent paralytic attacks. With the appearance of hyperthyroidism, however, the attacks of paralysis became severe and frequent. This suggests an intensification of the muscular disorder by excessive thyroid hormone secretion. The role of potassium in this complex is as yet uncertain, though the metabolic studies of McQuarrie and Zeigler showed that on a carbohydrate diet the patient with familial periodic paralysis has a greatly increased potassium requirement. Conversely, the onset of hyperthyroidism with its greater potassium requirement may be expected to help bring out an existent tendency to periodic paralysis. In other cases the weakness of hypopotassemia related to hyperthyroidism may be confused with periodic paralysis. During the attacks Hildebrand and Kepler found the serum potassium usually to be low, but in a number of cases intravenous potassium given during the paralytic period did not provide relief. Thyroidectomy almost always induces a partial or complete remission of the symptoms. The mechanism whereby thyroid hormone alters the serum potassium concentration and causes this muscle weakness is not understood. We have not seen reports of histologic studies of muscle in this combination of disorders and have not studied any specimens ourselves. The pathology of hypokalemic periodic paralysis is discussed in Chapter 12.

Exophthalmic Ophthalmoplegia. Exophthalmic ophthalmoplegia is another muscle disorder associated with hyperthyroidism. In our experience it is considerably more frequent than periodic paralysis and the general myasthenic varieties. Usually there is malignant exophthalmos with swelling of the eyelids, edema of the bulbar conjunctivae, injection of corneal vessels, and at times congestion of retinal veins and papilledema. Weakness of ocular muscles, most often those which elevate and converge the eyeballs, may occur as the malignant exophthalmos progresses, but in some cases it may develop with minimal degrees of proptosis. Diplopia is a troublesome symptom. Visual acuity is not compromised except after ulceration of the cornea.

As pointed out by Brain and Turnbull,[19] Means,[113] Dobyns,[44] and others, the relationship of the exophthalmos and ophthalmoplegia to hyperthyroidism is not clear. Some patients have hyperthyroidism with little or no exophthalmos. In others the two develop together, but the exophthalmos becomes more pronounced after medical or surgical treatment of the hyperthyroidism. In a few cases of exophthalmos the basal metabolic rate is normal throughout the illness. Dobyns and others suggested that the eye signs may be due to an excess of thyroid-stimulating hormone (TSH) from the pituitary gland. Injections of this substance produce a remarkable degree of exophthalmos in the guinea pig.[43, 132] Recent assays demonstrated the presence of TSH in the blood in most cases of Graves' disease. It is elevated in cases of malignant exophthalmos, particularly those in which the exophthalmos progresses after thyroidectomy. It may be reduced by the administration of thyroid hormone, and this is the basis for treating malignant exophthalmos by thyroxine and potassium iodide in combination. The cause of the exophthalmos is an increase in the volume of orbital fat behind the globe. This fat is edematous and infiltrated with lymphocytes, macrophages, and rare neutrophilic leukocytes. The orbital pressure becomes elevated, and according to Dobyns the venous drainage from the orbit is impaired. This accounts for the chemosis, edema of the eyelids, and papilledema. The strabismus and diplopia may result from displacement of the globe by retroorbital fat and from fatty and cellular infiltration of the ocular muscles.[123]

Naffziger[125] studied six cases of severe progressive exophthalmos of this kind. Grossly the orbital tissues were edematous, and the extra-ocular muscles were pale and edematous and often presented a "half-cooked" appearance. They were enlarged three to eight times their normal size in all cases, were firm and at times rubbery, and in advanced cases were hyalinized and gritty to the knife. Histologically there were varying degrees of degeneration of muscle fibers; the earliest lesions were said to be swelling of the fibers, which progressed to a loss of cross striations, accentuation of the longitudinal striation with fraying, and a change in the staining characteristics so as to resemble collagen. In the most advanced stage of the disease there was conversion of this fibrillar material into a fairly dense scar. In some muscles interstitial edema and round-cell infiltration, most marked about the small blood vessels, were found.

Anatomicopathologic examination of the ocular muscles was described by Rundle,[146] Miller *et al.*,[115] and Kroll and Kuwabara.[92] The fibers are separated by edema fluid, and the interstitial tissue is focally or diffusely infiltrated by lymphocytes. Lipocytes are increased in number. In the more advanced lesions the muscle fibers are atrophic and in places have degenerated. Fibrous tissue replaces them.

Brain and Turnbull,[19] Mulvaney,[123] and Dobyns[44] also reported the pathologic changes of ocular muscles in this disease. They observed swelling and edema of the muscles, infiltration of lymphocytes and other inflammatory cells, and increased amounts of fibrous connective tissue in the perimysium.[147] The increase in size may be two- to five-fold, and the fibers show degenerative changes—in contrast to thyrotoxicosis in which orbital fat deposition predominates. Pochin and Rundle[139] performed a chemical analysis of the muscles and other orbital tissues in simple exophthalmic goiter. They noted an increase in the amount of adipose tissue that normally lies between the muscle fibers in these muscles. The increase amounted to 85% both by histologic observations (Sudan III stain) and by chemical extraction methods. Naffziger[124] and Dobyns[44] stressed the similarity of the lesions of the eye muscles in exophthalmic ophthalmoplegia to those produced by Brooks[23] in the leg muscles of dogs after venous

occlusion. They proposed that the swelling of the orbital fat obstructed the venous drainage of the ocular muscles, with resultant fibrosis and infiltration of inflammatory cells. Extreme stretching of the eye muscles may be an additional factor.

Exophthalmic ophthalmoplegia may therefore be considered a thyrotropic affection of the muscles,[123] in contrast to the thyrotoxic myopathy mentioned above and to simple thyrotoxic exophthalmos. No satisfactory medical treatment has been developed, though often the condition tends to regress spontanously with time. In extremely severe cases in which vision is threatened, relief may be obtained only by decompression of the orbit, as practised by Naffziger. Some confusion has been caused by the presence of muscle atrophy of the shoulder girdles and neck in some cases, but we relate this to the concurrence of thyrotoxic myopathy. In other cases,[165] exophthalmic opthalmoplegia appeared or worsened after treatment of the hyperthyroidism. In the second case of McEachern and Ross,[111] exophthalmic ophthalmoplegia appeared after relief of a myasthenic syndrome by thyroidectomy. The interrelation of these syndromes is therefore complex and awaits further elucidation.

Luft's Hypermetabolic Myopathy. Closely related to the above is a curious state called Luft's hypermetabolic myopathy, where the basal metabolic rate exceeds 100+ without abnormality of the thyroid gland.[101] Luft's patient was a 35-year-old woman with muscular wasting and weakness, hyperhydrosis, polydipsia, polyphagia, and weight loss. The EMG changes were consistent with a myopathy. The urinary creatine levels ranged from 240–400 mg per day.

The important change in muscle was a great increase in the number and size of the mitochondria in the subsarcolemmal spaces and paranuclear regions. Many were altered and contained densely packed cristae in zigzag arrangement and rodlike inclusions. The outer membranes were thickened and reduplicated. The oxidative phosphorylation of these mitochondria when removed from the muscle fiber was "loosely coupled." It was concluded that the hypermetabolic state was caused by a defect in mitochondrial enzyme organization. This lesion contrasts to the degenerative changes in mitochondria, the autophagic vacuoles, lipid granules, and subsarcolemmal glycogen deposits observed by Engel in hyperthyroid muscle.

MUSCLE CHANGES IN HYPOTHYROIDISM

Several forms of muscle dysfunction are associated with hypothyroidism and clinical myxedema. Mollaret and Sigwald,[118] Mollaret and Rudaux,[117] Hesser,[70] and others described the association of true generalized muscle hypertrophy (hypertrophia musculorum vera) and hypothyroidism. The muscle symptoms included weakness; muscular aches, pains, and cramps; and slow relaxation—all of which was termed myotonia. This symptom complex of enlarged muscles, reduced strength, ready fatigue, and slowness of movements has for many years been recognized as one of the clinical manifestations of cretinism; in children it is known as Kocher-Debré-Semelaigne's syndrome and was first popularized by the latter two authors in 1935.[32] A similar clinical picture may occur in myxedematous children or adults and is called Hoffmann's syndrome. In the latter the movements are so slow as to suggest myotonia, though myotonia to percussion is absent and the typical prolongation of muscle contraction has not been substantiated by electromyographic study.[170] Persistent creatinuria is another feature of the disease. A unique feature of muscle contraction in myxedema, familiar to many clinicians, is decreased speed of the stretch reflexes. Lambert and his associates[93] investigated this phenomenon, concentrating on the ankle jerk and the biceps reflex in myxedema; they concluded that the slowness of response is due to an "abnormality of the contractile mechanism of the muscle" and not to an alteration of central or peripheral nervous conductivity. From this work it becomes clear that in the hypothyroid individual there is an alteration in muscle metabolism with a general slowing of the metabolic processes which bring about shortening and lengthening of the muscle fibers. The nature of these processes and the mechanism of action of thyroid hormone on muscle metabolism are not yet understood. Again, in

vitro studies of hypometabolic muscles done by Dennys and Hoffman[38] show them to contract normally but to relax more slowly owing to retardation of the factors involved in relaxing the muscle (uncoupling of filaments and restoration of calcium to vesicles of the triads).

Treatment with thyroid extract relieved the muscle symptoms in most of the cases, and they recurred when the treatment was discontinued. Muscle biopsies show a variety of minor changes of questionable significance. Hesser[70] found muscle fibers of average size with normal nuclei and cross striations. There was no inflammatory reaction or interstitial fibrous proliferation or fatty infiltration. However, in occasional fields he noted granular degeneration confined to groups of one or more fibers. A muscle biopsy from a cretin did not show these changes. Mollaret and Sigwald[118] reported slight inequality in the caliber of fibers with some dissociation of fibrils but no sarcolemmal changes or cellular infiltration of interstitial tissue. The most recent discussions on hypothyroid myopathy are those of Amyot,[2] Lanari,[94] Jesserer and Blacizek,[83] Norris and Panner,[127] Bergouignan et al.,[14] and Godet-Guillain and Fardeau.[65] The latter two studies are of particular interest because they include ultrastructural data. Norris and Panner found the sarcolemma and sarcoplasmic reticulum to be normal. In places myofilaments had degenerated (artifact?) and there were accumulations of glycogen. The mitochondria were deformed, and some contained electron-dense inclusions. There were rods throughout the sarcoplasm. The significance of these alterations is unknown. Feinberg and associates[54] described three cases of myxedema in association with myasthenia gravis. Apparently these are the only reported cases; because of the extreme rarity of this association, they concluded that it could be looked upon as a coincidence.

We have not examined skeletal muscles in cases of cretinism or prolonged myxedema. In one case presenting the clinical symptoms described above, Denny-Brown and Nevin found no abnormality in muscle biopsy or electromyogram. A volumetric increase in muscle fibers that are normal in all other respects (hypertrophia musculorum vera) has been described. In one reported case the basophilic change, known to occur in cardiac muscle of the myxedematous patient, was observed in skeletal muscle. The pathologic changes in this disease need further study.

MYOPATHY IN CUSHING'S DISEASE

In Cushing's original description of his syndrome, attention was called to muscular weakness, a condition which subsequent workers ascribed to either a general impairment of protein synthesis or to increased protein excretion (negative protein balance).

Two noteworthy reports appeared in recent years, one by Perkoff et al.[138] presenting a series of cases of myopathy occurring in the course of a therapeutically induced form of Cushing's disease, the other by Müller and Kugelberg[122] summarizing six cases of myopathy developing during the natural form of the disease.

Perkoff and his associates observed generalized muscle weakness, especially prominent in the thighs, developing several weeks or months after the initiation of long-term therapy with 17-corticosteroids (cortisone or prednisone). These substances had been given in an attempt to control some prexistent hematologic disease. The weakness was moderately severe, accompanied by some degree of atrophy and a diminution but not a loss of tendon reflexes, all of which coincided with the appearance of the well known features of Cushing's syndrome. Confirmation of the myopathic nature of the disorder was the absence of other neurologic signs, urinary excretion of large quantities of creatine (200–1000 mg per day), reduced urinary output of creatinine (less than 1000 mg per day), and muscle biopsy which disclosed an unconvincing vacuolation of fibers. Termination of steroid therapy resulted in a fairly prompt restoration of muscle power and normalization of creatine-creatinine excretion rates.

Müller and Kugelberg related the case histories of several clinical examples of "spontaneous" Cushing's disease. The myopathy appeared insidiously several years after the onset of the endocrine disorder, at a time when the other symptoms were severe. The maximal weakness was in the pelvic girdle and thighs, with lesser degrees of affection of lower legs and shoulders. No biochemical data were given. Muscle biopsy revealed degeneration of widely scattered fibers, hyalinization of fibers, and an increase in the number of fat cells within the muscle. Electromyograms demonstrated an alteration of motor units (decreased duration and voltage) without reduction in number. There was no evidence of fibrillation, fasciculation, or excessive muscular irritability. In three cases with adrenal tumors the myopathic disorder disappeared after the tumor was excised.

Reference was made in both these reports to the widespread necrosis of muscle fibers in experimental animals that received large amounts of ACTH or cortisone[49, 62, 64] (Chapter 3). It is our impression that the myopathy of Cushing's disease is not a necrotizing but rather an atrophic process; this was shown by Engel[52] and Engel[50] and Pearse.[133] The latter two investigators found a great increase in mitochondria and a degeneration of many of them. The transverse tubular system and sarcoplasmic reticulum and myofilaments were normal. Prineas et al.[141] noted an increase in lipofuscin, lipid droplets, and glycogen. This metabolic disorder is of obscure nature but bears resemblance to thyrotoxic myopathy, both with respect to distribution of muscle weakness and the tendency to the occurrence of necrosis of isolated muscle fibers.

MUSCLE DISEASE IN HYPERINSULINISM

On a number of occasions during the past several decades a syndrome consisting of sensory disturbances and muscular wasting was found to be associated with hyperinsulinism and its attendant hypoglycemia. Mulder et al.[121] presented a thorough report on this subject, reviewing in some detail all published cases as well as several of their own. Up to 1956 there were 20 patients in whom muscle atrophy was present in conjunction with hyperinsulinism.

Other reports of particular note were made by Levrat and Brette,[99] Barris,[10] and Tom and Richardson.[174]

The cause of the hyperinsulinism in all carefully investigated cases was a hyperfunctioning tumor of the pancreatic islets, solitary or multiple adenomas, and in one case a carcinoma of low-grade malignancy. Neuromuscular symptoms developed in relation to severe, prolonged, or recurrent bouts of hypoglycemia. Paresthesias were also reported, according to Mulder and his associates,[121] and were often the earliest symptoms; they were of the burning and tingling type and were distributed in the periphery of the extremities. However, objective sensory impairment occurred infrequently. Muscle atrophy was initially of the distal type and was present in 17 of the 20 reported cases. It was often accompanied by fascicular twitchings. Associated muscle weakness was reflected in an unsteady "slapping" gait and difficulty in performing fine manual movements. There was no tenderness of nerves. In two cases the muscle weakness was of a proximal limb-girdle distribution. The progression of the muscle atrophy and weakness and sensory disturbances was arrested by removing the pancreatic tumor, a matter of considerable practical interest. Electromyographic studies disclosed a marked reduction in the number of motor-unit potentials, and those that remained were of normal amplitude. There was no delay in nerve conduction time. Mulder *et al.*[121] concluded that these findings were compatible with either early peripheral neuropathy or affection of anterior horn cells. The former explanation is more plausible.

The muscle tissue was examined in only 2 of the 20 cases, and in each the findings described were not those of secondary neural atrophy. Levrat and Brette[99] reported a 31-year-old woman with carcinoma of the islets of Langerhans who suffered from repeated bouts of severe hypoglycemia. A notable feature of this case was muscle pain that eventually involved all extremities. A biopsy taken from a quadriceps muscle about 9 months after onset of the illness showed degeneration of muscle fibers in a spotty distribution. These fibers were said to show Zenker's degenera-

tion with nuclear proliferation and atrophy; the degenerating fibers were replaced by connective tissue. By using an iodine stain, a rather unreliable technic, it was found the glycogen was absent in most fibers and persisted in small amounts in only 5–10% of the fibers. Seven months after the biopsy was obtained, the patient died. Postmortem examination showed the psoas muscle to contain a great increase in fibrous connective tissue (fibrosis) which extended between the muscle fibers. Degenerating fibers were visible in many places, especially at the periphery of the sclerotic areas. Cellular infiltrations were not mentioned. In the second case (a 28-year-old woman who had had repeated episodes of hypoglycemic coma due to multiple adenomas of the pancreas and had developed a distal muscle weakness and atrophy reported by Mulder *et al.,*[121]) biopsy of the gastrocnemius revealed relatively normal muscle tissue except for scattered fibers of basophilic hue containing large nuclei with prominent nucleoli and collections of round cells around some intramuscular blood vessels.

The essential character of this muscle disease cannot be settled at this time; but on the basis of the usual clinical picture of distal muscular atrophy, fasciculations, and finding evidence of motor neuron damage in several cases,[121] the muscle disorder appears to be of the denervation type. However, the muscle biopsy findings in the only two available studies suggest the operation of another factor which leads to a primary myopathy of necrotizing type resembling that seen in other metabolic disorders. Further pathologic studies of this condition are needed.

The pathogenesis of this syndrome is believed to be related to recurrent severe bouts of hypoglycemia due to a hyperfunctioning adenoma of the pancreatic islets. If this is true, one might expect that repeated injections of insulin would evoke similar changes. In this connection, it is of interest that Ziegler[194] recently reported that paresthesias were recorded in 13 and weakness of the hand muscles in 2 of 22 patients in whom insulin coma had been induced as therapy for psychiatric illness. Another relevant observation is that of Tannenberg,[169] who noted pathologic changes in the skeletal muscles, myocardium, and liver of rabbits after repeated insulin injections in doses sufficient to cause shock. There were hydropic changes in the muscle fibers with disruption of myofibrils. Individual muscle fibers were necrotic; sparse infiltrations of neutrophilic leukocytes, lymphocytes, and macrophages as well as regeneration of the sarcolemmal nuclei were also observed. It was postulated on the basis of these experimental data that the combustion of carbohydrate is essential to the metabolism of muscle and that the fibers may be damaged by prolonged hypoglycemia.

MUSCLE DISORDER IN DIABETES MELLITUS

Neurologic syndromes in conjunction with diabetes mellitus are well recognized, and the pathologic changes are of the type discussed in Chapter 10. Several recent articles[61, 166] called attention to a primary motor disorder variously termed diabetic amyotrophy and asymmetrical motor neuropathy. As implied by the terms, there was weakness and wasting of the muscles of one or several extremities, usually involving first the distal ones. However, in several cases only the proximal or "girdle musculature" was affected on one or both sides of the body. Fasciculations, twitchings, and reflex changes were readily demonstrated. In several of the published cases the diabetes was of recent development or had been undetected prior to onset of the amyotrophy; in others diabetes mellitus was of long standing but had been poorly regulated. Electromyographic studies disclosed a pattern of findings suggestive of motor neuron disease; however, this was not certain in all cases. Careful control of the diabetes mellitus, usually with insulin and diet, resulted in partial or even complete reversal of signs and symptoms in the majority of cases.

We found only two sketchy notations[61] on the muscle pathology in this type of disorder; in each there was a combination of atrophy of muscle fibers with sarcolemmal nuclear proliferation and doubtful necrosis of muscle fibers. The changes illustrated could be interpreted to

be the result of neural atrophy. No mention was made of specific inflammatory cell infiltration.

MUSCLE SYNDROME IN HYPERPARATHYROIDISM, HYPERCALCEMIA, VITAMIN D DEFICIENCY AND HYPOPHOSPHATEMIC RICKETS

Vicale[180] presented the clinical details of three cases of hyperparathyroidism and one of osteomalacia due to renal tubular acidosis (a type of Milkman's syndrome) in which muscular weakness and fatigability were prominent complaints. In all three cases a multiplicity of neurologic diagnoses had been made before the underlying metabolic disorder was recognized. In essence the muscular syndrome consisted of a symmetrical proximal weakness and fatigability of limb muscles with discomfort on muscular effort, often amounting to pain, and an associated symmetrical and uniform atrophy of the affected muscles, which was marked in some instances. In addition there was a significant creatinuria. The deep tendon reflexes were hyperactive. There was no muscular tenderness, hypertonia, fasciculations or fibrillations, or abnormal reflexes. This writer observed that in 21 of 33 published cases of hyperparathyroidism and in all three of his own cases of renal tubular acidosis, muscular weakness and fatigability were prominent symptoms. He suggested that the muscular symptoms were due to a disorder of calcium metabolism similar in type to the potassium disturbance in periodic paralysis. Vitamin D, calcium, parathormone, and sodium bicarbonate have successfully restored muscle power in two cases reported by Shy and his collaborators.[162] No histologic studies of the skeletal muscles are available.

Richet *et al.*[145] presented a systematic study of the neuromuscular disorders associated with diseases of the parathyroid glands. They described a series of cases in which myosclerosis was demonstrated by muscle biopsy. By this term they referred to an increase in perivascular, endomysial, and perimysial connective tissue and atrophy of muscle fibers with an increase in sarcolemmal nuclei. The striations of the muscle fibers disappeared, and it was assumed that many of the muscle fibers had been converted to fibrous tissue. The clinical symptoms in their cases were variable. Some had typical scleroderma and Raynaud's phenomenon; others showed arthritis, or spasmophilia with Chvostek and Trousseau signs and prolonged chronaxie. One example of myotonia congenita was included. The evidence of parathyroid disease was not convincing. Blood calcium levels were often normal. Clinical improvement after parathyroidectomy was observed in several cases.

Cholod *et al.*[28] found subtle, nonspecific changes in biopsied quadriceps muscle in two cases of severe pelvic and thigh weakness due to tumors of parathyroid gland. Only a slight to moderate degree of simple atrophy of individual muscle fibers was noted by light microscopy. The muscle nuclei were relatively increased and formed rows. Under phase optics there were intermyofibrillar vacuoles and accumulated lipopigment, and in the spaces around some of the endomysial capillaries and venules there were small numbers of histiocytes containing an unidentified material. Under the electron microscope subsarcolemmal accumulations of lipopigment and vacuolation lay between the muscle nuclei and the plasma membrane. Myofibrils were separated by bands of proliferated mitochondria and vacuoles. Several regenerating muscle fibers were observed but no necrosis. The basement membranes of capillaries were thickened, and the author believes this might be the cause of the muscular lesions.

These changes are nonspecific and leave one in doubt as to the etiology and pathogenesis of the myopathy. Its reversibility when the parathyroid tumor is removed suggests a biochemical mechanism. Calcium levels correlate poorly with the myopathy. Possibly the low phosphorus is the key factor.

Vitamin D deficiency and hypocalcemia, whether from dietary lack or intestinal malabsorption, are also associated with muscle weakness but usually tetany is the more prominent feature. In contrast, in osteomalacia in adults and renal rickets in children there is muscle weakness without visible microscopic change in nerve or muscle. Vitamin D is believed to play a role in normal muscle activity possibly by regulating calcium activated ATPase or membrane phospholipids. A special form of hypophosphatemic rickets in children is that associated with bone tumors in adults again with normal levels of serum Ca. There is pronounced muscle weakness. In one of our cases[148] the weakness was so great that the child could not arise from the floor, climb stairs, or hold the arms above the head. A diagnosis of muscular dystrophy had been made. Finding osseous changes of mild rickets led to the discovery of the hypophosphatemia. The muscle biopsy by routine light microscopy showed no change. Improvement of the weakness was slight on vitamin D therapy. Later an ossifying hemangioma of bone was discovered in the radial bone; and when it was excised there was a complete cure, the strength returning to normal. Similar cases with bone tumors were reported by Salassa *et al.*[149] and others. It is tempting to believe that the low phosphate level interfered with either neuromuscular transmission or the contractile process.

MUSCLE HYPERTROPHY AND ATROPHY ASSOCIATED WITH DISEASES OF THE HYPOPHYSIS

Clinical medicine provides numerous illustrations of the influence of the pituitary body on striated muscle. The general hypertrophy of skeletal muscle in acromegalic patients is a well recognized clinical phenomenon. Maranon and Richet[103] commented upon the remarkable muscular power that many acromegalics exhibit during the early stages of their illness. They tell of two professional boxers, renowned for feats of strength, who had the characteristic features of acromegaly and great enlargement of the sella turcica on x rays. Obviously the hypophyseal tumor determined the muscular hypertrophy and great strength.

Muscular enfeeblement often ensues in the late stages of acromegaly. This may occur at a time when the total muscle bulk is still large.

Later the muscles undergo atrophy. Presumably the muscular asthenia results from a further growth of the hypophyseal tumor, which leads to hypopituitarism. Similar muscle weakness and atrophy are observed in Fröhlich's adiposogenital syndrome and in Simmonds' disease.

Little is known of the histologic changes of the striated muscle in these conditions. It is supposed that the muscle hypertrophy represents enlargement of the individual muscle fibers due to the activity of the pituitary factor which stimulates growth of viscera. This may be brought about by an excess of androgenic hormone.[131] Likewise in hypopituitarism small muscles indicate a lack of this growth hormone.[74] Marenzi,[104] Binet *et al.*,[15] and Shapiro and Zwarenstein[158] showed experimentally that hypophysectomy, or more specifically removal of the anterior lobe of the pituitary body, results in a progressive muscular asthenia accompanied by a disturbance of creatine metabolism and by diminution of certain chemical components of muscle such as phosphocreatine, total phosphorus, and glutathione.

Acromegaly and Fröhlich's syndrome have been observed in conjunction with syringomyelia, with the progressive muscular atrophy of Aran-Duchenne, and with muscular dystrophy. These unusual combinations of diseases were the subject of articles by Duchesneau[47] and Netter,[126] and examples were reported by Barker.[8] McCullagh and Hewlett[108] described a case of acromegaly and amyotrophic lateral sclerosis that was confirmed by autopsy. They also relate two other clinical cases in which the symptoms of amyotrophic lateral sclerosis were relieved by treatment of the acromegaly. In all probability some of these cases, particularly those with definite degeneration of motor nerve cells, are fortuitous concurrences of unrelated diseases, and others may be examples of thyrotoxic myopathy since hyperthyroidism not infrequently develops in acromegalic patients.

A number of authors reported examples of acromegaly and localized muscular atrophy. This may be limited to the arm muscles, interosseous muscles, thenar and hypothenar muscles, and leg muscles. Doubtless compression of nerve and spinal roots leading to denervation atrophy (carpal tunnel syndrome cervical spondylosis) accounts for some of

these changes. The large bibliography on this subject can be found in Netter's thesis.[126] Few patients have been examined postmortem, and the sporadic pathologic studies lack sufficient detail to furnish a clear idea as to the nature of the muscle disease. General statements to the effect that the muscles are atrophic, the fibers are of variable caliber with some degeneration, and the interstitial connective tissue and fat are increased are difficult to interpret.

CONTRACTURES OF THE LIMBS IN ADDISON'S DISEASE

A number of cases of Addison's disease with flexion contracture of the legs and arms have been reported. It is remarked that at some stage of the disease the legs gradually flex at the knees and ankles to a degree that at first interferes with locomotion and finally prevents standing or walking. Bartels[11] observed this condition, as did Thorn.[172]

We examined two such patients. One, a patient studied by VanderLaan and Adams,[178] was a middle-aged woman who was diagnosed as having Addison's disease within a few months after the onset of pigmentation and asthenia. When she was first admitted to the hospital for treatment with desoxycorticosterone it was noticed that her knees were flexed very slightly. During the succeeding months, even though there was symptomatic improvement from DOCA treatment, the legs flexed more and more until she could no longer stand. When examined 6 months later there was no swelling, deformity, pain, or x-ray changes in the knees, aside from slight osteoporosis. In attempts to extend the knees passively, a firm unyielding resistance was suddenly encountered. The hamstring muscles became taut. There was no tenderness of muscles and tendons. Although thin, the hamstring muscles contracted firmly, exhibited brisk normal tendon reflexes, and had a normal electromyogram. The quadriceps muscles were thin but otherwise normal. Biopsy of the biceps femoris disclosed normal-sized muscle fibers with perhaps more fraying artifact than is seen in most biopsies. There was no increase in connective tissue. Six months after treatment with ACTH the contractures were much less severe, and the legs were almost straight. The patient died unexpectedly about a year later, a few days after she had taken her first steps. At autopsy the knees were still flexed about 15–20°, and the strongest force would not straighten them. It was found that the knee joints were entirely normal, and that the contracture was due to shortening of either the muscles or the tendons of the biceps femoris, semitendinosus, and semimembranosus muscles. When the tendons were severed, the legs straightened to nearly 180°.

Microscopic sections of the sciatic nerves, muscles, and tendons disclosed no significant abnormalities. Transverse and longitudinal sections of the biceps femoris, semitendinosus, semimembranosus, and quadriceps were examined separately. There was no increase in collagen or elastic tissue in any of these muscles, and the muscle fibers were of natural appearance. The average diameter of the fibers of the biceps femoris was 35.5 μ, of the semitendinosus 39.6 μ, of the semimembranosus 34.5 μ, and of the quadriceps 38.4 μ. In the quadriceps muscle there was greater variation in size of the fibers and more sarcolemmal nuclei per fiber than in the other muscles; in some microscopic fields there were groups of 10–15 fibers averaging 26 μ. No alterations of blood vessels, perimysial connective tissue, or muscle spindles were noted. Two small infiltrations of lymphocytes and histiocytes were found in the semitendinosus muscle, one surrounding a small vessel in the perimysium and the other in the endomysium between several adjacent muscle fibers.

The other patient, and elderly black woman admitted because of recurrent episodes of coma and symptoms of Addison's disease, had the same type of flexion contractures of the legs, which had been present for 2–3 years. Roentgenograms of the knees showed narrowing of the joint spaces, hypertrophic spurs, and osteoporosis. At postmortem examination, which disclosed almost complete destruction of the anterior lobe of the pituitary body and atrophy of adrenal glands, the semimembranosus, semitendinosus, and biceps femoris muscles showed a type of change commonly seen in cachexia

and senility. The muscle fibers were extremely variable in size, ranging from 10 to 40 μ in width. The sarcolemmal nuclei in the small fibers were small and darkly stained and were increased in number. The striations of most of the fibers were well preserved. There was a definite increase in endomysial connective tissue (Fig. 11–15). The peripheral nerves and spinal cord showed no significant changes.

Our observations on flexion contractures of the legs in Addison's disease support the theory presented by Thorn[172] that this is *per primum* a tendinous and fascial contracture and not a disease of muscles or joints. Thorn postulated that the concentration of sodium, which is relatively high in normal tendon, accumulates in excessive quantities with a corresponding decrease in potassium. This he believes to be related in some manner to a shortening of tendons.

MYOGLOBINURIA

Myoglobin, or myohemoglobin (Chapter 2), is found in the urine in the following conditions: (1) spontaneous myoglobinuria of unknown etiology; (2) as a result of crush injuries to muscles and in intoxicated states (carbon monoxide poisoning, alcoholism); (3) following extreme muscular activity; (4) following ingestion of certain toxic agents; (5) in McArdle's disease and other metabolic myopathies (e.g., lipid storage-carnitine palmityl transferase deficiency) after excess of activity; (6) in severe polymyositis; and (7) in malignant hyperthermia. General reviews on the subject of myoglobinuria and on "myolysis" were written by Brass,[21] Kreutzer *et al.*,[91] and Köhler.[88]

IDIOPATHIC PAROXYSMAL MYOGLOBINURIA

An idiopathic form of recurrent myoglobinuria was first described in 1911 by Meyer-Betz.[114] The patient was a 13-year-old boy who from the age of 9 years suffered repeated episodes of

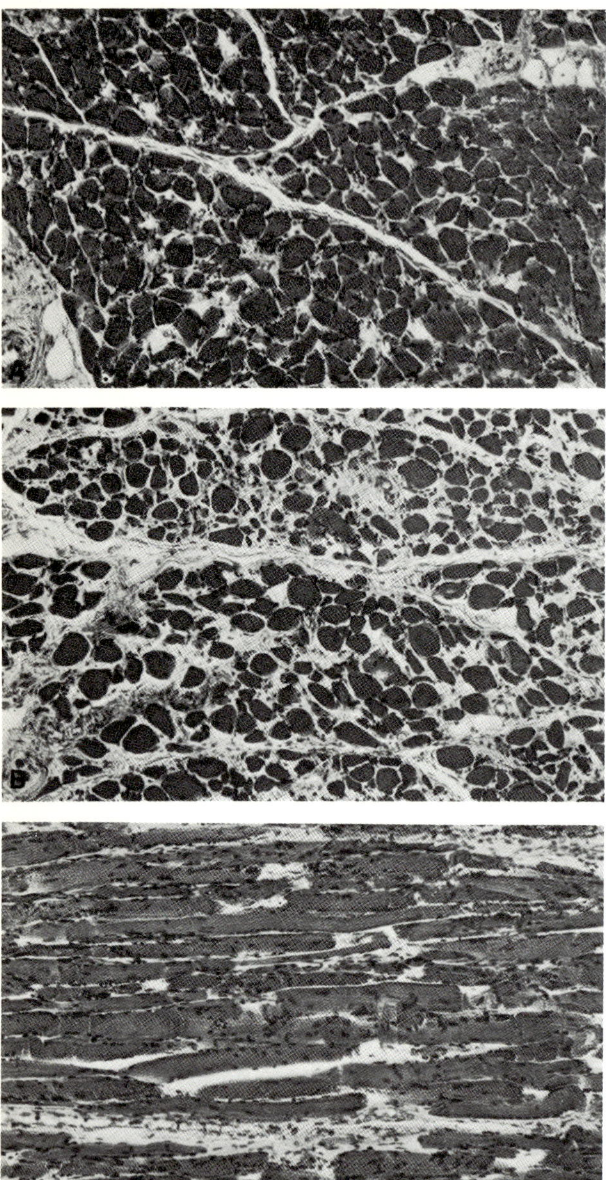

Fig. 11–15. Addison's disease with contractures of legs. Biceps femoris muscle. (A) Muscle fibers of normal appearance. (B) Another portion of same muscle showing atrophy of fibers and slight increase in connective tissue due presumably to associated senility and cachexia. (C) Higher magnification of atrophic fibers with slight nuclear proliferation. (H&E)

generalized or localized muscular cramping unrelated to use of the muscles, followed by the passage of reddish urine. Myoglobin was not specifically identified in the urine, but its presence was very likely from the history and a comparison with cases reported subsequently. Of further interest in this case was the gradual appearance of muscular atrophy (or poor muscular development) after a number of attacks and the patient's tendency to arise from a lying position in the same manner as does a dystrophic patient with paravertebral muscular weakness. Since this first report nearly a hundred cases have been recorded and in several recent studies[18, 88, 135, 144] more details of the pathology and pathophysiology of the disorder have been revealed. The actual cause of the disorder, however, remains unknown.

The onset of an attack is usually sudden, and the leading symptom is severe pain and cramping in the muscles, often followed by temporary muscle weakness or outright paralysis. Attacks occur under two circumstances, exertion and infection, which is the basis of a clinical division into two types.[90] In type I the onset is usually at puberty or during adolescence or early adult life; it occurs predominantly in males, and the family history is positive in one-third of cases. The symptoms follow heavy exertion within a few hours. In type II the onset is usually during childhood in association with an infection, and a positive family history is less often obtained. There is fever, leukocytosis, and often uremia. In both types the affected muscles are swollen and firm or "woody" in consistency, and they are tender to pressure and feel as though they are in spasm. The muscles of the lower extremities are most frequently affected, but other groups may be involved in some cases, especially in those patients whose attacks are initiated by the exertion of a particular group of muscles. Systemic symptoms such as anorexia, vomiting, pallor, abdominal pain, and fever, as well as actual shock and leukocytosis, have been reported. Within a few hours after muscle symptoms develop the urine becomes progressively discolored—first a faint pink and then within a few hours a deep reddish brown. In a number of cases oliguria or anuria ensued, apparently due to renal damage from myoglobin or other substances released from the injured muscle fibers; and several patients died in acute renal failure. Myoglobin pigment may continue to be present in the urine for 72 hours. After repeated attacks, weakness and wasting of affected muscles may or may not be permanent.

A number of patients have had recurrent attacks over many years, and one episode occurred after an interval of 40 years. Physical exertion has been the precipitating factor in about half the reported cases; usually the muscles exercised most strenuously are the ones maximally affected in the attack. In the other cases attacks evidently developed while the patient was at rest or upon his awakening after a night's sleep; and these are attacks that have been associated with an infection. Figure 11–16A was taken during the regenerative phase in a 5-year-old child whose myoglobinuria occurred during a respiratory infection. In families of apparent clinical muscular dystrophy some members have had myoglobinuria, but the possibility remains that these may have been instances of subclinical myoglobinuria with residual muscular damage mistaken for muscular dystrophy.

The sudden appearance of myoglobin in the urine following an attack of muscular cramping is a dramatic event that is rarely overlooked. The burgundy-red urine suggests, in addition to myoglobinuria, genitourinary hemorrhage, severe hemolytic disease, or less commonly porphyria. The differential diagnosis between these conditions can usually be made on the basis of a careful history or by the several simple laboratory tests listed in Table 11–2. Final identification of the disease can be made only by spectrophotometric analysis of the urine for myoglobin in either its oxidized, reduced, or metmyoglobin forms.[135]

Recent reports shed more light on the extent of the chemical and structural alterations that occur in this disease. In one case studied extensively by Pearson the patient, a 57-year-old male, had been strong and vigorous until the age of 20, when he first noted an initial episode of muscular cramping in the calves after a long hike. Subsequently cramps developed in any muscles subjected to heavy exercise, and after the age of 22 years the urine was red at these times. During the subsequent 35 years the patient noted repeated minor and at least 50 major episodes of muscular cramping, most of which were followed by the passage of red urine. In later years less and less exertion was required to initiate an attack. At the age of 57 the muscles were notably flabby but not particularly weak or atrophic. An episode of muscular cramping was evoked by repeatedly arising from a squatting position; cramps began within 20 sec, and in 40 sec he was powerless to arise and had to be lifted to the bed. The thigh muscles were stiff and hard. Serial collections of urine disclosed the initial appearance of a faint pink coloration in the urine 2 hours after exercise, with a subsequent increase in the color intensity to a deep red brown at 8 hours and beyond. The color of the blood serum remained normal. Postexercise examinations of urine and hematologic and blood chemistry revealed only minor variations from baseline values, except for urinary creatine levels, which rose sixfold in the postexercise period. The serum

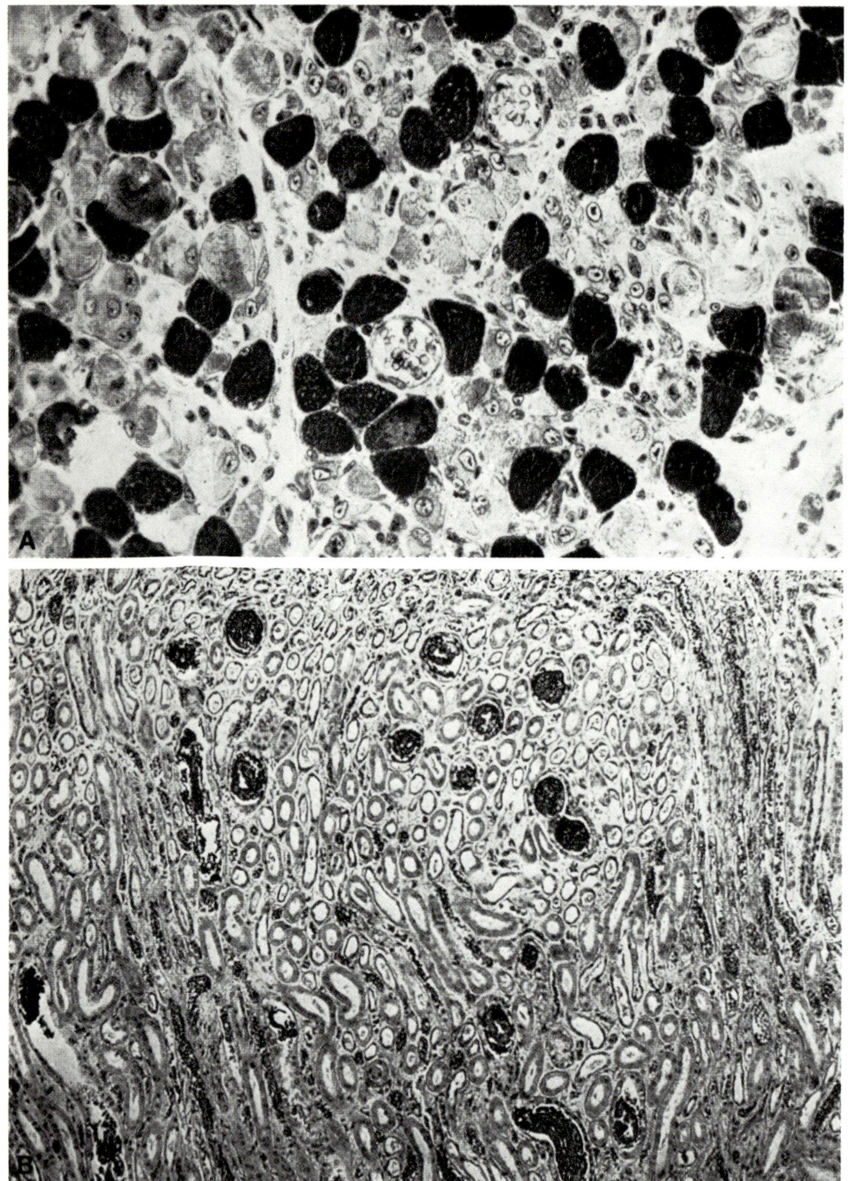

Fig. 11–16. (A) Transverse section of gastrocnemius muscle of a 5-year-old child taken 2 weeks after a first attack of idiopathic myoglobinuria. Dark fibers are undamaged, and pale ones with large sarcolemmal nuclei and few myofibrils are regenerating. (B) Section of kidney from patient with a fatal case of myoglobinuria. The dark casts in the collecting tubules are amorphous masses of myoglobin. (A, phosphotungstic hematoxylin; B, H&E)

levels of the enzymes glutamic oxalacetic transaminase, adolase, and phosphohexoisomerase rose to 10–20 times their baseline values within 8 hours and returned to normal in 96 hours.

On the basis of cases like this it may be said that paroxysmal myoglobinuria represents a syndrome in which the appearance of myoglobin in the urine is only one of a spectrum of biochemical derangements, albeit the most dramatic and useful change since it is an easily

TABLE 11–2. Laboratory Differential Diagnosis of Paroxysmal Pigmenturia

Type	Serum color	Urine color	Guaiac test on urine	Fluorescence of urine	Microscopy of urine
Myoglobinuria	Normal	Red to chocolate brown	Positive	None	Negative or rare RBC
Hemoglobinuria	Pink to red	Red to chocolate brown	Positive	None	Negative or rare RBC
Porphyria	Normal	Burgundy*	Negative	Present	Negative
GU hemorrhage	Normal	Red	Positive	None	Loaded RBC

* Color may develop only after exposure to light.

visible indicator of the severe necrosis of sarcoplasm. It should be added that Bowden et al.[18] found certain urinary amino acids (especially taurine) to be elevated after an attack, but again it seems likely that these are but other normal intracellular constituents released during the destruction of muscle fibers. In fact the myoglobin excretion itself may be a minor or secondary change but until the actual etiologic agent and pathogenesis of this condition have been ascertained it seems best to retain the descriptive term paroxysmal myoglobinuria.

The pathologic histologic changes of the muscle fibers after an attack were most carefully studied and recorded by Reiner et al.[144] in two cases. Their studies, made on biopsy material obtained on the eighth, tenth, and twentieth postattack days, revealed extensive damage of the contractile elements with all of the features of severe and widespread coagulative muscle necrosis similar to Zenker's degeneration. Hyaline necrosis and fragmentation of large segments of fibers were prominent, as was loss of cross striation and apparent decolorization of segments of fibers. Sequestrated necrotic fiber segments were often separated from the intact fiber by disintegrated sarcolemmal sheaths left with only a fine protein precipitate or a few phagocytic cells. Calcium impregnation was observed in some necrotic fibers, and a vacuolar degeneration of muscle fibers was seen occasionally. Figure 11–17 was prepared from the second case of Reiner et al.[144] Exudative cellular reactions were not observed; only a few lymphocytes and mononuclear histiocytes were seen. The histologic changes in the muscle appear to be the same in spontaneous "toxic" myoglobinuria (type I of Korein et al.[90] as in exertional myoglobinuria (type II). In both types the damaged fibers are scattered in haphazard fashion among healthy ones.[37] Type I can occur at any early age and is commonly fatal after a few attacks. Type II does not usually appear until after the age of 15 and in repeated attacks leads to proximal muscle atrophy (see also McArdle syndrome, Chapter 12).

We recently examined several cases. In one that ended fatally 5 days after the onset of muscular pains and weakness, the sarcoplasm of many fibers was hyalinized and fragmented. Myoglobin casts were found in the kidney tubules (Fig. 11–16B). In another case the muscle degeneration was followed by an extraodinary degree of proliferation of sarcolemmal nuclei and regeneration of sarcous substance (Fig. 11–16A). Brief descriptions of the pathologic changes were also made by Schaar et al.[151] and by Buchanan and Steiner.[24]

Speculations as to the etiology of type I myoglobinuria are few and so far have revolved around an analysis of myoglobin in these patients. Berenbaum et al.[12] suggested that perhaps an abnormal myoglobin molecule analogous to the defect in sickle cell disease and other abnormal hemoglobin disorders was the causative factor, but Prankerd[140] was not able to verify this; Prankerd observed no difference in electrophoretic mobilities of myoglobin extracted from muscle biopsy material from a typical patient with paroxysmal paralytic myoglobinuria and from a normal adult and a patient with sickle cell disease who served as controls. It is probable that the basic defect will eventually

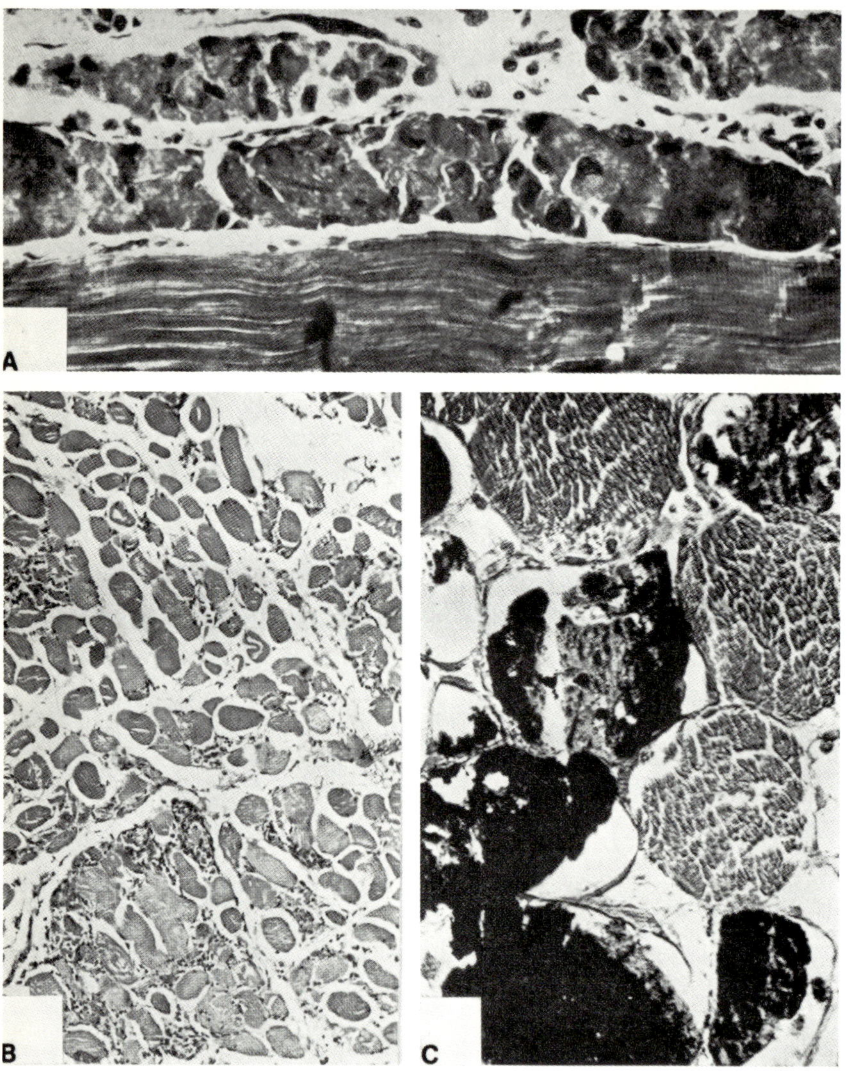

Fig. 11–17. Idiopathic paroxysmal myoglobinuria. (A) Phagocytosis of degenerating fibers. (B) Pattern of degenerating muscle fibers in transverse section. (C) Calcium deposit in degenerating sarcoplasm. (A,B, H&E; C, von Kossa)

prove to be a metabolic fault in another vital constituent of the muscle fiber, similar to the phosphorylase defect in exertional myoglobinuria.

CRUSH INJURIES OF MUSCLES

Another disorder characterized by the presence of myoglobin in the urine, intensively studied by Bywaters and associates,[25] is seen in some per-

sons suffering severe crush injuries of muscles. Such patients show evidence of renal failure and excreted a smoky-red urine that contain brownish casts of myoglobin. Anuria, due at least in part to blocking of the renal tubules by the muscle pigment, is the cause of death in some cases. The case described by deLangen[33] of a prisoner subjected to violent exercise, flogging, and exposure for several hours, with subsequent development of severe muscle aches, cramps, and tenderness, with reddish black urine spectroscopically positive for myoglobin, is probably another example of the same condition. Undoubtedly some of the reported cases of carbon monoxide poisoning and alcoholic myopathy are examples of crush injury in a comatose patient.

MARCH HEMOGLOBINURIA

In several cases of so-called march hemoglobinuria Bywaters observed only hemoglobin in the urine. However, it may be that some of the reported cases of march hemoglobinuria are in reality march myoglobinuria. Some patients with the anterior tibial syndrome (Chapter 9) died of anuria when attempts were made to mobilize the injured muscles.

HAFF DISEASE

The fourth condition associated with myoglobinuria is a peculiar malady called Haff disease, named after the locale in Germany where it was first described by Assmann *et al.*[6] in 1933 and others. Attacks of myoglobinuria followed ingestion of eels or fish poisoned by discarded waste products of nearby cellulose factories. The specific cause was thought to be two acids in the resinous waste products, specifically pimaric and abietic acids, although this was never proved. According to Biörck[16] an infectious element was not completely ruled out.

Myohemoglobin has been observed in the urine in animals, notably the horse, in a condition called equine paralytic myoglobinuria.[27] This appears to be comparable in many respects to spontaneous myoglobinuria in man.

In all these diseases myonecrosis has liberated myoglobin and other sarcoplasmic products and there is nothing otherwise specific about the muscle lesion.

POLYMYOSITIS

If unusually intense, polymyositis may cause such extensive necrosis that it produces a transitory myoglobinuria.

MYOPATHY AND MALIGNANT HYPERPYREXIA

Malignant hyperpyrexia is defined as a diffuse metabolic polymyopathy, occurring as an autosomal dominant trait, which renders the patient susceptible to certain anesthetic agents that induce widespread muscular rigidity, extreme hyperthermia, circulatory collapse, and death. Examples of hyperthermic anesthetic accidents are scattered widely in the medical literature, but it was not until the report by Denborough and Lovell[34] in 1960 that knowledge of the underlying malady became widely disseminated. By 1970 Britt and Kalow[22] collected 170 cases from the literature. For a systematic analysis of the problem and a review of recent writings on the subject the reader is referred to the article by Isaacs and Barlow.[81]

For a time it was believed that individuals susceptible to hyperthermic anesthetic catastrophes were entirely normal until the moment of the accident, but a more careful examination has shown that at least one group of them have certain musculoskeletal and myopathic abnormalites. As to the former, ptosis, strabismus, highly arched palate, dislocated patellae, and in a few kyphoscoliosis and abdominal hernias have been the most frequent. As to the latter,[86] a progressive congenital myopathy (in some instances not unlike that described by Barnes[9]) should be mentioned, in which the muscles were first hypertrophic and later pseudohypertrophic and atrophic. It has even been suggested that myotonia congenita may be associated with malignant hyperthemia; this is open to question, however, for as Isaacs and Barlow[81] remark, the review of a large series of patients with myotonia congenita and dystrophica (56 in all) who had surgical procedures requiring anesthesia had disclosed no such accidents.

The clinical picture of the hyperthermic episode is so dramatic as to be unforgettable. As halothane anesthesia is induced and suxemethonium is given for relaxation, the jaw muscles become tense rather than relaxed. Gradually the rigidity spreads to all the muscles, and soon thereafter the temperature rapidly rises to spectacular heights (up to 43°C). Coincidentally there is tachypnea, tachycardia, and finally failure of brainstem reflexes with circulatory collapse. If extreme hyperthermia is prevented the patient invariably survives. Apparently another closely related form of the disease consists of hyperthemia without muscular rigidity. Muscle fibers are destroyed during the attack, and marked increases in CPK and aldolase and myoglobinuria result. The latter carries its usual liability to renal failure. Muscular weakness and later atrophy may persist long after the episode has passed.

In many instances, unless other members of the family are known to have suffered an untoward anesthetic accident, the disease is discovered only when an anesthetic is first given. In such families, however, the disposed members can be detected by measuring CPK; only rarely are normal values found. The test may give false positive reactions, for when a normal population of subjects is tested the CPK is occasionally elevated and the EMG may be mildly myopathic.

The pathogenesis of this disease has been the topic of a number of investigations, some of which were extended to a study of a breed of pigs (Landrace) that also develops muscle spasm and hyperthermia when subjected to suxemethonium and halothane anesthesia or to great excitement. The lactic acid which accumulates in their muscles during the muscle spasm is said to cause their muscle tissue to be pale and soft. The spasm is believed to be a true contracture. In the pig there is a rapid depletion of ATP on exposure to halothane, and Venable[179] demonstrated a myopathy by electron microscopy in a susceptible strain of Poland-China pigs. In animals preliminary injection of curare blocks the triggering effect of suxemethonium but not the action of halothane. The suxemethonium-halothane sequence is known to increase oxygen consumption of muscle by 50–60% because of end-plate generation of nerve impulses. Suxemethonium causes intense muscle fasciculation. With halothane alone the increase in oxygen requirement is only 20%. Halothane leads also to the uncoupling of oxidative phosphorylation and accelerated glycolysis, which are necessary to restore ATP. This could explain both the contracture (Chapter 4) and increased heat production. The release of calcium into the

sarcoplasm from the cisternae occurs normally, but without ATP it reaccumulates in the sacroplasmic reticulum and the muscle cannot relax. The excised muscle from human subjects is abnormally susceptible also to caffeine contracture.

In sum, there is postulated an inborn error of metabolism which because of an enzymatic defect results in a depletion of ATP that renders the muscle incapable of maintaining calcium distribution. A primary defect in phosphodiesterase, the enzyme involved in the degradation of adenosine monophosphate (cyclic AMP) is possibly another factor. Damage to the sarcolemma releases muscle enzymes and potassium into the serum.

It is still an unsettled matter as to whether the hyperpyrexia is entirely secondary to the muscle spasm. On the face of it one could hardly believe that such an astronomical degree of fever could be produced in this way. In tetanus and other states of excessive muscle activity there is no such increase in body temperature. However, there is the possibility that the anesthetic might impede the heat regulatory mechanisms. The chief argument against this has been that the obvious sweating, flushing of skin, tachypnea, and tachycardia indicates these mechanisms to be working properly at least at lower grades of fever.

Segmental necrosis in large numbers of muscle fibers has been reported in patients who survived the hyperthermic crisis for a few days. However, if death is immediate such damage to muscle fibers cannot be distinguished from artifactual change. The usually myophagic response develops during the first 48–72 hours and is responsible for "mononuclear infiltrates." More puzzling are illustrations which show lymphocytic infiltrates.[81] Regeneration and restoration of injured segments follows. In some instances the myonecrosis affects several contiguous fibers in a small group as is true of many metabolic myopathies, raising the question of a circulatory factor. Fiber branching and group atrophy have also been observed and raised question of denervation but this can happen in any type of lesion that destroys groups of fibers and allows regeneration (Chapter 4). "Fiber type grouping" demonstrated in histochemical stains can be interpreted in the same way. No specific histochemical abnormality was found in muscle specimens obtained in susceptible individuals. Some degree of fiber loss, irregularity of fiber sizes, and increase in endomysial connective tissue are late stages in the reparative process. Myoglobin casts can be seen in renal tubules in fatal cases. Since these lesions do not differ from those of paroxysmal myglobinuria of Meyer-Betz, no additional illustrations are included.

GENERALIZED MYOSITIS OSSIFICANS

Two types of myositis ossificans are recognized. One is a localized form which appears in a single muscle or group of muscles after trauma, and the other is a progressive and widespread ossifying process in many muscles of the body unrelated to trauma. The localized traumatic type was discussed in Chapter 9 and is not considered further.

Generalized (progressive) myositis ossificans is a disease of unknown etiology. It affects the muscular system, usually in young growing children, adolescents, or young adults, and involves not only the interstitial tisue of the muscles but also the tendons, ligaments, fasciae, aponeuroses, and even the skin. In these structures there appear masses of fibrous tissue and bone, and the muscle fibers undergo atrophy and destruction secondary to pressure and inactivity. It is quite likely that this disorder is not primarily one of muscle substance but rather an affection of fibrous and collagenous connective tissue.

Although the cause of this condition is obscure, certain features are of particular interest. In about 70–75% of the cases there are associated congenital anomalies,[102, 112, 177] the most frequent of which is a failure of development of the great toes or thumbs and occasionally other digits, often referred to as microdactyly and supposedly due to synostosis or ankylosis of the interphalangeal joint and the absence of one phalanx. At times there is a family history of myositis ossificans or short thumbs. Exostoses, congenital hallux valgus, absence of the ear lobes, absence of the two upper incisors, and

spina bifida have also been observed in children with generalized ossifying myositis. In some cases the clinical picture at first is indistinguishable from that of a rheumatic infection with fever, joint pains, and swelling. A traumatic episode is said by some writers to be the initial event, but it becomes evident after reviewing a series of cases that traumatism occurs in a very small proportion of these, and as Mair[102] stated it probably does little more than determine the site at which a lesion may develop in an individual who already has the potentialities for abnormal bone formation.

The basic defect in this disorder is yet to be revealed, although it seems clear that it resides in the connective tissue. The observations of Wilkins *et al.*[187] that alkaline phosphatase activity in areas of heterotopic bone formation is higher than in the normal rib may be of considerable significance. The relationship between the antenatal events (i.e., microdactyly and other defects) and postnatal developments (i.e., various ankyloses and the heterotopic ossifying process) is at present unclear. Michelsohn, as mentioned by McKusick, suggested that the entire constellation of clinical abnormalities may be the expression of a single or closely related set of genetic factors operating in a different fashion at various stages in the development of the individual.

The disease is uncommon. Uehlinger[176] in 1936 was able to collect only 160 cases from the literature; since that time about 30 more have been reported. Nutt[128] found that 102 of 112 cases began before the age of 10 years. In rare instances it may be present at birth or as late as the 20th year. In many cases the patient attended medical clinics for one of the commonly associated congenital deformities prior to the onset of the disease.

The onset of the muscular disorder is characterized by the appearance of soft, fluctuant, or hard swellings anywhere on the body, but most commonly in the region of the neck or back. Wryneck deformity may develop, and during early childhood the differentiation from congenital torticollis or "sternomastoid tumor of infancy" may be difficult. In such a situation the finding of microdactyly can be a valuable diagnostic sign. The swellings develop spontaneously or follow minimal trauma. They vary in size and shape and after several days to weeks may disappear entirely or shrink slightly before becoming indurated and finally transformed into bony lumps. The masses are often covered by reddened skin. They are usually painless, although mild discomfort may occur. Occasionally, due to pressure on the overlying skin, the bony masses ulcerate through the skin and discharge a chalky white material. A succession of these lesions appears over a period of months or years.

Almost any of the striated muscles may be the seat of the ossifying process, but most commonly the dorsal aspect of the trunk and the proximal musculature of the limbs are affected. The plantar and palmar fascia may also be ossified, and exostoses are often found at the attachments of fibrous structures to bone, particularly in the occipital area of the skull or as an anterior calcaneal spur. Certain muscles are relatively immune from involvement. These include the tongue, larynx, diaphragm, and peri-neal muscles, as well as those of the eye and the cardiac muscle.

Progressive involvement of the back muscles and later those of the extremities, the abdominal wall, and lastly the intercostals usually occurs, in that order. As the disease progresses, entire muscles or groups of muscles may be converted into osseous tissue. The newly formed bones acquire various shapes, roughly conforming to the shape of the involved muscle. Calcified bridges between adjacent muscles and across joints result in ankylosis so that movements become more and more restricted. Along the back the bony masses assume the form of fattened angular plates, whereas in the extremities they are elongated and cylindrical. Plates of bone in the masseter muscle fix the jaw, and the vertebrae are ankylosed by bridges between the arches and the spinous process. The afflicted individual becomes a virtual "stone man." Usually the articular surfaces of the joints are not involved, although the joint capsule may be ossified. The cause of death is often a terminal

Fig. 11–18. Primary myositis ossificans. Biopsy from adductor muscle of thigh. All the usual changes in transformation of connective tissue to bone are illustrated. Fibrillar connective tissue (at left) merges into collagenous connective tissue. At center, homogeneous tissue resembles embryonic cartilage. At extreme right, darker tissues assumes the character of newly formed bone.

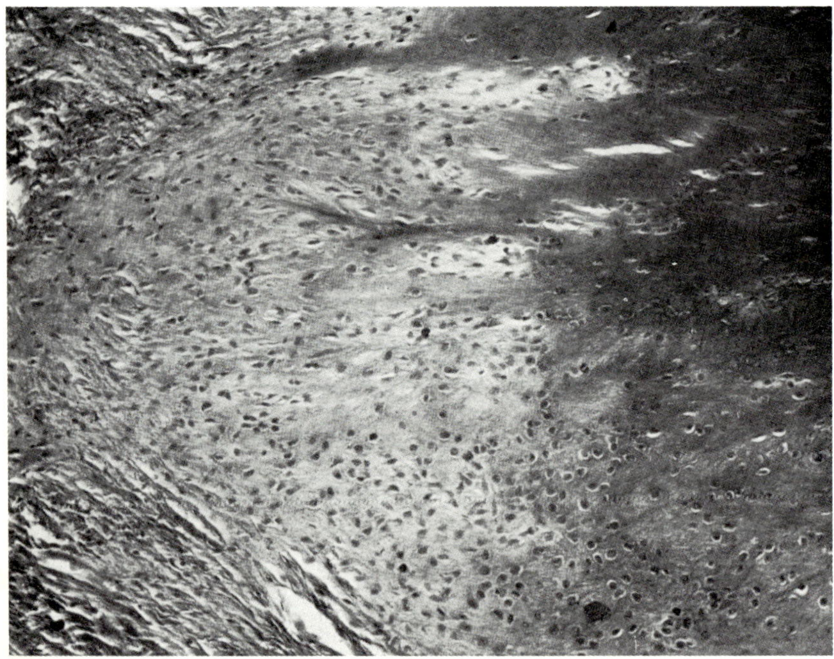

pulmonary infection secondary to progressive embarrassment of respiration from involvement of intercostal muscles.

The disease appears to begin as an interstitial myositis or fibrositis. Biopsy of the early indurated swellings reveals an extensive proliferation of interstitial cellular connective tissue. Very little inflammatory cell reaction or hemorrhage is present. Within the next few weeks the connective tissue becomes less cellular and forms reticular fibers and collagen, which then retracts and compresses muscle fibers and muscle bundles. The muscle fibers diminish in diameter, and some become fragmented or show Zenker's waxy degeneration. Sporadically attempts at regeneration, as evidenced by multinucleated muscle buds, may be seen. In the center of the dense mass of connective tissue the muscle fibers have usually disappeared completely. It is here that condensation of connective tissue seems to give rise to osteoid formation. Occasionally cartilage formation seems to be an intermediate step (Fig. 11–18). Spicules of bone are formed, and the ossification extends peripherally. The actual formation of bone is similar to the bone callus formed after fracture and does not differ from that described for traumatic myositis ossificans (Chapter 9). The newly formed bone is structurally and chemically similar to normal bone. One of the distinctive pathologic features is the persistence of intact muscle fibers within osseous tissue. This, however, does not distinguish the traumatic from the progressive type of myositis ossificans; indeed it is doubtful if the two can be set apart by a single microscopic section.

Since this disease appears to be a primary disorder of the fibrous connective tissues of muscles and not the muscle fibers, it seems more appropriate to term it interstitial ossifying myositis, or as Mair suggested fibrositis ossificans progressiva, or perhaps fibrodysplasia ossificans progressiva (McKusick).

REFERENCES

1. ADAMS RD: The giant muscle fiber: its place in myopathology. Modern Neurology, Boston, Little Brown, 1969, pp 225–241

2. AMYOT R: Myxoedéme et myotonie. Union Med Can 75:654–662, 1946

3. ANDRADE C: A peculiar form of peripheral neuropathy: familial atypical generalized amyloidosis with special involvement of peripheral nerves. Brain 75:408–427, 1952

4. ASBOE-HANSEN G, IVERSEN K, WICHMANN R: Malignant exophthalmos; muscular changes and thyrotrophin content in serum. Acta Endocrinol (Kbh) 11:376–399, 1952

5. ASKANAZY M: Pathologisch-anatomische beiträge zur kenntniss des morbus basedowii, insbesondere über die dobei auftretende muskelerkrankung. Dtsch Arch Klin Med 61:118–186, 1898

6. ASSMANN H, BIELENSTEIN H, HOBS H, et al: Beobachtungen und untersuchungen bei der haffkrankheit. Dtsch Med Wochenschr 59:122–126, 1933

7. BANDHUIN P, HERS HG, LOUBLT: An electron microscopic and chemical study of type II glycogenosis. Lab Invest 13:1139,

8. BARKER LF: Acromegaly with obesity, arterial hypertension, polyglobulin, eosinophilia, and paresis of the musculature of the lower trunk and thighs. Int Clin 4:37–44, 1933

9. BARNES S: A myopathic family with hypertrophic, pseudohypertrophic, atrophic and terminal stages. Brain 55:1–46, 1932

10. BARRIS RW: Pancreatic adenoma (hyperinsulinism) associated with neuromuscular disorders. Ann Intern Med 38:124–129, 1953

11. BARTELS EC: Personal communication, 1950

12. BERENBAUM MC, BIRCH CA, MORELAND JD: Paroxysmal myoglobinuria. Lancet 1:892–896, 1955

13. BERGEAT E: Fettbildung bei phosphorvergiftung. Munch Med Wochenschr 35:66–67, 1888

14. BERGUIGNON M, VITAL C, BATAILLE JM: Les myopathies hypothyroidiennes: aspects cliniques et histopathologiques. Presse Med 75:1551–1556, 1967

15. BINET L, KÉPINOV L, WELLER G: Le glutathione dans les tissus du chien hypophysoprive. C R Soc Biol 120:589–590, 1935

16. BIÖRCK G: On myoglobin and its occurrence in man. Acta Med Scand [Suppl] 133:216–226, 1949

17. BOMPAR F: Étude sur l'atrophie des muscles thoraciques chez les tuberculeux. J Med Bordeaux 17:121, 130, 149, 1887

18. BOWDEN DH, FRASER D, JACKSON SH, et al: Acute recurrent rhabdomyolysis (paroxysmal myohaemoglobinuria). Medicine (Ball) 35:335–353, 1956

19. BRAIN WR, TURNBULL HM: Exophthalmic opthalmoplegia. Q J Med 7:293–323, 1938

20. BRAMWELL B, MUIR R: A remarkable and hitherto undescribed muscular lesion occurring in sprue, with notes of a case of peculiar (myopathic) muscular atrophy in which somewhat similar changes were present. Q J Med 1:1–10, 1907

21. BRASS K: Ueber ein charackteristisches syndrom bei akuter schwerermyolyse. Frankfurt Z Pathol 58:387–442, 1944

22. BRITT BA, KALOW W: Malignant hyperthermia, etiology unknown. Can Anaesth Soc J 16:89–98, 1970

23. BROOKS B: Pathologic changes in muscle as a result of disturbances of circulation: an experimental study of Volkmann's ischemic paralysis. Arch Surg 5:188–216, 1922

24. BUCHANAN D, STEINER PE: Myoglobinuria with paralysis (Meyer-Betz disease). Arch Neurol Psychiatry 66:107–109, 1951

25. BYWATERS EGL, DELORY GE, RIMINGTON C, et al: Myohaemoglobin in urine in air raid casualties with crushing injuries. Biochem J 35:1164–1168, 1941

26. CARDIFF RD: A histochemical and electron microscopic study of skeletal muscle in a case of Pompe's disease (glycogenesis II). Pediatrics 27:249–259, 1966

27. CARLSTRÖM B: Ueber die ätiologie und pathogenese der kreuzlähme des pferdes. Scand Arch Physiol 61:161–224, 63:164–212, 1931

28. CHOLOD EJ, HAUST MD, HUDSON AJ, et al: Myopathy in primary familial hyperparathyroidism. Am J Med 48:700–707, 1970

29. CLEMENT DH, GODMAN GC: Glycogen disease resembling mongolism, cretinism and amyotonia congenita. J Pediatr 36:11–30, 1950

30. CORI GT: Biochemical aspects of glycogen deposition disease. Mod Probl Paediatr 3:344–358, 1957

31. DAHLIN DC: Amyloidosis. Proc Mayo Clin 24:637–648, 1949

32. DEBRE R, SEMELAIGNE G: Syndrome of diffuse muscular hypertrophy in infants causing athletic appearance: its connection with congenital myexedema. Am J Dis Child 50:1351–1361, 1935

33. DELANGEN CD: Myoglobin and myoglobinuria. Acta Med Scand 124:213–226, 1946

34. DENBOROUGH MA, LOVELL RRH: Anesthetic deaths in a family. Lancet 2:45, 1960

35. DENNY-BROWN D: Primary sensory neuropathy with muscular changes associated with carcinoma. J Neurol Neurosurg Psychiatry 11:73–87, 1948

36. DENNY-BROWN D: Hereditary sensory radicular neuropathy. J Neurol Neurosurg Psychiatry 14:237–252, 1951

37. DENNY-BROWN D: The nature of polymyositis and related diseases. Trans Coll Physicians (Phila) 28:14–29, 1960

38. DENNYS EH, HOFFMANN WW: An in vitro study of the biochemical changes produced by hypermetabolism and hypometabolism of skeletal muscle. Neurology (Minneap) 22:22–32, 1972

39. DI MAURO S, ROWLAND LP, DI MAURO P: Control of glycogen metabolism in human muscle: evidence from glycogen storage disease. Arch Neurol 25:534–540, 1970

40. DISANT'AGNESE PA, ANDERSEN DH, MASON HH: Glycogen storage disease of the heart. 2. Critical review of the literature. Pediatrics 6: 607–624, 1950

41. DISANT'AGNESE PA, ANDERSEN DH, MASON HH, et al.: Glycogen storage disease of the heart. 1. Report of 2 cases in siblings with chemical and pathologic studies. Pediatrics 6:402–424, 1950

42. DISANT'AGNESE PA, ANDERSON DH, METCALF KM: Glycogen storage disease of the muscles; report of a case with unusual features. J Pediatr 61:438–442, 1962

43. DOBYNS BM: Studies on exophthalmos produced by thyrotropic hormone. Surg Gynecol Obstet 82:717–722, 1946

44. DOBYNS BM: Present concepts of the pathologic physiology of exophthalmos. J Clin Endocrinol Metab 10:1202–1230, 1950

45. DOWLING GB: Scleroderma and dermatomyositis. Br J Dermatol 67:275–290, 1955

46. DREYFUS JC, ALEXANDRE E: Les phosphorylases et le système phosphorylasique en pathologie. Actualités de Pathologie Neuromusculaire. Edited by G Serratrice, H Roux. Paris, L'expansion Scientifique Française, 1971, pp 53–62

47. DUCHESNEAU G: Contribution à l'étude de l'acromégalie et en particulier d'une forme amyotrophique de cette maladie. Thèse, Lyons, 1891–1892

48. DURANTE G: Anatomie pathologique des muscles. Manuel d'Histologie Pathologique. Edited by B Cornil, L Ranvier. Paris, Félix Alcan, 1902, Vol 2, pp 1–477

49. ELLIS JT: Necrosis and regeneration of muscle in cortisone treated rabbits. Am J Pathol 32:993–1013, 1956

50. ENGEL AG: Electron microscopic observations in thyrotoxic and corticosteroid induced myopathies. Mayo Clin Proc 41:785–796, 1966

51. ENGEL AG: Acid maltase deficiency in adult life: morphology and biochemical data in 3 cases of a syndrome simulating other myopathies. Proceedings of the International Congress of Muscle Disease, Series 199. Edited by JN Walton, N Conal, G Scarlato. Amsterdam, Excerpta Medica, 1969, pp 236–245

52. ENGEL WK: Selective and nonselective susceptibility of fiber types: a new approach to human neuromuscular diseases. Arch Neurol 22:97–117, 1970

53. ERB W: Die Thomsen'sche krankheit (Myotonia congenita). Leipzig, Vogel, 1886, pp 1–128

54. FEINBERG WD, UNDERDAHL LO, EATON LM: Myasthenia gravis and myxedema. Mayo Clin Proc 32:299–305, 1957

55. FINDLEY JW, ADAMS W: Primary systemic amyloidosis simulating constrictive pericarditis with

steatorrhea and hyperesthesia. Arch Intern Med 81:342–351, 1948

56. FORBUS WD: Pathologic changes in voluntary muscle: degeneration and regeneration of the rectus abdominis in pneumonia. Arch Pathol 2:318–339, 1926

57. FORBUS WD: Pathologic changes in voluntary muscle: experimental study of degeneration and and regeneration of striated muscle with vital stains. Arch Pathol 2:486–499, 1926

58. FRAENKEL E: Ueber pathologische veranderungen der kehlkopfmuskeln bei phthisikern. Virchows Arch [Pathol Anat] 71:261–268, 1877

59. FRIEDMAN S, ASH R: Glycogen storage disease of the heart. J Pediatr 52:635–648, 1958

60. FRITZ E, RANVIER L, VERLIAC J: De la stéatose dans l'empoisonnement par le phosphore. Arch Gen Med 2:25–47, 1863

61. GARLAND H: Diabetic amyotrophy. Br Med J 2:1287–1290, 1955. Also Modern Trends in Neurology. Edited by D Williams. New York, Hoeber, 1957

62. GERMUTH FG, NEDZEL GA, OTTINGER B, et al: Anatomic and histologic changes in rabbits with experimental hypersensitivity treated with compound E and ACTH. Proc Soc Exp Biol Med 76:177–182, 1951

63. GIMLETTE TMD: The muscular lesion in hyperthyroidism. Br Med J 2:1143–1146, 1959

64. GLASER GH, STARK L: Excitability in experimental myopathy. Neurology (Minneap) 8:640–644, 1958

65. GODET-GUILLAIN J, FARDAEU M: Hypothyroid myopathy: histological and ultrastructural study of atrophic form. Proceedings of the International Congress of Muscle Disease, Series 199. Edited by JN Walton, N Conal, G Scarlato. Amsterdam, Excerpta Medica, 1969, pp 512–515

66. HABERSHON SO: Acute poisoning by phosphorus. Proc R Med Chir Soc 5:261–263, 1867

67. HARRISON RG: An experimental study of the relation of the nervous system to the developing musculature in the embryo of the frog. Am J Anat 3:197–220, 1904

68. HAYEM MG: Note sur les altérations musculaires qu'on observe dans les maladies chroniques. C R Soc Biol 1:69–78, 1875

69. HAYEM MG: Muscles (tissu), pathologie. Dict Encyclo Sc Med 10:697–770, 1876

70. HESSER FH: Hypertrophia musculorum vera (dystrophia musculorum hyperplastica) associated with hypothyroidism: a case study. Bull Hopkins Hosp 66:353–378, 1940

71. HILDEBRAND AG, KEPLER EJ: Familial periodic paralysis associated with exophthalmic goiter. J Nerv Ment Dis 94:713–721, 1941

72. HOCH FL: Thyrotoxicosis as a disease of mitochondria. N Engl J Med 266:446–455, 408–505, 1962

73. HOUGHTON JD, PARKER F: Personal communication, 1935

74. HOUSSAY BA: L'asthénie des crépauds sans hypophyse. C R Soc Biol 113:472–474, 1933

75. HOWARD CWH, CAMPBELL EDR, ROSS HB, et al: Electromyographic and histological findings in the muscles of patients with thyrotoxicosis. Q J Med 32:145–163, 1963

76. HUDGSON P, GARDNER-MEDWIN D, WARSFOLD M, et al: Adult myopathy from glycogen storage disease due to acid maltase deficiency. Brain 91:435–460, 1968

77. HUG G, GARANCIS JC, SCHUBERT WK, et al: Glycogen Storage disease: types II, III, VIII, IX; a biochemical and electron microscopic analysis. Am J Dis Child 111:457–474, 1966

78. HUG G, SCHUBERT WK: Lysosomes in type II glycogenesis; changes during administration of extract from Aspergillus niger. J Cell Biol 35: C1–C6, 1967

79. HUMPHREYS EM, KATO K: Glycogen-storage disease: thesaurismosis glycogenica (von Gierke). Am J Pathol 10:589–614, 1934

80. INGELMARK BE, GUSTAFSSON L: The aging of calf muscles in women, a morphological and roentgenological study. Acta Morphol Neerl Scand 1:173–187, 1957

81. ISAACS H, BARLOW MB: Malignant hyperpyrexia. J Neurol Neurosurg Psychiatry 36:228–243, 1973

82. JEGHERS HJ, HOUGHTON JD, FOLEY JA: Weil's disease: report of a case with postmortem observations and review of recent literature. Arch Pathol 20:447–476, 1935

83. JESSERER H, BLACIZEK O: Die hypothyreotische myopathie. Dtsch Med Wochenschr 79:1779–1783, 1954

84. JEWESBURY RC, TOPLEY WWC: On certain changes occurring in the voluntary muscles in general diseases. J Pathol 17:432–435, 1912–1913

85. KERNOHAN JW, WOLTMAN HW: Amyloid neuritis. Arch Neurol Psychiatry 47:132–140, 1942

86. KING JO, DENBOROUGH MA, ZAPF PW: Inheritance of malignant hyperpyrexia. Lancet 1:365–370, 1972

87. KLIPPEL M: Des Amyotrophies dans les Maladies Générales Chroniques et de leurs Relations avec les Lésions des Nerfs Peripheriques. Paris, Thèse, 1889

88. KÖHLER JH: Die myoglobinurien. Ergeb Inn Med Kinderheilkd 11:1–103, 1959

89. KOLETSKY S, STRECHER RM: Primary systemic amyloidosis: involvement of cardiac valves, joints and bone, with pathologic fracture of the femur. Arch Pathol 27:267–288, 1939

90. KOREIN J, CODDEN DR, MOWREY FH: The clinical

syndrome of paroxysmal paralytic myoglobinuria. Neurology (Minneap) 9:767–785, 1959

91. KREUTZER FL, STRAIT L, KERR WJ: Spontaneous myohemoglobinuria in man. Arch Intern Med 81:249–259, 1948

92. KROLL AG, KUWABARA T: Dysthyroid ocular myopathy. Arch Ophthalmol 76:244, 1966

93. LAMBERT EH, UNDERDAHL LO, BECKETT S, et al: A study of the ankle jerk in myxedema. J Clin Endocrinol Metab 11:1186–1205, 1951

94. LANARI A: Los musculos estriados en el hipotiroidismo. Medicina (B Aires) 7:197–214, 1947

95. LAURENT LPE: Acute thyrotoxic bulbar palsy. Lancet 1:87–88, 1944

96. LAYZER RC, ROWLAND LP, RAMEY HM: Muscle phosphofructokinase deficiency. Arch Neurol 17:512–523, 1967

97. LEHOCZKY T, HALASY M, SIMON G, et al: Glycogenic myopathy, a case of skeletal muscle glycogenosis in twins. J Neurol Sci 2:366, 1965

98. LELOIR LF, CARDINI CD: Biosynthesis of glycogen from uridine diphosphate glucose. J Am Chem Soc 79:6340–6342, 1957

99. LEVRAT M, BRETTE R: Cancer langerhansien du pancréas avec hypoglycémic douleurs musculaires et myosite dégénerative d'origine metabolique. Presse Med 56:530–531, 1948

100. LORENZ H: Die muskelerkrankungen. Handbuch der Speziellen Therapie Innerer Krankheiten. Edited by H Nothnagel. Vienna, Holder, 1904, Vol. XI, pp 1–727

101. LUFT R, IKKOS D, PALMIERI G, et al: A case of severe hypermetabolism of nonthyroid origin with a defect in the maintenance of mitochondrial respiratory control; a correlated clinical, biochemical and morphological study. J Clin Invest 41:1771, 1962

102. MAIR WF: Myositis ossificans progressiva. Edinburgh Med J 39:13–36, 69–92, 1932

103. MARANON G, RICHET C: Les syndromes neuromusculaires. Bull Acad Med 118:293–298, 1937

104. MARENZI AD: Chemical changes in the muscles of the hypophysectomized toad. Endocrinology 20:184–187, 1936

105. MARIN OS, DENNY-BROWN D: Changes in skeletal muscle associated with cachexia. Am J Pathol, 41:23–40, 1962

106. MARINESCO G: Maladies des muscles. Nouveau Traité de Medecine et de Thérapeutique, Edited by P Brouardel, A Gilbert, I Thinot. Paris, Masson et Cie, 1910, Vol 38, pp 1–172

107. MCARDLE B: Myopathy due to a defect in muscle glycogen breakdown. Clin Sci 10:13–35, 1951

108. MCCULLAGH EP, HEWLETT JS: Acromegaly associated with amyotrophic lateral sclerosis and acromegaly of the amyotrophic type; report of 3 cases. J Clin Endocrinol Metab 7:636–643, 1947

109. MCEACHERN D: Diseases and disorders of muscle function. Bull NY Acad Med 27: 3–23, 1951

110. MCEACHERN D, PARNELL JL: The relationship of hyperthyroidism to myasthenia gravis. J Clin Endocrinol Metab 8:842–850, 1948

111. MCEACHERN D, ROSS WD: Chronic thyrotoxic myopathy. Brain 65:181–192, 1942

112. MCKUSICK V: Heritable Disorders of Connective Tissue. St. Louis, Mosby, 1956, pp 184–195

113. MEANS JH: Hyperophthalmopathic Graves' disease. Ann Intern Med 23:779–789, 1945

114. MEYER-BETZ F: Beobachtungen au einem eigenartigen mit muskellähmungen verbundenen fall von hämoglobinurie. Dtsch Arch Klin Med 85–127, 1911

115. MILLER JE, VON HEUVEN W, WARD R: Surgical correction of hypotropias associated with thyroid dysfunction. Arch Ophthamol 74:509, 1965

116. MILLIKAN CH, HAINES SF: The thyroid gland in relation to neuromuscular disease. Arch Intern Med 92:5–39, 1953

117. MOLLARET P, RUDAUX P: Hypertrophie musculaire avec symptomes myotoniques et de constitution rapide, chez un hypothyroidienne latente. Soc Med Hop Paris 55:818–824, 1939

118. MOLLARET P, SIGWALD J: Hypertrophie musculaire generalisée de l'adulte à constitution rapide et myxoedéme fruste concomitants, cliniquement guérie par le traitement thyroidien. Rev Neurol (Paris) 71:513–547, 1939

119. MOMMAERTS WFHM, ILLINGWORTH B, PEARSON CM, et al.: A functional disorder of muscle associated with the absence of phosphorylase. Proc Natl Acad Sci USA 45:791–797, 1959

120. MORGAN HJ, WILLIAMS RH: Muscular atrophy and weakness in thyrotoxicosis (thyrotoxic myopathy: exophthalmic ophthalmoplegia). South Med J 33:261–268, 1940

121. MULDER DW, BASTRON JA, LAMBERT EH: Hyperinsulin neuronopathy. Neurology (Minneap) 6:627–235, 1956

122. MÜLLER R, KUGELBERG E: Myopathy in Cushing's syndrome. J Neurol Neurosurg Psychiatry 22:314–320, 1959

123. MULVANEY JH: The exophthalmos of hyperthyroidism. Am J Ophthalmol 27:589–612, 693–712, 820–832, 1944

124. NAFFZIGER HC: Discussion of paper by JL McCool and HC Naffziger. Trans Am Ophthamol Soc 30:103–124, 1932

125. NAFFZIGER HC: Pathologic changes in the orbit in progressive exophthalmos. Arch Ophthalmol 9:1–12, 1933

126. NETTER H: Les Atrophies Musculaires, Associées aux Affections Hypophysaires. Paris, Thèse, 1938, No. 374

127. NORRIS FH, PANNER BJ: Hypothyroid myopathy:

clinical, electromyographical and ultrastructural observations. Arch Neurol 14:574–589, 1966

128. NUTT JJ: Report of a case of myositis ossificans progressiva with bibliography. J Bone Joint Surg 5:344–359, 1923

129. OLINER, L, SHULMAN M, LARNER J: Myopathy associated with glycogen deposition resulting from generalized lack of amylo-1,6,glucosidase. Clin Res 9:243,1961

130. OLLIVIER A: Des Atrophies Musculaires. Thèse d'Agrégation. Paris, 1869

131. PAPANICOLAOU GN, FALK EA: General muscular hypertrophy induced by androgenic hormone. Science 87:238–239, 1938

132. PAULSON DL: Experimental exophthalmos and muscle degeneration induced by thyrotrophic hormone. Mayo Clin Proc 14:828–832, 1939

133. PEARCE GW: Electron microscopy of muscular dystrophy. Muscular Dystrophy in Man and Animals. Edited by GH Bourne, N Golasz. Basel, Karger, 1963, p 160

134. PEARSON CM: The incidence and type of pathologic alterations observed in muscles in a routine autopsy survey. Neurology (Minneap) 9:757–766, 1959

135. PEARSON CM, BECK WS, BLAHD WH: Idiopathic paroxysmal myoglobinuria; detailed study of a case including radioisotope and serum enzyme evaluations. Arch Intern Med 99:376–389, 1957

136. PEARSON CM, RIMER DG: Evidence for direct utilization of fructose in working muscle in man. Proc Soc Ex Biol Med 100:671–672, 1959

137. PEARSON CM, RIMER DG, MOMMAERTS WFHM: A metabolic myopathy due to absence of muscle phosphorylase. Am J Med 30:502–517, 1961

138. PERKOFF GT, SILBER R, TYLER FH, et al.: Myopathy due to the administration of therapeutic amounts of 17 hydroxycorticosteroids. Am J Med 26:891–898, 1959

139. POCHIN EE, RUNDLE FF: Deposition of adipose tissue between ocular muscle fibers in thyrotoxicosis. Clin Sci 8:89–95, 1949

140. PRANKERD TAJ: Electrophoretic properties of myoglobin and its character in sickle-cell disease and paroxysmal myoglobinuria. Br J Haematol 2:80–83, 1956

141. PRINEAS J, HALL R, BARWICK DD, et al: Myopathy associated with pigmentation following adrenalectomy for Cushing's syndrome. J Neuropathol Exp Neurol 28:598–621, 1969

142. REBEIZ JJ, ADAMS RD: Unpublished observations

143. RECANT L: Recent developments in the field of glycogen metabolism and the diseases of glycogen storage. Am J Med 19:610–619, 1955

144. REINER L, KONIKOFF N, ALTSCHULE MD, et al.: Idiopathic paroxysmal myoglobinuria; report of two cases and evaluation of the syndrome. Arch Intern Med 97:537–550, 1956

145. RICHET C, SOURDEL M, PERICOLA A: Syndromes parathyroidomusculaires: myopathies scléreuses liées a des troubles parathyroidiéns. J Med Fr 26:377–392, 1937

146. RUNDLE FF: Management of exophthalmos and related ocular changes in Graves' disease. Metabolism 61:36, 1957

147. RUNDLE FF, FINDLAY-JONES LR, NOAD KB: Malignant exophthalmos: quantitative analysis of orbital tissues. Australia Ann Med 2:128–135, 1953

148. SALAM MZ, ADAMS RD: Unpublished observation

149. SALASSA RM, JENNIFIER J, ARNAUD CD: Hypophosphatemic osteomalacia associated with nonendocrine tumors. N Engl J Med 283:65–70, 1970

150. SATAYOSHI E, KOWA H: A new myopathy due to glycolytic abnormalities. Trans Am Neurol Assoc 90:46–47, 1965

151. SCHAAR FE, LABREE JW, GLEASON DF: Paroxysmal myohemoglobinuria with fatal renal tubular injury. Cent Soc Clin Res Proc 22:71, 1949

152. SCHMID R, HAMMAKER L: Hereditary absence of muscle phosphorylase (McArdle's syndrome). N Engl J Med 264:223–225, 1961

153. SCHMID R, MAHLER R: Syndrome of muscular dystrophy with myoglobinuria: demonstration of a glycogenolytic defect in muscle. J Clin Invest 38:1040, 1959

154. SCHMID R, MAHLER R: Chronic progressive myopathy with myoglobinuria: demonstration of a glycogenolytic defect in the muscle. J Clin Invest 38:2044–2058, 1959

155. SCHMID R, ROBBINS PW, TRAUT RR: Glycogen synthesis in muscle lacking phosphorylase. Proc Natl Acad Sci USA 45:1236–1240, 1959

156. SCHNABEL R: Uber die neuromuskulare form der glykogenspeicherungskrankheit (neuromuscular form of glycogen storage disease). Virchows Arch [Pathol Anat] 331:287–313, 1958

157. SCHOTLAND DL, SPIRO D, ROWLAND LP, et al.: Ultrastructural studies of muscle in McArdle's disease. J Neuropathol Exp Neurol 24:629–644, 1965

158. SHAPIRO BG, ZWARENSTEIN H: On the relation of the pituitary gland to muscle creatine. Proc R Soc Edinburgh 56:164–168, 1936

159. Sheldon WH: Lesions of muscle in spirochetal jaundice (Weil's disease, spirochetosis icterohemorrhagica). Arch Intern Med 75:119–124, 1945

160. SHELDON WH: Leptospiral antigen demonstrated by the fluorescent antibody technic in human muscle lesions of leptospirosis icterohemor

rhagica. Proc Soc Exp Biol Med 84:165–167, 1953

161. SHY GM: Some metabolic and endocrinological aspects of disorders of striated muscle. Proc Assoc Res Nerv Ment Dis 38:274–317, 1960

162. SHY GM, MAGEE KR: A new congenital non-progressive myopathy. Brain 79:610–621, 1956

163. SILVER S, OSSERMAN KE: Hyperthyroidism and myasthenia gravis. J Mt Sinai Hosp 24:1214–1220, 1957

164. SMITH J, ZELLWEGER H, AFIFI AK: Muscular form of glycogenesis: type II (Pompé); report of a case with unusual features. Neurology (Minneap) 17:537–549, 1967

165. STRONG JA: Thyrotoxicosis with ophthalmoplegia, myopathy, Wolff-Parkinson-White syndrome, and pericardial friction. Lancet 1:959–961, 1949

166. SULLIVAN JF: The neuropathies of diabetes. Neurology (Minneap) 8:243–249, 1958

167. TAKEUCHI T: Histochemical demonstration of branching enzyme (amylo-1,4-1,6-transglucosidase) in animal tissues. J Histochem Cytochem 6:208–216, 1958

168. TALBOTT JH: Periodic paralysis: a clinical syndrome. Medicine (Balt) 20:85–143, 1941

169. TANNENBERG J: Pathological changes in the heart, skeletal musculature and liver in rabbits treated with insulin in shock dosage. Am J Pathol 15:25–54, 1939

170. THOMASEN E: Myotonia, Thomsen's Disease, Paramyotonia, Dystrophia Myotonica. Denmark, Universitetsforlaget i Aarhus, 1948, pp 1–251

171. THOMSON WHS, MACHAURIN JC, METCALF KM: Skeletal muscle glycogenesis: an investigation of 2 dissimilar cases. J Neurol Neurosurg Psychiatry 26:60–68, 1963

172. THORN GW: The Diagnosis and Treatment of Adrenal Insufficiency. Springfield, Ill, Charles C Thomas, 1949, pp 144–145

173. THORN GW, EDER HA: Studies on chronic thyrotoxic myopathy. J Med 1:583–601, 1946

174. TOM MI, RICHARDSON JC: Hypoglycemia from islet cell tumor of pancreas with amyotrophy and cerebrospinal nerve cell changes. J Neuropathol Exp Neurol 10:57–66, 1951

175. TOMLINSON BE, WALTON JN, REBEIZ JJ: Effects of aging and of cachexia upon skeletal muscle, a histopathological study. J Neurol Sci 9:321–346, 1969

176. UEHLINGER E: Myositis ossificans progressiva. Ergebn Med Strahlenforsch 7:175–220, 1936

177. VAN CREVELD S, SOETERS JM: Myositis ossificans progressiva. Am J Dis Child 62:1000–1013, 1941

178. VANDERLAAN WL, ADAMS RD: Observations of the contracture of addison's disease. Unpublished observation

179. VENABLE J: Electron microscopy of malignant hyperthermia in Poland-China pigs. First International Symposium on Malignant Hyperthermia, Toronto, 1971. Edited by RA Gordon, BA Britt, V Kalow. Springfield, Ill, Charles C Thomas, 1971, pp 541–567

180. VICALE CT: The diagnostic features of a muscular syndrome resulting from hyperparathyroidism, osteomalacia owing to renal tubular acidosis, and perhaps to related disorders of calcium metabolism. Trans Am Neurol Assoc 74:143–147, 1949

181. VIRCHOW R: Ueber entzündung und ruptur des M. rectus abdominalis. Wurzburger Berhandl 7:213, 1857

182. VOLKMANN R: Ueber die regeneration des quergestreiften muskelgeweben beim menschen und sängetier. Beitr Pathol Anat 12:233–332, 1893

183. VON MEYENBURG H: Die quergestreifte muskulatur. Handbuch der Speziellen Pathologischen Anatomie und Histologie. Edited by F Henke, O Lubarsch. Berlin, Springer, 1929, Vol IX, pp 229–509

184. WALDENSTRÖM J: Acute thyrotoxic encephalo- or myopathy, its cause and treatment. Acta Med Scand 121:251–294, 1945

185. WALDEYER W: Ueber die Veränderungen der quergestreiften muskeln bei der entzündung und dem typhusprozess, sowie über die regeneration derselben nach substanzdefecten. Virchows Arch [Pathol Anat] 34:473–514, 1865

186. WELLS HG: Waxy degeneration of the diaphragm; a factor in causing death in pneumonia and in other conditions. Arch Pathol 4:681–686, 1927

187. WILKINS WE, REGEN EM, CARPENTER GK: Phosphatase studies on biopsy tissue in progressive myositis ossificans, with report of case. Am J Dist Child 49:1219–1221, 1935

188. WILLIAMS RC: Dermatomyositis and malignancy: review of the literature. Ann Intern Med 50:1174–1181, 1959

189. WOHLFART G: Dystrophia myotonica and myotonia congenita, histopathologic studies with special reference to changes in muscle. J Neuropathol Exp Neurol 10:109–124, 1951

190. WOLBACH SB, FROTHINGHAM C: The influenza epidemic at Camp Devens in 1918: a study of the pathology of the fatal cases. Arch Intern Med 32:571–600, 1923

191. ZELLWEGER H, DARK A, HAIDAR GAA: Glycogen disease of skeletal muscle: report of two cases and review of literature. Pediatrics 15:715–732, 1955

192. ZENKER FA: Ueber die veränderungen der willkürlichen Muskeln in Typhus Abdominalis. Leipzig, Vogel, 1864

193. ZENKER FA: Sur les altérations des muscles

volontaires dans la fièvre typhoide. Arch Gen Med 6:143, 290, 1865

194. ZIEGLER DK: Minor neurologic signs and symptoms following insulin coma therapy. J Nerv Ment Dis 120:75–78, 1954

195. ZIERLER KL: Muscular wasting of obscure origin and the thyroid gland. Bull Hopkins Hosp 89: 263–280, 1951

196. ZOLLNER N: Inborn errors of metabolism. J Chronic Dis 10:6–26, 1959

chapter 12

DISORDERS OF THE NEURO-MUSCULAR AND MUSCULAR CONDUCTION APPARATUS

In an essay on the principles of myopathology[2] the author suggested that all disorders of muscle due primarily to nerve impulse control of contractility be placed in a separate category. In this group of diseases, unlike the neural atrophies and myopathies wherein there is usually close correspondence between the functional state and the pathologic findings in the muscle fibers, the conduction disorders result in the widest discrepancies. One must assume that the functional failure is due to an invisible block in impulse transmission in the terminal axonic branches, at the neuromuscular junction, or in the sarcolemma itself.

In the type of paralytic disorder to which reference is made, or its opposite state (spasm or myotonia), the lesion resides in parts of the neuromuscular apparatus that are not usually exposed by the conventional technics of light microscopy. Their study requires a special methodology that is not available in most laboratories.

MYASTHENIA GRAVIS

The symptoms of myasthenia gravis were described by Willis[219] in 1672. During the nineteenth century the disease was recognized by Erb[67] and others as a form of bulbar paralysis distinguished by its tendency to spontaneous recovery. Reports of fatal cases with negative autopsy findings by Wilks[218] in 1877, Oppenheim[153] in 1887, and Shaw[186] in 1890 led to the designation bulbar paralysis without anatomic changes, for which Jolly,[105] in describing a further case in 1895, proposed the name myasthenia gravis pseudoparalytica. Oppenheim and Jolly also described the progressive failure of the affected muscles to respond to repeated faradic stimulation, to which the term myasthenic reaction was applied. Buzzard[30] showed in 1905 that muscles thus paralzyed still reacted to stimulation by galvanic current. In 1893 Goldflam[89] described the daily variation of symptoms and the long remissions in some patients. Campbell and Bramwell[31] in 1900 and Oppenheim in 1901[154] each analyzed over 60 cases, and provided the classic descriptions.

Myasthenia gravis is a chronic disease characterized by a variable weakness worsened by the use of voluntary muscles, particularly those subserving ocular movements, facial expression, mastication, deglutition, and respiration. Cardiac and smooth muscles are not affected. Muscles of the neck, trunk, and limbs are usually involved, and respiratory difficulty occurs in severe cases. At first the weakness takes the form of a failure of contraction after any maintained effort, but in more severe phases of the disorder, the muscles first and most severely affected cease to contract even after prolonged rest. Since the disorder appears to bear an uncertain relationship to the natural phenomenon of muscular fatigue, except that both require a certain amount of preceding contraction and are ameliorated by rest, the use of the word fatigue in expressing the basic defect in myasthenia is imprecise. A complete definition of the disease should include the failure of muscles to contract strongly even after prolonged rest (if the disease is severe) and the capacity of anticholinesterase drugs to reverse the greater part of the weakness induced by use of the muscles but the retained ability of the weak muscle to respond to direct electrical stimulation.

Myasthenia gravis may pursue an intermittent or progressive course; the former is more often associated with lymphoid follicle hyperplasia of the thymus gland and the latter with thymic tumor. In intermittent myasthenia remissions may last as long as 15 years, though more often spontaneous improvement is partial and of only a few months duration. In 40% of the cases analyzed by Grob and Harvey[94] there was widespread muscular weakness within a month of onset, and in 34% weakness of the ocular muscles alone had been present for 1 month to 24 years. The benign course in the type of case in which myasthenia has been limited to the eye muscles is well established.

The onset of the disease may occur at any age, though it is relatively rare for the first signs to appear before age 10 and after age 70. The peak age of onset is 20 years, and females outnumber males by 3:1 in one type. In another type the peak incidence is in late adult life with male predominance and greater likelihood of thymoma. The first symptoms are most commonly a weakness of eye muscles (ptosis, paralytic strabismus, and diplopia) and at some phase of the disease these muscles are involved in over 90% of cases. The weakness extends to facial, jaw, lingual, pharyngeal, laryngeal, and posterior neck muscles in over 65% of cases. The onset of weakness of neck and shoulder muscles or those of the lower limbs is unusual, but these and in fact all of the muscles, even external sphincters of bladder and bowel, may be affected in the more advanced stages of the disease.

Variability of weakness with time of day and use of muscles is characteristic, as is the relative preservation of muscle bulk and retention of tendon reflexes. Smooth and cardiac muscle are invariably spared. Sensory and other neurologic functions escape the disease process, although rarely there is a mononeuropathy with sensory-motor-reflex loss; Alajouanine *et al.*[4]

reported two puzzling cases with ageusia and paresthesias in the perioral and lingual regions, which lessened and disappeared after Prostigmin therapy.

Although usually myasthenia gravis exists alone, unaccompanied by other diseases, there are a number of exceptions. The association with hyperthyroidism is more than fortuitous, as are combinations of myxedema, and lupus erythematosus. In the latter two settings the disease takes an unusual form, for it responds poorly to anticholinesterase drugs. Thymoma is rare before the age of 30 years and increases in frequency thereafter. It may be detected many years after the onset of the myasthenia or even after the myasthenia has disappeared. In some instances the thymic tumor is unaccompanied by myasthenia and the latter condition appears many years later. A high level of antinuclear antibodies is a helpful diagnostic test for thymoma. An association of thymoma, myasthenia gravis and polymyositis has also been recorded. The special type of myasthenia conjoined to bronchogenic carcinoma also differs in failing to respond to Prostigmin but being unusually sensitive still to curare (the latter, typical of myasthenia gravis); and guanidine HCl has had a strikingly beneficial effect (see below under Eaton-Lambert syndrome). A syndrome closely resembling myasthenia gravis has been recorded in states of malnutrition (Gerlier's syndrome and the Kubisagari of Japan) and was reviewed by Denny-Brown.[52] We have no experience with it.

The most remarkable feature of the paralysis is its relation to use of the muscles. After a few repetitions of contraction or while action is sustained for a few minutes, power fades noticeably and may as rapidly recover with rest. These phenomena can be put to use in the clinic and provide indubitable evidence of true myasthenic weakness. Noteworthy also is the fact that exhausted muscles do not ache or become tender as in natural physical fatigue, and the paresis of one group of muscles thus induced does not impair the contraction of others. As the disease advances muscles weakened by use take longer to recover and more of the musculature becomes continuously weakened. At this stage some additional exertion may precipitate a paralytic failure of the respiratory muscles, of which the critical state had not been suspected. Sudden death from failure of respiration is not infrequent, and the observation of respiratory tracings on a spirometer has shown the author that imminent failure may be anticipated by a lessening of maximal inspiration which may not have yet distressed the patient.

PATHOLOGIC CHANGES

LYMPHOCYTIC INFILTRATIONS (LYMPHORRHAGES OF BUZZARD)

Lymphorrhages were described for the first time in 1905 by Buzzard[30] who noted them to lie in relation to small venules and capillaries in the interstitial tissue of the affected muscles. A typical example appears in Figure 12–1 from an ocular muscle of one of our fatal cases. Such lymphorrhages are by no means a reliable index of myasthenia, for they may occur in a variety of other conditions; the author failed to find them in most of the muscles in some 30 fatal cases of the disease in the Pathology Department files at the Massachusetts General Hospital.

MYONECROSIS AND ATROPHY

Russell[177] observed both atrophy and segmental necrosis of single or small groups of muscle fibers in cases in the London City Hospital where Buzzard's material was still available. Such lesions, especially necrosis, were not present in all cases and were seen mainly in a few of the muscles commonly examined post mortem such as the upper esophagus and dia-

phragm. Mendelow and Genkins[141] confirmed these findings in 6 of 12 autopsied cases and agreed with Russell that myocardial changes of similar type are observed in 50% of cases. Once these lesions were brought to the author's attention a review of our 30 cases was undertaken. In this material, although the survey of the musculature was inadequate, such lesions were rare even in muscles virtually paralyzed before death. When myonecrosis, regeneration, and infiltrations of lymphocytes and other inflammatory cells are prominent, the picture is then indistinguishable from polymyositis. This combination of myasthenia gravis and polymyositis is usually associated with a thymoma. Some of the cases reported by Walton and Nattrass,[212] Sandifer,[180] Coërs,[37] and Rowland *et al.*[175] as myasthenic myopathy are probably of this type. While a definite Prostigmin response was obtained in some, in others the myasthenic reaction was not well described and had responded poorly to Prostigmin.

Several times in biopsy or autopsy material we observed a typical denervation atrophy of a muscle, and one can only assume that in some mysterious fashion the disease had implicated one or more peripheral nerves. Other instances of this have been documented by Fenichel.[71] Disuse atrophy can also be brought to light by use of a micrometer and apparently is more prominent in the muscle groups subject to a prolonged state of paralysis.

ALTERATIONS OF MOTOR END-PLATES

In the first edition of this book we illustrated a neuromuscular junction from one of Denny-Brown's cases in which the terminal axonic branches were thin and difficult to stain (Fig.

Fig. 12–1. Myasthenia gravis. Superior rectus muscle, showing a lymphorrhage. (hematoxylin, Van Gieson)

Fig. 12–2. Myasthenia gravis. Triceps muscle. Muscle exhibited extreme weakness during period immediately preceding death. Two motor nerve endings are shown. They are normal except for tenuousness of axis cylinder, the significance of which is doubtful. Absence of striation in upper figure is due only to lightness of staining. Blood corpuscles are stained darkly. (Gros-Bielschowsky method)

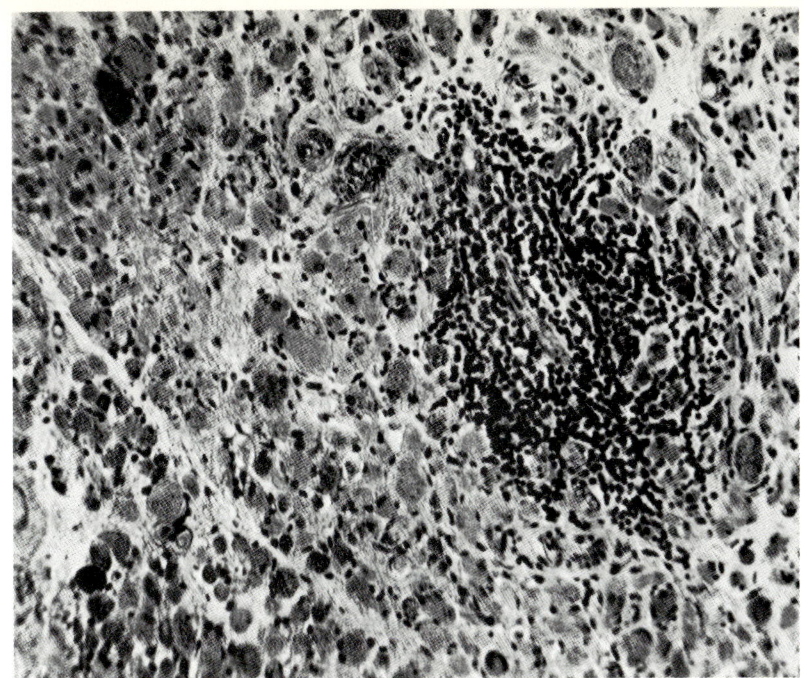

Figure 12–1.

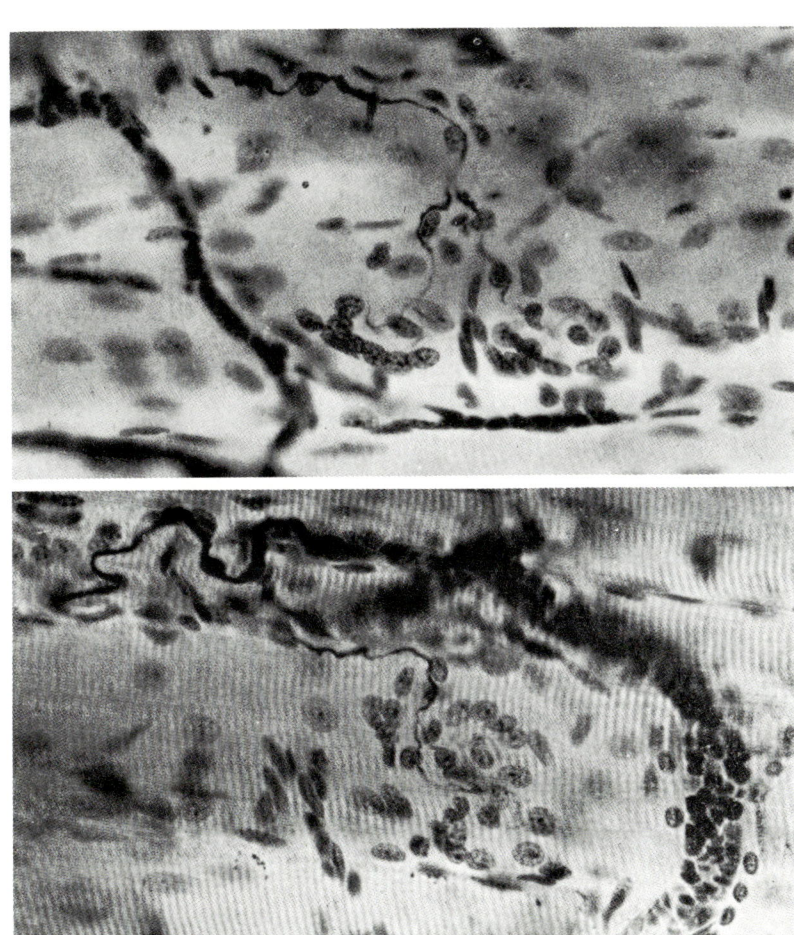

Figure 12–2.

12–2). At that time we thought the simplified neural apparatus to be a variation in the staining technic and not beyond the limits of normal. Soon thereafter, Coërs and Woolf[39] found that such simple, attenuated, terminal arborizations could be demonstrated with regularity by methylene blue staining of fresh biopsy material, even in the limb muscles when only ocular and bulbar muscles showed clinical myasthenic weakness. The subneural apparatus was also observed by Coërs to have an abnormally simplified pattern of cholinesterase, which was present in adequate quantities. In further studies Coërs[37] and Coërs and Desmedt[38] classified the changes in the end-plate (found in as many as 20% of those studies) as being of two types; they called one *dystrophic* (increased branching of the terminal arborization with the end-bulbs or knobs being distributed over a wider area of the muscle fiber than normally),

and the other *dysplastic* (fewer than the usual number of terminal branches and knobs). McDermot,[135] Bickerstaff and Woolf,[16] and Manalov and Haralanov,[128] who corroborated these findings, believe these two types of changes to be different aspects of a regenerative process, the purpose of which is to counteract the primary conductive defect of the disease.

Ultrastructural studies of human motor endplates in myasthenia gravis have proved instructive.[227] The presynaptic vesicles which contain the transmitter substance acetylcholine are present in normal numbers, tending to concentrate at points of thickening of the axolemma. The morphometric analyses of Santa *et al.*[182] show the surface area of the nerve terminals to be diminished by as much as 30–40%. The width of the synaptic clefts is increased (Fig. 12–3). The postsynaptic folds, especially the secondary and tertiary branches,

Fig. 12–3. End-plate from a patient with myasthenia gravis. The terminal axon contains abundant presynaptic vesicles, but the postsynaptic folds are wide and there are few secondary folds. The loose junctional sarcoplasm is filled with microtubules and ribosomes. The synaptic cleft (asterisk) is widened. The line indicates 1 μ. (Courtesy of Dr. Andrew Engel) (×39,600)

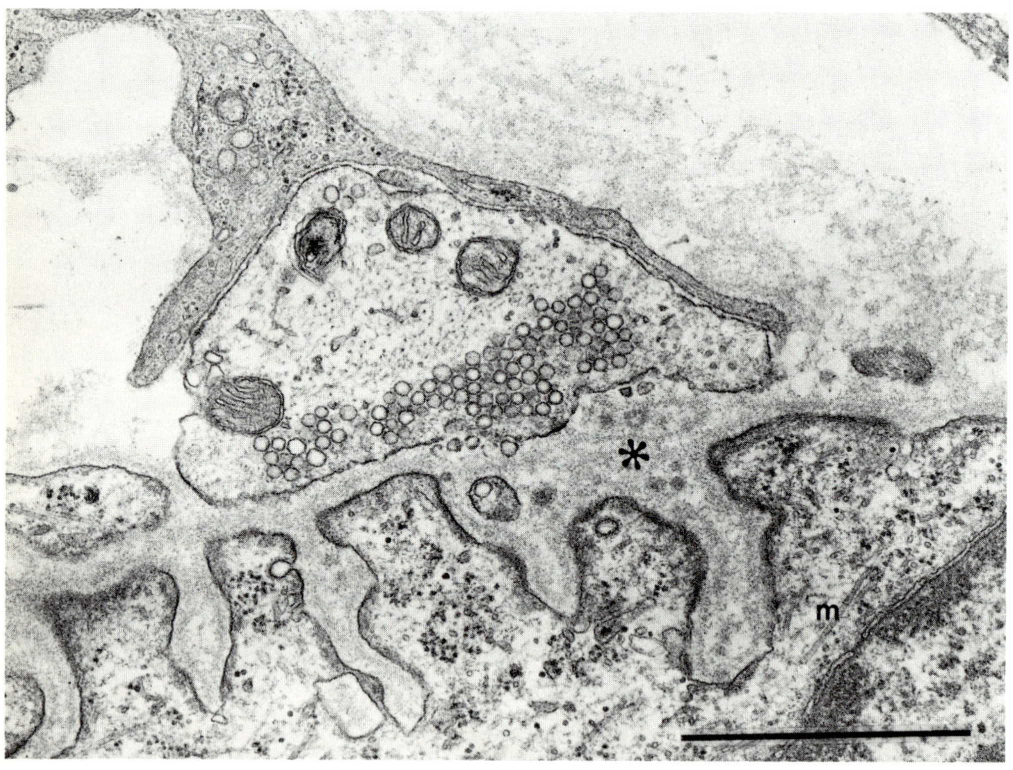

are atrophied, as had been noted earlier by Zacks *et al.*[229]

Since most of the specimens of Santa *et al.*[182] had come from patients (18 or 19) who had received long-term anticholinesterase medication, and since Lytle and Wellbrand[126] had found an increase in the extent of the post-synaptic cholinesterase zone in Prostigmin-treated rats, the obvious question is whether the junctional changes are therapeutically induced. Engel *et al.*[65] compared Prostigmin-treated rat motor end-plates with those of the human myasthenic and noted a difference. Postsynaptic folds were atrophic in both, but the presynaptic terminal area, vesicle counts, and postsynaptic area were not altered by Prostigmin. Further, the miniature motor end-plate potentials were much smaller in the human myasthenic than in the rat treated with Prostigmin.

PATHOLOGIC CHANGES IN THYMUS AND OTHER ORGANS

The association between myasthenia gravis and tumors or hyperplasia of lymph follicles in the thymus gland is well recognized. Weigert in 1901[213] made the first observation of the thymic tumor in myasthenia gravis, and Castleman and Norris[32] first drew attention to hyperplasia of the thymus. The tumors are of two cell types, the lymphoblastic and the larger pale cells of the Hassal's corpuscles. If the thymic tumor precedes the myasthenia, the latter eventually appears, sometimes after an interval of 10–15 years.

In a recent autopsy series approximately one-third (10 of 35 cases reported by Castleman and Norris,[32] 9 of 26 cases reported by Rowland *et al.*[175]) had a thymoma. In Rowland's series 6 of 26 patients had hyperplasia of the thymus, 3 a normal thymus, and in 8 no thymus was found. Of this series only 15 presented lymphorrhages and 3 myositis; myocarditis was present in 3. Removal of a thymoma has sometimes been followed by remarkable improvement. Removal of a thymus that presents only lymphoid hyperplasia with germinal follicles is more commonly followed by prolonged remission, but there is no doubt that severe relapse

may occur without the presence of thymus tissue. The disease has been reported to appear for the first time long after removal of a thymoma.[142, 170, 174] The severe and often rapid course of myasthenia associated with malignant thymoma and its poor prognosis in that circumstance were well reviewed by Morgan and Dudley.[145] Lymphocytic infiltrates may also be observed in the liver, thyroid gland, adrenals, and other organs,[5] and it appears probable that these are the result of a specific endocrine dysplasia.

ETIOLOGY

The cause and pathogenesis of myasthenia gravis have not been ascertained. There were many reasons originally offered for believing that it is a metabolic disorder secondary to some abnormality of the immune mechanisms of the body. The coincidence of disease of the thymus gland is so remarkable that the possibility that the thymus itself elaborates transformed cells or antibodies responsible for the myasthenic symptoms cannot easily be dismissed. Earlier attempts to isolate such a factor were not successful, though Torda and Wolff[205] claimed to have identified a factor that inhibits the synthesis of acetylcholine (ACh). Wilson and his associates[220, 221] demonstrated lessened twitch tension in nerve-muscle preparations as a result of exposure to serum of myasthenics and to extracts of thymus gland. Others[230] were unable to dissociate this effect from the potassium content of such extracts. Thymectomy, as practiced in a number of clinics and evaluated in the writings of Blalock,[19] Keynes,[111, 112] Castleman and Norris,[32] Eaton and Clagett,[60] Viets,[211] Grob and Harvey,[94] and others, has not totally relieved the myasthenic symptoms in many cases. Later follow-ups by Simpson[189] showed that the age of onset of myasthenia does not materially alter the prognosis after removal of the thymus. The prognosis for women was slightly poorer than that for men in patients not operated upon and better than for men after removal of the thymus.[184, 189] Severe and prolonged relapse can occur after complete removal of the thymus gland. The consensus at present is that in myasthenia gravis without thymic tumor remission following surgery is only slightly more frequent (30–40%) and slightly more complete than that which occurs naturally.

The close relationship of myasthenia to disease of the thyroid and adrenal glands has been investigated sporadically. Pathologic studies of the thyroid in cases of myasthenia reveal a higher incidence of disease, particularly hyperplasia, than normal.[138] Simpson[189] noted thyroid abnormalities to be twice as frequent in women (18%) as in men, whereas myositic changes were commoner in the men. Feinberg et al.[70] found hyperthyroidism present in 3% of myasthenics, and evidence of myxedema in a further 3%. Torda and Wolff[205] found an excess of an acidophilic substance in the anterior lobe of the pituitary gland. Mild Addison's disease may complicate myasthenia gravis, and in such a case Kane and Weed[107] found moderate atrophy of the adrenal glands, an enlarged thymus gland, and unusual hyperplasia of the pituitary. The thyroid gland showed mild lymphoid hyperplasia and a moderate increase in colloid. The relationship between hyperplasia of the thymus gland and these other endocrine changes is also obscure. Though Sloan[193] found hyperplasia of the thymus gland in three of seven cases of Addison's disease, myasthenia gravis had not developed. Thymic hyperplasia also occurs in hyperthyroidism and following castration. An interesting type of myasthenia gravis is seen in young women following failure of ovarian function; in two such cases observed by the author only partial remission followed thymectomy. One of these patients later succumbed in a severe relapse. The relationship of adrenal to thymus function and of both to myasthenia gravis is thus inconstant and in need of further clarification.

The significance of the changes in the pituitary gland are difficult to assess, for they are nonspecific. Beneficial effects of ACTH on the symptoms of myasthenia gravis were reported by Torda and Wolff.[205, 206] After receipt of 400 mg of this substance the myasthenia was worse temporarily and then underwent a partial remission, during which the requirement of Prostigmin was reduced by as much as 90%. Others have not obtained effects that were statistically significant. Cortisone was said to be less effective than ACTH.[206] Recently, however, there have been a series of papers documenting improvement in the myasthenic weakness and a decreased requirement of anticholinesterase drugs when prednisone was given. Growth hormone has not been effective in our patients. The effect of pregnancy in leading to remission of symptoms is often striking but is by no means constant.[77] No improvement has been obtained by using estrogen in treatment.[208] The causal relationship of the endocrine abnormalities so frequently yet inconstantly related to myasthenia gravis still remains obscure. In sum, one can conclude that except for thyrotoxicosis no real endocrine relationship has been shown and that the conjunction of thymic abnormalities may have an entirely different basis.

Further data supporting an autoimmune mechanism have slowly accumulated. The similarity of the thymus gland in myasthenia gravis to the thyroid gland in Hashimoto's thyroiditis was recognized early, and soon thereafter Strauss et al.[197] obtained evidence of a circulating antibody to muscle (but present in only about 75% of cases). Certainly the thymus gland influences various immune processes and immunoglobulin production. Simpson[189, 190] cites the variable course of the disease as being consistent with other autoimmune diseases, as are also the infiltrations of the target tissue (here muscle) with lymphocytes. Although the concept is attractive it is difficult to affirm or refute. Indeed there is no complete theory of the origin of autoimmune disease to date. Such a theory in reference to myasthenia gravis leaves unexplained the nature of the membrane events underlying the conduction failure and its localization predominantly in cranial muscles.

The primary defect in myasthenia gravis certainly relates to defective neuromuscular transmission, and the symptoms have obvious similarities to those of curare poisoning. Electromyographic studies of nerve-muscle transmission by Lindsley[125] and Harvey and Masland[99] provide further evidence of this similarity to the effects of curare. In the electromyogram the most characteristic abnormality is the irregularity and eventual loss of

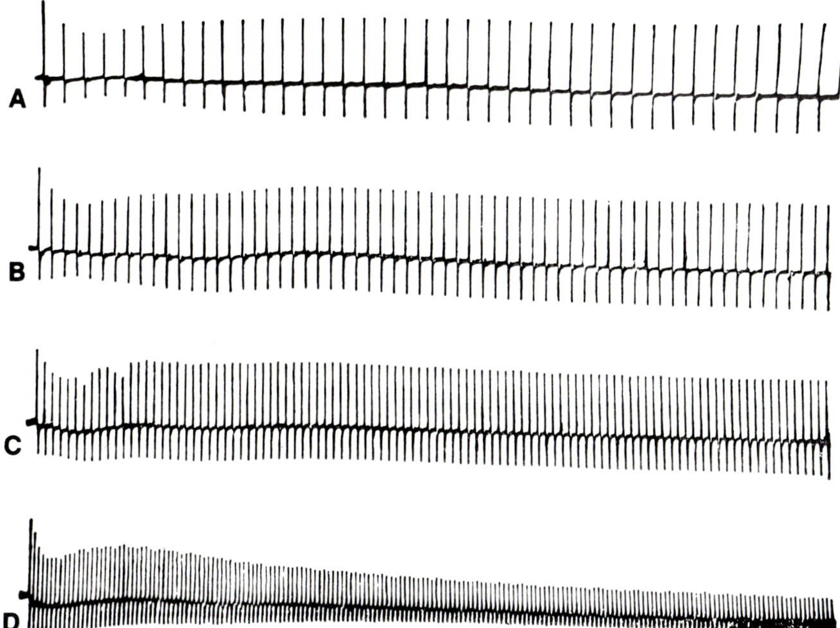

Fig. 12–4. Myasthenia gravis. Muscle action potentials in response to repetitive nerve stimulation. Initial fall in voltage is succeeded by a secondary decline. Stimulation frequencies: (A) 5/sec. (B) 10/sec. (C) 25/sec. (D) 50/sec. Response to initial stimulus measures 6.2 mV. (From Grob et al.[95])

voltage in the spikes of single units, presumed to be due to variation in the number of fibers contracting to each nerve impulse. When the nerve to the muscle is stimulated electrically by a rhythmic series of shocks, the response fails progressively—myasthenic reaction (Fig. 12–4). If a high frequency stimulus is used (e.g., 200–500 stimuli per second), an additional phenomenon is observed, i.e., a strong twitch to the first shock followed by only a poor contraction.[165] This is the Wedensky phenomenon, which means that the first impulse of the rapid series was the only one to excite many of the fibers. It is observed only at much higher frequency of stimulus in normal muscle and was shown by Wedensky to be due to the fact that the rate of stimulus is so fast that each subsequent stimulus falls into the relative refractory period of the one before, then exciting only a subnormal nerve impulse. The subnormal impulses, though insufficient to excite the muscle fiber, each set up yet another refractory period. The phenomenon therefore is one way of measuring the refractory period of the conducting system (nerve and neuro-

myal junction), and its occurrence at low frequencies in myasthenia gravis[165, 166, 176] and in curare poisoning[24] indicates that at some point near the nerve terminals (not necessarily in the end-plate) the refractory phase of the excitatory process is prolonged. All these abnormalities are reversed to some extent by neostigmine (Prostigmin) and other anticholinesterases, notably pyridostigmine (Mestinon), which has a more durable effect, and 3-hydroxyphenyl-dimethylethyl ammonium chloride (Edrophonium, Tensilon), which is potent, brief in action, and useful as a rapid test. Certain pyrophosphate compounds (TEPP and OMPA) that produce an irreversible inhibition of cholinesterases are also effective[95] but are too toxic for regular use. An excess of anticholinesterase leads to fasciculation, then further weakness, with excessive mucous and salivary secretions, miosis, and restlessness (the so-called cholinergic block).

The action of Prostigmin has led to formulation of a number of hypotheses of the mechanism of muscular failure in myasthenia gravis. These hypotheses, discussed at length in reviews by McEachern,[136, 137] Rowland *et al.,*[174, 175] and Simpson[189, 190] are: a deficiency or inadequate synthesis of ACh; an excess of cholinesterase, which destroys ACh; a reduced sensitivity of the muscle fiber to ACh; and the existence of some other substance with a curare-like action. All these hypotheses are unsatisfactory. The parenteral administration of ACh and other more stable cholines does not relieve the myasthenic symptoms, and cholinesterase does not increase either in the bloodstream or biopsied muscle. Myasthenic muscle is quite sensitive to an intraarterial injection of ACh, and the existence of a curare-like compound has not been established. Support for the presence of such a compound is derived from instances in which the newborn of a myasthenic mother exhibited myasthenic symptoms that responded to Prostigmin during the first weeks of neonatal life and then abruptly recovered, as if a toxic substance had been received through the placenta. Over 20 such cases are described by Mc-Keever[139] and others,[77, 84, 87, 94] but as pointed out by Simpson[189, 190] it could have some other basis than a circulating toxin. The occurrence of fatal myasthenia in the infant born of a myasthenic mother is well established.[198, 202] In rare cases the infant continues to manifest a congenital myasthenic syndrome.[123]

Examining each of these hypotheses—the first assumes the defect in myasthenia gravis is a defective production of ACh at the motor end-plates. The resemblance of myasthenia to the effect of curare, including the same incidence on muscles first affected by curare, led to the discovery that physostigmine and its analogue Prostigmin could correct the defect in transmission at the motor end-plate. It was known that these drugs were anticholinesterases, and their effect implied the possibility of a defect in the production or stability of ACh, which is closely related to transmission of excitation from nerve to muscle. It is now known that acetylcholine is produced continuously in very small quanta by the end-plate[108] and that the large amount produced by a nerve impulse lessens with each of a rapidly repeated series of impulses. There is evidently a large "safety factor" that ensures transmission, even though ACh production normally lessens in this way. In this sense a critical insufficiency develops in the myasthenic

state shown by Lambert and Elmquist[119] to be a decrease not in the number of ACh quanta but in their size. In distinction to conditions such as periodic paralysis in which the muscles most commonly active (respiratory, ocular) are less vulnerable owing to their more steady loss of potassium, these muscles are the most vulnerable in myasthenia. The selectivity of myasthenic paralysis for certain muscles may therefore relate only to their more continuous activity. An absence of miniature end-plate potentials and of their normal facilitation by potassium was demonstrated by Dahlbeck and his associates[45] in excised myasthenic muscle, providing direct evidence of a defect in release of transmitter substance.

The effect of Prostigmin in reversing the defect of neuromuscular transmission in myasthenia is most obvious in muscles that have been only recently and intermittently paralyzed. The anticholinesterase effect of Prostigmin does not wholly correct the fundamental defect in myasthenia gravis. Other anticholinesterases may have more prolonged but not more extensive effects. In severely affected patients dosage of Prostigmin large enough to produce generalized fascicular twitching may still overcome only some of the weakness. The twitching of the affected muscle then indicates the full Prostigmin effect at the end-plates; yet a defect in transmission remains. Moreover, such muscles do not undergo neural atrophy and may recover full function in a later natural remission of the disease. This phenomenon, which may be seen in every case of myasthenia gravis, implies that a defective amount of ACh is of only relative importance in the pathogenesis of myasthenia. Prostigmin likewise can reverse the initial stages of curarization, indicating that at this point the defect is a competitive block.

Decamethonium normally produces a depolarization block (not reversible by Prostigmin or ACh). Churchill-Davidson[34] found that in myasthenia gravis three to four times the normal dosage of decamethonium was required to lessen a test contraction to electrical stimulation of nerve. This pharmacologic resistance was found in muscles not yet affected by the disease and was still present in a full remission 3.5–5.0 years after thymectomy. This resistance to decamethonium was also found in three cases of lung carcinoma and one of thyrotoxicosis, but was absent in seven cases of dermatomyositis and in two of thyrotoxic myopathy.[35] When myasthenia is severe, however, this resistance to decamethonium disappears, and muscle weakness is immediately increased by a small fractional dose. At this stage transmission also fails to respond to Prostigmin, behaving as if the end-plate was partially or completely depolarized. In an intermediate stage decamethonium can cause a block in myasthenic muscle that can still be reversed by Prostigmin or Tensilon, whereas the decamethonium block in normal muscle cannot be so reversed. Churchill-Davidson concluded that these aspects of the conduction defect in myasthenia gravis point to some alteration in the end-plate membrane. In this connection the finding of simple linear end-plates and a corresponding pattern of cholinesterase in myasthenic muscles by Coërs[39] is of great interest. It has yet to be determined whether this is a fundamental structural change or an adaptation of the end-plate to the defect.

Grob *et al.*[95] approached the problem in another manner. They first showed that the defect in response of myasthenic muscle to repetitive nerve stimulation is complex. The action potentials of the second and subsequent stimuli lose voltage progressively during the first second or two, but there then occurs a delayed increase in action potential. This secondary increase is more prominent with rates of stimulation of over 25 per second (posttetanic facilitation) and with higher rates was earlier in onset and shorter in duration, being in any case followed by a second progressive decline (Fig. 12–4). These aspects of transmission were compared to the effect of intraarterial injection of ACh. After producing excitation, an excess of ACh has a depolarizing effect comparable to that of decamethonium but is very transient in normal and myasthenic muscle. This effect is seen as an initial depression of transmission that is less than in normal muscle but is followed in the myasthenic by a more prolonged, late depression than in normal subjects. These changes were interpreted to mean that some form of competitive block to ACh exists at the end-plates, resembling that produced by choline and worsened by some delayed product of

ACh, possibly choline. This aspect of the myasthenic defect is the basis of the "posttetanic exhaustion" of Desmedt,[56] which is not seen in curarized muscle. Churchill-Davidson and Richardson[35] reported a remarkable case of myasthenia in which progressive failure of Prostigmin to relieve symptoms disappeared if Prostigmin was omitted completely for a few days. Artificial respiration was maintained in the absence of Prostigmin treatment. On the hypothesis that some breakdown product of ACh such as choline was blocking the end-plates, moderate to large doses of *d*-tubocurarine were given the patient for brief intervals to block all production of ACh and choline. Six hours after the last dose of *d*-tubocurarine, a most remarkable recovery of muscular power occurred (without Prostigmin) that persisted for 10 days, following which Prostigmin had a dramatic effect for a time. Such an observation, supported by other instances in which cessation of Prostigmin and artificial respiration for intervals had been beneficial, is strongly in favor of the accumulation of a depolarizing breakdown product of ACh in myasthenia.

All these studies point to the presence of some defect in the mechanism of hydrolysis of ACh in myasthenia gravis, possibly a defect in the further breakdown or dispersion of choline, resulting in impairment of the membrane characteristics of the end-plate. The defect is for a long time competitive, which explains the striking effects of rest and activity, anticholinesterases, and curare. In this phase also the weaker effects of ephedrine and guanidine and conversely of quinine may be manifest. This prolonged stage of competitive defect in neuromuscular transmission therefore provides the Prostigmin test with relative validity. The importance of this conclusion is clear in studies of the myasthenic type of weakness found in some cases of malignant disease, notably pulmonary carcinoma. In such cases Eaton and Lambert[61] found only slight or equivocal response to Prostigmin but extreme sensitivity to tubocurarine. Further, the initial electrical response to stimulation was depressed, but posttetanic faciliation after rapid stimulation was remarkable in degree. The vulnerability of patients undergoing operations for carcinoma to relaxant drugs was noted independently.[44] Corresponding analysis of the myasthenia associated with thyroid disease has not yet been made, though it is considered likely that several distinct kinds of such defects may ultimately be differentiated.

In conclusion we would state that from the pathologic point of view, myasthenia gravis is a unique example of a neuromuscular disease caused by a disorder of the end-plate or neuromuscular junction. The changes in the muscle itself are unimpressive especially when examined in a large autopsy material where in the great majority of severely weakened muscles nothing at all is seen. Such inflammatory cell infiltrations and myonecrosis as do occur in some few muscles in occasional cases we are obliged to regard as an epiphenomenon, an associated polymyositis for example. The difficulty for the myopathologist is to visualize by reliable technics the ultraterminal nerve arborizations and the end-plate itself, which bear the brunt of the disease. The results of the methylene blue method and Kolle technic were already mentioned. Santa *et al.*[182] used an electron microscopic technic that is not practiced routinely. The importance of their findings relates to the fact that they provide strong support for Simpson's[189, 190] idea that a change in the geometry of the terminal nerve endings and receptor surface, caused in some manner by the thymus gland, may explain the reduced efficiency of ACh transmission.

Very recently, on both clinical[190] and experimental[150, 197] grounds, the hypothesis was advanced that myasthenia is a manifestation of an autoimmune disorder in which an antibody is produced against end-plate (receptor) protein. This antibody is assumed to compete with ACh in the end-plate. Further work is needed to clarify and corroborate this interesting suggestion.

MYASTHENIC-MYOPATHIC SYNDROME OF EATON AND LAMBERT

Eaton and Lambert and their colleagues in 1960[171] reported a special form of myasthenia associated with carcinoma, particularly the oat cell carcinoma of the bronchus. Preceding the clinical signs of the tumor, sometimes by many months or even several years, they noted the appearance of symptoms of myasthenia (i.e., weakness and fatigue) but in this instance associated with aching and wasting of the muscles of shoulders and thighs. Difficulty in elevating the arms, walking, and climbing stairs had called attention to the disorder. In contrast, ptosis, diplopia, dysphagia, and dysarthria were infrequent. They remarked on another feature not seen in myasthenia gravis, i.e., an augmentation of the power of contraction especially of shoulder and thigh muscles with each of the first half dozen attempts at voluntary contraction. In a sense this was a kind of inverse myasthenia, though true myasthenic weakening did become obvious as contraction continued. The tendon reflexes were impaired or absent altogether. Dry mouth, impotence, and paresthesias were included in the total clinical picture, as well as evidence of certain of the other paraneoplastic neurologic syndromes (polymyositis or dermatomyositis, polyneuritis, cerebellar degeneration, and multifocal leucoencephalitis). Since the original report many other examples have been seen and reported, and the original Mayo Clinic series enlarged. We have seen one or more cases each year. In addition there have been a few reports of cases of stiffness and mystonia associated with oat-cell carcinoma of the lung.

The electromyographic recordings confirm the early augmentation of power in successive contractions by a steplike increase in the voltage of the action potential upon nerve stimulation at tetanic rates. Supramaximal stimulation at rates below 10/sec results in progressive reduction in voltage similar to that of myasthenia gravis. Low-amplitude, brief ("myopathic") action potentials can also be recorded from weak muscles. In contrast to myasthenia gravis the strength of contraction is not greatly enhanced by Prostigmin or Tensilon, even though the muscles remain unusually sensitive to decamethonium, muscle relaxants, and tubocurarine.

The associated carcinoma (lung, breast, colon, stomach, prostate, etc.) if it does not clinically manifest within 2–3 years may never appear (the situation in one of five cases). In one of our patients an unexpected early death from pulmonary fibrosis and hypercapnia led to an autopsy at which time there was only a single carcinomatous lymph node in the mediastinum; the primary lesion could not be located. Boeck's sarcoid is sometimes accompanied by this myasthenic-myopathic syndrome in a few instances. Guanidine chloride corrects the muscle weakness by facilitating the release of ACh and is the treatment of choice.

The microscopic findings in muscles are relatively unimpressive. In biopsies of several cases diagnosed at the Massachusetts General Hospital there were no definite lesions. In autopsied cases the weak muscles tend to have rather thin fibers resembling more disuse atrophy than denervation. Necrosis of a few single fibers and myophagia or a few scattered lymphocytic-mononuclear infiltrates, when present, could not explain the myasthenia.

The electromyographic findings point to a disorder in the neuromuscular or sarcolemmal transmission of the nerve impulse, and an electron microscopic study of the motor end-plate revealed a widening of the synaptic cleft and reduction in the receptive surface, not unlike that of myasthenia gravis.

FAMILIAL PERIODIC PARALYSIS

Familial periodic paralysis is a rare and peculiar type of intermittent paralysis; because of its dramatic nature it could not have escaped the notice of ancient physicians, and indeed one finds reference to such illnesses in the papers of Musgrave and others in the early eighteenth century. The first clear descriptions, however, were given by Hartwig[97] in 1874 followed soon by authoritative reports from Westphal[214, 215] and Oppenheim.[152] Westphal succeeded in distinguishing the condition from hysteria, which had become a familiar state by then. Goldflam[20] recognized the remarkable vacuolation of the muscle fibers. The name given the disease was familial periodic paralysis. Wilson[221] favored Oddo's term intermittent myoplegia (now obsolete).

As the name implies the disease "runs in families," and its genetic pattern now establishes it as a mendelian autosomal dominant trait[15, 86, 151] though the degree of penetrance in some instances is incomplete. Sporadic cases without family history are described frequently, but in view of closely similar attacks now known to be associated with primary aldosteronism, thyroid disease, renal failure, and the ingestion of licorice such isolated cases must be carefully scrutinized. Enough nonfamilial cases are now on record, however, to indicate that the disease may appear as a spontaneous mutant.

Early in the twentieth century it was discovered that the administration of potassium salts had a beneficial effect, and finally Biemond noted that the serum potassium falls during attacks. This led to the designation hypokalemic paralysis. Study of further families, such as that reported by Schoenthal[183] and Wolf[224, 225] divulged the fact that not all were alike in this respect, and that in another type of disease the serum potassium could be elevated and its ingestion could produce an attack. Tyler et al.[207] and later Gamstorp[82] in describing other families concluded that there were two types of periodic paralysis. Finally Poskanzer and Kerr[162, 163] found a family with periodic paralysis whose serum potassium was within normal limits, thus demonstrating a third type. The second, hyperkalemic type is often accompanied by paramyotonia. In the Orient many cases of periodic paralysis are associated with hyperthyroidism. Renal failure and hyperaldosteronism in which high levels of serum potassium occur secondarily may also be attended by paralysis. Thus there has evolved a plethora of types that may be classified according to the headings listed in Table 12–1.[158, 159]

HYPOKALEMIC FAMILIAL PERIODIC PARALYSIS

Hypokalemic familial periodic paralysis usually appears between the ages of 14 and 20 years, although it has been reported as early as 8 and as late as 31 years. Males are affected more often than females (ratio 3:1). Attacks seem to appear less frequently during adulthood. Inheritance is by autosomal dominance.

TABLE 12–1. Classification of Periodic Paralysis

A. Primary periodic paralyses
   1. Hypokalemic (familial) periodic paralysis
   2. Hyperkalemic (familial) periodic paralysis
   3. Normokalemic (familial) periodic paralysis
   4. Paramyotonia of Von Eulenberg
B. Secondary periodic paralyses
   1. Endocrine dysfunctions
      a. Thyrotoxicosis
      b. Aldosteronism
   2. Renal disorders
      a. Renal tubular acidosis
      b. Diabetic acidosis
      c. Ureterosigmoidostomy
   3. Licorice intoxication

The onset of an attack is usually at night, often after a day of strenuous muscular activity. The patient awakens with all four extremities weak or paralyzed, and if the attack is severe even the cranial muscles are affected. He cannot move but is usually able to speak and breathe though sometimes with difficulty. The muscles are slack, electrically inexcitable, and do not even contract in a tendon reflex. The attack may last 2–3 hours to as long as 7 days, with gradual recovery. More brief abortive attacks occur, but the paralysis is then incomplete. The first muscles to be affected (ascertained when it begins during the day) are in the shoulders or back and thighs, and the weakness gradually spreads to involve the lower limbs, arms, and neck. Ptosis and dysphagia occur in severe attacks, and death from paralysis of respiratory muscles is not unknown.[110] Smooth muscle is spared but urination may be difficult. In some cases cardiac enlargement, with slow and irregular pulse, is prominent, as was pointed out in 1891 by Oppenheim.[153] An attack may be preceded and accompanied by excessive thirst and perspiration. There are no sensory changes.

The intervals between attacks vary from a few days to months or years. It has long been recognized that attacks can be precipitated by excessive carbohydrate ingestion and by the administration of insulin and adrenalin. The ability of the patient to "walk off" an incipient attack was noted by early writers. Between attacks the muscles are usually normal, but in some of the reported cases there was hyper-

trophy of the arm and leg muscles. Oppenheim,[152] Biemond and Daniels,[17] and Pearson[158, 159] observed some degree of atrophy during later life in the most frequently affected muscles of the legs and sometimes the arms.

Reviews of this disease were made by Talbot,[200] Gass *et al.,*[85] Grob *et al.,*[96] and McArdle.[133] An excellent discussion of the current concepts of intracellular inorganic ions and muscle action is given by Conway;[42] Smith[194] and Smith *et al.*[195] described an interesting animal homologue of human periodic paralysis.

The classic attacks related to sleep and carbohydrate ingestion are relieved by potassium and appear to be intimately related to carbohydrate metabolism. Serum potassium as low as 1.8 mEq/liter are found during the attack. The paralysis begins at 3.0–3.5 mEq/liter. Interestingly, in the hypokalemia induced by potassium deficiency in normal subjects[76] or by hyperthyroidism the onset of muscular weakness is at a much lower level. This is true also of the sporadic examples of paralysis with diabetic acidosis, after administration of desoxycorticosterone (DOCA), in Addison's disease, in primary aldosteronism, and in some cases of renal insufficiency.[27]

The enigma of periodic paralysis has been the reason for failure of muscular contraction comparable to the effect of potassium intoxication in the presence of a reduced serum potassium concentration. Empiric observations of the beneficial effect of administration of potassium salts during an attack were made as early as 1901 by Singer and Goodbody[192] and by Holzapple.[101] Biemond and Daniels[17] first recorded the reduced serum potassium level in 1934. Aitken and his associates[3] and Gammon[81] independently observed in 1937 that the fall in serum potassium was not associated with any increase in urinary output. This observation has since been documented many times.[6, 73, 82, 167] The possibility that potassium moves into skeletal muscle fibers is often suggested.[46] Zierler and Andres[231] and Grob *et al.*[96] clearly demonstrated in both spontaneous and induced attacks that potassium levels in venous blood in a limb dropped precipitously at the onset of an attack of weakness and rose to high levels in the recovery phase. The high A-V difference could be accounted for only on the basis of uptake

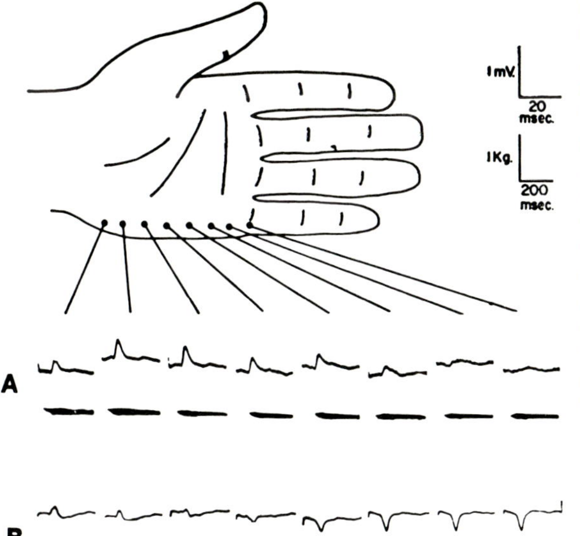

Fig. 12–5. Failure of propagation of action potentials along muscle fibers during an attack of periodic paralysis. A series of eight electrodes over hypothenar muscles record effect of single shock to ulnar nerve at elbow. End-plates lie under three proximal electrodes. (A) During an attack of paralysis, the tracing shows diminishing amplitide at increasing distances from end-plates. The line below each potential is a mechanical response, so weak that no change is seen at any point. (B) Recorded during start of recovery from attack, a small propagated potential is seen as an upward notch over end-plates and a downward deflection distally. (C) Recorded after complete recovery 2 hours later, full potential is propagated, and muscle contraction has returned. (From Grob et al.[95])

by the muscle cell. It was calculated that as much as 110 mEq potassium could move from extracellular fluid into muscle during an attack. Jantz[103] and Vastola and Bertrand[210] each demonstrated in one patient an increase of intracellular water and potassium in muscle, without a change in concentration. Shy and his associates[188] found only an increase of intracellular water, without an increase in potassium. Grob et al.[95] demonstrated an abnormally high uptake of potassium following ingestion of carbohydrate, or administration of glucose, insulin, and possibly epinephrine. During an attack the weak contractions of affected muscles resulted in release of four times as much potassium as from normal limb muscle. Cortisone tended to prevent the production of an attack following ingestion of glucose. No change in plasma concentration of sodium was detected during typical attacks of weakness.

No satisfactory explanation for the supposed movement of potassium into the muscle cell has yet been found. Zierler and Andres[231] noted no difference in the nocturnal movement of potassium between normal subjects and the patient suffering from periodic paralysis when free

from paralysis. Allott and McArdle[6] found a reduction in plasma inorganic phosphate during an attack, and McArdle and Merton[11, 133, 134] demonstrated a grossly abnormal uptake of potassium by muscle produced by ischemia when an attack of paralysis was not present. This was tentatively interpreted as indicating a partial block in the hexosephosphate stage of metabolism,[188] but Engel et al.[66] were unable to prove this hypothesis for the concentrations of intermediates of glycogen synthesis or of anionic phosphorylation were normal. Some such metabolic defect is clearly present. Its partial nature is shown by the effect of large amounts of potassium salts in counteracting the deleterious effect of high carbohydrate diet in patients suffering from periodic paralysis.[140]

The muscular weakness is associated with complete inexcitability of the affected muscle to either faradic or galvanic stimulus.[153] At the onset of an attack Grob et al.[96] found that a decline in strength of contraction precedes a loss of muscle action potential. Propagation of the muscle potential from the region of the end-plates is progressively decreased (Fig. 12–5), and during a severe attack only a reduced po-

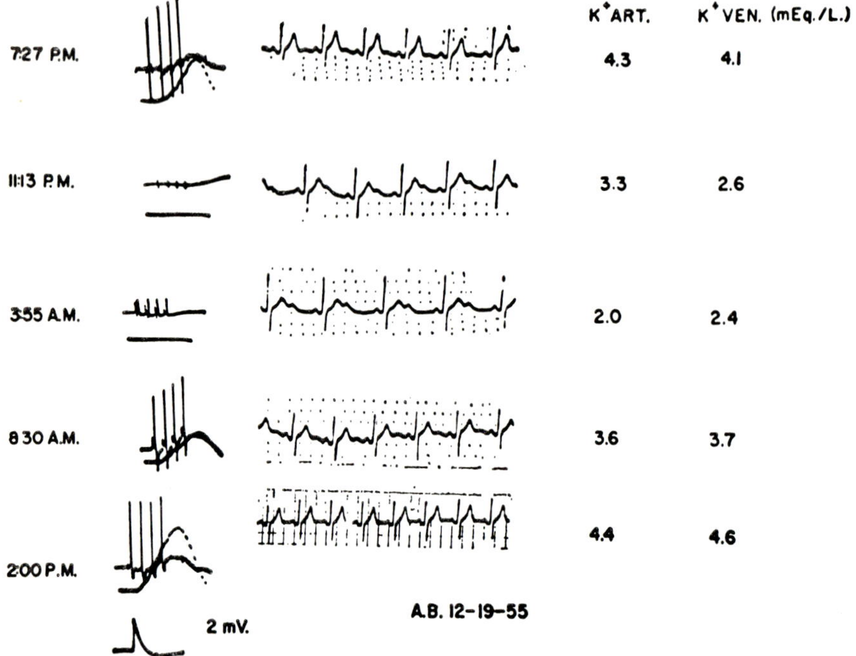

Fig. 12–6. Changes in muscle action potential resulting from four repeated shocks to the ulnar nerve, as well as the corresponding muscular contraction (lower line), electrocardiogram, and related potassium levels in artery and vein before, during, and after an attack of periodic paralysis induced by glucose. (From Grob et al.[96])

tential at the end-plate region may indicate arrival of the nerve impulse. The potentiating effect of rapidly repeated stimuli in causing momentary increase in size of potential (owing to loss of potassium with excitation) is still present (Fig. 12–6). Electrocardiographic changes began at the same plasma level (3.3 mEq/liter) as in normal subjects with potassium depletion and presented the usual positive afterpotential followed by prolongation of PR, QRS, and QT intervals and lowering of the T wave. The cardiac changes were much less pronounced than those in skeletal muscle and were evidently secondary to the inconstant change in serum potassium level. Similar changes were reported by van Buchem.[209]

Shy and his associates[187, 188] confirmed the maintenance of potassium concentration in the muscle cell during an attack, with large increases of intracellular water. They measured the polarization potential in 37 separate muscle fibers and found it to be within normal limits,[188] yet the muscle did not respond to direct stimu-

lation. This finding means that depolarization does not explain the paralysis, which is attributed by Shy et al.[188] to the increase of intracellular water. In electron micrographs of tissue taken from muscle during an attack, the sarcoplasmic reticulum was distended and vacuolar, with finely granular PAS-positive contents. A defect in the glycolytic cycle was postulated as the primary fault, with possible abnormal intermediaries of high molecular weight to account for the osmosis. The shifts in potassium and the loss of excitability remain to be explained. Transmission at the end-plate is not seriously affected, but Grob et al.[95] found that the late effects of ACTH and Prostigmin were lessened. Decamethonium could not reverse the paralysis. They likened the effect to that of an anodal (hyperpolarizing) current at the end-plate. Primary affection of the contractile process suggests a mechanism such as the diminished contractility of actomyosin associated with the increased intracellular potassium reported by Needham et al.[149] The relative immunity of respiratory and extraocular muscles in periodic paralysis is related to their more continuous activity. Grob *et al.*[95] clearly demonstrated the normal substantial movement of potassium in activity and in sleep. The striking immunity of respiratory muscles in periodic paralysis and the common onset in sleep must relate to these factors. Likewise, exercise of a muscle or limb during the period preceding paralysis may render these immune to weakness.

HISTOLOGY

The pathologic anatomy is based on muscle biopsies taken during and following attacks and on a few autopsies on subjects who died with paralysis. The gross and microscopic findings are believed to contribute relatively little to an understanding of the etiology and pathogenesis of the disturbance. Goldflam,[90, 91] in a comprehensive discussion of this subject, stated that on cross section the muscle fibers had a polygonal shape with rounded corners. The fibers were relatively thick and of similar size. Myofibrils were distinctly visible, more so than normally, creating a punctate appearance on the cut surface. In many fibers the Cohnheim's fields were thought to be more prominent than normal. The myofibrils were pressed apart by an amorphous material which gave the muscle fiber the appearance of a sieve. In several fibers there were spaces that appeared as round or oval vacuoles, usually in the center of the fibers (Fig. 12–7). Their content was mostly fluid but not stainable with carmine or eosin. The intramuscular connective tissue was not hyperplastic or infiltrated. The number of sarcolemmal nuclei was not increased, and the nuclei rarely penetrated the muscular substance. On longitudinal section the longitudinal and cross striations were distinct. The nerve bundles within the muscle appeared normal. The presence of vacuoles or unidentified droplets was confirmed by Crafts,[43] Bornstein,[21] Zabriski and Frantz.[228] Biemond and Daniels,[17] Allott and McArdle,[6] and Shy *et al.*[188] Tyler *et al.*[207] noted in addition a degeneration of isolated muscle fibers, an infiltration of mononuclear cells, and a proliferation of fibroblasts and sarcolemmal nuclei, but their cases were atypical. Sudan stains for fat and Best's carmine

stain for glycogen show these substances to be irregularly distributed within some of the vacuoles and adjacent parts of the fiber.

Phase optics, which in thin sections provide the least distorted view of the affected fibers, reveals not only the vacuoles but dense bodies of various shapes and myriads of tiny vesicles. Also parts of the fiber thought to be normal in routine hematoxylin and eosin sections are altered in some ill-defined way. The vacuoles themselves contain clusters of granules, tiny vesicles, and circular structures. In acid phosphatase stains these regions are reactive both in type I and type II fibers, as are some smaller regenerating fibers in their entirety. Von Kossa-positive mineral deposits (calcium) are seen in a few vacuoles.

Engel,[66] by electron microscopy, traced the origin of the vesicles from their earliest stage to the large distorting bubblelike structures that push all other organelles aside. Also he visualized the proliferation of t-tubules to form membrane networks. The longitudinal endoplasmic reticulum is dilated. The coalescence of the fine vesicles arising within and adjacent to the longitudinal and transverse sarcoplasmic reticulum is the natural evolution of the large vesicle. The content of vesicle and vacuole at high magnification varies from homogeneous amorphous material, dense bodies, and osmophilic rings to myelinlike structures. Glycogen particles, which usually surround the vesicles, sometimes penetrate them. The limiting membranes around many of the vacuoles, presumably derived from sarcoplasmic reticulum, are often ruptured with extension of adjacent sarcoplasm into the matrix of the vesicles. The proliferation of the t-system is believed to be a reaction to vesiculation. The vesicles and vacuoles are in communication with the surface of the fiber, i.e., are extracellular. An autophagic reaction degrades the substances within the vacuoles. The only part of the lesion not well visualized is the degeneration of segments of fibers and their regeneration.

This very complete study gives the best possible demonstration of the altered structure of the fiber yet offers no clue as to why vesicles and vacuoles form. The relation of the lesion to the electrical inexcitability of the fiber during attacks is equally obscure. Such changes are pres-

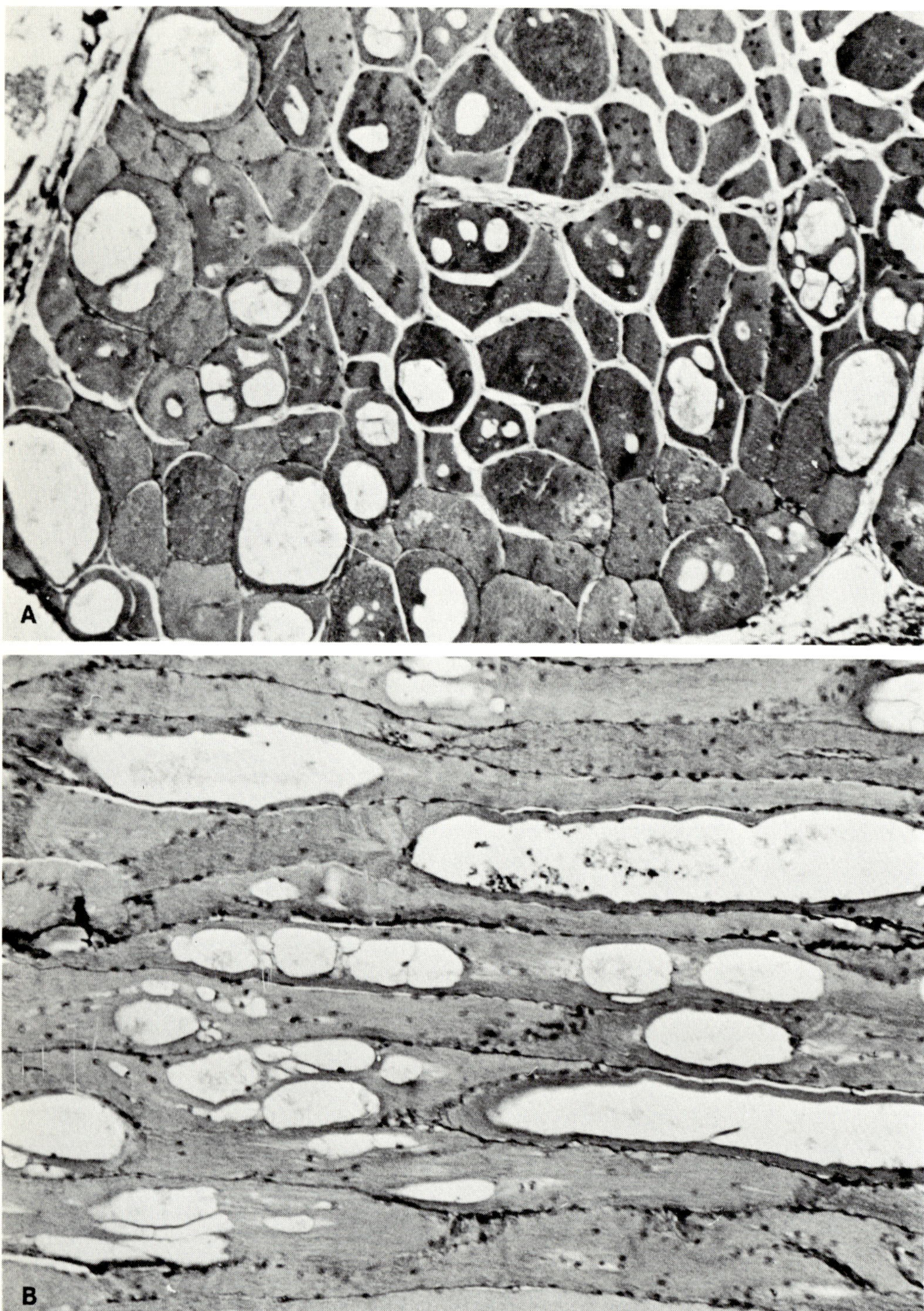

Fig. 12–7. Transverse (A) and longitudinal (B) sections of quadriceps muscle, showing an extreme degree of vacuolation of approximately half of the fibers. This 42-year-old patient has had numerous attacks of hypokalemic periodic paralysis during childhood and adolescence but none in the last 18 years. An attack was induced by a high carbohydrate meal and insulin. (A) Note the central nucleation of the majority of fibers and this variation in size. A fine precipitate is seen in some of the vacuoles, which vary in size. In longitudinal sections the large vacuoles, which are elliptical, swell the fiber to over 100 μ in diameter. (H&E)

ent between attacks when the muscle power is nearly normal and may be absent in some muscles that are virtually paralyzed. Therefore this lesion, which represents in purest form a disease of the intrinsic conducting mechanism of the fiber (sarcolemma, t-system, and SR), must as Engel says be "a delayed consequence of the paralytic attack" or only indirectly related to it.

HYPERKALEMIC PERIODIC PARALYSIS (ADYNAMIA HEREDITARIA AND PARAMYOTONIA)

In 1886 von Eulenberg[69] described in six generations of a family 26 cases of a disorder related to myotonia but differing from dystrophia myotonica and from congenital myotonia in several important respects. Unlike dystrophia myotonica, the condition was nonprogressive—indeed a considerable degree of improvement could occur during adult life—and it was not associated with muscular wasting. Inheritance was in the pattern of a single autosomal dominant, so that unaffected members of the family did not transmit the disease. Unlike those with Thomsen's disease, von Eulenberg's cases exhibited myotonia usually only in tongue and ocular muscles; spread to the face and distal extremities occurred only on cooling. Finally, the patients complained of attacks of flaccid limb weakness that could be precipitated by cold or appear spontaneously. Since von Eulenberg's account many other families suffering from this syndrome have been reported; summaries of the cases up to 1958 are given in the paper of Drager et al.[58] Yet in spite of all these reports most of the monographs on myotonia during the last 30 years noted only that all myotonia is worsened by cold, and they doubted the existence of paramyotonia as a separate entity.

Interest in the subject was renewed in relation to the study of intermittent attacks of weakness of limb muscles resembling periodic paralysis, but not related to sleep, ingestion of glucose, or fall of serum potassium, and not relieved by administration of potassium. In a study of such cases Gamstorp[83] noted the absence of muscle wasting. Although biopsy findings showed ringbinden and sarcolemmal masses in one of three cases, no myotonia was found on percussion of the thenar eminence or in electromyograms. Sixty-four of 94 cases showed an "unfavorable influence" of cold and were worse in winter. The serum potassium level was commonly raised somewhat in an attack, and attacks of weakness were precipitated by potassium administration. Gamstorp proposed the term adynamia episodica hereditaria and noted that the inheritance was of the single autosomal dominant type.

Drager et al.[58] more recently illuminated this subject by their report of a family with 30 affected members presenting all the criteria established by von Eulenberg for paramyotonia. These patients reported attacks of cramping and stiffness of muscles of the face and neck, and usually also of the fingers and hands, on exposure to cold. The eyelids closed, the facial expression became set, and if the cold continued the fingers clenched. Only exceptionally were the feet and ankles involved. Speech became slurred, and the tongue felt stiff after drinking cold liquids. The muscles of respiration and those supplied by the cranial nerves were not involved. The stiffness disappeared within 45 min of warming. In addition, weakness of proximal muscles of all four extremities with hyporeflexia occurred in some attacks. Such weakness disappeared more slowly on warming, taking as long as 3–4 days for complete recovery. Even a mild attack of weakness lasted several hours. Weakness could occur without stiffness in the extremities. Two members of the family described by Drager et al.[58] presented attacks of muscular weakness as the only symptom, not necessarily provoked by cold; this was described in three cases in the literature in addition to the series of Gamstorp.[83] In the series of Drager et al., the serum potassium was on the high side of normal levels during an attack, as in Gamstorp's cases. Intracellular potassium was not significantly raised. In an attack the affected muscle was refractory to both galvanic and faradic stimulation. Between attacks the muscle fibers showed exaggerated insertion potentials and short bursts of spontaneous activity, which were also observed in Gamstorp's cases. In the cases of Drager et al. myotonia on percussion of the tongue was universal, but

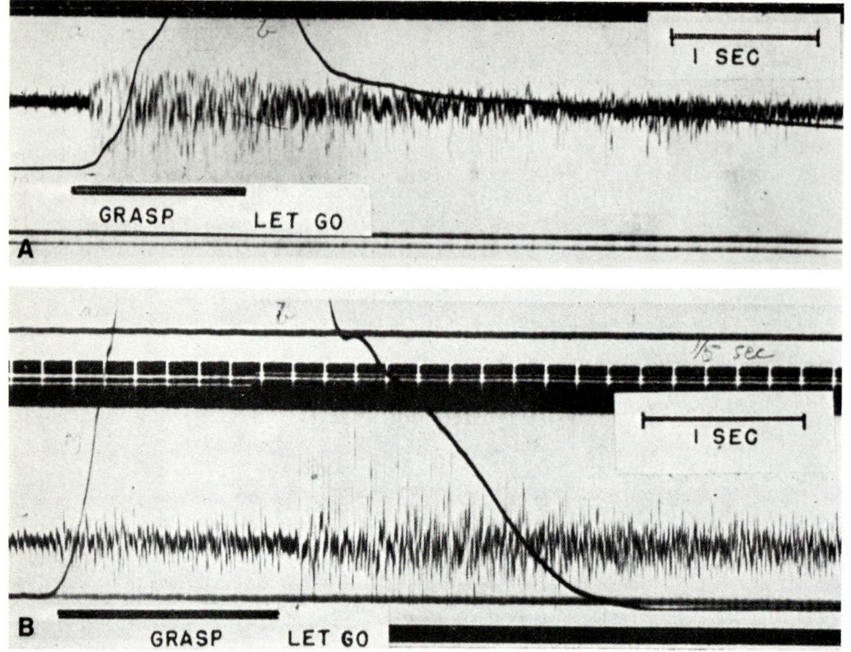

Fig. 12–8. Myotonia. *Upper tracing:* electromyogram of adductor pollicis, with recording of mechanical contraction of muscle (thin line). Patient grasped strongly and then attempted to relax. Full relaxation was reached 34 sec later. *Lower tracing:* mechanical record of a grasp of the hand and electromyogram of flexor sublimis digitorum. Note burst of large action potentials, though muscle had contracted feebly.

in other muscles it was provoked only by exposure to cold.

Thus Drager *et al* made a convincing argument for isolating paramyotonia as an entity distinct from congenital myotonia and dystrophia myotonica. These three syndromes could provisionally be viewed as a series in which myotonia is generalized as a spontaneous phenomenon in Thomsen's disease, limited usually to distal extremities, tongue, and face in myotonic dystrophy, and usually restricted to only the tongue in paramyotonia. Since the publication of the report by Drager *et al.* we have seen nine cases from four families with the typical syndrome as described by these authors. In all nine cases percussion myotonia was present only in the tongue, but in six myotonic spasm of orbicularis oculi muscles could always be elicited by closure of the eyelids; in another, myotonia of the levators could regularly be provoked by looking upward, and in yet another

by looking downward. The attacks of weakness were as described by Drager *et al.* and were associated with a rapid lessening of potential in successive contractions of some activated motor units of a type seen also in lesser degree in cases of dystrophia myotonica. Familial periodic paralysis in the past was reported in association with dystrophia myotonica by Bredemann[23] and by Stevens.[196] French and Kilpatrick[78] and we observed a similar family. It seems virtually certain that these are examples of Von Eulenberg's and not of Steinert's disease. The electrical change seen in Figure 12–8, in which a progressively smaller muscle potential follows a large stimulus potential, is a common finding in any type of myotonia. The attack of weakness therefore appears to be related to a progressive defect in polarization of the muscle fiber, unknown in type and independent of the phenomenon of myotonia itself. In our patients there has been no defect in plasma sodium level or in aldosterone excretion, and the essential change may have been in the muscle membrane characteristics.

Buchthal *et al.*[29] later reported on Gamstorp's syndrome noting the frequency of mechanical hyperexcitability of muscle fibers with trains of small potentials during attacks in their patients. There was also increased sensitivity to ACh (a prominent feature of myotonia). The rise of serum potassium was again noted but was less than 0.5 mEq/liter in 4 of 49 attacks. Artificial increase of the serum potassium levels of normal subjects to the highest levels observed in their cases did not induce weakness. A slight increase of serum sodium was also observed, together with positive sodium balances at the beginning of attacks in two subjects; this finding is of doubtful significance. Nevertheless, these authors still wish to maintain that adynamia episodica hereditaria is an entity distinct from paramyotonia.

The nature of the periodic brief attacks of muscular weakness of this type remains obscure. Shy[187] found on muscle biopsy that very few muscle fibers presented areas of vacuolation on cross section. Chemical analysis showed intrafibril loss of water and rise of potassium. The condition is certainly distinct from hypokalemic familial periodic paralysis, but whether it is identical with that observed in a large family by Tyler *et al.*[207] and in another by Poskanzer and Kerr[162, 163] is not established. This last family differed chiefly in the absence of myotonia and in the prominence of vacuolation in muscle fibers obtained by biopsy.

The consensus among myologists throughout the world now seems to be that the periodic adynamia and paramyotonia form a spectrum with episodic paralysis at one end and cold-sensitive myotonia at another, with attacks occurring at certain increases of serum potassium often within limits of normal. Most cases show both myotonia and periodic paralysis. Whether it is justifiable to separate hyperkalamic from normokalemic periodic paralysis, in consideration of the fact that the serum potassium does not correlate perfectly with muscle weakness in any of these conditions, cannot be settled until their mechanism is elucidated. It is noteworthy that myotonia has

only rarely been seen in individuals with hypokalemic periodic paralysis.

The attacks of paralysis in the Von Eulenberg-Gamstorp syndrome tend to begin earlier in life than those of the hypokalemic periodic paralysis and are also briefer and more frequent. In some families, as in the one reported by Bradley,[22] the attacks begin with pain, numbness, and tingling of legs. The weakness involves the muscles of legs and arms, with those of the upper parts of body and neck being less affected. Although weakening of the respiratory muscles is reported, death from asphyxiation or cardiac insufficiency must be extremely rare. A few patients remark upon diminished urinary output and increased thirst during the hours preceding the attack. The paralyzed muscles are sometimes firm and tender, and are inexcitable to stretch (loss of tendon reflex) and to electrical stimulus. Cold may induce either myotonia or paralysis or a mixture of the two, which probably explains the firmness of the muscles in some attacks. The serum CPK values occasionally rise slightly, but usually all serum enzymes are normal.

Attacks of this type are precipitated by cold and the ingestion of glucose has no effect. The glucose tolerance curves tend to be flat. Intraarterial injection of potassium may weaken only the muscles of that limb. Attacks may occur with serum potassium at 6.0 to 7.5 mEq/liter, and the critical level tends to be relatively constant for any given patient. The electromyogram during attacks shows diminished height and duration of amplitude of action potentials and increased numbers of polyphasic ones. The resting potentials of the muscle fibers are about 20 mV lower than normal (60–65 versus 85).

The biopsy reveals essentially the same charges as in hyperkalemic and normokalemic periodic paralysis except they are less pronounced. Interfibrillar glycogen is increased, and the sarcoplasmic reticulum is distended. The vacuoles are membrane-bound. In his thorough study of periodic paralysis Engel postulates an autophagic mechanism consequent to an unrecognized disorder of the electrically excitable membranes. The vacuoles are surrounded by membranes derived essentially from the t-system.

ALDOSTERONISM AND ITS RELATIONSHIP TO PERIODIC PARALYSIS

In 1955 Conn[40] drew attention to a clinical syndrome caused by excessive secretion of the mineralocorticoid aldosterone resulting from hyperplasia or tumor of the adrenal cortex. A loss of potassium from the body with increase in body sodium and metabolic alkalosis resulted in recurrent and severe muscular weakness, hypertension, paresthesias, tetany, excessive thirst, and polyuria. A number of cases were subsequently recognized, and by 1957 ten patients had been cured by removal of an adrenocortical tumor, and one had died from metastases. In many respects the clinical features of this syndrome of primary aldosteronism are similar to those of potassium-losing nephritis,[27] but the kidneys present only the tubular lesions of hypokalemia and hypertensive vascular changes.

In 1957 Conn and his associates[41] reported further studies on two unrelated young men, aged 19 and 24, with familial periodic paralysis. One of the patients was maintained on a rigid metabolic balance regimen for 11 months, the other for 30 days. It was found that attacks of weakness produced by glucose and insulin were preceded by great retention of sodium and accompanied by a sharp rise of urinary aldosterone. The urinary and serum levels of potassium fell sharply in a paralytic episode, but this "sequestration" of potassium was always preceded by intense sodium retention. The earliest return of muscle function was related to increased urinary sodium output, often when the serum potassium was still at a minimal value. When retention of body sodium was prevented by providing a diet low in sodium, attacks of paralysis could not be produced by glucose and insulin, or by potent mineralocorticoids. Muscle biopsies showed multiple vacuoles within muscle fibers. The vacuoles did not stain as fat or glycogen.

On the basis of these observations Conn and

his associates proposed that abnormally high concentrations of sodium within the muscle cell were responsible for family periodic paralysis and were related to an intermittent aldosteronism of unknown etiology. During attacks the potassium content of biopsy specimens of muscle was low and sodium content raised, and both levels were higher than in the intervals between attacks. These levels, repeated in five biopsies in one of the patients, were not greatly different from those found in primary aldosteronism due to tumor, except that the potassium was a little higher (81–85 mEq/kg) in familial periodic paralysis than in tumor cases (62–76 mEq/kg; normal, 91–107 mEq/kg). These values are sharply at variance with the increased muscle potassium found by Jantz[103] and Vastola and Bertrand[210] in family periodic paralysis. Jones et al.[106] more recently examined the sodium metabolism and aldosterone output in two cases of periodic paralysis, one familial, both with reduced serum potassium level in attacks, and both relieved by administration of potassium. In one patient an attack was induced without sodium retention. In the other patient sodium retention preceded the paralysis in one attack and accompanied it in two others, with associated potassium retention. In the first patient aldosterone excretion was high, in the second low, and in neither was there significant change before, during, or after an attack of paralysis. In the first patient paralysis was induced after a period on a low-sodium diet when he had lost 150 mEq sodium in 5 days. High retention of sodium by dietary increase plus aldosterone failed to produce paralysis. No muscle biopsies were obtainable. DeGraeff and Brocker[48] found sodium retention during attacks of family periodic paralysis but not necessarily preceding them. They were able to provoke attacks on a low-sodium diet.

From these studies and the observations made in earlier reports[6, 73] it can be concluded that some degree of sodium retention is common during attacks of family periodic paralysis but that it is by no means invariable and is not essential to the mechanism of paralysis. The sodium retention in Conn's striking cases was best seen when attacks were provoked by administration of a mineralocorticoid (2-methyl-

9-αfluorohydrocortisone) which itself has a sodium-retaining effect and possibly a direct muscular effect. Insulin and glucose produced a more striking retention of potassium than sodium, in line with previous work on family periodic paralysis. Van Buchem[209] described a case of primary aldosteronism with marked retention of sodium and serum potassium levels as low as 1.7 mEq/liter without paralysis, though fatigability was prominent. Analysis of muscle biopsy showed high sodium and low potassium content. He contrasted these findings to those of a typical case of periodic paralysis with normal aldosterone excretion. For these reasons we prefer to regard the carbohydrate fixation of potassium as the primary change in periodic paralysis, and intermittent aldosteronism as incidental, related to the presumed function of aldosterone as a compensatory regulator of electrolytes.

RENAL AND OTHER DISTURBANCES OF POTASSIUM (HYPOKALEMIA OR POTASSIUM LOSS; POTASSIUM INTOXICATION OR HYPERKALEMIA)

In his pioneer work on the effects of electrolytes on the heart in 1883, Ringer[169] found that either too low or too high a concentration of potassium ions caused the heart to cease beating. The effect could be antagonized to some extent by sodium and calcium ions. We already discussed the possible relationship of low serum potassium to family periodic paralysis. Low levels of serum potassium occur in the course of treatment of diabetic coma and spontaneously in chronic nephritis and a wide variety of other conditions, such as partial gastrectomy, primary aldosteronism, chronic enteritis, idiopathic steatorrhea, p-aminosalicylic acid therapy, and the use of ion-exchange resins. Conditions leading to the loss of large quantities of gastrointestinal fluids are a common cause.[62] Muscular paralysis in such states is rare, and more often apathy and muscular weakness with delirium, coarse muscular twitching, and tetany result. Electrocardiographic changes consisting of a progressive increase in the duration of systole, arrhythmias,

and finally arrest of the heart in systole are found. The QRS amplitude is increased, the QT interval prolonged, and the S-T segment depressed; there is also diminution, disappearance, or occasional inversion of the T wave. Severe paralysis of skeletal muscles is sometimes conjoined.[27, 57, 100] Respiratory difficulty with a peculiar "fish mouth" gasping may end in respiratory failure. The symptoms may be rapidly reversed if the serum potassium level is restored to normal, long before the total body potassium is replaced.[63] In the presence of renal insufficiency, however, there is great danger of causing a swing to hyperpotassemia by administration of potassium salts. Sinden et al.,[191] Elkinton and Tarail,[63] and Davidsen and Kjerulf-Jensen[47] discussed details of treatment.

The results of potassium intoxication (hyperkalemia) are more dramatic. Normally the potassium balance is maintained by an intake of just over 3.0 g a day and excretion of an equal amount. The normal kidney can excrete as much as 10.0 g within 8 hours, but when renal function is impaired any severe tissue trauma or a massive hemolytic reaction may liberate potassium from the tissues and lead to an accumulation of it in the serum. The serum potassium also becomes elevated in states of severe sodium chloride loss, such as salt-losing nephritis and diabetic coma. The electrocardiogram is a sensitive indicator of potassium intoxication, presenting first an elevation of the T wave, then a decrease in the size of R with an increase in the size of the S component, disappearance of the P wave and obliteration of the S-T segment, and finally a series of very large biphasic ST complexes.[74] These changes begin with serum potassium levels of 7–9 mEq/liter (27–35 mg/100 ml) and are fully developed at levels between 9.5 and 10.5 mEq/liter (37–41 mg/100 ml). Paralysis is seldom present and does not occur in animals dying of cardiac arrest or in some human cases with fatal outcome.[129] In the case of Sandkühler,[182] paralysis was present when the serum potassium level was 41.4 mg/100 ml; the patient recovered when the level was still 36.0 mg/100 ml. In a few cases, however, a most dramatic, severe, rapidly irreversible paralysis has been observed. Finch et al.[74] reported two such cases. Thickness of speech and difficulty in swallowing are followed by weakness and then paralysis of limbs, trunk, and neck muscles in the course of 1–2 hours. Only feeble movements of the fingers and toes and of the muscles supplied by the cranial nerves may be possible. Some subjective sensation of numbness, with impairment of sense of position and vibration and loss of tendon reflexes occurs. The heart becomes enlarged, the sounds indistinct, and the rhythm irregular. The skeletal muscles are flaccid and inert but may show twitches in response to direct percussion. Death results from cardiac arrest. Response to administration of calcium is slight, and the most effective treatment is administration of sodium chloride. Permanent recovery depends on controlling the underlying disease.

The paralysis of potassium intoxication is notable in its rapid onset, reversibility, and ascending progress. It is the most rapidly evolving type of Landry's ascending paralysis. The evolution and complete inactivity of the muscles, except on mechanical stimulation, resemble those of family periodic paralysis, in which, however, even mechanical stimulation eventually fails and cardiac disorder is usually minimal. The condition appears to result from inability of the end-plate mechanism to initiate an impulse in the sarcolemma, but direct evidence of this has not been obtained. The cardiac changes implicate the sarcolemma as the site of action rather than the end-plate. It is notable, however, that as in periodic paralysis, the muscles supplied by the cranial nerves are the last to be paralyzed, whereas in curare poisoning and myasthenia gravis the reverse is usually the case. This delicate difference in levels of susceptibility of neuromuscular transmission may result only from the effect of more constant muscular activity in diminishing potassium levels in the muscle cells.

CONGENITAL MYOTONIA (THOMSEN'S DISEASE)

Congenital myotonia was first brought to the attention of the medical profession in 1876 by Thomsen,[204] a Danish physician who suf-

fered from the disease himself. His description of tonic cramp during willed muscular movement leaves no doubt as to the condition about which he was writing. Over 20 members of his own family in four generations had suffered from it. He stressed its association with an inherited psychic disposition and was inclined to believe the disorder was nervous in origin. Strümpell[199] later (1881) assigned it the name myotonia congenita, and Westphal[214] in 1883 called it Thomsen's disease. Erb[67] presented a detailed description of three cases in 1886 and included a statement of the histologic findings in two of them in his classic monograph. In emphasizing the unique features of muscular excitability and hypertrophy, Erb first set the foundation of a muscular pathology for the disease. Jacoby[102] in the United States in 1887 and White[217] in England in 1889 confirmed these findings with excellent case reports. Of the many other case reports in the literature, those of Birt,[18] a Canadian physician and himself a sufferer, and Rosett,[172] who described a striking incidence in one family, may be mentioned as having both clinical and genealogic interest. Nissen[150] a grandnephew of Thomsen, continued the study of the original family and dealt particularly with the inheritance and the associated psychoses. Sanders[179] described a large family in the Netherlands. Thomasen[203] provided the most recent authoritative account of 19 living descendants of the family of Thomsen and 10 other cases, with a complete review of the syndrome. The association with psychosis is now believed to be fortuitous.

From the genetic studies of Becker and his colleagues[13] two forms of myotonia congenita are now recognized. In one, by far the most common, the inheritance is dominant, one parent and half the children being affected. The myotonia may begin at any period, even as late as adult life and is extremely variable in severity. About half of the patients are worsened by exposure to cold, but episodes of paralysis do not occur. Hypertrophy is absent or inconspicuous. In the other form the inheritance is autosomal recessive, and consanguinity is observed in 10–15% of such families. The sexes are equally affected. The myotonia has its onset at age 3–4 years and is severe. It begins in the limb muscles and ascends to involve those of the face. Later some weakness may be observed in the large muscles. These patients also may be worse in the cold. The family reported by Liebenam[122] is an example arising from coincidence of recessive genes.

In an earlier section on dystrophia myotonica we stressed the absence of dystrophic features in Thomsen's disease, which leads us to classify it separately as an inherited anomaly of muscular contraction and not as a dystrophy. De Jong[50] expressed a similar opinion after studying nearly 100 cases of both diseases. Maas and Paterson[127] recently reaffirmed their earlier contention that myotonia is but an aspect of dystrophia myotonica. They point out that in the course of time it has been possible to find some slight evidence of dystrophy (e.g., hollowing of the masseter muscles, an absent triceps reflex, some history of cataract in a relative, or a smallness of a testicle) in so many cases formerly accepted as myotonia congenita that Thomsen's disease has virtually ceased to exist. While we do not doubt that some meaning attaches to these slight signs patiently gathered by Maas in the course of an unrivaled collection of these syndromes, the striking differences in the natural history of myotonia congenita and dystrophia myotonica demand separate classification. Unequivocal dystrophic features imply an inevitably grave prognosis which is by no means attached to hereditary myotonia. Fragmentary evidence of possible dystrophic degeneration must be carefully weighed against the possibility of more banal coincidental cause. Instances have also been reported[33] in which myotonia with muscular hypertrophy persisted for years in one member of a family in which other members developed frank dystrophic features. Such exceptions indicate a close relationship between myotonia and dystrophy. Conceivably they could represent the antipodes of a spectrum with dystrophy at one end and myotonia at the other. At present one cannot refute the possibility that each is a separate disease.

The cardinal feature of Thomsen's disease is the phenomenon of myotonia, in combination with a more or less pronounced muscular hypertrophy. The myotonia usually manifests during childhood and in some instances may be traced back to the earliest months of life. The child may be late in learning to stand and walk because of it. More often the first difficulty in movement is noticed between the ages of 3 and 4. The muscular spasms become more intense during adolescence, and in typical families some members may not notice any spasms or other muscular disorder until after puberty, or indeed as late as the 23rd year. When the remainder of the family presents no advanced cases for study, such late onset should lead to suspicion that the disease is dystrophia myotonica. The complete absence of cataract or other dystrophic symptoms, the absence of progression in later years, and commonly a history of muscular hypertrophy since childhood indicate that the muscular abnormality has been present from infancy without having previously led to disabling spasms. It is therefore doubtful if "acquired myotonia" exists, for such cases usually turn out to be either true myotonia congenita, or dystrophia myotonica.

The degree of myotonia varies from case to case but is always more severe in the type with recessive inheritance and is also more severe than in myotonic dystrophy. It tends to affect all skeletal muscles, being especially prominent in the lower limbs; and in some cases it has been limited to these limbs. The patient complains of difficulty in initiating any movement after rest, and there is a painless stiffening of the limb with the first attempts to move after prolonged rest. The movement is slow in being recruited as well as delayed in cessation. With repetition of the effort the movement is gradually made more rapidly and relaxation is more prompt until after many trials it can be quickly and freely performed and is followed by natural relaxation. After a rest of some minutes, renewed attempts to repeat the movement again provoke some degree of muscular spasm, but prolonged rest is necessary before another movement initiates severe spasm. Loosening one set of muscles by repeated movement does not prevent the appearance of myotonia in another pattern of movement even though closely similar muscles or parts of muscles may be used in the latter. Thus the patient, having persisted in walking until his limbs are entirely free, nevertheless is immediately brought to a standstill by myotonic spasm when he begins to ascend a flight of stairs. The renewed myotonia is likewise loosened by continued climbing of the stairs, only to have further spasm slow the first steps when the patient again walks on a level surface. If the patient stumbles or trips when walking or running, the additional effort to maintain balance commonly results in a spasm of the whole trunk and limbs, so that he risks a fall to the ground as a rigid pillar.

In all but the mildest affections the upper limbs and face are also affected. The first grasp of the hand is made awkwardly and cannot be relaxed for 30–60 sec, during which time the muscles of the whole limb stand out in firm contraction. Small movements initiate myotonia only after prolonged rest, but any powerful effort sets up correspondingly severe spasm. Blinking may occur naturally, whereas a strong closure of the eyes may initiate a spasm that continues for more than a minute after attempted opening. A sudden movement such as a sneeze sets up a prolonged spasm of the muscles of the face, tongue, larynx, neck, and chest. Myotonia does not accompany the tendon reflex but has been reported to accompany more prolonged responses such as the abdominal and cremasteric reflexes.

Any skeletal muscle may be affected. Spasms of extraocular muscles may lead to convergent strabismus. Respiration may be embarrassed. There is no evidence that the visceral musculature is ever involved, and isolated instances of disorder of urination appear to be referable to myotonia of the striped external sphincter of the bladder. No cardiac abnormality has ever been reported. There is no evidence of endocrine disease. In distinction to Hoffmann's syndrome, myxedema is very rarely a feature of Thomsen's disease, and the myotonia is not affected by thyroid medication. There is no real resemblance to tetany.

The principle abnormality in any affected muscle is in the form of delayed relaxation. The same delay can readily be seen in the localized contraction set up by percussing the surface of the muscle with a percussion hammer, when the entire affected fasciculus (not just a transverse segment as in myoedema) remains tightened for many seconds before slowly relaxing. In myotonia congenita almost any muscle shows such percussion myotonia. Brief electrical stimulation through the skin also initiates a prolonged contraction, though a single shock fails to do so, as noted by Erb. Repeated electrical stimulation also abolishes the phenomenon for a time.

The muscles of patients suffering from Thomsen's disease are remarkably large and may give the patient a Herculean appearance. The hypertrophy affects particularly the muscles of the thighs, forearms, and shoulders but may involve the muscles of the neck and masseters. The tongue was not enlarged in the patients we studied. In complete relaxation the muscles have a natural consistency, but if myotonia is severe and persistent they feel firm and tense. Posture is not abnormal, and contractures do not occur. When the muscles have been freed of their myotonia by repeated movement, coordination and all other aspects of motor performance are normal. Muscular power may seem to be reduced, but this is related to the difficulty in initiating movements, possibly owing to an inability to relax antagonists. Power of resisting passive movement is strong in proportion to the size of the muscles, in marked contrast to dystrophia myotonica.

The only metabolic abnormality associated with myotonia congenita is an increase in creatine tolerance.[160, 161] This contrasts strongly with dystrophia myotonica, which like all muscular dystrophies reduces creatine tolerance. The explanation of this feature is that the hypertrophied muscles have greater ability to store creatine and convert it into creatinine in view of their correspondingly higher metabolism. No creatine is excreted in the urine, and the creatinine content of the urine falls in the high normal range. There is no change in the serum calcium level.

In the myogram of an affected muscle (Fig. 12–9) the tension of the muscle relaxes only very slowly. In the electromyogram this delay in the second half of relaxation is accompanied by very fine rapid electrical potentials, much smaller, faster, and more numerous than in normal motor unit discharges. These also accompany the delayed relaxation following electrical stimulation of the motor nerve below a novocain block. In the myotonia of goats the delayed relaxation with a long train of action

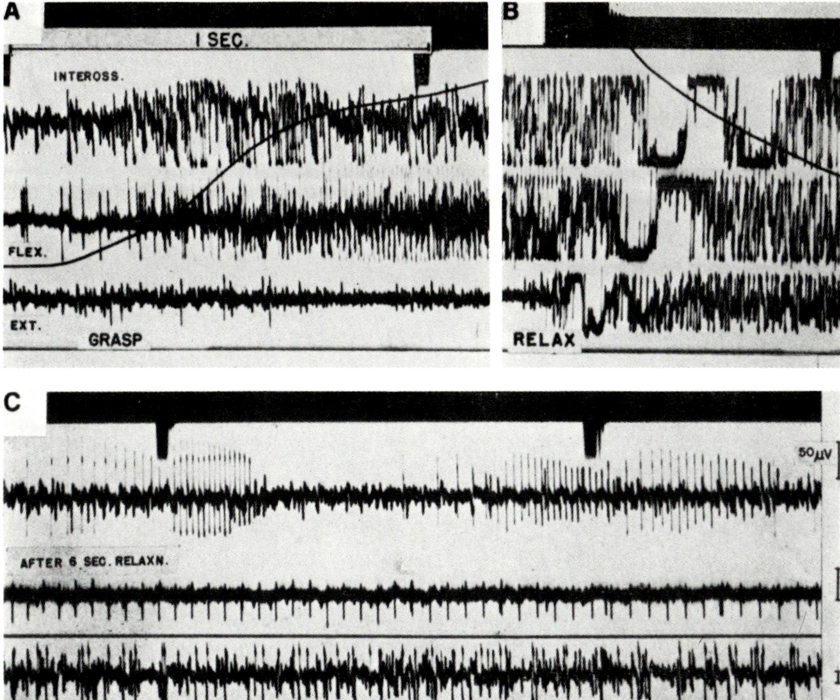

Fig. 12–9. Myotonia. Electromyograms from first interosseous muscle (Inteross.), flexor profundus (Flex.), and extensor communis (Ext.), with mechanical record of flexion of the fingers, (A) during the beginning of a grasp, (B) at commencement of relaxation (large slow fluctuations are due to movement of electrodes), and (C) 6 sec after relaxation, when myotonic rhythms are present in all three leads, and in addition, a few nerve action potentials in the extensor.

potentials has been found after curarization, and even after section and degeneration of the motor nerve.[26] The phenomenon is certainly muscular in origin, and the action potentials are of the same order of speed and size as those in muscular fibrillation, indicating that they represent independent incoordinate excitation of single muscle fibers. In critical electromyographic records it is clear that the fine muscle potentials developed during the preceding contraction may remain when the initiating discharge ceases (Figs. 12–10 and 12–11). When recording from single fibers Denny-Brown and Foley[54] found that one isolated excitation was unable to initiate the myotonic discharge, which appeared only after 2–20 repeated excitations depending upon the frequency of excitation (Figs. 12–8 and 12–2). It was possible to set up a minimal disturbance in this manner, and in this form myotonia con-

sisted of a rhythmic series of fibrillation potentials of diminishing size, corresponding to the estimated potential of a single muscle fiber (Fig. 12–8).

This isolation of the phenomenon of myotonia indicates that it is in the nature of an independent chemical contracture in a few of the fibers involved in the initiating excitation. It is not comparable to states of repetition of the initiating discharge following one excitation of the end-plate, as in veratrin contracture, but has the course and distribution of the cumulative effect of some by-product of the process of contraction.[54] The value of percussion in eliciting localized myotonic discharge is related to the brief but relatively intense repetitive excitation that occurs when a muscle is percussed. It is comparable to a brief tetanization of faradic shocks at a rate of 300 per second. After an interval of about 0.1 sec a number of the mus-

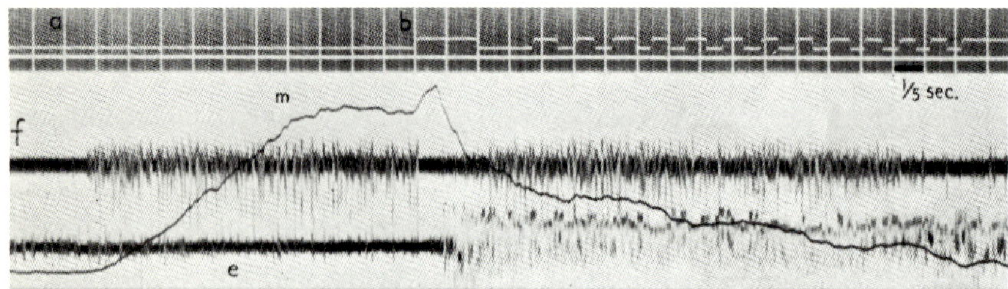

Fig. 12–10. Myotonia. Electromyograms from flexor profundus (f) and extensor communis (e) during a grasp from (a) to (b), and in relaxation following (b). Relatively insensitive electromyograms show myotonia as only a very small vibration, and large potentials starting 2/5 sec after beginning of relaxation are central neuronal discharge.

Fig. 12–11. Myotonia. Two leads were placed in a fasciculus of deltoid muscle about 0.5 cm apart. Fasciculus was stimulated rhythmically by 17 shocks in 0.5 sec, each of which gives a large deflection in the electromyogram. After the seventh shock a small train of action potentials at a rate of 136/sec appears near elecrode 1. Another gave two earlier beats. These abortive series are minimal myotonic contractions.

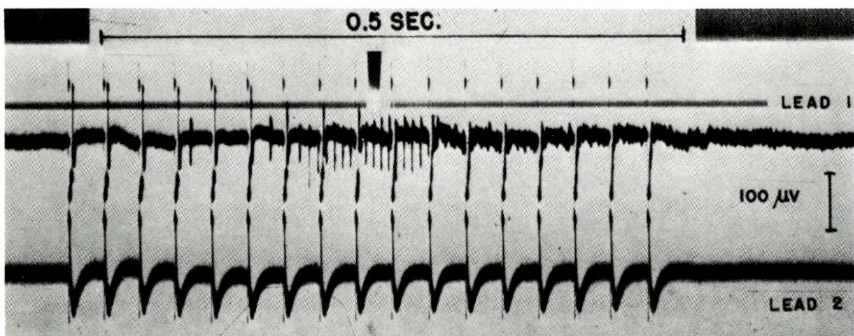

cle fibers in the excited fasciculus then exhibit a prolonged burst of myotonia (Fig. 12–13). The immediate contraction elicited by percussion differs in no way from that in normal muscle.

Although the trains of very small potentials of gradually diminishing frequency seen in myotonia have a striking appearance and a very characteristic "dive-bomber" sound when relayed to a loudspeaker, they differ only in intensity and in the circumstances of their appearance from other spontaneous activity of single muscle fibers. Their small amplitude make them difficult to pick up from surface electrodes, though when they are numerous and the skin thin they can be recognized as a high-frequency low-voltage discharge. Otherwise it could be possible that they result from irritation of irritable muscle by a needle electrode. It is likely that spontaneous series (Fig. 12–10) are the result of the distortion of the fibers by the needle electrode. Muscle fibers that are degenerating from any cause also show mechanical irritability, and insertion of the needle then provokes trains of potentials of similar aspect. Such insertion and electrode movement potentials are often seen in rapidly progressive types of muscular dystrophy, in myositis, and in denervation atrophy. The characteristic feature of myotonic potentials is therefore the requirement of some repetitive unit discharge in order to elicit them, and the large numbers that then appear. Conversely, in some cases of undoubted myotonia the characteristic po-

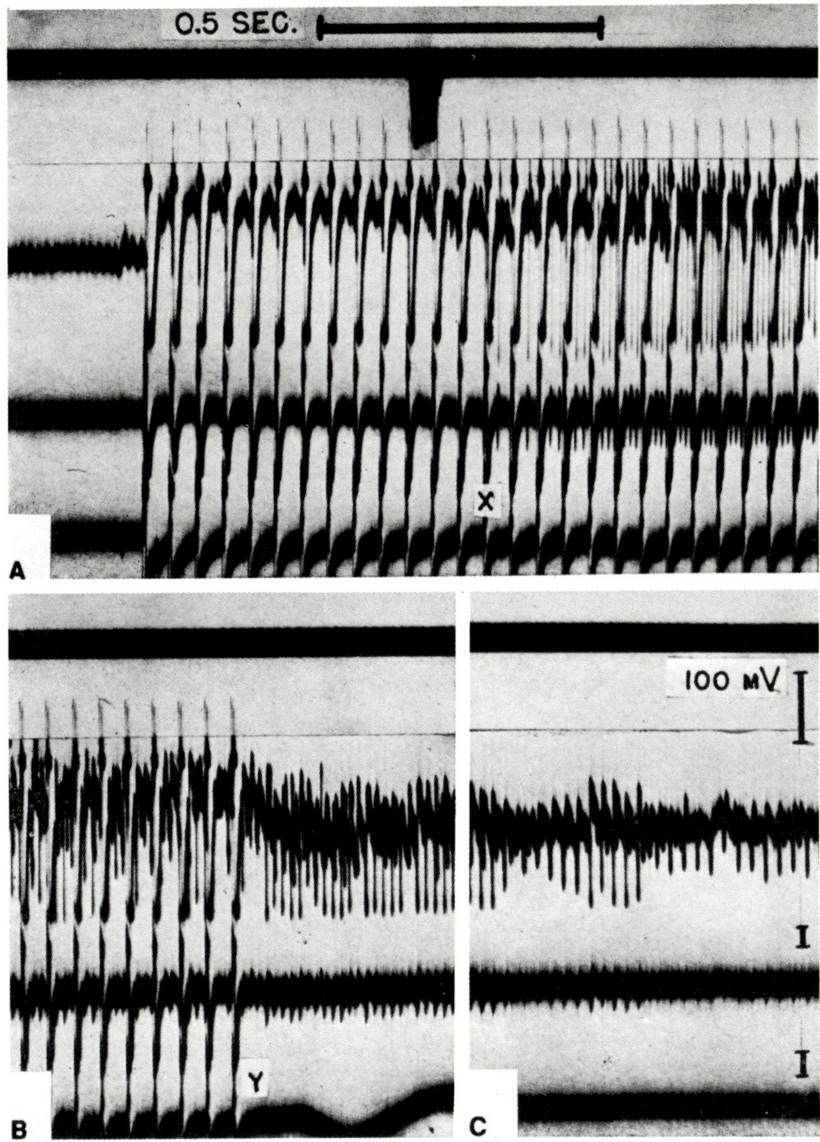

Fig. 12–12. Myotonia in single fibers. Arrangement was like that in Figure 12–8 but with (A) faster rate of stimulation. Myotonic discharge began at X and has been joined by two other series when seen 1 sec later at beginning of (B). At Y in (B) stimulation is stopped and a confusion of myotonic rhythms persists, still being present 5 sec later (C).

tentials may be difficult to demonstrate. In 1 such patient, for example, a young man of 25 with enormous muscles and complaints of stiffness in movement of the lower limbs after any prolonged rest, only doubtful myotonia could be produced by percussion, and several electromyographic recordings failed to reveal myotonia. Eventually it was found that myotonia potentials could be provoked with regularity only by intense contractions of less than 1 sec in duration. Yet his mother, who had had the same symptoms all her life without dystrophic change, showed intense percussion myotonia of many skeletal muscles and of the tongue. Since quinine can reduce the length and ease of elicitation of myotonia, it is assumed that other un-

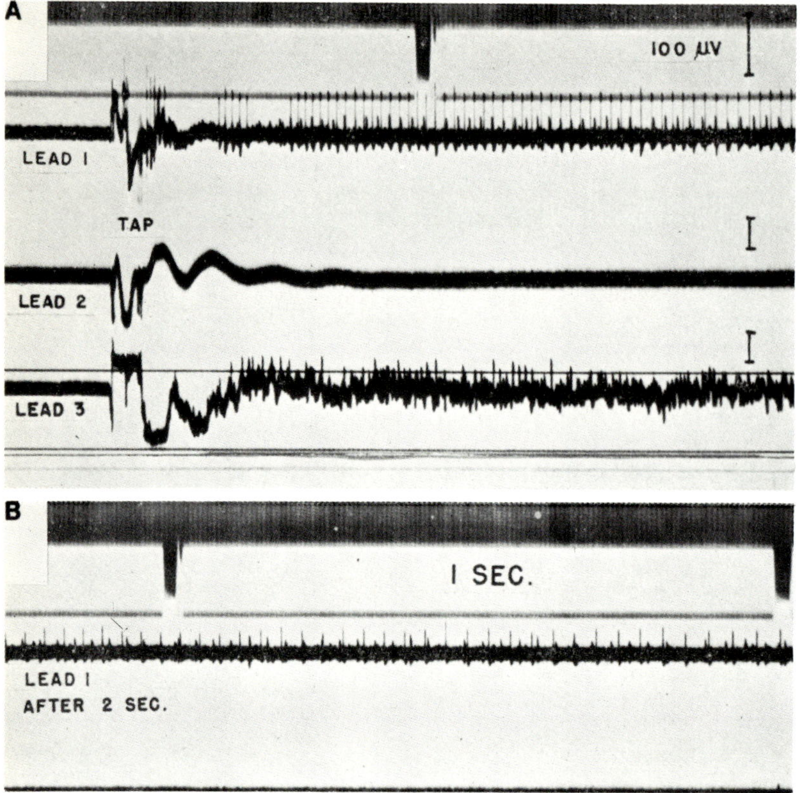

Fig. 12–13. As in Figure 12–8, with three electrodes. Percussion of other end of fasciculus (Tap) leads to momentary rhythmic discharge, followed by three trains of myotonia in lead 1, two of which are still present in B, 2 seconds later. In lead 2 there is no myotonia, and lead 3 records more confused rhythms in a nearby fasciculus.

known factors can also limit the phenomenon.

The phenomenon of myotonia, thus defined, is an independent by-product of the preceding contraction, affects only a small number of the muscle fibers initially excited, and is itself relatively weak in contractile force. Yet in Thomsen's disease, but also to a lesser and more localized extent in some cases of dystrophia myotonica, the observer is impressed by the active spasm that impedes the initiation of movement in the prime mover and at the cessation of effort involves whole groups of muscles not active in the initial effort. It may at first appear that the widespread afterspasm is an aspect of the effort to relax. With a cooperative patient, however, it is clear that simple relaxation of the grasp of the hand can alone set up an involuntary spasm in the forearm or whole limb, in flexors and extensors alike, more powerful than the original movement and incapable of modification by the patient. Such spasms are in fact the chief disability for they are responsible for the sudden immobilization of part or all of the musculature which temporarily disables the patient. Denny-Brown and Nevin[55] showed that this afterspasm in muscles remote from the prime mover is associated with an intense discharge of the large rhythmic action potentials of motor unit activity and is not necessarily preceded by motor innervation (Figs. 12–9 and 12–11). It is determined by activity of specific muscles in which muscular myotonia is well developed and can therefore be demonstrated most clearly in dystrophia myotonica. For example, when myotonia could be demonstrated only in the intrinsic muscles of the hand, such an afterspasm appeared in flexor carpi ulnaris only following a grasp of the hand and failed to appear after ulnar deviation. It did not appear at all in this muscle after a grasp of the fingers following Novocain block of the median and ulnar nerves at the wrist. Reflex afterspasm is easily fatigued by repetition, yet the peripheral myotonia still remains and can be elicited by percussion. Such muscular myotonia can then be seen to offer only feeble resistance to antagonistic muscles. In congenital myotonia the response to percussion is more powerful, but electromyographically combinations of peripheral and central action potentials also impede movement. From such observations Denny-Brown and Nevin concluded that the generalized afterspasms of Thomsen's disease (as distinct from the delay in relaxation of prime movers) are a reflex effect produced by an abnormal afferent proprioceptive discharge originating in myotonia of the prime movers. Though the disorder is primarily muscular, its reflex effects are therefore of considerable importance in the symptomatology. Such effects also afford along with hyperplasia of spindle fibers that the latter are affected by the disease. The high intensity of neuronal discharge thus caused in patients suffering from myotonia provides an adequate physiologic explanation for hypertrophy of the muscle fibers concerned.

Congenital myotonia has also been observed in certain strains of goats, known to farmers as "nervous," "stiff-legged," or fainting goats."[216] Such an animal becomes rigid and unable to move upon being startled after a rest of 15–30 min. The animal might fall to the ground, extremities stiff in extension for several seconds. The stiffness gradually subsides after repeated attempts to regain standing posture. Clark *et al.*[36] recognized the myotonic nature of the disorder, which is inherited and affects both male and female. They noted that the myotonia usually did not appear during the first month of life and disappeared during pregnancy.[114] Kolb found no pathologic changes in the muscles, which were only slightly larger than those of normal goats, an average width of 63 μ as compared to 58 μ, with the same range of fiber size. The number and distribution of sarcolemmal nuclei were unaltered. The myotonia of goats provides an opportunity for studying the

physiology and pharmacology of the nerve-muscle preparation. Kolb *et al.*[115] confirmed the effect of quinine and ionized calcium in lessening or abolishing the myotonia. Brown and Harvey[26] showed that the myotonia was not abolished by curare or by section and degeneration of the motor nerve. The ameliorating effect of quinine and calcium was demonstrated by closed arterial injection. Potassium and ACh excited myotonia when supplied directly to the muscle in this way. Lanari[120] also demonstrated the presence of myotonia to percussion and to direct electrical stimulation in human muscle curarized by intraarterial injection.

The effect of many substances on myotonia has been observed in patients suffering from congenital myotonia and dystrophia myotonica.[203] Wolf,[224] and Kennedy and Wolf[109, 110] discovered the beneficial effect of quinine, which is more useful in the treatment of congenital myotonia than of dystrophia myotonica, for in the latter disease weakness is the leading complaint. Calcium also lessens myotonia[178] probably by a central effect, as does procainamide.[88, 124] Passouant[157] found low levels of serum cholesterinase in myotonic dystrophy which were corrected by the administration of quinine and vitamin C. This observation, which seems incompatible with the transient relation of the disorder to localized activity, is probably only a reflection of the degree of associated atrophy.

In our studies of myotonic rhythms of single fibers we found that quinine, procainamide, and calcium lessened the duration of such bursts in single fibers, whereas ionized potassium and ACh provoked myotonia only in such concentrations as cause repetitive twitchings in normal muscle.[54] The electromyogram of a patient who no longer feels any myotonia as a result of quinine may still show many short bursts of the characteristic small action potentials during the first phase of relaxation. The abnormal discharge is still set up but is very brief. Quinine and calcium depress myotonia by virtue of their property of lengthening the refractory phase of excitable tissues in general. They do not interfere with the mechanism which sets up myotonia, and their effect is not of itself evidence that myotonia is an abnormal repetitive appearance of the initiating excitation, as postulated by Harvey[98] and others. Potassium and ACh evoke myotonia only because of their ability to excite contraction, which in turn can excite myotonia. Similarly the contracture which Lanari[121] produced by injecting ACh into the brachial artery can be interpreted as being secondary to the twitches which an injection of this size (0.02 g) sets up in normal muscle. There is no evidence of abnormality in the mechanism of action of these substances. The nature of the fundamental abnormality of myotonia is unknown. The fact that myotonia may be induced by certain substances such as 20–25 diazocholesterol, which are known to alter membrane resistance and to decrease Cl conductance, suggests that the defect lies in the sarcolemma. It is possible that it represents an alteration in the characteristics of the membrane of the muscle cell.

The effect of cold in increasing the disability is complained of by most patients suffering from Thomsen's disease and by a few suffering from dystrophia myotonica. Ravin[168] found no constant difference in the degree of observed or recorded myotonia when the limb was cooled, except that there was some greater slowness in initiation of movement. Such has also been our own experience. The difficulty is chiefly subjective. The validity of separating a group of cases as paramyotonia as done by von Eulenberg[69] was discussed above.

HISTOLOGIC CHANGES

In myotonia congenita (Thomsen's disease) biopsy does not reveal any abnormality other than hypertrophy of all fibers. In the beginning it was doubtful whether the increase in diameter was other than an expression of the intense and prolonged contraction of the living muscle. The condition is still present in muscle obtained postmortem,[49] however, and in life the muscles appear hypertrophied even in relaxation. The abnormal irritability of the muscle is evident for 10–15 min after its excision by biopsy. The threshold to stimulation is, however, not lowered, and the phenomenon is due only to the great delay in relaxation as compared with the

quick relaxation of the twitch following mechanical or electrical stimulation of normal muscle. If the muscle is fixed after allowing time for postmortem loss of contractility of the muscle fibers, the average diameter of muscle fibers in transverse section may be doubled,[102, 104, 217] or increased as much as threefold.[67] The increase is uniform throughout the muscle. In some fibers, as pointed out by Wohlfart,[223] a centrally placed nucleus is occasionally seen, but the long chains of such nuclei, which are characteristic of dystrophia myotonica, are not present. Some observers[67, 102] described a slight increase of endomysial connective tissue, but this has been denied by others. Jacoby[102] found no alteration in end-plates in sections using ordinary tissue stains. Coërs and Woolf[39] showed by the methylene blue method that the motor end-plates are extensively branched in the manner seen in hypertrophied muscle fibers in other conditions. The muscle spindles contain an excess of muscle fibers, sometimes numbering up to 30 or more. The proliferative changes seen in dystrophia myotonica are not seen in congenital myotonia.

Only two autopsy reports are available. Déjerine and Sottas[49] found no changes in the central nervous system at autopsy. Cell changes in the basal ganglia, hypothalamus, reticular substance, and dentate nuclei, noted by Foix and Nicolesco[75] in another case, do not appear to be of any significance as regards myotonia. The motor nuclei were intact.

The hypertrophied fibers are filled with myofibrils of natural appearance (Figs. 12–14 and 12–15), and no evidence of degeneration is observable. Contraction artifacts such as contraction bands are frequent in biopsy specimens (Fig. 12–14a). It is possible that these represent muscle fibers fixed while exhibiting the myotonic phenomenon. The series of small action potentials of diminishing size accompanying myotonia in one muscle fiber suggests that the abnormal contraction begins in a large extent of a muscle fiber and involves less and less of the fiber as it dies away. This indicates the presence of large contraction bands initially involving many sarcomeres and gradually diminishing to a small knot in the muscle. Ruptured or distorted fibers of the kind commonly

Fig. 12–14. Myotonia congenita. Sections from biopsy specimen of deltoid muscle. (phosphotungstic hematoxylin)

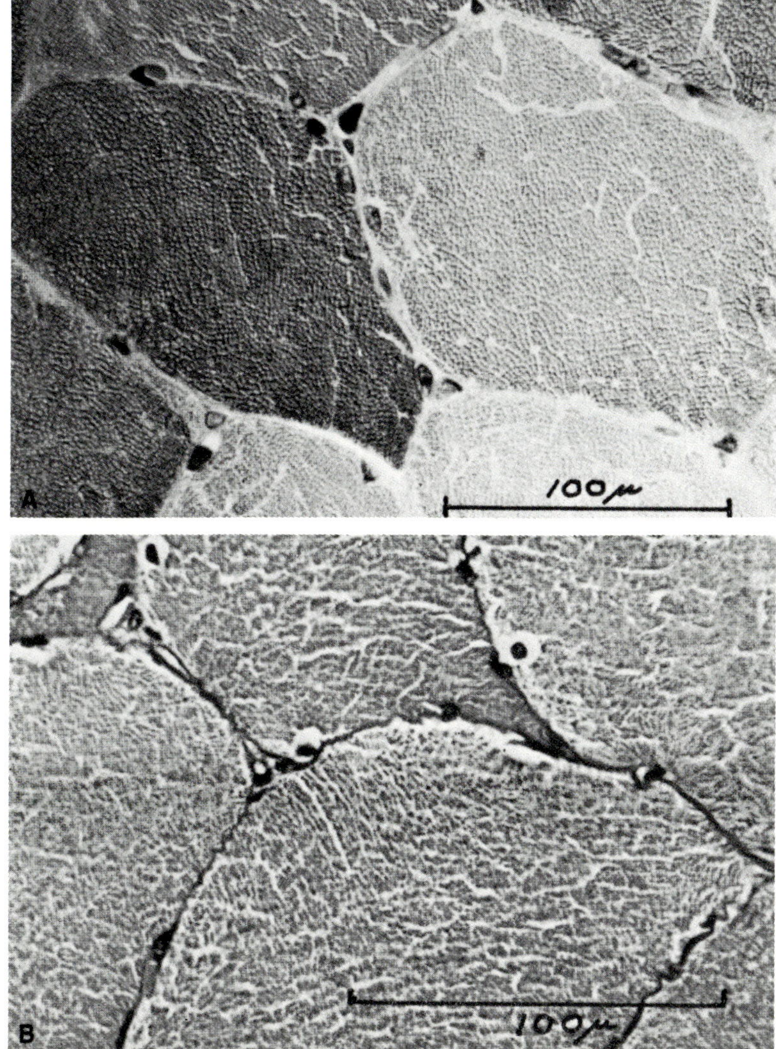

Fig. 12–15. Increased numbers of myofibrils in hypertrophied fibers. (A) Myotonia congenita. (B) For comparison, physiologic hypertrophy of gastrocnemius muscle in a patient with proximal polymyositis.

seen in dystrophia myotonica are not seen here, and their presence in that disease is therefore presumably related to the weakening of structure consequent upon dystrophic degeneration.

Schroeder and Adams[184] in excellent biopsied muscle from a patient with a dominant form of myotonia congenita were unable to find any definite abnormality in the conducting system or other structures under the electron microscope. The same is true of other forms of myotonia (Schwarz-Jampel syndrome, and neuromyotonia). Thus this peculiarity of the conducting membranes believed by Kohn[118] to be the result of a change in the pattern of phospholipids remains without morphologic counterpart.

HYPERTROPHIC MYOPATHY (HYPERTROPHIA MUSCULORUM VERA)

Aside from physiologic muscular hypertrophy and the enlargement of muscles that occasionally accompanies myxedema, other endocrine diseases, progressive muscular dystrophy, periodic paralysis, and Thomsen's disease, there is another entity: true muscular hypertrophy. Friedreich[79] and Auerbach[10] were the first to call attention to this condition, and additional cases were later contributed by Berger[14] and Fulda.[80] Talma[201] summarized the clinical and pathologic findings in 10 cases, 9 of which had been reported previously. Durante,[59] Woods,[226] and later Krabbe[116, 117] reviewed these same cases and added additional ones they had studied personally. Maxwell[131] reported one of the most recent examples of this condition.

The muscular hypertrophy usually begins during childhood, adolescence, or early adult life, though it has been seen in later adult life. Males are affected more than females. The muscular enlargement may be generalized or limited to half of the body, a limb, or a group of muscles. The musculature of the limbs is involved more often than that of the trunk, but any muscle may be affected, even the tongue. The onset of the muscular enlargement is insidious and the course progressive. Several muscles may be involved at the same time or suc-

cessively. The condition tends toward spontaneous arrest sooner or later. The strength of the hypertrophied muscles is usually increased, though in some instances weakness and excessive fatigue are present and the use of the affected limb is hampered by its large bulk and weight. The myotonic phenomenon was demonstrated in more than half of the reported cases. The tendon reflexes and electrical reactions are normal. Painful phenomena occur in many of the cases; these are vague, poorly localized, aching, drawing sensations for which no explanation is usually forthcoming. Objective sensory impairment is notably absent except in Talma's case in which sensation was lost in the arms. Vasomotor disturbances in the form of hyperhidrosis of a limb or one side of the head, or Horner's syndrome, have been reported on the side affected by the muscular disorder in several cases. Other neurologic disorders, except hemiathetosis in the questionable case of Pal[156] and epilepsy in those of Rybalktne (cited by Thomasen[203] and Krabbe[117]), have not been observed.

Other tissues do not participate in this hypertrophy. There is no enlargement of bones or trophic changes in the skin or subcutaneous tissues. In this respect it differs from the localized macrosomia or hypertrophy that accompanies some neuroectodermal disorders such as angiomatosis or neurofibromatosis.

The muscles in these cases are normal in gross appearance, except for their large size. Microscopically the only definite change is enlargement of the muscle fibers. When measurements have been made, as in Auerbach's case in which a biopsy of a biceps muscle was obtained, the fibers ranged from 96 to 200 μ in diameter (average 165 μ). The striations were completely preserved, as a rule, and both myofibrils and sarcoplasm are increased in the usual proportions. The sarcolemmal nuclei are said to be normal by some writers and increased slightly in size and number by others. Occasionally nuclei are found in the central portions of the fiber. Durante noted some of these nuclei to be surrounded by clear, granular basophilic sarcoplasm which in transverse sections took the form of rings or semicircular crescents of protoplasm. Longitudinal division of fibers, accounting for an increase in number as well as size of fibers, was claimed by Durante. Degenerative changes were not mentioned in most case reports and were a rare finding in others. The walls of the muscular arteries showed a slight thickening, incorrectly termed endarteritis and periarteritis by some writers. An excess of fat or connective tissue was not observed.

DELANGE'S SYNDROME

Another closely related form of true muscular hypertrophy was described by Bruck[28] and deLange[51] and is sometimes known in pediatric literature as deLange's syndrome. The latter writer drew attention to a clinical entity consisting of congenital muscular hypertrophy, extrapyramidal motor disturbances and mental deficiency. From the time of birth it was noted that all the muscles or particular muscle groups (e.g., deltoid, biceps, triceps, and pectoral) on both sides were unusually large and firm. The patients looked like infant wrestlers. In some instances the head was misshapen, being small or deformed. Retraction of the head and a pronounced rigidity of muscles which fluctuated in degree from time to time was found in all of deLange's cases and was interpreted as extrapyramidal rigidity since the tendon reflexes were not exaggerated. All patients died within a few months. The muscles were not studied in deLange's case, but the brain was carefully examined. Numerous cavities were found all through the cerebral white matter and in the thalami; there was polygyria and microgyria; the striatum and the corticospinal tracts had failed to develop properly. deLange rejected the notion that the muscle hypertrophy resulted from rigidity or athetosis occurring in utero but could offer no other explanation for its conjunction with extrapyramidal rigidity. She differentiates this generalized muscular hypertrophy from other localized congenital forms as reported by Kitaigorodskaja.[113]

VALIDITY OF "TRUE" HYPERTROPHY

Only the most tentative conclusions concerning true muscle hypertrophy can be drawn at this time, for the reported cases leave much to be desired both from the clinical and pathologic points of view. In the 20 or more cases reported, one looks in vain for a thorough pathologic study of the nervous system and muscles. The few remarks about morphology of the muscles have been based on examination of biopsies of single muscles; and in these, enlargement of muscle fibers has been the only constant finding. Very few of the patients were under observation for a sufficient time to allow determination of the clinical course of the disease. In many of the reported cases the muscular hypertrophy was associated with myotonia, and the syndrome has been said to differ from Thomsen's disease only in the absence of family history. Since cases of other hereditary diseases are known to occur sporadically, it is by no means clear that "acquired myotonia" with muscle hypertrophy differs from Thomsen's disease. Although it has been suggested that the myotonia is a consequence of muscle hypertrophy,[117] it is doubtful if this statement can be substantiated either by physiologic studies or clinical observation. Indeed the reverse seems more probable. The possibility that some of these cases are a form of progressive muscular dystrophy cannot be excluded, especially when it is remembered that in some genealogies a state of true muscular hypertrophy may precede weakness and muscular atrophy by some years, as in the interesting family of Barnes[12] and others the author has personally examined. Moreover, enlargement of the calf muscles in the typical Duchenne dystrophy involves a large element of hypertrophy, as was first pointed out by Erb. This diagnosis can be affirmed only by a long period of observation. In deLange's cases it is uncertain whether the enlargement and prominence of muscles is greater than that commonly seen in states or rigidity, double athetosis, and dystonia, and differs therefore from a physiologic hypertrophy.

The proposition of Krabbe[117] that true muscle hypertrophy may occur during recovery from peripheral neuropathy (postneuritic muscular hypertrophy) has not been established. Even though a number of the reported patients have had pains, in relatively few was paralysis, areflexia, or sensory loss demonstrated, either before the onset or during the development of muscular hypertrophy. We admit to the finding of hypertrophy of individual fibers, perhaps as a physiologic hypertrophy of residual sound fibers, in many forms of neural and spinal atrophy and muscle dystrophy but doubt whether this could account for a uniform enlargement of all the fibers within a muscle in the absence of gross mechanical defect. Reduplication of action potentials for a time following peripheral neuropathy is common, and this phenomenon conceivably could account for hypertrophy in such conditions.[53] Any patient with muscular hypertrophy must be scrutinized with great care to rule out myopathic disease such as myotonia congenita and progressive muscular dystrophy, though we believe that benign familial largeness of muscles is not uncommon.

LOCALIZED HYPERTROPHY

The configuration of particular muscles or groups of muscles is highly individualized, just as is the size of muscles in general. Some persons have unusually large calf, thigh, gluteal, or shoulder muscles, the basis of which has never been ascertained.

Hypertrophy of the masseters, which is a natural phenomenon in certain heavily jowled individuals, is an example. Moreover, habitual clenching and grinding of the teeth (bruxism) may cause work hypertrophy of the masseters, just as spasmodic torticollis leads to a "work hypertrophy" of the sternomastoid. In addition, there is a rare condition of masseter hypertrophy not dependent on overaction of such muscles. It may reach such proportions as to deform the face, requiring subtotal excision. The fibers of such muscles are enlarged, but it is not clear whether the condition is due to hypertrophy of individual fibers or hyperplasia.

OTHER DISEASES PRODUCING MUSCLE PARALYSIS AND SPASM

Muscle spasm and paralysis are the outstanding symptoms of a number of other disease states, but since these disorders are reversible and involve no demonstrable pathologic changes, they are referred to only briefly here.

TETANUS

Tetanus is an infectious disease caused by the toxin of the bacillus *Clostridium tetani* and is characterized by painful spasms of striated muscles. It usually follows an injury in which the tetanus bacilli are introduced into the wound. The symptoms begin after an incubation period of 1–2 weeks and consist of general restlessness and irritability followed by tightness of jaws, difficulty in chewing and swallowing, and stiffness of arms and legs. Later the masseters become firmly contracted so the mouth cannot be opened (trismus) and the neck and abdomen assume a boardlike rigidity. Generalized spasms initiated by noise or other sudden stimuli are accompanied by profuse sweating and cyanosis. Death occurs within a few days in about 15–30% of cases. Rarely the tetanus is at first localized to muscles in the region of the wound, and in such cases a localized contraction persists for some weeks after the period of intense spasms has passed.

The disease is entirely due to the effects of an exotoxin elaborated by the tetanus bacillus. This is one of the most powerful soluble poisons known; 0.22 mg is a fatal dose for man. The toxin has a great affinity for the nervous system and in man and monkey appears to reach it directly from the bloodstream. The cat and rat are more resistant, and even after moderately high dosage by inoculation the spasm remains localized to the general region of the segmental nerve supply of the region of inoculation. The spasm may then be delayed or prevented by section of the dorsal nerve roots to a region receiving a small but otherwise effective inoculation.[79] In such local tetanus the spasm begins within 36–72 hours and is at first manifested by a heightening of the stretch reflex and certain other local reflexes. Within 24 hours of its onset the affected motor neurons are discharging continually.

Our own experiments are in complete agreement with those of Acheson *et al.*[1] After 2–3 days of continuous spasm the affected muscles show a fixed shortening even if the motor nerve is then sectioned, a condition said to persist for several days, until the peripheral segment of the nerve undergoes wallerian degeneration. This is the contracture of tetanus, discussed earlier in relation to contracture in general. It appears likely that the contracture is related to a change at the neuromuscular junction. It is not associated with any remarkable histologic change, and the possibility of ultimate recovery shows the essential disorder to be reversible in nature.

There is no convincing neuropathologic change in this disease. The brain, spinal cord, and peripheral nerves appear to be normal. In experimental studies the motor nerve cells in the spinal cord reportedly show a diminution in the cytoplasmic and nuclear proteins. This cellular alteration is not evident in Nissl staining of human spinal cord. There are no convincing changes in skeletal muscles; many of those reported are probably agonal or post-mortem artifacts.

BOTULISM

Botulism is an acute bacterial intoxication caused by the toxin of *Clostridium botulinum.* The disease is characterized by weakness or paralysis of muscles and frequently death within 18–36 hours after ingestion of poisoned food. The earliest symptoms are lassitude and fatigue, and sometimes dizziness and headache. Vision becomes blurred owing to paralysis of accommodation. Diplopia due to weakness of ocular muscles, difficulty in swallowing and speaking, and generalized muscular weakness follow within a few hours. Death usually results from respiratory failure. Gastrointestinal symptoms are usually conspicuously absent; the patient is afebrile throughout the brief illness. The sensorium remains clear to the end, and there are no sensory disturbances. If recovery occurs, the return of the power of contraction of muscles is extremely slow.

This disease, like tetanus, is caused by an exotoxin elaborated in spoiled food by *Clostridium botulinum.* The toxin alone reproduces the essential features of the disease. It is by far the most potent soluble toxin known; 0.05 mg is fatal to man. The paralytic action of botulinus toxin was investigated by Ambache,[7] who demonstrated that appropriate doses in rabbits and cats produce a total paralysis of cholinergic motor fibers and affect little or not at all the adrenergic motor fibers and sensory fibers. When voluntary muscles are paralyzed by intramuscular injection of the toxin, stimulation of motor nerves no longer evoke muscle contractions, though the muscle still responds to intra-arterial injections of ACh. Masland[131] obtained electromyographic records of muscles locally paralyzed by this toxin and found that the initial response of the muscle to nerve stimulation was reduced. In this respect the electromyogram differs from that of a curarized muscle in which after a normal initial muscle response there is a rapid decline in action potential. The nerve conducts normally, and the muscle responds to direct stimulation. These and other findings suggest that botulinus toxin exerts its paralytic action by a selective peripheral effect on the terminals of the axon.

There are no definite gross or microscopic lesions in botulism. Although congestion of blood vessels within the central nervous system has been noted, there are no infiltrations of inflammatory cells or destruction of nerve cells. No convincing changes have been found in striated muscles other than atrophy.

SPIDER BITE

The "black widow spider" of the species *Latrodectus* produces a toxin that affects predominantly the neuromuscular system and may prove fatal. Within a few minutes after the bite the muscles of the abdomen, back, and lower extremities exhibit cramps and muscular spasms and finally an extremely painful rigidity. Bogen[20] observed in many cases urticaria, circulatory collapse, neusea, and vomiting, as well as the recurrent muscle spasm. If death does not occur, the symptoms usually subside in 24–48 hours, and recovery is complete. Presumably this heightened excitability of the neuromuscular apparatus is induced by a toxin injected at the time of the spider bite. Little or nothing is known of the nature or site of action of this toxin or of any histologic or histochemical changes produced by it in the peripheral nerve or in the muscle fibers. Greer[92] reviewed this subject.

TICK PARALYSIS

Certain species of ticks elaborate a toxin which if injected into humans causes a pronounced weakness of the muscles of the legs, arms, and respiratory and cranial musculature. The toxin is believed by Ross[173] to act on the central nervous system, since the paralyzed muscles continue to respond to faradic stimulation of the peripheral nerve. These experiments should be confirmed and extended, but the available data suggest that the tick toxin acts on central neural mechanisms much in the same way as does dithiobiuret, which Astwood and his associates[9] found induces a state of muscular paralysis without blocking neuromuscular transmission and without demonstrable histologic change in the peripheral nerves, spinal cord, and striated muscles. Murnaghan[146] produced evidence that tick paralysis is indeed peripheral; it is relieved by closed arterial injection of ACh and in other respects behaves like a defect in ACh liberation. Extracts of ticks do not cause paralysis. Evidently continuous production of the "toxin" is necessary for maintenance of paralysis, since rapid recovery follows removal of the tick. Other toxins such a cobra venom paralyze the neuromuscular apparatus more peripherally.

CRAMPS AND OTHER STATES OF MUSCULAR HYPERACTIVITY

MYOKYMIA AND PERSISTENT MUSCULAR ACTIVITY AND PATHOLOGIC CRAMP STATES

Cramps in leg muscles, especially during the night when the feet are cold, are commonplace. Nearly everyone has them sometimes. They are known to become more frequent after a day of strenuous activities, and also after sweating or salt depletion. They may interfere with work or prolonged athletic performance after heavy sweating. For reasons that are unclear, the tendency to cramp, whether at night or during exertion, increases in the majority of cases of motor system disease. In a cramp, the offending muscle hardens, the parts to which the contracted muscle is attached are fixated, and there is pain. Fasciculations usually become more prominent at times when one is disposed to cramp.

There is a group of patients whose principal complaint is that their muscles cramp after nearly every exertion. First one muscle and then another is seized by spasm, or all muscles may be involved for a period of time in a crisis of cramping. When this phenomenon is repeated again and again, the overly active muscles begin to hypertrophy. In the analysis of such patients it is found in some that ischemia of an arm or leg readily induces carpal or pedal spasms, as well as tingling, prickling paresthesias (Trousseau's sign). Also, tapping over a nerve produces a muscle contraction (in the face the Chvostek sign). The electromyogram during the cramp reveals bursts of high-voltage, fast-frequency action potentials (doublets and triplets) which persist during periods of attempted relaxation. These latter changes are ascribable to an instability of the polarization of the axonal membrane of the nerve fiber. This constellation of findings is known to be characteristic of tetany and is explained by the reduction in ionizable serum calcium. But in these cases of pathological cramp the calcium levels are normal. A feature common to both tetany and pseudotetany, if one may use such a term for pathological cramping, is that stimulation of muscle through its nerve at certain frequencies (10–15 cycles per second) will reproduce cramp in some but not in all the cases. When relaxed, the muscle is soft and electrically silent, except in the most severe degrees of tetany or pseudotetany, where bursts of muscle activity can be observed more or less continually.

The term *myokymia* has been applied to a constant rippling activity seen in the skin over muscles, due to the spontaneous contraction of one fascicle after another. Limited to one muscle like the orbicularis oculi or a hand muscle for a few hours or days it is always benign. This state is known to the laity as "live-flesh." But the condition may be universal and associated with severe cramps, hyperhydrosis and myotonic-like contractions, and then it is clearly pathological. There is difficulty at times in distinguishing myokymia from the pathological cramp syndrome. The abnormality in both is thought to be due to an instability of the membrane of the nerve fiber.

Isaacs and others have called attention to a closely related but unique syndrome of constant muscle activity simulating myotonia. It persists in sleep and general anesthesia (unlike myotonia), and is blocked by curare, depolarizing drugs and improved by diphenylhydantoin. There is no evidence, clinically or electrically, of excessive irritability of the muscle membrane. A burst of discharges is often followed by a brief period of electrical silence, after which the continuous muscle activity returns. In all these respects the condition differs from myokymia and from true myotonia. The only morphologic change is a slight thinning of scattered muscle fibers and relative increase in nuclei.

Closely related to neuromyotonia and myotonia is a third hereditary state of continuous muscle fiber activity in association with short stature and muscular hypertrophy (Schwartz-Jampel syndrome). Here the block or interruption of nerve diminishes but does not arrest the muscular activity, and percussion, needle insertion, and electrical stimulation reveal increased irritability of the muscle membrane. In this respect it resembles true myotonia, but differs in that the muscles are continuously active at rest. Both neural and muscular factors are postulated.

Unlike tetanus and the "stiff-man syndrome," wherein there is believed to be a disinhibition of alpha neurones (although in tetanus an additional myopathic factor has been postulated) these states of continuous muscular activity seem to involve an instability of the membranes of axons as well as muscle fibers. Their differentiation from one another is incomplete and the status of neuromuscular junctions is uncertain. Conditions in which there is overactivity of anterior horn cells, such as spasticity, rigidity, and dystonia, may be difficult to distinguish from these conditions of continuous overactivity of muscles, but usually there are other signs of disease of the central nervous system. However, we have observed a few patients in whom there is concurrence of the two conditions.

In none of these states of hyperactivity of muscle have the various investigators been able to define a consistent abnormality in the motor end-plate or the conducting mechanism of the muscle fibers. We observed spiraling of the superficial myofibrils in three of our five cases of pathologic cramp and pseudotetany (tetany without abnormality of serum calcium) but concluded that it was indicative only of hyperirritability of the biopsied muscle. No electron microscopic investigations have been undertaken.

"STIFF-MAN" SYNDROME

In 1956 Moersch and Woltmann[144] described 14 patients, seen over a period of 30 years, in whom the major symptoms had been a progressive stiffness and rigidity of the neck and abdominal muscles with superimposed muscular spasms. Ten of the patients were males and were severely affected. The onset of the disorder had been between the ages of 28 and 54 years, and the condition was followed over a period of 2–14 years without change. The proximal muscles of the limbs were involved in some cases and were the first to be affected in three. Neurologic examination disclosed only a fluctuating rigidity and spasm without changes in reflexes or any of the postural abnormalities of extrapyramidal diseases. The spasms could be precipitated by sudden stimulus or movement, but no heightening of reflex activity was demonstrated. There was no evidence of tetany or myotonia. An unknown reducing substance was found in the urine of four patients, but all other laboratory tests were negative. An autopsy limited to the brain and viscera in one case revealed no abnormality. Asher[8] described a similar condition in a woman who had had similar spasms for 15 years. A postmortem examination is said to have shown areas of atrophy and necrosis in muscle, with fibrous

tissue replacement and infiltrations of lymphocytes. Ischemia was thought to be the cause of
the necrosis. A further case described by Price
and Allott[164] showed no metabolic abnormality
except a raised level of inorganic phosphate in
the blood during muscle activity. The prevailing opinion is that this is a neurologic disorder
with disinhibition of alpha motor neurons in
the spinal cord, akin to that of tetanus.

CONCLUSION

The miscellaneous diseases of the muscular system illustrate the difficulty in classifying muscular disease in general. The gaps in knowledge
of etiology or even the site of the lesion in
many muscle diseases are so numerous that
classification is impossible. It is clear, however, that these more obscure forms of disease
reflect the great variety of independent metabolic processes that concern muscle tissue.

Most satisfying to the pathologist are those
diseases attended by a morphologic change in
the substance of the muscle fiber, such as
Zenker's vitreous necrosis or glycogen storage
disease. Most difficult are those diseases which
comprise a disorder of neuromuscular transmission, both at the motor end-plate (myasthenia gravis, botulism) and in the excitable
membrane of the muscle cell (periodic paralysis, hyperpotassemia, myotonia). Perplexing
also are the disturbances presumably of the
contractile mechanism itself, as occurs in adrenal insufficiency, and possibly in some hyperthyroid syndromes, myotonia, and hypertrophia
musculorum vera. Such a list not only illustrates the many different metabolic processes
in which muscle tissue is engaged, quite apart
from those underlying the contractile process
per se, and it stresses that contraction is only
one of the many biologic functions of muscle.
Only in a few disorders (e.g. phosphorylase
and acid maltase deficiency) has the specific
metabolic defect been clearly delineated.

We indeed considered the feasibility of describing these diseases in purely physiologic
terms. Further consideration, however, makes
it clear that such a plan, though attractive, is
premature and gives a misleading impression of
finality. Close inspection of the relevant data
concerning each of these diseases discloses the
extremely hypothetical nature of such grouping
and of its basis in current schemes of fundamental physiology. The chief purpose of this
chapter is well served if it stimulates the reader
and leads him to further study and investigation of the basis of pathophysiological muscle activity.

REFERENCES

1. ACHESON GH, RATNOFF OD, SCHOENBACH ER:
   Localized action on spinal cord of intramuscularly
   injected tetanus toxin. J Exp Med 75:465–480,
   1942

2. ADAMS RD: Thayer lectures. I. Principles of myopathology. II. Principles of clinical myology.
   Johns Hopkins Med J 131:24–63, 1972

3. AITKEN RS, ALLOTT EN, CASTLEDEN LIM, et al.:
   Observations on a case of familial periodic paralysis. Clin Sci 3:47–57, 1937

4. ALAJOUANINE T, CASTAIGNE P, NICK J, et al.: Sur
   l'existence de signes sensitifs et sensoriels au cours
   de la myasthénie. Rev Neurol (Paris) 96:242–248,
   1957

5. ALAJOURANINE T, HORNET T, THUREL R, et al.:
   Examen anatomique d'un cas de myasthenie a
   évolution intermittente pendant 27 ans. Rev.
   Neurol (Paris) 65:552–559, 559–564, 1936

6. ALLOTT EN, MCCARDLE B: Further observations on
   familial periodic paralysis. Clin Sci 3:229–239,
   1938

7. AMBACHE N: The peripheral action of C1 botulinum toxin. J Physiol (Lond) 108:127–141, 1949

8. ASHER R: A woman with the stiff-man syndrome.
   Br Med J 1:265–266, 1958

9. ASTWOOD EB, HUGHES AM, LUBIN M, et al.: The
   reversible paralysis of motor function in rats from
   chronic administration of dithiobiuret. Science
   102:196–197, 1945

10. AUERBACH L: Ein fall von Wahrer Muskelhyper-
    trophie. Virchows Arch [Pathol Anat] 53:397–
    417, 1871

11. BACQ: Quoted by McArdle[132]

12. BARNES S. Report of a myopathic family with
    hypertrophic, pseudohypertrophic, atrophic and
    terminal (distal in the upper extremities) stages.
    Brain 55:1–46, 1932

13. BECKER PE: Zur genetik der myotonien. Progres-
    sive Muskeldystrophie-Myotonie Myasthenie.
    Edited by E Kuhn. Berlin-Heidelberg, Springer-
    Verlag, 1966, pp 247–255

14. BERGER O: Zur aetiologie und pathologie der sogenannten muskelhypertrophie. Dtsch Arch Klin Med 9:363–396, 1872

15. BICKERSTOFF ER: Periodic paralysis. J Neurol Neurosurg Psychiatry 16:178–183, 1953

16. BICKERSTAFF ER, WOOLF AL: The intramuscular nerve endings in myasthenia gravis. Brain 83:10–23, 1960

17. BIEMOND A, DANIELS AP: Familial periodic paralysis and its transition into spinal muscular atrophy. Brain 57:91–108, 1934

18. BIRT A: A study of Thomsen's disease (congenital myotonia) by a sufferer from it. Montreal Med J 37:771–784, 1908

19. BLALOCK A: Thymectomy in the treatment of myasthenia gravis: report of 20 cases. J Thoracic Surg 13:316–339, 1944

20. BOGEN E: A study of spider poisoning. JAMA 84:1894–1896, 1926

21. BORNSTEIN M: Ueber die paroxysmale Lähmung (versuch einer theorie). Dtsch Z Nervenheilk 35:407–427, 1908

22. BRADLEY WG: Ultrastructural changes in adynamia episodica hereditaria and normokalemic familial periodic paralysis. Brain 92:379–390, 1969

23. BREDEMANN W: Ueber paroxysmale Lähmung bei dystrophischer myotonie. Artz Wochenschr 7:202–205, 1952

24. BRISCOE G: The antagonism between curarine and prostigmin and its relationship to the myasthenia problem. Lancet 1:469, 1936

25. BRISSET C: Paralysie periodique de Westphal; sept cas dans une famille. Rev Neurol (Paris) 88:45–56, 1953

26. BROWN GL, HARVEY AM: Congenital myotonia in the goat. Brain 62:341–363, 1939

27. BROWN MR, CURRENS JH, MARCHAND JF: Muscular paralysis and electrocardiographic abnormalities resulting from potassium loss in chronic nephritis. JAMA 124:545–549, 1944

28. BRUCK F: Ueber einen fall von congenitales makroglosse, combiniert mit allgemeiner wahrer muskelhypertrophie und idiotie. Dtsch Med Wochenschr 15:229–232, 1889

29. BUCHTHAL F, ENGBAEK L, GAMSTORP I: Paresis and hyperexcitability in adynamia episodica hereditaria. Neurology (Minneap) 8:347–351, 1958

30. BUZZARD EF: The clinical history and postmortem examination of five cases of myasthenia gravis. Brain 28:438–483, 1905

31. CAMPBELL H, BRAMWELL E: Myasthenia gravis. Brain 23:277–336, 1900

32. CASTLEMAN B, NORRIS EH: The pathology of the thymus in myasthenia gravis. Medicine (Balt) 28:27–58, 1949

33. CAUGHEY JE: Relationship of dystrophia myotonica (myotonic dystrophy) and myotonia congenita (Thomsen's disease). Neurology (Minneap) 8:470–476, 1958

34. CHURCHILL-DAVIDSON HC: Discussion on myasthenia gravis. Proc Soc Med 49:793–795, 1956

35. CHURCHILL-DAVIDSON HC, RICHARDSON AT: Myasthenic crisis: therapeutic use of d-tubocuràrine. Lancet 1:1221–1224, 1957

36. CLARK SL, LUTON FH, CUTLER JT: A form of congenital myotonia in goats. J Nerv Ment Dis 90:297–309, 1939

37. COËRS C: Sur quelques types cliniques et histologiques de polymyosite. Rev Belg Pathol Med Exp 25:369–383, 1956

38. COËRS C, DESMEDT JE: Mise en évidence d'une malformation caractéristique de la jonction neuromusculaire dans le myasthénie. Acta Neurol Belg 39:539–561, 1959

39. COËRS C. WOOLF AL: The Innervation of Muscle: A Biopsy Study. Springfield, Ill, Charles C Thomas, 1959

40. CONN JW: Primary aldosteronism, a new clinical syndrome. J Lab Clin Med 45:3–17, 661–664, 1955

41. CONN JW, LOUIS LH, FAJANS SS, et al.: Intermittent aldosteronism in periodic paralysis: dependence of attacks on retention of sodium, and failure to induce attacks by restriction of dietary sodium. Lancet 1:802–805, 1957

42. CONWAY EJ: Nature and significance of concentration relations of potassium and sodium ions in skeletal muscle. Physiol Rev 37:84–132, 1957

43. CRAFTS LM: A fifth case of family periodic paralysis. Am J Med Sci 119:651–661, 1900

44. CROFT PB: Abnormal responses to muscle relaxants in carcinomatous neuropathy. Br Med J 1:181–187, 1958

45. DAHLBÄCK O, ELMQVIST D, JOHNS TR, et al.: An electrophysiologic study of the neuromuscular junction in myasthenia gravis. J Physiol (Lond) 156:336–343, 1961

46. DANOWSKI TS, ELKINTON JR, BURROWS BA, et al.: Exchanges of sodium and potassium in familial periodic paralysis. J Clin Invest 27:65–73, 1948

47. DAVIDSEN HG, KJERULF-JENSEN K: Potassium replacement solutions. Lancet 2:17–18, 1951

48. DEGRAEFF J, BROCKER JSH: Intermittent aldosteronism in periodic paralysis. Lancet 1:1130, 1958

49. DÉJERINE J, SOTTAS J: Sur un cas de maladie de Thomsen, suivi d'autopsie. Rev Med 15:241–267, 1895

50. DE JONG JG: Dystrophia Myotonica, Paramyotonica and Myotonia Congenita. Assen, Netherlands, Van Gorcum, 1955, pp 1–255

51. DELANGE C: Congenital hypertrophy of muscles, extrapyramidal motor disturbances and mental deficiency: a clinical entity. Am J Dis Child 48:243–268, 1934.

52. DENNY-BROWN D: Neurological conditions resulting from prolonged and severe dietary restrictions. Medicine (Balt) 26:41–113, 1947

53. DENNY-BROWN D: Clinical problems in neuromuscular physiology. Am J Med 15:368–390, 1953

54. DENNY-BROWN D, FOLEY JM: Evidence of a chemical mediator in myotonia. Trans Assoc Am Physicians 62:187–191, 1949

55. DENNY-BROWN D, NEVIN S: The phenomenon of myotonia. Brain 64:1–18, 1941

56. DESMEDT JE: Nature of the defect of neuromuscular transmission in myasthenic patients: "post-tetanic exhaustion." Nature (Lond) 179:156–157, 1957

57. DIEFENBACH WOL, FISK SC, GILSON SB: Hypopotassemia following bilateral ureto-sigmoidoscopy. N Engl J Med 244:326–328, 1951

58. DRAGER GA, HAMMILL JF, SHY GM: Paramyotonia congenita. Arch Neurol Psychiatry 30:1–9, 1958

59. DURANTE G: Anatomie pathologique des muscles. Manuel d'Histologie Pathologique. Third edition. Edited by V Cornil, L Ranvier. Paris, Félix Alcan, 1902, pp 1–477

60. EATON LM, CLAGETT OT: Thymectomy in the treatment of myasthenia gravis. JAMA 142:963–967, 1950

61. EATON LM, LAMBERT EH: Electromyography and electric stimulation of nerves and diseases of motor unit: observations on myasthenic syndrome associated with malignant tumors. JAMA 163:1117–1124, 1957

62. ELIEL LP, PEARSON OH, RAWSON RW: Postoperative potassium deficit and metabolic alkalosis. N Engl J Med 243:472–478, 518–523, 1950

63. ELKINTON JR, TARAIL R: The present status of potassium therapy. Am J Med 9:200–207, 1950

64. ENGEL AG: Evolution and content of vacuoles in primary hypokalemic periodic paralysis. Mayo Clin Proc 45:774–814, 1970

65. ENGEL AG, LAMBERG EH, SANTA T: Study of long-term anticholinesterase therapy: effects on neuromuscular transmission and motor end-plate fine structure. Neurology (Minneap) 23:1273–1282, 1973

66. ENGEL WK, HOGENHUIS LAH: Genetically determined myopathies. Review of Current Concepts of Myopathies, Clinical Orthopedics and Related Research. Philadelphia, Lippincott, 1965, Vol 39, pp 34–42

67. ERB W: Zur Casuistik der bulbaren Lähmungen. Arch Psych Nervenkr 9:325–350, 1879

68. ERB W: Die Thomsen'sche Kraukheit (Myotonia congenita) Leipzig, Vogen, 1886, pp. 1–128

69. EULENBERG A: Ueber eine familiäre, durch 6 generationen verfolgbare form congenitaler paramyotonie. Neurol Zentralbl 5:265–272, 1886

70. FEINBERG WD, UNDERDAHL LO, EATON L: Myasthenia gravis and myxedema. Mayo Clin Proc 32:299–305, 1957

71. FENICHEL GM: Muscle lesions in myasthenia gravis. Ann NY Acad Sci 135:60–67, 1966

72. FERGUSON FR, HUTCHINSON EC, LIVERSEDGE LA: Myasthenia gravis: results of medical management. Lancet 2:636–639, 1955

73. FERREBEE JW, ATCHILEY DW, LOEB RF: A study of the electrolyte physiology in a case of familial periodic paralysis. J Clin Invest 17:504–505, 1938

74. FINCH CA, SAWYER CG, FLYNN JM: Clinical syndrome of potassium intoxication. Am J Med 1:337–352, 1946

75. FOIX C, NICOLESCO I: Note sur les altérations du système nerveux dans un cas de maladie de Thomsen. Soc Biol (Paris) 89:1095–1098, 1923

76. FOURMAN P: Depletion of potassium induced in man with exchange resin. Clin Sci 13:93–110, 1954

77. FRASER D, TURNER JWA: Myasthenia gravis and pregnancy. Lancet 2:417–419, 1953

78. FRENCH EB, KILPATRICK R: A variety of paramyotonia congenita. J Neurol Neurosurg Psychiatry 20:40–46, 1957

79. FREIDREICH N: Ueber congenitale halbseitige kopfhypertrophie. Virchows Arch [Pathol Anat] 28:474–481, 1863

80. FULDA F: Ein fall von wohrer muskelhypertrophie, nebst Bemerkungen über die Beziehungen der wahren hypertrophie zur pseudohypertrophie der muskeln. Dtsch Arch Klin Med 54:525–536, 1895

81. GAMMON GD: Relation of potassium to family periodic paralysis. Proc Soc Exp Biol Med 38:922–924, 1938

82. GAMMON GD, AUSTIN JH, BLITHE MD, et al.: The relation of potassium to family periodic paralysis. Am J Med Sci 197:326–332, 1939

83. GAMSTORP I: Adynamica episodica hereditaria. Acta Paediatr (Suppl) 108:1–126, 1956

84. GANS B, FORSDICK DH: Neonatal myasthenia gravis. Br Med J 1:314–316, 1953

85. GASS H, CHERKASKY M, SAVITSKY N: Potassium and periodic paralysis. Medicine (Balt) 27:105–137, 1948

86. GAUPP R, KALDEN O: Erblichkeitsuntersuchungen bei paroxysmaler Lähmung. Z Ges Neurol Psych 174:194–212, 1942

87. GEDDES AK, KIDD HM: Myasthenia gravis in the newborn child. Can Med J 64:152–156, 1951

88. GESCHWIND N, SIMPSON JA: Procaine amide in the treatment of myotonia. Brain 78:81–91, 1955

89. GOLDFLAM S: Ueber einen scheinbar heilbaren bulbärparalytischen symptomencomplex mit betheiligung der extremitäten. Dtsch Z Nervenheilk 4:312–352, 1893

90. GOLDFLAM S: Weitere mittheilung über die par-

oxysmale, familiäre Lähmung. Dtsch Z Nerven-heilk 7:1–31, 1895

91. GOLDFLAM S: Dritte mittheilung uber die par-oxysmale familiäre Lähmung. Dtsch Z Nerven-heilk 11:242–260, 1897

92. GREER WER: Arachnidism: effect of calcium gluconate in 6 cases. N Engl J Med 240:5–8,

93. GROB D: Course and management of myasthenia gravis. JAMA 153:529–532, 1953

94. GROB D, HARVEY AMCG: Abnormalities in neuro-muscular transmission, with special reference to myasthenia gravis. Am J Med 15:695–709, 1953

95. GROB D, JOHNS RJ, HARVEY AMCG: Studies in neuromuscular function. II: Effects of nerve stimulation in normal subjects and in patients with myasthenia gravis. Bull Hopkins Hosp 99:125–135, 1956. IV. Stimulating and depressant effects of acetylcholine and choline in patients with myasthenia gravis and their relationship to the defect of neuromuscular transmission. Bull Hopkins Hosp 99:153–181, 1956

96. GROB D, LILJESTRAND A, JOHNS RJ: Potassium movement in patients with familial periodic paralysis: relationship of the defect in muscle function. Am J Med 23:356–375, 1957

97. HARTWIG H: Ueber einen fall von intermittier-ender Paralysis spinalis. Inaugural dissertation, Halle, 1874

98. HARVEY AM: The actions of quinine on skeletal muscle. J Physiol (Lond) 95:45–67, 1939

99. HARVEY AM, MASLAND RL: The electromyogram in myasthenia gravis. Bull Hopkins Hosp 69:1–13, 1941

100. HOLLER JW: Potassium deficiency occurring dur-ing the treatment of diabetic acidosis. JAMA 131:1186–1189, 1946

101. HOLZAPPLE GE: Periodic paralysis. JAMA 45:1224–1231, 1905

102. JACOBY GW: Thomsen's disease. J Nerv Ment Dis 14:129–151, 1887

103. JANTZ H: Stoffwechseluntersuchungen bei par-oxysmaler Lähmung. Nervenarzt 18:360, 1947

104. JOHNSON W, MARSHALL G: Observations on Thomsen's disease. Q J Med 8:114–128, 1915

105. JOLLY F: Ueber myasthenia gravis pseudo-paralytica. Berl Klin Wochenschr 32:1–7, 1895

106. JONES RV, MCSWINEY RR, BROOKS RV: Periodic paralysis: sodium metabolism and aldosterone output in two cases. Lancet 1:177–181, 1959

107. KANE CA, WEED L: Myasthenia gravis associated with adrenocortical insufficiency. N Engl J Med 243:939–944, 1950

108. KATZ B, THESLEFF S: On the factors which de-termine the amplitude of the "miniature end-plate" potential. J Physiol (Lond) 137:267–278, 1957

109. KENNEDY F, WOLF A: Experiments with quinine

and Prostigmine in the treatment of myotonia and myasthenia. Arch Neurol Psychiatry 37:68–74, 1937

110. KENNEDY F, WOLF A: Quinine in myotonia and Prostigmine in myasthenia. JAMA 110:198–201, 1938

111. KEYNES G: The surgery of the thymus gland. Br J Surg 33:201–214, 1946

112. KEYNES G: The results of thymectomy in my-asthenia gravis. Br Med J 2:611–616, 1949

113. KITAIGORODSKAJA OD: Angeborene hypertrophie ins Kindesalter. Jahrb Klinderh 125:38–89, 1929

114. KOLB LC: Congenital myotonia in goats. Bull Hopkins Hosp 63:221–237, 1938

115. KOLB LC, HARVEY AM, WHITEHILL MR: A clinical study of myotonic dystrophy and myotonia con-genita with special reference to the therapeutic effect of quinine. Bull Hopkins Hosp 62:188–215, 1938

116. KRABBE KH: Les hypertrophies muscularies post-névritiques. Rev Neurol (Paris) 37:802–811, 1921

117. KRABBE KH: The myotonia acquista in relation to the postneuritic muscular hypertrophies. Brain 57:184–194, 1934

118. KUHN E: Myotonia congenita and dystrophia myotonica. Progressive Muskeldystrophie, Myo-tonie, Myasthenie. Edited by E Kuhn. Berlin-Heidelberg, Springer-Verlag, 1966

119. LAMBERT EH, ELMQUIST D: Quantal components of end-plate potentials in the myasthenic syn-drome. Trans Am Neurol Assoc 97:183, 1972

120. LANARI A: La contraccion miotonica en el hom-bre despues de curarization completa. Medicina (B Aires) 7:21–26, 1947

121. LANARI A: Miotonias. Buenos Aires, El Ateneo, 1951

122. LEIBENAM L: Zwillingspathologische beobachtung bei myotonia congenita (Thomsen'sche kran-kheit). Z Mensch Verer Konstitutionslchre 24:13–26, 1940

123. LELOIR LF, CARDINI CE: Biosynthesis of glycogen from uridine diphosphate glucose. J Am Chem Soc 79:6340–6342, 1957

124. LEYBURN P, WALTON JN: The treatment of myo-tonia: a controlled clinical trial. Brain 82:81–91, 1959

125. LINDSLEY DB: Myographic and electromyographic studies of myasthenia gravis. Brain 58:470–482, 1935

126. LYTLE RB, WELLBRAND WA: Increased synaptic area of neuromuscular junction in neostigmine treated rats (abstract). Anat Rec 166:399, 1970

127. MAAS O, PATERSON AS: Myotonia congenita, dys-trophia myotonica, and paramyotonia: reaffirma-tion of their identity. Brain 73:318–336, 1950

128. MANALOV S, HARALANOV H: Changes in the

myoneural synapses in myasthenia. Neurol Psychiatry 6:49–55, 1966

129. MARCHAND JF, FINCH CA: Fatal spontaneous potassium intoxication in patients with uremia. Arch Intern Med 73:384–390, 1944

130. MASLAND RL: The electromyogram in botulism. Dis Nerv System 8:355–356, 1947

131. MAXWELL IL: Hypertrophia musculorum vera. Br Md J 2:656, 1947

132. MCARDLE B: Myopathy due to a defect in muscle glycogen breakdown. Clin Sci 10:13–35, 1951

133. MCARDLE B: Familial periodic paralysis. Br Med Bull 12:226–229, 1956

134. MCARDLE B, MERTON PA: The behavior of radiopotassium in man. J Physiol (Lond) 116:51P–52P, 1952

135. MCDERMOT V: The changes in the motor endplate in myasthenia gravis. Brain 83:24–36, 1960

136. MCEACHERN D: The thymus in relation to myasthenia gravis. Medicine (Balt) 22:1–25, 1943

137. MCEACHERN D: Diseases and disorders of muscle function. Bull NY Acad Med 27:3–23, 1951

138. MCEACHERN D, PARNELL JL: The relationship of hyperthyroidism to myasthenia gravis. J Clin Endocrinol Metab 8:842–850, 1948

139. MCKEEVER GE: Myasthenia gravis in a mother and her newborn son. JAMA 147:320–322, 1951

140. MCQUARRIE I, ZIEGLER MR: Hereditary periodic paralysis: effects of fasting and of various types of diet on occurrence of paralytic attacks. Metabolism 1:129–144, 1952

141. MENDELOW H, GENKINS G: Studies in myasthenia gravis: cardiac and associated pathology. J Mt Sinai Hosp 21:218–225, 1954

142. MEYER A, HEYER-HEINE A, MONOD O, et al.: Myasthénie grave après exérèse chirurgicale d'une tumeur du thymus. Bull Soc Med Hop (Paris) 73:295–302, 1957

143. MEYER H, RANSOM F: Untersuchungen über den tetanus. Arch Exp Pathol Pharm 49:369–416, 1903

144. MOERSCH FP, WOLTMAN HW: Progressive fluctuating muscular rigidity ("stiff-man syndrome"): report of a case and some observations in 13 other cases. Mayor Clin Proc 31:421–427, 1956

145. MORGAN WL, DUDLEY HR: Malignant thymoma: report of a case and review of the literature. N Engl J Med 253:625–632, 1955

146. MURNAGHAN MF: Neuro-anatomical site in tick paralysis. Nature (Lond) 181:131, 1958

147. MYERS WA: Familial periodic paralysis traced in one family for five generations. Pa Med J 52:1060, 1949

148. NASTUK WL, Plescia OJ, Osserman KE: Changes in serum complement activity in patients with myasthenia gravis. Proc Soc Exp Biol Med 105:177–184, 1960

149. NEEDHAM J, SHE-CHANG SHE, NEEDHAM DM, et al.: Myosin birefringence and adenylpyrophosphate. Nature (Lond) 147:766–768, 1941

150. NISSEN K: *Beiträge zur kenntnis der Thomsen'*schen krankheit (myotonia congenita), mit besonderer berücksichtigung des heridären momentes und seinen beziehungen zu den mendelschen bererbungsregeln. Z Klin Med 97:58–93, 1923

151. OLIVER CP, MCQUARRIE I, ZIEGLER M: Hereditary periodic paralysis in family showing varied manifestations. Am J Dis Child 68:308–311, 1944

152. OPPENHEIM H: Ueber einen fall von chronischer progressiver bulbarparalyse ohne anatomischen befund. Virchow's Arch [Pathol Anat] 108:522–530, 1887

153. OPPENHEIM H: Neue mittheilungen über den von Professor Westphal beschriebende fall von periodischer Lähmung aller vier extremitaten. Charite-Annalen 16:350–372, 1891

154. OPPENHEIM H: Die Myasthenische Paralyse. Berlin, Karger, 1901

155. OSSERMAN KE, TENG P: Studies in myasthenia gravis: further progress with edrophonium (Tensilon), a rapid diagnostic test. JAMA 160:153–155, 1956

156. PAL J: Ueber einen fall von muskelhypertrophie mit nervösen symptomen. Wien Klin Wochenschr 2:195–197, 1889

157. PASSOUANT P: La Maladie de Steinert et le Problème de la Transmission Neuromusculaire. Montpellier, Charité Press, 1943

158. PEARSON CM, RIMER DG: Evidence for direct utilization of fructose in working muscle in man. Proc Soc Exp Biol Med 100:671–672, 1959

159. PEARSON CM, RIMER DG, MOMMAERTS WFHM: A metabolic myopathy due to absence of muscle phosphorylase. Am J Med 30:502–517, 1961

160. PONCHER H, WADE HW: Pathogenesis and treatment of myotonia congenita. Am J Dis Child 55:945–965, 1938

161. PONCHER H, WOODWARD H: Pathogenesis and treatment of myotonia congenita. Am J Dis Child 52:1065–1087, 1936

162. POSKANZER DC, KERR DNS: A third type of periodic paralysis with normokalemia and favourable response to sodium chloride. Am J Med 31:328–342, 1961

163. POSKANZER DC, KERR DNS: Periodic paralysis with response to spirolactone. Lancet 2:511–513, 1961

164. PRICE TML, ALLOTT EM: The stiff-man syndrome. Br Med J 2:682–685, 1958

165. PRITCHARD EAB: The occurrence of Wedensky inhibition in myasthenia gravis. J Physiol (Lond) 78:3, 1933

166. PRITCHARD EAB: The use of "Prostigmin" in the

treatment of myasthenia gravis. Lancet 1:432–435, 1935

167. PUDENZ RH, MCINTOSH JF, MCEACHERN D: The role of potassium in familial periodic paralysis. JAMA 111:2253–2258, 1938

168. RAVIN A: Studies in dystrophia myotonica: experimental studies in myotonia. Arch Neurol Psychiatry 43:649–668, 1940

169. RINGER S: A further contribution regarding the influence of the deficient constituents of the blood on the contraction of the heart. J Physiol (Lond) 4:29–42, 1883

170. ROE BB: Myasthenia gravis secondary to thymic neoplasm: report of a case in which symptoms developed six weeks after total thymectomy. J Thorac Surg 33:770–775, 1957

171. ROOKE ED, EATON LM, LAMBERT EH, et al.: Myasthenia and malignant intrathoracic tumor. Med Clin North Am 44:977, 1960

172. ROSETT J: A study of Thomsen's disease, based on 8 cases in a family exhibiting remarkable inheritance features in 3 generations. Brain 45:1–30, 1922

173. ROSS IC: An experimental study of thick paralysis in Australia. Parasitology 18:410–429, 1926

174. ROWLAND LP, ARANOW H, HOEFER FA: Myasthenia gravis appearing after removal of a thymoma. Neurology (Minneap) 7:584–588, 1957

175. ROWLAND LP, HOEFER PFA, ARANOW H JR, et al.: Fatalities in myasthenia gravis: a review of thirty-nine cases with twenty-six autopsies. Neurology (Minneap) 6:307–326, 1956

176. RUSSEL CK, ODOM G, MCEACHERN D: Physiological and chemical studies of neuromuscular disorders. Trans Am Neurol Assoc 64:120, 1938

177. RUSSELL DS: Histological changes in myasthenia gravis. J Pathol 65:279–289, 1953

178. RUSSELL WR, STEDMAN F: Observations on myotonia. Lancet 2:742–743, 1926

179. SANDERS J: Eine familie mit myotonia congenita (Thomsen'sche krankheit). Genetica 17:253–269, 1935

180. SANDIFER PH: The differential diagnosis of flaccid paralysis. Proc R Soc Med 48:186–189, 1955

181. SANDKÜHLER ST: Ueber paroxysmale Lähmung bei hyperkaliämie. Med Klin 52:208–210, 1957

182. SANTA T, ENGEL AG, LAMBERT EH: Histometric structure of neuromuscular junction understructure. Neurology (Minneap) 22:71–86, 1972

183. SCHOENTHAL L: Family periodic paralysis with review of literature. Am J Dis Child 48:799–813, 1934

184. SCHROEDER JM, ADAMS RD: The ultrastructural morphology of the muscle fiber in myotonic dystrophy. Acta Neuropathol 10:218–241, 1968

185. SCHWAB RS, LELAND CC: Sex and age in myasthenia as critical factors in incidence and remission. JAMA 153:1270–1273, 1953

186. SHAW LE: A case of bulbar paralysis without structural changes in the medulla. Brain 13:96–99, 1890

187. SHY GM: Some metabolic and endocrinological aspects of disorders of striated muscle. Proc Assoc Res Nerv Ment Dis 38:274–317, 1960

188. SHY GM, WANKO T, ROWLEY PT, et al.: Studies in familial periodic paralysis. Exp Neurol 3:53–121, 1961

189. SIMPSON JA: Discussion on myasthenia gravis: the value of thymectomy. Proc Soc Med 49:795–798, 1956

190. SIMPSON JA: Myasthenia gravis: a new hypothesis. Scott Med J 5:419–436, 1960

191. SINDEN RH, TULLIS JL, ROOT HF: Serum potassium levels in diabetic coma. N Engl J Med 240:502–505, 1949

192. SINGER HD, GOODBODY FW: A case of family periodic paralysis with a critical digest of the literature. Brain 24:257–285, 1901

193. SLOAN HE: Thymus in myasthenia gravis, with observations on normal anatomy and histology of thymus. Surgery 13:154–174, 1943

194. SMITH SG: An animal analogue of periodic paralysis. J Neuropathol Exp Neurol 9:172–178, 1950

195. SMITH SG, BLACK-SCHAFFER B, LASATER TE: Potassium deficiency syndrome in the rat and the dog: a description of the muscle changes in the potassium-depleted dog. Arch Pathol 49:185–199, 1950

196. STEVENS JR: Familial periodic paralysis, myotonia, progressive amyotrophy, and pes cavus in members of a single family. Arch Neurol Psychiatry 72:726–741, 1954

197. STRAUSS AJL, SEEGAL BC, HSU KC, et al.: Immunofluorescence demonstration of a muscle binding complement-fixing globulin fraction in myasthenia gravis. Proc Soc Exp Biol Med 105:184–191, 1960

198. STRICKFOOT FL, SCHAEFFER RL, BERGO HL: Myasthenia gravis occurring in an infant born of a myasthenic mother. JAMA 120:1207–1209, 1942

199. STRÜMPELL A: Tonische Krämpfe in Wilkürlich bewegten Muskeln (myotonia congenita). Berl Klin Wochenschr 18:119–121, 1881

200. TALBOT JH: Periodic paralysis: a clinical syndrome. Medicine (Balt) 20:85–143, 1941

201. TALMA S: Dystrophia muscularis hyperplastica (Wahre muskelhypertrophie). Dtsch Z Nervenheilk 2:197–209, 1892

202. TENG P, OSSERMANN KE: Studies in myasthenia gravis: neonatal and juvenile types. J Mt Sinai Hosp 23:711–727, 1956

203. THOMASEN E: Myotonia Thomsen's Disease, Paramyotonia, Dystrophia Myotonica. Denmark, Universitetsforlaget i Aarhus, 1948, pp 1–251

204. THOMSEN J: Tonische Krämpfe in willkürlich beweglichen muskeln in folge von erebterpsychischer disposition (ataxia muscularis?). Arch Psychiatr Nevenheilk 6:706–718, 1875–1876

205. TORDA C, WOLFF HG: Effects of adrenocorticotropic hormone on neuro-muscular function in patients with myasthenia gravis. Proc Soc Exp Biol Med 71:432–435. Also J Clin Invest 28:1228–1235, 1949

206. TORDA C, WOLFF HG: Effects of administration of the adrenocorticotropic hormone (ACTH) on patients with myasthenia gravis. Arch Neurol Psychiatry 66:163–170, 1951

207. TYLER FH, STEPHENS FE, GUNN FD, et al.: Studies in disorders of muscle. VII. Clinical manifestations and inheritance of a type of periodic paralysis without hypopotassemia. J Clin Invest 30:492–502, 1951

208. VACCA JB, KNIGHT WA: Estrogen therapy in myasthenia gravis: report of two cases. Mo Med 54:337–340, 1957

209. VAN BUCHEM FSP: The electrocardiogram and potassium metabolism: electrocardiographic abnormalities in primary aldosteronism and familial periodic paralysis. Am J Med 23:376–384, 1957

210. VASTOLA EF, BERTRAND CA: Intracellular water and potassium in periodic paralysis. Neurology (Minneap) 6:523–528, 1956

211. VIETS HR: Thymectomy in myasthenia gravis. Br Med J 1:139–147, 1950

212. WALTON JN, NATTRASS FJ: On the classification, natural history and treatment of the myopathies. Brain 77:169–231, 1954

213. WEIGERT C: Pathologisch-anatomischer beitrag zur Erb'schen Krankheit (myasthenia gravis). Neurol Zentralbl 20:597–601, 1901

214. WESTPHAL C: Demonstration zweier fälle von Thomsen'scher Krankheit. Berl Klin Wochenschr 20:153–155, 1883

215. WESTPHAL C: Ueber einen merkwurdigen fall van periodischer Lähmung aller vier extremitaten mit gleichszeitigen Erloschen der elektrischen Erregbarkeit wahrend der Lähmung. Berl Klin Wochenschr 22:489–491, 1885

216. WHITE JB, PLASKETT J: "Nervous," "stiff-legged," or "fainting goats." Amer Vet Rev 28:556, 1904

217. WHITE WH: On Thomsen's disease. Guys Hosp Rep 31:329–360, 1889

218. WILKS S: On cerebritis, hysteria and bulbar paralysis. Guys Hosp Rep 22:7–55, 1877

219. WILLIS T: De Anima Brutorum. Oxford, 1672, 4° pp 404–406. Also London, 1672, 8° pp 286–288

220. WILSON A: Myasthenia gravis. Glasgow Med J 33:381–394, 1952

221. WILSON A, OBRIST AR, WILSON H: Some effect of extracts of thymus glands removed from patients with myasthenia gravis. Lancet 2:368–371, 1953

222. WILSON SAK: Neurology. London, Edward Arnold, 1941

223. WOHLFART G: Dystrophia myotonica and myotonia congenita, histopathologic studies with special reference to changes in muscle. J Neuropathol Exp Neurol 10:109–124, 1951

224. WOLF A: Quinine: an effective form of treatment of myotonia: preliminary report of 4 cases. Arch Neurol Psychiatry 36:382–383, 1936

225. WOLF A: Effective use of thyroid in periodic paralysis. New York J Med 43:1951–1963, 1943

226. WOODS AH: Muscular hypertrophy with weakness. J Nerv Ment Dis 38:532–538, 1911

227. WOOLF AL: Morphology of the myasthenic neuromuscular junction. Ann NY Acad Sci 135:35–56, 1966

228. ZABRISKI EG, FRANTZ AM: Familial periodic paralysis: report of a case. Bull Neurol Inst NY 2:57–74, 1932

229. ZACKS SI, BAUER WC, BLUMBERG JM: The fine structure of the myasthenic neuromuscular junction. J Neuropathol Exp Neurol 21:335–347, 1962

230. ZACKS SI, COHEN RB: Myasthenia gravis. I. Importance of potassium in inhibitory action in normal and myasthenic thymius extracts. Proc Soc Exp Biol Med 90:601–606, 1955

231. ZIERLER KL, ANDRES R: Movement of potassium into skeletal muscle during spontaneous attack in family periodic paralysis. J Clin Invest 36:730–737, 1957

chapter 13

DIFFERENTIAL DIAGNOSIS AND METHODS OF PATHOLOGIC STUDY

The microscopic examination of skeletal muscle may provide information that is useful in the diagnosis, prognosis, and treatment of neuromuscular and general medical diseases. The customary procedure is to excise a small piece of an affected muscle and to prepare it for histologic study just as in a postmortem examination. This practice of muscle biopsy was introduced in 1868 by Duchenne[9] at about the same time Griesinger[13] first procured a surgical specimen from a case of muscular dystrophy. Duchenne devised a curved trocar with a punch at the side which he called a "muscle harpoon." This instrument was crude, and the technic was painful and not without hazard because of the danger of infection. The ethics of removing living tissue by a technic that carried an incalculable risk excited a lively debate among members of the medical profession, and there were some who opposed it. However, with the development of anesthesia and antisepsis the safety of the method was assured, and the critics were silenced. Nevertheless, the general usefulness of muscle biopsy was not appreciated, and only in recent years with the expansion of our knowledge of muscular disease have its advantages become well known.

Enthusiastic acceptance of biopsy as a way of acquiring knowledge about neuromuscular disease, especially by physicians not facile in thinking about disease in morphologic terms, has led to certain difficulties, particularly that of placing undue reliance on what may be seen in a biopsy specimen. Pathologists know that understanding a disease process is seldom pos-sible from an analysis of morphologic changes alone. Instead, concepts of disease depend always on a synthesis of genetic, clinical, physiologic, and biochemical data with morphologic ones. The latter gives only one dimension of a disease and often exposes only the advanced, irreversible stage. Tissue changes obtain significance only if interpreted in terms of the disease process. Sometimes absence of histologic change is as important as the demonstration of a particular lesion.

The deficiencies and inadequacies of the biopsy procedure must be appreciated. The variable topographies of disease within the musculature, the large number of artifacts created by the biopsy procedure, and the small number of fibers contained in a biopsy make it seem obvious that there is always risk of missing the lesion or wrongly interpreting an artifact. The procedure may then mislead the clinician more than it helps him. Experience teaches caution in the interpretation of biopsy pathology.

INDICATIONS FOR MUSCLE BIOPSY

The clinician is likely to resort to muscle biopsy for solving any one of five different clinical problems, as follows: (1) In any progressive muscular atrophy where it is desirable to know whether the disease is primarily neurogenic or myogenic; especially in the distal and atypical dystrophies and the proximal forms of motor system disease is the biopsy in a doubtful case likely to provide conclusive data on this point. (2) In patients suspected of having a localized or a diffuse inflammatory disease of muscle, the histologic examination may disclose the nature of the disease and at times the infective agent. (3) In cases of obscure fever and multiple visceral and cutaneous lesions where there is suspicion of a diffuse vascular or connective tissue disease that may be corroborated by muscle biopsy. (4) Following injury to the nerves and blood vessels of a limb where biopsy is sometimes necessary in order to assess the state of the muscle and intramuscular nerves, and from this information to direct further treatment and to determine the prognosis. (5) In a number of metabolic diseases where a sample of muscle may yield not only histologic but biochemical data referable to the underlying disease.

For the pathologist to be of assistance to the clinician, he must be apprised of the nature of the problem at hand and take steps to see that the muscle tissue is properly prepared so that the required data can be obtained.

DIAGNOSIS OF PROGRESSIVE ATROPHIC PARALYSIS

When the clinician is unable to decide whether a patient suffers from a spinal or neural atrophy or a muscular dystrophy, he is well advised in such a case to select for biopsy an accessible muscle that is definitely involved yet not completely atrophied. A small piece of superficial cutaneous nerve, such as a sensory branch, i.e., the sural nerve over the calf muscle, should also be obtained if possible. Formalin (Susa or Bouin's) fixation for both muscle and nerve are preferable because it is then possible to stain the sections for myelin sheaths and axis cylinders and to show general cellular detail as well. Hematoxylin and eosin or phloxine methylene blue are the most frequently used general tissue stains for muscle. Mann's methylene blue-eosin stain is preferred by some pathologists because of its more delicate portrayal of nuclear and cytoplasmic changes. Phosphotungstic acid hematoxylin (PTH) is excellent for the demonstration of myofibrils; and Lillie's allochrome, Masson's trichrome, and Wilder's

reticulum stain are useful in demonstrating the connective tissues. Sudan fat stains, Spielmeyer myelin, and Gros-Bielschowsky silver stains can be done on frozen sections of the nerve and muscle. The latter stain is also adaptable to celloidin sections. If paraffin is the embedding material, a hematoxylin and eosin and Bodian axis-cylinder stain can be used. Care must be exercised in obtaining both cross and longitudinal sections.

The histologic diagnosis of muscle atrophy secondary to disease of peripheral nerve or spinal motor neurons depends on finding groups of small atrophic muscle fibers with well preserved striations mixed with other groups of normal-appearing or hypertrophied muscle fibers. There is a slight increase in sarcolemmal nuclei in the wasted fibers, little or no increase in connective tissue, and less fat cell replacement than in dystrophy. Muscle fibers in groups, each showing different degrees of atrophy and each being of the same histochemical type, provide a clue as to the progressive character of the disease. Empty Schwann tubes in the nerve trunks or disappearance of medullated fibers denote a degeneration of motor nerve fibers, present or past. By contrast, in muscle dystrophy individual healthy muscle fibers are haphazardly mixed with very large and very small ones; there is loss and degeneration of individual muscle fibers, central nucleation, fibrous tissue replacement, and pronounced fatty infiltration. The intramuscular nerve trunks are normal.

Muscle dystrophy may be distinguished from chronic polymyositis by the less extensive degeneration of fibers, presence of hypertrophied fibers, the absence of inflammatory cell infiltration, and relative lack of muscle regeneration. In a dystrophy the presence of rows of centrally placed sarcolemmal nuclei in otherwise normal fibers is suggestive of the myotonic form of the disease, as are also ring spirals. Central nuclei should be carefully sought because this type of distal dystrophy is often confused clinically with a progressive neural or spinal atrophy.

Admittedly in some cases it is virtually impossible to distinguish between a muscular atrophy and dystrophy on the basis of a small biopsy specimen, especially if the disease has reached an advanced stage. The difficulty is due to the fact that in some cases of chronic neural atrophy there may be hypertrophied muscle fibers scattered among large fields of atrophied ones and interstitial fat cells in a pattern that resembles muscular dystrophy. Also degenerative changes (necrosis and myophagocytosis) and regeneration may eventually occur in larger fibers. However, we do not agree with Durante[8] and von Meyenburg[26] who doubt that atrophy can be distinguished from dystrophy. The criticism of Bowden and Gutmann[1] that Durante and others placed too much reliance on changes in the muscle fibers and ignored the intramuscular nerves is well taken. A muscle biopsy is in most instances a nerve biopsy as well, especially if taken from near the "motor point" (and the methylene blue of Woolf and Coërs applied). Bowden and Gutmann found nerve twigs in 133 of 140 muscle biopsies. By appropriate stains they were able to demonstrate partial or complete loss of nerve fibers in the form of empty Schwann tubes as well as regeneration of new nerve fibers. Another useful technic is histochemical staining which reveals large fields of fibers that are all of one type, indicating collateral regeneration of nerve twigs that have overtaken some of the denervated fibers. Central core (target fibers) are another frequent sign in denervation atrophy.[9]

The muscle atrophy of cachexia and senility may be confused with a neural muscular atrophy or dystrophy. Disuse atrophy in some cases differs, however, in being totally irregular. Small fibers are mixed with medium and normal-sized ones. There are no hypertrophic fibers, fibrosis, and lipomatosis, but in the majority of cases group atrophy, particularly of type II fibers, has been observed,[25] reflecting a denervative process. Motor unit atrophy is found in the calf muscles of more than 90% of persons over 65 years of age.

DIAGNOSIS OF LOCALIZED OR DIFFUSE INFLAMMATORY DISEASES OF MUSCLE

Zenker-acetic or Bouin's fixation with paraffin embedding is entirely satisfactory in the study of inflammatory diseases. Special care to obtain exact cross and longitudinal sections is not as important as in the study of muscular atrophy, but these should be obtained if possible. Apparatus for automatic fixation and embedding introduce difficulty in this respect.

Local suppurative processes associated with wounds seldom offer any problem in diagnosis that cannot be settled by bacterial cultures and microscopic examination of the stained exudate. On rare occasions the clinician seeks an explanation for a painful nodular focus within a skeletal muscle. The possible causes of such a lesion may be a solitary metastasis from some distant infection, e.g., staphylococcal or streptococcal septicemia, traumatic lesion, tumor nodule, or infarct due to vascular occlusion. Rarely the palpable lump is only a fascicle of muscle fibers in spasm or an uninterpretable one in which the admixture of small and large fibers and endomysial fibrosis resemble a dystrophic lesion. In such cases of these types the biopsy should be taken from the margin of the lesion and include some healthy as well as diseased tissue. Any exudate should be smeared and cultured, and if no exudate is present a small fragment of diseased muscle should be cultured. Necrosis of both interstitial and parenchymal elements, as in infarct necrosis, can usually be distinguished from inflammation by the extent of the destructive process, though the pseudoinflammatory reaction during the first 4–5 days of infarct necrosis may be a source of confusion. Traumatic necrosis can often be identified by hemorrhage, less complete destruction, and the relative paucity of inflammatory cells. Sometimes the differential diagnosis between Zenker's hyaline degeneration, infarct necrosis, and traumatic necrosis is extremely difficult, but usually the patchy involvement of single fibers and the preservation of the interstitial elements serve to differentiate Zenker's hyaline degeneration from infarct and traumatic necrosis. The final identification of a destructive pathologic process in muscle must sometimes be decided from clinical data alone.

Diagnosis of a diffuse or generalized inflammatory disease may present an especially difficult problem because the criteria of what constitutes a myositis are so poorly defined. A sharp distinction must be drawn between the myopathy of toxic and metabolic origin and myositis (Chapter 3, 7, and 9). Both classes of disease are usually generalized and may be either multifocal or diffuse. The only criterion that can be relied upon at the present time is the distribution of the pathologic process within the muscle substance. In an acute toxic myopathy it is usually possible to determine that the content of the muscle fiber is primarily affected and the endomysial connective tissue, blood vessels, and other interstitial structures are spared. Involvement of the sarcolemmal nuclei is variable; usually many of them survive, as shown by their enlargement and multiplication during the phase of regeneration. In acute myositis the supporting tissues as well as the muscle fibers are usually injured. In some diseases such as trichinosis and toxoplasmosis this is relatively slight, whereas in others such as muscle abscess or clostridial myositis it is pronounced.

The presence or absence of inflammatory cells is of some help in differentiating myopathy from myositis. They are almost invariably found in acute myositis, and their type is determined by the nature and duration of the infectious process. However, small numbers of inflammatory cells may be found in any acute myopathy, i.e., a few neutrophilic leukocytes may invade the necrotic portions of the muscle fiber, and there are sometimes a few collections of lymphocytes and mononuclear leukocytes in the interstitial tissues. The demonstration of a bacterium, virus, parasite, or fungus constitutes the only final proof of a true myositis. These two criteria—inflammatory cells and injury to the interstitial tissues and muscle fibers—are not absolute and infallible criteria, especially in diseases in which the etiologic agent has not been identified, as in dermato- and polymyositis.

There is probably no problem in the whole field of muscle pathology that is more difficult to solve than that of identifying a focal disseminated polymyositis. In surveying several sections of a muscle biopsy the pathologist finds foci of inflammatory cells, usually lymphocytes and mononuclear leukocytes, sometimes with a variable number of neutrophilic leukocytes, plasma cells, and occasionally eosinophils. These are usually situated in the supporting connective tissues, often near blood vessels.

Atrophy or degeneration of adjacent muscle fibers can be demonstrated.

This pathologic picture is nonspecific and is difficult or at times impossible to distinguish from focal interstitial polymyositis. In every case of this type there should be a careful search for the causative agent or some identifying feature of the pathologic process. Further sections from the same block of tissue may reveal a trichina or the minute bodies of *Toxoplasma*. Without demonstrating the parasite, the diagnosis of trichinosis cannot be established, though the finding of isolated degenerating muscle fibers invaded by eosinophilic leukocytes makes it probable. Calcified trichinae, on the other hand, are often found in the routine examination of muscle and of course would not explain any acute or subacute symptoms for which the biopsy was performed. Foci of inflammation and destruction of muscle fibers may occur in toxoplasmosis; the *Toxoplasma* can be seen within intact muscle fibers. Focal necrosis of inflammatory exudate and Langhan's giant cells are strongly suggestive of tuberculosis, a diagnosis that can be confirmed only by the demonstration of acid-fast bacilli. It may also occur in diffuse vascular disease, i.e., granulomatous polyarteritis and lupus erythematosus. A granulomatous lesion should always lead to a careful search for tubercle and leprous bacilli, spirochetes, fungi, or parasites. If the causative agent cannot be seen in the sections, other appropriate serologic studies and attempts to isolate the microbe from the muscle by bacterial cultures or animal inoculation should be attempted. The majority of the noncaseating granulomatous lesions in which no infective agent can be demonstrated prove to be due to Boeck's sarcoid.

Focal disseminated polymyositis of granulomatous and parasitic diseases can usually be differentiated from interstitial "nodules" and subacute or chronic diffuse polymyositis or dermatomyositis without difficulty. The cardinal pathologic feature of dermatomyositis is the widespread degeneration of muscle fibers. Some show phagocytosis, and others have disappeared and are replaced by fibrous tissue. The infiltrations of inflammatory cells are variable both in type and degree. They usually consist of lymphocytes, plasma cells, mononuclear leukocytes, and rarely neutrophils. Eosinophilic leukocytes are seldom found. There is no true arteritis though perivenous inflammatory cells are common. At times we have found it impossible to differentiate a focal disseminated polymyositis from a focal interstitial polymyositis. Indeed it must be admitted that there is no single microscopic feature that distinguishes mild recoverable forms of this group of disease from those that are progressive and fatal. The chronic stage is difficult to distinguish from muscular dystrophy except for the absence of hypertrophied fibers.

DIAGNOSIS OF DIFFUSE VASCULAR AND CONNECTIVE TISSUE DISEASE

Either Zenker or formalin fixation gives satisfactory results when looking for diffuse vascular and connective tissue disease. However, if formalin is used, celloidin embedding is better than paraffin because there is less shrinkage artifact. Stains for cellular detail such as hematoxylin and eosin, phosphotungstic acid hematoxylin, phloxine methylene blue, and Gomori trichrome demonstrate the lesions satisfactorily.

The so-called connective tissue diseases include rheumatic fever, lupus erythematosus, dermatomyositis, scleroderma, and polyarteritis nodosa. The classification of such a heterogeneous group of diseases in one category has been questioned by several workers and is not defended here. The point which concerns us is that the clinician is often hard-pressed to make an exact diagnosis and frequently must rely on biopsy of skin and muscle. Microscopic pathologic findings often permit accurate diagnosis, particularly in polyarteritis and dermatomyositis or polymyositis. There are two reliable histopathologic changes in polyarteritis nodosa: necrosis of the media or of all three coats of the vessel, often with aneurysm formation or thrombosis, and infiltration of the vessels and perivascular tissues by eosinophilic leukocytes and other inflammatory cells. In a series of 552 biopsies Wallace *et al.*[27] observed that necrotizing arteritis in all except one instance was due to polyarteritis nodosa.

Perivascular collections of neutrophilic leukocytes and other hematogenous cells without vessel damage are not sufficient for diagnosis, and as pointed out in the discussion of interstitial polymyositis they may be found in rheumatoid arthritis, lupus erythematosus, rheumatic fever, scleroderma, and a number of systemic infections. Infarct necrosis of muscle fibers, neural muscular atrophy, arterial lesions, and hemorrhage substantiate the diagnosis of polyarteritis nodosa. The diagnoses of idiopathic dermato- and polymyositis are established by the existence of extensive degeneration and regeneration of muscle fibers and a variable degree of inflammatory cell infiltration. Occasionally in scleroderma and lupus erythematosus the same type of muscle lesion may occur but usually is less extensive and constitutes a minor part of the syndrome.

Rheumatoid arthritis, rheumatic fever, lupus erythematosus, and scleroderma are characterized by an interstitial "nodular" polymyositis. Here the essential lesions are multiple, small foci of inflammatory cells (lymphocytes, plasma cells, histiocytes or epithelioid cells, and rare neutrophilic and eosinophilic leukocytes and mast cells) in the endomysium, perimysium, and epimysium, usually near small blood vessels. Damage to adjacent fibers occurs but is so slight as to appear secondary to the interstitial inflammation. No infective agent can be seen or isolated. We have been unable to distinguish between these diseases on the basis of the location or nature of the cellular infiltration. When we have found these lesions in muscle, we have reported them as interstitial polymyositis consistent with these several diseases and advised that the differential diagnosis depends on clinical and other laboratory data. The collagen changes which some workers regard as a characteristic feature of lupus erythematosus and scleroderma were not observed in our muscle biopsies. Langhan's giant cells in inflammatory foci suggest Boeck's sarcoid but may be seen in other diseases, particularly myasthenia gravis with thymoma (See Chapter 12).

Hyperplastic arterio- and arteriolosclerosis may be found in muscle specimens obtained by biopsy from patients suffering from malignant

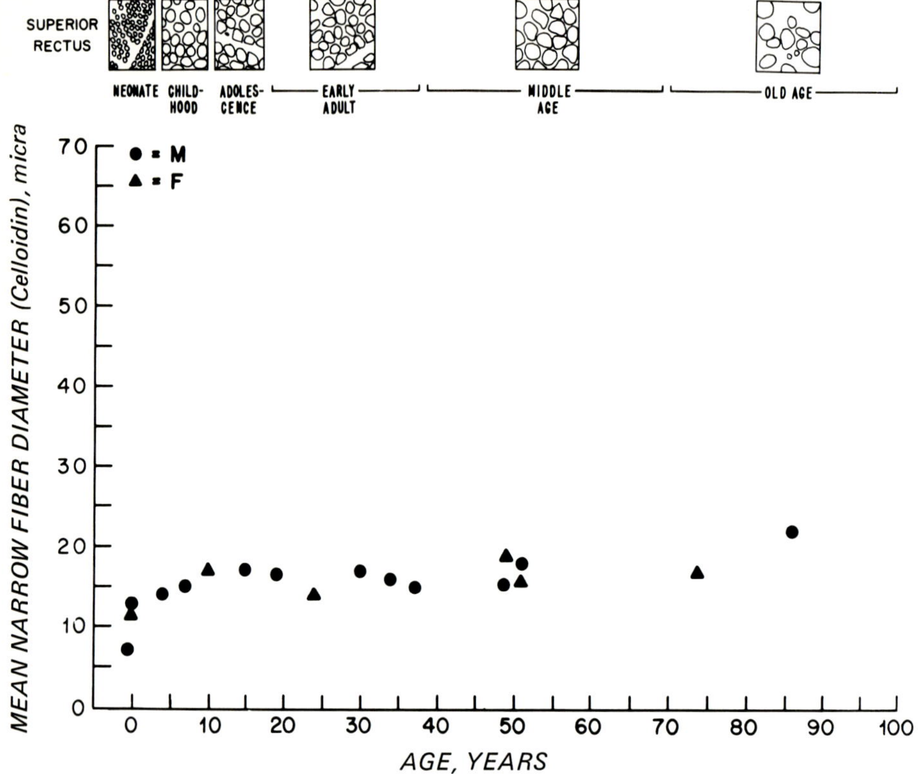

Fig. 13–1. Change of the mean narrow fiber diameter in the superior rectus muscle through age. The fiber size remains relatively constant from childhood throughout life. (Fiber tracings for each period, ×600)

Figures 13–1, 13–4, and 13–7 are included here for the convenience of the pathologist evaluating fiber sizes. They also appear in Chapter 1.

hypertension. Giant cell arteritis, which affect larger arteries, is seldom diagnosed from examining a piece of excised muscle.

DIFFERENTIATION OF DENERVATION ATROPHY AND ISCHEMIC DEGENERATION

When studying denervation in diseases of spinal cord and nerves, at least during the early weeks and months, one must resort to a biometric analysis of fiber size. The least transverse diameters of 200–500 fibers should be measured, this being the only easy measurement that can be made. Fibers showing vitreous change and retraction should not be included. Normative age standards for each muscle are needed for comparison since the fibers increase in diameter and volume with age. The growth curves for several of the commonly biopsied muscles are seen in Figures 13–1 through 13–7.[19]

An injury to an extremity may damage the nerves and blood vessels that supply the peripheral musculature. Muscle paralysis and atrophy result. It may be impossible, even knowing all the clinical data and the electrical reactions, to assess the state of the muscles. The usefulness of the biopsy procedure in determining the cause of atrophy or paralysis was discussed in an interesting and valuable paper by Bowden and Gutmann.[1]

The muscle tissue should be fixed in 10% neutral formalin for 2 days or more. If the piece is of sufficient size, it may be divided; half of it is then embedded in paraffin and the other half saved for frozen sections. A Bielschowsky stain and also Sudan black and Spielmeyer myelin stains are used on the frozen sections. Hematoxylin and eosin or vanGieson are adaptable to paraffin sections.

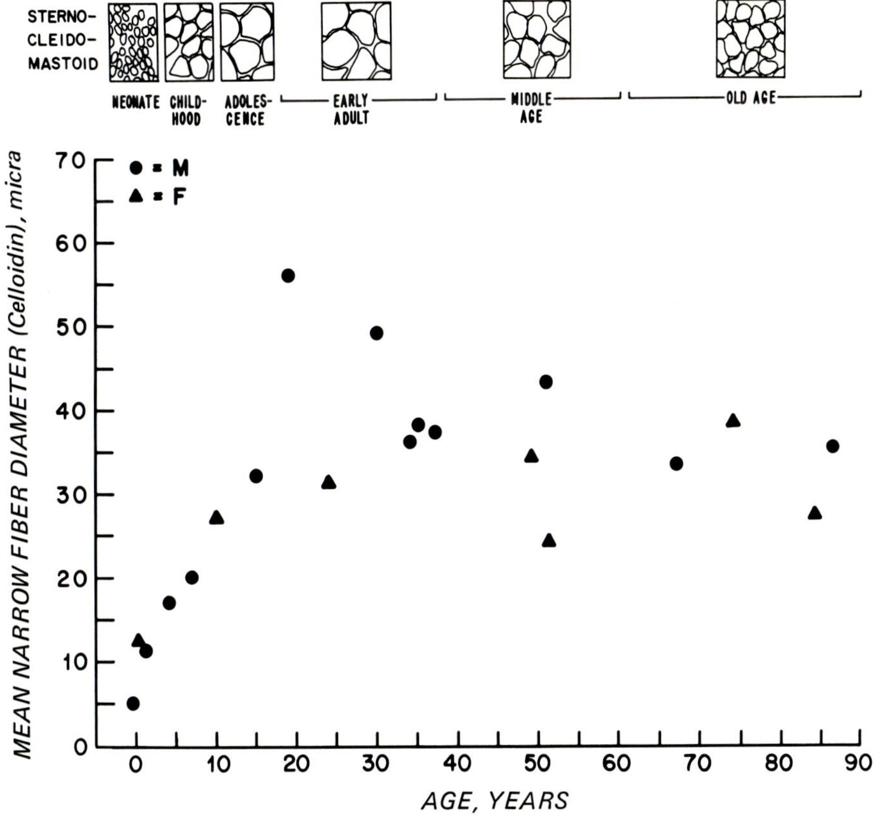

Fig. 13–2. Change of the mean narrow fiber diameter in the sternomastoid muscle through age. The periods of major fiber development occur after childhood and are maximum during adolescence, early adulthood, and middle age. (Fiber tracings for each age period, ×600)

During the first months after the denervative injury the muscle fibers become smaller with an apparent increase in nuclei. The cross striation is less well preserved. The blood vessels are dilated and congested. Nerves with empty Schwann tubes can be followed to the muscle fibers. Eventually the muscle fibers become minute threads of sarcoplasm with rows and clumps of dark nuclei. Fibrous tissue is not increased in amount during the first 6–12 months, but toward the end of the first year it is increased around the blood vessels and between the adjoining fibers. In human specimens from about 3 years onward there is almost complete replacement of the muscle by fibrous tissue and fat cells, only a few clusters of sarcolemmal nuclei remaining.

If all the motor units in a given muscle are in the same stage of atrophy and there are empty Schwann tubes in muscular nerves, it is likely that there has been complete interruption of the nerve. The presence of large numbers of thin, wavy, regenerating nerve fibers in a muscular nerve and large groups of muscle fibers of one histochemical type (loss of the normal checkerboard pattern) signifies that effective reinnervation is occurring. Regenerating nerve fibers of differing lengths and diameters suggest that a traumatic neuroma has impeded the regrowth of some nerve fibers and not others. There may then be disarrangement of the pattern of nerve supply. Regenerating axons may escape the nerve trunks, cross back and forth between the muscle fibers, and branch repeatedly at some distance from the motor end-plate unlike normal nerve fibers where the branching is obscured in the bundles of muscular nerves.

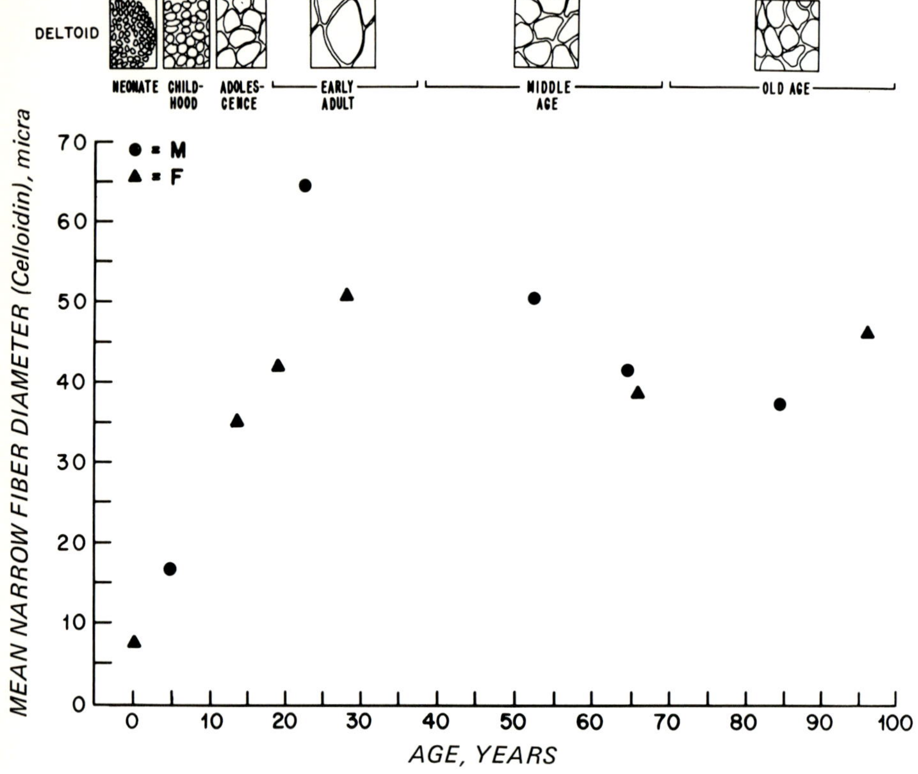

Fig. 13–3. Change of mean narrow fiber diameter in the deltoid muscle through age. There is a major increase of fiber size after childhood during adolescence and early adulthood, followed by a noticeable decrease during the middle and old age periods. (Fiber tracings of each age period, ×600)

Islands of necrotic muscle fibers surrounded by fibrous tissue and a peripheral cellular zone of phagocytes and muscle giant cells are the usual findings in the massive infarction resulting from occlusion of a major artery. Such lesions may be small and multiple, or an entire muscle may be involved. In small lesions only mixtures of hyaline degeneration and multinucleated young regenerating fibers may be seen. A diffuse increase in interstitial connective tissue with crowding of surviving muscle fibers suggests venous occlusion. Vascular lesions of this type together with denervation atrophy mean that the muscular nerves have also been damaged, either by trauma or ischemia.

These findings enable the clinician to determine whether reinnervation is proceeding as it should. After a delay of 15–20 days nerve fibers usually regenerate centrifugally—about 3 mm per day. Axonal growth precedes functional recovery. If reinnervation has not occurred within a year, there is little hope for recovery. If the muscles are fibrotic there can be no return of muscular contraction even though reinnervation occurs.

DIAGNOSIS OF METABOLIC DISEASES

Muscle biopsy has been employed with increasing frequency to study metabolic diseases, which explains in part our increasing knowledge of many of the metabolic activities of muscle. In such diseases as primary amyloidosis, the several glycogen storage diseases, and the group of disfigurative myopathies (central core, nemeline, mitochondrial, lipid, and myotubular), the diagnosis may actually be established by muscle biopsy. The tissue preferably

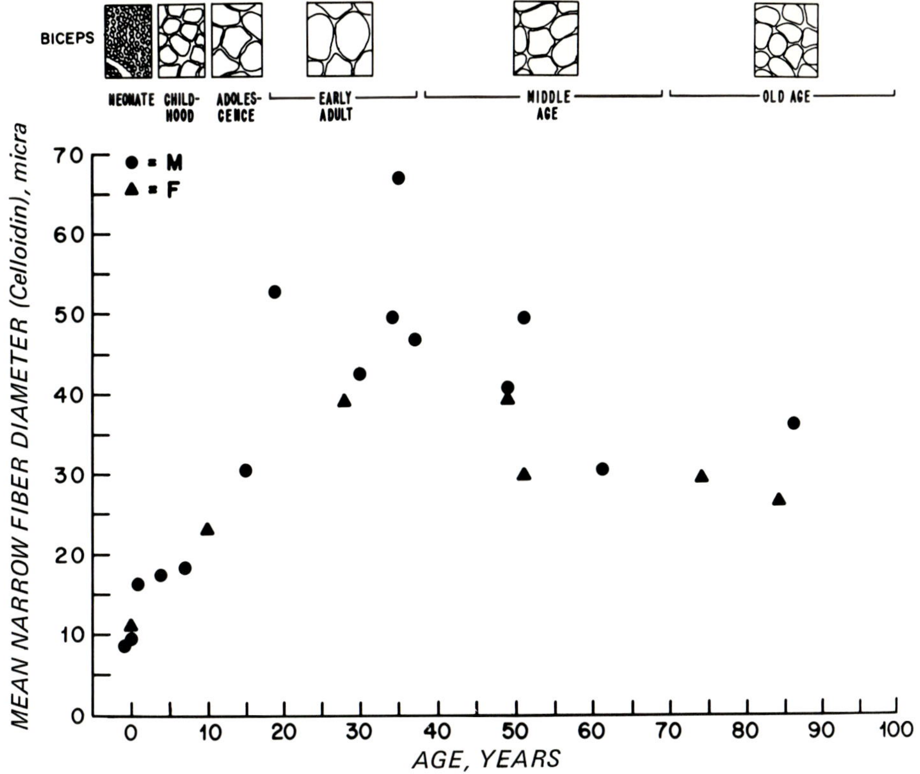

Fig. 13–4. Change of mean narrow fiber diameter in the biceps muscle through age. The growth curve approximates that of the deltoid. (Fiber tracings of each age period, ×600)

should be frozen or fixed in several different ways after preventing contraction during the biopsy by a special clamp or some other device. Absolute alcohol 95% or Rossman's fluid are used for good glycogen stains. Special histochemical technics are required to identify deposits of amyloid and glycogen. Freeze-drying technics are employed for histochemical fiber typing and osmium or glutaraldehyde for electron microscopy.

FREQUENCY OF PATHOLOGIC CHANGES IN ROUTINE AUTOPSY MATERIAL

Every pathologist must be concerned with the specificity of these individual pathologic changes. When any one or several are observed, can it be assumed that they represent an important disease of skeletal muscle? Or do they occur with such frequency in the course of the many medical and surgical diseases which terminate human life that they lack significance?

Clawson et al.,[5] Sokoloff et al.,[24] and Wallace et al.[27] attempted to answer these questions by examining muscle tissue from patients dying of trauma and other common fatal diseases. They discovered a distressingly large number of inflammatory cell infiltrations, atrophy, and degenerative changes in muscle fibers. Pearson conducted a similar study[20] in which microscopic sections were obtained from nine muscles (Table 13–1) in each of 110 successive autopsies. The patients were mostly males at a Veterans Hospital and as would be expected fell into the 50- to 70-year age bracket.

Some type of abnormality was found in the muscle fibers or adjacent connective tissues of one or more muscles in 54 of the 110 cases

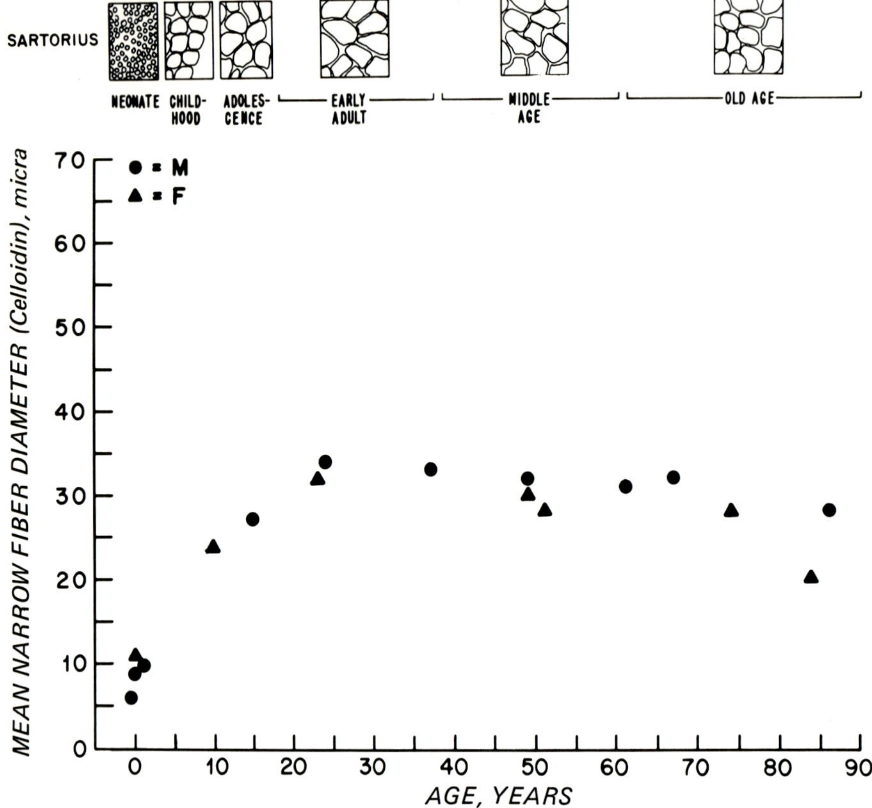

Fig. 13–5. Change of mean narrow fiber diameter in the sartorius muscle through age. The postpubertal growth curve is more gradual and lower than for any other muscle studied except the superior rectus. (Fiber tracings for each age period, ×600)

(49%) in Pearson's series. Even after the nonspecific alterations of muscle fibers such as cloudy swelling and the simple atrophy of disuse, cachexia, or senility were set aside, 213 of the 930 muscles examined (23%) exhibited some minor or major alteration. The diaphragm, rectus abdominis, deltoid, and iliopsoas were involved most frequently (Table 13–1), but each of the other muscles studied was found to be affected several times at least. The alterations that occurred with greatest frequency were focal or diffuse degenerative and regenerative changes in the muscle fibers, and less commonly interstitial or perivascular mononuclear cell foci (Table 13–2). Degeneration of muscle fibers, which was found in 85 muscles in 38 cases, was usually of the hyaline or Zenker's type (Fig. 13–8). Granular and vacuolar degeneration were decidedly less frequent. As a rule, these degenerative changes were restricted, involving only a single fiber or several fibers, and were best seen in longitudinal sections. Regenerative repair of previously

TABLE 13–1. Muscles Routinely Examined from 110 Autopsies: Incidence of Pathologic Change

Muscle	No. examined	No. affected
Quadriceps femoris	105	21
Iliopsoas	110	26
Rectus abdominis	110	31
Diaphragm	110	48
Intercostal	107	12
Pectoral	107	20
Deltoid	110	30
Sternocleidomastoid	104	22
Extraocular	67	3
Total	930	213

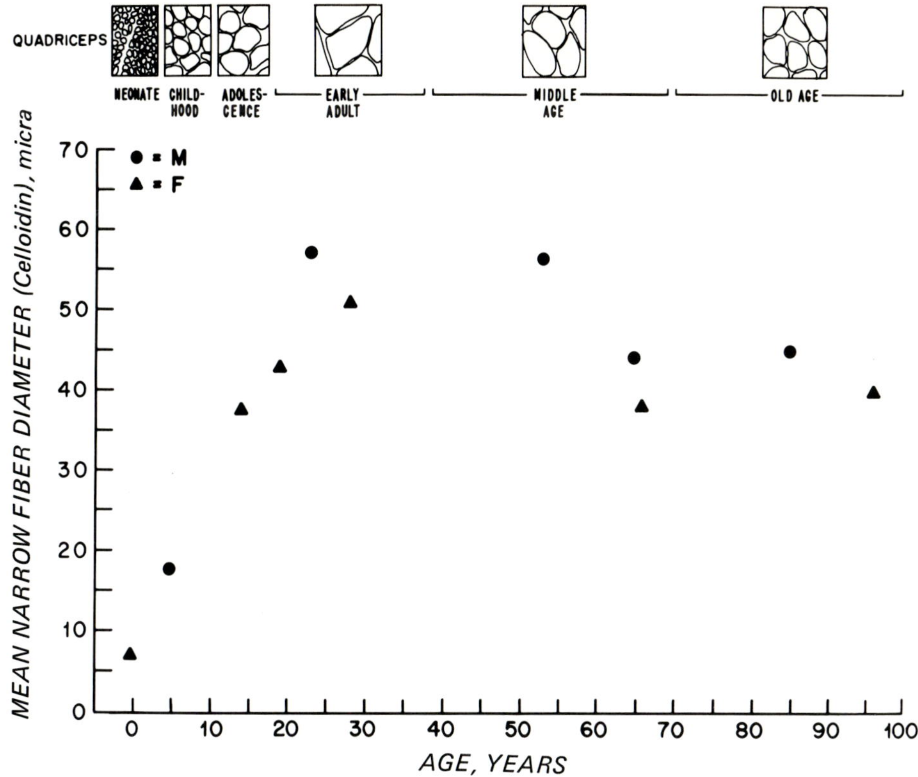

Fig. 13–6. Change of mean narrow fiber diameter in the quadriceps muscle through age. The growth curve closely parallels that of the deltoid and biceps muscle. (Fiber tracings for each age period, ×600)

damaged muscle was found in 54 muscles in 25 cases and seemed to be a continuing process in many serious diseases. Focal interstitial or perivascular collections of inflammatory cells were found in a wide variety of cases (Table 13–3), many of noninflammatory character. Al-though small, they were in other respects similar to those which accompany interstitial polymyositis (Chapter 7). In addition, two specific abnormalities occurred with notable frequency: trichina parasites and tumor cells. Trichina parasites were seen in 14 muscles from 5 cases and metastatic tumor nodules in 14 muscles from 6 cases. Even though there were 38 persons with malignancy in this series, it was somewhat surprising to find such a high incidence of metastatic tumors among them. The tumors were both carcinomas and lymphomas and had originated in various viscera. In the latter cases, degenerative and regenerative changes in muscle fibers were exceedingly prominent (Table 13–3), probably reflecting the frequent coincidence of neoplasia and polymyositis.

From this survey the following conclusions are justified: (1) Histologic alterations in muscle fibers or the supporting connective tissues

TABLE 13–2. Type and Incidence of Pathologic Changes in Sections of 930 Muscles

Type of alteration	No. of cases involved	No. of muscles affected
Degeneration of muscle fibers	38	85
Regenerative features	25	54
Focal interstitial myositis	7	8
Perivascular mononuclear cell foci	13	35
Trichina parasites	5	14
Metastatic tumor	6	14
Focal denervation atrophy	3	3
Total	97	213

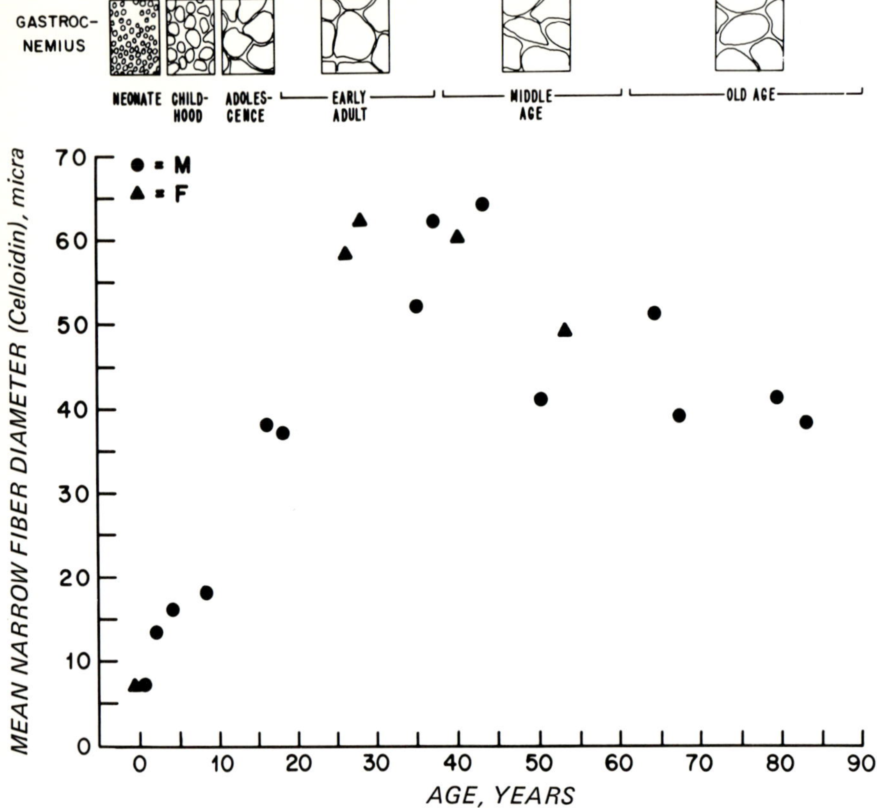

Fig. 13–7. Change of mean narrow fiber diameter in the gastrocnemius muscle through age. The growth curve closely parallels that of the deltoid, biceps, and quadriceps muscles. (Fiber tracings of each age period, ×600)

are common findings in the presence of serious systemic disease, and the chances of finding such abnormalities seem to increase with the size of the biopsy sample and the number of muscles studied. (2) No one of the above cellular changes is of itself characteristic of any disease; pathologic specificity (diagnosis) derives from the combinations of these changes, their probable duration, and their relationship to the clinical findings. (3) The occurrence of pathologic changes in striated muscle rises significantly when a visceral malignancy is present. (4) Degenerative and regenerative changes which occur in conjunction with a wide variety of systemic illnesses appear to represent phases of a continuous destruction and repair of muscle. Only if the destructive element predominates and is severe does it seem to contribute to muscular weakness. (5) It may be impossible to determine, in complicated diseases, which of several possible toxic, metabolic, and nutritional factors are responsible for the degeneration of muscle fibers (Fig. 13–8).

From surveys such as this it is evident that muscle biopsy may fail to elucidate a complex medical problem. Error may be committed in both directions—that of attaching undue importance to some minor tissue change or of excluding obvious disease because the specimen exhibits no pathologic abnormality. Experience with the biopsy procedure gives the pathologist and thoughtful clinician a proper perspective. They come to appreciate that the biopsy, sampling as it does but an infinitesimal part of the total musculature, is a relatively poor method of demonstrating the pathology of a muscular disease even when it is widespread. Negative findings obviously must carry less

TABLE 13–3. Major Disease and Pathologic Alteration in Muscle

Diagnosis	No. of Cases	Degeneration and regeneration (No.)	Interstitial myositis (No.)	Perivascular cellular foci (No.)
Malignancy	38	12	1	2
Congestive heart failure	9	4	2	1
Myocardial infarction	9	2	0	1
Hypertension	4	1	0	0
Cerebral vascular disease	9	2	0	0
Acute hemorrhage (ulcer; aorta)	4	2	0	0
Rheumatoid arthritis	6	5	2	2
Polyarteritis	1	0	0	1
Lupus erythematosus	1	1	0	1
Cirrhosis, decompensated	5	3	0	0
Diabetes mellitus	2	0	0	1
Pyelonephritis	2	0	0	0
Septicemia	1	1	1	1
Bronchopneumonia	3	2	0	1
Cor pulmonale	1	1	0	0
Uremia	2	1	0	0
Duodenal ulcer	2	1	0	1
Severe trauma	2	1	0	0
Myasthenia gravis	1	1	1	1
Miscellaneous conditions	8			
Total	110	40	7	13

weight than positive ones. Any given histopathologic change, it is soon realized, gains significance only when considered in the light of all the clinical, physiologic, and biochemical data. Microscopic alteration of normal muscular structure is but one aspect of a disease and seldom by itself can provide a complete picture.

OBTAINING AND PREPARING MUSCLE TISSUE FOR MICROSCOPIC STUDY

BIOPSY

Although biopsy of muscle can yield valuable information, the procedure is often unsuccessful for several reasons. First, the wrong muscle may be selected for biopsy. For example, in a patient with weakness and atrophy of forearm and hand muscles, the biopsy specimen may have been obtained from the deltoid or some other muscle because of its ready accessibility; or an unaffected leg muscle may be chosen for

biopsy even though all the muscle weakness is in the arms. Second, the tissue may be removed by blunt dissection, crushed with a clamp, or stretched so that most of the fibers are ruptured or distorted. Third, the specimen may be permitted to curl up or become twisted during fixation so that properly oriented sections are impossible. Fourth, the biopsy may be too small.

The following technic is recommended. A muscle is chosen that is actively involved in the disease. Normally strong muscles or those that are completely atrophied and fibrotic should not be selected. If possible a muscle is chosen that is readily accessible through a skin incision, such as the deltoid, biceps, triceps, quadriceps, peroneal, anterior tibial, or gastrocnemius, but this criterion alone should not determine the muscle to be examined. The specimen is taken from the point at which the nerve enters the muscle if there is interest also in the intramuscular nerves and motor endplates. Without this precaution only about half the specimens will contain neural elements. The operation is performed under local anesthesia, care being taken to avoid injection of the an-

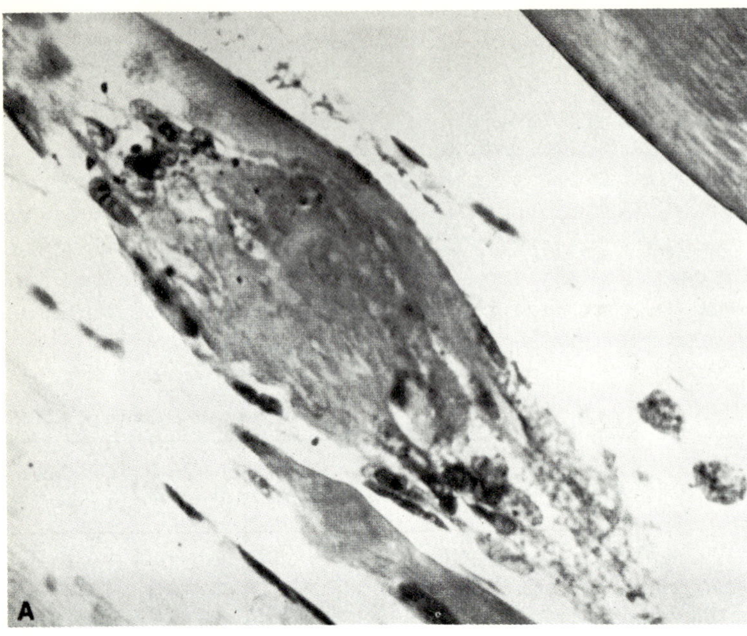

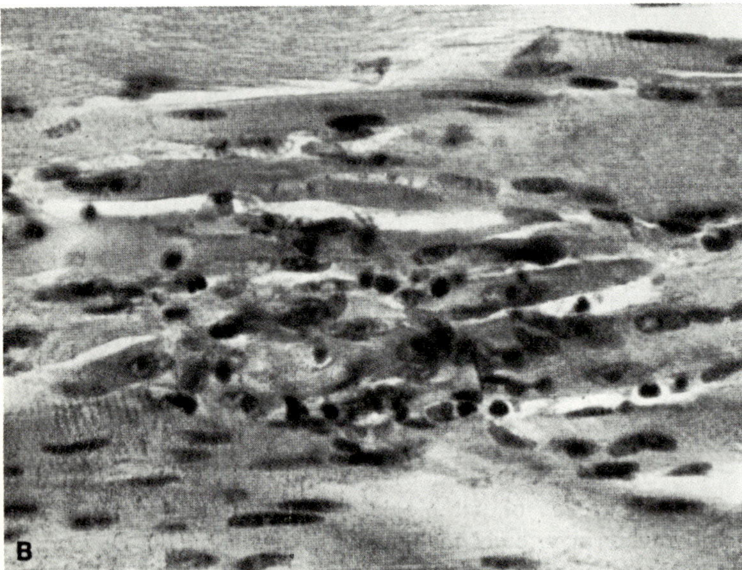

Fig. 13–8. (A) Zenker's hyaline degeneration in a single muscle fiber in the rectus abdominis muscle of a 55-year-old man with decompensated cirrhosis. Many such fibers were found in several muscles. Note beginning flocculation of hyalinized segment of muscle, early phagocytosis, and proliferation of sarcolemmal nuclei in margins of lesion. (B) Focus of thin basophilic regenerating fibers in deltoid muscle of a 67-year-old man with advanced uremia. Several intact fibers exhibit a granular appearance. (H&E; ×645)

esthetic into the part of the muscle selected for excision. After the body of the muscle to be biopsied is exposed under sterile technic, a block of muscle about 2.0 cm long, 1.0 cm wide, and 1.0 cm thick is excised cleanly with a sharp scalpel. The long dimension should be in the longitudinal axis of the muscle fasciculi. Crushing, tearing, or twisting the specimen must be avoided because artifacts are thereby produced. The specimen is laid on a piece of wood, straightened out to its original length. This usually leads to its firm adherence to the card. It is then either placed in a clean Petri dish and kept moist for 15–30 min before being placed in the fixative, or it is put directly into the fixative. A slight delay is best because the fixative may cause excessive contraction in muscle that is still excitable. Additional portions required for electron microscopy, histochemistry, or microchemistry estimations, are removed at this time. After fixation the block of muscle is cut either with a razor blade or sharp section knife so that the muscle fibers are in exact cross section in one piece and longitudinal section in the other.

From time to time special instruments have been devised for muscle biopsy, e.g., Duchenne's histologic harpoon and another instrument that was introduced by Shank and Hoagland[23] and Webster *et al.*[29] as an aid in procuring bits of muscle at regular intervals during treatment of a muscle disease. In general, the latter biopsy technics are advised only for special types of problems because of the small size of the specimen and the artifacts they produce. Meticulous attention to technic can provide excellent specimens, but the number of fibers examined is so small as to increase the sampling error.

One of the main sources of artifact is the sudden contraction of muscle at the moment of incision. Special clamps were devised to avoid this, and when applied before incision the contraction artifacts are thus prevented. Tying a stick to the surface of the muscle with a suture at each end and cutting around it is a practical way of avoiding contraction bands.

Dubowitz and Brooke's[7] recent monograph may be consulted for technical details of the biopsy procedure and histochemical staining techniques.

AUTOPSY

The standard autopsy technic permits examination of the thoracic, abdominal, and paravertebral muscles but not those of the extremities. In order to survey adequately the skeletal musculature it is necessary to make incisions in the arms and legs. The biceps as well as triceps muscles can be approached through an incision 12–15 cm long in the medial surface of the upper arm by dissecting anteriorly and posteriorly. This also affords access to brachial nerves. Similar incisions in the posteriorlateral surfaces of the thigh and lateral surfaces of the leg expose the quadriceps and hamstring muscles and the anterior tibial, peroneal, and gastrocnemius muscles, respectively. There is usually an objection to making a skin incision in the hands and, especially in women, in the

forearms. The intrinsic muscles of the feet can be examined through an incision in the plantar surfaces. The tongue can be removed with the block of pharyngeal and laryngeal tissues and the neck muscles by dissecting upward from the thoracic incision. The ocular muscles can be approached through the roof of the orbit from the cranial cavity.

The character and distribution of the muscle disease should determine which muscles are selected for examination. If the paralysis and atrophy involve only the muscles supplied by a single nerve, the examination can be limited to these muscles, though others should be studied for comparison. If the disease is generalized, the examination should include several muscles to see the earliest as well as the most advanced lesions. A complete examination should include an ocular muscle, tongue, sternomastoid, deltoid, biceps, triceps, a forearm muscle if possible, and paravertebral, intercostal, psoas, quadriceps, anterior tibial, and gastrocnemius muscle.

Exposure of the muscle should be adequate to permit satisfactory gross inspection. Blocks of affected muscles should be cleanly excised, placed on a piece of cardboard with the fibers as straight as possible, and labeled, just as in the muscle biopsy procedure described above. It is as important to avoid overstretching as it is to avoid twisting and overcontraction. Walton and Mastaglia[27] showed that histochemical demonstration of enzymes is possible many hours after death; hence blocks of tissue may be excised for freeze-drying.

Removing blocks of one of several muscles, the usual procedure, may still give an incomplete view of a pathologic process. Under certain circumstances (e.g., dystrophy or myopathy) it may be desirable to see the effect of a disease on whole muscles. Excision of muscles from origin to insertion is then performed and the specimen laid out straight and fixed in 10% neutral formalin. Transverse and longitudinally incised blocks, stained by hemotoxylin and eosin provides a more complete view of the lesion.

In most muscle diseases it is necessary to examine the peripheral and central nervous system. Several pieces of nerves, especially those supplying muscles known to have been paralyzed, should be carefully dissected without stretching and excised; they too are placed on cardboard and labeled. At least four or five of the lumbosacral dorsal root ganglia should be removed with the spinal cord. The nervous tissue is best fixed in 10% formalin, but a few pieces can be placed in Zenker's solution. From the formalin-fixed material it is possible to prepare satisfactory microscopic sections for studying motor cells of the spinal cord, the dorsal and ventral roots, the spinal ganglia, and the motor and sensory nerves.

FIXATION AND STAINING OF MUSCLE TISSUE

In considering the procedure of making microscopic sections of muscle, the objectives of the examination must be clearly in mind. The choice of fixative and stain are dictated largely by the purposes of the examination. If general cytologic detail is desired, formalin, Zenker's, or Bouin's fixative solutions followed by staining with iron alum hematoxylin after the method of Heidenhain or Häggquist, ordinary hematoxylin and eosin, or phloxine methylene blue and Verhoeff-Van Giesson picric acid are recommended, and the embedding should be in paraffin. Some pathologists prefer fixation of small pieces in Zenker's fixative without acetic acid for not longer than 24 hours and staining with eosin-methylene blue. Mallory's phosphotungstic acid hematoxylin and Masson's trichrome stain containing acid fuchsin can be used after these fixatives and give excellent details of the myofibrils and the striations. Gomori[12] recently described a rapid modification of the trichrome stain that gives excellent results in muscle tissue. If both cell and nerve fiber stains are contemplated, 10% neutral formalin is necessary, as this prepares the material for axis-cylinder and myelin stains and still permits satisfactory general cytologic study. Celloidin embedding has the advantage of lessening the shrinkage, which is often pronounced when formalin-fixed tissues are placed in paraffin. Frozen sections, so often utilized when these methods are applied to the nervous

tissues, can be obtained, and indeed all the routine tissue and histochemical stains can be done on muscle frozen in liquid N (Dubowitz and Brooke[7]).

More specialized technics[21] for the examination of muscle tissue include polarization microscopy (Chapter 1); fluorescence and phase microscopy; and special staining technics for histochemistry (particularly oxidative, phosphorylase, ATPase and PAS staining); for materials such as glycogen, fats, amino acids, nucleic acids, enzymes, and other fiber constituents. If the muscular nerves and motor endplates are to be examined, the nerve entering the muscle should be identified and followed for a short distance into the muscle. A block of muscle tissue with very small nerve fasciculus running through it is then selected for special fixation. The fixatives of choice are either citric or formic acid for Ranvier's and Cajal's gold chloride methods, or formalin for the Gros-Bielschowsky method. In the gold impregnation methods, which are generally unsatisfactory in human muscle (though Cajal's modification was successfully employed by Carey), the branching hypolemmal axons are rather thick, like the fronds of a fern, and are outlined by granules. In the Gros-Bielschowsky silver method the fine ramifications of the hypolemmal axon stain sharply, and a variety of counterstains can be used for studying nuclear changes. When the muscle is treated with methylene blue (Ehrlich) shortly before or immediately after death, the nerve fibers are also selectively stained. By this method, which is best suited for experimental work, there is variable coloration of the terminal branches of the axon. Successful staining of motor end-plates can be attained only by assiduous practice and requires specially fixed material. These procedures are well outlined in Lillie's[18] text on histopathologic technic, except for the Gros-Bielschowsky method the details of which may be found in Carleton and Leach's[4] text and the later editions of Lee.[17] The intravital staining of nerve endings by the biopsy method of Coërs and Woolf (Chapter 2) is particularly elegant and practical.

The effects of the various solutions and fixatives on striated muscle were examined closely by many of the early microscopists and more recently by Høncke.[14] Exsiccation after removal of the tissue may result in shrinkage. This is more evident, as a rule, in the transverse than the longitudinal microscopic sections.[21] The length of the A and I bands is usually not altered appreciably, although alcohol fixation, which has a dehydrating effect, causes more shrinkage of the A than the I band. When muscle tissue is immersed in hypotonic solutions or water, the fibers swell considerably and undergo a vitreous change with obliteration of the striations. Hypertonic solutions, on the other hand, have relatively little effect except to enhance the longitudinal striations and to induce numerous contraction bands.

Most of the fixatives in common use—alcohol, Zenker's or Bouin's solution, and formalin—cause shrinkage of muscle tissue, particularly if the fixative makes the tissue more acidic, which it may do either by being acidic itself or by combining with certain substances within the cells. As soon as living muscle tissue is immersed in an acidic fixative, the fibers are induced to brief contraction, during which their striations are invisible. Slow relaxation follows and according to Frank[11] and Hürthle,[15] fixation of living muscle in the contracted state is virtually impossible. Picric acid and mercuric chloride produce a turbidity of the muscle fibers due presumably to precipitation of proteins. Formalin causes an initial obscuration of the striations, but these again become visible and are well conserved in microscopic sections. The degree of shrinkage of the sarcomeres in formalin fixative was calculated by Høncke,[14] who found that both the A and I bands are reduced to a degree and in a proportion similiar to that in contraction. This is avoided to some extent by allowing the muscle to adhere to a piece of cardboard before fixation or tying it to a piece of wood. If a piece of muscle tissue is plunged directly into formalin, the connective tissue ultimately shrinks more than the muscle fibers, which then are wavy in outline.

If fixation is not attempted until the muscle fibers have lost irritability, these changes are lessened. Nevertheless, the strong chemical agents of the common fixatives do result in shrinkage and a certain amount of distortion. The myofibrils shrink; the differences between sarcoplasm and myofibrils are more evident; the cross striations become more clearly visible, and the nuclei become less discernible in hematoxylin and other routine stains.

COMMON ARTIFACTS IN MICROSCOPIC SECTIONS OF MUSCLE

The necessity of becoming familiar with the normal histologic characteristics of muscle before endeavoring to determine the existence of pathologic processes is of course too obvious to receive further comment. A practice of examining muscle tissue routinely in all types of disease serves to draw more clearly the boundaries between what is definitely pathologic and what is within the range of individual variation. Furthermore, the wider use of many different methods of fixation, embedding, and staining of muscle tissue should aid in recognizing artifacts resulting either from the biopsy procedure itself or from the method of processing the tissues. Sound judgment of the validity of the various histopathologic criteria of muscle disease can be acquired only through personal experience.

Although it is admitted that fixation and staining give rise to a number of artifacts that may be misleading, these are the only means whereby one can obtain information as to structural change in the main constituents of muscle tissue. The common artifacts can be recognized only after some experience with biopsy material. Hyalinization, granular change, and fibrillar splitting can all be produced by fixation. Excessive contraction with transverse bands throughout the muscle fibers may result from immersing irritable muscle in alcohol or an acidic fixative (Fig. 13–9). This tends to occur in biopsy specimens of the extremely irritable muscle of myotonia in spite of precautions. The corresponding irregular staining of fibers in transverse section can be very misleading. Stretching or crushing ruptures the sarcolemma. The sarcoplasm of the stretched fibers exhibits an altered staining reaction; e.g., in phloxine methylene blue the pulped sarcoplasm

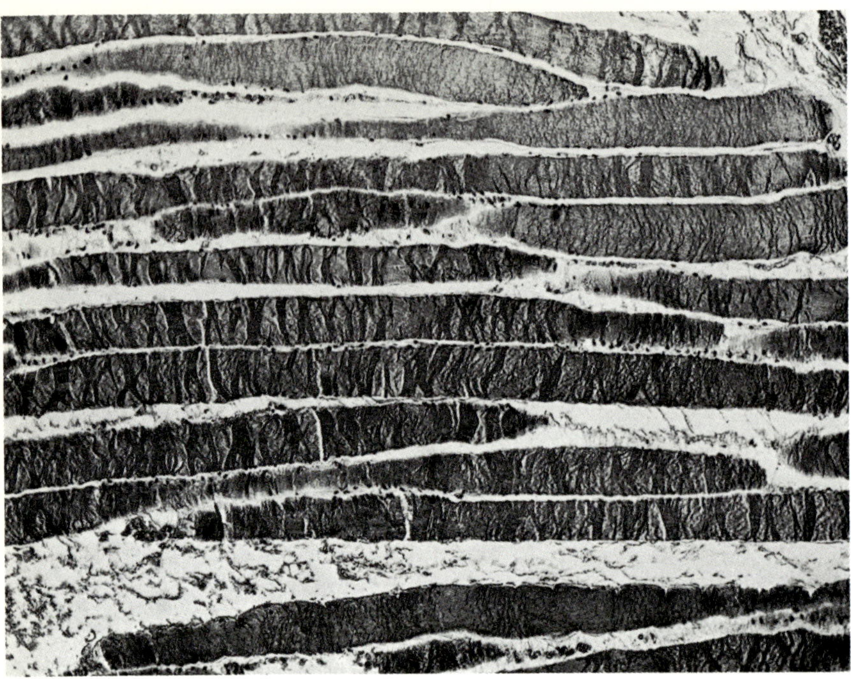

Fig. 13–9. Nageotte's change in biopsy of a trapezius muscle fixed in alcohol. (H&E; ×180)

may be blue or green instead of pink or red, and the transverse striations may disappear. Pressure ridges, section cracks, and free ends are familiar artifacts in muscle because of its brittleness after fixation (Fig. 13–10). Pressure ridges are caused by a dull knife encountering the tough, fixed muscle fiber in softer paraffin. The muscle fibers are compressed and pushed ahead of the knife and then suddenly cut through. This leaves ridges of darker staining condensed muscle tissue. Section cracks occur when the sharp knife blade encounters the brittle muscle tissue, well embedded. As the section is lifted from the block at the angle of the knife, it cracks at regular intervals. This happens more often in fragile, diseased muscle, which tends to be brittle (Fig. 13–10). Free ends are likely to occur in longitudinal sections of muscle fibers. Since the plane of section rarely coincides with the myofibrils or with connective tissue fibrils, these structures contain many cut ends in the length of the section. Such cut ends are apt to become free at their tips and to stick up or lie at an angle to the course of the fibers. Distortion and fissuration

of muscle fibers may result from rapid dehydration. These changes were discussed by Dahlgren,[6] Cancilla,[3] and Kakulas *et al.*[16] The latter investigators subjected fresh muscle to hypo- and hypertonic solutions and to all the standard fixatives as well as varied conditions of handling. Alcohol, FAB, acetic, formalin, Zenker's, Muller's, and Carnoy's fixatives with paraffin embedding caused the greatest degree of shrinkage of sarcoplasm from endomysium. Isotonic formalin, Susa fixative, and glutaraldehyde reduced the shrinkage. Moore *et al.*[19] with formalin fixation compared the degree of shrinkage with fresh freezing, fixed freezing, and paraffin and celloidin embedding, and found the ratio of fiber sizes to stand in the proportions of 10:9:8:7, respectively. Homogenization and vitreous change was common, particularly with Bouin's and Zenker's fixatives as well as with alcohol and lead acetate. It was increased by mechanical pressure on the unfixed tissue. Perfusion with glutaraldehyde and Heidenhain's Susa fixative gave different results; the organelles were much better preserved in the former. Dehydration before

Fig. 13–10. Biopsy section of muscle showing artifacts in form of altered staining reaction, fragmentation, and transverse fracturing of muscle fibers adjacent to area of resection. (phloxine, methylene blue)

complete fixation causes consistent vitreous change. It was concluded that for light microscopy the Susa, formalin, Flemming's Kultschitsky, and Perenyi's fixatives gave the best results.

Postmortem changes may also be a source of confusion, especially when they occur in conjunction with a muscle disease. The most frequently observed effect of postmortem autolysis is the failure of some fibers or groups of fibers to stain properly (Fig. 11–5). In hematoxylin and eosin such fibers are a pale pink or rose color. The outlines of the fiber are poorly defined, and it becomes increasingly difficult to distinguish between the muscle fibers and perimysial connective tissues. The sarcolemmal nuclei become dark and shrunken and may stain poorly with hematoxylin. In severe postmortem autolysis all the fibers may have this appearance (Fig. 4–12). With phosphotungstic acid hematoxylin the fibers are a grayish tan instead of blue, and the striations disappear. Some fibers are vacuolated.

All these artifacts can occur in healthy muscle tissue and of course are not accompanied by cellular reactions. When these same distortions occur in diseased muscle they must be disregarded. It is for this reason that in routine pathologic study alterations in the myofibrils or of their striations cannot solely be depended upon as reliable indications of disease states.

Another source of confusion results from a biopsy taken near the tendinous insertion of a muscle. At this point the fibers are tapering and in transverse sections are quite variable in size. Also some have a single, pale central nucleus. The mistake is to interpret this as regenerating muscle.

Near the motor point neurovascular bands traverse muscle fibers. In routine cell stains, which show only chains or clusters of nuclei, this may resemble a mononuclear infiltration.

SPECIAL PROBLEMS WITH BIOPSIES FOR HISTOCHEMICAL AND ELECTRON MICROSCOPIC METHODS

The introduction of histochemical stains in freeze-dried preparations and osmium and glutaraldehyde perfusion or immersion technics for electron microscopy has not averted artifacts. Instead these methods created new ones of different types. Of course when applied to biopsy specimens the distortions induced by the incision of irritable muscle are as prominent if not more so than with the preparative methods of light microscopy.

Freezing and drying fresh tissue before permitting it to thaw often results in the sarcoplasm having a finely vacuolated appearance owing to the formation of ice crystals. The sarcoplasm also swells so that the fibers are more closely opposed than with other methods of fixation. All measurements of transverse diameters made on such preparations are approximately 10% higher than with formalin-fixed frozen sections, as was mentioned above. Corrections must be made for these size differences according to the histologic technic used. The type I fibers are always slightly smaller than type II. In histochemical preparations the fine details of nuclear structure and of myofibrils are usually obscured. For example, in Brooke's elegant illustrations[2] only in his diphosphopyridine-reduced dehydrogenase (DPNP) preparation are the myofibrils seen and in many of the muscle fibers the central ones are poorly impregnated, distorted, or whorled. Dark coagulated masses are not infrequently seen in the sarcoplasm in ATPase preparations. If there are contraction bands the staining reaction of the entire contents of the fiber or only its central parts are altered, and there is a targetoid appearance that may simulate central core disease.

The difficulties in distinguishing artifactual and the specific and nonspecific modulations of structure in electron micrographs are obvious. Any critical pathologist is aware that the majority of such pictures in printed articles are of such poor quality that interpretations are virtually impossible. Muscle fibers with irregular, cleared zones, fraying and smudging of membranes, loss of tissue substance, swelling of mitochondria, rupture of endoplasmic reticulum have clearly been altered more by the effects of poor preparation than disease.[21] Indeed, one may use such alterations as criteria of the quality of the histologic technic. Moreover, there are a number of ultrastructural changes such as smudging or spreading of Z bands and poor staining or disruption of thin actin filaments of I bands, which at present are impossible to interpret. If combined with mitochondrial swelling and "edematous" appearance of the sarcoplasm it is likely that they are artifactual, though many electron microscopists attribute them to disease.

The idea that certain muscle diseases are the expression of specific changes in particular organelles is embodied in such new terms as mitochondrial myopathies or vacuolar myopathies. Indeed, it was a highly attractive notion that each disease began in or involved primarily one organelle. All subsequent experience with the electron microscope in myopathology, however, has served only to dispel this naive view of cellular pathology. Mitochondrial increase, enlargement, distortion, inclusions, etc. have been found in every disease so far studied. Even such intriguing structures as nemeline bodies (rod bodies) are now known to occur in tenotomized muscles and other diseases. Vacuolation may be due to several different diseases (Chapters 3 and 4), and the same may be said of a variety of peculiar cytoplasmic bodies, tubular structures, lipid inclusions, and autophagic vacuoles.

These remarks are not intended to disparage the contributions that have been and will be made by electron microscopy. We insist only that the same critique be applied to electron micrographs as are accorded to light microscopic preparations. Diseases of the muscle, like those of all cells, would hardly be expected to cause change in one organelle and spare others. Cells react in their totality, and one must think at the ultrastructural level in terms of both primary and secondary (reactive) effects. Whatever the morphologic changes prove to be, they must correspond in some meaningful way in light, phase, and histochemical preparations with other of the known attributes of disease. Electron microscopy is far from this level of accomplishment.

REFERENCES

1. BOWDEN REM, GUTMANN E: The clinical value of muscle biopsies. Lancet 1:768–771, 1945

2. BROOKE MH: The pathological interpretation of muscle histochemistry. The Striated Muscle. Monographs in Pathology. Edited by CM Pearson, FK Mostofi. Baltimore, Williams & Wilkins, 1973, Chap 8, pp 86–122

3. CANCILLA PA: Techniques of muscle biopsy and staining methods with particular emphasis on commonly encountered artifacts. The Striated Muscle. Monographs in Pathology. Edited by CM Pearson, FK Mostofi, Baltimore, Williams & Wilkins, 1973, Chap 2, pp 19–28

4. CARLETON HM, LEACH EH: Histological Technique for Normal Tissues, Morbid Changes and Identification of Parasites. Second edition. New York, Oxford University Press, 1938

5. CLAWSON BJ, NOBLE JF, LUFKIN NH: Nodular inflammatory and degenerative lesions of muscles from 450 autopsies. Arch Pathol 43:579–590, 1947

6. DAHLGREN U: Methods for preparing muscle and electric organ tissue. McClung's Handbook of Microscopical Technique. Third edition. New York, Hoeber, 1950, pp 329–345

7. DUBOWITZ and BROOKE MH: Muscle Biopsy: a modern approach. Philadelphia, WB Saunders Co, 1973

8. DURANTE G: Anatomie pathologique des muscles. Manuel d'Histologie Pathologique. Third edition. Edited by V Cornil, L Ranvier. Paris, Félix Alcan, 1902, pp 1–477

9. DUCHENNE GB: Récherches sur le paralysie musculaire pseudohypertrophie ou paralysie myosclerosique. Arch Gen Med 11:5–179, 1868

10. ENGEL WK, FOSTER JB, HUGHES BP, et al.: Central core disease—an investigation of a rare muscle cell abnormality. Brain 84:167–185, 1961

11. FRANK G: Das histologische bild der muskelkontraktion. Pfluegers Arch 218:37–53, 1928

12. GOMORI G: A rapid one-step trichrome stain. Am J Clin Pathol 20:661–664, 1950

13. GRIESINGER W: Ueber muskelhypertrophie. Arch Heilkunde 6:1–13, 1865

14. HØNCKE P: Investigations on the structure and function of living, isolated, cross striated muscle fibers of mammals. Acta Physiol Scand [Suppl 48] 15:1–230, 1947

15. HÜRTHLE K: Zur Kenntnis der Struktur des ruhenden und des tätigen Froschmuskels'. 1. Lässt sich die Struktur des Froschmuskels im Zustand der Verkürzung durch chemische Fixierungsmittel festhaltern? Pfluegers Arch 223:685–698, 1930

16. KAKULAS BA, ADAMS, MCGEE DA: The control of artefact in biopsied muscle. Proceedings of the V International Neuropathological Congress. International Congress Series 100. Excerpta Medica, Amsterdam, 1965, pp 632–639

17. LEE AB: The Microtomist's Vade-Mecum. Edited by JB Gatenby, AB Lee. Philadelphia, Blakiston, 1950

18. LILLIE RD: Histopathologic Technic. Philadelphia, Blakiston, 1948

19. MOORE MJ, REBEIZ JJ, HOLDEN M, et al.: Biometric analyses of normal skeletal muscle. Acta Neuropathol (Berl) 19:51–69, 1971

20. PEARSON CM: The incidence and type of pathologic alterations observed in muscle in a routine autopsy survey. Neurology (Minneap) 9:757–766, 1959

21. RUBNER M: Ueber die Wasserverbindung in Kolloiden mit besonderer Berücksichtigung des quergestreiften Muskels. Abhandl Preuss Akad Wissenschaften Phys-math K1 1–70, 1922

21. SCALZI HA, PRICE HM: Ultrastructural aspects of the normal mammalian muscle. The Striated Muscle. Edited by CM Pearson, FK Mostofi. Baltimore, Williams & Wilkins, 1973, Chap 11, pp 226–236

23. SHANK RE, HOAGLAND CL: A myotome for the biopsy of muscle. Science 98:592, 1943

24. SOKOLOFF L, WILENS SL, BUNIM JJ, et al.: Diagnostic value of histologic lesions of striated muscles in rheumatoid arthritis. Am J Med Sci 219:174–182, 1950

25. TOMLINSON BE, WALTON JM, REBEIZ JJ: Effects of ageing and cachexia upon skeletal muscle. J Neurol Sci 10:321, 1969

26. VON MEYENBERG H: Die quergestreifte Muskulatur. Handbuch der Speziellen Pathologischen Anatomie und Histologie. Edited by F Henke, O Lubarsch. Berlin, Springer, 1929, Vol 9, pp 299–509

27. WALLACE SL, LATTES R, RAGAN C: Diagnostic significance of the muscle biopsy. Am J Med 25:600–610, 1958

28. WALTON JN, MASTAGLIA FL: Histological and histochemical changes in skeletal muscle from cases of juvenile and early adult spinal muscular atrophy. J Neurol Sci 12:15, 1971

29. WEBSTER HDEF, PORRO RS, TOBIN W: Needle biopsy of skeletal muscle: phase and electron microscopy evaluation of its usefulness in the study of muscle disease. J Neuropathol Exp Neurol 28:129–242, 1969

INDEX